Texts in Applied Mathematics 23

Editors
J.E. Marsden
L. Sirovich
M. Golubitsky
W. Jäger

Advisor
G. Iooss

Springer
New York
Berlin
Heidelberg
Barcelona
Hong Kong
London
Milan
Paris
Singapore
Tokyo

Texts in Applied Mathematics

1. *Sirovich:* Introduction to Applied Mathematics.
2. *Wiggins:* Introduction to Applied Nonlinear Dynamical Systems and Chaos.
3. *Hale/Koçak:* Dynamics and Bifurcations.
4. *Chorin/Marsden:* A Mathematical Introduction to Fluid Mechanics, 3rd ed.
5. *Hubbard/West:* Differential Equations: A Dynamical Systems Approach: Ordinary Differential Equations.
6. *Sontag:* Mathematical Control Theory: Deterministic Finite Dimensional Systems, 2nd ed.
7. *Perko:* Differential Equations and Dynamical Systems, 2nd ed.
8. *Seaborn:* Hypergeometric Functions and Their Applications.
9. *Pipkin:* A Course on Integral Equations.
10. *Hoppensteadt/Peskin:* Mathematics in Medicine and the Life Sciences.
11. *Braun:* Differential Equations and Their Applications, 4th ed.
12. *Stoer/Bulirsch:* Introduction to Numerical Analysis, 2nd ed.
13. *Renardy/Rogers:* A First Graduate Course in Partial Differential Equations.
14. *Banks:* Growth and Diffusion Phenomena: Mathematical Frameworks and Applications.
15. *Brenner/Scott:* The Mathematical Theory of Finite Element Methods.
16. *Van de Velde:* Concurrent Scientific Computing.
17. *Marsden/Ratiu:* Introduction to Mechanics and Symmetry, 2nd ed.
18. *Hubbard/West:* Differential Equations: A Dynamical Systems Approach: Higher-Dimensional Systems.
19. *Kaplan/Glass:* Understanding Nonlinear Dynamics.
20. *Holmes:* Introduction to Perturbation Methods.
21. *Curtain/Zwart:* An Introduction to Infinite-Dimensional Linear Systems Theory.
22. *Thomas:* Numerical Partial Differential Equations: Finite Difference Methods.
23. *Taylor:* Partial Differential Equations: Basic Theory.
24. *Merkin:* Introduction to the Theory of Stability of Motion.
25. *Naber:* Topology, Geometry, and Gauge Fields: Foundations.
26. *Polderman/Willems:* Introduction to Mathematical Systems Theory: A Behavioral Approach.
27. *Reddy:* Introductory Functional Analysis with Applications to Boundary-Value Problems and Finite Elements.
28. *Gustafson/Wilcox:* Analytical and Computational Methods of Advanced Engineering Mathematics.
29. *Tveito/Winther:* Introduction to Partial Differential Equations: A Computational Approach.
30. *Gasquet/Witomski:* Fourier Analysis and Applications: Filtering, Numerical Computation, Wavelets.

(continued after index)

Michael E. Taylor

Partial Differential Equations

Basic Theory

With 37 Illustrations

Springer

Michael E. Taylor
Department of Mathematics
University of North Carolina
Chapel Hill, NC 27599
USA

Series Editors

Jerrold E. Marsden
Control and Dynamical Systems, 104–44
California Institute of Technology
Pasadena, CA 91125
USA

L. Sirovich
Division of Applied Mathematics
Brown University
Providence, RI 02912
USA

M. Golubitsky
Department of Mathematics
University of Houston
Houston, TX 77204-3476
USA

W. Jäger
Department of Applied Mathematics
Universität Heidelberg
Im Neuenheimer Feld 294
69120 Heidelberg, Germany

Mathematics Subject Classification (1991): 35-01, 35J05, 35Exx, 35G10

Library of Congress Cataloging-in-Publication Data
Taylor, Michael Eugene, 1946–
Partial differential equations / Michael E. Taylor.
p. cm. – (Texts in applied mathematics; 23)
Includes bibliographical references and index.
ISBN 0-387-94654-3 (hc : alk. paper).
1. Differential equations, Partial. I. Title. II. Series:
Texts in applied mathematics (Springer-Verlag New York Inc.) ; v. 23.
QA1.A647
[QA374]
510 s–dc20
[515′.353] 95-54104

Printed on acid-free paper.

Production coordinated by Chernow Editorial Services, Inc. and managed by Bill Imbornoni; manufacturing supervised by Joseph Quatela.
Camera-ready copy prepared from the author's AMS-TeX files.
Printed and bound by Sheridan Books, Ann Arbor, MI.
Printed in the United States of America.

9 8 7 6 5 4 3 2 (Corrected second printing, 1999)

ISBN 0-387-94654-3 Springer-Verlag New York Berlin Heidelberg SPIN 10728799

To my wife and daughter, Jane Hawkins and Diane Taylor

Series Preface

Mathematics is playing an ever more important role in the physical and biological sciences, provoking a blurring of boundaries between scientific disciplines and a resurgence of interest in the modern as well as the classical techniques of applied mathematics. This renewal of interest, both in research and teaching, has led to the establishment of the series: *Texts in Applied Mathematics (TAM).*

The development of new courses is a natural consequence of a high level of excitement on the research frontier as newer techniques, such as numerical and symbolic computer systems, dynamical systems, and chaos, mix with and reinforce the traditional methods of applied mathematics. Thus, the purpose of this textbook series is to meet the current and future needs of these advances and encourage the teaching of new courses.

TAM will publish textbooks suitable for use in advanced undergraduate and beginning graduate courses, and will complement the *Applied Mathematical Sciences (AMS)* series, which will focus on advanced textbooks and research level monographs.

Contents

Introduction

Partial differential equations is a many-faceted subject. Created to describe the mechanical behavior of objects such as vibrating strings and blowing winds, it has developed into a body of material that interacts with many branches of mathematics, such as differential geometry, complex analysis, and harmonic analysis, as well as a ubiquitous factor in the description and elucidation of problems in mathematical physics.

This work is intended to provide a course of study of some of the major aspects of PDE. It is addressed to readers with a background in the basic introductory graduate mathematics courses in American universities: elementary real and complex analysis, differential geometry, and measure theory.

Chapter 1 provides background material on the theory of ordinary differential equations (ODE). This includes both very basic material—on topics such as the existence and uniqueness of solutions to ODE and explicit solutions to equations with constant coefficients and relations to linear algebra—and more sophisticated results—on flows generated by vector fields, connections with differential geometry, the calculus of differential forms, stationary action principles in mechanics, and their relation to Hamiltonian systems. We discuss equations of relativistic motion as well as equations of classical Newtonian mechanics. There are also applications to topological results, such as degree theory, the Brouwer fixed-point theorem, and the Jordan-Brouwer separation theorem. In this chapter we also treat scalar first-order PDE, via Hamilton-Jacobi theory.

Chapters 2 through 6 constitute a survey of basic linear PDE. Chapter 2 begins with the derivation of some partial differential equations of continuum mechanics in a fashion similar to the derivation of ODE in mechanics in Chapter 1, via variational principles. We obtain equations for vibrating strings and membranes. Further material in Chapter 2 centers around the Laplace operator, which on Euclidean space $\mathbb{R}^n$ is

$$\Delta = \frac{\partial^2}{\partial x_1^2} + \cdots + \frac{\partial^2}{\partial x_n^2}, \tag{1}$$

and the linear wave equation,

$$\frac{\partial^2 u}{\partial t^2} - \Delta u = 0. \tag{2}$$

We also consider the Laplace operator on a general Riemannian manifold and the wave equation on a general Lorentz manifold. We discuss basic consequences of Green's formula, including energy conservation and finite propagation speed for solutions to linear wave equations. We also discuss Maxwell's equations for electromagnetic fields and their relation with special relativity. Before we can establish general results on the solvability of these equations, it is necessary to develop some analytical techniques. This is done in the next couple of chapters.

Chapter 3 is devoted to Fourier analysis and the theory of distributions. These topics are crucial for the study of linear PDE. We give a number of basic applications to the study of linear PDE with constant coefficients. Among these applications are results on harmonic and holomorphic functions in the plane, including a short treatment of elementary complex function theory. We derive explicit formulas for solutions to Laplace and wave equations on Euclidean space, and also the heat equation,

$$\frac{\partial u}{\partial t} - \Delta u = 0. \tag{3}$$

We also produce solutions on certain subsets, such as rectangular regions, using the method of images. We include material on the discrete Fourier transform, germane to the discrete approximation of PDE, and on the fast evaluation of this transform, the FFT. Chapter 3 is the first chapter to make extensive use of functional analysis. Basic results on this topic are compiled in Appendix A, Outline of Functional Analysis.

Sobolev spaces have proven to be a very effective tool in the existence theory of PDE, and in the study of regularity of solutions. In Chapter 4 we introduce Sobolev spaces and study some of their basic properties. We start with $H^k(\mathbb{R}^n)$, which consists of L^2 functions whose derivatives of order $\le k$ (defined in a distributional sense, in Chapter 3) belong to $L^2(\mathbb{R}^n)$, when k is a positive integer. We also replace k by a general real number s.

Chapter 5 is devoted to the study of the existence and regularity of solutions to linear elliptic PDE, on bounded regions. We begin with the Dirichlet problem for the Laplace operator,

$$\Delta u = f \text{ on } \Omega, \qquad u = g \text{ on } \partial\Omega, \tag{4}$$

and then treat the Neumann problem and various other boundary problems, including some that apply to electromagnetic fields. We also study general boundary problems for linear elliptic operators, giving a condition that guarantees regularity and solvability (perhaps given a finite number of linear conditions on the data). Also in Chapter 5 are some applications to other areas, such as a proof of the Riemann mapping theorem, first for smooth, simply connected domains in the complex plane $\mathbb{C}$, then, after a treatment of the Dirichlet problem for the Laplace operator on domains with rough boundary, for general, simply connected domains in $\mathbb{C}$. We also develop Hodge theory and apply it to DeRham cohomology, extending the study of topological applications of differential forms begun in Chapter 1.

In Chapter 6, the last chapter of this volume, we study linear evolution equations, in which there is a "time" variable t, and initial data are given at $t = 0$. We discuss the heat and wave equations. We also treat Maxwell's equations, for an electromagnetic field, and more general hyperbolic systems. We prove the Cauchy-Kowalewsky theorem, in the linear case, establishing the local solvability of the Cauchy initial-value problem for general linear PDE with analytic coefficients, and analytic data, as long as the initial surface is "noncharacteristic." Also in Chapter 6 we treat geometrical optics, providing approximations to solutions of wave equations whose initial data either are highly oscillatory or possess simple singularities, such as a jump across a smooth hypersurface.

In addition to Chapters 1 through 6, there are two appendices. Appendix A provides an outline of functional analysis. It gives definitions and basic properties of Banach and Hilbert spaces (of which L^p-spaces and Sobolev spaces are examples), Fréchet spaces (such as $C^\infty(\mathbb{R}^n)$), and other locally convex spaces (such as spaces of distributions). It discusses some basic facts about bounded linear operators, including some special properties of compact operators, and also considers certain classes of unbounded operators. This functional analytical material plays a major role in the development of PDE from Chapter 3 on. Appendix B gives definitions and basic properties of manifolds and vector bundles. It also discusses some elementary properties of Lie groups.

This book is the first of a three-volume treatment of PDE. The second volume covers a selection of more advanced topics in linear PDE. It consists of Chapters 7 through 12 and Appendix C. Chapter 7 deals with pseudodifferential operators. This class of operators includes both differential operators and parametrices of elliptic operators, that is, inverses modulo smoothing operators. Chapter 8 is devoted to spectral theory, particularly for self-adjoint elliptic operators. In Chapter 9 we study the scattering of waves by a compact obstacle K in $\mathbb{R}^3$. This scattering theory is to some degree an extension of the spectral theory of the Laplace operator on $\mathbb{R}^3 \setminus K$, with the Dirichlet boundary condition. Chapter 10 is devoted to the Atiyah-Singer index theorem. In Chapter 11 we study Brownian motion, described mathematically by Wiener measure on the space of continuous paths in $\mathbb{R}^n$. This provides a probabilistic approach to diffusion, and it both uses and provides new tools for the analysis of the heat equation. In Chapter 12 we tackle the $\bar{\partial}$-Neumann problem, a boundary problem for an elliptic operator (essentially the Laplace operator) on a domain $\Omega \subset \mathbb{C}^n$, which is very important in the theory of functions of several complex variables. Appendix C contains material of a differential geometric nature, particularly a discussion of curvature.

The third and final volume of this work is devoted to nonlinear PDE. It contains Chapters 13 through 18. Chapter 13 contains a development of function space and operator theory, for use in nonlinear analysis. This includes the theory of L^p-Sobolev spaces and Hölder spaces, among other things. Chapter 14 is devoted to nonlinear elliptic PDE. In Chapter 15, we treat nonlinear parabolic equations. Nonlinear hyperbolic equations are studied in Chapter 16. We include a study of weak solutions to nonlinear hyperbolic systems, with shocks. Another topic covered in Chapter 16 is the Cauchy-Kowalewsky theorem, in the nonlinear case. We use a method introduced by P. Garabedian to transform the Cauchy problem for an analytic equation into a symmetric hyperbolic system. In Chapter 17 we study incompressible fluid flow. This is governed by the Euler equation in the absence of viscosity, and by the Navier-Stokes equation in the presence of viscosity. Chapter 18, the last chapter in this work, is devoted to Einstein's gravitational equations.

Acknowledgments

I have had the good fortune to teach at least one course relevant to the material of this book, almost every year since 1976. These courses led to many course notes, and I am grateful to many colleagues at Rice University, SUNY at Stony Brook, the California Institute of Technology, and the University of North Carolina, for the supportive atmospheres at these institutions. Also, a number of individuals provided valuable advice on various portions of the manuscript, as it grew over the years. I particularly want to thank Florin David, Martin Dindos, David Ebin, Frank Jones, Anna Mazzucato, Richard Melrose, James Ralston, Jeffrey Rauch, Santiago Simanca, and James York. The final touches were put on the manuscript while I was visiting the Institute for Mathematics and its Applications, at the University of Minnesota, which I thank for its hospitality and excellent facilities.

During the time I spent preparing this book, I received research support from the National Science Foundation.

Finally, I would like to acknowledge the impact on my studies of my senior thesis and Ph.D. thesis advisors, Edward Nelson and Heinz Cordes.

1

Basic Theory of ODE and Vector Fields

Introduction

This chapter examines basic topics in the field of ordinary differential equations (ODE), as it has developed from the era of Newton into modern times. This is closely tied to the development of a number of concepts in advanced calculus. We begin with a brief discussion of the derivative of a vector-valued function of several variables as a linear map. We then establish in §2 the fundamental local existence and uniqueness of solutions to ODE, of the form

$$\frac{dy}{dt} = F(t, y), \quad y(t_0) = y_0, \tag{0.1}$$

where $F(t, y)$ is continuous in both arguments and Lipschitz in y, and y takes values in $\mathbb{R}^k$. The proof uses a nice tool known as the contraction mapping principle; next we use this principle to establish the inverse and implicit function theorems in §3. After a discussion of constant-coefficient linear equations, in which we recall the basic results of linear algebra, in §4, we treat variable-coefficient linear ODE in §5, emphasizing a result known as Duhamel's principle, and then use this to examine smooth dependence on parameters for solutions to nonlinear ODE in §6.

The first six sections have a fairly purely analytic character and present ODE from a perspective similar to that seen in introductory courses. It is expected that the reader has seen much of this material before. Beginning in §7, the material begins to acquire a geometrical flavor as well. This section interprets solutions to (0.1) in terms of a flow generated by a vector field. The next two sections examine the Lie derivative of vector fields and some of its implications for ODE. While we initially work on domains in $\mathbb{R}^n$, here we begin a transition to global constructions, involving working on manifolds and hence making use of concepts that are invariant under changes of coordinates. By the end of §13, this transition is complete. Appendix B, at the end of this volume, collects some of the basic facts about manifolds which are useful for such an approach to analysis.

Physics is a major source of differential equations, and in §10 we discuss some of the basic ODE arising from Newton's force law, converting the resulting second-order ODE to first-order systems known as Hamiltonian systems. The study of

Hamiltonian vector fields is a major focus for the subsequent sections in this chapter. In §11 we deal with an apparently disjoint topic, the equations of geodesics on a Riemannian manifold. We introduce the covariant derivative as a tool for expressing the geodesic equations, and later show that these equations can also be cast in Hamiltonian form. In §12 we study a general class of variational problems, giving rise to both the equations of mechanics and the equations of geodesics, all expressible in Hamiltonian form.

In §13 we develop the theory of differential forms, one of E. Cartan's great contributions to analysis. There is a differential operator, called the exterior derivative, acting on differential forms. In beginning courses in multivariable calculus, one learns of div, grad, and curl as the major first-order differential operators; from a more advanced perspective, it is reasonable to think of the Lie derivative, the covariant derivative, and the exterior derivative as filling this role. The relevance of differential forms to ODE has many roots, but its most direct relevance for Hamiltonian systems is through the symplectic form, discussed in §14.

Results on Hamiltonian systems are applied in §15 to the study of first-order nonlinear PDE for a single unknown. The next section studies "completely integrable" systems, reversing the perspective, to apply solutions to certain nonlinear PDE to the study of Hamiltonian systems. These two sections comprise what is known as Hamilton-Jacobi theory. In §17 we make a further study of integrable systems arising from central force problems, particularly the one involving the gravitational attraction of two bodies, the solution to which was Newton's triumph. Section 18 gives a brief relativistic treatment of the equations of motion arising from the electromagnetic force, which ushered in Einstein's theory of relativity.

In §19 we apply material from §13 on differential forms to some topological results, such as the Brouwer fixed-point theorem, the study of the degree of a map between compact oriented manifolds, and the Jordan-Brouwer separation theorem. We apply the degree theory in §20 to a study of the index of a vector field, which reflects the behavior of its critical points. Other applications, and extensions, of results on degree theory and index theory in §§19–20 can be found in Appendix C and in Chapters 5 and 10. Also the Brouwer fixed-point theorem will be extended to the Leray-Schauder fixed-point theorem, and applied to problems in nonlinear PDE, in Chapter 14.

The appendix at the end of this chapter discusses the existence and uniqueness of solutions to (0.1) when F satisfies a condition weaker than Lipschitz in y. Results established here are applicable to the study of ideal fluid flow, as will be seen in Chapter 17.

1. The derivative

Let $\mathcal{O}$ be an open subset of $\mathbb{R}^n$, and let $F : \mathcal{O} \to \mathbb{R}^m$ be a continuous function. We say that F is differentiable at a point $x \in \mathcal{O}$, with derivative L, if $L : \mathbb{R}^n \to \mathbb{R}^m$

is a linear transformation such that, for small $y \in \mathbb{R}^n$,

$$F(x+y) = F(x) + Ly + R(x,y), \tag{1.1}$$

with

$$\frac{\|R(x,y)\|}{\|y\|} \to 0 \text{ as } y \to 0. \tag{1.2}$$

We denote the derivative at x by $DF(x) = L$. With respect to the standard bases of $\mathbb{R}^n$ and $\mathbb{R}^m$, $DF(x)$ is simply the matrix of partial derivatives,

$$DF(x) = \left(\frac{\partial F_j}{\partial x_k}\right), \tag{1.3}$$

so that, if $v = (v_1, \dots, v_n)$ (regarded as a column vector), then

$$DF(x)v = \left(\sum_k \frac{\partial F_1}{\partial x_k} v_k, \dots, \sum_k \frac{\partial F_m}{\partial x_k} v_k\right). \tag{1.4}$$

It will be shown that F is differentiable whenever all the partial derivatives exist and are *continuous* on $\mathcal{O}$. In such a case we say that F is a C^1-function on $\mathcal{O}$. In general, F is said to be C^k if all its partial derivatives of order $\le k$ exist and are continuous.

In (1.2) we can use the *Euclidean* norm on $\mathbb{R}^n$ and $\mathbb{R}^m$. This norm is defined by

$$\|x\| = \left(x_1^2 + \cdots + x_n^2\right)^{1/2} \tag{1.5}$$

for $x = (x_1, \dots, x_n) \in \mathbb{R}^n$. Any other norm would do equally well. Some basic results on the Euclidean norm are derived in §4.

More generally, the definition of the derivative given by (1.1) and (1.2) extends to a function $F : \mathcal{O} \to Y$, where $\mathcal{O}$ is an open subset of X, and X and Y are Banach spaces. Basic material on Banach spaces appears in Appendix A, Functional Analysis. In this case, we require L to be a bounded linear map from X to Y. The notion of differentiable function in this context is useful in the study of nonlinear PDE.

We now derive the *chain rule* for the derivative. Let $F : \mathcal{O} \to \mathbb{R}^m$ be differentiable at $x \in \mathcal{O}$, as above; let U be a neighborhood of $z = F(x)$ in $\mathbb{R}^m$; and let $G : U \to \mathbb{R}^k$ be differentiable at z. Consider $H = G \circ F$. We have

$$\begin{aligned} H(x+y) &= G(F(x+y)) \\ &= G\big(F(x) + DF(x)y + R(x,y)\big) \\ &= G(z) + DG(z)\big(DF(x)y + R(x,y)\big) + R_1(x,y) \\ &= G(z) + DG(z)DF(x)y + R_2(x,y), \end{aligned} \tag{1.6}$$

with

$$\frac{\|R_2(x,y)\|}{\|y\|} \to 0 \text{ as } y \to 0.$$

Thus $G \circ F$ is differentiable at x, and

$$D(G \circ F)(x) = DG(F(x)) \cdot DF(x). \tag{1.7}$$

This result works equally well if $\mathbb{R}^n$, $\mathbb{R}^m$, and $\mathbb{R}^k$ are replaced by general Banach spaces.

Another useful remark is that, by the fundamental theorem of calculus, applied to $\varphi(t) = F(x+ty)$,

$$F(x+y) = F(x) + \int_0^1 DF(x+ty)y\,dt, \tag{1.8}$$

provided F is C^1. For a typical application, see (6.6).

A closely related application of the fundamental theorem of calculus is that if we assume that $F : \mathcal{O} \to \mathbb{R}^m$ is differentiable in each variable separately, and that each $\partial F/\partial x_j$ is continuous on $\mathcal{O}$, then

$$\begin{aligned} F(x+y) &= F(x) + \sum_{j=1}^n \big[F(x+z_j) - F(x+z_{j-1})\big] \\ &= F(x) + \sum_{j=1}^n A_j(x,y)y_j, \end{aligned} \tag{1.9}$$

$$A_j(x,y) = \int_0^1 \frac{\partial F}{\partial x_j}\big(x + z_{j-1} + ty_je_j\big)\,dt,$$

where $z_0 = 0$, $z_j = (y_1, \dots, y_j, 0, \dots, 0)$, and $\{e_j\}$ is the standard basis of $\mathbb{R}^n$. Now (1.9) implies that F is differentiable on $\mathcal{O}$, as we stated beneath (1.4). As is shown in many calculus texts, by using the mean value theorem instead of the fundamental theorem of calculus, one can obtain a slightly sharper result. We leave the reconstruction of this argument to the reader.

We now describe two convenient notations to express higher-order derivatives of a C^k-function $f : \Omega \to \mathbb{R}$, where $\Omega \subset \mathbb{R}^n$ is open. In the first, let J be a k-tuple of integers between 1 and n; $J = (j_1, \dots, j_k)$. We set

$$f^{(J)}(x) = \partial_{j_k} \cdots \partial_{j_1} f(x), \quad \partial_j = \frac{\partial}{\partial x_j}. \tag{1.10}$$

Also, we set $|J| = k$, the total order of differentiation. As will be seen in the exercises, $\partial_i\partial_j f = \partial_j\partial_i f$, provided $f \in C^2(\Omega)$. Hence, if $f \in C^k(\Omega)$, then $\partial_{j_k} \cdots \partial_{j_1} f = \partial_{\ell_k} \cdots \partial_{\ell_1} f$ whenever $\{\ell_1, \dots, \ell_k\}$ is a permutation of $\{j_1, \dots, j_k\}$. Thus, another convenient notation to use is the following. Let α be an n-tuple of nonnegative integers, $\alpha = (\alpha_1, \dots, \alpha_n)$. Then we set

$$f^{(\alpha)}(x) = \partial_1^{\alpha_1} \cdots \partial_n^{\alpha_n} f(x), \quad |\alpha| = \alpha_1 + \cdots + \alpha_n. \tag{1.11}$$

Note that if $|J| = |\alpha| = k$ and $f \in C^k(\Omega)$, then

$$f^{(J)}(x) = f^{(\alpha)}(x), \quad \text{with } \alpha_i = \#\{\ell : j_\ell = i\}. \tag{1.12}$$

Correspondingly, there are two expressions for monomials in x:

$$x^J = x_{j_1} \cdots x_{j_k}, \quad x^\alpha = x_1^{\alpha_1} \cdots x_n^{\alpha_n}, \tag{1.13}$$

and $x^J = x^\alpha$, provided J and α are related as in (1.12). Both of these notations are called "multi-index" notations.

We now derive Taylor's formula with remainder for a smooth function $F : \Omega \to \mathbb{R}$, making use of these multi-index notations. We will apply the one-variable formula,

$$\varphi(t) = \varphi(0) + \varphi'(0)t + \frac{1}{2}\varphi''(0)t^2 + \cdots + \frac{1}{k!}\varphi^{(k)}(0)t^k + r_k(t), \tag{1.14}$$

with

$$r_k(t) = \frac{1}{k!}\int_0^t (t-s)^k \varphi^{(k+1)}(s)\, ds, \tag{1.15}$$

given $\varphi \in C^{k+1}(I)$, $I = (-a, a)$. Let us assume that $0 \in \Omega$ and that the line segment from 0 to x is contained in Ω. We set $\varphi(t) = F(tx)$ and apply (1.14) and (1.15) with $t = 1$. Applying the chain rule, we have

$$\varphi'(t) = \sum_{j=1}^n \partial_j F(tx)x_j = \sum_{|J|=1} F^{(J)}(tx)x^J. \tag{1.16}$$

Differentiating again, we have

$$\varphi''(t) = \sum_{|J|=1,|K|=1} F^{(J+K)}(tx)x^{J+K} = \sum_{|J|=2} F^{(J)}(tx)x^J, \tag{1.17}$$

where, if $|J| = k$ and $|K| = \ell$, we take $J + K = (j_1, \dots, j_k, k_1, \dots, k_\ell)$. Inductively, we have

$$\varphi^{(k)}(t) = \sum_{|J|=k} F^{(J)}(tx)x^J. \tag{1.18}$$

Hence, from (1.14), with $t = 1$,

$$F(x) = F(0) + \sum_{|J|=1} F^{(J)}(0)x^J + \cdots + \frac{1}{k!}\sum_{|J|=k} F^{(J)}(0)x^J + R_k(x),$$

or, more briefly,

$$F(x) = \sum_{|J|\le k} \frac{1}{|J|!} F^{(J)}(0)x^J + R_k(x), \tag{1.19}$$

where

$$R_k(x) = \frac{1}{k!}\sum_{|J|=k+1}\Big(\int_0^1 (1-s)^k F^{(J)}(sx)\, ds\Big)x^J. \tag{1.20}$$

This gives Taylor's formula with remainder for $F \in C^{k+1}(\Omega)$, in the J-multi-index notation.

We also want to write the formula in the α-multi-index notation. We have

$$\sum_{|J|=k} F^{(J)}(tx)x^J = \sum_{|\alpha|=k} \nu(\alpha)F^{(\alpha)}(tx)x^\alpha, \tag{1.21}$$

where

$$\nu(\alpha) = \#\{J : \alpha = \alpha(J)\}, \tag{1.22}$$

and we define the relation $\alpha = \alpha(J)$ to hold provided (1.12) holds or, equivalently, provided $x^J = x^\alpha$. Thus, $\nu(\alpha)$ is uniquely defined by

$$\sum_{|\alpha|=k} \nu(\alpha)x^\alpha = \sum_{|J|=k} x^J = (x_1 + \cdots + x_n)^k. \tag{1.23}$$

One sees that, if $|\alpha| = k$, then $\nu(\alpha)$ is equal to the product of the number of combinations of k objects, taken α_1 at a time, times the number of combinations of $k - \alpha_1$ objects, taken α_2 at a time, and so on, times the number of combinations of $k - (\alpha_1 + \cdots + \alpha_{n-1})$ objects, taken α_n at a time. Thus

$$\nu(\alpha) = \binom{k}{\alpha_1}\binom{k-\alpha_1}{\alpha_2}\cdots\binom{k-\alpha_1-\cdots-\alpha_{n-1}}{\alpha_n} = \frac{k!}{\alpha_1!\alpha_2!\cdots\alpha_n!}. \tag{1.24}$$

In other words, for $|\alpha| = k$,

$$\nu(\alpha) = \frac{k!}{\alpha!}, \quad \text{where } \alpha! = \alpha_1!\cdots\alpha_n!. \tag{1.25}$$

Thus, the Taylor formula (1.19) can be rewritten as

$$F(x) = \sum_{|\alpha|\le k} \frac{1}{\alpha!} F^{(\alpha)}(0)x^\alpha + R_k(x), \tag{1.26}$$

where

$$R_k(x) = \sum_{|\alpha|=k+1} \frac{k+1}{\alpha!}\Bigl(\int_0^1 (1-s)^k F^{(\alpha)}(sx)\, ds\Bigr)x^\alpha. \tag{1.27}$$

Exercises

1. Let $M_{n\times n}$ be the space of complex $n \times n$ matrices, and let $\det : M_{n\times n} \to \mathbb{C}$ denote the determinant. Show that if I is the identity matrix, then

$$D\,\det(I)B = \operatorname{Tr} B,$$

i.e.,

$$\frac{d}{dt}\det(I + tB)|_{t=0} = \operatorname{Tr} B.$$

2. If $A(t) = \bigl(a_{jk}(t)\bigr)$ is a curve in $M_{n\times n}$, use the expansion of $(d/dt)\det A(t)$ as a sum of n determinants, in which the rows of $A(t)$ are successively differentiated, to show that, for $A \in M_{n\times n}$,

$$D\,\det(A)B = \operatorname{Tr}\bigl(\operatorname{Cof}(A)^t \cdot B\bigr),$$

where $\operatorname{Cof}(A)$ is the cofactor matrix of A.

3. Suppose $A \in M_{n\times n}$ is invertible. Using

$$\det(A + tB) = (\det A)\ \det(I + tA^{-1}B),$$

show that

$$D\,\det(A)B = (\det A)\operatorname{Tr}(A^{-1}B).$$

Comparing the result of Exercise 2, deduce Cramer's formula:

$$(\det A)A^{-1} = \mathrm{Cof}(A)^t. \tag{1.28}$$

4. Identify $\mathbb{R}^2$ and $\mathbb{C}$ via $z = x+iy$. Then multiplication by i on $\mathbb{C}$ corresponds to applying

$$J = \begin{pmatrix} 0 & -1 \\ 1 & 0 \end{pmatrix}.$$

Let $\mathcal{O} \subset \mathbb{R}^2$ be open, and let $f : \mathcal{O} \to \mathbb{R}^2$ be C^1. Say $f = (u, v)$. Regard $Df(x, y)$ as a 2×2 real matrix. One says f is *holomorphic*, or complex analytic, provided the Cauchy-Riemann equations hold:

$$\frac{\partial u}{\partial x} = \frac{\partial v}{\partial y}, \quad \frac{\partial u}{\partial y} = -\frac{\partial v}{\partial x}. \tag{1.29}$$

Show that this is equivalent to the condition

$$Df(x, y)J = J\,Df(x, y).$$

Generalize to $\mathcal{O}$ open in $\mathbb{C}^m$, $f : \mathcal{O} \to \mathbb{C}^n$.

5. If $R(x)$ is a C^∞-function near the origin in $\mathbb{R}^n$, satisfying $R(0) = 0$ and $DR(0) = 0$, show that there exist smooth functions $r_{jk}(x)$ such that

$$R(x) = \sum r_{jk}(x)x_j x_k.$$

(*Hint*: Using (1.8), write $R(x) = \Phi(x)x$, $\Phi(x) = \int_0^1 DR(tx)dt$, since $R(0) = 0$. Then $\Phi(0) = DR(0) = 0$, so (1.8) can be applied again, to give $\Phi(x) = \Psi(x)x$.)

6. If f is C^1 on a region in $\mathbb{R}^2$ containing $[a, b] \times \{y\}$, show that

$$\frac{d}{dy}\int_a^b f(x, y)\,dx = \int_a^b \frac{\partial f}{\partial y}(x, y)\,dx.$$

(*Hint*: Show that the left side is equal to

$$\lim_{h\to 0} \int_a^b \frac{1}{h}\int_0^h \frac{\partial f}{\partial y}(x, y+s)\,ds\,dx.)$$

7. Suppose $F : \mathcal{O} \to \mathbb{R}^m$ is a C^2-function. Applying the fundamental theorem of calculus, first to

$$G_j(x) = F(x + he_j) - F(x)$$

(as a function of h) and then to

$$H_{jk}(x) = G_j(x + he_k) - G_j(x),$$

where $\{e_j\}$ is the standard basis of $\mathbb{R}^n$, show that if $x \in \mathcal{O}$ and h is small, then

$$F(x + he_j + he_k) - F(x + he_k) - F(x + he_j) + F(x) = \int_0^h\int_0^h \frac{\partial}{\partial x_k}\frac{\partial F}{\partial x_j}(x + se_j + te_k)\,ds\,dt.$$

Similarly, show that this quantity is equal to

$$\int_0^h\int_0^h \frac{\partial}{\partial x_j}\frac{\partial F}{\partial x_k}(x + se_j + te_k)\,dt\,ds.$$

Deduce that

$$\frac{\partial}{\partial x_k}\frac{\partial F}{\partial x_j}(x) = \frac{\partial}{\partial x_j}\frac{\partial F}{\partial x_k}(x).$$

(*Hint*: Use Exercise 6.)
Arguments that use the mean value theorem instead of the fundamental theorem of calculus can be found in many calculus texts.

2. Fundamental local existence theorem for ODE

The goal of this section is to establish the existence of solutions to an ODE:

$$\frac{dy}{dt} = F(t, y), \quad y(t_0) = y_0. \tag{2.1}$$

We will prove the following fundamental result.

Theorem 2.1. *Let $y_0 \in \mathcal{O}$, an open subset of $\mathbb{R}^n$, $I \subset \mathbb{R}$ an interval containing t_0. Suppose F is continuous on $I \times \mathcal{O}$ and satisfies the following Lipschitz estimate in y:*

$$\|F(t, y_1) - F(t, y_2)\| \le L\|y_1 - y_2\|, \tag{2.2}$$

for $t \in I$, $y_j \in \mathcal{O}$. Then the equation (2.1) has a unique solution on some t-interval containing t_0.

To begin the proof, we note that the equation (2.1) is equivalent to the integral equation

$$y(t) = y_0 + \int_{t_0}^{t} F(s, y(s))\, ds. \tag{2.3}$$

Existence will be established via the Picard iteration method, which is the following. Guess $y_0(t)$, e.g., $y_0(t) = y_0$. Then set

$$y_k(t) = y_0 + \int_{t_0}^{t} F\big(s, y_{k-1}(s)\big)\, ds. \tag{2.4}$$

We aim to show that, as $k \to \infty$, $y_k(t)$ converges to a (unique) solution of (2.3), at least for t close enough to t_0. To do this, we will use the following tool, known as the *contraction mapping principle.*

Theorem 2.2. *Let X be a complete metric space, and let $T : X \to X$ satisfy*

$$dist(Tx, Ty) \le r\, dist(x, y), \tag{2.5}$$

for some $r < 1$. (We say that T is a contraction.) Then T has a unique fixed point x. For any $y_0 \in X$, $T^k y_0 \to x$ as $k \to \infty$.

Proof. Pick $y_0 \in X$, and let $y_k = T^k y_0$. Then

$$\text{dist}(y_{k+1}, y_k) \le r^k \,\text{dist}(y_1, y_0),$$

so

$$\begin{aligned}\text{dist}(y_{k+m}, y_k) &\le \text{dist}(y_{k+m}, y_{k+m-1}) + \cdots + \text{dist}(y_{k+1}, y_k) \\ &\le \left(r^k + \cdots + r^{k+m-1}\right) \text{dist}(y_1, y_0) \\ &\le r^k\left(1 - r\right)^{-1} \text{dist}(y_1, y_0).\end{aligned} \tag{2.6}$$

It follows that (y_k) is a Cauchy sequence, so it converges; $y_k \to x$. Since $Ty_k = y_{k+1}$ and T is continuous, it follows that $Tx = x$, that is, x is a fixed point. The uniqueness of the fixed point is clear from the estimate $\text{dist}(Tx, Tx') \le r\,\text{dist}(x, x')$, which implies $\text{dist}(x, x') = 0$ if x and x' are fixed points. This completes the proof.

Tackling the solvability of (2.3), we look for a fixed point of T, defined by

$$(Ty)(t) = y_0 + \int_{t_0}^{t} F(s, y(s))\, ds. \tag{2.7}$$

Let

$$X = \left\{u \in C(J, \mathbb{R}^n) : u(t_0) = y_0,\ \sup_{t \in J} \|u(t) - y_0\| \le K\right\}. \tag{2.8}$$

Here $J = [t_0 - \varepsilon, t_0 + \varepsilon]$, where ε will be chosen, sufficiently small, below. K is picked so $\{y : \|y - y_0\| \le K\}$ is contained in $\mathcal{O}$, and we also suppose $J \subset I$. Then there exists an M such that

$$\sup_{s \in J, \|y - y_0\| \le K} \|F(s, y)\| \le M. \tag{2.9}$$

Then, provided

$$\varepsilon \le \frac{K}{M}, \tag{2.10}$$

we have

$$T : X \to X. \tag{2.11}$$

Now, using the Lipschitz hypothesis (2.2), we have, for $t \in J$,

$$\|(Ty)(t) - (Tz)(t)\| \le \int_{t_0}^{t} L\|y(s) - z(s)\|\, ds \le \varepsilon\, L \sup_{s \in J} \|y(s) - z(s)\|, \tag{2.12}$$

assuming y and z belong to X. It follows that T is a contraction on X provided one has

$$\varepsilon < \frac{1}{L}, \tag{2.13}$$

in addition to the hypotheses above. This proves Theorem 2.1.

In view of the lower bound on the length of the interval J on which the existence theorem works, it is easy to show that the only way a solution can fail to be globally defined, that is, to exist for all $t \in I$, is for $y(t)$ to "explode to infinity" by leaving every compact set $K \subset \mathcal{O}$, as $t \to t_1$, for some $t_1 \in I$.

We remark that the local existence proof given above works if $\mathbb{R}^n$ is replaced by any Banach space.

Often one wants to deal with a higher-order ODE. There is a standard method of reducing an nth-order ODE

$$y^{(n)}(t) = f(t, y, y', \dots, y^{(n-1)}) \tag{2.14}$$

to a first-order system. One sets $u = (u_0, \dots, u_{n-1})$, with

$$u_0 = y, \quad u_j = y^{(j)}, \tag{2.15}$$

and then

$$\frac{du}{dt} = \big(u_1, \dots, u_{n-1}, f(t, u_0, \dots, u_{n-1})\big) = g(t, u). \tag{2.16}$$

If y takes values in $\mathbb{R}^k$, then u takes values in $\mathbb{R}^{kn}$.

If the system (2.1) is nonautonomous, that is, if F explicitly depends on t, it can be converted to an autonomous system (one with no explicit t-dependence) as follows. Set $z = (t, y)$. We then have

$$z' = (1, y') = (1, F(z)) = G(z). \tag{2.17}$$

Sometimes this process destroys important features of the original system (2.1). For example, if (2.1) is linear, (2.17) might be nonlinear. Nevertheless, the trick of converting (2.1) to (2.17) has some uses.

Many systems of ODE are difficult to solve explicitly. One very basic class of ODE can be solved explicitly, in terms of integrals, namely the single first-order linear ODE:

$$\frac{dy}{dt} = a(t)y + b(t), \quad y(0) = y_0, \tag{2.18}$$

where $a(t)$ and $b(t)$ are continuous real- or complex-valued functions. Set

$$A(t) = \int_0^t a(s)\, ds. \tag{2.19}$$

Then (2.18) can be written as

$$e^{A(t)} \frac{d}{dt}\big(e^{-A(t)} y\big) = b(t), \tag{2.20}$$

which yields

$$y(t) = e^{A(t)} y_0 + e^{A(t)} \int_0^t e^{-A(s)} b(s)\, ds. \tag{2.21}$$

Compare this result with formulas (4.42) and (5.8), in subsequent sections of this chapter.

Exercises

1. Solve the initial-value problem

$$y' = y^2, \quad y(0) = a,$$

given $a \in \mathbb{R}$. On what t-interval is the solution defined?

2. Under the hypotheses of Theorem 2.1, if y solves (2.1) for $t \in [T_0, T_1]$, and $y(t) \in K$, compact in $\mathcal{O}$, for all such t, prove that $y(t)$ extends to a solution for $t \in [S_0, S_1]$, with $S_0 < T_0$, $T_1 > T_0$, as stated beneath (2.13).
3. Let M be a compact, smooth surface in $\mathbb{R}^n$. Suppose $F : \mathbb{R}^n \to \mathbb{R}^n$ is a smooth map (vector field) such that, for each $x \in M$, $F(x)$ is tangent to M, that is, the line $\gamma_x(t) = x + tF(x)$ is tangent to M at x, at $t = 0$. Show that if $x \in M$, then the initial-value problem

$$y' = F(y), \quad y(0) = x$$

has a solution for all $t \in \mathbb{R}$, and $y(t) \in M$ for all t.
(*Hint*: Locally, straighten out M to be a linear subspace of $\mathbb{R}^n$, to which F is tangent. Use uniqueness. Material in §3 will help do this local straightening.)
Reconsider this problem after reading §7.
4. Show that the initial-value problem

$$\frac{dx}{dt} = -x(x^2 + y^2), \quad \frac{dy}{dt} = -y(x^2 + y^2), \quad x(0) = x_0, \quad y(0) = y_0$$

has a solution for all $t \geq 0$, but not for all $t < 0$, unless $(x_0, y_0) = (0, 0)$.

3. Inverse function and implicit function theorems

We will use the contraction mapping principle to establish the inverse function theorem, which together with its corollary, the implicit function theorem, is a fundamental result in multivariable calculus. First we state the inverse function theorem.

Theorem 3.1. *Let F be a C^k-map from an open neighborhood Ω of $p_0 \in \mathbb{R}^n$ to $\mathbb{R}^n$, with $q_0 = F(p_0)$. Suppose the derivative $DF(p_0)$ is invertible. Then there is a neighborhood U of p_0 and a neighborhood V of q_0 such that $F : U \to V$ is one-to-one and onto, and $F^{-1} : V \to U$ is a C^k-map. (One says that $F : U \to V$ is a diffeomorphism.)*

Proof. Using the chain rule, it is easy to reduce to the case $p_0 = q_0 = 0$ and $DF(p_0) = I$, the identity matrix, so we suppose this has been done. Thus,

$$F(u) = u + R(u), \quad R(0) = 0, \quad DR(0) = 0. \tag{3.1}$$

For v small, we want to solve

$$F(u) = v. \tag{3.2}$$

This is equivalent to $u + R(u) = v$, so let

$$T_v(u) = v - R(u). \tag{3.3}$$

Thus, solving (3.2) is equivalent to solving

$$T_v(u) = u. \tag{3.4}$$

We look for a fixed point $u = K(v) = F^{-1}(v)$. Also, we want to prove that $DK(0) = I$, that is, that $K(v) = v + r(v)$, with $r(v) = o(\|v\|)$. If we succeed in doing this, it follows easily that, for general x close to 0,

$$DK(x) = \Big(DF\big(K(x)\big)\Big)^{-1},$$

and a simple inductive argument shows that K is C^k if F is C^k. Now consider

$$T_v : X_v \longrightarrow X_v, \tag{3.5}$$

with

$$X_v = \{u \in \Omega : \|u - v\| \le A_v\}, \tag{3.6}$$

where we set

$$A_v = \sup_{\|w\| \le 2\|v\|} \|R(w)\|. \tag{3.7}$$

We claim that (3.5) holds if $\|v\|$ is sufficiently small. To prove this, note that $T_v(u) - v = -R(u)$, so we need to show that, provided $\|v\|$ is small, $u \in X_v$ implies $\|R(u)\| \le A_v$. But indeed, if $u \in X_v$, then $\|u\| \le \|v\| + A_v$, which is $\le 2\|v\|$ if $\|v\|$ is small, so then

$$\|R(u)\| \le \sup_{\|w\| \le 2\|v\|} \|R(w)\| = A_v;$$

this establishes (3.5).

Note that if $\|v\|$ is small enough, the map (3.5) is a contraction map, so there exists a unique fixed point $u = K(v) \in X_v$. Also note that since $u \in X_v$,

$$\|K(v) - v\| \le A_v = o(\|v\|). \tag{3.8}$$

Hence, the inverse function theorem is proved.

Thus, if DF is invertible on the domain of F, then F is a local diffeomorphism, although stronger hypotheses are needed to guarantee that F is a global diffeomorphism onto its range. Here is one result along these lines.

Proposition 3.2. *If $\Omega \subset \mathbb{R}^n$ is open and convex, $F : \Omega \to \mathbb{R}^n$ is C^1, and the symmetric part of $DF(u)$ is positive-definite for each $u \in \Omega$, then F is one-to-one on Ω.*

Proof. Suppose that $F(u_1) = F(u_2)$, where $u_2 = u_1 + w$. Consider $\varphi : [0, 1] \to \mathbb{R}$, given by

$$\varphi(t) = w \cdot F(u_1 + tw).$$

Thus $\varphi(0) = \varphi(1)$, so $\varphi'(t_0)$ must vanish for some $t_0 \in (0, 1)$, by the mean value theorem. But $\varphi'(t) = w \cdot DF(u_1 + tw)w > 0$, if $w \neq 0$, by the hypothesis on DF. This shows that F is one-to-one.

We can obtain the following implicit function theorem as a consequence of the inverse function theorem.

Theorem 3.3. *Suppose U is a neighborhood of $x_0 \in \mathbb{R}^k$, V is a neighborhood of $z_0 \in \mathbb{R}^\ell$, and*

$$F : U \times V \longrightarrow \mathbb{R}^\ell \tag{3.9}$$

is a C^k-map. Assume $D_z F(x_0, z_0)$ is invertible; say $F(x_0, z_0) = u_0$. Then the equation $F(x, z) = u_0$ defines $z = f(x, u_0)$ for x near x_0, with f a C^k-map.

Proof. Consider $H : U \times V \to \mathbb{R}^k \times \mathbb{R}^\ell$ defined by

$$H(x, z) = \bigl(x, F(x, z)\bigr). \tag{3.10}$$

We have

$$DH = \begin{pmatrix} I & D_x F \\ 0 & D_z F \end{pmatrix}. \tag{3.11}$$

Thus $DH(x_0, z_0)$ is invertible, so $J = H^{-1}$ exists and is C^k, by the inverse function theorem. It is clear that $J(x, u_0)$ has the form

$$J(x, u_0) = \bigl(x, f(x, u_0)\bigr), \tag{3.12}$$

and f is the desired map.

As in §2, we remark that the inverse function theorem generalizes. One can replace $\mathbb{R}^n$ by any Banach space and the proof of Theorem 3.1 given above extends with no change. Such generalizations are useful in nonlinear PDE, as we will see in Chapter 14.

Exercises

1. Suppose that $F : U \to \mathbb{R}^n$ is a C^2-map, U is open in $\mathbb{R}^n$, $p \in U$, and $DF(p)$ is invertible. With $q = F(p)$, define a map N on a neighborhood of p by

$$N(x) = x + DF(x)^{-1}\bigl(q - F(x)\bigr). \tag{3.13}$$

Show that there exists $\varepsilon > 0$ and $C < \infty$ such that, for $0 \le r < \varepsilon$,

$$\|x - p\| \le r \Longrightarrow \|N(x) - p\| \le C\, r^2.$$

Conclude that if $\|x_1 - p\| \le r$, with $r < \min(\varepsilon, 1/2C)$, then $x_{j+1} = N(x_j)$ defines a sequence converging very rapidly to p. This is the basis of *Newton's method*, for solving $F(p) = q$ for p.

(*Hint*: Write $x = p + y$, $F(x) = F(p) + DF(x)y + R$, with R given as in (1.27), with $k = 2$. Then $N(x) = p + \tilde{y}$, $\tilde{y} = -DF(x)^{-1}R$.)

2. Applying Newton's method to $f(x) = 1/x$, show that you get a fast approximation to division using only addition and multiplication.
(*Hint*: Carry out the calculation of $N(x)$ in this case and notice a "miracle.")
3. Identify $\mathbb{R}^{2n}$ with $\mathbb{C}^n$ via $z = x + iy$, as in Exercise 4 of §1. Let $U \subset \mathbb{R}^{2n}$ be open, and let $F : U \to \mathbb{R}^{2n}$ be C^1. Assume that $p \in U$ and $DF(p)$ is invertible. If $F^{-1} : V \to U$ is given as in Theorem 3.1, show that F^{-1} is holomorphic provided F is.
4. Let $\mathcal{O} \subset \mathbb{R}^n$ be open. We say that a function $f \in C^\infty(\mathcal{O})$ is *real analytic* provided that, for each $x_0 \in \mathcal{O}$, we have a convergent power-series expansion

$$f(x) = \sum_{\alpha \geq 0} \frac{1}{\alpha!} f^{(\alpha)}(x_0)(x - x_0)^\alpha, \tag{3.14}$$

valid in a neighborhood of x_0. Show that we can let x be complex in (3.14), and obtain an extension f to a neighborhood of $\mathcal{O}$ in $\mathbb{C}^n$. Show that the extended function is holomorphic, that is, satisfies the Cauchy-Riemann equations.
Remark. It can be shown that, conversely, any holomorphic function has a power-series expansion. See (2.30) of Chapter 3 for one such proof. For the next exercise, assume this to be known.
5. Let $\mathcal{O} \subset \mathbb{R}^n$ be open, $p \in \mathcal{O}$, and $f : \mathcal{O} \to \mathbb{R}^n$ be real analytic, with $Df(p)$ invertible. Take $f^{-1} : V \to U$ as in Theorem 3.1. Show f^{-1} is real analytic.
(*Hint*: Consider a holomorphic extension $F : \Omega \to \mathbb{C}^n$ of f, and apply Exercise 3.)

4. Constant-coefficient linear systems; exponentiation of matrices

Let A be an $n \times n$ matrix, real or complex. We consider the linear ODE

$$y' = Ay, \quad y(0) = y_0. \tag{4.1}$$

In analogy to the scalar case, we can produce the solution in the form

$$y(t) = e^{tA} y_0, \tag{4.2}$$

where we define the matrix exponential

$$e^{tA} = \sum_{k=0}^{\infty} \frac{t^k}{k!} A^k. \tag{4.3}$$

We will establish estimates implying the convergence of this infinite series for all real t, indeed for all complex t. Then term-by-term differentiation is valid and gives (4.1). To discuss convergence of (4.3), we need the notion of the *norm* of a matrix. This is a special case of results discussed in Appendix A, Functional Analysis.

If $u = (u_1, \dots, u_n)$ belongs to $\mathbb{R}^n$ or to $\mathbb{C}^n$, set, as in (1.5),

$$\|u\| = \left(|u_1|^2 + \cdots + |u_n|^2\right)^{1/2}. \tag{4.4}$$

Then, if A is an $n \times n$ matrix, set

$$\|A\| = \sup\{\|Au\| : \|u\| \le 1\}. \tag{4.5}$$

The norm (4.4) possesses the following properties:

$$\|u\| \ge 0, \quad \|u\| = 0 \text{ if and only if } u = 0, \tag{4.6}$$

$$\|cu\| = |c|\,\|u\|, \quad \text{for real or complex } c, \tag{4.7}$$

$$\|u + v\| \le \|u\| + \|v\|. \tag{4.8}$$

The last property, known as the *triangle inequality*, follows from Cauchy's inequality:

$$|(u, v)| \le \|u\| \cdot \|v\|, \tag{4.9}$$

where the inner product is $(u, v) = u_1\overline{v}_1 + \cdots + u_n\overline{v}_n$. To deduce (4.8) from (4.9), just square both sides of (4.8). To prove (4.9), use $(u - v, u - v) \ge 0$ to get

$$2\,\mathrm{Re}\,(u, v) \le \|u\|^2 + \|v\|^2.$$

Then replace u by $e^{i\theta}u$ to deduce

$$2|(u, v)| \le \|u\|^2 + \|v\|^2.$$

Next, replace u by tu and v by $t^{-1}v$, to get

$$2|(u, v)| \le t^2\|u\|^2 + t^{-2}\|v\|^2,$$

for any $t > 0$. Picking t so that $t^2 = \|v\|/\|u\|$, we have Cauchy's inequality (4.9).

Given (4.6)–(4.8), we easily get

$$\begin{aligned} &\|A\| \ge 0, \\ &\|cA\| = |c|\,\|A\|, \\ &\|A + B\| \le \|A\| + \|B\|. \end{aligned} \tag{4.10}$$

Also, $\|A\| = 0$ if and only if $A = 0$. The fact that $\|A\|$ is the smallest constant K such that $\|Au\| \le K\|u\|$ gives

$$\|AB\| \le \|A\| \cdot \|B\|. \tag{4.11}$$

In particular,

$$\|A^k\| \le \|A\|^k. \tag{4.12}$$

This makes it easy to check the convergence of the power-series (4.3).

Power-series manipulations can be used to establish the identity

$$e^{sA}e^{tA} = e^{(s+t)A}. \tag{4.13}$$

Another way to prove this is as follows. Regard t as fixed; denote the left side of (4.13) as $X(s)$ and the right side as $Y(s)$. Then differentiation with respect to s gives, respectively,

$$\begin{aligned} X'(s) &= AX(s), \quad X(0) = e^{tA}, \\ Y'(s) &= AY(s), \quad Y(0) = e^{tA}, \end{aligned} \tag{4.14}$$

so the uniqueness of solutions to the ODE implies $X(s) = Y(s)$ for all s. We note that (4.13) is a special case of the following.

Proposition 4.1. *$e^{t(A+B)} = e^{tA}e^{tB}$ for all t, if and only if A and B commute.*

Proof. Let

$$Y(t) = e^{t(A+B)}, \quad Z(t) = e^{tA}e^{tB}. \tag{4.15}$$

Note that $Y(0) = Z(0) = I$, so it suffices to show that $Y(t)$ and $Z(t)$ satisfy the same ODE, to deduce that they coincide. Clearly,

$$Y'(t) = (A+B)Y(t). \tag{4.16}$$

Meanwhile,

$$Z'(t) = Ae^{tA}e^{tB} + e^{tA}Be^{tB}. \tag{4.17}$$

Thus we get the equation (4.16) for $Z(t)$ provided we know that

$$e^{tA}B = Be^{tA} \text{ if } AB = BA. \tag{4.18}$$

This follows from the power-series expansion for e^{tA}, together with the fact that

$$A^kB = BA^k, \quad \forall\, k \geq 0, \text{ if } AB = BA. \tag{4.19}$$

For the converse, if $Y(t) = Z(t)$ for all t, then $e^{tA}B = Be^{tA}$, by (4.17), and hence, taking the t-derivative, $e^{tA}AB = BAe^{tA}$; setting $t = 0$ gives $AB = BA$.

If A is in diagonal form,

$$A = \begin{pmatrix} a_1 & & \\ & \ddots & \\ & & a_n \end{pmatrix}, \tag{4.20}$$

then clearly

$$e^{tA} = \begin{pmatrix} e^{ta_1} & & \\ & \ddots & \\ & & e^{ta_n} \end{pmatrix}. \tag{4.21}$$

The following result makes it useful to diagonalize A in order to compute e^{tA}.

Proposition 4.2. *If K is an invertible matrix and $B = KAK^{-1}$, then*

$$e^{tB} = K\, e^{tA}\, K^{-1}. \tag{4.22}$$

Proof. This follows from the power-series expansion (4.3), given the observation that

$$B^k = K\, A^k\, K^{-1}. \tag{4.23}$$

In view of (4.20)–(4.22), it is convenient to record a few standard results about eigenvalues and eigenvectors here. Let A be an $n \times n$ matrix over F, $F = \mathbb{R}$ or $\mathbb{C}$. An eigenvector of A is a nonzero $u \in F^n$ such that

$$Au = \lambda u, \tag{4.24}$$

for some $\lambda \in F$. Such an eigenvector exists if and only if $A - \lambda I : F^n \to F^n$ is not invertible, that is, if and only if

$$\det(A - \lambda I) = 0. \tag{4.25}$$

Now (4.25) is a polynomial equation, so it always has a complex root. This proves the following.

Proposition 4.3. *Given an $n \times n$ matrix A, there exists at least one (complex) eigenvector u.*

Of course, if A is real, and we know there is a real root of (4.25) (e.g., if n is odd), then a real eigenvector exists. One important class of matrices guaranteed to have real eigenvalues is the class of self-adjoint matrices. The adjoint of an $n \times n$ complex matrix is specified by the identity $(Au, v) = (u, A^*v)$.

Proposition 4.4. *If $A = A^*$, then all eigenvalues of A are real.*

Proof. $Au = \lambda u$ implies

$$\lambda \|u\|^2 = (\lambda u, u) = (Au, u) = (u, Au) = (u, \lambda u) = \overline{\lambda} \|u\|^2. \tag{4.26}$$

Hence $\lambda = \overline{\lambda}$, if $u \neq 0$.

We now establish the following important result.

Theorem 4.5. *If $A = A^*$, then there is an orthonormal basis of $\mathbb{C}^n$ consisting of eigenvectors of A.*

Proof. Let u_1 be one unit eigenvector; $Au_1 = \lambda u_1$. Existence is guaranteed by Proposition 4.3. Let $V = (u_1)^\perp$ be the orthogonal complement of the linear span of u_1. Then dim V is $n - 1$ and

$$A : V \to V, \quad \text{if } A = A^*. \tag{4.27}$$

The result follows by induction on n.

Corollary 4.6. *If $A = A^t$ is a real symmetric matrix, then there is an orthonormal basis of $\mathbb{R}^n$ consisting of eigenvectors of A.*

Proof. By Proposition 4.4 and the remarks following Proposition 4.3, there is one unit eigenvector $u_1 \in \mathbb{R}^n$. The rest of the proof is as above.

The proofs of the last four results rest on the fact that every nonconstant polynomial has a complex root. This is the fundamental theorem of algebra. A proof is given in §19 (Exercise 5), and another after Corollary 4.7 of Chapter 3. An alternative approach to Proposition 4.3 when $A = A^*$, yielding Proposition 4.4–Corollary 4.6, is given in one of the exercises at the end of this section.

Given an ODE in upper triangular form,

$$\frac{dy}{dt} = \begin{pmatrix} a_{11} & * & * \\ & \ddots & * \\ & & a_{nn} \end{pmatrix} y, \tag{4.28}$$

you can solve the last ODE for y_n, as it is just $dy_n/dt = a_{nn}y_n$. Then you get a single nonhomogeneous ODE for y_{n-1}, which can be solved as demonstrated in (2.18)–(2.21), and you can continue inductively to solve. Thus, it is often useful to be able to put an $n \times n$ matrix A in upper triangular form, with respect to a convenient choice of basis. We will establish two results along these lines. The first is due to Schur.

Theorem 4.7. *For any $n \times n$ matrix A, there is an orthonormal basis $u_1, \dots, u_n$ of $\mathbb{C}^n$ with respect to which A is in upper triangular form.*

This result is equivalent to the following proposition.

Proposition 4.8. *For any A, there is a sequence of vector spaces V_j of dimension j, contained in $\mathbb{C}^n$, with*

$$V_n \supset V_{n-1} \supset \cdots \supset V_1 \tag{4.29}$$

and

$$A : V_j \longrightarrow V_j. \tag{4.30}$$

To see the equivalence, if we are granted (4.29)–(4.30), pick $u_n \perp V_{n-1}$, a unit vector, then pick $u_{n-1} \in V_{n-1}$ such that $u_{n-1} \perp V_{n-2}$, and so forth. Meanwhile, Proposition 4.8 is a simple inductive consequence of the following result.

Lemma 4.9. *For any matrix A acting on V_n, there is a linear subspace V_{n-1}, of codimension 1, such that $A : V_{n-1} \to V_{n-1}$.*

Proof. Use Proposition 4.3, applied to A^*. There is a vector v_1 such that $A^*v_1 = \lambda v_1$. Let $V_{n-1} = (v_1)^\perp$. This completes the proof of the lemma, hence of Theorem 4.7.

Let us look more closely at what you can say about solutions to an ODE that has been put in the form (4.28). As mentioned, we can obtain y_j inductively by solving nonhomogeneous scalar ODE

$$\frac{dy_j}{dt} = a_{jj} y_j + b_j(t), \tag{4.31}$$

where $b_j(t)$ is a linear combination of $y_{j+1}(t), \ldots, y_n(t)$, and the formula (2.21) applies, with $A(t) = a_{jj}t$. We have $y_n(t) = Ce^{a_{nn}t}$, so $b_{n-1}(t)$ is a multiple of $e^{a_{nn}t}$. If $a_{n-1,n-1} \neq a_{nn}$, $y_{n-1}(t)$ will be a linear combination of $e^{a_{nn}t}$ and $e^{a_{n-1,n-1}t}$, but if $a_{n-n,n-1} = a_{nn}$, $y_{n-1}(t)$ may be a linear combination of $e^{a_{nn}t}$ and $te^{a_{nn}t}$. Further integration will involve $\int p(t)e^{\alpha t}dt$, where $p(t)$ is a polynomial. That no other sort of function will arise is guaranteed by the following result.

Lemma 4.10. *If $p(t) \in \mathcal{P}_n$, the space of polynomials of degree $\leq n$, and $\alpha \neq 0$, then*

$$\int p(t)e^{\alpha t}\,dt = q(t)e^{\alpha t} + C, \tag{4.32}$$

for some $q(t) \in \mathcal{P}_n$.

Proof. The map $p = Tq$ defined by $(d/dt)(q(t)e^{\alpha t}) = p(t)e^{\alpha t}$ is a map on $\mathcal{P}_n$; in fact, we have

$$Tq(t) = \alpha q(t) + q'(t). \tag{4.33}$$

It suffices to show that $T : \mathcal{P}_n \to \mathcal{P}_n$ is invertible. But $D = d/dt$ is *nilpotent* on $\mathcal{P}_n$; $D^{n+1} = 0$. Hence

$$T^{-1} = \alpha^{-1}(I + \alpha^{-1}D)^{-1} = \alpha^{-1}\big(I - \alpha^{-1}D + \cdots + \alpha^{-n}(-D)^n\big).$$

Note that this gives a neat formula for the integral (4.32). For example,

$$\begin{aligned} \int t^n e^{-t}\,dt &= -(t^n + nt^{n-1} + \cdots + n!)e^{-t} + C \\ &= -n!\Big(1 + t + \frac{1}{2}t^2 + \cdots + \frac{1}{n!}t^n\Big)e^{-t} + C. \end{aligned} \tag{4.34}$$

This could also be established by integration by parts and induction. Of course, when $\alpha = 0$ in (4.32), the result is different; $q(t)$ is a polynomial of degree $n+1$.

Now the implication for the solution to (4.28) is that all the components of $y(t)$ are products of polynomials and exponentials. By Theorem 4.7, we can draw the same conclusion about the solution to $dy/dt = Ay$ for any $n \times n$ matrix A. We can formally state the result as follows.

Proposition 4.11. *For any $n \times n$ matrix A,*

$$e^{tA}v = \sum e^{\lambda_j t}v_j(t), \tag{4.35}$$

where $\{\lambda_j\}$ is the set of eigenvalues of A and $v_j(t)$ are $\mathbb{C}^n$-valued polynomials. All the $v_j(t)$ are constant when A is diagonalizable.

To see that the λ_j are the eigenvalues of A, note that in the upper triangular case only the exponentials $e^{a_{jj}t}$ arise, and in that case the eigenvalues are precisely the diagonal elements.

If we let $\mathcal{E}_\lambda$ denote the space of $\mathbb{C}^n$-valued functions of the form $V(t) = e^{\lambda t} v(t)$, where $v(t)$ is a $\mathbb{C}^n$-valued polynomial, then $\mathcal{E}_\lambda$ is invariant under the action of both d/dt and A, hence of $d/dt - A$. Hence, if a sum $V_1(t) + \cdots + V_k(t)$, $V_j(t) \in \mathcal{E}_{\lambda_j}$ (with λ_js distinct), is annihilated by $d/dt - A$, so is each term in this sum.

Therefore, if (4.35) is a sum over the distinct eigenvalues λ_j of A, it follows that each term $e^{\lambda_j t} v_j(t)$ is annihilated by $d/dt - A$ or, equivalently, is of the form $e^{tA} w_j$, where $w_j = v_j(0)$. This leads to the following conclusion. Set

$$G_\lambda = \{v \in \mathbb{C}^n : e^{tA} v = e^{t\lambda} v(t),\ v(t) \text{ polynomial}\}. \tag{4.36}$$

Then $\mathbb{C}^n$ has a direct-sum decomposition

$$\mathbb{C}^n = G_{\lambda_1} + \cdots + G_{\lambda_k}, \tag{4.37}$$

where $\lambda_1, \ldots, \lambda_k$ are the distinct eigenvalues of A. Furthermore, each G_{λ_j} is invariant under A, and

$$A_j = A|_{G_{\lambda_j}} \text{ has exactly one eigenvalue, } \lambda_j. \tag{4.38}$$

This last statement holds because $e^{tA} v$ involves only the exponential $e^{\lambda_j t}$, when $v \in G_{\lambda_j}$. We say that G_{λ_j} is the *generalized eigenspace* of A, with eigenvalue λ_j. Of course, G_{λ_j} contains $\ker(A - \lambda_j I)$. Now $B_j = A_j - \lambda_j I$ has only 0 as an eigenvalue. It is subject to the following result.

Lemma 4.12. *If $B : \mathbb{C}^k \to \mathbb{C}^k$ has only 0 as an eigenvalue, then B is nilpotent; in fact,*

$$B^m = 0 \text{ for some } m \le k. \tag{4.39}$$

Proof. Let $W_j = B^j(\mathbb{C}^k)$; then $\mathbb{C}^k \supset W_1 \supset W_2 \supset \cdots$ is a sequence of finite-dimensional vector spaces, each invariant under B. This sequence must stabilize, so for some m, $B : W_m \to W_m$ bijectively. If $W_m \ne 0$, B has a nonzero eigenvalue.

We next discuss the famous Jordan normal form of a complex $n \times n$ matrix. The result is the following.

Theorem 4.13. *If A is an $n \times n$ matrix, then there is a basis of $\mathbb{C}^n$ with respect to which A becomes a direct sum of blocks of the form*

$$\begin{pmatrix} \lambda_j & 1 & & \\ & \lambda_j & \ddots & \\ & & \ddots & 1 \\ & & & \lambda_j \end{pmatrix}. \tag{4.40}$$

In light of the decomposition (4.37) and Lemma 4.12, it suffices to establish the Jordan normal form for a nilpotent matrix B. Given $v_0 \in \mathbb{C}^k$, let m be the smallest integer such that $B^m v_0 = 0$; $m \le k$. If $m = k$, then $\{v_0, Bv_0, \dots, B^{m-1}v_0\}$ gives a basis of $\mathbb{C}^k$, putting B in Jordan normal form. We then say v_0 is a *cyclic* vector for B, and $\mathbb{C}^k$ is generated by v_0. We call $\{v_0, \dots, B^{m-1}v_0\}$ a *string*.

We will have a Jordan normal form precisely if we can write $\mathbb{C}^k$ as a direct sum of cyclic subspaces. We establish that this can be done by induction on the dimension.

Thus, inductively, we can suppose that $W_1 = B(\mathbb{C}^k)$ is a direct sum of cyclic subspaces, so W_1 has a basis that is a union of strings, let's say a union of d strings $\{v_j, Bv_j, \dots, B^{\ell_j}v_j\}$, $1 \le j \le d$. In this case, $\ker B \cap W_1 = N_1$ has dimension d, and the vectors $B^{\ell_j}v_j$, $1 \le j \le d$, span N_1. Furthermore, each v_j has the form $v_j = Bw_j$ for some $w_j \in \mathbb{C}^k$.

Now $\dim \ker B = k - r \ge d$, where $r = \dim W_1$. Let $\{z_1, \dots, z_{k-r-d}\}$ span a subspace of $\ker B$ complementary to N_1. Then the strings $\{w_j, v_j = Bw_j, \dots, B^{\ell_j}v_j\}$, $1 \le j \le d$, and $\{z_1\}, \dots, \{z_{k-r-d}\}$ generate cyclic subspaces whose direct sum is $\mathbb{C}^k$, giving the Jordan normal form.

The argument above is part of an argument of Filippov. In fact, Filippov's proof contains a further clever twist, enabling one to prove Theorem 4.13 without using the decomposition (4.37). However, since we got this decomposition almost for free as a byproduct of the ODE analysis in Proposition 4.11, this author decided to make use of it. See Strang [Str] for Filippov's proof.

We have seen how constructing e^{tA} solves the equation (4.1). We can also use it to solve a nonhomogeneous equation, of the form

$$y' = Ay + b(t), \quad y(0) = y_0. \tag{4.41}$$

Direct calculation shows that the solution is given by

$$y(t) = e^{tA}y_0 + \int_0^t e^{(t-s)A}\, b(s)\, ds. \tag{4.42}$$

Note how this partially generalizes the formula (2.21). This formula is a special case of Duhamel's principle, which will be discussed further in §5.

We remark that the definition of e^{tA} by power series (4.3) extends to the case where A is a bounded linear operator on a Banach space. In that case, e^{tA} furnishes the simplest sort of example of a one-parameter group of operators. Compare §9 in Appendix A, Functional Analysis, for a further discussion of semigroups of operators. A number of problems in PDE amount to exponentiating various *unbounded* operators. The discussion of eigenvalues, eigenvectors, and normal forms above relies heavily on finite dimensionality, although a good deal of it carries over to compact operators on infinite-dimensional Banach and Hilbert spaces; see §6 of Appendix A. Also, there is a somewhat more subtle extension of Theorem 4.5 for general self-adjoint operators on a Hilbert space, which is discussed in §1 of Chapter 8.

Exercises

1. In addition to the operator norm $\|A\|$ of an $n \times n$ matrix, defined by (4.5), we consider the Hilbert-Schmidt norm $\|A\|_{\mathrm{HS}}$, defined by

$$\|A\|_{\mathrm{HS}}^2 = \sum_{j,k} |a_{jk}|^2,$$

if $A = (a_{jk})$. Show that

$$\|A\| \le \|A\|_{\mathrm{HS}}.$$

(*Hint*: If $r_1, \dots, r_n$ are the rows of A, then for $u \in \mathbb{C}^n$, Au has entries $r_j \cdot u$, $1 \le j \le n$. Use Cauchy's inequality (4.9) to estimate $|r_j \cdot u|^2$.)

Show also that

$$\sum_j |a_{jk}|^2 \le \|A\|^2 \text{ for each } k,$$

and hence

$$\|A\|_{\mathrm{HS}}^2 \le n\|A\|^2.$$

(*Hint*: $\|A\| \ge \|Ae_k\|$ for each standard basis vector e_k.)

2. Show that, in analogy with (4.11), we have

$$\|AB\|_{\mathrm{HS}} \le \|A\|_{\mathrm{HS}} \|B\|_{\mathrm{HS}}.$$

Indeed, show that

$$\|AB\|_{\mathrm{HS}} \le \|A\| \cdot \|B\|_{\mathrm{HS}},$$

where the first factor on the right is the operator norm $\|A\|$.

3. Let X be an $n \times n$ matrix. Show that

$$\det e^X = e^{Tr\, X}.$$

(*Hint*: Use a normal form.)

Let M_n denote the space of complex $n \times n$ matrices. If $A \in M_n$ and $\det A = 1$, we say that $A \in \mathrm{SL}(n, \mathbb{C})$. If $X \in M_n$ and $\mathrm{Tr}\, X = 0$, we say that $X \in \mathrm{sl}(n, \mathbb{C})$.

4. Let $X \in \mathrm{sl}(2, \mathbb{C})$. Suppose X has eigenvalues $\{\lambda, -\lambda\}$, $\lambda \ne 0$. Such an X can be diagonalized, so we know that there exist matrices $Z_j \in M_2$ such that

$$e^{tX} = Z_1\, e^{t\lambda} + Z_2\, e^{-t\lambda}.$$

Evaluating both sides at $t = 0$, and the t-derivative at $t = 0$, show that $Z_1 + Z_2 = I$, $\lambda Z_1 - \lambda Z_2 = X$, and solve for Z_1, Z_2. Deduce that

$$e^{tX} = (\cosh t\lambda)I + \lambda^{-1}(\sinh t\lambda)X.$$

5. Define holomorphic functions $C(z)$ and $S(z)$ by

$$C(z) = \cosh\sqrt{z}, \quad S(z) = \frac{\sinh\sqrt{z}}{\sqrt{z}}.$$

Deduce from Exercise 4 that, for $X \in \mathrm{sl}(2, \mathbb{C})$,

$$e^X = C(-\det X)I + S(-\det X)X.$$

Show that this identity is also valid when 0 is an eigenvalue of X.

6. Rederive the formula above for e^X, $X \in \mathrm{sl}(2, \mathbb{C})$, by using the power series for e^X together with the identity

$$X^2 = -(\det X)I, \quad X \in \mathrm{sl}(2, \mathbb{C}).$$

The next set of exercises examines the derivative of the map

$$\mathrm{Exp} : M_n \to M_n, \quad \mathrm{Exp}(X) = e^X.$$

7. Set $U(t, s) = e^{t(X+sY)}$, where X and Y are $n \times n$ matrices, and set $U_s = \partial U / \partial s$. Show that U_s satisfies

$$\frac{\partial U_s}{\partial t} = (X + sY)U_s + YU, \quad U_s(0, s) = 0.$$

8. Use Duhamel's principle, formula (4.42), to show that

$$U_s(t, s) = \int_0^t e^{(t-\tau)(X+sY)}\, Y\, e^{\tau(X+sY)}\, d\tau.$$

Deduce that

$$\frac{d}{ds} e^{X+sY}\Big|_{s=0} = e^X \int_0^1 e^{-\tau X} Y e^{\tau X}\, d\tau. \tag{4.43}$$

9. Given $X \in M_n$, define $\mathrm{ad}\, X \in \mathrm{End}(M_n)$, that is,

$$\mathrm{ad}\, X : M_n \to M_n,$$

by

$$\mathrm{ad}\, X(Y) = XY - YX.$$

Show that

$$e^{-tX} Y e^{tX} = e^{-t\, \mathrm{ad}\, X}\, Y.$$

(*Hint*: If $V(t)$ denotes either side, show that $dV/dt = -(\mathrm{ad}\, X)V$, $V(0) = Y$.)

10. Deduce from Exercise 8 that

$$\frac{d}{ds} e^{X+sY}\Big|_{s=0} = e^X\, \Xi(\mathrm{ad}\, X)Y, \tag{4.44}$$

where $\Xi(z)$ is the entire holomorphic function

$$\Xi(z) = \int_0^1 e^{-\tau z}\, d\tau = \frac{1 - e^{-z}}{z}. \tag{4.45}$$

The operator $\Xi(\mathrm{ad}\, X)$ is defined in the following manner. For any $L \in \mathrm{End}(\mathbb{C}^m) = M_m$, any function $F(z)$ holomorphic on $|z| < a$, with $a > \|L\|$, define $F(L)$ by power series:

$$F(L) = \sum_{n=0}^{\infty} f_n L^n, \quad \text{where } F(z) = \sum_{n=0}^{\infty} f_n z^n. \tag{4.46}$$

For further material on holomorphic functions of operators, see §5 in Appendix A.

11. With $\mathrm{Exp} : M_n \to M_n$ as defined above, describe the set of matrices X such that the transformation $D\, \mathrm{Exp}(X)$ is not invertible.

12. Let $A : \mathbb{R}^n \to \mathbb{R}^n$ be symmetric, and let $Q(x) = (Ax, x)$. Let $v_1 \in S^{n-1} = \{x \in \mathbb{R}^n : |x| = 1\}$ be a point where $Q\big|_{S^{n-1}}$ assumes a maximum. Show that v_1 is an eigenvector of A.
(*Hint*: Show that $\nabla Q(v_1)$ is parallel to $\nabla E(v_1)$, where $E(x) = (x, x)$.)
Use this result to give an alternative proof of Corollary 4.6. Extend this argument to establish Theorem 4.5.

5. Variable-coefficient linear systems of ODE: Duhamel's principle

Let $A(t)$ be a continuous, $n \times n$ matrix-valued function of $t \in I$. We consider the general linear, homogeneous ODE

$$\frac{dy}{dt} = A(t)y, \quad y(0) = y_0. \tag{5.1}$$

The general theory of §2 gives local solutions. We claim that the solutions here exist for all $t \in I$. This follows from the discussion after the proof of Theorem 2.1, together with the following *estimate* on the solution to (5.1).

Proposition 5.1. *If $\|A(t)\| \le M$ for $t \in I$, then the solution to (5.1) satisfies*

$$\|y(t)\| \le e^{M|t|}\|y_0\|. \tag{5.2}$$

It suffices to prove this for $t \ge 0$. Then $z(t) = e^{-Mt}y(t)$ satisfies

$$z' = C(t)z, \quad z(0) = y_0, \tag{5.3}$$

with $C(t) = A(t) - M$. Hence $C(t)$ satisfies

$$\text{Re}\,(C(t)u, u) \le 0, \quad \text{for all } u \in \mathbb{C}^n. \tag{5.4}$$

Thus (5.2) is a consequence of the following energy estimate, which is of independent interest.

Proposition 5.2. *If z solves (5.3) and if (5.4) holds for $C(t)$, then*

$$\|z(t)\| \le \|z(0)\|, \quad \textit{for } t \ge 0.$$

Proof. We have

$$\begin{aligned}\frac{d}{dt}\|z(t)\|^2 &= (z'(t), z(t)) + (z(t), z'(t)) \\ &= 2\,\text{Re}\,(C(t)z(t), z(t)) \\ &\le 0.\end{aligned} \tag{5.5}$$

Thus we have global existence for (5.1). There is a matrix-valued function $S(t, s)$ such that the unique solution to (5.1) satisfies

$$y(t) = S(t, s)y(s). \tag{5.6}$$

Using this solution operator, we can treat the nonhomogeneous equation

$$y' = A(t)y + b(t), \quad y(0) = y_0. \tag{5.7}$$

Indeed, direct calculation yields

$$y(t) = S(t, 0)y_0 + \int_0^t S(t, s)b(s)\, ds. \tag{5.8}$$

This identity is known as *Duhamel's principle.*

Next we prove an identity that might be called the "noncommutative fundamental theorem of calculus."

Proposition 5.3. *If $A(t)$ is a continuous matrix function and $S(t, 0)$ is defined as above, then*

$$S(t, 0) = \lim_{n\to\infty} e^{(t/n)A((n-1)t/n)} \cdots e^{(t/n)A(0)}, \tag{5.9}$$

where there are n factors on the right.

Proof. To prove this at $t = T$, divide the interval $[0, T]$ into n equal parts. Set $y = S(t, 0)y_0$, and define $z_n(t)$ by $z_n(0) = y_0$ and

$$z_n' = A(jT/n)z_n, \quad \text{for } t \in \big(jT/n, (j+1)T/n\big), \tag{5.10}$$

requiring continuity across each endpoint of these intervals. We see that

$$z_n' = A(t)z_n + R_n(t), \tag{5.11}$$

with

$$\|R_n(t)\| \le \delta_n \|z_n(t)\|, \quad \delta_n \to 0 \text{ as } n \to \infty. \tag{5.12}$$

Meanwhile we see that $\|z_n(t)\| \le C_T\|y_0\|$ on $[0, T]$. We want to compare $z_n(t)$ and $y(t)$. We have

$$\frac{d}{dt}(z_n - y) = A(t)(z_n - y) + R_n(t); \quad z_n(0) - y(0) = 0. \tag{5.13}$$

Hence Duhamel's principle gives

$$z_n(t) - y(t) = \int_0^t S(t, s)R_n(s)\, ds, \tag{5.14}$$

and since we have an a priori bound $\|S(t, s)\| \le K$ for $|s|, |t| \le T$, we get

$$\|z_n(t) - y(t)\| \le KTC_T\delta_n\|y_0\| \to 0 \text{ as } n \to \infty,\ |t| \le T. \tag{5.15}$$

In particular, $z_n(T) \to y(T)$ as $n \to \infty$. Since $z_n(T)$ is given by the right side of (5.9) with $t = T$, this proves (5.9).

Exercises

1. Let $A(t)$ and $X(t)$ be $n \times n$ matrices satisfying
$$\frac{dX}{dt} = A(t)X.$$
We form the *Wronskian* $W(t) = \det X(t)$. Show that W satisfies the ODE
$$\frac{dW}{dt} = a(t)W, \quad a(t) = \operatorname{Tr} A(t).$$
(*Hint*: Use Exercise 2 of §1 to write $dW/dt = \operatorname{Tr}(\operatorname{Cof}(X)^t dX/dt)$, and use Cramer's formula, $(\det X)X^{-1} = \operatorname{Cof}(X)^t$. *Alternative*: Write $X(t+h) = e^{hA(t)}X(t) + O(h^2)$ and use Exercise 3 of §4 to write $\det e^{hA(t)} = e^{ha(t)}$, hence $W(t+h) = e^{ha(t)}W(t) + O(h^2)$.)
2. Let $u(t) = \|y(t)\|^2$, for a solution y to (5.1). Show that
$$u' \le M(t)u(t), \tag{5.16}$$
provided $\|A(t)\| \le M(t)/2$. Such a differential inequality implies the integral inequality
$$u(t) \le A + \int_0^t M(s)u(s)\, ds, \quad t \ge 0, \tag{5.17}$$
with $A = u(0)$. The following is a *Gronwall inequality*; namely, if (5.17) holds for a real-valued function u, then provided $M(s) \ge 0$, we have, for $t \ge 0$,
$$u(t) \le Ae^{N(t)}, \quad N(t) = \int_0^t M(s)\, ds. \tag{5.18}$$
Prove this. Note that the quantity dominating $u(t)$ in (5.18) is equal to U, solving $U(0) = A,\ dU/dt = M(t)U(t)$.
3. Generalize the Gronwall inequality of Exercise 2 as follows. Assume $F(t,u)$ and $\partial_u F(t,u)$ are continuous, let U be a real-valued solution to
$$U' = F(t,U), \quad U(0) = A, \tag{5.19}$$
and let u satisfy the integral inequality
$$u(t) \le A + \int_0^t F(s, u(s))\, ds. \tag{5.20}$$
Then prove that
$$u(t) \le U(t), \quad \text{for } t \ge 0, \tag{5.21}$$
provided $\partial F/\partial u \ge 0$. Show that this continues to hold if we replace (5.19) by
$$U(t) \ge A + \int_0^t F(s, U(s))\, ds. \tag{5.19a}$$
(*Hint*: Set $v = u - U$. Then (5.19a) and (5.20) imply
$$v(t) \le \int_0^t \big[F(s,u(s)) - F(s,U(s))\big]\, ds = \int_0^t M(s)v(s)\, ds,$$
where
$$M(s) = \int_0^1 F_u\big(s, \tau u(s) + (1-\tau)U(s)\big)\, d\tau.$$
Thus (5.17) applies, with $A = 0$.)

4. Let $x(t)$ be a smooth curve in $\mathbb{R}^3$; assume it is parameterized by arc length, so $T(t) = x'(t)$ has unit length; $T(t) \cdot T(t) = 1$. Differentiating, we have $T'(t) \perp T(t)$. The *curvature* is defined to be $\kappa(t) = \|T'(t)\|$. If $\kappa(t) \neq 0$, we set $N(t) = T'/\|T'\|$, so

$$T' = \kappa N,$$

and N is a unit vector orthogonal to T. We define $B(t)$ by

$$B = T \times N. \tag{5.22}$$

Note that (T, N, B) form an orthonormal basis of $\mathbb{R}^3$ for each t, and

$$T = N \times B, \quad N = B \times T. \tag{5.23}$$

By (5.22) we have $B' = T \times N'$. Deduce that B' is orthogonal to both T and B, hence parallel to N. We set

$$B' = -\tau N,$$

for smooth $\tau(t)$, called the *torsion*.

5. From $N' = B' \times T + B \times T'$ and the formulas for T' and B' given in Exercise 4, deduce the following system, called the *Frenet-Serret formula*:

$$\begin{aligned} T' &= \qquad\quad \kappa N \\ N' &= -\kappa T \qquad\qquad + \tau B \\ B' &= \qquad\quad -\tau N \end{aligned} \tag{5.24}$$

Form the 3×3 matrix

$$A(t) = \begin{pmatrix} 0 & -\kappa & 0 \\ \kappa & 0 & -\tau \\ 0 & \tau & 0 \end{pmatrix}, \tag{5.25}$$

and deduce that the 3×3 matrix $F(t)$ whose columns are T, N, B,

$$F = (T, N, B),$$

satisfies the ODE

$$F' = F\,A(t).$$

6. Derive the following *converse* to the Frenet-Serret formula. Let $T(0)$, $N(0)$, and $B(0)$ be an orthonormal set in $\mathbb{R}^3$, such that $B(0) = T(0) \times N(0)$; let $\kappa(t)$ and $\tau(t)$ be given smooth functions; and solve the system (5.24). Show that there is a unique curve $x(t)$ such that $x(0) = 0$ and $T(t)$, $N(t)$, and $B(t)$ are associated to $x(t)$ by the construction in Exercise 4, so in particular the curve has curvature $\kappa(t)$ and torsion $\tau(t)$.
(*Hint*: To prove that (5.22) and (5.23) hold for all t, consider the next exercise.)

7. Let $A(t)$ be a smooth, $n \times n$ real matrix function that is *skew-adjoint* for all t (of which (5.25) is an example). Suppose $F(t)$ is a real $n \times n$ matrix function satisfying

$$F' = F\,A(t).$$

If $F(0)$ is an orthogonal matrix, show that $F(t)$ is orthogonal for all t.
(*Hint*: Set $J(t) = F(t)^* F(t)$. Show that $J(t)$ and $J_0(t) = I$ both solve the initial-value problem

$$J' = [J, A(t)], \quad J(0) = I.)$$

8. Let $U_1 = T$, $U_2 = N$ and $U_3 = B$, and set $\omega(t) = \tau T + \kappa B$. Show that (5.24) is equivalent to $U_j' = \omega \times U_j$, $1 \le j \le 3$.
9. Suppose τ and κ are constant. Show that ω is constant, so $T(t)$ satisfies the constant-coefficient ODE

$$T'(t) = \omega \times T(t).$$

Note that $\omega \cdot T(0) = \tau$. Show that after a translation and rotation, $x(t)$ takes the form

$$\gamma(t) = \big(\lambda^{-2}\kappa \cos \lambda t, \lambda^{-2}\kappa \sin \lambda t, \lambda^{-1}\tau t\big), \quad \lambda^2 = \kappa^2 + \tau^2.$$

6. Dependence of solutions on initial data and on other parameters

We consider how a solution to an ODE depends on the initial conditions. Consider a nonlinear system

$$y' = F(y), \quad y(0) = x. \tag{6.1}$$

As noted in §2, we can consider an autonomous system, such as (6.1), without loss of generality. Suppose $F : U \to \mathbb{R}^n$ is smooth, $U \subset \mathbb{R}^n$ open; for simplicity we assume U is convex. Say $y = y(t, x)$. We want to examine smoothness in x.

Note that *formally* differentiating (6.1) with respect to x suggests that $W = D_x y(t, x)$ satisfies an ODE called the *linearization* of (6.1):

$$W' = DF(y)W, \quad W(0) = I. \tag{6.2}$$

In other words, $w(t, x) = D_x y(t, x) w_0$ satisfies

$$w' = DF(y)w, \quad w(0) = w_0. \tag{6.3}$$

To justify this, we want to compare $w(t)$ and

$$z(t) = y_1(t) - y(t) = y(t, x + w_0) - y(t, x). \tag{6.4}$$

It would be convenient to show that z satisfies an ODE similar to (6.3). Indeed, $z(t)$ satisfies

$$z' = F(y_1) - F(y) = \Phi(y_1, y)z, \quad z(0) = w_0, \tag{6.5}$$

where

$$\Phi(y_1, y) = \int_0^1 DF\big(\tau y_1 + (1 - \tau)y\big)\, d\tau. \tag{6.6}$$

If we assume that

$$\|DF(u)\| \le M, \quad \text{for } u \in U, \tag{6.7}$$

then the solution operator $S(t, 0)$ of the linear ODE $d/dt - B(t)$, with $B(y) = \Phi(y_1(t), y(t))$, satisfies a bound $\|S(t, 0)\| \le e^{|t|M}$ as long as $y(t)$ and $y_1(t)$ belong to U. Hence

$$\|y_1(t) - y(t)\| \le e^{|t|M} \|w_0\|. \tag{6.8}$$

This establishes that $y(t,x)$ is *Lipschitz* in x.

To continue, since $\Phi(y,y) = DF(y)$, we rewrite (6.5) as

$$z' = \Phi(y+z,y)z = DF(y)z + R(y,z), \quad w(0) = w_0, \tag{6.9}$$

where

$$F \in C^1(U) \Longrightarrow \|R(y,z)\| = o(\|z\|) = o(\|w_0\|). \tag{6.10}$$

Now comparing (6.9) with (6.3), we have

$$\frac{d}{dt}(z-w) = DF(y)(z-w) + R(y,z), \quad (z-w)(0) = 0. \tag{6.11}$$

Then Duhamel's principle yields

$$z(t) - w(t) = \int_0^t S(t,s)R\big(y(s),z(s)\big)\,ds, \tag{6.12}$$

so by the bound $\|S(t,s)\| \le e^{|t-s|M}$ and (6.10), we have

$$z(t) - w(t) = o(\|w_0\|). \tag{6.13}$$

This is precisely what is required to show that $y(t,x)$ is differentiable with respect to x, with derivative $W = D_x y(t,x)$ satisfying (6.2). We state our first result.

Proposition 6.1. *If $F \in C^1(U)$, and if solutions to (6.1) exist for $t \in (-T_0, T_1)$, then for each such t, $y(t,x)$ is C^1 in x, with derivative $D_x y(t,x) = W(t,x)$ satisfying (6.2).*

So far we have shown that $y(t,x)$ is both Lipschitz and differentiable in x, but the continuity of $W(t,x)$ in x follows easily by comparing the ODEs of the form (6.2) for $W(t,x)$ and $W(t,x+w_0)$, in the spirit of the analysis of (6.11).

If F possesses further smoothness, we can obtain higher differentiability of $y(t,x)$ in x by the following trick. *Couple* (6.1) and (6.2), to get an ODE for (y,W):

$$\begin{aligned} y' &= F(y), \\ W' &= DF(y)W, \end{aligned} \tag{6.14}$$

with initial conditions

$$y(0) = x, \quad W(0) = I. \tag{6.15}$$

We can reiterate the preceeding argument, getting results on $D_x(y,W)$, that is, on $D_x^2 y(t,x)$, and continue, proving:

Proposition 6.2. *If $F \in C^k(U)$, then $y(t,x)$ is C^k in x.*

Similarly, we can consider dependence of the solution to a system of the form

$$\frac{dy}{dt} = F(\tau, y), \quad y(0) = x \tag{6.16}$$

on a parameter τ, assuming F is smooth jointly in τ, y. This result can be deduced from the previous one by the following trick: Consider the ODE

$$y' = F(z, y),\ z' = 0; \quad y(0) = x,\ z(0) = \tau. \tag{6.17}$$

Thus we get smoothness of $y(t, \tau, x)$ in (τ, x). As one special case, let $F(\tau, y) = \tau F(y)$. In this case $y(t_0, \tau, x) = y(\tau t_0, 1, x)$, so we can improve the conclusion of Proposition 6.2 to the following:

$$F \in C^k(U) \Longrightarrow y \in C^k \text{ jointly in } (t, x). \tag{6.18}$$

It is also true that if F is analytic, then one has the analytic dependence of solutions on parameters, especially on t, so that power-series techniques work in that case. One approach to the proof of this is given in the exercises below, and another at the end of §9.

Exercises

1. Let Ω be open in $\mathbb{R}^{2n}$, identified with $\mathbb{C}^n$, via $z = x + iy$. Let $X : \Omega \to \mathbb{R}^{2n}$ have components $X = (a_1, \dots, a_n, b_1, \dots, b_n)$, where $a_j(x, y)$ and $b_j(x, y)$ are real-valued. Denote the solution to $du/dt = X(u)$, $u(0) = z$ by $u(t, z)$. Assume $f_j(z) = a_j(z) + ib_j(z)$ is holomorphic in z, that is, its derivative commutes with J, acting on $\mathbb{R}^{2k} = \mathbb{C}^k$ as multiplication by i. Show that, for each t, $u(t, z)$ is holomorphic in z, that is, $D_z u(t, z)$ commutes with J.
(*Hint*: Use the linearized equation (6.2) to show that $K(t) = [W(t), J]$ satisfies the ODE
$$K' = DX(z)K, \quad K(0) = 0.)$$
2. If $\mathcal{O} \subset \mathbb{R}^n$ is open and $F : \mathcal{O} \to \mathbb{R}^n$ is real analytic, show that the solution $y(t, x)$ to (6.1) is real analytic in x.
(*Hint*: With $F = (a_1, \dots, a_n)$, take holomorphic extensions $f_j(z)$ of $a_j(x)$ and use Exercise 1.)
Using the trick leading to (6.18), show that $y(t, x)$ is real analytic jointly in (t, x).

In the next set of problems, consider a linear ODE of the form

$$A(x)\frac{du}{dx} = B(x)u, \quad 0 < x < 1, \tag{6.19}$$

where we assume that the $n \times n$ matrix functions A and B have holomorphic extensions to $\Delta = \{z \in \mathbb{C} : |z| < 1\}$, such that $\det A(z) = 0$ at $z = 0$, but at no other point of Δ. We say $z = 0$ is a singular point. Let $u_1(x), \dots, u_n(x)$ be n linearly independent solutions to (6.19), obtained, for example, by specifying u at $x = 1/2$.

3. Show that each u_j has a unique holomorphic extension to the universal covering surface $\mathcal{M}$ of $\Delta \setminus 0$, and show that there are $c_{jk} \in \mathbb{C}$ such that
$$u_j(e^{2\pi i}x) = \sum_k c_{jk}\, u_k(x), \quad 0 < x < 1.$$

4. Suppose the matrix $C = (c_{jk})$ is diagonalizable, with eigenvalues $\lambda_\ell \in \mathbb{C},\ 1 \le \ell \le n$. Show that there is a basis of solutions v_ℓ to (6.19) such that

$$v_\ell(e^{2\pi i}x) = \lambda_\ell\, v_\ell(x),$$

and hence, picking $\alpha_\ell \in \mathbb{C}$ such that $e^{2\pi i\alpha_\ell} = \lambda_\ell$,

$$v_\ell(x) = x^{\alpha_\ell} w_\ell(x); \quad w_\ell \text{ holomorphic on } \Delta \setminus 0.$$

5. Suppose $\|A(z)^{-1}B(z)\| \le K|z|^{-1}$. Show that $\|v_\ell(z)\| \le C|z|^{-K}$. Deduce that each $w_\ell(z)$ has at most a pole at $z = 0$; hence, shifting α_ℓ by an integer, we can assume that w_ℓ is holomorphic on Δ. (*Hint*: Recall the statement of Gronwall's inequality, in Exercises 2 and 3 of §5.)
6. Suppose that instead of C being diagonalizable, it has the Jordan normal form

$$\begin{pmatrix} \lambda & 1 \\ 0 & \lambda \end{pmatrix}$$

(in case $n = 2$). What can you say? Generalize.
7. If $a(z)$ and $b(z)$ are holomorphic on Δ, convert

$$x^2u''(x) + xa(x)u'(x) + b(x)u(x) = 0$$

to a first-order system to which Exercises 3–6 apply. (*Hint.* Take $v = xu'$ rather than $v = u'$.)

The next set of exercises deals with certain small perturbations of the system $\dot{x} = -y,\ \dot{y} = x$, whose solution curves are circles centered at the origin.

8. Let $x = x_\varepsilon(t),\ y = y_\varepsilon(t)$ solve

$$\dot{x} = -y + \varepsilon(x^2 + y^2), \quad \dot{y} = x,$$

with initial data $x(0) = 1,\ y(0) = 0$. Knowing smooth dependence on ε, find ODEs for the coefficients $x_j(t),\ y_j(t)$ in power-series expansions

$$x(t) = x_0(t) + \varepsilon x_1(t) + \varepsilon^2 x_2(t) + \cdots, \quad y(t) = y_0(t) + \varepsilon y_1(t) + \varepsilon^2 y_2(t) + \cdots.$$

9. Making use of the substitution $\xi(t) = -x(-t),\ \eta(t) = y(-t)$, show that, for fixed initial data and ε sufficiently small, the orbits of the ODE in Exercise 8 are periodic.
10. Show that, for ε small, the period of the orbit in Exercise 8 is a smooth function of ε. Compute the first three terms in its power-series expansion.

7. Flows and vector fields

Let $U \subset \mathbb{R}^n$ be open. A vector field on U is a smooth map

$$X : U \longrightarrow \mathbb{R}^n. \tag{7.1}$$

Consider the corresponding ODE:

$$y' = X(y), \quad y(0) = x, \tag{7.2}$$

with $x \in U$. A curve $y(t)$ solving (7.2) is called an integral curve of the vector field X. It is also called an *orbit*. For fixed t, write

$$y = y(t, x) = \mathcal{F}_X^t(x). \tag{7.3}$$

The locally defined $\mathcal{F}_X^t$, mapping (a subdomain of) U to U, is called the *flow* generated by the vector field X.

The vector field X defines a differential operator on scalar functions, as follows:

$$\mathcal{L}_X f(x) = \lim_{h\to 0} h^{-1}\big[f(\mathcal{F}_X^h x) - f(x)\big] = \frac{d}{dt} f(\mathcal{F}_X^t x)\big|_{t=0}. \tag{7.4}$$

We also use the common notation

$$\mathcal{L}_X f(x) = Xf, \tag{7.5}$$

that is, we apply X to f as a first-order differential operator.

Note that if we apply the chain rule to (7.4) and use (7.2), we have

$$\mathcal{L}_X f(x) = X(x)\cdot \nabla f(x) = \sum a_j(x)\frac{\partial f}{\partial x_j}, \tag{7.6}$$

if $X = \sum a_j(x)e_j$, with $\{e_j\}$ the standard basis of $\mathbb{R}^n$. In particular, using the notation (7.5), we have

$$a_j(x) = Xx_j. \tag{7.7}$$

In the notation (7.5),

$$X = \sum a_j(x)\frac{\partial}{\partial x_j}. \tag{7.8}$$

We note that X is a *derivation*, that is, a map on $C^\infty(U)$, linear over $\mathbb{R}$, satisfying

$$X(fg) = (Xf)g + f(Xg). \tag{7.9}$$

Conversely, any derivation on $C^\infty(U)$ defines a vector field, namely, has the form (7.8), as we now show.

Proposition 7.1. *If X is a derivation on $C^\infty(U)$, then X has the form (7.8).*

Proof. Set $a_j(x) = Xx_j$, $X^\# = \sum a_j(x)\partial/\partial x_j$, and $Y = X - X^\#$. Then Y is a derivation satisfying $Yx_j = 0$ for each j; we aim to show that $Yf = 0$ for all f. Note that whenever Y is a derivation,

$$1\cdot 1 = 1 \Rightarrow Y\cdot 1 = 2Y\cdot 1 \Rightarrow Y\cdot 1 = 0,$$

that is, Y annihilates constants. Thus, in this case Y annihilates all polynomials of degree ≤ 1.

Now we show that $Yf(p) = 0$ for all $p \in U$. Without loss of generality, we can suppose $p = 0$, the origin. Then, by (1.8), we can take $b_j(x) = \int_0^1 (\partial_j f)(tx)\,dt$, and write

$$f(x) = f(0) + \sum b_j(x)x_j.$$

It immediately follows that Yf vanishes at 0, so the proposition is proved.

If U is a manifold, it is natural to regard a vector field X as a section of the tangent bundle of U, as explained in Appendix B. Of course, the characterization given in Proposition 7.1 makes good invariant sense on a manifold.

A fundamental fact about vector fields is that they can be "straightened out" near points where they do not vanish. To see this, suppose a smooth vector field X is given on U such that, for a certain $p \in U$, $X(p) \neq 0$. Then near p there is a hypersurface M that is nowhere tangent to X. We can choose coordinates near p so that p is the origin and M is given by $\{x_n = 0\}$. Thus, we can identify a point $x' \in \mathbb{R}^{n-1}$ near the origin with $x' \in M$. We can define a map

$$\mathcal{F} : M \times (-t_0, t_0) \longrightarrow U \tag{7.10}$$

by

$$\mathcal{F}(x', t) = \mathcal{F}_X^t(x'). \tag{7.11}$$

This is C^∞ and has surjective derivative and so by the inverse function theorem is a local diffeomorphism. This defines a new coordinate system near p, in which the flow generated by X has the form

$$\mathcal{F}_X^s(x', t) = (x', t + s). \tag{7.12}$$

If we denote the new coordinates by $(u_1, \dots, u_n)$, we see that the following result is established.

Theorem 7.2. *If X is a smooth vector field on U with $X(p) \neq 0$, then there exists a coordinate system $(u_1, \dots, u_n)$ centered at p (so $u_j(p) = 0$) with respect to which*

$$X = \frac{\partial}{\partial u_n}. \tag{7.13}$$

We now make some elementary comments on vector fields in the plane. Here the object is to find the integral curves of

$$f(x, y)\frac{\partial}{\partial x} + g(x, y)\frac{\partial}{\partial y}, \tag{7.14}$$

that is, to solve

$$x' = f(x, y), \quad y' = g(x, y). \tag{7.15}$$

This implies

$$\frac{dy}{dx} = \frac{g(x, y)}{f(x, y)}, \tag{7.16}$$

or, written in differential-form notation (which will be discussed more thoroughly in §13),

$$g(x, y)\, dx - f(x, y)\, dy = 0. \tag{7.17}$$

Suppose we manage to find an explicit solution to (7.16):

$$y = \varphi(x), \quad x = \psi(y). \tag{7.18}$$

Often it is not feasible to do so, but ODE texts frequently give methods for doing so in some cases. Then the original system becomes

$$x' = f(x, \varphi(x)), \quad y' = g(\psi(y), y). \tag{7.19}$$

In other words, we have reduced ourselves to integrating vector fields on the line. We have

$$\begin{aligned} \int \left[f(x, \varphi(x))\right]^{-1} dx &= t + C_1, \\ \int \left[g(\psi(y), y)\right]^{-1} dy &= t + C_2. \end{aligned} \tag{7.20}$$

If (7.18) can be explicitly achieved, it may be that one integral or the other in (7.20) is easier to evaluate. With either x or y solved as a function of t, the other is determined by (7.18).

One case when the planar vector field can be integrated explicitly (locally) is when there is a smooth u, with nonvanishing gradient, explicitly given, such that

$$Xu = 0, \tag{7.21}$$

where X is the vector field (7.14). One says u is a *conserved quantity*. In such a case, let w be any smooth function such that (u, w) form a local coordinate system. In this coordinate system,

$$X = b(u, w)\frac{\partial}{\partial w} \tag{7.22}$$

by (7.7), so

$$Xv = 1, \tag{7.23}$$

with

$$v(u, w) = \int_{w_0}^{w} b(u, s)^{-1}\, ds, \tag{7.24}$$

and the local coordinate system (u, v) linearizes X.

Exercises

1. Suppose $h(x, y)$ is homogeneous of degree 0, that is, $h(rx, ry) = h(x, y)$, so $h(x, y) = k(x/y)$. Show that the ODE

$$\frac{dy}{dx} = h(x, y)$$

is changed to a separable ODE for $u = u(x)$, if $u = y/x$.

2. Using Exercise 1, discuss constructing the integral curves of a vector field
$$X = f(x, y)\frac{\partial}{\partial x} + g(x, y)\frac{\partial}{\partial y}$$
when $f(x, y)$ and $g(x, y)$ are homogeneous of degree a, that is,
$$f(rx, ry) = r^a\, f(x, y) \ \text{ for } r > 0,$$
and similarly for g.
3. Describe the integral curves of
$$(x^2 + y^2)\frac{\partial}{\partial x} + xy\frac{\partial}{\partial y}.$$
4. Describe the integral curves of
$$A(x, y)\frac{\partial}{\partial x} + B(x, y)\frac{\partial}{\partial y}$$
when $A(x, y) = a_1x + a_2y + a_3$, $B(x, y) = b_1x + b_2y + b_3$.
5. Let $X = f(x, y)(\partial/\partial x) + g(x, y)(\partial/\partial y)$ be a vector field on a disc $\Omega \subset \mathbb{R}^2$. Suppose that div $X = 0$, that is, $\partial f/\partial x + \partial g/\partial y = 0$. Show that a function $u(x, y)$ such that
$$\frac{\partial u}{\partial x} = g, \quad \frac{\partial u}{\partial y} = -f$$
is given by a line integral. Show that $Xu = 0$, and hence integrate X. Reconsider this problem after reading §13.
6. Find the integral curves of the vector field
$$X = (2xy + y^2 + 1)\frac{\partial}{\partial x} + (x^2 + 1 - y^2)\frac{\partial}{\partial y}.$$
7. Show that
$$\operatorname{div}(e^v X) = e^v(\operatorname{div} X + Xv).$$
Hence, if X is a vector field on $\Omega \subset \mathbb{R}^2$, as in Exercise 5, show that you can integrate X if you can construct a function $v(x, y)$ such that $Xv = -\operatorname{div} X$. Construct such v if either
$$\frac{\operatorname{div} X}{f(x, y)} = \varphi(x) \quad \text{or} \quad \frac{\operatorname{div} X}{g(x, y)} = \psi(y).$$
For now, we define div $X = \partial X_1/\partial x_1 + \cdots + \partial X_n/\partial x_n$. See Chapter 2, §2, for another definition.
8. Find the integral curves of the vector field
$$X = 2xy\frac{\partial}{\partial x} + (x^2 + y^2 - 1)\frac{\partial}{\partial y}.$$

Let X be a vector field on $\mathbb{R}^n$, with a critical point at 0, that is, $X(0) = 0$. Suppose that for $x \in \mathbb{R}^n$ near 0,

$$X(x) = Ax + R(x), \quad \|R(x)\| = O(\|x\|^2), \tag{7.25}$$

where A is an $n \times n$ matrix. We call Ax the linearization of X at 0.
9. Suppose all the eigenvalues of A have negative real part. Construct a quadratic polynomial $Q : \mathbb{R}^n \to [0, \infty)$, such that $Q(0) = 0$, $\left(\partial^2 Q/\partial x_j \partial x_k\right)$ is positive-definite,

and for any integral curve $x(t)$ of X as in (7.25),

$$\frac{d}{dt} Q(x(t)) < 0 \ \text{ if } t \geq 0,$$

provided $x(0) = x_0 (\neq 0)$ is close enough to 0. Deduce that for small enough C, if $\|x_0\| \leq C$, then $x(t)$ exists for all $t \geq 0$ and $x(y) \to 0$ as $t \to \infty$.
(*Hint*: Take $Q(x) = \langle x, x \rangle$, using Exercise 10 below.)

10. Let A be an $n \times n$ matrix, all of whose eigenvalues λ_j have *negative* real part. Show that there exists a Hermitian inner product $\langle \, , \rangle$ on $\mathbb{C}^n$ such that Re $\langle Au, u \rangle < 0$ for nonzero $u \in \mathbb{C}^n$. (*Hint*: Put A in Jordan normal form, but with εs instead of 1s above the diagonal, where ε is small compared with $|\text{Re}\, \lambda_j|$.)

8. Lie brackets

If $F : V \to W$ is a diffeomorphism between two open domains in $\mathbb{R}^n$, or between two smooth manifolds, and Y is a vector field on W, we define a vector field $F_\# Y$ on V so that

$$\mathcal{F}^t_{F_\# Y} = F^{-1} \circ \mathcal{F}^t_Y \circ F, \tag{8.1}$$

or equivalently, by the chain rule,

$$F_\# Y(x) = \big(DF^{-1}\big)\big(F(x)\big) Y\big(F(x)\big). \tag{8.2}$$

In particular, if $U \subset \mathbb{R}^n$ is open and X is a vector field on U defining a flow $\mathcal{F}^t$, then for a vector field Y, $\mathcal{F}^t_\# Y$ is defined on most of U, for $|t|$ small, and we can define the Lie derivative,

$$\mathcal{L}_X Y = \lim_{h \to 0} h^{-1}\big(\mathcal{F}^h_\# Y - Y\big) = \frac{d}{dt} \mathcal{F}^t_\# Y\big|_{t=0}, \tag{8.3}$$

as a vector field on U.

Another natural construction is the operator-theoretic bracket:

$$[X, Y] = XY - YX, \tag{8.4}$$

where the vector fields X and Y are regarded as first-order differential operators on $C^\infty(U)$. One verifies that (8.4) defines a derivation on $C^\infty(U)$, hence a vector field on U. The basic elementary fact about the Lie bracket is the following.

Theorem 8.1. *If X and Y are smooth vector fields, then*

$$\mathcal{L}_X Y = [X, Y]. \tag{8.5}$$

Proof. Let us first verify the identity in the special case

$$X = \frac{\partial}{\partial x_1}, \quad Y = \sum b_j(x) \frac{\partial}{\partial x_j}.$$

Then $\mathcal{F}^t_\# Y = \sum b_j(x + te_1)\, \partial/\partial x_j$, so $\mathcal{L}_X Y = \sum (\partial b_j / \partial x_1)\, \partial / \partial x_j$, and a straightforward calculation shows that this is also the formula for $[X, Y]$, in this case.

Now we verify (8.5) in general, at any point $x_0 \in U$. First, if X is nonvanishing at x_0, we can choose a local coordinate system so the example above gives the identity. By continuity, we get the identity (8.5) on the closure of the set of points x_0, where $X(x_0) \neq 0$. Finally, if x_0 has a neighborhood where $X = 0$, clearly $\mathcal{L}_X Y = 0$ and $[X, Y] = 0$ at x_0. This completes the proof.

Corollary 8.2. *If X and Y are smooth vector fields on U, then*

$$\frac{d}{dt}\mathcal{F}^t_{X\#}Y = \mathcal{F}^t_{X\#}[X, Y], \tag{8.6}$$

for all t.

Proof. Since locally $\mathcal{F}^{t+s}_X = \mathcal{F}^s_X \mathcal{F}^t_X$, we have the same identity for $\mathcal{F}^{t+s}_{X\#}$, which yields (8.6) upon taking the s-derivative.

We make some further comments about cases when one can explicitly integrate a vector field X in the plane, exploiting "symmetries" that may be apparent. In fact, suppose one has in hand a vector field Y such that

$$[X, Y] = 0. \tag{8.7}$$

By (8.6), this implies $\mathcal{F}^t_{Y\#}X = X$ for all t; this connection will be pursued further in the next section. Suppose that one has an explicit hold on the flow generated by Y, so one can produce explicit local coordinates (u, v) with respect to which

$$Y = \frac{\partial}{\partial u}. \tag{8.8}$$

In this coordinate system, write $X = a(u, v)\partial/\partial u + b(u, v)\partial/\partial v$. The condition (8.7) implies $\partial a/\partial u = 0 = \partial b/\partial u$, so in fact we have

$$X = a(v)\frac{\partial}{\partial u} + b(v)\frac{\partial}{\partial v}. \tag{8.9}$$

Integral curves of (8.9) satisfy

$$u' = a(v), \quad v' = b(v) \tag{8.10}$$

and can be found explicitly in terms of integrals; one has

$$\int b(v)^{-1}\, dv = t + C_1 \tag{8.11}$$

and then

$$u = \int a(v(t))\, dt + C_2. \tag{8.12}$$

More generally than (8.7), we can suppose that, for some constant c,

$$[X, Y] = cX, \tag{8.13}$$

which by (8.6) is the same as

$$\mathcal{F}^t_{Y\#}X = e^{-ct}X. \tag{8.14}$$

An example would be

$$X = f(x,y)\,\frac{\partial}{\partial x} + g(x,y)\,\frac{\partial}{\partial y}, \tag{8.15}$$

where f and g satisfy "homogeneity" conditions of the form

$$f(r^a x, r^b y) = r^{a-c} f(x,y), \quad g(r^a x, r^b y) = r^{b-c} g(x,y), \tag{8.16}$$

for $r > 0$; in such a case one can take explicitly

$$\mathcal{F}_Y^t(x,y) = (e^{at}x, e^{bt}y). \tag{8.17}$$

Now, if one again has (8.8) in a local coordinate system (u, v), then X must have the form

$$X = e^{cu}\Big[a(v)\frac{\partial}{\partial u} + b(v)\frac{\partial}{\partial v}\Big], \tag{8.18}$$

which can be explicitly integrated, since

$$u' = e^{cu}a(v), \ v' = e^{cu}b(v) \Longrightarrow \frac{du}{dv} = \frac{a(v)}{b(v)}. \tag{8.19}$$

The hypothesis (8.13) implies that the linear span (over $\mathbb{R}$) of X and Y is a two-dimensional, solvable Lie algebra. Sophus Lie devoted a good deal of effort to examining when one could use constructions of solvable Lie algebras of vector fields to integrate vector fields explicitly; his investigations led to his foundation of what is now called the theory of Lie groups.

Exercises

1. Verify that the bracket (8.4) satisfies the "Jacobi identity"

$$[X, [Y, Z]] - [Y, [X, Z]] = [[X, Y], Z],$$

i.e.,

$$[\mathcal{L}_X, \mathcal{L}_Y]Z = \mathcal{L}_{[X,Y]}Z.$$

2. Find the integral curves of

$$X = (x + y^2)\,\frac{\partial}{\partial x} + y\,\frac{\partial}{\partial y}$$

using (8.16).

3. Find the integral curves of

$$X = (x^2y + y^5)\,\frac{\partial}{\partial x} + (x^2 + xy^2 + y^4)\,\frac{\partial}{\partial y}.$$

9. Commuting flows; Frobenius's theorem

Let $G : U \to V$ be a diffeomorphism. Recall from §8 the action on vector fields:

$$G_\# Y(x) = DG(y)^{-1} Y(y), \quad y = G(x). \tag{9.1}$$

As noted there, an alternative characterization of $G_\# Y$ is given in terms of the flow it generates. One has

$$\mathcal{F}_Y^t \circ G = G \circ \mathcal{F}_{G_\# Y}^t. \tag{9.2}$$

The proof of this is a direct consequence of the chain rule. As a special case, we have the following

Proposition 9.1. *If $G_\# Y = Y$, then $\mathcal{F}_Y^t \circ G = G \circ \mathcal{F}_Y^t$.*

From this, we derive the following condition for a pair of flows to commute. Let X and Y be vector fields on U.

Proposition 9.2. *If X and Y commute as differential operators, that is,*

$$[X, Y] = 0, \tag{9.3}$$

then locally $\mathcal{F}_X^s$ and $\mathcal{F}_Y^t$ commute; in other words, for any $p_0 \in U$, there exists a $\delta > 0$ such that for $|s|, |t| < \delta$,

$$\mathcal{F}_X^s \mathcal{F}_Y^t p_0 = \mathcal{F}_Y^t \mathcal{F}_X^s p_0. \tag{9.4}$$

Proof. By Proposition 9.1, it suffices to show that $\mathcal{F}_{X\#}^s Y = Y$. This clearly holds at $s = 0$. But by (8.6), we have

$$\frac{d}{ds} \mathcal{F}_{X\#}^s Y = \mathcal{F}_{X\#}^s [X, Y],$$

which vanishes if (9.3) holds. This finishes the proof.

We have stated that given (9.3), the identity (9.4) holds locally. If the flows generated by X and Y are not complete, this can break down globally. For example, consider $X = \partial/\partial x_1$, $Y = \partial/\partial x_2$ on $\mathbb{R}^2$, which satisfy (9.3) and generate commuting flows. These vector fields lift to vector fields on the universal covering surface $\tilde{M}$ of $\mathbb{R}^2 \setminus (0, 0)$, which continue to satisfy (9.3). The flows on $\tilde{M}$ do not commute globally. This phenomenon does not arise, for example, for vector fields on a compact manifold.

We now consider when a family of vector fields has a multidimensional integral manifold. Suppose $X_1, \dots, X_k$ are smooth vector fields on U which are linearly independent at each point of a k-dimensional surface $\Sigma \subset U$. If each X_j is tangent to Σ at each point, Σ is said to be an integral manifold of $(X_1, \dots, X_k)$.

Proposition 9.3. *Suppose $X_1, \dots, X_k$ are linearly independent at each point of U and $[X_j, X_\ell] = 0$ for all j, ℓ. Then, for each $x_0 \in U$, there is a k-dimensional integral manifold of $(X_1, \dots, X_k)$ containing x_0.*

Proof. We define a map $F : V \to U$, V a neighborhood of 0 in $\mathbb{R}^k$, by

$$F(t_1, \dots, t_k) = \mathcal{F}_{X_1}^{t_1} \cdots \mathcal{F}_{X_k}^{t_k} x_0. \tag{9.5}$$

Clearly, $(\partial/\partial t_1)F = X_1(F)$. Similarly, since $\mathcal{F}_{X_j}^{t_j}$ all commute, we can put any $\mathcal{F}_{X_j}^{t_j}$ first and get $(\partial/\partial t_j)F = X_j(F)$. This shows that the image of V under F is an integral manifold containing x_0.

We now derive a more general condition guaranteeing the existence of integral submanifolds. This important result is due to Frobenius. We say $(X_1, \dots, X_k)$ is *involutive* provided that, for each j, ℓ, there are smooth $b_m^{j\ell}(x)$ such that

$$[X_j, X_\ell] = \sum_{m=1}^{k} b_m^{j\ell}(x) X_m. \tag{9.6}$$

The following is Frobenius's theorem.

Theorem 9.4. *If $(X_1, \dots, X_k)$ are C^∞ vector fields on U, linearly independent at each point, and the involutivity condition (9.6) holds, then through each x_0 there is, locally, a unique integral manifold Σ, of dimension k.*

We will give two proofs of this result. First, let us restate the conclusion as follows. There exist local coordinates $(y_1, \dots, y_n)$ centered at x_0 such that

$$\operatorname{span}(X_1, \dots, X_k) = \operatorname{span}\left(\frac{\partial}{\partial y_1}, \dots, \frac{\partial}{\partial y_k}\right). \tag{9.7}$$

First proof. The result is clear for $k = 1$. We will use induction on k. So let the set of vector fields $X_1, \dots, X_{k+1}$ be linearly independent at each point and involutive. Choose a local coordinate system so that $X_{k+1} = \partial/\partial u_1$. Now let

$$Y_j = X_j - (X_j u_1) \frac{\partial}{\partial u_1} \quad \text{for } 1 \le j \le k, \quad Y_{k+1} = \frac{\partial}{\partial u_1}. \tag{9.8}$$

Since in $(u_1, \dots, u_n)$ coordinates, no $Y_1, \dots, Y_k$ involves $\partial/\partial u_1$, neither does any Lie bracket, so

$$[Y_j, Y_\ell] \in \operatorname{span}(Y_1, \dots, Y_k), \quad j, \ell \le k.$$

Thus $(Y_1, \dots, Y_k)$ is involutive. The induction hypothesis implies that there exist local coordinates $(y_1, \dots, y_n)$ such that

$$\operatorname{span}(Y_1, \dots, Y_k) = \operatorname{span}\left(\frac{\partial}{\partial y_1}, \dots, \frac{\partial}{\partial y_k}\right).$$

Now let

$$Z = Y_{k+1} - \sum_{\ell=1}^{k} (Y_{k+1} y_\ell) \frac{\partial}{\partial y_\ell} = \sum_{\ell > k} (Y_{k+1} y_\ell) \frac{\partial}{\partial y_\ell}. \tag{9.9}$$

Since, in the $(u_1, \dots, u_n)$ coordinates, $Y_1, \dots, Y_k$ do not involve $\partial/\partial u_1$, we have

$$[Y_{k+1}, Y_j] \in \operatorname{span}(Y_1, \dots, Y_k).$$

Thus $[Z, Y_j] \in \operatorname{span}(Y_1, \dots, Y_k)$ for $j \le k$, while (9.9) implies that $[Z, \partial/\partial y_j]$ belongs to the span of $(\partial/\partial y_{k+1}, \dots, \partial/\partial y_n)$, for $j \le k$. Thus we have

$$\left[Z, \frac{\partial}{\partial y_j}\right] = 0, \quad j \le k.$$

Proposition 9.3 implies span $(\partial/\partial y_1, \dots, \partial/\partial y_k, Z)$ has an integral manifold through each point, and since this span is equal to the span of $X_1, \dots, X_{k+1}$, the first proof is complete.

Second proof. Let $X_1, \dots, X_k$ be C^∞ vector fields, linearly independent at each point and satisfying the condition (9.6). Choose an $(n-k)$-dimensional surface $\mathcal{O} \subset U$, transverse to $X_1, \dots, X_k$. For V a neighborhood of the origin in $\mathbb{R}^k$, define $\Phi : V \times \mathcal{O} \to U$ by

$$\Phi(t_1, \dots, t_k, x) = \mathcal{F}_{X_1}^{t_1} \cdots \mathcal{F}_{X_k}^{t_k} x. \tag{9.10}$$

We claim that, for x fixed, the image of V in U is a k-dimensional surface Σ tangent to each X_j, at each point of Σ. Note that since $\Phi(0, \dots, t_j, \dots, 0, x) = \mathcal{F}_{X_j}^{t_j} x$, we have

$$\frac{\partial}{\partial t_j} \Phi(0, \dots, 0, x) = X_j(x), \quad x \in \mathcal{O}. \tag{9.11}$$

To establish the claim, it suffices to show that $\mathcal{F}_{X_j \#}^{t} X_\ell$ is a linear combination with coefficients in $C^\infty(U)$ of $X_1, \dots, X_k$. This is accomplished by the following:

Lemma 9.5. *Suppose $[Y, X_j] = \sum_\ell \lambda_{j\ell}(x) X_\ell$, with smooth coefficients $\lambda_{j\ell}(x)$. Then $\mathcal{F}_{Y\#}^t X_j$ is a linear combination of $X_1, \dots, X_k$, with coefficients in $C^\infty(U)$.*

Proof. Denote by Λ the matrix $(\lambda_{j\ell})$, and let $\Lambda(t) = \Lambda(t, x) = (\lambda_{j\ell}(\mathcal{F}_Y^t x))$. Now let $A(t) = A(t, x)$ be the unique solution to the ODE

$$A'(t) = \Lambda(t) A(t), \quad A(0) = I. \tag{9.12}$$

Write $A = (\alpha_{j\ell})$. We claim that

$$\mathcal{F}_{Y\#}^t X_j = \sum_\ell \alpha_{j\ell}(t, x) X_\ell. \tag{9.13}$$

This formula will prove the lemma. Indeed, we have

$$\begin{aligned}\frac{d}{dt}(\mathcal{F}_Y^t)_\# X_j &= (\mathcal{F}_Y^t)_\#[Y, X_j] \\ &= (\mathcal{F}_Y^t)_\# \sum_\ell \lambda_{j\ell} X_\ell \\ &= \sum_\ell (\lambda_{j\ell} \circ \mathcal{F}_Y^t)(\mathcal{F}_{Y\#}^t X_\ell).\end{aligned}$$

Uniqueness of the solution to (9.12) gives (9.13), and we are done.

This completes the second proof of Frobenius's theorem.

Exercises

1. Let Ω be open in $\mathbb{R}^{2n}$, identified with $\mathbb{C}^n$ via $z = x + iy$. Let

$$X = \sum \Big[a_j(x, y) \frac{\partial}{\partial x_j} + b_j(x, y) \frac{\partial}{\partial y_j} \Big]$$

be a vector field on Ω, where $a_j(x, y)$ and $b_j(x, y)$ are real-valued. Form $f_j(z) = a_j(z) + ib_j(z)$. Consider the vector field

$$Y = JX = \sum_j \Big[-b_j(x, y) \frac{\partial}{\partial x_j} + a_j(x, y) \frac{\partial}{\partial y_j} \Big].$$

Show that X and Y commute, that is, $[X, Y] = 0$, provided $f(z)$ is holomorphic, namely if the Cauchy-Riemann equations hold:

$$\frac{\partial a_j}{\partial x_k} = \frac{\partial b_j}{\partial y_k}, \quad \frac{\partial a_j}{\partial y_k} = -\frac{\partial b_j}{\partial x_k}.$$

2. Assuming $f_j(z) = a_j(z) + ib_j(z)$ are holomorphic, show that, for $z \in \Omega$,

$$z(t, s) = \mathcal{F}_X^t \mathcal{F}_Y^s z$$

satisfies $\partial z/\partial s = J \partial z/\partial t$, and hence that $z(t, s)$ is holomorphic in $t + is$.
3. Suppose $a_j(x)$ are real analytic (and real-valued) on $\mathcal{O} \subset \mathbb{R}^n$. Let $X = \sum a_j(x) \partial/\partial x_j$. Show that, for $x \in \mathcal{O}$, $x(t) = \mathcal{F}_X^t x$ is real analytic in t (for t near 0), by applying Exercises 1 and 2.
Compare the proof of this indicated in Exercise 2 of §6.
4. Discuss the *uniqueness* of integral manifolds arising in Theorem 9.4.
5. Let A_j be smooth $m \times m$ matrix-valued functions on $\mathcal{O} \subset \mathbb{R}^n$. Suppose the operators $L_j = \partial/\partial x_j + A_j(x)$, acting on functions with values in $\mathbb{R}^m$, all commute, $1 \le j \le n$. If $p \in \mathcal{O}$, show that there is a solution in a neighborhood of p to

$$L_j u = 0, \quad 1 \le j \le n,$$

with $u(p) \in \mathbb{R}^m$ prescribed.

10. Hamiltonian systems

Hamiltonian systems arise from classical mechanics. As a most basic example, consider the equations of motion that arise from Newton's law $F = ma$, where the force F is given by

$$F = -\operatorname{grad} V(x), \tag{10.1}$$

with V the potential energy. We get the ODE

$$m\,\frac{d^2x}{dt^2} = -\frac{\partial V}{\partial x}. \tag{10.2}$$

We can convert this into a first-order system for (x, ξ), where

$$\xi = m\,\frac{dx}{dt} \tag{10.3}$$

is the momentum. We have

$$\frac{dx}{dt} = \frac{\xi}{m}, \quad \frac{d\xi}{dt} = -\frac{\partial V}{\partial x}. \tag{10.4}$$

Now consider the total energy

$$f(x, \xi) = \frac{1}{2m}\,|\xi|^2 + V(x). \tag{10.5}$$

Note that $\partial f/\partial \xi = \xi/m$ and $\partial f/\partial x = \partial V/\partial x$. Thus (10.4) is of the form

$$\frac{dx_j}{dt} = \frac{\partial f}{\partial \xi_j}, \quad \frac{d\xi_j}{dt} = -\frac{\partial f}{\partial x_j}. \tag{10.6}$$

Hence we're looking for the integral curves of the vector field

$$H_f = \sum_{j=1}^{n} \left[\frac{\partial f}{\partial \xi_j}\frac{\partial}{\partial x_j} - \frac{\partial f}{\partial x_j}\frac{\partial}{\partial \xi_j}\right]. \tag{10.7}$$

For smooth $f(x, \xi)$, we call H_f, defined by (10.7), a Hamiltonian vector field. Note that, directly from (10.7),

$$H_f f = 0. \tag{10.8}$$

A useful notation is the Poisson bracket, defined by

$$\{f, g\} = H_f g. \tag{10.9}$$

One verifies directly from (10.7) that

$$\{f, g\} = -\{g, f\}, \tag{10.10}$$

generalizing (10.8). Also, a routine calculation verifies that

$$[H_f, H_g] = H_{\{f,g\}}. \tag{10.11}$$

As noted at the end of §7, if X is a vector field in the plane and we explicitly have a function u with nonvanishing gradient such that $Xu = 0$, then X can be explicitly integrated. These comments apply to $X = H_f$, $u = f$, when H_f is a planar Hamiltonian vector field. We can rephrase this description as follows. If $x \in \mathbb{R}$, $\xi \in \mathbb{R}$, then integral curves of

$$x' = \frac{\partial f}{\partial \xi}, \quad \xi' = -\frac{\partial f}{\partial x} \tag{10.12}$$

lie on a level set

$$f(x, \xi) = E. \tag{10.13}$$

Suppose that locally this set is described by

$$x = \varphi(\xi) \text{ or } \xi = \psi(x). \tag{10.14}$$

Then we have one of the following ODEs:

$$x' = f_\xi(x, \psi(x)) \text{ or } \xi' = -f_x(\varphi(\xi), \xi), \tag{10.15}$$

and hence we have

$$\int f_\xi\big(x, \psi(x)\big)^{-1} dx = t + C \tag{10.16}$$

or

$$-\int f_x\big(\varphi(\xi), \xi\big)^{-1} d\xi = t + C'. \tag{10.17}$$

Thus, solving (10.12) is reduced to a quadrature, that is, a calculation of an explicit integral, (10.16) or (10.17).

If the planar Hamiltonian vector field H_f arises from describing motion in a force field on a line, via Newton's laws given in (10.2), so that

$$f(x, \xi) = \frac{1}{2m}\xi^2 + V(x), \tag{10.18}$$

then the second curve in (10.14) is

$$\xi = \pm\big[(2m)\big(E - V(x)\big)\big]^{1/2}, \tag{10.19}$$

and the formula (10.16) becomes

$$\pm\Big(\frac{m}{2}\Big)^{1/2} \int \big[E - V(x)\big]^{-1/2} dx = t + C, \tag{10.20}$$

defining x implicitly as a function of t.

In some cases, the integral in (10.20) can be evaluated by elementary means. This includes the trivial case of a constant force, where $V(x) = cx$, and also the case of the "harmonic oscillator" or linearized spring, where $V(x) = cx^2$. It also includes the case of the motion of a rocket in space, along a line through the center of a planet, where $V(x) = -K/|x|$. This gravitational attraction problem for motion in several-dimensional space will be studied further in §§16 and 17. The case $V(x) = -K \cos x$ arises in the analysis of the pendulum (see (12.38)). In that case, (10.20) is an elliptic integral, rather than one that arises in first-year calculus.

For Hamiltonian vector fields in higher dimensions, more effort is required to understand the resulting flows. The notion of complete integrability provides a method of constructing explicit solutions in some cases, as will be discussed in §§16 and 17.

Hamiltonian vector fields arise in the treatment of many problems in addition to those derived from Newton's laws in Cartesian coordinates. In §11 we study the equations of geodesics and then show how they can be transformed to Hamiltonian systems. In §12 this is seen to be a special case of a broad class of variational problems, which lead to Hamiltonian systems, and which also encompass classical mechanics. This variational approach has many convenient features, such as allowing an easy formulation of the equations of motion in arbitrary coordinate systems, a theme that will be developed in a number of subsequent sections.

Exercises

1. Verify that $[H_f, H_g] = H_{\{f,g\}}$.
2. Demonstrate that the Poisson bracket satisfies the Jacobi identity

$$\{f, \{g, h\}\} - \{g, \{f, h\}\} = \{\{f, g\}, h\}. \tag{10.21}$$

   (*Hint*: Use Exercise 1 above and Exercise 1 of §8.)
3. Identifying y and ξ, show that a planar vector field $X = f(x, y)(\partial/\partial x) + g(x, y)(\partial/\partial y)$ is Hamiltonian if and only if $\operatorname{div} X = 0$.
   Reconsider Exercise 5 in §7.
4. Show that

$$\frac{d}{dt} g(x, \xi) = \{f, g\}$$

   on an orbit of H_f.
5. If $X = \sum X_j(x)\partial/\partial x_j$ is a vector field on $U \subset \mathbb{R}^n$, associate to X a function on $U \times \mathbb{R}^n \approx T^*U$:

$$s_X(x, \xi) = \langle X, \xi \rangle = \sum \xi_j X_j(x). \tag{10.22}$$

   Show that

$$s_{[X,Y]} = \{s_X, s_Y\}. \tag{10.23}$$

11. Geodesics

Here we define the concept of a geodesic on a region with a Riemannian metric (more generally, a Riemannian manifold). A Riemannian metric on $\Omega \subset \mathbb{R}^n$ is specified by $g_{jk}(x)$, where (g_{jk}) is a positive-definite, smooth, $n \times n$ matrix-valued function on Ω. If $U = \sum u^j(x)\partial/\partial x_j$ and $V = \sum v^j(x)\partial/\partial x_j$ are two vector fields on Ω, their inner product is the smooth scalar function

$$\langle U, V \rangle = g_{jk}(x)\, u^j(x) v^k(x), \tag{11.1}$$

using the summation convention (i.e., summing over repeated indices). If Ω is a manifold, a Riemannian metric is an inner product on each tangent space $T_x\Omega$, given in local coordinates by (11.1). Thus, (g_{jk}) gives rise to a tensor field of type $(0, 2)$, that is, a section of the bundle $\otimes^2 T^*\Omega$.

If $\gamma(t)$, $a \le t \le b$, is a smooth curve on Ω, its length is

$$L = \int_a^b \|\gamma'(t)\|\, dt = \int_a^b \Big[g_{jk}(\gamma(t))\gamma_j'(t)\gamma_k'(t)\Big]^{1/2}\, dt. \tag{11.2}$$

A curve γ is said to be a geodesic if, for $|t_1 - t_2|$ sufficiently small, $t_j \in [a, b]$, the curve $\gamma(t)$, $t_1 \le t \le t_2$, has the shortest length of all smooth curves in Ω from $\gamma(t_1)$ to $\gamma(t_2)$.

We derive the ODE for a geodesic. We start with the case where Ω has the metric induced from a diffeomorphism $\Omega \to S$, S a hypersurface in $\mathbb{R}^{n+1}$; we will identify Ω and S here. This short computation will serve as a guide for the general case.

So let $\gamma_0(t)$ be a smooth curve in S $(a \le t \le b)$, joining p and q. Suppose $\gamma_s(t)$ is a smooth family of such curves. We look for a condition guaranteeing that $\gamma_0(t)$ has minimum length. Since the length of a curve is independent of its parameterization, we may additionally suppose that

$$\|\gamma_0'(t)\| = c_0, \quad \text{constant, for } a \le t \le b. \tag{11.3}$$

Let N denote a field of normal vectors to S. Note that

$$V = \frac{\partial}{\partial s}\gamma_s(t) \perp N. \tag{11.4}$$

Also, any vector field $V \perp N$ over the image of γ_0 can be obtained by some variation γ_s of γ_0, provided $V = 0$ at p and q. Recall that we are assuming $\gamma_s(a) = p$, $\gamma_s(b) = q$. If $L(s)$ denotes the length of γ_s, we have

$$L(s) = \int_a^b \|\gamma_s'(t)\|\, dt, \tag{11.5}$$

and hence

$$\begin{aligned} L'(s) &= \frac{1}{2}\int_a^b \|\gamma_s'(t)\|^{-1}\frac{\partial}{\partial s}\big(\gamma_s'(t), \gamma_s'(t)\big)\, dt \\ &= \frac{1}{c_0}\int_a^b \Big(\frac{\partial}{\partial s}\gamma_s'(t), \gamma_s'(t)\Big)\, dt, \quad \text{at } s = 0. \end{aligned} \tag{11.6}$$

Using the identity

$$\frac{d}{dt}\Big(\frac{\partial}{\partial s}\gamma_s(t), \gamma_s'(t)\Big) = \Big(\frac{\partial}{\partial s}\gamma_s'(t), \gamma_s'(t)\Big) + \Big(\frac{\partial}{\partial s}\gamma_s(t), \gamma_s''(t)\Big), \tag{11.7}$$

together with the fundamental theorem of calculus, in view of the fact that

$$\frac{\partial}{\partial s}\gamma_s(t) = 0, \quad \text{at } t = a \text{ and } b, \tag{11.8}$$

we have

$$L'(s) = -\frac{1}{c_0}\int_a^b \big(V(t), \gamma_s''(t)\big)\,dt, \quad \text{at } s = 0. \tag{11.9}$$

Now, if γ_0 were a geodesic, we would have

$$L'(0) = 0, \tag{11.10}$$

for all such variations. In other words, we must have $\gamma_0''(t) \perp V$ for all vector fields V tangent to S (and vanishing at p and q), and hence

$$\gamma_0''(t) \| N. \tag{11.11}$$

This vanishing of the tangential curvature of γ_0 is the usual geodesic equation for a hypersurface in $\mathbb{R}^{n+1}$.

We proceed to derive from (11.11) an ODE in standard form. Suppose S is defined locally by $u(x) = C$, $\nabla u \neq 0$. Then (11.11) is equivalent to

$$\gamma_0''(t) = K \nabla u(\gamma_0(t)), \tag{11.12}$$

for a scalar K that remains to be determined. But the condition that $u(\gamma_0(t)) = C$ implies

$$\gamma_0'(t) \cdot \nabla u(\gamma_0(t)) = 0,$$

and differentiating this gives

$$\gamma_0''(t) \cdot \nabla u(\gamma_0(t)) = -\gamma_0'(t) \cdot D^2 u(\gamma_0(t)) \cdot \gamma_0'(t), \tag{11.13}$$

where $D^2 u$ is the matrix of second-order partial derivatives of u. Comparing (11.12) and (11.13) gives K, and we obtain the ODE

$$\gamma_0''(t) = -\big|\nabla u(\gamma_0(t))\big|^{-2} \Big[\gamma_0'(t) \cdot D^2 u(\gamma_0(t)) \cdot \gamma_0'(t)\Big] \nabla u(\gamma_0(t)) \tag{11.14}$$

for a geodesic γ_0 lying in S.

We now want to parallel (11.6)–(11.11), to provide the ODE for a geodesic on Ω with a general Riemannian metric. As before, let $\gamma_s(t)$ be a one-parameter family of curves satisfying $\gamma_s(a) = p$, $\gamma_s(b) = q$, and (11.3). Then

$$V = \frac{\partial}{\partial s}\gamma_s(t)\Big|_{s=0} \tag{11.15}$$

is a vector field defined on the curve $\gamma_0(t)$, vanishing at p and q, and a general vector field of this sort could be obtained by a variation $\gamma_s(t)$. Let

$$T = \gamma_s'(t). \tag{11.16}$$

With the notation of (11.1), we have, parallel to (11.6),

$$\begin{aligned} L'(s) &= \int_a^b V\langle T, T\rangle^{1/2}\,dt \\ &= \frac{1}{2c_0}\int_a^b V\langle T, T\rangle\,dt, \quad \text{at } s = 0. \end{aligned} \tag{11.17}$$

Now we need a generalization of $(\partial/\partial s)\gamma_s'(t)$ and of the formula (11.7). One natural approach involves the notion of a *covariant derivative*.

If X and Y are vector fields on Ω, the covariant derivative $\nabla_X Y$ is a vector field on Ω. The following properties are to hold: We assume that $\nabla_X Y$ is additive in both X and Y, that

$$\nabla_{fX} Y = f \nabla_X Y, \tag{11.18}$$

for $f \in C^\infty(\Omega)$, and that

$$\nabla_X (fY) = f\nabla_X Y + (Xf)Y \tag{11.19}$$

(i.e., ∇_X acts as a derivation). The operator ∇_X is required to have the following relation to the Riemannian metric:

$$X\langle Y, Z\rangle = \langle \nabla_X Y, Z\rangle + \langle Y, \nabla_X Z\rangle. \tag{11.20}$$

One further property, called the "zero torsion condition," will uniquely specify ∇:

$$\nabla_X Y - \nabla_Y X = [X, Y]. \tag{11.21}$$

If these properties hold, one says that ∇ is a "Levi-Civita connection." We have the following existence result.

Proposition 11.1. *Associated with a Riemannian metric is a unique Levi-Civita connection, given by*

$$\begin{aligned} 2\langle \nabla_X Y, Z\rangle =& X\langle Y, Z\rangle + Y\langle X, Z\rangle - Z\langle X, Y\rangle \\ &+ \langle [X, Y], Z\rangle - \langle [X, Z], Y\rangle - \langle [Y, Z], X\rangle. \end{aligned} \tag{11.22}$$

Proof. To obtain the formula (11.22), cyclically permute X, Y, and Z in (11.20) and take the appropriate alternating sum, using (11.21) to cancel out all terms involving ∇ but two copies of $\langle \nabla_X Y, Z\rangle$. This derives the formula and establishes uniqueness. On the other hand, if (11.22) is taken as the definition of $\nabla_X Y$, then verification of the properties (11.18)–(11.21) is a routine exercise.

We can resume our analysis of (11.17), which becomes

$$L'(s) = \frac{1}{c_0} \int_a^b \langle \nabla_V T, T\rangle \, dt, \quad \text{at } s = 0. \tag{11.23}$$

Since $\partial/\partial s$ and $\partial/\partial t$ commute, we have $[V, T] = 0$ on γ_0, and (11.21) implies

$$L'(s) = \frac{1}{c_0} \int_a^b \langle \nabla_T V, T\rangle \, dt, \quad \text{at } s = 0. \tag{11.24}$$

The replacement for (11.7) is

$$T\langle V, T\rangle = \langle \nabla_T V, T\rangle + \langle V, \nabla_T T\rangle, \tag{11.25}$$

so, by the fundamental theorem of calculus,

$$L'(0) = -\frac{1}{c_0}\int_a^b \langle V, \nabla_T T\rangle\, dt. \tag{11.26}$$

If this is to vanish for all smooth vector fields over γ_0, vanishing at p and q, we must have

$$\nabla_T T = 0. \tag{11.27}$$

This is the geodesic equation for a general Riemannian metric.

If $\Omega \subset \mathbb{R}^n$ carries a Riemannian metric $g_{jk}(x)$ and a corresponding Levi-Civita connection, the *Christoffel symbols* $\Gamma^k{}_{ij}$ are defined by

$$\nabla_{D_i} D_j = \sum_k \Gamma^k{}_{ji} D_k, \tag{11.28}$$

where $D_k = \partial/\partial x_k$. The formula (11.22) implies

$$g_{k\ell}\Gamma^\ell{}_{ij} = \frac{1}{2}\Big[\frac{\partial g_{jk}}{\partial x_i} + \frac{\partial g_{ik}}{\partial x_j} - \frac{\partial g_{ij}}{\partial x_k}\Big]. \tag{11.29}$$

We can rewrite the geodesic equation (11.27) for $\gamma_0(t) = x(t)$ as follows. With $x = (x_1, \dots, x_n)$ and $T = (\dot{x}^1, \dots, \dot{x}^n)$, we have

$$0 = \sum_\ell \nabla_T(\dot{x}^\ell D_\ell) = \sum_\ell \big[\ddot{x}^\ell D_\ell + \dot{x}^\ell \nabla_T D_\ell\big]. \tag{11.30}$$

In view of (11.28), this becomes

$$\ddot{x}^\ell + \dot{x}^j\, \dot{x}^k\, \Gamma^\ell{}_{jk} = 0 \tag{11.31}$$

(with the summation convention). The standard existence and uniqueness theory applies to this system of second-order ODE. We will call any smooth curve satisfying the equation (11.27), or equivalently (11.31), a geodesic. Shortly we will verify that such a curve is indeed locally length-minimizing. Note that if $T = \gamma'(t)$, then $T\langle T, T\rangle = 2\langle \nabla_T T, T\rangle$; so if (11.27) holds, $\gamma(t)$ automatically has constant speed.

For a given $p \in \Omega$, the exponential map

$$\mathrm{Exp}_p : U \longrightarrow \Omega \tag{11.32}$$

is defined on a neighborhood U of $0 \in \mathbb{R}^n = T_p\Omega$ by

$$\mathrm{Exp}_p(v) = \gamma_v(1), \tag{11.33}$$

where $\gamma_v(t)$ is the unique constant-speed geodesic satisfying

$$\gamma_v(0) = p, \quad \gamma_v'(0) = v. \tag{11.34}$$

Note that $\mathrm{Exp}_p(tv) = \gamma_v(t)$. It is clear that Exp_p is well defined and C^∞ on a sufficiently small neighborhood U of $0 \in \mathbb{R}^n$, and its derivative at 0 is the identity. Thus, perhaps shrinking U, we have that Exp_p is a diffeomorphism of U onto a neighborhood $\mathcal{O}$ of p in Ω. This provides what is called an exponential coordinate system, or a normal coordinate system. Clearly, the geodesics through p are the lines through the origin in this coordinate system. We claim that in this coordinate system

$$\Gamma^{\ell}{}_{jk}(p) = 0. \tag{11.35}$$

Indeed, since the line through the origin in any direction $aD_j + bD_k$ is a geodesic, we have

$$\nabla_{(aD_j+bD_k)}(aD_j + bD_k) = 0, \quad \text{at } p, \tag{11.36}$$

for all $a, b \in \mathbb{R}$ and all j, k. This implies

$$\nabla_{D_j} D_k = 0, \quad \text{at } p \text{ for all } j, k, \tag{11.37}$$

which implies (11.35). We note that (11.35) implies $\partial g_{jk}/\partial x_\ell = 0$ at p, in this exponential coordinate system. In fact, a simple manipulation of (11.29) gives

$$\frac{\partial g_{jk}}{\partial x_\ell} = g_{mk}\Gamma^m{}_{j\ell} + g_{mj}\Gamma^m{}_{k\ell}. \tag{11.38}$$

As a consequence, a number of calculations in differential geometry can be simplified by working in exponential coordinate systems.

We now establish a result, known as the *Gauss lemma*, which implies that a geodesic is locally length-minimizing. For a small, let $\Sigma_a = \{v \in \mathbb{R}^n : \|v\| = a\}$, and let $S_a = \operatorname{Exp}_p(\Sigma_a)$.

Proposition 11.2. *Any unit-speed geodesic through p hitting S_a at $t = a$ is orthogonal to S_a.*

Proof. If $\gamma_0(t)$ is a unit-speed geodesic, $\gamma_0(0) = p$, $\gamma_0(a) = q \in S_a$, and $V \in T_q\Omega$ is tangent to S_a, there is a smooth family of unit-speed geodesics, $\gamma_s(t)$, such that $\gamma_s(0) = p$ and $(\partial/\partial s)\gamma_s(a)\big|_{s=0} = V$. Using (11.24) and (11.25) for this family, with $0 \le t \le a$, since $L(s)$ is constant, we have

$$0 = \int_0^a T\langle V, T\rangle\, dt = \langle V, \gamma_0'(a)\rangle,$$

which proves the proposition.

Though a geodesic is locally length-minimizing, it need not be globally length-minimizing. There are many simple examples of this, some of which are discussed in the exercises.

We next consider a "naive" alternative to the calculations (11.17)–(11.31), not bringing in the notion of covariant derivative, in order to compute $L'(0)$ when $L(s)$ is given by

$$L(s) = \int_a^b \Big[g_{jk}\big(x_s(t)\big)\dot{x}_s^j(t)\,\dot{x}_s^k(t)\Big]^{1/2}\, dt. \tag{11.39}$$

We use the notation $T^j = \dot{x}_0^j(t)$, $V^j = (\partial/\partial s)x_s^j(t)|_{s=0}$. Calculating in a spirit similar to that of (11.6), we have (with $x = x_0$)

$$L'(0) = \frac{1}{c_0} \int_a^b \Big[g_{jk} \frac{\partial}{\partial s} \dot{x}_s^j(t)\Big|_{s=0} T^k + \frac{1}{2} V^j \frac{\partial g_{k\ell}}{\partial x_j} T^k T^\ell \Big] \, dt. \tag{11.40}$$

Now, in analogy with (11.7), and in place of (11.25), we can write

$$\frac{d}{dt}\Big(g_{jk}(x(t)) V^j T^k\Big) = g_{jk} \frac{\partial}{\partial s} \dot{x}_s^j(t)\Big|_{s=0} T^k + g_{jk} V^j \ddot{x}^k(t) + T^\ell \frac{\partial g_{jk}}{\partial x_\ell} V^j T^k. \tag{11.41}$$

Thus, by the fundamental theorem of calculus,

$$L'(0) = -\frac{1}{c_0} \int_a^b \Big[g_{jk} V^j \ddot{x}^k + T^\ell \frac{\partial g_{jk}}{\partial x_\ell} V^j T^k - \frac{1}{2} V^j \frac{\partial g_{k\ell}}{\partial x_j} T^k T^\ell \Big] \, dt, \tag{11.42}$$

and the stationary condition $L'(0) = 0$ for all variations of the form described before implies

$$g_{jk} \ddot{x}^k(t) = -\Big(\frac{\partial g_{jk}}{\partial x_\ell} - \frac{1}{2} \frac{\partial g_{k\ell}}{\partial x_j} \Big) T^k T^\ell. \tag{11.43}$$

Symmetrizing the quantity in parentheses with respect to k and ℓ yields the ODE (11.31), with ${\Gamma^\ell}_{jk}$ given by (11.29).

Of the two derivations for the equations of (constant-speed) geodesics given in this section, the latter is a bit shorter and more direct. On the other hand, the slight additional complication of the first derivation paid for the introduction of the notion of covariant derivative, a fundamental object in differential geometry. As we will see in the next section, the methods of the second derivation are very flexible; there we consider a class of extremal problems, containing the problem of geodesics, and also containing problems giving rise to the equations of classical physics, via the stationary action principle.

We now show that the geodesic flow equations can be transformed to a Hamiltonian system. Let (g^{jk}) denote the matrix inverse of (g_{jk}), and relate $v \in \mathbb{R}^n$ to $\xi \in \mathbb{R}^n$ by

$$\xi_j = g_{jk}(x) v_k, \quad \text{i.e.,}\ v_j = g^{jk}(x) \xi_k. \tag{11.44}$$

Define $f(x, \xi)$ on $\Omega \times \mathbb{R}^n$ by

$$f(x, \xi) = \frac{1}{2} g^{jk}(x) \xi_j \xi_k, \tag{11.45}$$

as before using the summation convention. For a manifold M, (11.44) is a local coordinate expression of the Riemannian metric tensor, providing an isomorphism of $T_x M$ with $T_x^* M$, and (11.45) defines half the square norm on T^*M. Then the integral curves $(x(t), \xi(t))$ of H_f satisfy

$$\dot{x}_\ell = g^{\ell k}(x) \xi_k, \quad \dot{\xi}_\ell = -\frac{1}{2} \frac{\partial g^{jk}}{\partial x_\ell} \xi_j \xi_k. \tag{11.46}$$

If we differentiate the first equation and plug in the second one for $\dot{\xi}_k$, we get

$$\ddot{x}_\ell = \sum\Big[-\frac{1}{2}g^{\ell j}\frac{\partial g^{ik}}{\partial x_j} + g^{kj}\frac{\partial g^{i\ell}}{\partial x_j}\Big]\xi_i\xi_k, \tag{11.47}$$

and using $\xi_j = \sum g_{jk}(x)\dot{x}_k$, straightforward manipulations yield the geodesic equation (11.31), with $\Gamma^\ell{}_{jk}$ given by (11.29).

We now describe a relatively noncomputational approach to the result just obtained. Identifying (x, v)-space and (x, ξ)-space via (11.44), let Y be the resulting vector field on (x, ξ)-space defined by the geodesic flow. The result we want to reestablish is that Y and H_f coincide at an arbitrary point $(x_0, \xi_0) \in \Omega \times \mathbb{R}^n$. We will make use of an exponential coordinate system centered at x_0; recall that in this coordinate system the geodesics through x_0 become precisely the lines through the origin. (Of course, geodesics through nearby points are not generally straight lines in this coordinate system.) In such a coordinate system, we can arrange $g^{jk}(x_0) = \delta^{jk}$ and, by (11.35), $(\partial g^{jk}/\partial x_\ell)(x_0) = 0$. Thus, if $\xi_0 = (a_1, \dots, a_n)$, using (11.46) we have

$$H_f(x_0, \xi_0) = \sum a_k \frac{\partial}{\partial x_k} = Y(x_0, \xi_0) \tag{11.48}$$

in this coordinate system. The identity of H_f and Y at (x_0, ξ_0) is independent of the coordinate system used, so our result is again established. Actually, there is a little cheat here. We have not shown that H_f is defined independently of the choice of coordinates on Ω. This will be established in §14; see (14.15)–(14.19).

In the next section there will be a systematic approach to converting variational problems to Hamiltonian systems.

Exercises

1. Suppose $\text{Exp}_p : B_a \to M$ is a diffeomorphism of $B_a = \{v \in T_pM : \|v\| \le a\}$ onto its image, $\mathcal{B}$. Use the Gauss lemma to show that, for each $q \in \mathcal{B}$, $q = \text{Exp}(w)$, the curve $\gamma(t) = \text{Exp}(tw)$, $0 \le t \le 1$, is the unique shortest path from p to q. If Exp_p is defined on B_a but is *not* a diffeomorphism, show that this conclusion does not hold.
2. Let M be a connected Riemannian manifold. Define $d(p, q)$ to be the infimum of lengths of smooth curves from p to q. Show that this makes M a metric space.
3. Let $p, q \in M$, and suppose there exists a *Lipschitz* curve $\gamma : [a, b] \to M$, $\gamma(a) = p$, $\gamma(b) = q$, parameterized by arc length, of length equal to $d(p, q)$. Show that γ is a C^∞-curve. (*Hint*: Make use of Exercise 1.)
4. Let M be a connected Riemannian manifold that, with the metric of Exercise 2, is compact. Show that any $p, q \in M$ can be joined by a geodesic of length $d(p, q)$. (*Hint*: Let $\gamma_k : [0, 1] \to M$, $\gamma_k(0) = p$, $\gamma_k(1) = q$ be constant-speed curves of lengths $\ell_k \to d(p, q)$. Use Ascoli's theorem to produce a Lipschitz curve of length $d(p, q)$ as a uniform limit of a subsequence of these.)
5. Try to extend the result of Exercise 4 to the case where M is assumed to be *complete*, rather than compact.
6. Verify that the definition of ∇_X given by (11.22) does indeed provide a Levi-Civita connection, having properties (11.18)–(11.21).

(*Hint*: For example, if you interchange the roles of Y and Z in (11.22), and add it to the resulting formula for $2\langle Y, \nabla_X Z\rangle$, you can cancel all the terms on the right side except $X\langle Y, Z\rangle + X\langle Z, Y\rangle$; this gives (11.20).)

12. Variational problems and the stationary action principle

The calculus of variations consists of the study of stationary points (e.g., maxima and minima) of a real-valued function that is defined on some space of functions. Here, we let M be a region in $\mathbb{R}^n$, or more generally an n-dimensional manifold, fix two points $p, q \in M$ and an interval $[a, b] \subset \mathbb{R}$, and consider a space of functions $\mathcal{P}$ consisting of smooth curves $u : [a, b] \to M$ satisfying $u(a) = p,\ u(b) = q$. We consider functions $I : \mathcal{P} \to \mathbb{R}$ of the form

$$I(u) = \int_a^b F\big(u(t), \dot{u}(t)\big)\, dt. \tag{12.1}$$

Here $F(x, v)$ is a smooth function on the tangent bundle TM, or perhaps on some open subset of TM. By definition, the condition for I to be stationary at u is that

$$\frac{d}{ds} I(u_s)\big|_{s=0} = 0 \tag{12.2}$$

for any smooth family u_s of elements of $\mathcal{P}$ with $u_0 = u$. Note that

$$\frac{d}{ds} u_s(t)\big|_{s=0} = w(t) \tag{12.3}$$

defines a tangent vector to M at $u(t)$, and precisely those tangent vectors $w(t)$ vanishing at $t = a$ and at $t = b$ arise from making some variation of u within $\mathcal{P}$.

As in the last section, we can compute the left side of (12.2) by differentiating under the integral, and obtaining a formula for this involves considering t-derivatives of w. Recall the two approaches to this taken in §11. Here we will emphasize the second approach, since the data at hand do not generally pick out some distinguished covariant derivative on M. Thus we work in local coordinates on M. Since any smooth curve on M can be enclosed by a single coordinate patch, this involves no loss of generality. Then, given (12.3), we have

$$\frac{d}{ds} I(u_s)\big|_{s=0} = \int_a^b \big[F_x(u, \dot{u})w + F_v(u, \dot{u})\dot{w}\big]\, dt. \tag{12.4}$$

Integrating the last term by parts and recalling that $w(a)$ and $w(b)$ vanish, we see that this is equal to

$$\int_a^b \Big[F_x(u, \dot{u}) - \frac{d}{dt} F_v(u, \dot{u})\Big] w\, dt. \tag{12.5}$$

It follows that the condition for u to be stationary is precisely that u satisfy the equation

$$\frac{d}{dt} F_v(u, \dot{u}) - F_x(u, \dot{u}) = 0, \tag{12.6}$$

a second-order ODE, called *Lagrange's equation*. Written more fully, it is

$$F_{vv}(u,\dot{u})\ddot{u} + F_{vx}(u,\dot{u})\dot{u} - F_x(u,\dot{u}) = 0, \tag{12.7}$$

where F_{vv} is the $n \times n$ matrix of second-order v-derivatives of $F(x, v)$, acting on the vector $\ddot{u}$, etc. This is a nonsingular system as long as $F(x, v)$ satisfies the condition

$$F_{vv}(x,v) \text{ is invertible,} \tag{12.8}$$

as an $n \times n$ matrix, for each $(x, v) = (u(t), \dot{u}(t))$, $t \in [a, b]$.

The ODE (12.6) suggests a particularly important role for

$$\xi = F_v(x,v). \tag{12.9}$$

Then, for $(x, v) = (u, \dot{u})$, we have

$$\dot{\xi} = F_x(x,v), \quad \dot{x} = v. \tag{12.10}$$

We claim that this system, in (x, ξ)-coordinates, is in *Hamiltonian* form. Note that (x, ξ) gives a local coordinate system under the hypothesis (12.8), by the inverse function theorem. In other words, we will produce a function $E(x, \xi)$ such that (12.10) is the same as

$$\dot{x} = E_\xi, \quad \dot{\xi} = -E_x, \tag{12.11}$$

so the goal is to construct $E(x, \xi)$ such that

$$E_x(x,\xi) = -F_x(x,v), \quad E_\xi(x,\xi) = v, \tag{12.12}$$

when $v = v(x, \xi)$ is defined by inverting the transformation

$$(x,\xi) = (x, F_v(x,v)) = \lambda(x,v). \tag{12.13}$$

If we set

$$E^b(x,v) = E(\lambda(x,v)), \tag{12.14}$$

then (12.12) is equivalent to

$$E^b_x(x,v) = -F_x + vF_{vx}, \quad E^b_v(x,v) = v\,F_{vv}, \tag{12.15}$$

as follows from the chain rule. This calculation is most easily performed using differential forms, details on which can be found in the next section; in the differential form notation, our task is to find $E^b(x, v)$ such that

$$dE^b = (-F_x + vF_{vx})\,dx + vF_{vv}\,dv. \tag{12.16}$$

It can be seen by inspection that this identity is satisfied by

$$E^b(x,v) = F_v(x,v)v - F(x,v). \tag{12.17}$$

Thus the ODE (12.7) describing a stationary point for (12.1) has been converted to a first-order Hamiltonian system, in the (x, ξ)-coordinates, given the hypothesis (12.8) on F_{vv}. In view of (12.13), one often writes (12.17) informally as

$$E(x,\xi) = \xi \cdot v - F(x,v).$$

We make some observations about the transformation λ of (12.13). If $v \in T_xM$, then $F_v(x, v)$ acts naturally as a linear functional on T_xM. In other words, $\xi = F_v(x, v)$ is naturally regarded as an element of T_x^*M, in the cotangent bundle of M; it makes invariant sense to regard

$$\lambda : TM \longrightarrow T^*M \tag{12.18}$$

(if F is defined on all of TM). This map is called the *Legendre transformation*. As we have already noted, the hypothesis (12.8) is equivalent to the statement that λ is a local diffeomorphism.

As an example, suppose M has a Riemannian metric g and

$$F(x, v) = \frac{1}{2} g(v, v).$$

Then the map (12.18) is the identification of TM and T^*M associated with "lowering indices," using the metric tensor g_{jk}. A straightforward calculation gives, in this case, $E(x, \xi)$ equal to half the natural square norm on cotangent vectors. On the other hand, the function $F(x, v) = \sqrt{g(v, v)}$ *fails* to satisfy the hypothesis (12.8). Since this is the integrand for arc length, it is important to incorporate this case into our analysis. Recall from the previous section that obtaining equations for a geodesic involves parameterizing a curve by arc length. We now look at the following more general situation.

We say $F(x, v)$ is homogeneous of degree r in v if $F(x, cv) = c^r F(x, v)$ for $c > 0$. Thus $\sqrt{g(v, v)}$ above is homogeneous of degree 1. When F is homogeneous of degree 1, hypothesis (12.8) is never satisfied. Furthermore, $I(u)$ is independent of the parameterization of a curve in this case; if $\sigma : [a, b] \to [a, b]$ is a diffeomorphism (fixing a and b), then $I(u) = I(\tilde{u})$ for $\tilde{u}(t) = u(\sigma(t))$. Let us look at a function $f(x, v)$ related to $F(x, v)$ by

$$f(x, v) = \psi(F(x, v)), \quad F(x, v) = \varphi(f(x, v)). \tag{12.19}$$

Given a family u_s of curves as before, we can write

$$\begin{aligned} \frac{d}{ds} I(u_s)|_{s=0} = \int_a^b \Big[& \varphi'(f(u, \dot{u})) f_x(u, \dot{u}) \\ & - \frac{d}{dt}\{\varphi'(f(u, \dot{u})) f_v(u, \dot{u})\}\Big] w \, dt. \end{aligned} \tag{12.20}$$

If u satisfies the condition

$$f(u, \dot{u}) = c, \tag{12.21}$$

with c constant, this is equal to

$$c' \int_a^b \big[f_x(u, \dot{u}) - (d/dt) f_v(u, \dot{u})\big] w \, dt, \tag{12.22}$$

with $c' = \varphi'(c)$. Of course, setting

$$J(u) = \int_a^b f(u, \dot{u}) \, dt, \tag{12.23}$$

we have

$$\frac{d}{ds} J(u_s)\big|_{s=0} = \int_a^b \Big[f_x(u,\dot{u}) - \frac{d}{dt} f_v(u,\dot{u})\Big] w\, dt. \tag{12.24}$$

Consequently, if u satisfies (12.21), then u is stationary for I if and only if u is stationary for J (provided $\varphi'(c) \neq 0$).

It is possible that $f(x, v)$ satisfies (12.8) even though $F(x, v)$ does not, as the case $F(x, v) = \sqrt{g(v, v)}$ illustrates. Note that

$$f_{v_j v_k} = \psi'(F) F_{v_j v_k} + \psi''(F) F_{v_j} F_{v_k}.$$

Let us specialize to the case $\psi(F) = F^2$, so $f(x, v) = F(x, v)^2$ is homogeneous of degree 2. If F is convex in v and $(F_{v_j v_k})$, a positive-semidefinite matrix, annihilates only radial vectors, and if $F > 0$, then $f(x, v)$ is strictly convex (i.e., f_{vv} is positive-definite), and hence (12.8) holds for $f(x, v)$. This is the case when $F(x, v) = \sqrt{g(v, v)}$ is the arc length integrand.

If $f(x, v) = F(x, v)^2$ satisfies (12.8), then the stationary condition for (12.23) is that u satisfy the ODE

$$f_{vv}(u,\dot{u})\ddot{u} + f_{vx}(u,\dot{u})\dot{u} - f_x(u,\dot{u}) = 0,$$

a nonsingular ODE for which we know there is a unique local solution, with $u(a) = p$, $\dot{u}(a)$ given. We will be able to say that such a solution is also stationary for (12.1) once we know that (12.21) holds, that is, $f(u, \dot{u})$ is constant. Indeed, if $f(x, v)$ is homogeneous of degree 2, then $f_v(x, v)v = 2f(x, v)$, and hence

$$e^b(x, v) = f_v(x, v)v - f(x, v) = f(x, v). \tag{12.25}$$

But since the equations for u take Hamiltonian form in the coordinates $(x, \xi) = (x, f_v(x, v))$, it follows that $e^b(u(t), \dot{u}(t))$ is constant for u stationary, so (12.21) does hold in this case.

There is a general principle, known as the *stationary action principle*, or Hamilton's principle, for producing equations of mathematical physics. In this set-up, the state of a physical system at a given time is described by a pair (x, v), position and velocity. One has a *kinetic energy* function $T(x, v)$ and a *potential energy* function $V(x, v)$, determining the dynamics, as follows. Form the difference

$$L(x, v) = T(x, v) - V(x, v), \tag{12.26}$$

known as the *Lagrangian*. Hamilton's principle states that a path $u(t)$ describing the evolution of the state in this system is a stationary path for the *action integral*

$$I(u) = \int_a^b L(u, \dot{u})\, dt. \tag{12.27}$$

In many important cases, the potential $V = V(x)$ is velocity independent and $T(x, v)$ is a quadratic form in v; say $T(x, v) = (1/2)v \cdot G(x)v$ for a symmetric matrix $G(x)$. In that case, we consider

$$L(x, v) = \frac{1}{2} v \cdot G(x)v - V(x). \tag{12.28}$$

Thus we have

$$\xi = L_v(x, v) = G(x)v, \tag{12.29}$$

and the conserved quantity (12.17) becomes

$$\begin{aligned} E^b(x, v) &= v \cdot G(x)v - \left[\frac{1}{2}v \cdot G(x)v - V(x)\right] \\ &= \frac{1}{2}v \cdot G(x)v + V(x), \end{aligned} \tag{12.30}$$

which is the *total energy* $T(x, v) + V(x)$. Note that the nondegeneracy condition is that $G(x)$ be invertible (in physical problems, $G(x)$ is typically positive-definite, but see (18.20)); assuming this, we have

$$E(x, \xi) = \frac{1}{2}\xi \cdot G(x)^{-1}\xi + V(x), \tag{12.31}$$

whose Hamiltonian vector field defines the dynamics. Note that, in this case, Lagrange's equation (12.6) takes the form

$$\frac{d}{dt}\big[G(u)\dot{u}\big] = \frac{1}{2}\dot{u} \cdot G_x(u)\dot{u} - V_x(u), \tag{12.32}$$

which can be rewritten as

$$\ddot{u} + \Gamma\dot{u}\dot{u} + G(u)^{-1}V_x(u) = 0, \tag{12.33}$$

where $\Gamma\dot{u}\dot{u}$ is a vector whose ℓth component is $\Gamma^\ell{}_{jk}\dot{u}^j\dot{u}^k$, with $\Gamma^\ell{}_{jk}$ the connection coefficients defined by (11.29) with $(g_{jk}) = G(x)$. In other words, (12.33) generalizes the geodesic equation for the Riemannian metric $(g_{jk}) = G(x)$, which is what would arise in the case $V = 0$.

We refer to [Ar] and [Go] for a discussion of the relation of Hamilton's principle to other formulations of the laws of Newtonian mechanics, but we will briefly illustrate it here with a couple of examples.

Consider the basic case of motion of a particle in Euclidean space $\mathbb{R}^n$, in the presence of a force field of potential type $F(x) = -\operatorname{grad} V(x)$, as in the beginning of §10. Then

$$T(x, v) = \frac{1}{2}m|v|^2, \quad V(x, v) = V(x). \tag{12.34}$$

This is of course the special case of (12.28) with $G(x) = mI$, and the ODE satisfied by stationary paths for (12.27) hence has the form

$$m\ddot{u} + V_x(u) = 0, \tag{12.35}$$

precisely the equation (10.2) expressing Newton's law $F = ma$.

Next we consider one example where Cartesian coordinates are not used, namely the motion of a pendulum (Fig. 12.1). We suppose a mass m is at the end of a (massless) rod of length ℓ, swinging under the influence of gravity. In this case, we can express the potential energy as

$$V(\theta) = -mg\ell\cos\theta, \tag{12.36}$$

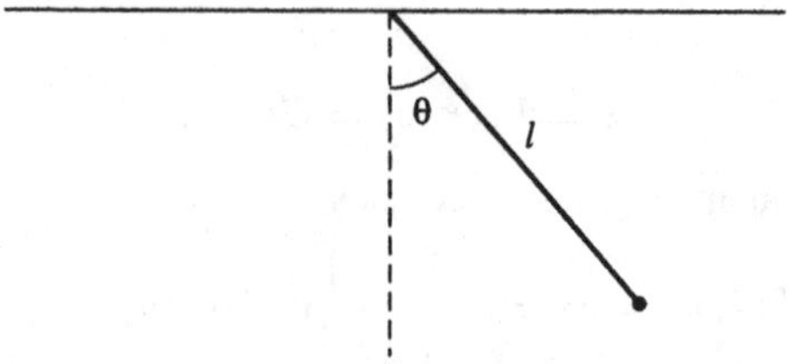

FIGURE 12.1

where θ is the angle the rod makes with the downward vertical ray, and g denotes the strength of gravity. The speed of the mass at the end of the pendulum is $\ell|\dot{\theta}|$, so the kinetic energy is

$$T(\theta, \dot{\theta}) = \frac{1}{2} m\ell^2 |\dot{\theta}|^2. \tag{12.37}$$

In this case we see that Hamilton's principle leads to the ODE

$$\ell\ddot{\theta} + g \sin\theta = 0, \tag{12.38}$$

describing the motion of a pendulum.

Next we consider a very important physical problem that involves a *velocity-dependent* force, leading to a Lagrangian of a form different from (12.28), namely the (nonrelativistic) motion of a charged particle (with charge e) in an electromagnetic field (E, B). One has Newton's law

$$m\frac{dv}{dt} = F, \tag{12.39}$$

where $v = dx/dt$ and F is the *Lorentz force*, given by

$$F = e(E + v \times B). \tag{12.40}$$

Certainly F here is not of the form $-\nabla V(x)$. To construct a replacement for the potential V, one makes use of two of Maxwell's equations for E and B:

$$\text{curl } E = -\frac{\partial B}{\partial t}, \quad \text{div } B = 0, \tag{12.41}$$

in units where the speed of light is 1. We will return to Maxwell's equations later on. As we will show in §18, these equations imply the existence of a real-valued $\varphi(t, x)$ and a vector-valued $A(t, x)$ such that

$$B = \text{curl } A, \quad E = -\text{ grad } \varphi - \frac{\partial A}{\partial t}. \tag{12.42}$$

Given these quantities, we set

$$V(x, v) = e(\varphi - A \cdot v), \tag{12.43}$$

and use the Lagrangian $L = T - V$, with $T = (1/2)m|v|^2$. We have

$$L_v = mv + eA, \quad L_x = -e\varphi_x + e \text{ grad}(A \cdot v).$$

Consequently, $(d/dt)L_v = m\, dv/dt + e\partial A/\partial t + eA_x v$. Using (12.42), we can obtain

$$\frac{d}{dt}L_v - L_x = m\,\frac{dv}{dt} - e(E + v \times \text{curl } A), \tag{12.44}$$

showing that Lagrange's equation

$$\frac{d}{dt}L_v - L_x = 0 \tag{12.45}$$

is indeed equivalent to (12.39)–(12.40).

If the electromagnetic field varies with t, then the Lagrangian L produced by (12.43) has explicit t-dependence:

$$L = L(t, x, v). \tag{12.46}$$

The equation (12.45) is still the stationary condition for the integral

$$I(u) = \int_a^b L\big(t, u(t), \dot{u}(t)\big)\, dt, \tag{12.47}$$

as in (12.6). Of course, instead of (12.7), we have

$$L_{vv}(t, u, \dot{u})\ddot{u} + L_{vx}(t, u, \dot{u})\dot{u} - L_x(t, u, \dot{u}) + L_{tv}(t, u, \dot{u}) = 0. \tag{12.48}$$

Finally, we note that for this Lorentz force the Legendre transformation (12.13) is given by

$$(x, \xi) = (x, mv + eA), \tag{12.49}$$

and hence the Hamiltonian function $E(x, \xi)$ as in (12.11) is given by

$$E(x, \xi) = \frac{1}{2m}|\xi - eA|^2 + e\varphi. \tag{12.50}$$

A treatment of the *relativistic* motion of a charged particle in an electromagnetic field (which in an important sense is cleaner than the nonrelativistic treatment) is given in §18.

Hamilton's principle can readily be extended to produce *partial differential equations*, describing the motion of continua, such as vibrating strings, moving fluids, and numerous other important phenomena. Some of these results will be discussed in the beginning of Chapter 2, and others in various subsequent chapters.

We end this section by noting that Lagrange's equation (12.6) depends on the choice of a coordinate system. We can write down an analogue of (12.6), which depends on a choice of Riemannian metric on M, but not on a coordinate system.

Thus, let M be a Riemannian manifold, and denote by ∇ the Levi-Civita connection constructed in §11. If we have a family of curves in TM, that is, a map

$$u : I \times I \longrightarrow M, \quad u = u(t, s), \tag{12.51}$$

with velocity $u_t : I \times I \to TM$, we can write

$$I(s) = \int_a^b F\big(u_t(t, s)\big)\, dt, \tag{12.52}$$

for a given $F : TM \to \mathbb{R}$. We have

$$I'(s) = \int_a^b DF\big(u_t(t,s)\big)\partial_s u_t \, dt. \tag{12.53}$$

Note that $DF(u_t)$ acts on $\partial_s u_t \in T_{u_t}(TM)$. Now, given $v \in TM$, we can write

$$T_v(TM) = V_v(TM) \oplus H_v(TM). \tag{12.54}$$

Here the "vertical" space $V_v(TM)$ is simply $T_v(T_{\pi(v)}M)$, where $\pi : TM \to M$ is the usual projection. The "horizontal" space $H_v(TM)$ is a complementary space, isomorphic to $T_{\pi(v)}M$, defined as follows.

For any smooth curve γ on M, such that $\gamma(0) = x = \pi(v)$, let $V(t) \in T_{\gamma(t)}M$ be given by parallel translation of v along γ, that is, if $T = \gamma'(t)$, V solves $\nabla_T V = 0$, $V(0) = v$. Thus $V(t)$ is a curve in TM, and $V(0) = v$. The map $\gamma'(0) \mapsto V'(0)$ is an injective linear map of $T_{\pi(v)}M$ into $T_v(TM)$, whose range we call $H_v(TM)$. One might compare the construction in §6 of Appendix C, Connections and Curvature. Thus we have both the decomposition (12.54) and the isomorphisms

$$V_v(TM) \approx T_{\pi(v)}M, \quad H_v(TM) \approx T_{\pi(v)}M. \tag{12.55}$$

The first isomorphism is canonical. The second isomorphism is simply the restriction of $D\pi : T_v(TM) \to T_{\pi(v)}M$ to the subspace $H_v(TM)$.

The splitting (12.54) gives

$$DF(v)(\partial_s u_t) = \langle F_v(v), (\partial_s u_t)_{\text{vert}}\rangle + \langle F_x(v), (\partial_s u_t)_{\text{horiz}}\rangle, \tag{12.56}$$

where we use this to define

$$F_v(v) \in T_{\pi(v)}M \approx V_v(TM), \quad F_x(v) \in T_{\pi(v)}M \approx H_v(TM). \tag{12.57}$$

If we set $v = u_t$, $w = u_s$, we have

$$I'(s) = \int_a^b \Big[\big\langle F_v(u_t), \nabla_v w\big\rangle + \big\langle F_x(v), w\big\rangle\Big]\, dt. \tag{12.58}$$

Parallel to (11.24)–(11.26), we have

$$\int_a^b \big\langle F_v(u_t), \nabla_v w\big\rangle\, dt = -\int_a^b \big\langle \nabla_v F_v(u_t), w\big\rangle\, dt, \tag{12.59}$$

where to apply ∇_v we regard $F_v(u_t)$ as a vector field defined over the curve $t \mapsto u(t,s)$ in M. Hence the stationary condition that $I'(0) = 0$ for all variations of $u(t) = u(t,0)$ takes the form

$$\nabla_{\dot u} F_v(\dot u) - F_x(\dot u) = 0. \tag{12.60}$$

Note that if $v(s)$ is a smooth curve in TM, with $\pi(v(s)) = u(s)$ and $u'(s) = w(s)$, then, under the identification in (12.55),

$$v'(s)_{\text{vert}} = \nabla_w v, \quad v'(s)_{\text{horiz}} = w. \tag{12.61}$$

Then, for smooth $F : TM \to \mathbb{R}$,

$$\frac{d}{ds} F\big(v(s)\big) = \big\langle F_v(v), \nabla_w v\big\rangle + \big\langle F_x(v), w\big\rangle. \tag{12.62}$$

In particular,

$$F(v) = \langle v, v\rangle \Longrightarrow F_v(v) = 2v \text{ and } F_x(v) = 0. \tag{12.63}$$

Thus, for this function $F(v)$, the Lagrange equation (12.60) becomes the geodesic equation $\nabla_v v = 0$, as expected. If, parallel to (12.28), we take $L(v) = (1/2)\langle v, v\rangle - V(x),\ x = \pi(v)$, then

$$L_v(v) = v, \quad L_x(v) = -\operatorname{grad} V(x), \tag{12.64}$$

where grad $V(x)$ is the vector field on M defined by $\langle \operatorname{grad} V(x), W\rangle = \mathcal{L}_W V(x)$. The Lagrange equation becomes

$$\nabla_{\dot u}\dot u + \operatorname{grad} V(u) = 0, \tag{12.65}$$

in agreement with (12.33).

Exercises

1. Suppose that, more generally than (12.28), we have a Lagrangian of the form

$$L(x, v) = \frac{1}{2} v \cdot G(x)v + A(x) \cdot v - V(x).$$

   Show that (12.30) continues to hold, that is,

$$E^b(x, v) = \frac{1}{2} v \cdot G(x)v + V(x),$$

   and that the Hamiltonian function becomes, in place of (12.31),

$$E(x, \xi) = \frac{1}{2}(\xi - A(x)) \cdot G(x)^{-1}(\xi - A(x)) + V(x).$$

   Work out the modification to (12.33) when the extra term $A(x) \cdot v$ is included. Relate this to the discussion of the motion in an electromagnetic field in (12.39)–(12.50).
2. Work out the differential equations for a planar double pendulum, in the spirit of (12.36)–(12.38). See Fig. 12.2. (*Hint*: To compute kinetic and potential energy, think of the plane as the complex plane, with the real axis pointing down. The position of particle 1 is $\ell_1 e^{i\theta_1}$ and that of particle 2 is $\ell_1 e^{i\theta_1} + \ell_2 e^{i\theta_2}$.)
3. After reading §18, show that the identity $\mathcal{F} = d\mathcal{A}$ in (18.19) implies the identity (12.42), with $\mathcal{A} = \varphi\, dx_0 + \sum_{j\ge 1} A_j\, dx_j$.
4. If $A(x)$ is a vector field on $\mathbb{R}^3$ and v is a constant vector, show that

$$\operatorname{grad}(v \cdot A) = \nabla_v A + v \times \operatorname{curl} A.$$

   Use this to verify (12.44). How is the formula above modified if $v = v(x)$ is a function of x? Reconsider this last question after looking at the exercises following §8 of Chapter 5.

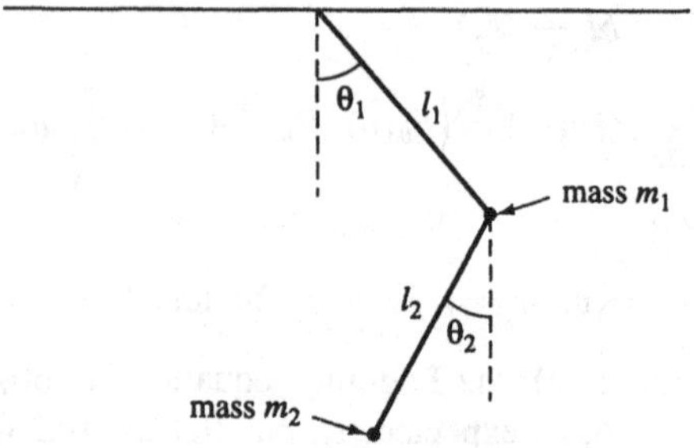

FIGURE 12.2

5. The statement before (12.4)—that any smooth curve $u(s)$ on M can be enclosed by a single coordinate patch—is not strictly accurate, as the curve may have self-intersections. Give a more precise statement.

13. Differential forms

It is very desirable to be able to make constructions that depend as little as possible on a particular choice of coordinate system. The calculus of differential forms, whose study we now take up, is one convenient set of tools for this purpose.

We start with the notion of a 1-form. It is an object that is integrated over a curve; formally, a 1-form on $\Omega \subset \mathbb{R}^n$ is written

$$\alpha = \sum_j a_j(x)\, dx_j. \tag{13.1}$$

If $\gamma : [a, b] \to \Omega$ is a smooth curve, we set

$$\int_\gamma \alpha = \int_a^b \sum a_j\big(\gamma(t)\big)\gamma_j'(t)\, dt. \tag{13.2}$$

In other words,

$$\int_\gamma \alpha = \int_I \gamma^*\alpha, \tag{13.3}$$

where $I = [a, b]$ and $\gamma^*\alpha = \sum_j a_j(\gamma(t))\gamma_j'(t)$ is the *pull-back* of α under the map γ. More generally, if $F : \mathcal{O} \to \Omega$ is a smooth map ($\mathcal{O} \subset \mathbb{R}^m$ open), the pull-back $F^*\alpha$ is a 1-form on $\mathcal{O}$ defined by

$$F^*\alpha = \sum_{j,k} a_j(F(y)) \frac{\partial F_j}{\partial y_k}\, dy_k. \tag{13.4}$$

The usual change of variable for integrals gives

$$\int_{\gamma} \alpha = \int_{\sigma} F^* \alpha \tag{13.5}$$

if γ is the curve $F \circ \sigma$.

If $F : \mathcal{O} \to \Omega$ is a diffeomorphism, and

$$X = \sum b^j(x) \frac{\partial}{\partial x_j} \tag{13.6}$$

is a vector field on Ω, recall that we have the vector field on $\mathcal{O}$:

$$F_\# X(y) = \big(DF^{-1}(p)\big) X(p), \quad p = F(y). \tag{13.7}$$

If we define a pairing between 1-forms and vector fields on Ω by

$$\langle X, \alpha \rangle = \sum_j b^j(x) a_j(x) = b \cdot a, \tag{13.8}$$

a simple calculation gives

$$\langle F_\# X, F^* \alpha \rangle = \langle X, \alpha \rangle \circ F. \tag{13.9}$$

Thus, a 1-form on Ω is characterized at each point $p \in \Omega$ as a linear transformation of *vectors* at p to $\mathbb{R}$.

More generally, we can regard a k-form α on Ω as a k-multilinear map on vector fields:

$$\alpha(X_1, \dots, X_k) \in C^\infty(\Omega); \tag{13.10}$$

we impose the further condition of antisymmetry:

$$\alpha(X_1, \dots, X_j, \dots, X_\ell, \dots, X_k) = -\alpha(X_1, \dots, X_\ell, \dots, X_j, \dots, X_k). \tag{13.11}$$

We use a special notation for k-forms: If $1 \le j_1 < \cdots < j_k \le n$, $j = (j_1, \dots, j_k)$, we set

$$\alpha = \sum_j a_j(x)\, dx_{j_1} \wedge \cdots \wedge dx_{j_k}, \tag{13.12}$$

where

$$a_j(x) = \alpha(D_{j_1}, \dots, D_{j_k}), \quad D_j = \frac{\partial}{\partial x_j}. \tag{13.13}$$

More generally, we assign meaning to (13.12) summed over all k-indices $(j_1, \dots, j_k)$, where we identify

$$dx_{j_1} \wedge \cdots \wedge dx_{j_k} = (\operatorname{sgn} \sigma)\, dx_{j_{\sigma(1)}} \wedge \cdots \wedge dx_{j_{\sigma(k)}}, \tag{13.14}$$

σ being a permutation of $\{1, \dots, k\}$. If any $j_m = j_\ell$ ($m \neq \ell$), then (13.14) vanishes. A common notation for the statement that α is a k-form on Ω is

$$\alpha \in \Lambda^k(\Omega). \tag{13.15}$$

In particular, we can write a 2-form β as

$$\beta = \sum b_{jk}(x)\, dx_j \wedge dx_k \tag{13.16}$$

and pick coefficients satisfying $b_{jk}(x) = -b_{kj}(x)$. According to (13.12) and (13.13), if we set $U = \sum u_j(x)\, \partial/\partial x_j$ and $V = \sum v_j(x)\, \partial/\partial x_j$, then

$$\beta(U, V) = 2\sum b_{jk}(x) u^j(x) v^k(x). \tag{13.17}$$

If b_{jk} is not required to be antisymmetric, one gets $\beta(U, V) = \sum (b_{jk} - b_{kj}) u^j v^k$.

If $F : \mathcal{O} \to \Omega$ is a smooth map as above, we define the pull-back $F^*\alpha$ of a k-form α, given by (13.12), to be

$$F^*\alpha = \sum_j a_j\big(F(y)\big)(F^*dx_{j_1}) \wedge \cdots \wedge (F^*dx_{j_k}), \tag{13.18}$$

where

$$F^*dx_j = \sum_\ell \frac{\partial F_j}{\partial y_\ell}\, dy_\ell, \tag{13.19}$$

the algebraic computation in (13.18) being performed using the rule (13.14). Extending (13.9), if F is a diffeomorphism, we have

$$(F^*\alpha)(F_\# X_1, \dots, F_\# X_k) = \alpha(X_1, \dots, X_k) \circ F. \tag{13.20}$$

If $B = (b_{jk})$ is an $n \times n$ matrix, then, by (13.14),

$$\begin{aligned} &\Big(\sum_k b_{1k}\, dx_k\Big) \wedge \Big(\sum_k b_{2k}\, dx_k\Big) \wedge \cdots \wedge \Big(\sum_k b_{nk}\, dx_k\Big) \\ &= \Big(\sum_\sigma (\operatorname{sgn} \sigma) b_{1\sigma(1)} b_{2\sigma(2)} \cdots b_{n\sigma(n)}\Big)\, dx_1 \wedge \cdots \wedge dx_n \\ &= \big(\det B\big)\, dx_1 \wedge \cdots \wedge dx_n, \end{aligned} \tag{13.21}$$

Hence, if $F : \mathcal{O} \to \Omega$ is a C^1-map between two domains of dimension n, and $\alpha = A(x)\, dx_1 \wedge \cdots \wedge dx_n$ is an n-form on Ω, then

$$F^*\alpha = \det DF(y)\, A(F(y))\, dy_1 \wedge \cdots \wedge dy_n. \tag{13.22}$$

Comparison with the change-of-variable formula for multiple integrals suggests that one has an intrinsic definition of $\int_\Omega \alpha$ when α is an n-form on Ω, $n = \dim \Omega$. To implement this, we need to take into account that $\det DF(y)$ rather than $|\det DF(y)|$ appears in (13.21). We say that a smooth map $F : \mathcal{O} \to \Omega$ between two open subsets of $\mathbb{R}^n$ *preserves orientation* if $\det DF(y)$ is everywhere positive. The object called an "orientation" on Ω can be identified as an equivalence class of nowhere-vanishing n-forms on Ω, where two such forms are equivalent if one is a multiple of another by a positive function in $C^\infty(\Omega)$; the standard orientation on $\mathbb{R}^n$ is determined by $dx_1 \wedge \cdots \wedge dx_n$. If S is an n-dimensional surface in $\mathbb{R}^{n+k}$, an orientation on S can also be specified by a nowhere-vanishing form $\omega \in \Lambda^n(S)$. If such a form exists, S is said to be orientable. The equivalence class of positive multiples $a(x)\omega$ is said to consist of "positive" forms. A smooth map $\psi : S \to M$ between oriented n-dimensional surfaces preserves orientation provided $\psi^*\sigma$ is positive on S whenever $\sigma \in \Lambda^n(M)$ is positive. If S is oriented, one can choose coordinate charts that are all orientation-preserving. Surfaces that cannot be oriented also exist.

If $\mathcal{O}, \Omega$ are open in $\mathbb{R}^n$ and $F : \mathcal{O} \to \Omega$ is an orientation-preserving diffeomorphism, we have

$$\int_{\mathcal{O}} F^*\alpha = \int_{\Omega} \alpha. \tag{13.23}$$

More generally, if S is an n-dimensional manifold with an orientation, say the image of an open set $\mathcal{O} \subset \mathbb{R}^n$ by $\varphi : \mathcal{O} \to S$, carrying the natural orientation of $\mathcal{O}$, we can set

$$\int_S \alpha = \int_{\mathcal{O}} \varphi^*\alpha \tag{13.24}$$

for an n-form α on S. If it takes several coordinate patches to cover S, define $\int_S \alpha$ by writing α as a sum of forms, each supported on one patch.

We need to show that this definition of $\int_S \alpha$ is independent of the choice of coordinate system on S (as long as the orientation of S is respected). Thus, suppose $\varphi : \mathcal{O} \to U \subset S$ and $\psi : \Omega \to U \subset S$ are both coordinate patches, so that $F = \psi^{-1} \circ \varphi : \mathcal{O} \to \Omega$ is an orientation-preserving diffeomorphism. We need to check that if α is an n-form on S, supported on U, then

$$\int_{\mathcal{O}} \varphi^*\alpha = \int_{\Omega} \psi^*\alpha. \tag{13.25}$$

To see this, first note that, for any form α of any degree,

$$\psi \circ F = \varphi \Longrightarrow \varphi^*\alpha = F^*\psi^*\alpha. \tag{13.26}$$

It suffices to check this for $\alpha = dx_j$. Then $\psi^*\, dx_j = \sum(\partial\psi_j/\partial x_\ell)\, dx_\ell$, by (13.14), so

$$F^*\psi^*\, dx_j = \sum_{\ell,m} \frac{\partial F_\ell}{\partial x_m}\frac{\partial \psi_j}{\partial x_\ell}\, dx_m, \quad \varphi^*\, dx_j = \sum_m \frac{\partial \varphi_j}{\partial x_m}\, dx_m; \tag{13.27}$$

but the identity of these forms follows from the chain rule:

$$D\varphi = (D\psi)(DF) \Longrightarrow \frac{\partial \varphi_j}{\partial x_m} = \sum_\ell \frac{\partial \psi_j}{\partial x_\ell}\frac{\partial F_\ell}{\partial x_m}. \tag{13.28}$$

Now that we have (13.26), we see that the left side of (13.25) is equal to

$$\int_{\mathcal{O}} F^*(\psi^*\alpha), \tag{13.29}$$

which is equal to the right side of (13.25), by (13.23). Thus the integral of an n-form over an oriented n-dimensional surface is well defined.

Having discussed the notion of a differential form as something to be integrated, we now consider some operations on forms. There is a *wedge product*, or exterior product, characterized as follows. If $\alpha \in \Lambda^k(\Omega)$ has the form (13.12), and if

$$\beta = \sum_i b_i(x)\, dx_{i_1} \wedge \cdots \wedge dx_{i_\ell} \in \Lambda^\ell(\Omega), \tag{13.30}$$

define

$$\alpha \wedge \beta = \sum_{j,i} a_j(x) b_i(x)\, dx_{j_1} \wedge \cdots \wedge dx_{j_k} \wedge dx_{i_1} \wedge \cdots \wedge dx_{i_\ell} \tag{13.31}$$

in $\Lambda^{k+\ell}(\Omega)$. A special case of this arose in (13.18)–(13.21). We retain the equivalence (13.14). It follows easily that

$$\alpha \wedge \beta = (-1)^{k\ell} \beta \wedge \alpha. \tag{13.32}$$

In addition, there is an *interior product* if $\alpha \in \Lambda^k(\Omega)$ with a vector field X on Ω, producing $\iota_X \alpha = \alpha \rfloor X \in \Lambda^{k-1}(\Omega)$, defined by

$$(\alpha \rfloor X)(X_1, \dots, X_{k-1}) = \alpha(X, X_1, \dots, X_{k-1}). \tag{13.33}$$

Consequently, if $\alpha = dx_{j_1} \wedge \cdots \wedge dx_{j_k},\ D_i = \partial/\partial x_i$, then

$$\alpha \rfloor D_{j_\ell} = (-1)^{\ell-1} dx_{j_1} \wedge \cdots \wedge \widehat{dx}_{j_\ell} \wedge \cdots \wedge dx_{j_k}, \tag{13.34}$$

where $\widehat{dx}_{j_\ell}$ denotes removing the factor dx_{j_ℓ}. Furthermore,

$$i \notin \{j_1, \dots, j_k\} \Longrightarrow \alpha \rfloor D_i = 0.$$

If $F : \mathcal{O} \to \Omega$ is a diffeomorphism and α, β are forms and X a vector field on Ω, it is readily verified that

$$F^*(\alpha \wedge \beta) = (F^*\alpha) \wedge (F^*\beta), \quad F^*(\alpha \rfloor X) = (F^*\alpha) \rfloor (F_\# X). \tag{13.35}$$

We make use of the operators $\wedge_k$ and ι_k on forms:

$$\wedge_k \alpha = dx_k \wedge \alpha, \quad \iota_k \alpha = \alpha \rfloor D_k. \tag{13.36}$$

There is the following useful *anticommutation relation*:

$$\wedge_k \iota_\ell + \iota_\ell \wedge_k = \delta_{k\ell}, \tag{13.37}$$

where $\delta_{k\ell}$ is 1 if $k = \ell$, 0 otherwise. This is a fairly straightforward consequence of (13.34). We also have

$$\wedge_j \wedge_k + \wedge_k \wedge_j = 0, \quad \iota_j \iota_k + \iota_k \iota_j = 0. \tag{13.38}$$

From (13.37) and (13.38) one says that the operators $\{\iota_j, \wedge_j : 1 \le j \le n\}$ generate a "Clifford algebra." For more on this, see Chapter 10.

Another important operator on forms is the *exterior derivative*:

$$d : \Lambda^k(\Omega) \longrightarrow \Lambda^{k+1}(\Omega), \tag{13.39}$$

defined as follows. If $\alpha \in \Lambda^k(\Omega)$ is given by (13.12), then

$$d\alpha = \sum_{j,\ell} \frac{\partial a_j}{\partial x_\ell}\, dx_\ell \wedge dx_{j_1} \wedge \cdots \wedge dx_{j_k}. \tag{13.40}$$

Equivalently,

$$da = \sum_{\ell=1}^{n} \partial_\ell \wedge_\ell \alpha, \tag{13.41}$$

where $\partial_\ell = \partial/\partial x_\ell$ and $\wedge_\ell$ is given by (13.36). The antisymmetry $dx_m \wedge dx_\ell = -dx_\ell \wedge dx_m$, together with the identity $\partial^2 a_j/\partial x_\ell \partial x_m = \partial^2 a_j/\partial x_m \partial x_\ell$, implies

$$d(d\alpha) = 0, \tag{13.42}$$

for any differential form α. We also have a product rule:

$$d(\alpha \wedge \beta) = (d\alpha) \wedge \beta + (-1)^k \alpha \wedge (d\beta), \quad \alpha \in \Lambda^k(\Omega),\ \beta \in \Lambda^j(\Omega). \tag{13.43}$$

The exterior derivative has the following important property under pull-backs:

$$F^*(d\alpha) = dF^*\alpha, \tag{13.44}$$

if $\alpha \in \Lambda^k(\Omega)$ and $F : \mathcal{O} \to \Omega$ is a smooth map. To see this, extending (13.43) to a formula for $d(\alpha \wedge \beta_1 \wedge \cdots \wedge \beta_\ell)$ and using this to apply d to $F^*\alpha$, we have

$$\begin{aligned} dF^*\alpha = & \sum_{j,\ell} \frac{\partial}{\partial x_\ell}\big(a_j \circ F(x)\big)\, dx_\ell \wedge \big(F^* dx_{j_1}\big) \wedge \cdots \wedge \big(F^* dx_{j_k}\big) \\ & + \sum_{j,\nu} (\pm) a_j\big(F(x)\big)\big(F^* dx_{j_1}\big) \wedge \cdots \wedge d\big(F^* dx_{j_\nu}\big) \wedge \cdots \wedge \big(F^* dx_{j_k}\big). \end{aligned} \tag{13.45}$$

Now

$$d\big(F^* dx_i\big) = \sum_{j,\ell} \frac{\partial^2 F_i}{\partial x_j \partial x_\ell}\, dx_j \wedge dx_\ell = 0,$$

so only the first sum in (13.45) contributes to $dF^*\alpha$. Meanwhile,

$$F^* d\alpha = \sum_{j,m} \frac{\partial a_j}{\partial x_m}\big(F(x)\big)\, \big(F^* dx_m\big) \wedge \big(F^* dx_{j_1}\big) \wedge \cdots \wedge \big(F^* dx_{j_k}\big), \tag{13.46}$$

so (13.44) follows from the identity

$$\sum_{\ell} \frac{\partial}{\partial x_\ell}\big(a_j \circ F(x)\big)\, dx_\ell = \sum_m \frac{\partial a_j}{\partial x_m}\big(F(x)\big)\, F^* dx_m, \tag{13.47}$$

which in turn follows from the chain rule.

If $d\alpha = 0$, we say α is *closed*; if $\alpha = d\beta$ for some $\beta \in \Lambda^{k-1}(\Omega)$, we say α is *exact*. Formula (13.42) implies that every exact form is closed. The converse is not always true globally. Consider the multivalued angular coordinate θ on $\mathbb{R}^2 \setminus (0, 0)$; $d\theta$ is a single-valued, closed form on $\mathbb{R}^2 \setminus (0, 0)$ that is not globally exact. As we will see shortly, every closed form is locally exact.

First we introduce another important construction. If $\alpha \in \Lambda^k(\Omega)$ and X is a vector field on Ω, generating a flow $\mathcal{F}_X^t$, the *Lie derivative* $\mathcal{L}_X\alpha$ is defined to be

$$\mathcal{L}_X \alpha = \frac{d}{dt}\big(\mathcal{F}_X^t\big)^* \alpha\big|_{t=0}. \tag{13.48}$$

Note the formal similarity to the definition (8.2) of $\mathcal{L}_X Y$ for a vector field Y. Recall the formula (8.4) for $\mathcal{L}_X Y$. The following is not only a computationally convenient formula for $\mathcal{L}_X \alpha$, but also an identity of fundamental importance.

Proosition 13.1. *We have*

$$\mathcal{L}_X \alpha = d(\alpha \rfloor X) + (d\alpha) \rfloor X. \tag{13.49}$$

Proof. First we compare both sides in the special case $X = \partial/\partial x_\ell = D_\ell$. Note that

$$\left(\mathcal{F}^t_{D_\ell}\right)^* \alpha = \sum_j a_j(x + te_\ell)\, dx_{j_1} \wedge \cdots \wedge dx_{j_k},$$

so

$$\mathcal{L}_{D_\ell} \alpha = \sum_j \frac{\partial a_j}{\partial x_\ell}\, dx_{j_1} \wedge \cdots \wedge dx_{j_k} = \partial_\ell \alpha. \tag{13.50}$$

To evaluate the right side of (13.49) with $X = D_\ell$, use (13.41) to write this quantity as

$$d(\iota_\ell \alpha) + \iota_\ell d\alpha = \sum_{j=1}^n \left(\partial_j \wedge_j \iota_\ell + \iota_\ell \partial_j \wedge_j\right) \alpha. \tag{13.51}$$

Using the commutativity of ∂_j with $\wedge_j$ and with ι_ℓ, and the anticommutation relations (13.37), we see that the right side of (13.51) is $\partial_\ell \alpha$, which coincides with (13.50). Thus the proposition holds for $X = \partial/\partial x_\ell$.

Now we can prove the proposition in general, for a smooth vector field X on Ω. It is to be verified at each point $x_0 \in \Omega$. If $X(x_0) \neq 0$, choose a coordinate system about x_0 so that $X = \partial/\partial x_1$, and use the calculation above. This shows that the desired identity holds on the set of points $\{x_0 \in \Omega : X(x_0) \neq 0\}$, and by continuity it holds on the closure of this set. However, if $x_0 \in \Omega$ has a neighborhood on which X vanishes, it is clear that $\mathcal{L}_X \alpha = 0$ near x_0 and also $\alpha \rfloor X$ and $d\alpha \rfloor X$ vanish near x_0. This completes the proof.

The identity (13.49) can furnish a formula for the exterior derivative in terms of Lie brackets, as follows. By (8.4) and (13.49), we have, for a k-form ω,

$$(\mathcal{L}_X \omega)(X_1, \dots, X_k) = X \cdot \omega(X_1, \dots, X_k) - \sum_j \omega(X_1, \dots, [X, X_j], \dots, X_k). \tag{13.52}$$

Now (13.49) can be rewritten as

$$\iota_X d\omega = \mathcal{L}_X \omega - d\iota_X \omega. \tag{13.53}$$

This implies

$$(d\omega)(X_0, X_1, \dots, X_k) = (\mathcal{L}_{X_0} \omega)(X_1, \dots, X_k) - (d\iota_{X_0} \omega)(X_1, \dots, X_k). \tag{13.54}$$

We can substitute (13.52) into the first term on the right in (13.54). In case ω is a 1-form, the last term is easily evaluated; we get

(13.55) $$(d\omega)(X_0, X_1) = X_0 \cdot \omega(X_1) - X_1 \cdot \omega(X_0) - \omega([X_0, X_1]).$$

More generally, we can tackle the last term on the right side of (13.54) by the same method, using (13.53) with ω replaced by the $(k-1)$-form $\iota_{X_0}\omega$. In this way we inductively obtain the formula

(13.56)
$$(d\omega)(X_0, \dots, X_k) = \sum_{\ell=0}^{k} (-1)^\ell X_\ell \cdot \omega(X_0, \dots, \widehat{X}_\ell, \dots, X_k) + \sum_{0\le\ell<j\le k} (-1)^{j+\ell} \omega([X_\ell, X_j], X_0, \dots, \widehat{X}_\ell, \dots, \widehat{X}_j, \dots, X_k).$$

Note that from (13.48) and the property $\mathcal{F}_X^{s+t} = \mathcal{F}_X^s \mathcal{F}_X^t$ it easily follows that

(13.57) $$\frac{d}{dt} (\mathcal{F}_X^t)^* \alpha = \mathcal{L}_X (\mathcal{F}_X^t)^* \alpha = (\mathcal{F}_X^t)^* \mathcal{L}_X \alpha.$$

It is useful to generalize this. Let F_t be any smooth family of diffeomorphisms from M to $F_t(M) \subset M$. Define vector fields X_t on $F_t(M)$ by

(13.58) $$\frac{d}{dt} F_t(x) = X_t(F_t(x)).$$

Then it easily follows that, for $\alpha \in \Lambda^k M$,

(13.59) $$\begin{aligned} \frac{d}{dt} F_t^* \alpha &= F_t^* \mathcal{L}_{X_t} \alpha \\ &= F_t^* \big[d(\alpha \rfloor X_t) + (d\alpha) \rfloor X_t \big]. \end{aligned}$$

In particular, if α is *closed*, then if F_t are diffeomorphisms for $0 \le t \le 1$,

(13.60) $$F_1^* \alpha - F_0^* \alpha = d\beta, \quad \beta = \int_0^1 F_t^* (\alpha \rfloor X_t)\, dt.$$

Using this, we can prove the celebrated *Poincaré lemma.*

Theorem 13.2. *If B is the unit ball in $\mathbb{R}^n$, centered at* 0, *$\alpha \in \Lambda^k(B)$, $k > 0$, and $d\alpha = 0$, then $\alpha = d\beta$ for some $\beta \in \Lambda^{k-1}(B)$.*

Proof. Consider the family of maps $F_t : B \to B$ given by $F_t(x) = tx$. For $0 < t \le 1$, these are diffeomorphisms, and the formula (13.59) applies. Note that

$$F_1^* \alpha = \alpha, \quad F_0^* \alpha = 0.$$

Now a simple limiting argument shows that (13.60) remains valid, so $\alpha = d\beta$ with

(13.61) $$\beta = \int_0^1 F_t^* (\alpha \rfloor V) t^{-1}\, dt,$$

where $V = r\partial/\partial r = \sum x_j\, \partial/\partial x_j$. Since $F_0^* = 0$, the apparent singularity in the integrand is removable.

Since in the proof of the theorem we dealt with F_t such that F_0 was not a diffeomorphism, we are motivated to generalize (13.60) to the case where $F_t : M \to N$ is a smooth family of maps, not necessarily diffeomorphisms. Then (13.58) does not work to define X_t as a vector field, but we do have

$$\frac{d}{dt} F_t(x) = Z(t,x); \quad Z(t,x) \in T_{F_t(x)}N. \tag{13.62}$$

Now in (13.60) we see that

$$F^*(\alpha \rfloor X_t)(Y_1, \dots, Y_{k-1}) = \alpha\big(F_t(x)\big)\big(X_t, DF_t(x)Y_1, \dots, DF_t(x)Y_{k-1}\big),$$

and we can replace X_t by $Z(t,x)$. Hence, in this more general case, if α is closed, we can write

$$F_1^*\alpha - F_0^*\alpha = d\beta, \quad \beta = \int_0^1 \gamma_t\, dt, \tag{13.63}$$

where, at $x \in M$,

$$\gamma_t(Y_1, \dots, Y_{k-1}) = \alpha\big(F_t(x)\big)\big(Z(t,x), DF_t(x)Y_1, \dots, DF_t(x)Y_{k-1}\big). \tag{13.64}$$

For an alternative approach to this homotopy invariance, see Exercise 7.

A basic result in the theory of differential forms is the generalized *Stokes formula*:

Proposition 13.3. *Given a compactly supported $(k-1)$-form β of class C^1 on an oriented k-dimensional manifold $\overline{M}$ (of class C^2) with boundary ∂M, with its natural orientation,*

$$\int_M d\beta = \int_{\partial M} \beta. \tag{13.65}$$

The orientation induced on ∂M is uniquely determined by the following requirement. If

$$\overline{M} = \mathbb{R}^k_- = \{x \in \mathbb{R}^k : x_1 \le 0\}, \tag{13.66}$$

then $\partial M = \{(x_2, \dots, x_k)\}$ has the orientation determined by $dx_2 \wedge \cdots \wedge dx_k$.

Proof. Using a partition of unity and invariance of the integral and the exterior derivative under coordinate transformations, it suffices to prove this when $\overline{M}$ has the form (13.66). In that case, we will be able to deduce (13.65) from the fundamental theorem of calculus. Indeed, if

$$\beta = b_j(x)\, dx_1 \wedge \cdots \wedge \widehat{dx_j} \wedge \cdots \wedge dx_k, \tag{13.67}$$

with $b_j(x)$ of bounded support, we have

$$d\beta = (-1)^{j-1} \frac{\partial b_j}{\partial x_j} \, dx_1 \wedge \cdots \wedge dx_k. \tag{13.68}$$

If $j > 1$, we have

$$\int_M d\beta = \int \left\{ \int_{-\infty}^{\infty} \frac{\partial b_j}{\partial x_j} \, dx_j \right\} dx' = 0, \tag{13.69}$$

and also $\kappa^* \beta = 0$, where $\kappa : \partial M \to \overline{M}$ is the inclusion. On the other hand, for $j = 1$, we have

$$\begin{aligned} \int_M d\beta &= \int \left\{ \int_{-\infty}^{0} \frac{\partial b_1}{\partial x_1} \, dx_1 \right\} dx_2 \cdots dx_k \\ &= \int b_1(0, x') \, dx' \\ &= \int_{\partial M} \beta. \end{aligned} \tag{13.70}$$

This proves Stokes' formula (13.65).

It is useful to allow singularities in ∂M. We say a point $p \in \overline{M}$ is a *corner* of dimension ν if there is a neighborhood $\overline{U}$ of p in $\overline{M}$ and a C^2-diffeomorphism of $\overline{U}$ onto a neighborhood of 0 in

$$K = \{x \in \mathbb{R}^k : x_j \leq 0, \text{ for } 1 \leq j \leq k - \nu\}, \tag{13.71}$$

where k is the dimension of M. If M is a C^2-manifold and every point $p \in \partial M$ is a corner (of some dimension), we say $\overline{M}$ is a C^2-manifold with corners. In such a case, ∂M is a locally finite union of C^2-manifolds with corners. The following result extends Proposition 13.3.

Proposition 13.4. *If $\overline{M}$ is a C^2-manifold of dimension k, with corners, and β is a compactly supported $(k-1)$-form of class C^1 on $\overline{M}$, then (13.65) holds.*

Proof. It suffices to establish this when β is supported on a small neighborhood of a corner $p \in \partial M$, of the form $\overline{U}$ described above. Hence it suffices to show that (13.65) holds whenever β is a $(k-1)$-form of class C^1, with compact support on K in (13.71); and we can take β to have the form (13.67). Then, for $j > k - \nu$,

(13.69) still holds, while for $j \le k - \nu$, we have, as in (13.70),
(13.72)

$$\int_K d\beta = (-1)^{j-1} \int \Big\{ \int_{-\infty}^0 \frac{\partial b_j}{\partial x_j}\, dx_j \Big\}\, dx_1 \cdots \widehat{dx_j} \cdots dx_k$$

$$= (-1)^{j-1} \int b_j(x_1, \dots, x_{j-1}, 0, x_{j+1}, \dots, x_k)\, dx_1 \cdots \widehat{dx_j} \cdots dx_k$$

$$= \int_{\partial K} \beta.$$

The reason we required $\overline{M}$ to be a manifold of class C^2 (with corners) in Propositions 13.3 and 13.4 is the following. Due to the formulas (13.18)–(13.19) for a pull-back, if β is of class C^j and F is of class C^ℓ, then $F^*\beta$ is generally of class C^μ, with $\mu = \min(j, \ell - 1)$. Thus, if $j = \ell = 1$, $F^*\beta$ might be only of class C^0, so there is not a well-defined notion of a differential form of class C^1 on a C^1-manifold, though such a notion is well defined on a C^2-manifold. This problem can be overcome, and one can extend Propositions 13.3 and 13.4 to the case where $\overline{M}$ is a C^1-manifold (with corners) and β is a $(k-1)$-form with the property that both β and $d\beta$ are continuous. We will not go into the details. Substantially more sophisticated generalizations are given in [Fed].

Exercises

1. If $F : U_0 \to U_1$ and $G : U_1 \to U_2$ are smooth maps and $\alpha \in \Lambda^k(U_2)$, (13.26) implies

$$(G \circ F)^*\alpha = F^*(G^*\alpha) \ \text{ in } \Lambda^k(U_0).$$

In the special case that $U_j = \mathbb{R}^n$, F and G are linear maps, and $k = n$, show that this identity implies

$$\det(GF) = (\det F)(\det G).$$

2. If α is a closed form and β is exact, show that $\alpha \wedge \beta$ is exact. (*Hint*: Use (13.43).)

Let $\Lambda^k(\mathbb{R}^n)$ denote the space of k-forms (13.12) with constant coefficients. If $T : \mathbb{R}^m \to \mathbb{R}^n$ is linear, then T^* preserves this class of spaces; we denote the map

$$\Lambda^k T^* : \Lambda^k \mathbb{R}^n \longrightarrow \Lambda^k \mathbb{R}^m.$$

Similarly, replacing T by T^* yields

$$\Lambda^k T : \Lambda^k \mathbb{R}^m \longrightarrow \Lambda^k \mathbb{R}^n.$$

3. Show that $\Lambda^k T$ is uniquely characterized as a linear map from $\Lambda^k\mathbb{R}^m$ to $\Lambda^k\mathbb{R}^n$ that satisfies

$$(\Lambda^k T)(v_1 \wedge \cdots \wedge v_k) = (Tv_1) \wedge \cdots \wedge (Tv_k), \quad v_j \in \mathbb{R}^m.$$

4. If $\{e_1, \dots, e_n\}$ is the standard orthonormal basis of $\mathbb{R}^n$, define an inner product on $\Lambda^k\mathbb{R}^n$ by declaring an orthonormal basis to be

$$\{e_{j_1} \wedge \cdots \wedge e_{j_k} : 1 \le j_1 < \cdots < j_k \le n\}.$$

Show that if $\{u_1, \dots, u_n\}$ is any other orthonormal basis of $\mathbb{R}^n$, then the set

$$\{u_{j_1} \wedge \cdots \wedge u_{j_k} : 1 \le j_1 < \cdots < j_k \le n\}$$

is an orthonormal basis of $\Lambda^k \mathbb{R}^n$.

5. Let F be a vector field on U, open in $\mathbb{R}^3$, $F = \sum_1^3 f_j(x)\, \partial/\partial x_j$. Consider the 1-form $\varphi = \sum_1^3 f_j(x)\, dx_j$. Show that $d\varphi$ and curl F are related in the following way:

$$\text{curl } F = \sum_1^3 g_j(x) \frac{\partial}{\partial x_j},$$

$$d\varphi = g_1(x)\, dx_2 \wedge dx_3 + g_2(x)\, dx_3 \wedge dx_1 + g_3(x)\, dx_1 \wedge dx_2.$$

6. If F and φ are related as in Exercise 5, show that curl F is uniquely specified by the relation

$$d\varphi \wedge \alpha = \langle \text{curl } F, \alpha \rangle \omega$$

for all 1-forms α on $U \subset \mathbb{R}^3$, where $\omega = dx_1 \wedge dx_2 \wedge dx_3$ is the volume form.

7. Suppose $f_0, f_1 : X \to Y$ are smoothly homotopic maps, via $\Phi : X \times \mathbb{R} \to Y$, $\Phi(x, j) = f_j(x)$. Let $\alpha \in \Lambda^k Y$ be closed. Apply (13.60) to $\tilde{\alpha} = \Phi^* \alpha \in \Lambda^k(X \times \mathbb{R})$, with $F_t(x, s) = (x, s + t)$, to obtain $\tilde{\beta} \in \Lambda^{k-1}(X \times \mathbb{R})$ such that $F_1^* \tilde{\alpha} - \tilde{\alpha} = d\tilde{\beta}$, and from there produce $\beta \in \Lambda^{k-1}(X)$ such that $f_1^* \alpha - f_0^* \alpha = d\beta$.
(*Hint*: Use $\beta = \iota^* \tilde{\beta}$, where $\iota(x) = (x, 0)$.)

For the next set of exercises, let Ω be a planar domain, $X = f(x, y)\, \partial/\partial x + g(x, y)\, \partial/\partial y$ a nonvanishing vector field on Ω. Consider the 1-form $\alpha = g(x, y)\, dx - f(x, y)\, dy$.

8. Let $\gamma : I \to \Omega$ be a smooth curve, $I = (a, b)$. Show that the image $C = \gamma(I)$ is the image of an integral curve of X if and only if $\gamma^* \alpha = 0$. Consequently, with slight abuse of notation, one describes the integral curves by $g\, dx - f\, dy = 0$. If α is exact (i.e., $\alpha = du$,) conclude that the level curves of u are the integral curves of X.
9. A function φ is called an integrating factor if $\tilde{\alpha} = \varphi \alpha$ is exact (i.e., if $d(\varphi \alpha) = 0$, provided Ω is simply connected). Show that an integrating factor always exists, at least locally. Show that $\varphi = e^v$ is an integrating factor if and only if $Xv = -$ div X. Reconsider Exercise 7 in §7. Find an integrating factor for $\alpha = (x^2 + y^2 - 1)\, dx - 2xy\, dy$.
10. Let Y be a vector field that you know how to linearize (i.e., conjugate to $\partial/\partial x$) and suppose $\mathcal{L}_Y \alpha = 0$. Show how to construct an integrating factor for α. Treat the more general case $\mathcal{L}_X \alpha = c\alpha$ for some constant c. Compare the discussion in §8 of the situation where $[X, Y] = cX$.

14. The symplectic form and canonical transformations

Recall from §10 that a Hamiltonian vector field on a region $\Omega \subset \mathbb{R}^{2n}$, with coordinates $\zeta = (x, \xi)$, is a vector field of the form

$$H_f = \sum_{j=1}^n \left[\frac{\partial f}{\partial \xi_j} \frac{\partial}{\partial x_j} - \frac{\partial f}{\partial x_j} \frac{\partial}{\partial \xi_j} \right]. \tag{14.1}$$

We want to gain an understanding of Hamiltonian vector fields, free from coordinates. In particular, we ask the following question. Let $F : \mathcal{O} \to \Omega$ be a diffeomorphism, and let H_f be a Hamiltonian vector field on Ω. Under what condition on F is $F_\# H_f$ a Hamiltonian vector field on $\mathcal{O}$?

A central object in this study is the *symplectic form*, a 2-form on $\mathbb{R}^{2n}$ defined by

$$\sigma = \sum_{j=1}^{n} d\xi_j \wedge dx_j. \tag{14.2}$$

Note that if

$$U = \sum \Big[u^j(\zeta) \frac{\partial}{\partial x_j} + a^j(\zeta) \frac{\partial}{\partial \xi_j} \Big], \quad V = \sum \Big[v^j(\zeta) \frac{\partial}{\partial x_j} + b^j(\zeta) \frac{\partial}{\partial \xi_j} \Big],$$

then

$$\sigma(U, V) = \sum_{j=1}^{n} \big[-u^j(\zeta) b^j(\zeta) + a^j(\zeta) v^j(\zeta) \big]. \tag{14.3}$$

In particular, σ satisfies the following nondegeneracy condition: If U has the property that, for some $(x_0, \xi_0) \in \mathbb{R}^{2n}$, $\sigma(U, V) = 0$ at (x_0, ξ_0) for all vector fields V, then U must vanish at (x_0, ξ_0). The relation between the symplectic form and Hamiltonian vector fields is as follows:

Proposition 14.1. *The vector field H_f is uniquely determined by the identity*

$$\sigma \rfloor H_f = -df. \tag{14.4}$$

Proof. The content of the identity is

$$\sigma(H_f, V) = -Vf, \tag{14.5}$$

for any smooth vector field V. If V has the form used in (14.3), then that identity gives

$$\sigma(H_f, V) = -\sum_{j=1}^{n} \Big[\frac{\partial f}{\partial \xi_j} b^j(\zeta) + \frac{\partial f}{\partial x_j} v^j(\zeta) \Big],$$

which coincides with the right side of (14.5). In view of the nondegeneracy of σ, the proposition is proved. Note the special case

$$\sigma(H_f, H_g) = \{f, g\}. \tag{14.6}$$

The following is an immediate corollary.

Proposition 14.2. *If $\mathcal{O}, \Omega$ are open in $\mathbb{R}^{2n}$, and $F : \mathcal{O} \to \Omega$ is a diffeomorphism preserving σ, that is, satisfying*

$$F^* \sigma = \sigma, \tag{14.7}$$

then for any $f \in C^\infty(\Omega)$, $F_\# H_f$ is Hamiltonian on Ω and

$$F_\# H_f = H_{F^* f}, \tag{14.8}$$

where $F^* f(y) = f(F(y))$.

A diffeomorphism satisfying (14.7) is called a *canonical transformation*, or a symplectic transformation. Let us now look at the condition on a vector field X on Ω that the flow $\mathcal{F}_X^t$ generated by X preserve σ for each t. There is a simple general condition in terms of the Lie derivative for a given form to be preserved.

Lemma 14.3. *Let* $\alpha \in \Lambda^k(\Omega)$. *Then* $(\mathcal{F}_X^t)^*\alpha = \alpha$ *for all* t *if and only if*

$$\mathcal{L}_X\alpha = 0.$$

Proof. This is an immediate consequence of (13.57).

Recall the formula (13.49):

$$\mathcal{L}_X\alpha = d(\alpha \rfloor X) + (d\alpha)\rfloor X. \tag{14.9}$$

We apply it in the case where $\alpha = \sigma$ is the symplectic form. Clearly, (14.2) implies

$$d\sigma = 0, \tag{14.10}$$

so

$$\mathcal{L}_X\sigma = d(\sigma \rfloor X). \tag{14.11}$$

Consequently, $\mathcal{F}_X^t$ preserves the symplectic form σ if and only if $d(\sigma \rfloor X) = 0$ on Ω. In view of Poincaré's lemma, at least locally, one has a smooth function $f(x, \xi)$ such that

$$\sigma \rfloor X = df, \tag{14.12}$$

provided $d(\sigma \rfloor X) = 0$. Any two f's satisfying (14.12) must differ by a constant, and it follows that such f exists globally provided Ω is simply connected. In view of Proposition 14.1, (14.12) is equivalent to the identity

$$X = -H_f. \tag{14.13}$$

In particular, we have established the following result.

Proposition 14.4. *The flow generated by a Hamiltonian vector field* H_f *preserves the symplectic form* σ.

It follows a fortiori that the flow $\mathcal{F}^t$ generated by a Hamiltonian vector field H_f leaves invariant the $2n$-form

$$v = \sigma \wedge \cdots \wedge \sigma \quad (n \text{ factors}),$$

which provides a *volume form* on Ω. That this volume form is preserved is known as a theorem of Liouville. This result has the following refinement. Let S be a level surface of the function f; suppose f is nondegenerate on S. Then we can define a $(2n-1)$-form w on S (giving rise to a volume element on S) which is also invariant under the flow $\mathcal{F}^t$, as follows. Let X be any vector field on Ω such that $Xf = 1$ on S, and define

$$w = j^*(v\rfloor X), \tag{14.14}$$

where $j : S \hookrightarrow \Omega$ is the natural inclusion. We claim this is well defined.

Lemma 14.5. *The form (14.14) is independent of the choice of X, as long as $Xf = 1$ on S.*

Proof. The difference of two such forms is $j^*(v\rfloor Y_1)$, where $Y_1 f = 0$ on S, that is, Y_1 is tangent to S. Now this form, acting on vectors $Y_2, \dots, Y_{2n}$, all tangent to S, is merely $(j^*v)(Y_1, \dots, Y_{2n})$; but obviously $j^*v = 0$ since $\dim S < 2n$.

We can now establish the invariance of the form w on S.

Proposition 14.6. *The form (14.14) is invariant under the flow $\mathcal{F}^t$ on S.*

Proof. Since v is invariant under $\mathcal{F}^t$, we have

$$\begin{aligned}\mathcal{F}^{t*}w &= j^*(\mathcal{F}^{t*}v\rfloor \mathcal{F}^t_\# X)\\ &= j^*(v\rfloor \mathcal{F}^t_\# X)\\ &= w + j^*\big(v\rfloor (\mathcal{F}^t_\# X - X)\big).\end{aligned}$$

Since $\mathcal{F}^{t*}f = f$, we see that $(\mathcal{F}^t_\# X)f = 1 = Xf$, so the last term vanishes, by Lemma 14.5, and the proof is complete.

Let $\mathcal{O} \subset \mathbb{R}^n$ be open; we claim that the symplectic form σ is well defined on $T^*\mathcal{O} = \mathcal{O} \times \mathbb{R}^n$, in the following sense. Suppose $g : \mathcal{O} \to \Omega$ is a diffeomorphism (i.e., a coordinate change). The map this induces from $T^*\mathcal{O}$ to $T^*\Omega$ is

$$G(x, \xi) = \big(g(x), \big((Dg)^t\big)^{-1}(x)\xi\big) = (y, \eta). \tag{14.15}$$

Our invariance result is

$$G^*\sigma = \sigma. \tag{14.16}$$

In fact, a stronger result is true. We can write

$$\sigma = d\kappa, \quad \kappa = \sum_j \xi_j\, dx_j, \tag{14.17}$$

where the 1-form κ is called the *contact form*. We claim that

$$G^*\kappa = \kappa, \tag{14.18}$$

which implies (14.16), since $G^*d\kappa = dG^*\kappa$. To see (14.18), note that

$$dy_j = \sum_k \frac{\partial g_j}{\partial x_k}\, dx_k, \quad \eta_j = \sum_\ell H_{j\ell}\xi_\ell,$$

where $(H_{j\ell})$ is the matrix of $\big((Dg)^t\big)^{-1}$, that is, the inverse matrix of $(\partial g_\ell/\partial x_j)$. Hence

$$\begin{aligned}\sum_j \eta_j\,dy_j &= \sum_{j,k,\ell} \frac{\partial g_j}{\partial x_k} H_{j\ell}\xi_\ell\,dx_k \\ &= \sum_{k,\ell} \delta_{k\ell}\xi_\ell\,dx_k \\ &= \sum_k \xi_k\,dx_k,\end{aligned} \tag{14.19}$$

which establishes (14.18).

As a particular case, a vector field Y on $\mathcal{O}$, generating a flow $\mathcal{F}_Y^t$ on $\mathcal{O}$, induces a flow $\mathcal{G}_Y^t$ on $T^*\mathcal{O}$. Not only does this flow preserve the symplectic form; in fact, $\mathcal{G}_Y^t$ is generated by the Hamiltonian vector field H_Φ, where

$$\Phi(x,\xi) = \langle Y(x),\xi\rangle = \sum_j \xi_j v^j(x) \tag{14.20}$$

if $Y = \sum v^j(x)\,\partial/\partial x_j$.

The symplectic form given by (14.2) can be regarded as a special case of a general symplectic form, which is a closed, nondegenerate 2-form on a domain (or manifold) Ω. Often such a form ω arises naturally, in a form not a priori looking like (14.2). It is a theorem of *Darboux* that locally one can pick coordinates in such a fashion that ω does take the standard form (14.2). We present a short proof, due to J. Moser, of that theorem.

To start, pick $p \in \Omega$, and consider $B = \omega(p)$, a nondegenerate, antisymmetric, bilinear form on the vector space $V = T_p\Omega$. It is a simple exercise in linear algebra that if one has such a form, then dim V must be even, say $2n$, and V has a basis $\{e_j, f_j : 1 \le j \le n\}$ such that

$$B(e_j, e_\ell) = B(f_j, f_\ell) = 0, \quad B(e_j, f_\ell) = \delta_{j\ell}, \tag{14.21}$$

for $1 \le j,\ell \le n$. Using such a basis to impose linear coordinates (x,ξ) on a neighborhood of p, taken to the origin, we have $\omega = \omega_0 = \sum d\xi_j \wedge dx_j$ at p. Thus Darboux' theorem follows from:

Proposition 14.7. *If ω and ω_0 are closed, nondegenerate 2-forms on Ω, and $\omega = \omega_0$ at $p \in \Omega$, then there is a diffeomorphism G_1 defined on a neighborhood of p, such that*

$$G_1(p) = p \ \text{ and } \ G_1^*\omega = \omega_0. \tag{14.22}$$

Proof. For $t \in [0,1]$, let

$$\omega_t = (1-t)\omega_0 + t\omega = \omega_0 + t\alpha, \quad \alpha = \omega - \omega_0. \tag{14.23}$$

Thus $\alpha = 0$ at p, and α is a closed 2-form. We can therefore write

$$\alpha = d\beta \tag{14.24}$$

on a neighborhood of p, and if β is given by the formula (13.61) in the proof of the Poincaré lemma, we have $\beta = 0$ at p. Since for each t, $\omega_t = \omega$ at p, we see that each ω_t is nondegenerate on some common neighborhood of p, for $t \in [0, 1]$.

Our strategy will be to produce a smooth family of local diffeomorphisms G_t, $0 \le t \le 1$, such that $G_t(p) = p$, $G_0 = id.$, and such that $G_t^*\omega_t$ is independent of t, hence $G_t^*\omega_t = \omega_0$. G_t will be specified by a time-varying family of vector fields, via the ODE

$$\frac{d}{dt} G_t(x) = X_t(G_t(x)), \quad G_0(x) = x. \tag{14.25}$$

We will have $G_t(p) = p$ provided $X_t(p) = 0$. To arrange for $G_t^*\omega_t$ to be independent of t, note that, by the product rule,

$$\frac{d}{dt} G_t^*\omega_t = G_t^*\mathcal{L}_{X_t}\omega_t + G_t^*\frac{d\omega_t}{dt}. \tag{14.26}$$

By (14.23), $d\omega_t/dt = \alpha = d\beta$, and by Proposition 13.1,

$$\mathcal{L}_{X_t}\omega_t = d(\omega_t \rfloor X_t) \tag{14.27}$$

since ω_t is closed. Thus we can write (14.26) as

$$\frac{d}{dt} G_t^*\omega_t = G_t^* d(\omega_t \rfloor X_t + \beta). \tag{14.28}$$

This vanishes provided X_t is defined to satisfy

$$\omega_t \rfloor X_t = -\beta. \tag{14.29}$$

Since ω_t is nondegenerate near p, this does indeed uniquely specify a vector field X_t near p, for each $t \in [0, 1]$, which vanishes at p, since $\beta = 0$ at p. The proof of Darboux' theorem is complete.

Exercises

1. Do the linear algebra exercise stated before Proposition 14.7, as a preparation for the proof of Darboux' theorem.
2. On $\mathbb{R}^2$, identify (x, ξ) with (x, y), so the symplectic form is $\sigma = dy \wedge dx$. Show that

$$X = f\frac{\partial}{\partial x} + g\frac{\partial}{\partial y} \quad \text{and} \quad \alpha = g\,dx - f\,dy$$

are related by

$$\alpha = \sigma \rfloor X.$$

Reconsider Exercises 8–10 of §13 in light of this.

3. Show that the volume form w on the level surface S of f, given by (14.14), can be characterized as follows. Let S_h be the level set $\{f(x, \xi) = c + h\}$, $S = S_0$. Given any vector field X transversal to S, any open set $\mathcal{O} \subset S$ with smooth boundary, let $\widetilde{\mathcal{O}}_h$ be the thin set sandwiched between S and S_h, lying on orbits of X through $\mathcal{O}$. Then, with

$v = \sigma \wedge \cdots \wedge \sigma$ the volume form on Ω,

$$\int_{\mathcal{O}} w = \lim_{h \to 0} \frac{1}{h} \int_{\tilde{\mathcal{O}}_h} v.$$

4. A manifold $M \subset \mathbb{R}^{2n}$ is said to be *coisotropic* if, for each $p \in M$, the tangent space T_pM *contains* its symplectic annihilator

$$T_p^\sigma = \{w \in \mathbb{R}^{2n} : \sigma(v, w) = 0 \text{ for all } v \in T_pM\}.$$

It is said to be *Lagrangian* if $T_pM = T_p^\sigma$ for all $p \in M$. If M is coisotropic, show that it is naturally foliated by manifolds $\{N_q\}$ such that, for $p \in N_q$, $T_pN_q = T_p^\sigma$. (*Hint*: Apply Frobenius's theorem.)

15. First-order, scalar, nonlinear PDE

This section is devoted to a study of PDE of the form

$$F(x, u, \nabla u) = 0, \tag{15.1}$$

for a real-valued $u \in C^\infty(\Omega)$, $\dim \Omega = n$, given $F(x, u, \xi)$ smooth on $\Omega \times \mathbb{R} \times \mathbb{R}^n$, or some subdomain thereof. We study local solutions of (15.1) satisfying

$$u|_S = v, \tag{15.2}$$

where S is a smooth hypersurface of Ω, $v \in C^\infty(S)$. The study being local, we suppose S is given by $x_n = 0$. Pick a point $x_0 \in S \subset \mathbb{R}^n$, and set $\zeta_0 = (\partial v/\partial x_1, \dots, \partial v/\partial x_{n-1})$ at x_0. Assume

$$\begin{aligned} &F(x_0, v(x_0), (\zeta_0, \tau_0)) = 0, \\ &\frac{\partial F}{\partial \xi_n} \neq 0 \text{ at this point.} \end{aligned} \tag{15.3}$$

We call this the *noncharacteristic* hypothesis on S. We look for a solution to (15.1) near x_0.

In the paragraph above, ∇u denotes the n-tuple $(\partial u/\partial x_1, \dots, \partial u/\partial x_n)$. In view of the material in §§13 and 14, one should be used to the idea that the 1-form $du = \sum(\partial u/\partial x_j)\, dx_j$ has an invariant meaning. As we will see later, a Riemannian metric on Ω then associates to du a vector field, denoted grad u.

Thus, we will rephrase (15.1) as

$$F(x, u, du) = 0. \tag{15.4}$$

We think of F as being defined on $T^*\Omega \times \mathbb{R}$, or some open subset of this space. The first case we will treat is the case

$$F(x, du) = 0. \tag{15.5}$$

This sort of equation is known as an *eikonal* equation. From the treatment of (15.5), we will be able to deduce a treatment of the general case (15.4), using a device known as Jacobi's trick.

The equation (15.5) is intimately connected with the theory of Hamiltonian systems. We will use this theory to construct a surface Λ in $\mathbb{R}^{2n}$, of dimension n, the graph of a function $\xi = \Xi(x)$, which ought to be the graph of du for some smooth u. Thus our first goal is to produce a geometrical description of when

$$\Lambda = \text{graph of } \xi = \Xi(x) \tag{15.6}$$

is the graph of du for some smooth u.

Proposition 15.1. *The surface (15.6) is locally the graph of du for some smooth u if and only if*

$$\frac{\partial \Xi_j}{\partial x_k} = \frac{\partial \Xi_k}{\partial x_j}, \quad \forall\, j, k. \tag{15.7}$$

Proof. This follows from the Poincaré lemma, since (15.7) is the same as the condition that $\sum \Xi_j(x)\, dx_j$ be closed.

The next step is to produce the following geometrical restatement.

Proposition 15.2. *The surface Λ of (15.6) is the graph of du (locally) if and only if $\sigma(X, Y) = 0$ for all vectors X, Y tangent to Λ, where σ is the symplectic form.*

If Λ satisfies this condition, and $\dim \Lambda = n$, we say Λ is a *Lagrangian* surface.

Proof. We may as well check $\sigma(X_j, X_k)$ for some specific set $X_1, \dots, X_n$ of linearly independent vector fields, tangent to Λ. Thus, take

$$X_j = \frac{\partial}{\partial x_j} + \sum_\ell \frac{\partial \Xi_\ell}{\partial x_j} \frac{\partial}{\partial \xi_\ell}. \tag{15.8}$$

In view of the formula (14.3), we have

$$\sigma(X_j, X_k) = \frac{\partial \Xi_k}{\partial x_j} - \frac{\partial \Xi_j}{\partial x_k}, \tag{15.9}$$

so the result follows from Proposition 15.1.

To continue our pursuit of the solution to (15.5), we next specify a surface Σ, of dimension $n-1$, lying over $S = \{x_n = 0\}$, namely, with $\partial_j v = \partial v / \partial x_j$,

$$\Sigma = \{(x, \xi) : x_n = 0,\ \xi_j = \partial_j v,\ \text{ for } 1 \le j \le n-1,\ F(x, \xi) = 0\}. \tag{15.10}$$

The noncharacteristic hypothesis implies, by the implicit function theorem, that (with $x' = (x_1, \dots, x_{n-1})$), the equation

$$F(x', 0; \partial_1 v, \dots, \partial_{n-1} v, \tau) = 0$$

implicitly defines $\tau = \tau(x')$, so (15.10) defines a smooth surface of dimension $n-1$ through the point $(x_0, (\zeta_0, \tau_0))$.

We now define Λ to be the union of the integral curves of the Hamiltonian vector field H_F through Σ. Note that the noncharacteristic hypothesis implies that H_F has a nonvanishing $\partial/\partial x_n$ component over S, so Λ is a surface of dimension n, and is the graph of a function $\xi = \Xi(x)$, at least for x close to x_0 (Fig. 15.1). Since F is constant on integral curves of H_F, it follows that $F = 0$ on Λ.

Theorem 15.3. *The surface Λ constructed above is locally the graph of du, for a solution u to*

$$F(x, du) = 0, \quad u|_S = v. \tag{15.11}$$

Proof. We will show that Λ is Lagrangian. So let X, Y be vector fields tangent to Λ at (x, ξ) in $\Lambda \subset \mathbb{R}^{2n}$. We need to examine $\sigma(X, Y)$. First suppose $x \in S$ (i.e., $(x, \xi) \in \Sigma$). Then we may decompose X and Y into $X = X_1 + X_2,\ Y = Y_1 + Y_2$, with X_1, Y_1 tangent to Σ and X_2, Y_2 multiples of H_F at (x, ξ). It suffices to show that $\sigma(X_1, Y_1) = 0$ and $\sigma(X_1, Y_2) = 0$. Since Σ, regarded simply as projecting over $\{x_n = 0\}$, is the graph of a gradient, Proposition 15.2 implies $\sigma(X_1, Y_1) = 0$. On the other hand, $\sigma(X_1, Y_2)$ is a multiple of $\sigma(X_1, H_F) = \langle X_1, dF\rangle = X_1 F$. Since X_1 is tangent to Σ and $F = 0$ on Σ, $X_1 F = 0$.

Thus we know that $\sigma(X, Y) = 0$ if X and Y are tangent to Λ at a point in Σ. Suppose now that X and Y are tangent to Λ at a point $\mathcal{F}^t(x, \xi)$, where $(x, \xi) \in \Sigma$ and $\mathcal{F}^t$ is the flow generated by H_F. We have

$$\sigma(X, Y) = \big(\mathcal{F}^{t*}\sigma\big)(\mathcal{F}^t_{\#}X, \mathcal{F}^t_{\#}Y).$$

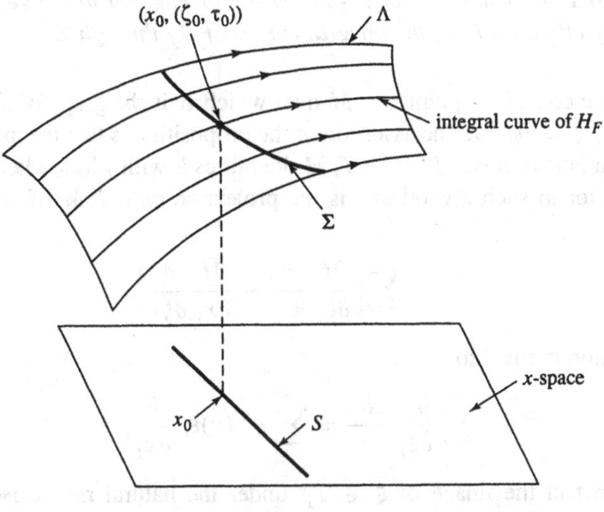

FIGURE 15.1

Now $\mathcal{F}^t_{\#}X$ and $\mathcal{F}^t_{\#}Y$ are tangent to Λ at $(x, \xi) \in \Sigma$. We use the important fact that the flow generated by H_F leaves the symplectic form invariant to conclude that

$$\sigma(X, Y) = \sigma(\mathcal{F}^t_{\#}X, \mathcal{F}^t_{\#}Y) = 0.$$

This shows that Λ is Lagrangian.

Thus Λ is the graph of du for some smooth u, uniquely determined up to an additive constant. Pick $x_0 \in S$ and set $u(x_0) = v(x_0)$. We see that, on S, $\partial u/\partial x_j = \partial v/\partial x_j$ for $1 \le j \le n-1$, so this forces $u|_S = v$. We have seen that $F = 0$ on Λ, so we have solved (15.11).

An important example of an eikonal equation is

$$|d\varphi|^2 = 1 \tag{15.12}$$

on a Riemannian manifold, with metric tensor g_{jk}. In local coordinates, (15.12) is

$$\sum_{j,k} g^{jk}(x) \frac{\partial \varphi}{\partial x_j} \frac{\partial \varphi}{\partial x_k} = 1, \tag{15.13}$$

where, as before, (g^{jk}) is the matrix inverse to (g_{jk}). We want to give a geometrical description of solutions to this equation. Let φ be specified on a hypersurface $S \subset M$; $\varphi|_S = \psi$. Assume that $|d\psi| < 1$ on S. Then there are two possible sections of T^*M over S, giving the graphs of $d\varphi$ over S. Pick one of then; call it Σ. As we have seen, the graph of $d\varphi$ is the flow-out Λ of Σ, via the flow generated by H_f, with $f(x, \xi) = (1/2)|\xi|^2 = (1/2)\sum g^{jk}(x)\xi_j\xi_k$, that is, via the "geodesic flow" on T^*M. The projections onto M of the integral curves of H_f in T^*M are geodesics on M. The geometrical description of φ arises from the following result.

Proposition 15.4. *The level surfaces of φ are orthogonal to the geodesics that are the projections on M of the integral curves of H_f through Σ.*

Proof. If we consider a point $x \in M$ over which Λ is the graph of $d\varphi$, we have $(x, \xi) \in \Lambda$, $\xi = d\varphi(x)$. The assertion of the proposition is that the metric tensor, inducing an isomorphism $T^*_xM \approx T_xM$, identifies ξ with $\gamma'(t)$, where $\gamma'(t)$, the tangent vector to such a geodesic, is the projection onto T_xM of H_f at (x, ξ). Since

$$H_f = \sum \Big[\frac{\partial f}{\partial \xi_j} \frac{\partial}{\partial x_j} - \frac{\partial f}{\partial x_j} \frac{\partial}{\partial \xi_j} \Big], \tag{15.14}$$

this projection is equal to

$$\sum \frac{\partial f}{\partial \xi_j} \frac{\partial}{\partial x_j} = \sum g^{jk}(x)\xi_k \frac{\partial}{\partial x_j}, \tag{15.15}$$

which is in fact the image of $\xi \in T^*_x$ under the natural metric isomorphism $T^*_xM \approx T_xM$. This proves the proposition.

We can restate it this way. The metric isomorphism $T^*M \approx TM$ produces from the 1-form $d\varphi$, the *gradient vector field* grad φ. In local coordinates, with $d\varphi = \sum(\partial\varphi/\partial x_j)\,dx_j$, we have

$$\text{grad}\,\varphi = \sum g^{jk}(x)\frac{\partial\varphi}{\partial x_j}\frac{\partial}{\partial x_k}. \tag{15.16}$$

Thus, the content of the last proposition is the following:

Corollary 15.5. *If $\gamma(t)$ is the geodesic of unit speed that is the projection on M of an integral curve of H_f through Σ, then*

$$\text{grad}\,\varphi(x) = \gamma'(t), \quad \text{at } x = \gamma(t). \tag{15.17}$$

Suppose, for example, that for an initial condition on φ we take $\varphi = c$ (constant) on the surface S. Then, near S, the other level sets of φ are described as follows. For $p \in S$, let $\gamma_p(t)$ be the unit-speed geodesic through p, so $\gamma_p(0) = p$, orthogonal to S, going in one of two possible directions, corresponding to a choice of one of two possible Σs, as mentioned above. Then

$$\varphi(x) = c + t, \quad \text{at } x = \gamma_p(t). \tag{15.18}$$

This gives a very geometrical picture of solutions to (15.12).

On flat Euclidean space, where geodesics are just straight lines, these formulas become quite explicit. Suppose, for example, that we want to solve $|d\varphi|^2 = 1$ on $\mathbb{R}^n$ (i.e., $\sum(\partial\varphi/\partial x_j)^2 = 1$), and we prescribe

$$\varphi = 0 \text{ on a surface } S \text{ defined by } \psi(x) = 0, \tag{15.19}$$

where $\psi(x)$ is given. Then it is clear that, for $|t|$ not too large, φ is defined by

$$\varphi\big(x + t|\nabla\psi(x)|^{-1}\nabla\psi(x)\big) = t, \quad \text{for } x \in S. \tag{15.20}$$

For small a, $I = (-a, a)$, the map

$$\Psi : S \times I \longrightarrow \mathbb{R}^n \tag{15.21}$$

given by

$$\Psi(x, t) = x + t|\nabla\psi(x)|^{-1}\nabla\psi(x) \tag{15.22}$$

is a diffeomorphism, but simple examples show that this can break down for large $|t|$.

Having solved the special sort of first-order PDE known as the eikonal equation, we now tackle the general case (15.1)–(15.2), subject to the condition (15.3). We use a method, called Jacobi's trick, of defining u implicitly by

$$V(x, u(x)) = 0 \tag{15.23}$$

and producing a PDE for V of the eikonal type. Indeed (15.23) gives, with $V = V(x, z)$,

$$\nabla_x V + V_z \nabla u = 0, \text{ or } \nabla u = -V_z^{-1}\nabla_x V, \tag{15.24}$$

so set

(15.25) $$g(x, z, \xi, \zeta) = F(x, z, -\zeta^{-1}\xi).$$

Our equation for V is hence $F(x, z, -V_z^{-1}\nabla_x V) = 0$, or

(15.26) $$g(x, z, \nabla_{x,z} V) = 0.$$

This is of eikonal type. Our initial condition is

(15.27) $$V = z - v \text{ on } x_n = 0.$$

This gives $V_z \neq 0$ locally, so by the implicit function theorem, (15.23) defines a function $u(x)$, which solves the system (15.1)–(15.2).

Exercises

1. Let X be a vector field on a region Ω, generating a flow $\mathcal{F}^t$, which we will assume is defined everywhere. Consider the *linear* PDE

(15.28) $$\frac{\partial u}{\partial t} = Xu, \quad u(0, x) = f(x).$$

Show that a solution is given by

$$u(t, x) = f(\mathcal{F}^t x).$$

Show that the equation

(15.29) $$\frac{\partial u}{\partial t} = Xu + g(t, x), \quad u(0, x) = f(x)$$

is solved by

$$u(t, x) = f(\mathcal{F}^t x) + \int_0^t g\big(s, \mathcal{F}^{t-s} x\big)\, ds,$$

and that

(15.30) $$\frac{\partial u}{\partial t} = Xu + a(t, x)u, \quad u(0, x) = f(x)$$

is solved by

$$u(t, x) = e^{\left(\int_0^t a(s, \mathcal{F}^{t-s} x) ds\right)} f(\mathcal{F}^t x).$$

(*Hint*: The solution to (15.28) is constant on integral curves of $\partial/\partial t - X$ in $\mathbb{R} \times \Omega$. Apply Duhamel's principle to (15.29). Then find $A(t, x)$ such that (15.30) is equivalent to

$$e^{-A}\Big(\frac{\partial}{\partial t} - X\Big)(e^A u) = 0.)$$

2. A PDE of the form

$$\frac{\partial u}{\partial t} + \sum_{j=1}^n a_j(x, u) \frac{\partial u}{\partial x_j} = 0,$$

for a real-valued $u = u(t, x)$, is a special case of a *quasilinear* equation. Show that if we set $u(0, x) = v(x) \in C^\infty(\mathbb{R}^n)$, then there is a unique smooth solution in a neighborhood of $\{0\} \times \mathbb{R}^n$ in $\mathbb{R}^{n+1}$, and $u(t,x)$ has the following property. For each $x_0 \in \mathbb{R}^n$, consider the vector field

$$V_{x_0} = \frac{\partial}{\partial t} + \sum_{j=1}^{n} a_j(x, v(x_0)) \frac{\partial}{\partial x_j}.$$

Then $u(t,x)$ is equal to $v(x_0)$ on the integral curve of V_{x_0} through $(0, x_0)$. Considering the example

$$u_t + u u_x = 0, \quad u(0,x) = e^{-x^2},$$

show that this smooth solution can cease to exist globally, due to two such lines crossing.

3. Work out explicitly the solution to

$$\left(\frac{\partial \varphi}{\partial x}\right)^2 + \left(\frac{\partial \varphi}{\partial y}\right)^2 = 1,$$

satisfying $\varphi(x,y) = 0$ on the parabola $y = x^2$, and $\partial\varphi/\partial y > 0$ there, using (15.19) and (15.20). Write a computer program to graph the level curves of φ. How does the solution break down?

4. The group of dilations of T^*M, defined (in local coordinates) by $D(r)(x,\xi) = (x, r\xi)$, is generated by a vector field ϑ on T^*M, which we call the natural radial vector field. Show that ϑ is uniquely specified by the identity

$$\sigma(\vartheta, X) = \langle X, \kappa \rangle,$$

when X is a vector field on T^*M, and $\kappa = \sum \xi_j \, dx_j$ is the contact form (14.17).

5. Suppose Λ is a submanifold of T^*M of dimension $n = \dim M$, with $\iota : \Lambda \hookrightarrow T^*M$. Show that Λ is Lagrangian if and only if $\iota^*\kappa$ is a closed 1-form on Λ (hence locally exact). If Λ is Lagrangian, relate $\iota^*\kappa = df$ on Λ to du, in the context of Proposition 15.1.

6. Suppose Λ is a Lagrangian submanifold of T^*M, transverse to ϑ. Define a subbundle $\mathcal{V}$ of $T\Lambda$ by

$$\mathcal{V}_{(x,\xi)} = (\vartheta)^\sigma \cap T_{(x,\xi)}\Lambda,$$

where $(\vartheta)^\sigma$ is the set of vectors $v \in T_{(x,\xi)}T^*M$ such that $\sigma(\vartheta, v) = 0$. Show that $\mathcal{V}$ is an *integrable* subbundle of $T\Lambda$, that is, that Frobenius's theorem applies to $\mathcal{V}$, giving a foliation of Λ. If Λ is the graph of du, $u \in C^\infty(M)$, show that the inverse image, under $\pi : \Lambda \to M$, of the level sets of u gives the leaves of this foliation of Λ.

16. Completely integrable Hamiltonian systems

Here we will examine the consequences of having n "conservation laws" for a Hamiltonian system with n degrees of freedom. More precisely, suppose $\mathcal{O}$ is a region in $\mathbb{R}^{2n}$, with coordinates (x,ξ) and symplectic form $\sigma = \sum_{j=1}^n d\xi_j \wedge dx_j$, or more generally $\mathcal{O}$ could be a symplectic manifold of dimension $2n$. Suppose we have n functions $u_1, \dots, u_n$, in involution, that is,

$$\{u_j, u_k\} = 0, \quad 1 \le j, k \le n. \tag{16.1}$$

The function $u_1 = F$ could be the energy function whose Hamiltonian vector field we want to analyze, and $u_2, \dots, u_n$ auxiliary functions, constructed to reflect conservation laws. We give some examples shortly. In case one has n such functions, with linearly independent gradients, one is said to have a *completely integrable* system.

Our goal here will be to show that in such a case the flows generated by the H_{u_j} can be constructed *by quadrature*. We define the last concept as follows. Given a collection of functions $\{u_j\}$, a map is said to be constructed by quadrature if it is produced by a composition of the following operations:

(i) elementary algebraic manipulation,
(ii) differentiation,
(iii) integration,
(iv) constructing inverses of maps.

To begin the study of a completely integrable system, given (16.1), consider, for a given $p \in \mathbb{R}^n$, the level set

$$M_p = \{(x, \xi) \in \mathcal{O} : u_j(x, \xi) = p_j\}. \tag{16.2}$$

Assuming the u_j have linearly independent gradients, each nonempty M_p is a manifold of dimension n. Note that each vector field H_{u_j} is tangent to M_p, by (16.1), and therefore $\{H_{u_j} : 1 \le j \le n\}$ *spans* the tangent space to M_p at each point. Since $\sigma(H_{u_j}, H_{u_k}) = \{u_j, u_k\}$, we conclude from (16.1) that

$$\text{each } M_p \text{ is Lagrangian.} \tag{16.3}$$

If we make the "generic" hypothesis

$$\pi : M_p \to \mathbb{R}^n \text{ is a local diffeomorphism,} \tag{16.4}$$

where $\pi(x, \xi) = x$, then M_p is the graph of a closed 1-form Ξ_p (depending smoothly on p); note that $\Xi_p(x)$ is constructed by inverting a map, one of the operations involved in construction by quadrature. Furthermore, Ξ_p being closed, we can construct a smooth function $\varphi(x, p)$ such that

$$M_p \text{ is the graph of } x \mapsto d_x\varphi(x, p). \tag{16.5}$$

The function $\varphi(x, p)$ is constructed from Ξ_p by an integration, another ingredient in construction by quadrature. Note that a statement equivalent to (16.5) is that φ simultaneously satisfies the eikonal equations

$$u_j\big(x, d_x\varphi(x, p)\big) = p_j, \quad 1 \le j \le n. \tag{16.6}$$

Consider now the following maps:

$$\begin{array}{ccc} (x, p) & \xrightarrow{F_1} & \big(d_p\varphi(x, p), p\big) \\ & & \Big\downarrow c \\ (x, p) & \xrightarrow{F_2} & \big(x, d_x\varphi(x, p)\big). \end{array} \tag{16.7}$$

Since $F_2(x, p) = (x, \Xi_p(x))$, it is clear that F_2 is a local diffeomorphism under our hypotheses. This implies that the matrix

$$\frac{\partial^2 \varphi}{\partial p_j \partial x_k} \tag{16.8}$$

is invertible, which hence implies that F_1 is a local diffeomorphism (by the inverse function theorem). Hence C is locally defined, as a diffeomorphism:

$$C\big(d_p\varphi(x, p), p\big) = \big(x, d_x\varphi(x, p)\big). \tag{16.9}$$

Write $C(q, p) = (x, \xi)$. Note that

$$\begin{aligned} F_2^* \sum d\xi_j \wedge dx_j &= \sum_{j,k} \frac{\partial^2 \varphi}{\partial p_k \partial x_j} dp_k \wedge dx_j \\ &= F_1^* \sum dp_j \wedge dq_j, \end{aligned} \tag{16.10}$$

so

$$C^*\Big(\sum d\xi_j \wedge dx_j\Big) = \sum dp_j \wedge dq_j, \tag{16.11}$$

that is, C preserves the symplectic form. One says C is a canonical transformation with *generating function* $\varphi(x, p)$. Now conjugation by C takes the Hamiltonian vector fields H_{u_j} on (x, ξ)-space to the Hamiltonian vector fields $H_{\tilde{u}_j}$ on (q, p)-space, with

$$\tilde{u}_j(q, p) = u_j \circ C(q, p) = p_j,$$

in view of (16.6). Thus

$$H_{\tilde{u}_j} = \frac{\partial}{\partial q_j}, \tag{16.12}$$

so C conjugates the flows generated by H_{u_j} to simple straight-line flows. This provides the construction of the H_{u_j}-flows by quadrature.

Note that if $\mathcal{O}$ has dimension 2, one needs only one function u_1. Thus the construction above generalizes the treatment of Hamiltonian systems on $\mathbb{R}^2$ given in §10. In fact, the approach given above, specialized to $n = 1$, is closer to the analysis in §10 than it might at first appear. Using notation as in §10, let $u_1 = f,\ p_1 = E$, so

$$M_E = \{(x, \xi) : f(x, \xi) = E\}$$

is the graph of $\xi = \psi(x, E) = d_x\varphi(x, E)$, with

$$\varphi(x, E) = \int \psi(x, E)\, dx.$$

Note that $f(x, \psi(x, E)) = E \Rightarrow f_\xi \psi_E = 1$, so

$$d_E\varphi(x, E) = \int f_\xi\big(x, \psi(x, E)\big)^{-1}\, dx, \tag{16.13}$$

and $\mathcal{C}$ maps $(\int f_\xi^{-1} dx, E)$ to $(x, \psi(x, \xi))$. To say $\mathcal{C}$ conjugates H_f to $H_E = \partial/\partial q$ (in (q, E) coordinates) is to say that under the time-t Hamiltonian flow, $\int f_\xi^{-1} dx$ is augmented by t; but this is precisely the content of (10.16), namely,

$$\int f_\xi\big(x, \psi(x, E)\big)^{-1}\, dx = t + C(E). \tag{16.14}$$

We also note that, for the purpose of linearizing H_{u_1}, it suffices to have $\varphi(x, p)$, satisfying only the eikonal equation

$$u_1\big(x, d_x\varphi(x, p)\big) = p_1, \tag{16.15}$$

such that the matrix (16.8) is invertible. The existence of $u_2, \dots, u_n$, which together with u_1 are in involution, provides a way to construct $\varphi(x, p)$, but any other successful attack on (16.15) is just as satisfactory. Integrating H_{u_1} by perceiving solutions to (16.15) is the essence of the Hamilton-Jacobi method.

We now look at some examples of completely integrable Hamiltonian systems. First we consider geodesic flow on a two-dimensional surface of revolution $M^2 \subset \mathbb{R}^3$. Note that T^*M^2 is four-dimensional, so we want u_1 and u_2, in involution. The function u_1 is, of course, the energy funtion $u_1 = (1/2)\sum g^{jk}(x)\xi_j\xi_k$; as we have seen, H_{u_1} generates the geodesic flow. Our function u_2 will arise from the group of rotations R_θ of M^2 about its axis of symmetry, $\theta \in \mathbb{R}/2\pi\mathbb{Z}$. This produces a group $\mathcal{R}_\theta$ of canonical transformations of T^*M^2, generated by a Hamiltonian vector field $X = H_{u_2}$, with $u_2(x, \xi) = \langle \partial/\partial\theta, \xi\rangle$. Since R_θ is a group of isometries of M^2, $\mathcal{R}_\theta$ preserves u_1 (i.e., $Xu_1 = 0$), or equivalently, $\{u_2, u_1\} = 0$. We have our pair of functions in involution. Thus geodesics on such a surface of revolution can be constructed by quadrature.

Another important class of completely integrable Hamiltonian systems is provided by motion in a central force field in the plane $\mathbb{R}^2$. In other words, let $x(t)$, a path in $\mathbb{R}^2$, satisfy

$$\ddot{x} = -\nabla V(x), \quad V(x) = v(|x|). \tag{16.16}$$

The Hamiltonian system is

$$\dot{x} = \nabla_\xi F, \quad \dot{\xi} = -\nabla_x F, \tag{16.17}$$

with

$$F(x, \xi) = \frac{1}{2}|\xi|^2 + v(|x|). \tag{16.18}$$

We take $u_1 = F$ and look for u_2, in involution. Again u_2 arises from a group of rotations, this time rotations of $\mathbb{R}^2$ about the origin. The method we have given by which a vector field on Ω produces a Hamiltonian vector field on $T^*\Omega$ yields the formula

$$(16.19)\qquad \begin{aligned} u_2(x,\xi) &= \left\langle \frac{\partial}{\partial\theta}, \xi \right\rangle \\ &= \left\langle -x_2 \frac{\partial}{\partial x_1} + x_1 \frac{\partial}{\partial x_2}, \xi \right\rangle \\ &= x_1\xi_2 - x_2\xi_1. \end{aligned}$$

This is the "angular momentum." The symmetry of $V(x)$ implies that the group of rotations on $T^*\mathbb{R}^2$ generated by H_{u_2} preserves $F = u_1$, that is,

$$(16.20)\qquad \{u_1, u_2\} = 0,$$

a fact that is also easily verified from (16.18) and (16.19) by a computation. This expresses the well-known law of conservation of angular momentum. It also establishes the complete integrability of the general central force problem on $\mathbb{R}^2$. We remark that, for the general central force problem in $\mathbb{R}^n$, conservation of angular momentum forces any path to lie in a plane, so there is no loss of generality in studying planar motion.

The case

$$(16.21)\qquad V(x) = -\frac{K}{|x|} \quad (K > 0)$$

of the central force problem is called the Kepler problem. It gives Newton's description of a planet traveling about a massive star, or of two celestial bodies revolving about their center of mass. We will give a direct study of central force problems, with particular attention to the Kepler problem, in the next section.

These examples of completely integrable systems have been based on only the simplest of symmetry considerations. For many other examples of completely integrable systems, see [Wh].

We have dealt here only with the local behavior of completely integrable systems. There is also an interesting "global" theory, which among other things studies the distinction between the regular behavior of completely integable systems on the one hand and varieties of "chaotic behavior" exhibited by (globally) nonintegrable systems on the other. The reader can find out more about this important topic (begun in [Poi]) in [Mos], [TS], [Wig], and references given therein.

Exercises

1. Let $u_1(x,\xi) = (1/2)|\xi|^2 - |x|^{-1}$ be the energy function for the Kepler problem (with $K = 1$), and let $u_2(x,\xi)$ be given by (16.19). Set

$$v_j(x,\xi) = x_j|x|^{-1} - x_j|\xi|^2 + (x\cdot\xi)\xi_j, \quad j = 1, 2.$$

(v_1, v_2) is called the Lenz vector. Show that the following Poisson bracket relations hold:

$$\{u_1, v_j\} = 0, \quad j = 1, 2,$$
$$\{u_2, v_j\} = \pm v_j,$$
$$\{v_1, v_2\} = 2u_1 u_2.$$

Also show that

$$v_1^2 + v_2^2 - 2u_1 u_2^2 = 1.$$

2. Deduce that the Kepler problem is integrable in several different ways. Can you relate this to the fact that all bounded orbits are periodic?

In Exercises 3–5, suppose a given M_p, as in (16.2), is compact, and du_j, $1 \le j \le n$ are linearly independent at each point of M_p.

3. Show that there is an $\mathbb{R}^n$-action on M_p, defined by $\Phi(t)(\zeta) = \mathcal{F}_1^{t_1} \cdots \mathcal{F}_n^{t_n} \zeta$, for $t = (t_1, \dots, t_n)$, $\zeta \in M_p$, where $\mathcal{F}_j^s$ is the flow generated by H_{u_j}. Show that $\Phi(t+s)\zeta = \Phi(t)\Phi(s)\zeta$.
4. Show that $\mathbb{R}^n$ acts transitively on M_p, that is, given $\zeta \in M_p$, $\mathcal{O}(\zeta) = \{\Phi(t)\zeta : t \in \mathbb{R}^n\}$ is all of M_p. (*Hint*: Use the linear independence to show $\mathcal{O}(\zeta)$ is open. Then, if ζ_1 is on the boundary of $\mathcal{O}(\zeta)$ in M_p, show that $\mathcal{O}(\zeta_1) \cap \mathcal{O}(\zeta) \neq \emptyset$.)
5. Fix $\zeta_0 \in M_p$ and let $\Gamma = \{t \in \mathbb{R}^n : \Phi(t)\zeta_0 = \zeta_0\}$. Show that M_p is diffeomorphic to $\mathbb{R}^n / \Gamma$ and that this is a torus.
6. If $u_1 = F$ can be extended to a completely integrable system in two different ways, with the setting of Exercises 3–5 applicable in each case, then phase space may be foliated by tori in two different ways. Hence intersections of various tori will be invariant under H_F. How does this relate to Exercise 2?

17. Examples of integrable systems; central force problems

In the last section it was noted that central force problems give rise to a class of completely integrable Hamiltonian systems with two degrees of freedom. Here we will look at this again, from a more elementary point of view. We look at a class of Hamiltonians on a region in $\mathbb{R}^4$, of the form

$$F(y, \eta) = F(y_1, \eta_1, \eta_2), \tag{17.1}$$

that is, with no y_2-dependence. Thus Hamilton's equations take the form

$$\dot{y}_j = \frac{\partial F}{\partial \eta_j}, \quad \dot{\eta}_1 = -\frac{\partial F}{\partial y_1}, \quad \dot{\eta}_2 = 0. \tag{17.2}$$

In particular, η_2 is constant on any orbit, say

$$\eta_2 = L. \tag{17.3}$$

This, in addition to F, provides the second conservation law implying integrability; note that $\{F, \eta_2\} = 0$. If $F(y_1, \eta_1, L) = E$ on an integral curve, we write this relation as

$$\eta_1 = \psi(y_1, L, E). \tag{17.4}$$

We can now pursue an analysis that is a variant of that described by (10.14)–(10.20). The first equation in (17.2) becomes

$$\dot{y}_1 = F_{\eta_1}\big(y_1, \psi(y_1, L, E), L\big), \tag{17.5}$$

with solution given implicitly by

$$\int F_{\eta_1}\big(y_1, \psi(y_1, L, E), L\big)^{-1} dy_1 = t + C. \tag{17.6}$$

Once one has $y_1(t)$, then one has

$$\eta_1(t) = \psi(y_1(t), L, E), \tag{17.7}$$

and then the remaining equation in (17.2) becomes

$$\dot{y}_2 = F_{\eta_2}\big(y_1(t), \eta_1(t), L\big), \tag{17.8}$$

which is solved by an integration.

We apply this method to the central force problem, with

$$F(x, \xi) = \frac{1}{2}|\xi|^2 + v(|x|), \quad x \in \mathbb{R}^2. \tag{17.9}$$

Use of polar coordinates is clearly suggested, so we set

$$y_1 = r,\ y_2 = \theta; \quad x_1 = r\cos\theta,\ x_2 = r\sin\theta. \tag{17.10}$$

In these coordinates, the Euclidean metric $dx_1^2 + dx_2^2$ becomes $dr^2 + r^2 d\theta^2$, so, as in (12.31), the function F becomes

$$F(y, \eta) = \frac{1}{2}\big(\eta_1^2 + y_1^{-2}\eta_2^2\big) + v(y_1). \tag{17.11}$$

We see that the first pair of ODEs in (17.2) takes the form

$$\dot{r} = \eta_1, \quad \dot{\theta} = Lr^{-2}, \tag{17.12}$$

where L is the constant value of η_2 along an integral curve, as in (17.3). The last equation, rewritten as

$$r^2\dot{\theta} = L, \tag{17.13}$$

expresses conservation of angular momentum. The remaining ODE in (17.2) becomes

$$\dot{\eta}_1 = L^2 r^{-3} - v'(r). \tag{17.14}$$

Note that differentiating the first equation of (17.12) and using (17.14) gives

$$\ddot{r} = L^2 r^{-3} - v'(r), \tag{17.15}$$

an equation that can be integrated by the methods described in (10.12)–(10.20). We will not solve (17.15) by this means here, though (17.15) will be used below, to produce (17.23). For now, we instead use (17.4)–(17.6). In the present case, (17.4) takes the form

$$\eta_1 = \pm\big[2E - 2v(r) - L^2 r^{-2}\big]^{1/2}, \tag{17.16}$$

and since $F_{\eta_1} = \eta_1$, (17.6) takes the form

$$\pm \int \left[2Er^2 - 2r^2 v(r) - L^2\right]^{-1/2} r\, dr = t + C. \tag{17.17}$$

In the case of the Kepler problem (16.21), where $v(r) = -K/r$, the resulting integral

$$\pm \int \left(2Er^2 + 2Kr - L^2\right)^{-1/2} r\, dr = t + C \tag{17.18}$$

can be evaluated using techniques of first-year calculus, by completing the square in $2Er^2 + 2Kr - L^2$. Once $r = r(t)$ is given, the equation (17.13) provides an integral formula for $\theta = \theta(t)$.

One of the most remarkable aspects of the analysis of the Kepler problem is the demonstration that orbits all lie on some conic section, given in polar coordinates by

$$r\left[1 + e\,\cos(\theta - \theta_0)\right] = ed, \tag{17.19}$$

where e is the "eccentricity." We now describe the famous, elementary but ingenious trick used to demonstrate this. The method involves producing a differential equation for r in terms of θ, from (17.13) and (17.15). More precisely, we produce a differential equation for u, defined by

$$u = r^{-1}. \tag{17.20}$$

By the chain rule,

$$\frac{dr}{dt} = -r^2 \frac{du}{dt} = -r^2 \frac{du}{d\theta}\frac{d\theta}{dt} = -L\frac{du}{d\theta}, \tag{17.21}$$

in light of (17.13). Differentiating this with respect to t gives

$$\frac{d^2 r}{dt^2} = -L\frac{d}{dt}\frac{du}{d\theta} = -L\frac{d^2 u}{d\theta^2}\frac{d\theta}{dt} = -L^2 u^2 \frac{d^2 u}{d\theta^2}, \tag{17.22}$$

again using (17.13). Comparing this with (17.15), we get $-L^2u^2(d^2u/d\theta^2) = L^2u^3 - v'(1/u)$ or, equivalently,

$$\frac{d^2 u}{d\theta^2} + u = (Lu)^{-2} v'\left(\frac{1}{u}\right). \tag{17.23}$$

In the case of the Kepler problem, $v(r) = -K/r$, the right side becomes the constant K/L^2, so in this case (17.23) becomes the *linear* equation

$$\frac{d^2 u}{d\theta^2} + u = \frac{K}{L^2}, \tag{17.24}$$

with general solution

$$u(\theta) = A\cos(\theta - \theta_0) + \frac{K}{L^2}, \tag{17.25}$$

which is equivalent to the formula (17.19) for a conic section.

For more general central force problems, the equation (17.23) is typically not linear, but it is of the form treatable by the method of (10.12)–(10.20).

Exercises

1. Solve explicitly $w''(t) = -w(t)$, for w taking values in $\mathbb{R}^2 = \mathbb{C}$. Show that $|w(t)|^2 + |w'(t)|^2 = 2E$ is constant for each orbit.
2. For $w(t)$ taking values in $\mathbb{C}$, define a new curve by
$$Z(\tau) = w(t)^2, \quad \frac{d\tau}{dt} = |w(t)|^2.$$
Show that if $w''(t) = -w(t)$, then
$$Z''(\tau) = -4E\,\frac{Z(\tau)}{|Z(\tau)|^3},$$
that is, $Z(\tau)$ solves the Kepler problem.
3. Analyze the flow of H_F, for F of the form (17.1), in a manner more directly parallel to the approach in §16, in a spirit similar to (16.13) and (16.14). Note that, with $u_1 = F,\ u_2 = \eta_2,\ p_1 = E,\ p_2 = L$, the canonical transformation C of (16.9) is defined by
$$C\Bigl(\int F_{\eta_1}^{-1}\,dy_1,\, y_2 - \int F_{\eta_1}^{-1}F_{\eta_2}\,dy_1;\, E, L\Bigr) = \bigl(y_1, y_2; \psi(y_1, L, E), L\bigr),$$
where the first integrand is $F_{\eta_1}\bigl(y_1, \psi(y_1, L, E), L\bigr)^{-1}$, and so on.
4. Analyze the equation (17.23) for $u(\theta)$ in the following cases.

   (a) $v(r) = -K/r^2$,
   (b) $v(r) = Kr^2$,
   (c) $v(r) = -K/r + \varepsilon r^2$.

   Show that, in case (c), $u(\theta)$ is typically not periodic in θ.
5. Consider motion on a surface of revolution, under a force arising from a rotationally invariant potential. Show that you can choose coordinates (r, θ) so that the metric tensor is $ds^2 = dr^2 + \beta(r)^{-1}\,d\theta^2$, and then you get a Hamiltonian system of the form (17.2) with
$$F(y_1, \eta_1, \eta_2) = \frac{1}{2}\eta_1^2 + \frac{1}{2}\beta(y_1)\eta_2^2 + v(y_1),$$
where $y_1 = r,\ y_2 = \theta$. Show that, parallel to (17.16) and (17.17), you get
$$\dot r = \pm\bigl[2E - 2v(r) - L^2\beta(r)\bigr]^{1/2}.$$
Show that $u = 1/r$ satisfies
$$\frac{du}{d\theta} = \mp\frac{u^2}{L\beta(1/u)}\Bigl[2E - 2v\Bigl(\frac{1}{u}\Bigr) - L^2\beta\Bigl(\frac{1}{u}\Bigr)\Bigr]^{1/2}.$$

18. Relativistic motion

Mechanical systems considered in previous sections were formulated in the Newtonian framework. The description of a particle moving subject to a force was given in terms of a curve in *space* (with a positive-definite metric), parameterized by *time*. In the relativistic set-up, one has not space and time as separate entities, but rather *spacetime*, provided with a metric of Lorentz signature. In particular, Minkowski spacetime is $\mathbb{R}^4$ with inner product

$$\langle x, y\rangle = -x_0 y_0 + \sum_{j=1}^{3} x_j y_j, \tag{18.1}$$

given $x = (x_0, \dots, x_3),\ y = (y_0, \dots, y_3)$. The behavior of a particle moving in a force field is described by a curve in spacetime, which is *timelike*, that is, its tangent vector T satisfies $\langle T, T\rangle < 0$. We parameterize the curve not by time, but by arc length, so we consider a curve $x(\tau)$ satisfying

$$\langle u(\tau), u(\tau)\rangle = -1, \quad u(\tau) = x'(\tau). \tag{18.2}$$

The parameter τ is often called "proper time," and $u(\tau)$ the "4-velocity." Such a curve $x(\tau)$ is sometimes called a "world line."

Relativistic laws of physics are to be formulated in a manner depending only on the Lorentz metric (18.1), but contact is made with the Newtonian picture by using the product decomposition $\mathbb{R}^4 = \mathbb{R} \times \mathbb{R}^3$, writing $x = (t, x_s),\ t = x_0$, and $x_s = (x_1, x_2, x_3)$. The "3-velocity" is $v = dx_s/dt$. Then

$$u = \gamma(1, v), \tag{18.3}$$

where, by (18.2),

$$\gamma = \frac{dt}{d\tau} = \left(1 - |v|^2\right)^{-1/2}, \tag{18.4}$$

with $|v|^2 = v_1^2 + v_2^2 + v_3^2$. In the limit of small velocities, γ is close to 1.

The particle whose motion is to be described is assumed to have a constant "rest mass" m_0, and then the "4-momentum" is defined to be

$$p = m_0 u. \tag{18.5}$$

In terms of the decomposition (18.3),

$$p = (m_0\gamma, m_0\gamma v), \tag{18.6}$$

where $m_0 v$ is the momentum in Newtonian theory. The replacement for Newton's equation $m_0 dv/dt = f$ is

$$\frac{dp}{d\tau} = F, \tag{18.7}$$

the right side being the "Minkowski 4-force."

Newtonian theory and Einstein's relativity are related as follows. Define m by $m = m_0\gamma$ and, using (18.6) and (18.7), write

$$F = \left(\frac{dm}{d\tau}, \frac{d(mv)}{d\tau}\right) = \left(\frac{dm}{d\tau}, \gamma\frac{d(mv)}{dt}\right). \tag{18.8}$$

Then we identify $f_C = d(mv)/dt$ as the "classical force" and write the last expression as $(f^0, \gamma f_C)$. If (18.2) is to hold, we require $f^0 = \gamma f_C \cdot v$ (the dot product in Euclidean $\mathbb{R}^3$), so

$$F = \gamma(f_C \cdot v, f_C). \tag{18.9}$$

With this correspondence, the equation (18.7) yields Newton's equation in the small velocity limit.

Since the 4-velocity has constant length, by (18.2), the Minkowski 4-force F must satisfy

$$\langle F, u \rangle = 0. \tag{18.10}$$

It follows that in relativity one cannot have velocity-independent forces. The simplest situation compatible with (18.10) is for F to be linear in u, say

$$F(x, u) = \tilde{\mathcal{F}}(x)u, \tag{18.11}$$

where for each $x \in \mathbb{R}^4$, $\tilde{\mathcal{F}}(x)$ is a linear transformation on $\mathbb{R}^4$; in other words, $\tilde{\mathcal{F}}$ is a tensor field of type (1, 1). The condition (18.10) holds provided $\tilde{\mathcal{F}}$ is skew-adjoint with respect to the Lorentz inner product:

$$\langle \tilde{\mathcal{F}}u, w \rangle = -\langle u, \tilde{\mathcal{F}}w \rangle. \tag{18.12}$$

Equivalently, if we consider the related tensor $\mathcal{F}$ of type (0, 2),

$$\mathcal{F}(u, w) = \langle u, \tilde{\mathcal{F}}w \rangle, \tag{18.13}$$

then $\mathcal{F}$ is antisymmetric, that is, $\mathcal{F}$ is a 2-form. In index notation, $\mathcal{F}_{jk} = h_{j\ell}\mathcal{F}^{\ell}{}_k$, where h_{jk} defines the Lorentz metric.

The electromagnetic field is of this sort. The classical force exerted by an electric field E and a magnetic field B on a particle with charge e is the Lorentz force

$$f_L = e(E + v \times B), \tag{18.14}$$

as in (12.40). Using this in (18.9) gives, for $u = (u^0, v)$,

$$\tilde{\mathcal{F}}u = e(E \cdot v, Eu^0 + v \times B). \tag{18.15}$$

Consequently the 2-form $\mathcal{F}$ is $\mathcal{F}(u, w) = e \sum \mathcal{F}_{\mu\nu} u_\mu w_\nu$ with

$$(\mathcal{F}_{\mu\nu}) = \begin{pmatrix} 0 & -E_1 & -E_2 & -E_3 \\ E_1 & 0 & B_3 & -B_2 \\ E_2 & -B_3 & 0 & B_1 \\ E_3 & B_2 & -B_1 & 0 \end{pmatrix}. \tag{18.16}$$

In relativity it is this 2-form which is called the electromagnetic field.

To change notation slightly, let us denote by $\mathcal{F}$ the 2-form described by (18.16), namely, with $t = x_0$,

$$\mathcal{F} = \sum_{j=1}^{3} E_j \, dx_j \wedge dt + B_1 \, dx_2 \wedge dx_3 + B_2 \, dx_3 \wedge dx_1 + B_3 \, dx_1 \wedge dx_2. \tag{18.17}$$

Thus the force in (18.11) is now denoted by $e\tilde{\mathcal{F}}u$.

We can construct a Lagrangian giving the equation of motion (18.7), (18.11), in a fashion similar to (12.44). The part of Maxwell's equations for the electromagnetic field recorded as (12.41) is equivalent to the statement that

$$d\mathcal{F} = 0. \tag{18.18}$$

Thus we can find a 1-form $\mathcal{A}$ on Minkowski spacetime such that

$$\mathcal{F} = d\mathcal{A}. \tag{18.19}$$

Then we can set

$$L(x, u) = \frac{1}{2} m_0 \langle u, u \rangle + e \langle \mathcal{A}, u \rangle, \tag{18.20}$$

and the force law $dp/d\tau = e\widetilde{\mathcal{F}}(x)u$ is seen to be equivalent to

$$\frac{d}{d\tau} L_u - L_x = 0. \tag{18.21}$$

See Exercise 3 below. In this case, the Legendre transform (12.13) becomes, with $u^b = (-u^0, u^1, u^2, u^3)$,

$$(x, \xi) = (x, m_0 u^b + e\mathcal{A}), \tag{18.22}$$

and we get the Hamiltonian system

$$\frac{dx}{d\tau} = E_\xi, \quad \frac{d\xi}{d\tau} = -E_x, \tag{18.23}$$

with

$$E(x, \xi) = \frac{1}{2m_0} \langle \xi - e\mathcal{A}, \xi - e\mathcal{A} \rangle. \tag{18.24}$$

Exercises

1. Consider a constant electromagnetic field of the form

$$E = (1, 0, 0), \quad B = 0.$$

Work out the solution to Newton's equation

$$m \frac{dv}{dt} = e(E + v \times B), \quad v = \frac{dx}{dt},$$

for the path $x = x(t)$ in $\mathbb{R}^3$ of a particle of charge e, mass m, moved by the Lorentz force arising from this field. Then work out the solution to the relativistic equation

$$m_0 \frac{du}{d\tau} = e(E \cdot v, Eu^0 + v \times B),$$

with $u = (u^0, v)$ (having square norm -1), $u = dx/d\tau$, for the path in $\mathbb{R}^4$ of a particle of charge e, rest mass m_0, moved by such an electromagnetic field. Compare the results. Do the same for

$$E = 0, \quad B = (1, 0, 0).$$

2. Take another look at Exercise 3 in §12.
3. Show that taking (18.20) for the Lagrangian implies that Lagrange's equation (18.21) is equivalent to the force law $dp/d\tau = e\tilde{\mathcal{F}}u$, on Minkowski spacetime.
(*Hint*: To compute L_x, use
$$d\langle \mathcal{A}, u\rangle = -(d\mathcal{A})\rfloor u + \mathcal{L}_u \mathcal{A},$$
regard u as independent of x, and note that $d\mathcal{A}/d\tau = \nabla_u \mathcal{A} = \mathcal{L}_u \mathcal{A}$, in that case.) Compare Exercise 4 in §12.
4. Verify formula (18.16) for $\mathcal{F}_{\mu\nu}$. Show that the matrix for $\tilde{\mathcal{F}}$ has the same form, except all E_j carry plus signs.
5. An alternative sign convention for the Lorentz metric on Minkowski spacetime is to replace (18.1) by $\langle x, y\rangle = x_0 y_0 - \sum_{j\geq 1} x_j y_j$. Show that this leads to a sign change in (18.16). What other sign changes arise?
6. Suppose a 1-form $\mathcal{A}$ is given, satisfying (18.19), on a general four-dimensional Lorentz manifold M. Let $L : TM \to \mathbb{R}$ be given by (18.20). Use the set-up described in (12.51)–(12.65) to derive equations of motion, extending the Lorentz force law from Minkowski spacetime to any Lorentz 4-manifold.
(*Hint*: In analogy with (12.64), show that L_v is given by
$$L_v = m_0 u + e\mathcal{A}^{\#},$$
where $\mathcal{A}^{\#}$ is the vector field corresponding to $\mathcal{A}$ via the metric (by raising indices). Taking a cue from Exercise 3, show that L_x satisfies
$$L_x = e\tilde{\mathcal{F}}u + e\nabla_u \mathcal{A}^{\#}.$$
Deduce that the equation
$$m_0 \nabla_u u = e\tilde{\mathcal{F}}u$$
is the stationary condition for this Lagrangian.)

19. Topological applications of differential forms

Differential forms are a fundamental tool in calculus. In addition, they have important applications to topology. We give a few here, starting with simple proofs of some important topological results of Brouwer.

Proposition 19.1. *There is no continuous retraction $\varphi : B \to S^{n-1}$ of the closed unit ball B in $\mathbb{R}^n$ onto its boundary S^{n-1}.*

In fact, it is just as easy to prove the following more general result. The approach we use is adapted from [Kan].

Proposition 19.2. *If $\overline{M}$ is a compact, oriented manifold with nonempty boundary ∂M, there is no continuous retraction $\varphi : \overline{M} \to \partial M$.*

Proof. A retraction φ satisfies $\varphi \circ j(x) = x$, where $j : \partial M \hookrightarrow \overline{M}$ is the natural inclusion. By a simple approximation, if there were a continuous retraction there would be a smooth one, so we can suppose φ is smooth.

Pick $\omega \in \Lambda^{n-1}(\partial M)$ to be the volume form on ∂M, endowed with some Riemannian metric ($n = \dim M$), so $\int_{\partial M} \omega > 0$. Now apply Stokes' theorem to $\alpha = \varphi^*\omega$. If φ is a retraction, $j^*\varphi^*\omega = \omega$, so we have

$$\int_{\partial M} \omega = \int_M d\varphi^*\omega. \tag{19.1}$$

But $d\varphi^*\omega = \varphi^* d\omega = 0$, so the integral (19.1) is zero. This is a contradiction, so there can be no retraction.

A simple consequence of this is the famous Brouwer fixed-point theorem.

Theorem 19.3. *If $F : B \to B$ is a continuous map on the closed unit ball in $\mathbb{R}^n$, then F has a fixed point.*

Proof. We are claiming that $F(x) = x$ for some $x \in B$. If not, define $\varphi(x)$ to be the endpoint of the ray from $F(x)$ to x, continued until it hits $\partial B = S^{n-1}$. It is clear that φ would be a retraction, contradicting Proposition 19.1.

We next show that an even-dimensional sphere cannot have a smooth nonvanishing vector field.

Proposition 19.4. *There is no smooth nonvanishing vector field on S^n if $n = 2k$ is even.*

Proof. If X were such a vector field, we could arrange it to have unit length, so we would have $X : S^n \to S^n$, with $X(v) \perp v$ for $v \in S^n \subset \mathbb{R}^{n+1}$. Thus there is a unique unit-speed geodesic γ_v from v to $X(v)$, of length $\pi/2$. Define a smooth family of maps $F_t : S^n \to S^n$ by $F_t(v) = \gamma_v(t)$. Thus $F_0(v) = v$, $F_{\pi/2}(v) = X(v)$, and $F_\pi = A$ would be the *antipodal map*, $A(v) = -v$. By (13.63), we deduce that $A^*\omega - \omega = d\beta$ is exact, where ω is the volume form on S^n. Hence, by Stokes' theorem,

$$\int_{S^n} A^*\omega = \int_{S^n} \omega. \tag{19.2}$$

On the other hand, it is straightforward that $A^*\omega = (-1)^{n+1}\omega$, so (19.2) is possible only when n is odd.

Note that an important ingredient in the proof of both Proposition 19.2 and Proposition 19.4 is the existence of n-forms on a compact, oriented, n-dimensional manifold M which are not exact (though of course they are closed). We next establish the following important counterpoint to the Poincaré lemma.

Proposition 19.5. *If M is a compact, connected, oriented manifold of dimension n and $\alpha \in \Lambda^n M$, then $\alpha = d\beta$ for some $\beta \in \Lambda^{n-1}(M)$ if and only if*

$$\int_M \alpha = 0. \tag{19.3}$$

We have already discussed the necessity of (19.3). To prove the sufficiency, we first look at the case $M = S^n$.

In that case, any n-form α is of the form $a(x)\omega$, $a \in C^\infty(S^n)$, ω the volume form on S^n, with its standard metric. The group $G = \mathrm{SO}(n+1)$ of rotations of $\mathbb{R}^{n+1}$ acts as a transitive group of isometries on S^n. In Appendix B, Manifolds, Vector Bundles, and Lie Groups, we construct the integral of functions over $\mathrm{SO}(n+1)$, with respect to Haar measure.

As noted in Appendix B, we have the map $\mathrm{Exp} : \mathrm{Skew}(n+1) \to \mathrm{SO}(n+1)$, giving a diffeomorphism from a ball $\mathcal{O}$ about 0 in $\mathrm{Skew}(n+1)$ onto an open set $U \subset \mathrm{SO}(n+1) = G$, a neighborhood of the identity. Since G is compact, we can pick a finite number of elements $\xi_j \in G$ such that the open sets $U_j = \{\xi_j g : g \in U\}$ cover G. Pick $\eta_j \in \mathrm{Skew}(n+1)$ such that $\mathrm{Exp}\,\eta_j = \xi_j$. Define $\Phi_{jt} : U_j \to G$ for $0 \le t \le 1$ by

$$\Phi_{jt}\big(\xi_j\,\mathrm{Exp}(A)\big) = (\mathrm{Exp}\,t\eta_j)(\mathrm{Exp}\,tA), \quad A \in \mathcal{O}. \tag{19.4}$$

Now partition G into subsets Ω_j, each of whose boundaries has content zero, such that $\Omega_j \subset U_j$. If $g \in \Omega_j$, set $g(t) = \Phi_{jt}(g)$. This family of elements of $SO(n+1)$ defines a family of maps $F_{gt} : S^n \to S^n$. Now, as in (13.60), we have

$$\alpha = g^*\alpha - d\kappa_g(\alpha), \quad \kappa_g(\alpha) = \int_0^1 F_{gt}^*(\alpha \rfloor X_{gt})\,dt, \tag{19.5}$$

for each $g \in \mathrm{SO}(n+1)$, where X_{gt} is the family of vector fields on S^n generated by F_{gt}, as in (13.58). Therefore,

$$\alpha = \int_G g^*\alpha\,dg - d\int_G \kappa_g(\alpha)\,dg. \tag{19.6}$$

Now the first term on the right is equal to $\overline{\alpha}\omega$, where $\overline{\alpha} = \int a(g\cdot x)dg$ is a constant; in fact, the constant is

$$\overline{\alpha} = \frac{1}{\mathrm{Vol}\,S^n}\int_{S^n} \alpha. \tag{19.7}$$

Thus, in this case, (19.3) is precisely what serves to make (19.6) a representation of α as an exact form. This finishes the case $M = S^n$.

For a general compact, oriented, connected M, proceed as follows. Cover M with open sets $\mathcal{O}_1, \dots, \mathcal{O}_K$ such that each $\overline{\mathcal{O}}_j$ is diffeomorphic to the closed unit ball in $\mathbb{R}^n$. Set $U_1 = \mathcal{O}_1$, and inductively enlarge each $\mathcal{O}_j$ to U_j, so that $\overline{U}_j$ is also diffeomorphic to the closed ball, and such that $U_{j+1} \cap U_j \ne \emptyset$, $1 \le j < K$. You can do this by drawing a simple curve from $\overline{\mathcal{O}}_{j+1}$ to a point in U_j and thickening it. Pick a smooth partition of unity φ_j, subordinate to this cover.

Given $\alpha \in \Lambda^n M$, satisfying (19.3), take $\tilde{\alpha}_j = \varphi_j \alpha$. Most likely $\int \tilde{\alpha}_1 = c_1 \neq 0$, so take $\sigma_1 \in \Lambda^n M$, with compact support in $U_1 \cap U_2$, such that $\int \sigma_1 = c_1$. Set $\alpha_1 = \tilde{\alpha}_1 - \sigma_1$, and redefine $\tilde{\alpha}_2$ to be the old $\tilde{\alpha}_2$ plus σ_1. Make a similar construction using $\int \tilde{\alpha}_2 = c_2$, and continue. When you are done, you have

$$\alpha = \alpha_1 + \cdots + \alpha_K, \tag{19.8}$$

with α_j compactly supported in U_j. By construction,

$$\int \alpha_j = 0, \tag{19.9}$$

for $1 \le j < K$. But then (19.3) implies $\int \alpha_K = 0$ too.

Now pick $p \in S^n$ and define smooth maps

$$\psi_j : M \longrightarrow S^n, \tag{19.10}$$

which map U_j diffeomorphically onto $S^n \setminus p$ and map $M \setminus U_j$ to p. There is a unique $v_j \in \Lambda^n S^n$, with compact support in $S^n \setminus p$, such that $\psi^* v_j = \alpha_j$. Clearly

$$\int_{S^n} v_j = 0,$$

so by the case $M = S^n$ of Proposition 19.5 already established, we know that $v_j = dw_j$ for some $w_j \in \Lambda^{n-1} S^n$, and then

$$\alpha_j = d\beta_j, \quad \beta_j = \psi_j^* w_j. \tag{19.11}$$

This concludes the proof.

We can sharpen and extend some of the topological results given above, using the notion of the degree of a map between compact, oriented surfaces. Let X and Y be compact, oriented, n-dimensional surfaces. We want to define the degree of a smooth map $F : X \to Y$. To do this, assume Y is connected. We pick $\omega \in \Lambda^n Y$ such that

$$\int_Y \omega = 1. \tag{19.12}$$

We want to define

$$\operatorname{Deg}(F) = \int_X F^* \omega. \tag{19.13}$$

The following result shows that $\operatorname{Deg}(F)$ is indeed well defined by this formula. The key argument is an application of Proposition 19.5.

Lemma 19.6. *The quantity (19.13) is independent of the choice of ω, as long as (19.12) holds.*

Proof. Pick $\omega_1 \in \Lambda^n Y$ satisfying $\int_Y \omega_1 = 1$, so $\int_Y \omega - \omega_1 = 0$. By Proposition 19.5, this implies

$$\omega - \omega_1 = d\alpha, \quad \text{for some } \alpha \in \Lambda^{n-1} Y. \tag{19.14}$$

Thus

$$\int_X F^*\omega - \int_X F^*\omega_1 = \int_X dF^*\alpha = 0, \tag{19.15}$$

and the lemma is proved.

The following is a most basic property.

Proposition 19.7. *If F_0 and F_1 are homotopic, then $Deg(F_0) = Deg(F_1)$.*

Proof. As noted in Exercise 7 of §13, if F_0 and F_1 are homotopic, then $F_0^*\omega - F_1^*\omega$ is exact, say $d\beta$, and of course $\int_X d\beta = 0$.

We next give an alternative formula for the degree of a map, which is very useful in many applications. A point $y_0 \in Y$ is called a regular value of F provided that, for each $x \in X$ satisfying $F(x) = y_0$, $DF(x) : T_x X \to T_{y_0} Y$ is an isomorphism. The easy case of Sard's theorem, discussed in Appendix B, implies that *most* points in Y are regular. Endow X with a volume element ω_X, and similarly endow Y with ω_Y. If $DF(x)$ is invertible, define $JF(x) \in \mathbb{R}\setminus 0$ by $F^*(\omega_Y) = JF(x)\omega_X$. Clearly the *sign* of $JF(x)$ (i.e., sgn $JF(x) = \pm 1$), is independent of the choices of ω_X and ω_Y, as long as they determine the given orientations of X and Y.

Proposition 19.8. *If y_0 is a regular value of F, then*

$$Deg(F) = \sum \{sgn\ JF(x_j) : F(x_j) = y_0\}. \tag{19.16}$$

Proof. Pick $\omega \in \Lambda^n Y$, satisfying (19.12), with support in a small neighborhood of y_0. Then $F^*\omega$ will be a sum $\sum \omega_j$, with ω_j supported in a small neighborhood of x_j, and $\int \omega_j = \pm 1$ as sgn $JF(x_j) = \pm 1$.

The following result is a powerful tool in degree theory.

Proposition 19.9. *Let $\overline{M}$ be a compact, oriented manifold with boundary. Assume that dim $M = n+1$. Given a smooth map $F : \overline{M} \to Y$, let $f = F\big|_{\partial M} : \partial M \to Y$. Then*

$$Deg(f) = 0.$$

Proof. Applying Stokes' theorem to $\alpha = F^*\omega$, we have

$$\int_{\partial M} f^*\omega = \int_M dF^*\omega.$$

But $dF^*\omega = F^*d\omega$, and $d\omega = 0$ if $\dim Y = n$, so we are done.

An easy corollary of this is another proof of Brouwer's no-retraction theorem. Compare the proof of Proposition 19.2.

Corollary 19.10. *If $\overline{M}$ is a compact, oriented manifold with nonempty boundary ∂M, then there is no smooth retraction $\varphi : \overline{M} \to \partial M$.*

Proof. Without loss of generality, we can assume that $\overline{M}$ is connected. If there were a retraction, then $\partial M = \varphi(\overline{M})$ must also be connected, so Proposition 19.9 applies. But then we would have, for the map $id. = \varphi\big|_{\partial M}$, the contradiction that its degree is both 0 and 1.

For another application of degree theory, let X be a compact, smooth, oriented hypersurface in $\mathbb{R}^{n+1}$, and set $\Omega = \mathbb{R}^{n+1} \setminus X$. (Assume $n \geq 1$.) Given $p \in \Omega$, define

$$F_p : X \longrightarrow S^n, \quad F_p(x) = \frac{x - p}{|x - p|}. \tag{19.17}$$

It is clear that $\mathrm{Deg}(F_p)$ is constant on each connected component of Ω. It is also easy to see that, when p crosses X, $\mathrm{Deg}(F_p)$ jumps by ± 1. Thus Ω has at least two connected components. This is most of the smooth case of the Jordan-Brouwer separation theorem:

Theorem 19.11. *If X is a smooth, compact, oriented hypersurface of $\mathbb{R}^{n+1}$, which is connected, then $\Omega = \mathbb{R}^{n+1} \setminus X$ has exactly two connected components.*

Proof. Since X is oriented, it has a smooth, global, normal vector field. Use this to separate a small collar neighborhood C of X into two pieces; $C \setminus X = C_0 \cup C_1$. The collar C is diffeomorphic to $[-1, 1] \times X$, and each C_j is clearly connected. It suffices to show that any connected component $\mathcal{O}$ of Ω intersects either C_0 or C_1. Take $p \in \partial\mathcal{O}$. If $p \notin X$, then $p \in \Omega$, which is open, so p cannot be a boundary point of any component of Ω. Thus $\partial\mathcal{O} \subset X$, so $\mathcal{O}$ must intersect a C_j. This completes the proof.

Let us note that, of the two components of Ω, exactly one is unbounded, say Ω_0, and the other is bounded; call it Ω_1. Then we claim that if X is given the orientation it gets as $\partial\Omega_1$,

$$p \in \Omega_j \Longrightarrow \mathrm{Deg}(F_p) = j. \tag{19.18}$$

Indeed, for p very far from X, $F_p : X \to S^n$ is not onto, so its degree is 0. And when p crosses X, from Ω_0 to Ω_1, the degree jumps by $+1$.

For a simple closed curve in $\mathbb{R}^2$, this result is the smooth case of the Jordan curve theorem. That special case of the argument given above can be found in [Sto].

We remark that, with a bit more work, one can show that any compact, smooth hypersurface in $\mathbb{R}^{n+1}$ is orientable. For one proof, see Appendix B to Chapter 5.

The next application of degree theory is useful in the study of closed orbits of planar vector fields. Let C be a simple, smooth, closed curve in $\mathbb{R}^2$, parameterized by arc length, of total length L. Say C is given by $x = \gamma(t),\ \gamma(t+L) = \gamma(t)$. Then we have a unit tangent field to C, $T(\gamma(t)) = \gamma'(t)$, defining

$$T : C \longrightarrow S^1. \tag{19.19}$$

Proposition 19.12. *For T given by (19.19), we have*

$$Deg(T) = 1. \tag{19.20}$$

Proof. Pick a tangent line ℓ to C such that C lies on one side of ℓ, as in Fig. 19.1. Without changing $\mathrm{Deg}(T)$, you can flatten out C a little, so it intersects ℓ along a line segment, from $\gamma(L_0)$ to $\gamma(L) = \gamma(0)$, where we take $L_0 = L - 2\varepsilon,\ L_1 = L - \varepsilon$.

Now T is close to the map $T_s : C \to S^1$, given by

$$T_s\big(\gamma(t)\big) = \frac{\gamma(t+s) - \gamma(t)}{|\gamma(t+s) - \gamma(t)|}, \tag{19.21}$$

for any $s > 0$ small enough; hence T and T_s are homotopic, for small positive s. It follows that T and T_s are homotopic for all $s \in (0, L)$. Furthermore, we can even let $s = s(t)$ be any continuous function $s : [0, L] \to (0, L)$ such that $s(0) = s(L)$. In particular, T is homotopic to the map $V : C \to S^1$, obtained from (19.21) by taking

$$s(t) = L_1 - t, \quad \text{for } t \in [0, L_0],$$

and $s(t)$ going monotonically from $L_1 - L_0$ to L_1, for $t \in [L_0, L]$. Note that

$$V\big(\gamma(t)\big) = \frac{\gamma(L_1) - \gamma(t)}{|\gamma(L_1) - \gamma(t)|}, \quad 0 \le t \le L_0.$$

The parts of V over the ranges $0 \le t \le L_0$ and $L_0 \le t \le L$, respectively, are illustrated in Figures 19.1 and 19.2. We see that V maps the segment of C from $\gamma(0)$ to $\gamma(L_0)$ into the lower half of the circle S^1, and it maps the segment of C from $\gamma(L_0)$ to $\gamma(L)$ into the upper half of the circle S^1. Therefore, V (hence T) is homotopic to a one-to-one map of C onto S^1, preserving orientation, and (19.20) is proved.

The material of this section can be cast in the language of deRham cohomology, which we now define. Let M be a smooth manifold. A smooth k-form u is said to be *exact* if $u = dv$ for some smooth $(k-1)$-form v, and *closed* if $du = 0$. Since $d^2 = 0$, every exact form is closed:

$$\mathcal{E}^k(M) \subset \mathcal{C}^k(M), \tag{19.22}$$

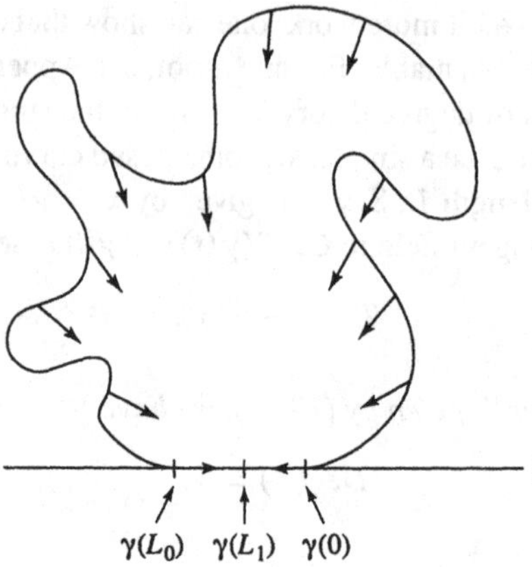

FIGURE 19.1

where $\mathcal{E}^k(M)$ and $\mathcal{C}^k(M)$ denote respectively the spaces of exact and closed k-forms. The deRham cohomology groups are defined as quotient spaces:

$$\mathcal{H}^k(M) = \mathcal{C}^k(M)/\mathcal{E}^k(M). \tag{19.23}$$

There are no nonzero (-1)-forms, so $\mathcal{E}^0(M) = 0$. A 0-form is a real-valued function, and it is closed if and only if it is constant on each connected component of M, so

$$\mathcal{H}^0(M) \approx \mathbb{R}^\nu, \quad \nu = \#\text{ connected components of } M. \tag{19.24}$$

An immediate consequence of Proposition 19.5 is the following:

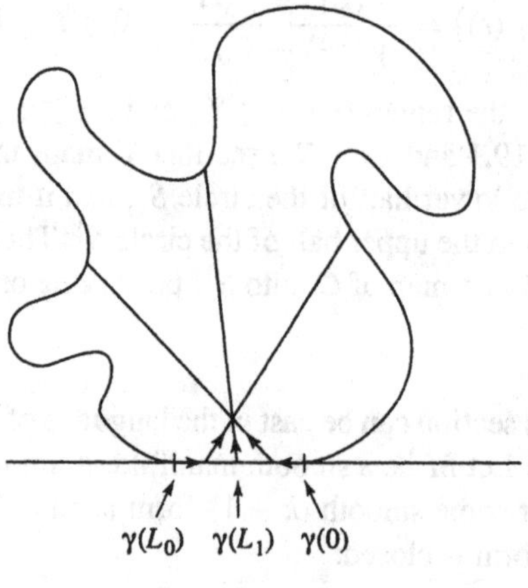

FIGURE 19.2

Proposition 19.13. *If M is a compact, connected, oriented manifold of dimension n, then*

$$\mathcal{H}^n(M) \approx \mathbb{R}. \tag{19.25}$$

Via the pull-back of forms, a smooth map $F : X \to Y$ between two manifolds induces maps on cohomology:

$$F^* : \mathcal{H}^j(Y) \longrightarrow \mathcal{H}^j(X). \tag{19.26}$$

If X and Y are both compact, connected, oriented, and of dimension n, then we have $F^* : \mathcal{H}^n(Y) \to \mathcal{H}^n(X)$, and, via the isomorphism $\mathcal{H}^n(X) \approx \mathbb{R} \approx \mathcal{H}^n(Y)$ arising from integration of n-forms, this map is simply multiplication by Deg F.

The subject of deRham cohomology plays an important role in material we develop later, such as Hodge theory, in Chapter 5, and index theory, in Chapter 10.

Exercises

1. Show that the identity map $I : X \to X$ has degree 1.
2. Show that if $F : X \to Y$ is not onto, then $\text{Deg}(F) = 0$.
3. If $A : S^n \to S^n$ is the antipodal map, show that $\text{Deg}(A) = (-1)^{n-1}$.
4. Show that the homotopy invariance property given in Proposition 19.7 can be deduced as a corollary of Proposition 19.9. (*Hint*: Take $\overline{M} = X \times [0, 1]$.)
5. Let $p(z) = z^n + a_{n-1}z^{n-1} + \cdots + a_1 z + a_0$ be a polynomial of degree $n \geq 1$. Show that if we identify $S^2 \approx \mathbb{C} \cup \{\infty\}$, then $p : \mathbb{C} \to \mathbb{C}$ has a unique continuous extension $\widetilde{p} : S^2 \to S^2$, with $\widetilde{p}(\infty) = \infty$. Show that

$$\text{Deg}\, \widetilde{p} = n.$$

   Deduce that $\widetilde{p} : S^2 \to S^2$ is onto, and hence that $p : \mathbb{C} \to \mathbb{C}$ is onto. In particular, each nonconstant polynomial in z has a complex root.
   This result is the fundamental theorem of algebra.

20. Critical points and index of a vector field

A *critical point* of a vector field V is a point at which V vanishes. Let V be a vector field defined on a neighborhood $\mathcal{O}$ of $p \in \mathbb{R}^n$, with a single critical point, at p. Then, for any small ball B_r about p, $B_r \subset \mathcal{O}$, we have a map

$$V_r : \partial B_r \to S^{n-1}, \quad V_r(x) = \frac{V(x)}{|V(x)|}. \tag{20.1}$$

The degree of this map is called the *index* of V at p, denoted $\text{ind}_p(V)$; it is clearly independent of r. If V has a finite number of critical points, then the index of V is defined to be

$$\text{Index}(V) = \sum \text{ind}_{p_j}(V). \tag{20.2}$$

If $\psi : \mathcal{O} \to \mathcal{O}'$ is an orientation-preserving diffeomorphism, taking p to p and V to W, then we claim that

$$\text{ind}_p(V) = \text{ind}_p(W). \tag{20.3}$$

In fact, $D\psi(p)$ is an element of $GL(n, \mathbb{R})$ with positive determinant, so it is homotopic to the identity, and from this it readily follows that V_r and W_r are homotopic maps of $\partial B_r \to S^{n-1}$. Thus one has a well-defined notion of the index of a vector field with a finite number of critical points on any oriented manifold M.

A vector field V on $\mathcal{O} \subset \mathbb{R}^n$ is said to have a nondegenerate critical point at p provided $DV(p)$ is a nonsingular $n \times n$ matrix. The following formula is convenient.

Proposition 20.1. *If V has a nondegenerate critical point at p, then*

$$\textit{ind}_p(V) = \textit{sgn det } DV(p). \tag{20.4}$$

Proof. If p is a nondegenerate critical point, and we set $\psi(x) = DV(p)x$, $\psi_r(x) = \psi(x)/|\psi(x)|$, for $x \in \partial B_r$, it is readily verified that ψ_r and V_r are homotopic, for r small. The fact that $\text{Deg}(\psi_r)$ is given by the right side of (20.4) is an easy consequence of Proposition 19.8.

The following is an important global relation between index and degree.

Proposition 20.2. *Let $\overline{\Omega}$ be a smooth bounded region in $\mathbb{R}^{n+1}$. Let V be a vector field on $\overline{\Omega}$, with a finite number of critical points p_j, all in the interior Ω. Define $F : \partial\Omega \to S^n$ by $F(x) = V(x)/|V(x)|$. Then*

$$\textit{Index}(V) = \textit{Deg}(F). \tag{20.5}$$

Proof. If we apply Proposition 19.9 to $\overline{M} = \overline{\Omega} \setminus \bigcup_j B_\varepsilon(p_j)$, we see that $\text{Deg}(F)$ is equal to the sum of degrees of the maps of $\partial B_\varepsilon(p_j)$ to S^n, which gives (20.5).

Next we look at a process of producing vector fields in higher-dimensional spaces from vector fields in lower-dimensional spaces.

Proposition 20.3. *Let W be a vector field on $\mathbb{R}^n$, vanishing only at 0. Define a vector field V on $\mathbb{R}^{n+k}$ by $V(x, y) = \big(W(x), y\big)$. Then V vanishes only at $(0, 0)$. Then we have*

$$\textit{ind}_0 W = \textit{ind}_{(0,0)} V. \tag{20.6}$$

Proof. If we use Proposition 19.8 to compute degrees of maps, and choose $y_0 \in S^{n-1} \subset S^{n+k-1}$, a regular value of W_r, and hence also for V_r, this identity follows.

We turn to a more sophisticated variation. Let X be a compact, oriented, n-dimensional submanifold of $\mathbb{R}^{n+k}$, W a (tangent) vector field on X with a finite number of critical points p_j. Let $\overline{\Omega}$ be a small tubular neighborhood of X, $\pi : \overline{\Omega} \to X$ mapping $z \in \overline{\Omega}$ to the nearest point in X. Let $\varphi(z) = \text{dist}(z, X)^2$. Now define a vector field V on $\overline{\Omega}$ by

$$V(z) = W(\pi(z)) + \nabla\varphi(z). \tag{20.7}$$

Proposition 20.4. *If $F : \partial\Omega \to S^{n+k-1}$ is given by $F(z) = V(z)/|V(z)|$, then*

$$Deg(F) = Index(W). \tag{20.8}$$

Proof. We see that all the critical points of V are points in X that are critical for W, and, as in Proposition 20.3, $\text{Index}(W) = \text{Index}(V)$. But Proposition 20.2 implies that $\text{Index}(V) = \text{Deg}(F)$.

Since $\varphi(z)$ is increasing as one moves away from X, it is clear that, for $z \in \partial\Omega$, $V(z)$ points out of $\overline{\Omega}$, provided it is a sufficiently small tubular neighborhood of X. Thus $F : \partial\Omega \to S^{n+k-1}$ is homotopic to the *Gauss map*

$$N : \partial\Omega \longrightarrow S^{n+k-1}, \tag{20.9}$$

given by the outward-pointing normal. This immediately gives the next result.

Corollary 20.5. *Let X be a compact oriented manifold in $\mathbb{R}^{n+k}$, $\overline{\Omega}$ a small tubular neighborhood of X, and $N : \partial\Omega \to S^{n+k-1}$ the Gauss map. If W is a vector field on X with a finite number of critical points, then*

$$Index(W) = Deg(N). \tag{20.10}$$

Clearly, the right side of (20.10) is independent of the choice of W. Thus any two vector fields on X with a finite number of critical points have the same index, that is, $\text{Index}(W)$ is an invariant of X. This invariant is denoted by

$$\text{Index}(W) = \chi(X), \tag{20.11}$$

and is called the *Euler characteristic* of X. See the exercises for more results on $\chi(X)$. A different definition of $\chi(X)$ is given in Chapter 5. These two definitions are related in §8 of Appendix C, Connections and Curvature.

Exercises

In Exercises 1–3, V is a vector field on a region $\Omega \subset \mathbb{R}^2$. A nondegenerate critical point p of a vector field V is said to be a *source* if the real parts of the eigenvalues of $DV(p)$ are all positive, a *sink* if they are all negative, and a *saddle* if they are all either positive or negative, and there exist some of each sign. Such a critical point is called a *center* if all orbits of V close to p are closed orbits, which stay near p; this requires all the eigenvalues of $DV(p)$ to be purely imaginary.

1. Let V have a nondegenerate critical point at p. Show that

$$\begin{aligned} p \text{ saddle} &\Longrightarrow \operatorname{ind}_p(V) = -1, \\ p \text{ source} &\Longrightarrow \operatorname{ind}_p(V) = 1, \\ p \text{ sink} &\Longrightarrow \operatorname{ind}_p(V) = 1, \\ p \text{ center} &\Longrightarrow \operatorname{ind}_p(V) = 1. \end{aligned}$$

2. If V has a closed orbit γ, show that the map $T : \gamma \to S^1$, $T(x) = V(x)/|V(x)|$, has degree $+1$. (*Hint*: Use Proposition 19.8.)
3. If V has a closed orbit γ whose inside $\mathcal{O}$ is contained in Ω, show that V must have at least one critical point in $\mathcal{O}$, and that the sum of the indices of such critical points must be $+1$. (*Hint*: Use Proposition 20.2.)
   If V has exactly one critical point in $\mathcal{O}$, show that it cannot be a saddle.
4. Let M be a compact, oriented surface. Given a triangulation of M, within each triangle construct a vector field, vanishing at seven points as illustrated in Fig. 20.1, with the vertices as attractors, the center as a repeller, and the midpoints of each side as saddle points. Fit these together to produce a smooth vector field X on M. Show directly that

$$\operatorname{Index}(X) = V - E + F,$$

where

$$V = \#\text{ vertices}, \quad E = \#\text{ edges}, \quad F = \#\text{ faces},$$

in the triangulation.
5. More generally, construct a vector field on an n-simplex so that when a compact, oriented, n-dimensional manifold M is triangulated into simplices, one produces a vector field X on M such that

$$\operatorname{Index}(X) = \sum_{j=0}^{n} (-1)^j \nu_j, \tag{20.12}$$

where ν_j is the number of j-simplices in the triangulation, namely, $\nu_0 = \#$ vertices, $\nu_1 = \#$ edges, ..., $\nu_n = \#$ of n-simplices. (See Fig. 20.2 for a picture of a 3-simplex, with its faces (i.e., 2-simplices), edges, and vertices labeled.)

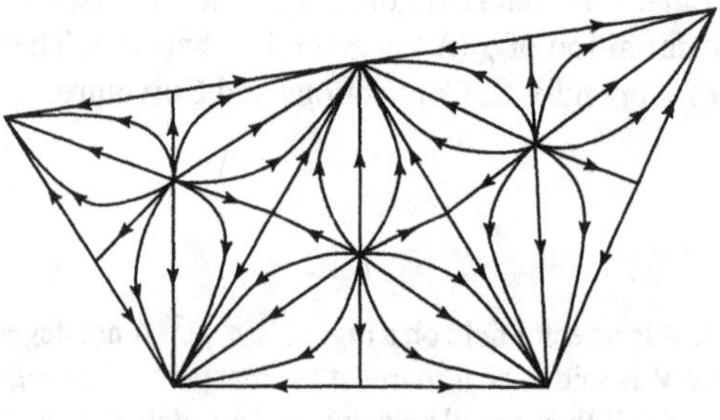

FIGURE 20.1

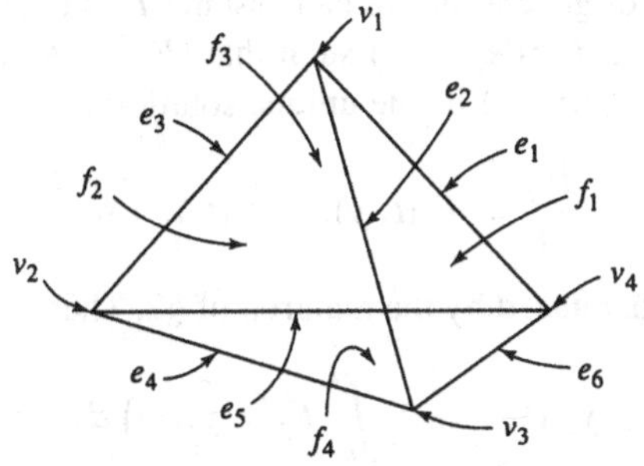

FIGURE 20.2

The right side of (20.12) is one definition of $\chi(M)$. As we have seen, the left side of (20.12) is independent of the choice of X, so it follows that the right side is independent of the choice of triangulation.

6. Let M be the sphere S^n, which is homeomorphic to the boundary of an $(n+1)$-simplex. Computing the right side of (20.12), show that

$$\chi(S^n) = 2 \text{ if } n \text{ even}, \quad 0 \text{ if } n \text{ odd}. \tag{20.13}$$

Conclude that if n is even, there is no smooth nowhere-vanishing vector field on S^n, thus obtaining another proof of Proposition 19.4.

7. With $X = S^n \subset \mathbb{R}^{n+1}$, note that the manifold $\partial\Omega$ in (20.9) consists of two copies of S^n, with opposite orientations. Compute the degree of the map N in (20.9) and (20.10), and use this to give another derivation of (20.13), granted (20.11).
8. Consider the vector field R on S^2 generating rotation about an axis. Show that R has two critical points, at the "poles." Classify the critical points, compute Index(R), and compare the $n = 2$ case of (20.13).
9. Show that the computation of the index of a vector field X on a manifold M is independent of orientation and that Index(X) can be defined when M is not orientable.

A. Nonsmooth vector fields

Here we establish properties of solutions to the ODE

$$\frac{dy}{dt} = F(t, y), \quad y(t_0) = x_0 \tag{A.1}$$

of a sort done in §§2–6, under weaker hypotheses than those used there; in particular, we do not require F to be Lipschitz in y. For existence, we can assume considerably less:

Proposition A.1. *Let $x_0 \in \mathcal{O}$, an open subset of $\mathbb{R}^n$, $I \subset \mathbb{R}$ an interval containing t_0. Assume F is continuous on $I \times \mathcal{O}$. Then the equation (A.1) has a solution on some t-interval containing t_0.*

Proof. Without loss of generality, we can assume F is bounded and continuous on $\mathbb{R} \times \mathbb{R}^n$. Take $F_j \in C^\infty(\mathbb{R} \times \mathbb{R}^n)$ such that $|F_j| \le K$ and $F_j \to F$ locally uniformly, and let $y_j \in C^\infty(\mathbb{R})$ be the unique solution to

$$\frac{dy_j}{dt} = F_j(t, y), \quad y_j(t_0) = x_0, \tag{A.2}$$

whose existence is guaranteed by the material of §2. Thus

$$y_j(t) = x_0 + \int_{t_0}^{t} F_j\big(s, y_j(s)\big)\, ds. \tag{A.3}$$

Now

$$|F_j| \le K \Longrightarrow |y_j(t') - y_j(t)| \le K|t' - t|. \tag{A.4}$$

Hence, by Ascoli's theorem (see Proposition 6.2 in Appendix A, Functional Analysis) the sequence (y_j) has a subsequence (y_{j_ν}) which converges locally uniformly: $y_{j_\nu} \to y$. It follows immediately that

$$y(t) = x_0 + \int_{t_0}^{t} F\big(s, y(s)\big)\, ds, \tag{A.5}$$

so y solves (A.1).

Under the hypotheses of Proposition A.1, a solution to (A.1) may not be unique. The following family of examples illustrates the phenomenon. Take $a \in (0, 1)$ and consider

$$\frac{dy}{dt} = |y|^a, \quad y(0) = 0. \tag{A.6}$$

Then one solution on $[0, \infty)$ is given by

$$y_0(t) = (1 - a)^{1/(1-a)}\, t^{1/(1-a)}, \tag{A.7}$$

and another is given by

$$y_*(t) = 0.$$

Note that, for any $\varepsilon > 0$, the problem $dy/dt = |y|^a$, $y(0) = \varepsilon$ has a unique solution on $t \in [0, \infty)$, and $\lim_{\varepsilon \to 0} y_\varepsilon(t) = y_0(t)$. Understanding this provides the key to the following uniqueness result, due to W. Osgood.

Let $\omega : \mathbb{R}^+ \to \mathbb{R}^+$ be a modulus of continuity, i.e., $\omega(0) = 0$, ω is continuous, and increasing. We may as well assume ω is bounded and C^∞ on $(0, \infty)$.

Proposition A.2. *In the setting of Proposition A.1, assume F is continuous on $I \times \mathcal{O}$ and that*

$$|F(t, y_1) - F(t, y_2)| \le \omega\big(|y_1 - y_2|\big), \tag{A.8}$$

for all $t \in I$, $y_j \in \mathcal{O}$. Then solutions to (A.1) (with range in $\mathcal{O}$) are unique, provided

$$\int_0^1 \frac{ds}{\omega(s)} = \infty. \tag{A.9}$$

Proof. If $y_1(t)$ and $y_2(t)$ are two solutions to (A.1), then

$$y_1(t) - y_2(t) = \int_{t_0}^{t} \{F(s, y_1(s)) - F(s, y_2(s))\}\, ds. \tag{A.10}$$

Let us set $\theta(t) = |y_1(t) - y_2(t)|$. Hence, by (A.8), for $t \geq t_0$,

$$\theta(t) \leq \int_{t_0}^{t} \omega(\theta(s))\, ds. \tag{A.11}$$

In particular, for each $\varepsilon > 0$, $\theta(t) \leq \int_{t_0}^{t} \omega(\theta(s)+\varepsilon)\, ds$. Since we are assuming ω is smooth on $(0, \infty)$, we can apply the Gronwall inequality, derived in (5.19)–(5.21), to deduce that

$$\theta(t) \leq \varphi_\varepsilon(t), \quad \forall\, t \geq t_0,\ \varepsilon > 0, \tag{A.12}$$

where φ_ε is uniquely defined on $[t_0, \infty)$ by

$$\varphi_\varepsilon'(t) = \omega(\varphi_\varepsilon(t) + \varepsilon), \quad \varphi_\varepsilon(t_0) = 0. \tag{A.13}$$

Thus

$$\int_0^{\varphi_\varepsilon(t)} \frac{d\zeta}{\omega(\zeta + \varepsilon)} = t - t_0. \tag{A.14}$$

Now the hypothesis (A.9) implies

$$\lim_{\varepsilon \searrow 0} \varphi_\varepsilon(t) = 0, \quad \forall\, t \geq t_0, \tag{A.15}$$

so we have $\theta(t) = 0$, for all $t \geq t_0$. Similarly, one shows $\theta(t) = 0$, for $t \leq t_0$, and uniqueness is proved.

An important example to which Proposition A.2 applies is

$$\omega(s) = s \log \frac{1}{s}, \quad s \leq \frac{1}{2}. \tag{A.16}$$

This arises in the study of ideal fluid flow, as will be seen in Chapter 17.

A similar argument establishes continuous dependence on initial data. If

$$\frac{dy_j}{dt} = F(t, y_j), \quad y_j(t_0) = x_j, \tag{A.17}$$

then

$$y_1(t) - y_2(t) = x_1 - x_2 + \int_{t_0}^{t} \{F(s, y_1(s)) - F(s, y_2(s))\}\, ds, \tag{A.18}$$

so $\theta_{12}(t) = |y_1(t) - y_2(t)|$ satisfies

$$\theta_{12}(t) \le |x_1 - x_2| + \int_{t_0}^{t} \omega\big(\theta_{12}(s)\big)\,ds. \tag{A.19}$$

An argument similar to that used above gives (for $t \ge t_0$)

$$\theta_{12}(t) \le \vartheta(|x_1 - x_2|, t), \tag{A.20}$$

where, for $a > 0,\ t \ge t_0,\ \vartheta(a,t)$ is the unique solution to

$$\partial_t \vartheta = \omega(\vartheta), \quad \vartheta(a, t_0) = a, \tag{A.21}$$

that is,

$$\int_a^{\vartheta(a,t)} \frac{d\zeta}{\omega(\zeta)} = t - t_0. \tag{A.22}$$

Again, the hypothesis (A.9) implies

$$\lim_{a \searrow 0} \vartheta(a,t) = 0, \quad \forall\, t \ge t_0. \tag{A.23}$$

By (A.20), we have

$$|y_1(t) - y_2(t)| \le \vartheta\big(|x_1 - x_2|, t\big), \tag{A.24}$$

for all $t \ge t_0$, and a similar argument works for $t \le t_0$.

References

[AM] R. Abraham and J. Marsden, *Foundations of Mechanics*, Benjamin/Cummings, Reading, Mass., 1978.

[Ar] V. Arnold, *Mathematical Methods of Classical Mechanics*, Springer-Verlag, New York, 1978.

[Ar2] V. Arnold, *Geometrical Methods in the Theory of Ordinary Differential Equations*, Springer-Verlag, New York, 1983.

[Bir] G. D. Birkhoff, *Dynamical Systems*, AMS Colloq. Publ., Vol. 9, Providence, R.I., 1927.

[Car] C. Caratheodory, *Calculus of Variations and Partial Differential Equations of the First Order*, Holden-Day, San Francisco, 1965.

[Fed] H. Federer, *Geometric Measure Theory*, Springer-Verlag, New York, 1969.

[Go] H. Goldstein, *Classical Mechanics*, Addison-Wesley, New York, 1950.

[GS] V. Guillemin and S. Sternberg, *Symplectic Techniques in Physics*, Cambridge Univ. Press, Cambridge, 1984.

[Hal] J. Hale, *Ordinary Differential Equations*, Wiley, New York, 1969.

[Har] P. Hartman, *Ordinary Differential Equations*, Baltimore, 1973.

[HS] M. Hirsch and S. Smale, *Differential Equations, Dynamical Systems, and Linear Algebra*, Academic Press, New York, 1974.

[Ja] J. Jackson, *Classical Electrodynamics*, J. Wiley, New York, 1962.

[Kan] Y. Kannai, An elementary proof of the no-retraction theorem, *Amer. Math. Monthly* 88(1981), 264–268.

[Lef] S. Lefschetz, *Differential Equations, Geometric Theory*, J. Wiley, New York, 1957.

[LS] L. Loomis and S. Sternberg, *Advanced Calculus*, Addison-Wesley, New York, 1968.

[Mos] J. Moser, *Stable and Random Motions in Dynamical Systems*, Princeton Univ. Press, Princeton, N.J., 1973.

[NS] V. Nemytskii and V. Stepanov, *Qualitative Theory of Differential Equations*, Dover, New York, 1989.

[Poi] H. Poincaré, *Les Méthodes Nouvelles de la Mécanique Céleste*, Gauthier-Villars, 1899.

[Poo] W. Poor, *Differential Geometric Structures*, McGraw-Hill, New York, 1981.

[Rin] W. Rindler, *Essential Relativity*, Springer, New York, 1977.

[Spi] M. Spivak, *A Comprehensive Introduction to Differential Geometry*, Vols. 1–5, Publish or Perish Press, Berkeley, 1979.

[Stb] S. Sternberg, *Lectures on Differential Geometry*, Prentice Hall, New Jersey, 1964.

[Sto] J. J. Stoker, *Differential Geometry*, Wiley-Interscience, New York, 1969.

[Str] G. Strang, *Linear Algebra and its Applications*, Harcourt Brace Jovanovich, San Diego, 1988.

[TS] J. M. Thompson and H. Stewart, *Nonlinear Dynamics and Chaos*, J. Wiley, New York, 1986.

[Wh] E. Whittaker, *A Treatise on the Analytical Dynamics of Particles and Rigid Bodies*, Dover, New York, 1944.

[Wig] S. Wiggins, *Introduction to Applied Dynamical Systems and Chaos*, Springer-Verlag, New York, 1990.

2

The Laplace Equation and Wave Equation

Introduction

In this chapter we introduce the central linear partial differential equations of the second order, the Laplace equation

$$\Delta u = f \tag{0.1}$$

and the wave equation

$$\left(\frac{\partial^2}{\partial t^2} - \Delta\right)u = f. \tag{0.2}$$

For flat Euclidean space $\mathbb{R}^n$, the Laplace operator is defined by

$$\Delta u = \frac{\partial^2 u}{\partial x_1^2} + \cdots + \frac{\partial^2 u}{\partial x_n^2}. \tag{0.3}$$

The wave equation arose early in the history of continuum mechanics, in a mathematical description of the motion of vibrating strings and membranes. We discuss this in §1. The analysis, based on an appropriate version of Hamilton's stationary action principle, generally produces nonlinear partial differential equations, of a sort that will be studied more in Chapters 14 through 16. The wave equation described by (0.2), which is linear, arises as a "linearized" PDE, describing such vibratory motion, as will be seen in §1.

In this chapter we consider the Laplace operator on a general Riemannian manifold and emphasize concepts defined in a coordinate-independent fashion. Also, more generally than the wave equation (0.2) on the Cartesian product of a spatial region with the time axis, we consider natural generalizations defined on a manifold endowed with a Lorentz metric.

Before defining the Laplace operator on Riemannian manifolds, we devote two sections to some first-order operators. In §2 we discuss the divergence operator applied to vector fields, and in §3 we generalize the operations of covariant derivative and divergence from vector fields to tensor fields. These concepts play important roles in the study of the Laplace and wave equations.

In §4 we define the Laplace operator acting on real- (or complex-) valued functions on a Riemannian manifold M, and in §5 we write down the wave equation for functions on $\mathbb{R} \times M$ and discuss energy conservation. In §6 we extend energy identities in a way that leads to proofs of results on finite propagation speed for solutions to such a wave equation.

In §7 we extend the notion of the wave equation from $\mathbb{R} \times M$ to a general Lorentz manifold. We extend the notion of energy conservation. To a solution of the wave equation is associated a second-order tensor field, the "stress-energy tensor," and the law of conservation of energy can be expressed as the vanishing of the divergence of this field, as is shown in §7. One can pass from such a "local" conservation law to an integral conservation law via the divergence theorem, for a certain class of Lorentz manifolds, namely those with a timelike Killing field. We derive the phenomenon of "finite propagaton speed" for solutions to the wave equation as a consequence of such a conservation law.

In §8 we consider a more general class of hyperbolic equations. To solutions we can still associate a tensor with some of the properties of a stress-energy tensor, but the energy conservation law may not hold, and instead we look for "energy estimates."

The Stokes formula used in §2 to derive the divergence theorem is a special case of a more general Stokes-type formula, which we discuss in §9. This more general formula is used in §10 to produce a variant of Green's formula for the Laplace operator acting on differential forms. In these sections we also make use of the notion of the "principal symbol" of a differential operator, as an invariantly defined function on the cotangent bundle.

In §11 we look at Maxwell's equations for the electromagnetic field. We show how they can be manipulated to yield the wave equation. This mathematical fact will be further exploited in Chapter 6. We deal with Maxwell's equations in the framework of relativity and work with the electromagnetic field on a general Lorentz 4-manifold.

Though we discuss some qualitative properties of solutions to the Laplace equation and the wave equation, such as Green's identities and finite propagation speed (in the case of the wave equation), we do not tackle the question of existence of solutions in this chapter, except for the very simplest case, namely the $n = 1$ case of (0.2), treated in §1. In the case of such equations on flat Euclidean space, Fourier analysis provides an adequate tool to construct and analyze solutions, and this will be developed in the next chapter. Then functional analytical methods, centered on the theory of Sobolev spaces, will be developed in Chapter 4 and applied in subsequent chapters. As we will see in Chapter 6, energy estimates, such as those derived in §8 of this chapter, in concert with Sobolev space theory, form the principal tools for existence theorems for linear hyperbolic equations. Existence of solutions to nonlinear hyperbolic equations, which requires somewhat more subtle analysis, will be studied in Chapter 16.

1. Vibrating strings and membranes

The problem of describing the motion of a vibrating string was one of the earliest problems of continuum mechanics, producing a partial differential equation. Such a PDE can be derived by a procedure similar to that described in §12 of Chapter 1, using a stationary action principle. To carry this out, we need formulas for the kinetic energy and the potential energy of a vibrating string.

Suppose our string is vibrating in $\mathbb{R}^k$; say its ends are tied down at two points, the origin 0 and a vector $Le_1 \in \mathbb{R}^k$, of length L. We suppose the string is uniform, of mass density m (i.e., total mass mL). The motion of the string is described by a function $u = u(t, x)$, $t \in \mathbb{R}$, $x \in [0, L]$, taking values in $\mathbb{R}^k$ and satisfying $u(t, 0) = 0$, $u(t, L) = Le_1$ for all t. Then the kinetic energy at time t is given by

$$T(t) = \frac{m}{2} \int_0^L |u_t(t, x)|^2 \, dx, \tag{1.1}$$

and the integral $\int_{t_0}^{t_1} T(t)\, dt$ is given by

$$J_0(u) = \frac{m}{2} \iint\limits_{I\times\Omega} |u_t(t, x)|^2 \, dx\, dt, \tag{1.2}$$

where $I = (t_0, t_1)$, $\Omega = (0, L)$.

As for the potential energy at a given time t, we will use the law that the potential energy in a small piece of string is a function of the degree that the string has been stretched, namely,

$$V(t) = \int_0^L f\big(u_x(t, x)\big)\, dx \tag{1.3}$$

for a function

$$f : \mathbb{R}^k \longrightarrow \mathbb{R}. \tag{1.4}$$

This is known as Hooke's law. The case of an "ideal" string (where the force exerted by a small piece of string is proportional to the amount by which it has been stretched) is

$$f(y) = \sigma(|y| - a)^2, \tag{1.5}$$

where the unstretched string has length $aL < L$ and $\sigma > 0$ is a given constant. The term accompanying (1.2) in the expression for the action is

$$J_1(u) = \iint\limits_{I\times\Omega} f\big(u_x(t, x)\big)\, dx\, dt. \tag{1.6}$$

The stationary condition according to Hamilton's principle is

$$\frac{d}{ds}(J_0 - J_1)(u + sv)\Big|_{s=0} = 0, \tag{1.7}$$

for all $v \in C_0^\infty(I \times \Omega, \mathbb{R}^k)$. A simple computation gives

$$\begin{aligned} \frac{d}{ds} J_0(u+sv)\big|_{s=0} &= \iint_{I\times\Omega} m u_t v_t \, dx\, dt \\ &= -\iint m v u_{tt} \, dx\, dt, \end{aligned} \tag{1.8}$$

where the last identity is obtained by integration by parts. Furthermore, also integrating by parts, we have

$$\begin{aligned} \frac{d}{ds} J_1(u+sv)\big|_{s=0} &= \iint f'\big(u_x(t,x)\big) \cdot v_x(t,x) \, dx\, dt \\ &= -\iint \Big\{ \frac{\partial}{\partial x} f'\big(u_x(t,x)\big) \Big\} \cdot v(t,x)\, dx\, dt. \end{aligned} \tag{1.9}$$

Note that

$$\frac{\partial}{\partial x} f'(u_x(t,x)) = f''(u_x)u_{xx}, \tag{1.10}$$

where $f''(y)$ is the $k \times k$ matrix valued function of second-order partial derivatives of $f : \mathbb{R}^k \to \mathbb{R}$, and u_{xx} takes values in $\mathbb{R}^k$. In other words,

$$\frac{d}{ds} J_1(u+sv)\big|_{s=0} = -\iint_{I\times\Omega} f''(u_x)u_{xx} \cdot v \, dx\, dt. \tag{1.11}$$

Combining (1.8) and (1.11), we see that the stationary condition (1.7) is equivalent to the partial differential equation

$$m u_{tt} - f''(u_x)u_{xx} = 0. \tag{1.12}$$

If $f(y)$ is a second-order polynomial in y, that is, of the form

$$f(y) = a + b \cdot y + Ay \cdot y, \tag{1.13}$$

where $a \in \mathbb{R}$, $b \in \mathbb{R}^k$, and A is a real, symmetric, $k \times k$ matrix, then $f''(y) = 2A$, and the PDE (1.12) becomes

$$m u_{tt} - 2A u_{xx} = 0. \tag{1.14}$$

The example (1.5) does not satisfy this condition, and the resulting PDE is not linear. Let us rewrite this PDE, setting

$$u(t,x) = xe_1 + w(t,x), \tag{1.15}$$

so that $w(t,0) = 0$ and $w(t,L) = 0$ in $\mathbb{R}^k$. Then

$$J_1(u) = K_1(w) = \iint \varphi(w_x)\, dx\, dt, \tag{1.16}$$

where $\varphi : \mathbb{R}^k \to \mathbb{R}$ is given by

$$\varphi(y) = f(e_1 + y), \tag{1.17}$$

and the corresponding PDE for w is

$$mw_{tt} - \varphi''(w_x)w_{xx} = 0. \tag{1.18}$$

The *linearization* of this equation is, by definition, obtained by replacing $\varphi(y)$ by its quadratic part, that is, by the terms of order ≤ 2 in its power series about $y = 0$:

$$\varphi_0(y) = a_0 + b_0 \cdot y + \frac{1}{2} A_0 y \cdot y, \tag{1.19}$$

where $a_0 = \varphi(0) = f(e_1)$, $b_0 = \varphi'(0) = f'(e_1)$, and $A_0 = \varphi''(0) = f''(e_1)$. For one reason why the term "linearization" is appropriate, see Exercise 4 at the end of this section. If φ is replaced by φ_0 in (1.16), the stationary condition yields the linear PDE

$$mw_{tt} - A_0 w_{xx} = 0 \quad \big(A_0 = \varphi''(0)\big). \tag{1.20}$$

In the case of an ideal string (1.5), this linearized PDE is readily computed to be

$$mw_{tt} - 2\sigma(I - aP)w_{xx} = 0, \tag{1.21}$$

where P is the orthogonal projection of $\mathbb{R}^k$ onto the orthogonal complement of e_1. (Compare the calculations (1.43)–(1.47) and (1.51)–(1.55) below.) Recall that we are assuming $0 < a < 1$.

For this linear equation, we can write $w = w^b + w^\#$, where w^b is parallel to e_1 and $w^\#$ is orthogonal to e_1. The equation (1.21) decouples, and we have

$$mw^b_{tt} - 2\sigma w^b_{xx} = 0 \tag{1.22}$$

as the equation for the *longitudinal* wave w^b and

$$mw^\#_{tt} - 2\sigma(1 - a)w^\#_{xx} = 0, \tag{1.23}$$

as the equation for the *transverse* wave $w^\#$. Both of these equations are cases (with different values of c) of the wave equation

$$v_{tt} - c^2 v_{xx} = 0. \tag{1.24}$$

Here c is identified with the *propagation speed* for solutions to (1.24), for the following reason. Namely, for any C^2-functions f_j of one variable,

$$v(t, x) = f_1(x + ct) + f_2(x - ct) \tag{1.25}$$

is a solution to (1.24). Conversely, the general solution to (1.24) on $(t, x) \in \mathbb{R} \times \mathbb{R}$, satisfying the *initial conditions*

$$v(0, x) = g(x), \quad v_t(0, x) = h(x), \tag{1.26}$$

can be expressed in the form (1.25). Indeed, a solution to (1.24) in the form (1.25) satisfies these initial conditions if and only if

$$f_1(x) + f_2(x) = g(x) \ \text{ and } \ cf_1'(x) - cf_2'(x) = h(x). \tag{1.27}$$

This implies $f_1'(x) + f_2'(x) = g'(x)$, so we can solve algebraically for f_1' and f_2'; thus we can set

$$(1.28)\qquad \begin{aligned} f_1(x) &= \frac{1}{2}g(x) + \frac{1}{2c}\int_0^x h(s)\,ds, \\ f_2(x) &= \frac{1}{2}g(x) - \frac{1}{2c}\int_0^x h(s)\,ds. \end{aligned}$$

That the solution (1.25) so produced is the only solution to (1.24) satisfying the initial conditions (1.26) is a special case of a uniqueness result proved in §5.

One can arrange that the *boundary condition*

$$(1.29)\qquad v(t,0) = v(t,L) = 0$$

be satisfied by taking g and h that satisfy

$$(1.30)\qquad g(s) = g(s+2L) = -g(-s), \quad h(s) = h(s+2L) = -h(-s).$$

This is a special case of the *method of images*, discussed further in Chapter 3, §7.

Whenever one has the linear equation (1.14), if A is a positive-definite matrix, one can diagonalize A and construct solutions as above. Constructing solutions for the equation (1.12), or (1.18) in the nonlinear case, is much more difficult; Chapter 16 gives some results for this problem.

Now we look at the higher-dimensional case, of a vibrating membrane. Let Ω be some open region in $\mathbb{R}^n$. We consider vibrations of Ω in $\mathbb{R}^k$, with $k \ge n$. Define the inclusion $j : \mathbb{R}^n \hookrightarrow \mathbb{R}^k$ by

$$j(x_1, \dots, x_n) = (x_1, \dots, x_n, 0, \dots, 0).$$

This time suppose the boundary of Ω is tied down. The motion of the membrane is described by a function $u = u(t,x)$, $t \in \mathbb{R}$, $x \in \overline{\Omega}$, taking values in $\mathbb{R}^k$ and satisfying $u(t,x) = j(x)$ for $x \in \partial\Omega$. We suppose the membrane is of a uniform substance, with mass density m. The kinetic energy at a given time t is then

$$(1.31)\qquad T(t) = \frac{m}{2}\int_\Omega |u_t(t,x)|^2\,dx,$$

parallel to (1.1), and the integral $\int_{t_0}^{t_1} T(t)\,dt = J_0(u)$ is again given by (1.2), with Ω now an n-dimensional domain. As for the potential energy, we will again work under the hypothesis that it is a function of the "stretching" of the membrane, of the form

$$(1.32)\qquad V(t) = \int_\Omega f\big(u_x(t,x)\big)\,dx,$$

where, for each $(t,x) \in \mathbb{R} \times \Omega$,

$$(1.33)\qquad u_x(t,x) \in \mathcal{L}(T_x\Omega, \mathbb{R}^k) \approx \mathcal{L}(\mathbb{R}^n, \mathbb{R}^k)$$

is the x-derivative, and

$$(1.34)\qquad f : \mathcal{L}(\mathbb{R}^n, \mathbb{R}^k) \longrightarrow \mathbb{R}$$

is a given smooth function. Again $\int_{t_0}^{t_1} V(t)\,dt = J_1(u)$ is given by (1.6), the stationary action principle takes the form (1.7), and the variation of $J_0(u)$ is given by (1.8). The variation of $J_1(u)$ is also given by a formula of the form (1.11). More precisely, if we set

$$f = f(y), \quad y = \big(y_{\nu j}\big) \in \mathcal{L}(\mathbb{R}^n, \mathbb{R}^k), \tag{1.35}$$

then (1.11) holds, with the interpretation

$$f''(u_x)u_{xx} \cdot v = \sum_{\mu,\nu=1}^{k} \sum_{i,j=1}^{n} \frac{\partial^2 f(u_x)}{\partial y_{\mu i} \partial y_{\nu j}} u^{\mu}_{x_i x_j} v^{\nu}, \tag{1.36}$$

where $u = (u^1, \dots, u^k)$, $v = (v^1, \dots, v^k) \in \mathbb{R}^k$. With this notation, the PDE obtained for u is again of the form (1.12).

As in (1.15)–(1.17), we can concentrate on the deviation of u from the map $j : \Omega \to \mathbb{R}^k$. Set

$$u(t,x) = j(x) + w(t,x), \tag{1.37}$$

so the boundary condition becomes $w(t,x) = 0$ for $x \in \partial\Omega$; then the PDE for w is of the form (1.18), again interpreted as in (1.36), with

$$\varphi(y) = f(j + y), \tag{1.38}$$

for $y \in \mathcal{L}(\mathbb{R}^n, \mathbb{R}^k)$. As before, we have the linearized PDE

$$m w_{tt} - A w_{xx} = 0, \quad A = \varphi''(0), \tag{1.39}$$

where, for $w = (w^1, \dots, w^k)$,

$$\big(A w_{xx}\big)^{\nu} = \sum_{\mu=1}^{k} \sum_{i,j=1}^{n} \frac{\partial^2 \varphi(0)}{\partial y_{\mu i} \partial y_{\nu j}} w^{\mu}_{x_i x_j}. \tag{1.40}$$

We can regard A as defining a symmetric bilinear map

$$A : \mathcal{L}(\mathbb{R}^n, \mathbb{R}^k) \times \mathcal{L}(\mathbb{R}^n, \mathbb{R}^k) \longrightarrow \mathbb{R}. \tag{1.41}$$

There are a number of different forms the potential energy function $f(y)$ can take, depending on the physical properties of the membrane. In a number of models, one has $f(y) = \psi(y^* y)$, a function invariant under conjugating $y^* y$ by an orthogonal $n \times n$ matrix. These models have the form

$$f(y) = \Psi\big(\mathrm{Tr}\, g_1(y^* y), \dots, \mathrm{Tr}\, g_K(y^* y)\big), \tag{1.42}$$

where $g_\ell : \mathbb{R} \to \mathbb{R}$ is smooth and, for a self adjoint matrix $z = y^* y$, $g_\ell(z)$ is defined by the spectral representation; $g_\ell(z) v_j = g_\ell(\lambda_j) v_j$ for v_j in the λ_j-eigenspace of z. There is no loss in generality in assuming $g_\ell(1) = 0$.

To compute the linearized PDE when $f(y)$ is given by (1.42), start with

$$\begin{aligned} g_\ell\big((j^*+y^*)(j+y)\big) &= g_\ell(I+j^*y+y^*j+y^*y) \\ &= g_\ell(1)I + g_\ell'(1)(j^*y+y^*j+y^*y) \\ &\quad + \frac{1}{2}g_\ell''(1)(j^*y+y^*j)^2 + O(\|y\|^3). \end{aligned} \tag{1.43}$$

If $(1/2)\tau = \operatorname{Tr} j^*y = \operatorname{Tr} y^*j$, $\sigma = \operatorname{Tr} y^*y$, and $\gamma = \operatorname{Tr}(j^*y+y^*j)^2$, we obtain

$$\begin{aligned} \varphi(y) = f(j+y) &= \Psi(0) + \sum \partial_\ell\Psi(0)\big[g_\ell'(1)(\tau+\sigma) + \frac{1}{2}g_\ell''(1)\gamma\big] \\ &\quad + \sum \frac{1}{2}\partial_\ell\partial_m\Psi(0)g_\ell'(1)g_m'(1)\tau^2 + O(\|y\|^3). \end{aligned} \tag{1.44}$$

Thus the purely quadratic part, which yields the linearized PDE, is

$$\begin{aligned} \varphi_0(y) &= \sum_\ell \partial_\ell\Psi(0)\big[g_\ell'(1)\operatorname{Tr} y^*y + \frac{1}{2}g_\ell''(1)\operatorname{Tr}(j^*y+y^*j)^2\big] \\ &\quad + \sum_{\ell,m} \frac{1}{2}\partial_\ell\partial_m\Psi(0)g_\ell'(1)g_m'(1)\big[\operatorname{Tr}(j^*y+y^*j)\big]^2 \\ &= A\operatorname{Tr} y^*y + B\operatorname{Tr}(j^*y+y^*j)^2 + C\big(\operatorname{Tr}(j^*y+y^*j)\big)^2. \end{aligned} \tag{1.45}$$

As in the case of the linearized equations of the vibrating string, the resulting linear PDE decouples into an equation for the components of w orthogonal to the space $\mathbb{R}^n \subset \mathbb{R}^k$ in which Ω sits and an equation for the components of w parallel to this space. For the orthogonal component $w^\#$, since $j^*w^\# = 0$ in this case, we can replace $\varphi_0(y)$ by

$$\varphi^\#(y) = A\operatorname{Tr} y^*y, \quad y \in \mathcal{L}(\mathbb{R}^n, \mathbb{R}^{k-n}). \tag{1.46}$$

In this case, we have

$$\frac{\partial^2\varphi^\#}{\partial y_{\mu i}\partial y_{\nu j}} = 2A\delta_{ij}\delta_{\mu\nu}. \tag{1.47}$$

Hence the linearized equation for the orthogonal (or transverse) wave is

$$m w^\#_{tt} - 2A\Delta w^\# = 0, \tag{1.48}$$

where Δ is the Laplace operator on $\mathbb{R}^n$:

$$\Delta v(x) = \frac{\partial^2 v}{\partial x_1^2} + \cdots + \frac{\partial^2 v}{\partial x_n^2}. \tag{1.49}$$

If $A > 0$, we can rewrite (1.48) in the form

$$v_{tt} - c^2\Delta v = 0. \tag{1.50}$$

The equation (1.50) is typically called "the wave equation." As in (1.24), c is the propagation speed for waves satisfying (1.50); we will discuss this further in §6.

The construction of solutions to (1.50), satisfying initial conditions of the form (1.26), is not as elementary for $n > 1$ as the construction for $n = 1$ given by (1.25)–(1.28). In Chapter 3, we will give a construction, valid for $\Omega = \mathbb{R}^n$, using Fourier analysis. A symmetry trick similar to (1.30) will work if Ω is a rectangular solid in $\mathbb{R}^n$, though not for general bounded regions Ω. The existence and uniqueness of solutions to the wave equation (1.50) for such more general Ω are proven in Chapter 6.

The equation for the components of w parallel to the plane $\mathbb{R}^n$ of $\Omega \subset \mathbb{R}^k$, in this case, has a somewhat different form, as we now compute. Note that this case is the same as considering the entire linearized PDE for the case $k = n$. Then j is the identity map, so the linearization is of the form (1.39)–(1.40), with $\varphi(y)$ replaced by

$$\begin{aligned}\varphi^b(y) &= A \operatorname{Tr} y^*y + B \operatorname{Tr}(y+y^*)^2 + C\big(\operatorname{Tr}(y+y^*)\big)^2 \\ &= (A+2B)\operatorname{Tr} y^*y + 2B \operatorname{Tr} y^2 + 4C\big(\operatorname{Tr} y\big)^2,\end{aligned} \tag{1.51}$$

since $\operatorname{Tr} y^*y = \operatorname{Tr} yy^*$ and $\operatorname{Tr} y^2 = \operatorname{Tr} (y^*)^2$, for a real $n \times n$ matrix y. If we denote the sum of the three terms on the last line in (1.51) by

$$\psi_0(y) + \psi_1(y) + \psi_2(y),$$

then, as in (1.47),

$$\frac{\partial^2 \psi_0}{\partial y_{\mu i}\partial y_{\nu j}} = (2A+4B)\delta_{ij}\delta_{\mu\nu}. \tag{1.52}$$

Also, a brief computation gives

$$\frac{\partial^2 \psi_1}{\partial y_{\mu i}\partial y_{\nu j}} = 4B\delta_{\mu j}\delta_{\nu i} \tag{1.53}$$

and

$$\frac{\partial^2 \psi_2}{\partial y_{\mu i}\partial y_{\nu j}} = 8C\delta_{\mu i}\delta_{\nu j}. \tag{1.54}$$

Now, when φ is replaced by ψ_0, the differential operator of the form (1.40) is $(2A+4B)\Delta$, similar to the computation giving (1.48). When φ is replaced by $\psi_1 + \psi_2$, the differential operator becomes

$$\begin{aligned}\big(\mathcal{L}w\big)^\nu &= 4B \sum_{\mu,i,j=1}^{n} \delta_{\mu j}\delta_{\nu i} w^\mu_{x_i x_j} + 8C \sum_{\mu,i,j=1}^{n} \delta_{\mu i}\delta_{\nu j} w^\mu_{x_i x_j} \\ &= (4B+8C)\sum_j w^j_{x_\nu x_j}.\end{aligned} \tag{1.55}$$

We can write this as

$$\mathcal{L}w = (4B+8C)\ \text{grad div}\ w, \tag{1.56}$$

where the *divergence* of the vector field $w = (w^1, \dots, w^n)$ is

$$\text{div}\, w = \sum_j \frac{\partial w^j}{\partial x_j}, \tag{1.57}$$

and, as before, the gradient of a real-valued function on $\mathbb{R}^n$ is

$$\text{grad}\, u = \left(\frac{\partial u}{\partial x_1}, \dots, \frac{\partial u}{\partial x_n}\right). \tag{1.58}$$

Thus the linearized PDE for vibration in the plane of Ω is

$$m w_{tt} - (2A + 4B)\Delta w - (4B + 8C)\, \text{grad div}\, w = 0. \tag{1.59}$$

The situation where $k = n$ represents a vibrating elastic solid, and the equation (1.59) is known as the equation of *linear elasticity*.

In linear elasticity it is common to linearize about an unstrained state. One writes (1.59) as

$$m w_{tt} - \mu \Delta w - (\lambda + \mu)\, \text{grad div}\, w = 0;$$

$\mu = 2A + 4B$ and $\lambda = 8C$ are called Lamé constants. For more on this, see [MH].

We will concentrate primarily on linear equations in this chapter, indeed, on scalar equations like (1.50). Methods of Chapter 16 will yield results on nonlinear equations of the form (1.12), in any number of x-variables, under a "hyperbolicity" assumption, which is that, for some $C > 0$,

$$\sum_{\mu,\nu=1}^{k} \sum_{i,j=1}^{n} \frac{\partial^2 f(y)}{\partial y_{\mu i}\, \partial y_{\nu j}}\, \xi_i \xi_j \tau_\mu \tau_\nu \geq C|\xi|^2 |\tau|^2, \tag{1.60}$$

for $\xi \in \mathbb{R}^n$, $\tau \in \mathbb{R}^k$. A sufficient, though not necessary, condition for this to hold is that f be a strongly convex function of y. For example (in the case $k = n$), (1.60) holds for

$$f(y) = a\, \text{Tr}\, y^* y + b\, \text{Tr}\, y^2 \tag{1.61}$$

whenever $a > \max(0, -b)$, but such f is strongly convex only if $a > |b|$.

The notions of divergence, gradient, and Laplacian given above are for the case of Euclidean space $\mathbb{R}^n$. All these notions extend to more general Riemannian manifolds. The Laplacian will be defined in such a way as to generalize the identity

$$\int_{\mathbb{R}^n} (\Delta u) v\, dx = -\int_{\mathbb{R}^n} \text{grad}\, u \cdot \text{grad}\, v\, dx, \tag{1.62}$$

for $u, v \in C_0^\infty(\mathbb{R}^n)$, which follows from the definition (1.49) by integration by parts. A further identity that generalizes to the case of Riemannian manifolds is

$$\Delta u = \text{div grad}\, u, \tag{1.63}$$

which for a real-valued function on $\mathbb{R}^n$ follows immediately from the definitions of div, grad, and Δ given above.

We will discuss extensions of these concepts to Riemannian manifolds in the next few sections, starting with the notion of divergence in §2. Then we will derive a number of properties of solutions to wave equations, in §§5–8, and also discuss an extension of the wave equation (1.50) from the case $\mathbb{R} \times \mathbb{R}^n$ to Lorentz manifolds. The problem of proving existence of solutions will be tackled only in later chapters.

We will state here more precisely what the basic existence problem is. In the case of one of the wave equations produced above, say

$$\frac{\partial^2 u}{\partial t^2} - \Delta u = 0, \tag{1.64}$$

we desire to find u satisfying this PDE, given *initial conditions*

$$u(0, x) = f(x), \quad u_t(0, x) = g(x). \tag{1.65}$$

If $\partial\Omega \neq 0$, we also need to impose a boundary condition. There is in particular the *Dirichlet condition*

$$u(t, x) = 0, \quad \text{for } x \in \partial\Omega, \tag{1.66}$$

in the case of a membrane tied down along $\partial\Omega$, as discussed above. There are other boundary conditions that arise in other situations, such as the Neumann boundary condition described in §5, and others mentioned in subsequent chapters. We also can replace (1.64) and (1.66) by nonhomogeneous equations, that is, replace the zeros on the right by given functions.

In this section we have concentrated on evolution equations, involving motion with the passage of time. It is also of interest to study stationary problems, where there is no time dependence. In other words, one looks for stationary points for

$$J(u) = \int_\Omega f(u_x(x))\, dx. \tag{1.67}$$

Thus one obtains a PDE of the form

$$f''(u_x)u_{xx} = 0, \tag{1.68}$$

interpreted via (1.36), as the stationary condition for $J(u)$. In the case $f(u_x) = |u_x|^2$, this becomes the Laplace equation

$$\Delta u = 0. \tag{1.69}$$

A typical boundary condition is the nonhomogeneous Dirichlet condition

$$u = \psi \text{ on } \partial\Omega. \tag{1.70}$$

The existence of a solution to this will follow from results of Chapter 5.

Exercises

1. Compare the formulas (1.22) and (1.23) for longitudinal and transverse waves. For a piano wire, a is very close to 1. What does this imply about the relative propagation

speeds of longitudinal and transverse waves along a piano wire? Which type of waves produce audible sounds?

2. For a function f appearing in (1.60), to be strongly convex means

$$\sum_{\mu,\nu}\sum_{i,j}\frac{\partial^2 f(y)}{\partial y_{\mu i}\partial y_{\nu j}}\lambda_{\mu i}\lambda_{\nu j}\geq C_0|\lambda|^2, \tag{1.71}$$

where $|\lambda|^2=\sum_{\mu,i}|\lambda_{\mu i}|^2$. Show that this estimate implies (1.60). Prove the statements made about $f(y)=a\ \mathrm{Tr}\ y^*y+b\ \mathrm{Tr}\ y^2$ after (1.61).

3. Suppose more generally that $f(y)=a\ \mathrm{Tr}\ y^*y+b\ \mathrm{Tr}\ y^2+c(\mathrm{Tr}\ y)^2$. For what values of a, b, and c is f strongly convex? For what values of a, b, and c does one have the strong ellipticity condition (1.60)?

4. The following exercise relates to the choice of the word "linearization" in describing the relation between the equations (1.12) and (1.20). For $\Omega\subset\mathbb{R}^n$, bounded with smooth boundary, define

$$F: C^2(\overline{\Omega},\mathbb{C}^k)\to C(\overline{\Omega},\mathbb{C}^k)$$

by

$$F(u)=f''(u_x)u_{xx},$$

the right side defined by (1.36). Assume f is C^∞. Show that F is differentiable, as a map between Banach spaces, and that

$$DF(j)w=Lw,$$

where $Lw=Aw_{xx},\ A=f''(j)$, as defined by (1.40).

5. If $u=u(t,x)$ is a *real-valued* function on $\mathbb{R}\times\Omega$, show that the PDE for u giving the stationary condition for the function (1.67) can be written in the form

$$\mathrm{div}\ f_p(u_x)=0, \tag{1.72}$$

where, if $f=f(p)=f(p_1,\dots,p_n)$, then $f_p(u_x)$ is the vector field with components $(\partial f/\partial p_j)(u_x)$. Compare (5.39).

2. The divergence of a vector field

Let M be an n-dimensional manifold, provided with a volume form $\omega\in\Lambda^n M$. Let X be a vector field on M. Then the divergence of X, denoted div X, is a function on M that measures the rate of change of the volume form under the flow generated by X. Thus it is defined by

$$\mathcal{L}_X\omega=(\mathrm{div}\ X)\omega. \tag{2.1}$$

Here, $\mathcal{L}_X$ denotes the Lie derivative. In view of the general formula $\mathcal{L}_X\alpha=d\alpha\rfloor X+d(\alpha\rfloor X)$, derived in Chapter 1, since $d\omega=0$ for any n-form ω on M, we have

$$(\mathrm{div}\ X)\omega=d(\omega\rfloor X). \tag{2.2}$$

If $M=\mathbb{R}^n$, with the standard volume element

$$\omega=dx_1\wedge\cdots\wedge dx_n, \tag{2.3}$$

and if

$$X = \sum X^j(x) \frac{\partial}{\partial x_j}, \tag{2.4}$$

then

$$\omega \rfloor X = \sum_{j=1}^{n} (-1)^{j-1} X^j(x)\, dx_1 \wedge \cdots \wedge \widehat{dx_j} \wedge \cdots \wedge dx_n. \tag{2.5}$$

Hence, in this case, (2.2) yields the formula used in (1.57):

$$\operatorname{div} X = \sum_{j=1}^{n} \partial_j X^j, \tag{2.6}$$

where we use the notation

$$\partial_j f = \frac{\partial f}{\partial x_j}. \tag{2.7}$$

Suppose now that M is an oriented manifold endowed with a Riemannian metric $g_{jk}(x)$. Then M carries a natural volume element ω, determined by the condition that, if one has a coordinate system in which $g_{jk}(p_0) = \delta_{jk}$, then $\omega(p_0) = dx_1 \wedge \cdots \wedge dx_n$. This condition produces the following formula, in any oriented coordinate system:

$$\omega = \sqrt{g}\, dx_1 \wedge \cdots \wedge dx_n, \tag{2.8}$$

where

$$g = \det(g_{jk}). \tag{2.9}$$

In order to derive (2.8), note that if coordinates y are related to x linearly, that is, $y_j = \sum A_{jk} x_k$, then

$$\sum dy_j^2 = \sum_{j,k,\ell} A_{jk} A_{j\ell}\, dx_k\, dx_\ell = \sum g_{k\ell}\, dx_k\, dx_\ell,$$

with

$$g_{k\ell} = \sum_j A_{\ell j} A_{jk},$$

provided $A = (A_{jk})$ is symmetric. Now construct A as the positive-definite square root of the positive-definite matrix $G = \big(g_{jk}(x_0)\big)$. In other words, if $\{v_j\}$ is an orthonormal basis of $\mathbb{R}^n$ with $Gv_j = c_j v_j$, set $Av_j = c_j^{1/2} v_j$. The transformation law for $\Lambda^n A$ on $\Lambda^n \mathbb{R}$ gives

$$\begin{aligned} dy_1 \wedge \cdots \wedge dy_n &= (\det A)\, dx_1 \wedge \cdots \wedge dx_n \\ &= \sqrt{g(x_0)}\, dx_1 \wedge \cdots \wedge dx_n, \end{aligned}$$

from which the formula (2.8) follows.

We now compute div X when the volume element on M is given by (2.8). We have

$$\omega \rfloor X = \sum_j (-1)^{j-1} X^j \sqrt{g}\, dx_1 \wedge \cdots \wedge \widehat{dx_j} \wedge \cdots \wedge dx_n \tag{2.10}$$

and hence

$$d(\omega \rfloor X) = \partial_j(\sqrt{g} X^j)\, dx_1 \wedge \cdots \wedge dx_n. \tag{2.11}$$

Here, as below, we use the summation convention. Hence the formula (2.2) gives

$$\operatorname{div} X = g^{-1/2} \partial_j (g^{1/2} X^j). \tag{2.12}$$

We next derive a result known as the divergence theorem, as a consequence of Stokes' formula, proved in Chapter 1. Recall that Stokes' formula for differential forms is

$$\int_M d\alpha = \int_{\partial M} \alpha, \tag{2.13}$$

for an $(n-1)$-form on M, assumed to be a smooth, compact, oriented manifold with boundary. If $\alpha = \omega \rfloor X$, the formula (2.2) gives

$$\int_M (\operatorname{div} X)\omega = \int_{\partial M} \omega \rfloor X. \tag{2.14}$$

This is one form of the divergence theorem. We will produce an alternative expression for the integrand on the right before stating the result formally.

Given that ω is the volume form for M determined by a Riemannian metric, we can write the interior product $\omega \rfloor X$ in terms of the volume element ω_∂ on ∂M, with its induced Riemannian metric, as follows. Pick normal coordinates on M, centered at $p_0 \in \partial M$, such that ∂M is tangent to the hyperplane $\{x_n = 0\}$ at $p_0 = 0$. Then it is clear that, at p_0,

$$j^*(\omega \rfloor X) = \langle X, \nu \rangle \omega_\partial, \tag{2.15}$$

where ν is the unit vector normal to ∂M, pointing out of M and $j : \partial M \hookrightarrow M$ is the natural inclusion. The two sides of (2.15), which are both defined in a coordinate-independent fashion, are hence equal on ∂M, and the identity (2.14) becomes

$$\int_M (\operatorname{div} X)\omega = \int_{\partial M} \langle X, \nu \rangle \omega_\partial. \tag{2.16}$$

Finally, we adopt the following common notation: we denote the volume element on M by dV and that on ∂M by dS, obtaining the *divergence theorem*:

Theorem 2.1. *If M is a compact manifold with boundary, X a smooth vector field on M, then*

$$\int_M (\operatorname{div} X)\, dV = \int_{\partial M} \langle X, \nu \rangle\, dS, \tag{2.17}$$

where ν is the unit outward-pointing normal to ∂M.

The only point left to mention here is that M need not be orientable. Indeed, we can treat dV and dS as measures and note that all objects in (2.17) are independent of a choice of orientation. To prove the general case, just use a partition of unity supported on orientable pieces.

The definition of the divergence of a vector field given by (2.1), in terms of how the flow generated by the vector field magnifies or diminishes volumes, is a good geometrical characterization, explaining the use of the term "divergence." There are other characterizations of the divergence operation, of a more analytical flavor, which are also quite useful. Here is one.

Proposition 2.2. *The divergence operation is the negative of the adjoint of the gradient operation on vector fields; if X is a vector field and u a function on M, one compactly supported on the interior of M, then*

$$(X, \operatorname{grad} u)_{L^2(M)} = -(\operatorname{div} X, u)_{L^2(M)}. \tag{2.18}$$

The asserted integral identity here is

$$\int_M \langle X, \operatorname{grad} u \rangle\, dV(x) = -\int_M (\operatorname{div} X) u\, dV(x),$$

provided either u or X has compact support in the interior of M. Note that

$$\langle X, \operatorname{grad} u \rangle = \langle X, du \rangle = Xu.$$

In fact, we will use the divergence theorem to obtain a more general result, in which neither u or X is required to vanish on ∂M. We apply (2.17) with X replaced by uX. We have the following "derivation" identity:

$$\operatorname{div} uX = u \operatorname{div} X + \langle du, X \rangle = u \operatorname{div} X + Xu, \tag{2.19}$$

which follows easily from the formula (2.12). The divergence theorem immediately gives the following result.

Proposition 2.3. *If M is a smooth, compact manifold with boundary, u a smooth function, X a smooth vector field on M, then*

$$\int_M (\operatorname{div} X) u\, dV + \int_M Xu\, dV = \int_{\partial M} \langle X, \nu \rangle u\, dS. \tag{2.20}$$

We can also express the adjoint of the differential operator X, defined by

$$\int_M (X^* u)\overline{v}\, dV = \int_M u(X\overline{v})\, dV, \tag{2.21}$$

for $v \in C_0^\infty(\overset{\circ}{M})$, using the divergence, as follows:

Proposition 2.4. *If X is a smooth vector field on M, then*

$$X^* u = -Xu - (\mathrm{div}\, X)u. \tag{2.22}$$

This is equivalent to the statement that

$$\int_M \big[(Xu)v + u(Xv)\big]\, dV = -\int_M (\mathrm{div}\, X)uv\, dV, \tag{2.23}$$

for $u, v \in C_0^\infty(\overset{\circ}{M})$. In fact, from (2.20) we can obtain the following more general result.

Proposition 2.5. *If u and v are smooth functions and X a smooth vector field on a compact manifold M with boundary, then*

$$\int_M \big[(Xu)v + u(Xv)\big]\, dV = -\int_M (\mathrm{div}\, X)uv\, dV + \int_{\partial M} \langle X, \nu\rangle uv\, dS. \tag{2.24}$$

Proof. Replace u by uv in (2.20) and use the derivation identity $X(uv) = (Xu)v + u(Xv)$.

Exercises

1. Given a Hamiltonian vector field

$$H_f = \sum_{j=1}^{n} \Big[\frac{\partial f}{\partial \xi_j}\frac{\partial}{\partial x_j} - \frac{\partial f}{\partial x_j}\frac{\partial}{\partial \xi_j}\Big],$$

calculate $\mathrm{div}\, H_f$ directly from (2.6).

2. If M is a smooth domain in $\mathbb{R}^2$, apply the divergence theorem (2.17) to the vector field $X = g\partial/\partial x - f\partial/\partial y$ to deduce Green's formula:

$$\int_{\partial M} f\, dx + g\, dy = \iint_M \Big(\frac{\partial g}{\partial x} - \frac{\partial f}{\partial y}\Big)\, dx\, dy.$$

3. Show that the identity (2.19) for $\mathrm{div}\,(uX)$ follows from (2.2) and

$$du \wedge (\omega \rfloor X) = (Xu)\omega.$$

Prove this identity, for any n-form ω on M^n. What happens if ω is replaced by a k-form, $k < n$?

4. Relate Exercise 3 to the calculations

$$\mathcal{L}_{uX}\alpha = u\mathcal{L}_X\alpha + du \wedge (\iota_X\alpha) \tag{2.25}$$

and

$$du \wedge (\iota_X\alpha) = -\iota_X(du \wedge \alpha) + (Xu)\alpha, \tag{2.26}$$

valid for any k-form α. The last identity follows from (13.37) of Chapter 1; compare with formula (10.27) of this chapter.

5. Show that

$$\operatorname{div}[X, Y] = X(\operatorname{div} Y) - Y(\operatorname{div} X).$$

3. The covariant derivative and divergence of tensor fields

The covariant derivative of a vector field on a Riemannian manifold was introduced in Chapter 1, §11, in connection with the study of geodesics. We will briefly recall this concept here and relate the divergence of a vector field to the covariant derivative, before generalizing these notions to apply to more general tensor fields. A still more general setting for covariant derivatives is discussed in Appendix C.

If X and Y are vector fields on a Riemannian manifold M, then $\nabla_X Y$ is a vector field on M, the covariant derivative of Y with respect to X. We have the properties

$$\nabla_{(fX)}Y = f\nabla_X Y \tag{3.1}$$

and

$$\nabla_X(fY) = f\nabla_X Y + (Xf)Y, \tag{3.2}$$

the latter being the *derivation property*. Also, ∇ is related to the metric on M by

$$Z\langle X, Y\rangle = \langle \nabla_Z X, Y\rangle + \langle X, \nabla_Z Y\rangle, \tag{3.3}$$

where $\langle X, Y\rangle = g_{jk}X^jY^k$ is the inner product on tangent vectors. The Levi-Civita connection on M is uniquely specified by (3.1)–(3.3) and the torsion free property:

$$\nabla_X Y - \nabla_Y X = [X, Y]. \tag{3.4}$$

There is the explicit defining formula (derived already in (11.22) of Chapter 1)

$$\begin{aligned} 2\langle \nabla_X Y, Z\rangle &= X\langle Y, Z\rangle + Y\langle X, Z\rangle - Z\langle X, Y\rangle \\ &\quad + \langle [X, Y], Z\rangle - \langle [X, Z], Y\rangle - \langle [Y, Z], X\rangle, \end{aligned} \tag{3.5}$$

which follows from cyclically permuting X, Y, and Z in (3.3) and combining the results, exploiting (3.4) to cancel out all covariant derivatives but one. Another way of writing this is the following. If

$$X = X^k D_k, \quad D_k = \frac{\partial}{\partial x_k} \text{ (summation convention)}, \tag{3.6}$$

then

$$\nabla_{D_j} X = X^k{}_{;j}\, D_k, \tag{3.7}$$

with

$$X^k{}_{;j} = \partial_j X^k + \sum_\ell \Gamma^k{}_{\ell j} X^\ell, \tag{3.8}$$

where the "connection coefficients" are given by the formula

$$\Gamma^\ell{}_{jk} = \frac{1}{2} g^{\ell\mu} \Big[\frac{\partial g_{j\mu}}{\partial x_k} + \frac{\partial g_{k\mu}}{\partial x_j} - \frac{\partial g_{jk}}{\partial x_\mu} \Big], \tag{3.9}$$

equivalent to (3.5). We also recall that $\partial g_{k\mu}/\partial x_j$ can be recovered from $\Gamma^\ell{}_{jk}$:

$$\frac{\partial g_{k\mu}}{\partial x_j} = g_{\ell\mu}\Gamma^\ell{}_{jk} + g_{\ell k}\Gamma^\ell{}_{j\mu}. \tag{3.10}$$

The divergence of a vector field has an important expression in terms of the covariant derivative.

Proposition 3.1. *Given a vector field X with components X^k as in (3.6),*

$$\text{div } X = X^j{}_{;j}. \tag{3.11}$$

Proof. This can be deduced from our previous formula for div X,

$$\begin{aligned} \text{div } X &= g^{-1/2}\partial_j (g^{1/2} X^j) \\ &= \partial_j X^j + (\partial_j \log g^{1/2}) X^j. \end{aligned} \tag{3.12}$$

One way to see this is the following. We can think of ∇X as defining a tensor field of type $(1, 1)$:

$$(\nabla X)(Y) = \nabla_Y X. \tag{3.13}$$

Then the right side of (3.11) is the trace of such a tensor field:

$$X^j{}_{;j} = \text{Tr}\, \nabla X. \tag{3.14}$$

This is clearly defined independently of any choice of coordinate system. If we choose an exponential coordinate system centered at a point $p \in M$, then $g_{jk}(p) = \delta_{jk}$ and $\partial g_{jk}/\partial x_\ell = 0$ at p, so (3.12) gives div $X = \partial_j X^j$ at p, in this coordinate system, while the right side of (3.11) is equal to $\partial_j X^j + \Gamma^j{}_{\ell j} X^\ell = \partial_j X^j$ at p. This proves the identity (3.11).

The covariant derivative can be applied to forms, and other tensors, by requiring ∇ to be a derivation. On scalar functions, set

$$\nabla_X u = Xu. \tag{3.15}$$

For a 1-form α, $\nabla_X \alpha$ is characterized by the identity

$$\langle Y, \nabla_X \alpha \rangle = X\langle Y, \alpha \rangle - \langle \nabla_X Y, \alpha \rangle. \tag{3.16}$$

Denote by $\mathfrak{X}(M)$ the space of smooth vector fields on M, and by $\Lambda^1(M)$ the space of smooth 1-forms; each of these is a module over $C^\infty(M)$. Generally, a tensor field of type (k, j) defines a map (with j factors of $\mathfrak{X}(M)$ and k of $\Lambda^1(M)$)

$$F : \mathfrak{X}(M) \times \cdots \times \mathfrak{X}(M) \times \Lambda^1(M) \times \cdots \times \Lambda^1(M) \longrightarrow C^\infty(M), \tag{3.17}$$

which is linear in each factor, over the ring $C^\infty(M)$. A vector field is of type $(1, 0)$ and a 1-form is of type $(0, 1)$. The covariant derivative $\nabla_X F$ is a tensor of the same type, defined by

(3.18)

$$\begin{aligned}(\nabla_X F)(Y_1, \dots, Y_j, \alpha_1, \dots, \alpha_k) = X \cdot \bigl(F(Y_1, \dots, Y_j, \alpha_1, \dots, \alpha_k)\bigr) \\ - \sum_{\ell=1}^{j} F(Y_1, \dots, \nabla_X Y_\ell, \dots, Y_j, \alpha_1, \dots, \alpha_k) \\ - \sum_{\ell=1}^{k} F(Y_1, \dots, Y_j, \alpha_1, \dots, \nabla_X \alpha_\ell, \dots, \alpha_k),\end{aligned}$$

where $\nabla_X \alpha_\ell$ is uniquely defined by (3.16). We can naturally consider ∇F as a tensor field of type $(k, j + 1)$:

$$(\nabla F)(X, Y_1, \dots, Y_j, \alpha_1, \dots, \alpha_k) = (\nabla_X F)(Y_1, \dots, Y_j, \alpha_1, \dots, \alpha_k). \tag{3.19}$$

For example, if Z is a vector field, ∇Z is a vector field of type $(1, 1)$, as already anticipated in (3.13). Hence it makes sense to consider the tensor field $\nabla(\nabla Z)$, of type $(1, 2)$. For vector fields X and Y, we define the *Hessian* $\nabla^2_{(X,Y)} Z$ to be the vector field characterized by

$$\langle \nabla^2_{(X,Y)} Z, \alpha \rangle = (\nabla \nabla Z)(X, Y, \alpha). \tag{3.20}$$

Since, by (3.19), if $F = \nabla Z$, we have

$$F(Y, \alpha) = \langle \nabla_Y Z, \alpha \rangle, \tag{3.21}$$

and, by (3.18),

$$(\nabla_X F)(Y, \alpha) = X \cdot \bigl(F(Y, \alpha)\bigr) - F(\nabla_X Y, \alpha) - F(Y, \nabla_X \alpha), \tag{3.22}$$

it follows by substituting (3.21) into (3.22) and using (3.16) that

$$\nabla^2_{(X,Y)} Z = \nabla_X \nabla_Y Z - \nabla_{(\nabla_X Y)} Z; \tag{3.23}$$

this is a useful formula for the Hessian of a vector field.

More generally, for any tensor field F, of type (j, k), the Hessian $\nabla^2_{(X,Y)} F$, also of type (j, k), is defined in terms of the tensor field $\nabla^2 F = \nabla(\nabla F)$, of type $(j, k + 2)$, by the same type of formula as (3.20), and we have

$$\nabla^2_{(X,Y)} F = \nabla_X (\nabla_Y F) - \nabla_{(\nabla_X Y)} F, \tag{3.24}$$

by an argument similar to that for (3.23).

The metric tensor g is of type $(0, 2)$, and the identity (3.3) is equivalent to

$$\nabla_X g = 0 \tag{3.25}$$

for all vector fields X (i.e., to $\nabla g = 0$). In index notation, this means

$$g_{jk;\ell} = 0 \ \text{ or, equivalently, } \ g^{jk}{}_{;\ell} = 0. \tag{3.26}$$

We also note that the zero torsion condition (3.4) implies

$$u_{;j;k} = u_{;k;j} \tag{3.27}$$

when u is a smooth scalar function, with second covariant derivative $\nabla\nabla u$, a tensor field of type $(0, 2)$. It turns out that analogous second-order derivatives of a vector field differ by a term arising from the curvature tensor; this point is discussed in Appendix C, Connections and Curvature.

We have seen an expression for the divergence of a vector field in terms of the covariant derivative. We can use this latter characterization to provide a general notion of divergence of a tensor field. If T is a tensor field of type (k, j), with components

$$T_\alpha{}^\beta = T_{\alpha_1\cdots\alpha_j}{}^{\beta_1\cdots\beta_k} \tag{3.28}$$

in a given coordinate system, then div T is a tensor field of type $(k-1, j)$, with components

$$T_{\alpha_1\cdots\alpha_j}{}^{\beta_1\cdots\beta_{k-1}\ell}{}_{;\ell}. \tag{3.29}$$

In view of the special role played by the last index, the divergence of a tensor field T is mainly interesting when T has some symmetry property. In §7 we will introduce the stress-energy tensor, a symmetric second-order covariant tensor; raising indices produces a symmetric second-order tensor field of type $(2, 0)$, whose divergence is an important object.

In view of (3.11), we know that a vector field X generates a volume-preserving flow if and only if $X^j{}_{;j} = 0$. Complementing this, we investigate the condition that the flow generated by X consists of isometries, that is, the flow leaves the metric g invariant, or equivalently

$$\mathcal{L}_X g = 0. \tag{3.30}$$

For vector fields U and V, we have

$$\begin{aligned}(\mathcal{L}_X g)(U, V) &= -\langle \mathcal{L}_X U, V\rangle - \langle U, \mathcal{L}_X V\rangle + X\langle U, V\rangle \\ &= \langle \nabla_X U - \mathcal{L}_X U, V\rangle + \langle U, \nabla_X V - \mathcal{L}_X V\rangle \\ &= \langle \nabla_U X, V\rangle + \langle U, \nabla_V X\rangle,\end{aligned} \tag{3.31}$$

where the first identity follows from the derivation property of $\mathcal{L}_X$, the second from the metric property (3.3) expressing $X\langle U, V\rangle$ in terms of covariant derivatives, and the third from the zero torsion condition (3.4). If U and V are coordinate vector fields $D_j = \partial/\partial x_j$, we can write this identity as

$$(\mathcal{L}_X g)(D_j, D_k) = g_{k\ell} X^\ell{}_{;j} + g_{j\ell} X^\ell{}_{;k}. \tag{3.32}$$

Thus X generates a group of isometries (one says X is a *Killing field*) if and only if

$$g_{k\ell} X^\ell{}_{;j} + g_{j\ell} X^\ell{}_{;k} = 0. \tag{3.33}$$

This takes a slightly shorter form for the covariant field

$$X_j = g_{jk}X^k. \tag{3.34}$$

We state formally the consequence, which follows immediately from (3.33) and the vanishing of the covariant derivatives of the metric tensor.

Proposition 3.2. *X is a Killing vector field if and only if*

$$X_{k;j} + X_{j;k} = 0. \tag{3.35}$$

Generally, half the quantity on the left side of (3.35) is called the *deformation tensor* of X. If we denote by ξ the 1-form $\xi = \sum X_j\, dx_j$, the deformation tensor is the *symmetric part* of $\nabla\xi$, a tensor field of type $(0, 2)$. It is also useful to identify the antisymmetric part, which is naturally regarded as a 2-form.

Proposition 3.3. *We have*

$$d\xi = \frac{1}{2}\sum_{j,k}(X_{j;k} - X_{k;j})\, dx_k \wedge dx_j. \tag{3.36}$$

Proof. By definition,

$$d\xi = \frac{1}{2}\sum_{j,k}(\partial_k X_j - \partial_j X_k)\, dx_k \wedge dx_j, \tag{3.37}$$

and the identity with the right side of (3.36) follows from the symmetry $\Gamma^\ell{}_{jk} = \Gamma^\ell{}_{kj}$.

There is a useful generalization of the concept of a Killing field, namely a *conformal* Killing field, which is a vector field X whose flow consists of conformal diffeomorphisms of M, that is, preserves the metric tensor up to a scalar factor:

$$\mathcal{F}_X^{t*} g = \alpha(t, x) g \Longleftrightarrow \mathcal{L}_X g = \lambda(x) g. \tag{3.38}$$

Note that the trace of $\mathcal{L}_X g$ is 2 div X, by (3.32), so the last identity in (3.38) is equivalent to $\mathcal{L}_X g = (2/n)(\text{div } X) g$ or, with $(1/2)\mathcal{L}_X g = \text{Def } X$,

$$\text{Def } X - \frac{1}{n}(\text{div } X) g = 0 \tag{3.39}$$

is the equation of a conformal Killing field.

To end this section, and prepare for subsequent material, we note that concepts developed so far for Riemannian manifolds, that is, manifolds with positive-definite metric tensors, have extensions to indefinite metric tensors, including *Lorentz* metrics.

A Riemannian metric tensor produces a symmetric isomorphism

$$G : T_x M \longrightarrow T_x^* M, \tag{3.40}$$

which is positive. More generally, a symmetric isomorphism (3.40) corresponds to a nondegenerate metric tensor. Such a tensor has a well defined signature (j, k), $j + k = n = \dim M$; at each $x \in M$, T_xM has a basis $\{e_1, \dots, e_n\}$ of mutually orthogonal vectors such that $\langle e_1, e_1\rangle = \cdots = \langle e_j, e_j\rangle = 1$, while $\langle e_{j+1}, e_{j+1}\rangle = \cdots = \langle e_n, e_n\rangle = -1$. If $j = 1$ (or $k = 1$), we say M has a Lorentz metric.

The concepts discussed in this section in the Riemannian case, such as the covariant derivative, all extend with little change to the general nondegenerate case. We will see this in use, in the Lorentz case, in §7.

Exercises

1. Let φ be a tensor field of type $(0, k)$ on a Riemannian manifold, endowed with its Levi-Civita connection. Show that
$$(\mathcal{L}_X\varphi - \nabla_X\varphi)(U_1, \dots, U_k) = \sum_j \varphi(U_1, \dots, \nabla_{U_j}X, \dots, U_k).$$
How does this generalize (3.31)?

2. Recall the formula (13.56) of Chapter 1, when ω is a k-form:
$$\begin{aligned}(d\omega)(X_0, \dots, X_k) = &\sum_{j=0}^{k}(-1)^j X_j \cdot \omega(X_0, \dots, \widehat{X}_j, \dots, X_k)\\ &+ \sum_{0\le\ell<j\le k} (-1)^{j+\ell}\omega([X_\ell, X_j], X_0, \dots, \widehat{X}_\ell, \dots, \widehat{X}_j, \dots, X_k).\end{aligned}$$
Show that the last double sum can be replaced by
$$\begin{aligned}&-\sum_{\ell<j}(-1)^j\omega(X_0, \dots, \nabla_{X_j}X_\ell, \dots, \widehat{X}_j, \dots, X_k)\\ &-\sum_{\ell>j}(-1)^j\omega(X_0, \dots, \widehat{X}_j, \dots, \nabla_{X_j}X_\ell, \dots, X_k).\end{aligned}$$

3. Using Exercise 2 and the expansion of $(\nabla_{X_j}\omega)(X_0, \dots, \widehat{X}_j, \dots, X_k)$ via the derivation property, show that
$$(d\omega)(X_0, \dots, X_k) = \sum_{j=0}^{k}(-1)^j(\nabla_{X_j}\omega)(X_0, \dots, \widehat{X}_j, \dots, X_k). \tag{3.41}$$
Note that this generalizes Proposition 3.3.

4. Prove the identity
$$\frac{\partial \log\sqrt{g}}{\partial x_j} = \sum_\ell \Gamma^\ell{}_{\ell j}.$$
Use either the identity (3.11), involving the divergence, or the formula (3.9) for $\Gamma^\ell{}_{jk}$. Which is easier?

5. Show that the characterization (3.17) of a tensor field of type (k, j) is equivalent to the condition that F be a section of the vector bundle $(\otimes^j T^*) \otimes (\otimes^k T)$ or, equivalently, of the bundle $\mathrm{Hom}\,(\otimes^j T, \otimes^k T)$. Think of other variants.

6. The operation $X_j = g_{jk}X^k$ is called *lowering indices*. It produces a 1-form (section of T^*M) from a vector field (section of TM), implementing the isomorphism (3.38). Similarly, one can raise indices:

$$Y^j = g^{jk}Y_k,$$

producing a vector field from a 1-form, that is, implementing the inverse isomorphism. Define more general operations raising and lowering indices, passing from tensor fields of type (j,k) to other tensor fields, of type (ℓ,m), with $\ell+m=j+k$. One says that these tensor fields are associated to each other via the metric tensor.
7. Using (3.16), show that if $\alpha = a_k(x)\,dx_k$ (summation convention), then $\nabla_{D_j}\alpha = a_{k;j}\,dx_k$, with

$$a_{k;j} = \partial_j a_k - \sum_\ell \Gamma^\ell{}_{kj} a_\ell.$$

Compare with (3.8). Use this to verify that (3.36) and (3.37) are equal. Work out a corresponding formula for $\nabla_{D_\ell} T$ when T is a tensor field of type (j,k), as in (3.28).
8. Using the formula (3.23) for the Hessian, show that, for vector fields X, Y, Z on M,

$$\big(\nabla^2_{(X,Y)} - \nabla^2_{(Y,X)}\big)Z = \big([\nabla_X, \nabla_Y] - \nabla_{[X,Y]}\big)Z.$$

Denoting this by $R(X,Y,Z)$, show that it is linear in each of its three arguments over the ring $C^\infty(M)$, for example, $R(X,Y,fZ) = f\,R(X,Y,Z)$ for $f \in C^\infty(M)$. Discussion of $R(X,Y,Z)$ as the *curvature tensor* is given in Appendix C, Connections and Curvature.
9. Verify (3.24). For a function u, to show that $\nabla^2_{(X,Y)}u = \nabla^2_{(Y,X)}u$, use the special case

$$\nabla^2_{(X,Y)}u = XYu - (\nabla_X Y)\cdot u$$

of (3.24). Note that this is an invariant formulation of (3.27). Show that

$$\nabla^2_{(X,Y)}u = \frac{1}{2}(\mathcal{L}_V g)(X,Y), \quad V = \text{grad}\, u.$$

10. Let ω be the *volume form* of an oriented Riemannian manifold M. Show that $\nabla_X\omega = 0$ for all vector fields X.
11. Let X be a vector field on a Riemannian manifold M. Show that the formal adjoint of ∇_X, acting on vector fields, is

$$\nabla_X^* Y = -\nabla_X Y - (\text{div}\, X)Y. \tag{3.42}$$

12. Show that the formal adjoint of $\mathcal{L}_X$, acting on vector fields, is

$$\mathcal{L}_X^* Y = -\mathcal{L}_X Y - (\text{div}\, X)Y - 2\,\text{Def}(X)Y, \tag{3.43}$$

where $\text{Def}(X)$ is a tensor field of type $(1,1)$, given by

$$\frac{1}{2}(\mathcal{L}_X g)(Z,Y) = g(Z, \text{Def}(X)Y), \tag{3.44}$$

g being the metric tensor.
13. With div defined by (3.29) for tensor fields, show that

$$\text{div}\,(X\otimes Y) = (\text{div}\, Y)X + \nabla_Y X. \tag{3.45}$$

14. If X, Y, and Z have compact support, show that

$$(Z, \text{div}\,(X\otimes Y))_{L^2} = -(\nabla_Y Z, X)_{L^2}.$$

15. If $\gamma(s)$ is a unit-speed geodesic on a Riemannian manifold M, $\gamma'(s) = T(s)$, and X is a vector field on M, show that

$$\frac{d}{ds}\langle T(s), X(\gamma(s))\rangle = \frac{1}{2}(\mathcal{L}_X g)(T, T). \tag{3.46}$$

Deduce that if X is a Killing field, then $\langle T, X\rangle$ is constant on γ. Relate this to the conservation law for geodesic flow on a surface of revolution, discussed in Chapter 1, §16. (*Hint*: Show that the left side of (3.46) is equal to $\langle T, \nabla_T X\rangle$.)

16. If we define Def: $C^\infty(M, T) \to C^\infty(M, S^2T^*)$ by $\operatorname{Def}(X) = (1/2)\mathcal{L}_X g$, show that

$$\operatorname{Def}^* u = -\operatorname{div} u,$$

where $(\operatorname{div} u)^j = u^{jk}{}_{;k}$, as in (3.29).

4. The Laplace operator on a Riemannian manifold

We define the Laplace operator on a Riemannian manifold M, with metric g_{jk}, in a way that naturally generalizes the characterizations of the Laplace operator on Euclidean space, given by (1.49), (1.62), and (1.63). Taking (1.62) as fundamental, we define the Laplace operator Δ on M to be the second-order differential operator satisfying

$$-(\Delta u, v) = (du, dv) = (\operatorname{grad} u, \operatorname{grad} v), \tag{4.1}$$

for $u, v \in C_0^\infty(M)$. Here the left side is

$$-\int_M (\Delta u)\overline{v}\, dV, \tag{4.2}$$

where dV is the natural volume element, given in local coordinates by $\sqrt{g}dx_1 \cdots dx_n$. The right side of (4.1), for u and v supported in a coordinate patch, is

$$\begin{aligned}\int \langle du, dv\rangle\, dV &= \int g^{jk}(\partial_j u)(\partial_k \overline{v})\sqrt{g}\, dx \\ &= -\int \overline{v}\partial_k\big(g^{1/2}g^{jk}\,\partial_j u\big)g^{-1/2}g^{1/2}\, dx,\end{aligned} \tag{4.3}$$

integrating by parts, so we see that Δ is given in local coordinates by

$$\Delta u = g^{-1/2}\,\partial_j\big(g^{jk}g^{1/2}\,\partial_k u\big). \tag{4.4}$$

Soon we will see how to modify (4.1) when u and v do not vanish on ∂M, in case M is a compact Riemannian manifold with boundary.

We now show that (1.63) generalizes, that is, we have

$$\Delta u = \operatorname{div}\operatorname{grad} u. \tag{4.5}$$

In fact, in view of the formula

$$\operatorname{div} X = g^{-1/2}\,\partial_j(g^{1/2}X^j)$$

derived in (2.12), together with

$$X^j = g^{jk}\partial_k u, \quad \text{for } X = \text{ grad } u,$$

we see that (4.5) follows directly from the local coordinate formula (4.4). Note that the identity

$$(X, \text{grad } v)_{L^2} = -(\text{div } X, v)_{L^2}, \tag{4.6}$$

proved in (2.18), when applied to $X = \text{grad } u$, also gives (4.5) directly.

Applying the refinement (2.20) of (4.6) gives us important identities due to Green. Let us use the notation

$$\frac{\partial u}{\partial \nu} = \langle \text{grad } u, \nu\rangle \tag{4.7}$$

for the normal component of grad u; $\partial u/\partial \nu$ is called the *normal derivative* of u. If we exploit (2.20) with $X = \text{grad } \overline{v}$, we get the identity (4.8) below; if we interchange u and $\overline{v}$ and subtract the resulting expression from (4.8), we obtain (4.9). This provides a proof of *Green's identities*:

Proposition 4.1. *If M is a compact Riemannian manifold with boundary, then for $u, v \in C^\infty(M)$, we have*

$$-(u, \Delta v)_{L^2} = (du, dv) - \int_{\partial M} u\Big(\frac{\partial \overline{v}}{\partial \nu}\Big)\, dS \tag{4.8}$$

and

$$(\Delta u, v) - (u, \Delta v) = \int_{\partial M} \Big[\Big(\frac{\partial u}{\partial \nu}\Big)\overline{v} - u\Big(\frac{\partial \overline{v}}{\partial \nu}\Big)\Big]\, dS. \tag{4.9}$$

Next we express the Laplace operator in terms of covariant derivatives. As we have seen,

$$\text{div } X = X^j{}_{;j}.$$

If we set $X = \text{grad } u$, we obtain

$$\Delta u = g^{jk} u_{;j;k}, \tag{4.10}$$

using the fact that $g^{jk}{}_{;\ell} = 0$. Here, $\sum u_{;j;k} dx_k \otimes dx_j$ is a tensor field of type $(0, 2)$, which is the same as $\nabla^2 u$. Recall that $\nabla^2 F$ is a tensor field of type $(j, k+2)$ whenever F is a tensor field of type (j, k). The formula (4.10) can be rewritten as

$$\Delta u = \text{ Tr}_g \nabla^2 u, \tag{4.11}$$

where Tr_g denotes the trace of $\nabla^2 u(x)$, as a quadratic form on $T_x M$, in terms of the quadratic form given by the metric tensor g. In other words, we can define a tensor field $H(u)$, of type $(1, 1)$, by

$$\langle H(u)X, Y\rangle = (\nabla^2 u)(X, Y), \tag{4.12}$$

and $\mathrm{Tr}_g \nabla^2 u = \mathrm{Tr}\, H(u)$.

Since the Laplace operator is defined in a coordinate-independent manner on a Riemannian manifold, it is clear that if $F : M \to M$ is a diffeomorphism and $F^* : C^\infty(M) \to C^\infty(M)$ is defined by $F^*u(x) = u(F(x))$, then F^* commutes with the Laplace operator provided F is an isometry. Thus, if X is a vector field on M, X commutes with Δ provided the flow $\mathcal{F}_X^t$ generated by X consists of isometries. This result has a converse.

Proposition 4.2. *A vector field X commutes with Δ if and only if X generates a group of isometries.*

The proof rests on a computation of independent interest. In fact, a manipulation of (4.10), which we leave to the reader, yields the general identity

$$\begin{aligned} [\Delta, X]u &= (X^{j;k} + X^{k;j})u_{;j;k} + (X^{j;k} + X^{k;j})_{;j} u_{;k} \\ &= g^{-1/2}\partial_j\big(g^{1/2}(X^{j;k} + X^{k;j})\,\partial_k u\big). \end{aligned} \tag{4.13}$$

Thus $[\Delta, X] = 0$ if and only if $X^{j;k} + X^{k;j} = 0$, which is equivalent to the condition (3.35) for a Killing field.

Exercises

1. If $u \in C^\infty(M)$, $X = \mathrm{grad}\, u$, the condition that X generates a volume-preserving flow is that $\Delta u = 0$. What PDE on u is equivalent to the statement that X is a Killing field?
2. Verify formula (4.13) for $[\Delta, X]$. Show that it has the invariant formulation

$$\frac{1}{2}[\Delta, X]u = \langle \mathrm{Def}(X), \nabla^2 u\rangle + \langle \mathrm{div}\,\mathrm{Def}(X), du\rangle = \mathrm{div}\big(\mathrm{Def}(X)\cdot du\big), \tag{4.14}$$

   in terms of the deformation tensor $\mathrm{Def}(X)$, with components $(1/2)(X^{j;k} + X^{k;j})$, that is, the type $(2, 0)$ analogue of the tensor field of type $(1, 1)$ given by (3.42), or the tensor field of type $(0, 2)$ equal to half of (3.35).
3. Show that the Laplace operator $\Delta = \partial^2/\partial x_1^2 + \cdots + \partial^2/\partial x_n^2$ on $\mathbb{R}^n$ has the following expressions in various coordinate systems:

   (a) Polar coordinates on $\mathbb{R}^2$: $x_1 = r\cos\theta$, $x_2 = r\sin\theta$.

$$\Delta = \frac{\partial^2}{\partial r^2} + \frac{1}{r}\frac{\partial}{\partial r} + \frac{1}{r^2}\frac{\partial^2}{\partial\theta^2}. \tag{4.15}$$

   (b) Spherical polar coordinates on $\mathbb{R}^3$: $x_1 = \rho\sin\varphi\sin\theta$, $x_2 = \rho\sin\varphi\cos\theta$, $x_3 = \rho\cos\varphi$.

$$\Delta = \frac{\partial^2}{\partial\rho^2} + \frac{2}{\rho}\frac{\partial}{\partial\rho} + \frac{1}{\rho^2\sin\varphi}\Big(\frac{\partial^2}{\partial\theta^2} + \sin\varphi\frac{\partial^2}{\partial\varphi^2} + \cos\varphi\frac{\partial}{\partial\varphi}\Big). \tag{4.16}$$

   (c) Spherical polar coordinates on $\mathbb{R}^n$: $x = r\omega$, $\omega \in S^{n-1}$.

$$\Delta = \frac{\partial^2}{\partial r^2} + \frac{n-1}{r}\frac{\partial}{\partial r} + \frac{1}{r^2}\Delta_S, \tag{4.17}$$

   where Δ_S is the Laplace operator on the unit sphere S^{n-1}. (Compare (4.19) below.) (*Hint*: Express the Euclidean metric tensor $ds^2 = dx_1^2 + \cdots + dx_n^2$ in these coordinates.)

4. Let N be a Riemannian manifold, of dimension $n-1$. Denote by $C(N)$ the cone with base N, that is, the space $\mathbb{R}^+ \times N$, with Riemannian metric

$$g = dr^2 + r^2 g_N. \tag{4.18}$$

Show that the Laplace operator on $C(N)$ is of the form

$$\Delta = \frac{\partial^2}{\partial r^2} + \frac{n-1}{r}\frac{\partial}{\partial r} + \frac{1}{r^2}\Delta_N, \tag{4.19}$$

where Δ_N is the Laplace operator on the base N. Apply this to the expression of the Laplace operator Δ on $\mathbb{R}^n$, in polar coordinates, with $N = S^{n-1}$.
5. Show that, in local coordinates,

$$\Delta u = g^{jk}\,\partial_j\partial_k u - g^{jk}\Gamma^\ell{}_{jk}\,\partial_\ell u.$$

5. The wave equation on a product manifold and energy conservation

The analysis of vibrating membranes in Euclidean space has important extensions to studies of vibrating manifolds. We will start with a fairly general situation, specializing quickly to models that give rise to "the wave equation"

$$\frac{\partial^2 u}{\partial t^2} - \Delta u = 0, \tag{5.1}$$

for $u = u(t,x)$, a scalar function on $\mathbb{R} \times M$, where Δ is the Laplace operator on M defined in §4.

We consider vibrations of one manifold M within another, N. Suppose these manifolds are endowed with Riemannian metric tensors g and h, respectively. The vibration is described by a map

$$u : \mathbb{R} \times M \longrightarrow N. \tag{5.2}$$

In §1 we dealt with the special case where M is a bounded region in $\mathbb{R}^n$ and $N = \mathbb{R}^k$. Now we allow M to be a compact manifold with boundary. We again use a stationary action principle to produce equations governing the vibration. The appropriate expression for "kinetic energy" is

$$T(t) = \frac{1}{2}\int_M m(x)|u_t(t,x)|^2\,dV, \tag{5.3}$$

where dV is the natural volume element on M and $m(x) > 0$ is a given "mass density." The velocity $u_t(t,x)$ takes values in T_yN, with $y = u(t,x)$, and the square-norm in the integrand in (5.3) is given by the metric tensor h;

$$|u_t|^2 = h(u, u_t, u_t) \tag{5.4}$$

if $h(y, v, w)$ denotes the inner product of v and w in T_yN.

The form that we will consider for the potential energy is the following generalization of (1.3):

$$V(t) = \int_M f(x, u(t,x), u_x(t,x))\, dV, \tag{5.5}$$

where

$$u_x(t,x) \in \mathcal{L}(T_xM, T_{u(t,x)}N), \tag{5.6}$$

and f is a smooth, real-valued function defined on the bundle $\mathcal{L}$ over $M \times N$ with fiber over (x, y) given by $\mathcal{L}(T_xM, T_yN)$:

$$f = f(x, y, A), \quad A \in \mathcal{L}(T_xM, T_yN). \tag{5.7}$$

In particular, one has examples analogous to (1.42), that is,

$$f(x, y, A) = \Psi\big(\operatorname{Tr} g_1(A^*A), \dots, \operatorname{Tr} g_K(A^*A)\big), \tag{5.8}$$

where $A^* \in \mathcal{L}(T_yN, T_xM)$ is the adjoint of A, defined using the inner products on T_xM and T_yN defined by their Riemannian metrics. The $g_\ell(A^*A)$ are defined as described below (1.42). Many interesting cases of this sort arise naturally, including

$$f(x, y, A) = \operatorname{Tr} A^*A. \tag{5.9}$$

Applying the stationary action principle will yield for u a second-order system of PDE of a form that generalizes (1.12). We look here at the details for a special case.

Namely, take $N = \mathbb{R}$, and suppose $f(x, y, A)$ is independent of $y \in \mathbb{R}$. In other words, we consider a potential energy of the form

$$V(t) = \int_M f(x, u_x(t,x))\, dV, \tag{5.10}$$

where $u_x(t,x) \in T_x^*M$ and $f = f(x, \xi)$ is a smooth, real-valued function defined on T^*M, or perhaps on some open subset. In that case, the stationary condition for $(J_0 - J_1)(u) = \int_{t_0}^{t_1} \big[T(t) - V(t)\big]\, dt$ is derived from the following calculations. First, as in (1.8),

$$\frac{d}{ds} J_0(u + sv)\big|_{s=0} = -\iint m u_{tt} v \, dV\, dt, \tag{5.11}$$

provided $v \in C_0^\infty(I \times \overset{\circ}{M})$, $I = (t_0, t_1)$. Here $\overset{\circ}{M}$ denotes the interior of M. Furthermore, for such v,

$$\frac{d}{ds} J_1(u + sv)\big|_{s=0} = \iint f_\xi(x, u_x) \cdot v_x \, dV\, dt, \tag{5.12}$$

where, in local coordinates,

$$f_\xi(x, u_x) \cdot v_x = \sum_j \frac{\partial f}{\partial \xi_j} \frac{\partial v}{\partial x_j}. \tag{5.13}$$

If v is supported in a coordinate patch, in which $dV = \sqrt{g}dx$, we can integrate by parts and write

$$(5.14)\quad \frac{d}{ds} J_1(u+sv)\big|_{s=0} = -\iint \sum_j g^{-1/2}\, \partial_{x_j}\big(g^{1/2} f_{\xi_j}(x,u_x)\big) v\sqrt{g}\, dx\, dt.$$

Thus we get the following PDE for u, in a local coordinate system:

$$(5.15)\qquad mu_{tt} - g^{-1/2}\partial_{x_j}\big(g^{1/2} f_{\xi_j}(x,u_x)\big) = 0,$$

using the summation convention. Written out more fully, this is

$$(5.16)\quad mu_{tt} - \Big[f_{\xi_j\xi_k}(x,u_x)u_{x_jx_k} + f_{\xi_jx_j}(x,u_x) + \frac{1}{2}g^{-1}(\partial_{x_j}g) f_{\xi_j}(x,u_x)\Big] = 0.$$

An invariant formulation of this PDE is given in the exercises.

The choice of $f(x,\xi)$ that produces a wave equation of the form (5.1) is that of a constant times the Riemannian metric on covariant vectors:

$$(5.17)\qquad f(x,\xi) = \sigma\, g(x,\xi,\xi) = \sigma\, g^{jk}\, \xi_j\xi_k,$$

with σ a positive constant. In that case, (5.15) becomes

$$(5.18)\qquad mu_{tt} - \sigma\Delta u = 0$$

in view of the local coordinate formula

$$(5.19)\qquad \Delta u = g^{-1/2}\, \partial_j(g^{1/2} g^{jk}\partial_k u)$$

derived in §4. If m is a constant, this is of the form (5.1) provided $\sigma = m$, which could be arranged by a rescaling of the t-variable.

Other choices of $f(x,\xi)$ arise naturally in the study of vibrating membranes, choices that lead to nonlinear PDE. We will return to this in Chapter 16, but for now we concentrate on the linear case (5.18), until the very end of this section where we make a few brief comments on nonlinear problems.

Let us redo the calculation of the variation of $J_1(u)$ in an invariant fashion, when $f(x,\xi)$ is given by (5.17), so

$$(5.20)\qquad J_1(u) = \sigma \iint_{I\times M} |d_x u|^2\, dV\, dt.$$

We have, for $v \in C_0^\infty(I\times \overset{\circ}{M})$,

$$(5.21)\qquad \frac{d}{ds} J_1(u+sv)\big|_{s=0} = 2\sigma \iint \langle d_x u, d_x v\rangle\, dV\, dt,$$

and Green's formula (4.8) shows that this is equal to

$$(5.22)\qquad -2\sigma \iint (\Delta u) v\, dV\, dt,$$

since the boundary integral vanishes in this case. Again the stationary condition for $(J_0 - J_1)(u)$ is seen to be the wave equation (5.18).

As in (1.26), it is typical to specify *initial conditions*, of the form

$$u(0,x) = f(x), \quad u_t(0,x) = g(x). \tag{5.23}$$

If $\partial M \neq \emptyset$, we also need to specify a *boundary condition* for u. One typical condition is

$$u(t,x) = 0, \quad \text{for } x \in \partial M. \tag{5.24}$$

This is known as the *Dirichlet* boundary condition for u. It models a vibrating drum head that is firmly attached to its boundary. Tying down the boundary provides a justification for considering only variations v that vanish on $I \times \partial M$ in the specification of the stationary condition above. Another natural physical problem is to describe vibrations of M when the boundary is allowed to move freely. Then we should allow any $v \in C^\infty(\overline{I} \times M)$ that vanishes at $t = t_0$ and $t = t_1$, as a variation. The formula (5.11) for the variation of $J_0(u)$ continues to hold, and so does (5.21), but an application of Green's formula to (5.21) now yields

$$\frac{d}{ds} J_1(u+sv)\big|_{s=0} = -2\sigma \iint\limits_{I\times M} (\Delta u) v \, dV \, dt + 2\sigma \iint\limits_{I\times\partial M} v \frac{\partial u}{\partial \nu} \, dS \, dt. \tag{5.25}$$

If we do apply this to the subclass of $v \in C_0^\infty(I \times \overset{\circ}{M})$, we see that the wave equation (5.18) must still be satisfied for u to be a stationary point. Now, granted that u satisfies (5.18), we hence have

$$\frac{d}{ds} (J_0 - J_1)(u+sv)\big|_{s=0} = -2\sigma \iint\limits_{I\times\partial M} v \frac{\partial u}{\partial \nu} \, dS \, dt, \tag{5.26}$$

for all $v \in C^\infty(\overline{I} \times M)$ that vanish at $t = t_0$ and at $t = t_1$. This yields the following boundary condition for freely vibrating M:

$$\frac{\partial u}{\partial \nu} = 0, \quad \text{for } x \in \partial M. \tag{5.27}$$

This is known as the *Neumann* boundary condition for u. Another situation it models is the propagation of small-amplitude sound waves in a region bounded by a hard wall.

Since we have introduced the kinetic energy and the potential energy, we should look at the total energy. In the case when (5.17) gives the potential energy, if we take $m = 1$ and $\sigma = 1/2$, the total energy is

$$E(t) = \frac{1}{2} \int\limits_M \left[|u_t(t,x)|^2 + \langle d_x u, d_x u \rangle \right] dV(x). \tag{5.28}$$

We aim to establish the energy conservation law

$$E(t) = \text{const.} \tag{5.29}$$

whenever u is a sufficiently smooth solution to the wave equation (5.1), assuming that u satisfies either the Dirichlet condition (5.24) or the Neumann condition

(5.27) on ∂M. In fact, we have

$$\frac{dE}{dt} = \int_M \left[u_t u_{tt} + \langle d_x u_t, d_x u\rangle\right] dV. \tag{5.30}$$

We want to factor u_t out of the integrand, so we integrate by parts the last term in (5.30), using Green's identity to get

$$\frac{dE}{dt} = \int_M u_t(u_{tt} - \Delta u)\, dV + \int_{\partial M} u_t \frac{\partial u}{\partial \nu}\, dS. \tag{5.31}$$

The right side of (5.31) vanishes provided u satisfies the wave equation and either the Dirichlet or Neumann boundary condition. This proves the energy conservation law (5.29), equivalent to

$$\int_M \left[|u_t(t,x)|^2 + \langle d_x u, d_x u\rangle\right] dV = \int_M \left[|g(x)|^2 + \langle d_x f, d_x f\rangle\right] dV, \tag{5.32}$$

given the initial conditions (5.23).

We continue briefly the discussion of stationary problems from the end of §1. These problems do not involve t-dependence, that is, they arise via describing critical points for a function

$$J(u) = \int_M f\big(x, u(x), u_x(x)\big)\, dV, \tag{5.33}$$

with

$$f = f(x, y, A), \quad A \in \mathcal{L}(T_x M, T_y N). \tag{5.34}$$

If $N = \mathbb{R}$ and $f(x, y, \xi) = f(x, \xi)$ is given by (5.17), then the PDE obtained as the stationary condition for $J(u)$ is

$$\Delta u = 0, \tag{5.35}$$

involving the Laplace operator (5.19). A typical boundary condition is the nonhomogeneous Dirichlet condition

$$u = \psi \text{ on } \partial M. \tag{5.36}$$

Another is the nonhomogeneous Neumann condition

$$\frac{\partial u}{\partial \nu} = \varphi \text{ on } \partial M. \tag{5.37}$$

These will be studied in Chapter 5.

There are also very important nonlinear problems arising from the problem of finding stationary points, particularly extrema, of (5.33). We mention in particular the choice (5.9) for $f(x, y, A)$, namely, $\operatorname{Tr} A^*A$. Maps $u : M \to N$ critical for such $J(u)$ are called *harmonic maps*. In case $N = \mathbb{R}^k$, these are just functions whose components are harmonic in the sense of (5.35), but for a nonflat Riemannian manifold N, one gets a nonlinear problem. For example, as seen in Chapter 1, for

$M = I \subset \mathbb{R}$, one gets the geodesic equation. Harmonic maps will be studied in Chapter 14, by variational methods, and in Chapter 15, via techniques involving nonlinear parabolic PDE.

Exercises

1. For $J_1(u) = \int_M f(x, u_x)dV$ as in (5.10), $f : T^*M \to \mathbb{R}$, demonstrate the invariant formula

$$\frac{d}{ds} J_1(u + sv)\big|_{s=0} = \int_M \langle A_f(x, u_x), v_x \rangle \, dV,$$

where $A_f : T^*M \to TM$ is given by

$$A_f(x, \xi) = D\pi(x, \xi) H_f, \tag{5.38}$$

H_f being the Hamiltonian vector field of f, and $\pi : T^*M \to M$ the natural projection. For fixed t, $u_x = d_x u$ is a 1-form on M. Consequently, $A_f(x, u_x)$ is a vector field on M.

2. In the context of Exercise 1, show that the resulting PDE (5.15) has the invariant description

$$mu_{tt} - \operatorname{div} A_f(x, u_x) = 0. \tag{5.39}$$

Compare (1.72).

3. Show that (under an appropriate nondegeneracy hypothesis) maps of the form A_f invert Legendre transformations $\lambda : TM \to T^*M$, discussed in §12 of Chapter 1.
(*Hint*: Using (12.9)–(12.18), consider the Legendre transform associated to the function $F(x, v)$ on TM defined implicitly by

$$F\big(x, f_\xi(x, \xi)\big) = f(x, \xi) - \xi \cdot f_\xi(x, \xi)$$

or, in the notation used above,

$$F\big(A_f(x, \xi)\big) = f(x, \xi) - \langle A_f(x, \xi), \xi \rangle.)$$

6. Uniqueness and finite propagation speed

We study some properties of solutions to the wave equation on $\mathbb{R} \times M$:

$$u_{tt} - \Delta u = 0, \tag{6.1}$$

with initial conditions

$$u(0, x) = f(x), \quad u_t(0, x) = g(x), \tag{6.2}$$

and boundary condition either the Dirichlet condition or the Neumann condition, if $\partial M \neq \emptyset$. We leave aside for the present the issue of the existence of solutions, for arbitrarily given f and g. We examine the uniqueness; u is assumed sufficiently smooth. If u_1 and u_2 solve (6.1) with initial data f_j, g_j, then $u_1 - u_2$ solves (6.1) with initial data $f = f_1 - f_2$, $g = g_1 - g_2$. To establish uniqueness, it suffices to show that if $f = g = 0$, then the solution $u = 0$ for all t. But by energy conservation, we have, for all t,

$$\int_M \left[u_t^2 + \langle d_x u, d_x u\rangle\right] dV = \int_M \left[|g|^2 + \langle d_x f, d_x f\rangle\right] dV = 0. \tag{6.3}$$

Thus u is constant. Since $u(0, x) = 0$, we conclude that $u = 0$ everywhere. This establishes uniqueness.

A closer look at how Green's formula enters into this argument will produce both a generalization of the notion of energy conservation and a localization of this uniqueness theorem to a result implying finite propagation speed for solutions to the wave equation. Note that the identity (5.31) can be written as

$$E(t_2) - E(t_1) = \int_{t_1}^{t_2} \int_M u_t(u_{tt} - \Delta u)\, dV\, dt + \int_{t_1}^{t_2} \int_{\partial M} u_t \frac{\partial u}{\partial \nu}\, dS\, dt. \tag{6.4}$$

In particular, for u satisfying either the Dirichlet or Neumann condition on ∂M, with $\Omega = [t_1, t_2] \times M$, we have

$$\begin{aligned} &\int_\Omega u_t(u_{tt} - \Delta u)\, dV\, dt = \\ &\frac{1}{2} \int_{\{t=t_2\}} \left[|u_t|^2 + |d_x u|^2\right] dV - \frac{1}{2} \int_{\{t=t_1\}} \left[|u_t|^2 + |d_x u|^2\right] dV. \end{aligned} \tag{6.5}$$

Next we want to look at the left side of (6.5) when Ω is a more general sort of region in $\mathbb{R} \times M$ than a product region $[t_1, t_2] \times M$.

First, we assume for simplicity that Ω does not intersect $\mathbb{R} \times \partial M$. We suppose $\partial\Omega$ consists of two smooth surfaces, Σ_1 and Σ_2, as indicated in Fig. 6.1. We denote by Ω_t the intersection of Ω with $\{t\} \times M \subset \mathbb{R} \times M$. Now, making use of formula (2.19), we have

$$\begin{aligned} \int_\Omega u_t(u_{tt} - \Delta u)\, dV\, dt &= \int_\Omega \frac{\partial}{\partial t}\left(\frac{1}{2}u_t^2\right) dV\, dt + \int_\Omega \langle d_x u_t, d_x u\rangle\, dV\, dt \\ &\quad - \int_\Omega \operatorname{div}_x(u_t \operatorname{grad}_x u)\, dV\, dt. \end{aligned} \tag{6.6}$$

Note that

$$\langle d_x u_t, d_x u\rangle = \frac{1}{2}\frac{\partial}{\partial t}\langle d_x u, d_x u\rangle. \tag{6.7}$$

Applying the fundamental theorem of calculus to the first two integrals on the right side of (6.6), and the divergence theorem to the last integral, we get

$$\int_\Omega u_t(u_{tt} - \Delta u)\, dV\, dt = \frac{1}{2}\int_{\partial\Omega} \left[u_t^2 + \langle d_x u, d_x u\rangle\right]\omega - \int \int_{\partial\Omega_t} u_t \frac{\partial u}{\partial \nu_x}\, dS_t\, dt. \tag{6.8}$$

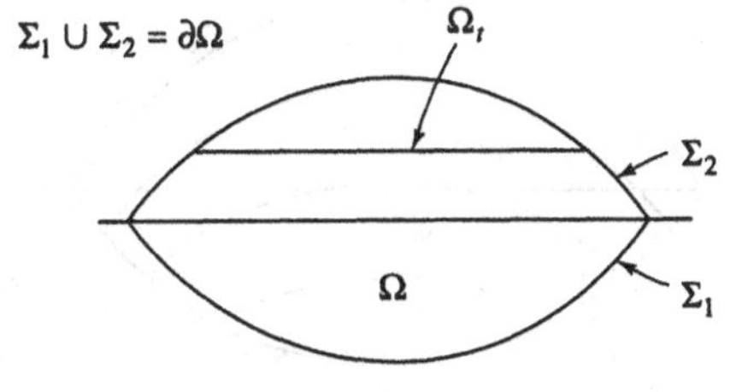

FIGURE 6.1

Both integrals on the right side of (6.8) are integrals over $\partial\Omega$. Here ω is the volume form on M, thought of as an n-form on $\mathbb{R} \times M$, pulled back to $\partial\Omega$, and dS_t is the natural surface measure on $\partial\Omega_t$, thought of as a surface in M. We want to express both ω and $dS_t\, dt$ in terms of the natural surface measure on $\partial\Omega$, induced from the inclusion $\partial\Omega \subset \mathbb{R} \times M$, endowed with the natural product Riemannian metric. Indeed, we easily obtain

$$\omega = N_t\, dS, \quad dS_t\, dt = |N_x|\, dS, \tag{6.9}$$

where $N = (N_t, N_x)$ is the outward unit normal to $\Sigma \subset \mathbb{R} \times M$. Hence (6.8) becomes

$$\int_\Omega u_t(u_{tt} - \Delta u)\, dV\, dt = \frac{1}{2} \int_{\partial\Omega} \Big\{ \big[u_t^2 + |d_x u|^2\big] N_t - 2u_t \frac{\partial u}{\partial \nu_x} |N_x| \Big\}\, dS. \tag{6.10}$$

Thus, if u satisfies the wave equation in Ω, we see that

$$\begin{aligned} &\int_{\Sigma_2} \Big\{ \big[u_t^2 + |d_x u|^2\big] |N_t| - 2u_t \frac{\partial u}{\partial \nu_x} |N_x| \Big\}\, dS \\ &= \int_{\Sigma_1} \Big\{ \big[u_t^2 + |d_x u|^2\big] |N_t| + 2u_t \frac{\partial u}{\partial \nu_x} |N_x| \Big\}\, dS. \end{aligned} \tag{6.11}$$

This is a useful "energy identity" provided the integrands are positive-definite quadratic forms in $du = (u_t, d_x u)$. Note that Cauchy's inequality implies

$$2\Big| u_t \frac{\partial u}{\partial \nu_x} \Big| \le u_t^2 + |d_x u|^2. \tag{6.12}$$

Thus the integrands have the desired property, provided

$$|N_x| < |N_t|. \tag{6.13}$$

Definition. *A surface $\Sigma \subset \mathbb{R} \times M$ is called spacelike provided its normal $N = (N_t, N_x)$ satisfies (6.13). A vector satisfying (6.13) is called timelike.*

Clearly any surface $t = \text{const.}$ is spacelike, as is a small perturbation of such a surface. Suppose $\Omega \subset \mathbb{R} \times M$ is bounded by spacelike surfaces Σ_1 and Σ_2 and

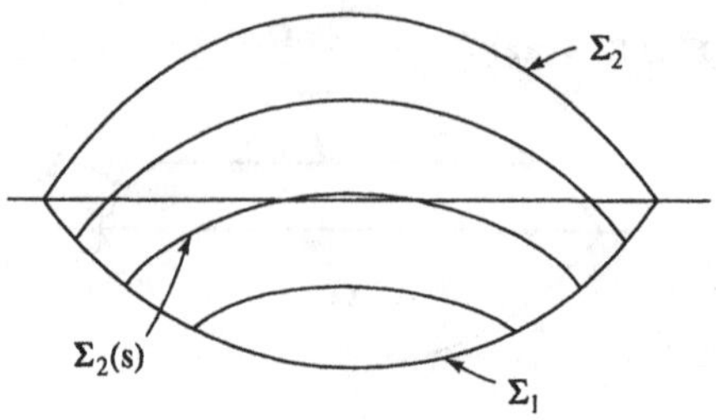

FIGURE 6.2

furthermore is swept out by spacelike surfaces $\Sigma_2(s)$, as in Fig.6.2. We call Ω a domain of influence for its lower boundary Σ_1.

Theorem 6.1. *Suppose $\Omega \subset \mathbb{R} \times M$ is a domain of influence for its lower boundary Σ_1. If u solves the wave equation $u_{tt} - \Delta u = 0$ on $\mathbb{R} \times M$, and if u and $du = (u_t, d_x u)$ vanish on Σ_1, then u vanishes throughout Ω.*

Proof. The energy identity implies that du vanishes on each $\Sigma_2(s)$; hence du vanishes on Ω, so u is constant on Ω. Since $u = 0$ on Σ_1, this constant is 0.

One interpretation of this theorem is that it shows that signals propagate at speed at most 1. In other words, in the special case $\Sigma_1 = \{t = 0\}$, if $u(0, x) = f(x)$ and $u_t(0, x) = g(x)$ vanish on some open set $\mathcal{O} \subset M$, then the solution to the wave equation vanishes on $\{(t, x) : x \in \mathcal{O},\ \text{dist}(x, \partial\mathcal{O}) > |t|\}$.

A slight variation of the argument above treats the case when $\partial\Omega$ consists of three parts, Σ_1 and Σ_2, both spacelike as above, and a part in $\mathbb{R} \times \partial M$, provided the solution u to $u_{tt} - \Delta u = 0$ satisfies the Dirichlet or Neumann boundary condition.

Exercises

1. Use (1.24)–(1.28) to write out the explicit solution to the initial value problem (6.1)–(6.2) in case $\Delta = \partial^2/\partial x^2$ on $\mathbb{R}$, and explicitly observe finite propagation speed in this case.
2. Extend the finite propagation speed argument of Theorem 6.1 to the case where M has a boundary, on which either the Dirichlet or Neumann boundary condition is imposed.
3. Consider the equations of linear elasticity, derived in (1.59), $Lu = 0$, where

$$Lu = mu_{tt} - \mu\Delta u - (\lambda + \mu) \text{ grad div } u.$$

Suppose $\mu > 0$, $\lambda + 2\mu > 0$, $m > 0$. For each $(t, x) \in \mathbb{R} \times M$, $u(t, x) \in T_x M$. Take $M = \mathbb{R}^n$. Let Ω be a region in $\mathbb{R} \times M$ of the form depicted in Fig. 6.1. Perform an integration by parts of

$$\int_\Omega u_t \cdot Lu \, dV \, dt,$$

along the lines of (6.6)–(6.10), to derive an identity similar to (6.11). What geometrical conditions should be placed on Σ_1 and Σ_2, replacing the "spacelike" condition (6.13), in order to ensure that the resulting integrands are positive-definite quadratic forms in $\nabla u = (u_t, \nabla_x u)$? Derive a finite propagation speed result.

7. Lorentz manifolds and stress-energy tensors

The analysis of the wave equation in the last section made strong use of the fact that we were working with $\partial^2/\partial t^2 - \Delta$ on a product $\mathbb{R} \times M$. We will take a deeper look at the notion of energy, which will produce concepts that are important in the study of the wave equation on more general Lorentz manifolds.

For starters, we will stick with the product case $\mathbb{R} \times M$, M a Riemannian manifold. This has a natural structure of a Lorentz manifold, with metric

$$h = -dt^2 + g. \tag{7.1}$$

Contrast this with the Riemannian metric $dt^2 + g$ on $\mathbb{R} \times M$ we considered in the last section. In coordinates, h_{jk} has the form

$$(h_{jk}) = \begin{pmatrix} -1 & 0 \\ 0 & g_{\mu\nu} \end{pmatrix}. \tag{7.2}$$

The *stress-energy tensor* T associated with u is supposed to be a symmetric, second order tensor such that, if Z is a unit timelike vector (representing the "world line" of an observer), then $T(Z, Z)$ gives the observed energy density. The energy density $(1/2)u_t^2 + (1/2)\langle d_x u, d_x u\rangle$ encountered before specifies

$$T_{00} = \frac{1}{2}u_t^2 + \frac{1}{2}\langle d_x u, d_x u\rangle = u_t^2 + \frac{1}{2}\langle du, du\rangle, \tag{7.3}$$

where

$$\langle du, du\rangle = h^{jk}\, \partial_j u\, \partial_k u \tag{7.4}$$

is the Lorentz square-length of du. If we expect that T is constructed in a "natural" manner from du and the metric tensor h, we are led to require

$$T(Z, Z) = \langle Z, du\rangle^2 + \frac{1}{2}\langle du, du\rangle \ \text{ whenever } \ \langle Z, Z\rangle = -1.$$

If $\langle Z, Z\rangle = -z^2$, this leads to $T(Z, Z) = \langle Z, du\rangle^2 - (1/2)\langle du, du\rangle\langle Z, Z\rangle$, and polarizing this identity gives

$$T(Z, W) = \langle Z, du\rangle\langle W, du\rangle - \frac{1}{2}\langle du, du\rangle\langle Z, W\rangle. \tag{7.5}$$

This should hold for all vectors Z, W. Equivalently, we write

$$T = du \otimes du - \frac{1}{2}\langle du, du\rangle h. \tag{7.6}$$

We call (7.6) the stress-energy tensor associated to a wave $u = u(t, x)$. See the exercises for more on the construction of T.

More generally, let Ω be any Lorentz manifold, with metric tensor, of signature $(n, 1)$, denoted h. The "Laplacian" in this metric is defined by

$$\Box u = |h|^{-1/2}\,\partial_j(h^{jk}|h|^{1/2}\,\partial_k u) = h^{jk}u_{;j;k}, \tag{7.7}$$

in analogy with the formula for the Laplace operator on a Riemannian manifold. Here, $|h| = |\det(h_{jk})|$. The wave equation on a general Lorentz manifold is

$$\Box u = 0. \tag{7.8}$$

In this more general context, it is still meaningful to assign to u the tensor T, defined by (7.5) and (7.6). We continue to call T the stress-energy tensor. We have the following important result.

Proposition 7.1. *For a solution to (7.8) on a general Lorentz manifold Ω, the stress-energy tensor has vanishing divergence, that is,*

$$T^{jk}{}_{;k} = 0. \tag{7.9}$$

More generally, for any u,

$$T^{jk}{}_{;k} = u^{;j}\Box u. \tag{7.10}$$

Proof. This is a straightforward calculation. We have

$$T^{jk} = u^{;j}u^{;k} - \frac{1}{2}h^{jk}h^{\mu\nu}u_{;\mu}u_{;\nu}, \tag{7.11}$$

where $u^{;j} = h^{jk}u_{;k}$ denotes the gradient. Hence, using $h^{jk}{}_{;\ell} = 0$, we obtain

$$\begin{aligned} T^{jk}{}_{;k} &= u^{;j}{}_{;k}u^{;k} + u^{;j}u^{;k}{}_{;k} - \frac{1}{2}h^{jk}h^{\mu\nu}u_{;\mu;k}u_{;\nu} - \frac{1}{2}h^{jk}h^{\mu\nu}u_{;\mu}u_{;\nu;k} \\ &= u^{;j}\Box u + u^{;j}{}_{;k}u^{;k} - h^{jk}h^{\mu\nu}u_{;\mu;k}u_{;\nu} \\ &= u^{;j}\Box u + u^{;j}{}_{;k}u^{;k} - u_{;\mu}{}^{;j}u^{;\mu}. \end{aligned}$$

Since, as we have seen, $u_{;j;k} = u_{;k;j}$, we obtain (7.10), and the proposition follows.

We have seen that the divergence theorem applies to reduce the integral $\int_\Omega (\text{div } X)\, dV$ to a boundary integral, when X is a vector field; in particular, when X is a divergence-free vector field, it yields that a certain boundary integral is zero or, equivalently, that integrals over two parts of $\partial\Omega$ are equal in magnitude. However, T is not a divergence-free vector field; it is a second-order tensor field. In general vanishing of div T will not lead to integral conservation laws. It will, however, in the following case.

Suppose a Lorentz manifold Ω has a timelike Killing field Z, that is, a timelike vector field whose flow preserves the metric tensor h. As derived in the Riemannian case, the condition for the metric to be preserved is

$$Z_{j;k} + Z_{k;j} = 0, \quad Z_j = h_{jk}Z^k. \tag{7.12}$$

Here, "timelike" means that $h(Z, Z) < 0$. This means Z lies inside the light cone determined by the Lorentz metric.

Lemma 7.2. *If T^{jk} is divergence free and Z^k is a Killing field, then*

$$X^j = T^{jk} Z_k \text{ is divergence free.} \tag{7.13}$$

Proof. We have

$$X^j{}_{;j} = T^{jk}{}_{;j} Z_k + T^{jk} Z_{k;j}.$$

Now the symmetry of T^{jk} implies $T^{jk}{}_{;j} = 0$ and

$$T^{jk} Z_{k;j} = \frac{1}{2} T^{jk}(Z_{k;j} + Z_{j;k}) = 0,$$

assuming (7.12) holds. This proves the lemma.

We denote the vector (7.13) by

$$X = \widetilde{T} Z. \tag{7.14}$$

Suppose $\mathcal{O}$ is a region in the Lorentz manifold Ω, bounded by two surfaces Σ_1 and Σ_2, as in Fig. 7.1.

By (2.14), we have

$$\begin{aligned} 0 = \int_{\mathcal{O}} (\operatorname{div} \widetilde{T} Z)\, dV &= \int_{\Sigma_1 \cup \Sigma_2} \omega \rfloor (\widetilde{T} Z) \\ &= \int_{\Sigma_1} \langle \widetilde{T} Z, \nu_1 \rangle\, dS - \int_{\Sigma_2} \langle \widetilde{T} Z, \nu_2 \rangle\, dS, \end{aligned} \tag{7.15}$$

where ν_j is the unit vector, normal to Σ_j, with respect to the Lorentz metric h, pointing in the same "forward" direction as Z. The last identity in (7.15) holds in analogy with (2.15). We make the hypothesis as before, that Σ_1 and Σ_2 are spacelike (i.e., ν_j are timelike), so it makes sense to specify that they lie inside the

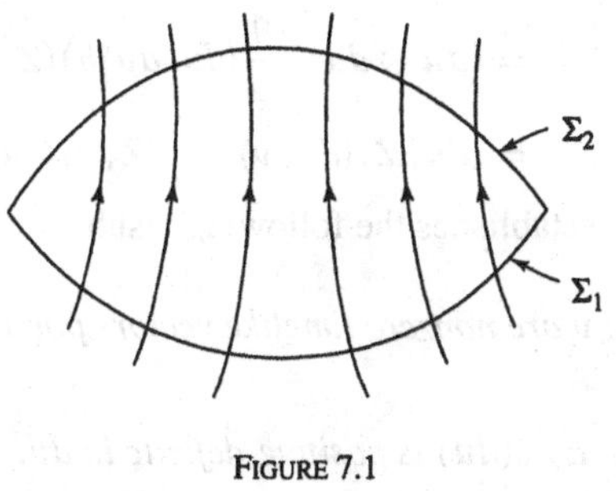

FIGURE 7.1

forward light cone. Equation (7.15) is equivalent to

$$\int_{\Sigma_2} T(Z, \nu_2)\, dS = \int_{\Sigma_1} T(Z, \nu_1)\, dS. \tag{7.16}$$

The volume element dS on Σ_j is determined here by the Riemannian metric on Σ_j, induced by restricting the Lorentz metric h to tangent vectors to Σ_j.

Again we seek to guarantee that the integrand in (7.16), which is a quadratic form in du for T given by (7.5), is positive-definite. In order to check this at a point $p_0 \in \partial\mathcal{O}$, choose a coordinate system such that

$$\big(h_{jk}(p_0)\big) = \begin{pmatrix} -1 & 0 \\ 0 & I \end{pmatrix}, \quad \nu(p_0) = (1, 0, \dots, 0)^t \quad (\nu = \nu_1 \text{ or } \nu_2), \tag{7.17}$$

which is always possible. Suppose $Z(p_0) = (Z^0, Z^1, \dots, Z^n)$. The condition that $Z(p_0)$ belong to the forward light cone is

$$Z^0 > 0, \quad (Z^0)^2 > (Z^1)^2 + \cdots + (Z^n)^2. \tag{7.18}$$

Now, if we set $M = \widetilde{T}\nu$, then, at p_0,

$$M^0 = -\frac{1}{2}\big[(\partial_0 u)^2 + (\partial_1 u)^2 + \cdots + (\partial_n u)^2\big], \quad M^j = (\partial_0 u)(\partial_j u), \tag{7.19}$$

if T is given by (7.5). Consequently, at p_0,

$$\begin{aligned} T(Z, \nu) &= \langle Z, M\rangle = -Z^0 M^0 + \sum_{j=1}^n Z^j M^j \\ &= \frac{1}{2} Z^0\big[(\partial_0 u)^2 + \cdots + (\partial_n u)^2\big] + \sum_{j=1}^n Z^j (\partial_0 u)(\partial_j u). \end{aligned} \tag{7.20}$$

The positive definiteness of this quadratic form in $(\partial_0 u, \dots, \partial_n u)$ follows immediately from Cauchy's inequality, granted (7.18). This definiteness calculation does not use the hypothesis that Z is a Killing field, of course. For positive definiteness of $T(Z, \nu)$ in du, it suffices that Z and ν both be nonzero timelike vectors inside the forward light cone.

In order to emphasize that the dependence of $T(Z, \nu)$ on du has fundamental significance, we adopt the following notation. Set

$$\begin{aligned} E_{Z,\nu}(du) &= T(Z, \nu) \\ &= \Big(du \otimes du - \frac{1}{2}\langle du, du\rangle h\Big)(Z, \nu) \\ &= \langle du, Z\rangle\langle du, \nu\rangle - \tfrac{1}{2}\langle Z, \nu\rangle\langle du, du\rangle. \end{aligned} \tag{7.21}$$

The calculation above establishes the following result.

Lemma 7.3. *If Z and ν are nonzero timelike vectors pointing inside the forward light cone, then*

$$E_{Z,\nu}(du) \text{ is positive-definite in } du.$$

Note that the identity (7.16) is

$$\int_{\Sigma_2} E_{Z,\nu_2}(du)\, dS = \int_{\Sigma_1} E_{Z,\nu_1}(du)\, dS. \tag{7.22}$$

It follows that if $\mathcal{O}$, as in Fig. 7.1, is swept out by spacelike surfaces, as in Fig. 7.2, then the same argument as given in §6 leads to the uniqueness result: $\Box u = 0$ in $\mathcal{O}$, u and $du = 0$ on Σ_1 imply $u = 0$ in $\mathcal{O}$, provided Ω has a timelike Killing field Z. This gives finite propagation speed for solutions to the wave equation on such a Lorentz manifold.

If a Lorentz manifold Ω has no timelike Killing field, which is typical, then natural energy identities such as (7.22) do not arise. However, there are *inequalities* involving the stress-energy tensor, that are powerful enough to imply the local uniqueness (finite propagation speed) of solutions to the wave equation $\Box u = 0$ on a general Lorentz manifold. In the next section we will establish this as a special case of a more general result on hyperbolic equations.

Exercises

1. If M is a Lorentz manifold, $S \subset M$ a hypersurface (codimension 1), show that S is spacelike if and only if the metric tensor restricted to S is positive-definite. In the product case (7.1), show that the definitions of "spacelike" given in this section and the previous one are equivalent.
2. On $\mathbb{R}^{n+1}$, with coordinates $(x_0, \dots, x_n)$, place the Lorentz inner product

$$\langle u, v\rangle = -u_0 v_0 + u_1 v_1 + \cdots + u_n v_n.$$

Show that $A : \mathbb{R}^{n+1} \to \mathbb{R}^{n+1}$, defined by

$$A(u_0, u_1, u_2, \dots, u_n) = (u_1, u_0, u_2, \dots, u_n)$$

is *skew-adjoint* for the Lorentz metric (i.e., $\langle Au, v\rangle = -\langle u, Av\rangle$), and hence the group $\mathcal{F}(t) = e^{tA}$ preserves the Lorentz metric.
3. Consider the hyperboloids

$$M = M_s = \{x \in \mathbb{R}^{n+1} : \langle x, x\rangle = s\}.$$

Show that M_s is spacelike if and only if $s < 0$.

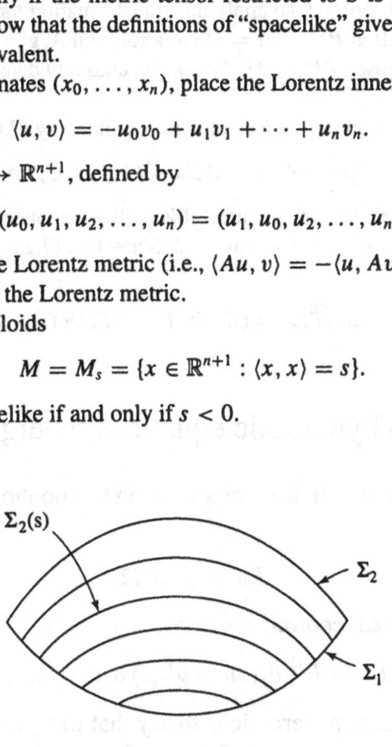

FIGURE 7.2

4. If $s > 0$ and M_s is as in Exercise 3, show that M_s gets a Lorentz metric, induced from $\mathbb{R}^{n+1}$. Show that the group $\mathcal{F}(t)$ of Exercise 2 leaves M_s invariant and its generator is a timelike Killing field on M_s.
5. We consider a general approach to constructing a second-order tensor of the form
$$T^{jk} = A^{jk\ell m} u_{;\ell} u_{;m},$$
where $A^{jk\ell m}$ is a tensor field of type (4, 0), such that the conclusion (7.10) of Proposition 7.1 holds. Let us assume that $\nabla A = 0$. Show that
$$T^{jk}{}_{;k} = B^{jk\ell m} u_{;k;\ell} u_{;m},$$
where
$$B = P^{23} P^{34} A.$$
Here, $P^{\mu\nu}$ denotes the operation on tensors of type (4, 0) of symmetrizing with respect to the μth and νth indices, for example, $(P^{23} C)^{jk\ell m} = (1/2)[C^{jk\ell m} + C^{j\ell km}]$. Consequently, (7.10) holds provided
$$P^{23} P^{34} A = H, \quad H^{jk\ell m} = h^{jm} h^{k\ell}.$$
6. Show that $P^{\mu\nu}$ are all projections of the same rank and H belongs to the range of P^{23}. Show that $\operatorname{Ker} P^{23} \cap R(P^{34}) = 0$ and hence
$$P^{23} : R(P^{34}) \longrightarrow R(P^{23}) \text{ is an isomorphism.}$$
(*Hint*: If $B \in \operatorname{Ker} P^{23} \cap R(P^{34})$, show that $B^{jk\ell m} = -B^{jmk\ell}$. $(k\ \ell\ m) \mapsto (m\ k\ \ell)$ is a cyclic permutation of order 3, so apply this transformation three times.)
7. Deduce that the equation $P^{23} P^{34} A = H$ has a solution A, given uniquely, mod Ker P^{34}, and hence that the tensor $T^{jk} = A^{jk\ell m} u_{;\ell} u_{;m}$ is uniquely determined by the conditions set in Exercise 5.
8. Show that, for general smooth scalar u, with T defined by (7.6), then
$$\operatorname{div} \widetilde{T} Z = (Zu)\Box u + \langle T, \operatorname{Def}(Z)\rangle, \tag{7.23}$$
where Def(Z) is the deformation tensor of Z, with components $(1/2)(Z_{j;k} + Z_{k;j})$ and $\langle T, V\rangle = T^{jk} V_{jk}$. This implies Lemma 7.2. Show that (7.23) follows from the general identity
$$\operatorname{div}(\widetilde{T} Z) = \langle Z, \operatorname{div} T\rangle + \langle T, \operatorname{Def} Z\rangle. \tag{7.24}$$

8. More general hyperbolic equations; energy estimates

In this section we derive estimates for a solution to a nonhomogeneous hyperbolic equation of the form
$$Lu = f \ \text{ in } \Omega, \tag{8.1}$$
where L is given in local coordinates by
$$Lu = h^{jk}\, \partial_j \partial_k u + b^j(x)\, \partial_j u + c(x) u. \tag{8.2}$$
By definition, to say L is hyperbolic is to say that (h^{jk}) is a symmetric matrix of signature $(n, 1)$, if $\dim \Omega = n + 1$. One can then use the inverse matrix (h_{jk}) to

define a Lorentz metric on Ω, and in view of the formula (7.7), we can write (8.2) as

$$Lu = \Box u + Xu, \tag{8.3}$$

for some first-order differential operator X on Ω.

Suppose $\mathcal{O} \subset \Omega$ is bounded by two surfaces Σ_1 and Σ_2, both spacelike. As at the end of §7, we suppose that $\mathcal{O}$ is swept out by spacelike surfaces. Specifically, we suppose that there is a smooth function on a neighborhood of $\overline{\mathcal{O}}$, which in fact we denote by t, such that dt is timelike, and set

$$\mathcal{O}(s) = \overline{\mathcal{O}} \cap \{t \le s\}, \quad \Sigma_2(s) = \overline{\mathcal{O}} \cap \{t = s\}.$$

We suppose $\mathcal{O}$ is swept out by $\Sigma_2(s)$, $s_0 \le s \le s_1$, as illustrated in Fig. 8.1, with $\Sigma_2 = \Sigma_2(s_1)$. Also set

$$\Sigma_1^b(s) = \Sigma_1 \cap \{t \le s\}.$$

As in (7.15), the divergence theorem implies

$$\int\limits_{\Sigma_2(s)} E_{Z,\nu_2}(du)\, dS = \int\limits_{\Sigma_1^b(s)} E_{Z,\nu_1}(du)\, dS - \int\limits_{\mathcal{O}(s)} (\operatorname{div} \widetilde{T} Z)\, dV, \tag{8.4}$$

where $E_{Z,\nu}(du)$ is defined by (7.21) and T by (7.5), though at this point it is not physically meaningful in general to think of T as the stress-energy tensor. Here ν_1 is the forward-pointing unit normal to Σ_1, with respect to the Lorentz metric, and ν_2 is the normalization of grad t, the vector field obtained from dt via the Lorentz metric. Z is any timelike vector field; we will set $Z = \nu_2$. Note that Lemma 7.3 applies to the integrands $E_{Z,\nu_j}(du)$.

We no longer have $\operatorname{div} \widetilde{T} Z = 0$, but we can *estimate* this quantity, as follows. First,

$$\operatorname{div} \widetilde{T} Z = T^{jk}{}_{;k} h_{j\ell} Z^\ell + T^{jk} h_{j\ell} Z^\ell{}_{;k} = \langle \operatorname{div} T, Z \rangle + \langle T, \nabla Z \rangle. \tag{8.5}$$

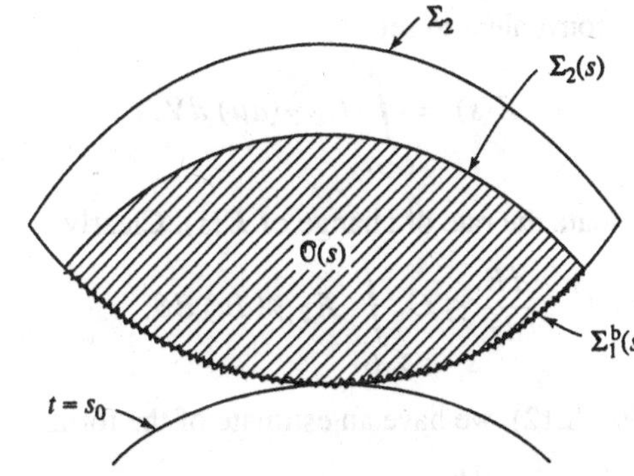

FIGURE 8.1

The term $\langle T, \nabla Z\rangle$ is a quadratic form in du, and hence, by Lemma 7.3, we have an estimate

$$|\langle T, \nabla Z\rangle| \le K\, E_{Z,Z}(du). \tag{8.6}$$

As for the first term on the right side of (8.5), (7.10) implies

$$\operatorname{div} T = (\operatorname{grad} u)\Box u. \tag{8.7}$$

If u satisfies $Lu = f$, this implies

$$\operatorname{div} T = (\operatorname{grad} u)(f - Xu). \tag{8.8}$$

Cauchy's inequality together with Lemma 7.3 gives an estimate

$$|\langle \operatorname{div} T, Z\rangle| \le K\, E_{Z,Z}(du) + K|u|^2 + K|f|^2. \tag{8.9}$$

Consequently, (8.4) yields the estimate

$$\int\limits_{\Sigma_2(s)} E_{Z,Z}(du)\, dS \le \int\limits_{\Sigma_1^b(s)} E_{Z,\nu_1}(du)\, dS + K \int\limits_{\mathcal{O}(s)} \big[2E_{Z,Z}(du) + |u|^2 + |f|^2\big]\, dV. \tag{8.10}$$

Suppose that u satisfies the following initial conditions on Σ_1:

$$u = g, \quad du = \omega \ \text{ on } \Sigma_1. \tag{8.11}$$

We want to estimate the left side of (8.10) in terms of f, g, and ω. Our first goal will be to deive a variant of (8.10) without the $|u|^2$ term. We can work on the term $\int_{\mathcal{O}(s)} |u|^2\, dV$ on the right side of (8.10) as follows. An easy consequence of the fundamental theorem of calculus, Cauchy's inequality, and Lemma 7.3 gives

$$\int\limits_{\mathcal{O}(s)} |u|^2\, dV \le C \int\limits_{\Sigma_1^b(s)} |g|^2\, dS + C \int\limits_{\mathcal{O}(s)} E_{Z,Z}(du)\, dV, \tag{8.12}$$

which can be applied to (8.10).

At this point, it is convenient to set

$$E(s) = \int\limits_{\mathcal{O}(s)} E_{Z,Z}(du)\, dV. \tag{8.13}$$

We will want to estimate the rate of change of $E(s)$. Clearly,

$$\frac{dE}{ds} \le C \int\limits_{\Sigma_2(s)} E_{Z,Z}(du)\, dS, \tag{8.14}$$

and hence, by (8.10)–(8.12), we have an estimate of the form

$$\frac{dE}{ds} \le C E(s) + F(s), \tag{8.15}$$

where

$$F(s) = C \int_{\Sigma_1} \left[E_{Z,Z}(\omega) + |g|^2 \right] dS + C \int_{\mathcal{O}(s)} |f|^2 \, dV. \tag{8.16}$$

Note that (8.15) is equivalent to

$$\frac{d}{ds}\left(e^{-Cs} E(s)\right) \le e^{-Cs} F(s), \tag{8.17}$$

and since $E(s_0) = 0$, we have

$$e^{-Cs} E(s) \le \int_{s_0}^{s} e^{-Cr} F(r) \, dr. \tag{8.18}$$

In view of (8.16), this establishes the following "energy estimate."

Proposition 8.1. *If u solves the hyperbolic equation $Lu = f$ of the form (8.3), with initial data (8.11) on Σ_1, and if $\mathcal{O}(s)$ satisfies the geometrical hypotheses made above and illustrated in Fig. 8.1, then*

$$\int_{\mathcal{O}(s)} E_{Z,Z}(du) \, dV \le C(s - s_0) \int_{\Sigma_1} \left[|g|^2 + |\omega|^2 \right] dS + C \int_{\mathcal{O}(s)} |f|^2 \, dV, \tag{8.19}$$

for $s \in [s_0, s_1]$.

In particular, if g and ω vanish on Σ_1 and f vanishes on $\mathcal{O}$, then (8.19) implies $du = 0$ on $\mathcal{O}$, so u is constant on $\mathcal{O}$, that constant being $g = 0$. This gives the local uniqueness (finite propagation speed) for solutions to the homogeneous hyperbolic equation $Lu = 0$, extending the result of §7.

We note that, using (8.10) and (8.12), we deduce from (8.19) that

$$\int_{\Sigma_2(s)} E_{Z,Z}(du) \, dS \le C \int_{\Sigma_1} \left[|g|^2 + |\omega|^2 \right] dS + C \int_{\mathcal{O}(s)} |f|^2 \, dV. \tag{8.20}$$

Exercises

1. Prove the estimate

$$(1 - \varepsilon) \int_0^1 |u(s)|^2 \, ds \le |u(0)|^2 + C_\varepsilon \int_0^1 |u'(s)|^2 \, ds.$$

   What is the best value of C_ε that will work?
2. Give a detailed proof of the estimate (8.12).
3. Sharpen the estimate (8.19) to

$$\int_{\mathcal{O}(s)} E_{Z,Z}(du) \, dV \le C(s - s_0) \int_{\Sigma_1} \left[|g|^2 + |\omega|^2 \right] dS + C(s - s_0) \int_{\mathcal{O}(s)} |f|^2 \, dV, \tag{8.21}$$

   under the hypotheses of Proposition 8.1. (*Hint*: Use (8.18) more carefully.)

4. Work out generalizations of the energy estimates (8.10)–(8.19) when u satisfies the *semilinear* PDE

$$\square u = f(x, u, du). \tag{8.22}$$

Formulate and prove a finite propagation speed result in this case.
(*Hint*: Given solutions u_1 and u_2 to (8.22), derive a *linear* PDE for $w = u_1 - u_2$, to get the finite propagation speed result.)

9. The symbol of a differential operator and a general Green-Stokes formula

Let P be a differential operator of order m on a manifold M; P could operate on sections of a vector bundle. In local coordinates, P has the form

$$Pu(x) = \sum_{|\alpha|\le m} p_\alpha(x) D^\alpha u(x), \tag{9.1}$$

where $D^\alpha = D_1^{\alpha_1} \cdots D_n^{\alpha_n}$, $D_j = (1/i)\,\partial/\partial x_j$. The coefficients $p_\alpha(x)$ could be matrix valued. The homogeneous polynomial in $\xi \in \mathbb{R}^n$ ($n = \dim M$),

$$p_m(x, \xi) = \sum_{|\alpha|=m} p_\alpha(x)\xi^\alpha, \tag{9.2}$$

is called the *principal symbol* (or just the symbol) of P. We want to give an intrinsic characterization, which will show that $p_m(x, \xi)$ is well defined on the cotangent bundle of M. For a smooth function ψ, a simple calculation, using the product rule and chain rule of differentiation, gives

$$P\big(u(x)e^{i\lambda\psi}\big) = \big[p_m(x, d\psi)u(x)\lambda^m + r(x, \lambda)\big]e^{i\lambda\psi}, \tag{9.3}$$

where $r(x, \lambda)$ is a polynomial of degree $\le m - 1$ in λ. In (9.3), $p_m(x, d\psi)$ is evaluated by substituting $\xi = (\partial\psi/\partial x_1, \dots, \partial\psi/\partial x_n)$ into (9.2). Thus the formula

$$p_m(x, d\psi)u(x) = \lim_{\lambda\to\infty} \lambda^{-m} e^{-i\lambda\psi} P\big(u(x)e^{i\lambda\psi}\big) \tag{9.4}$$

provides an intrinsic characteristization of the symbol of P as a function on T^*M. We also use the notation

$$\sigma_P(x, \xi) = p_m(x, \xi). \tag{9.5}$$

If

$$P : C^\infty(M, E_0) \longrightarrow C^\infty(M, E_1), \tag{9.6}$$

where E_0 and E_1 are smooth vector bundles over M, then, for each $(x, \xi) \in T^*M$,

$$p_m(x, \xi) : E_{0x} \longrightarrow E_{1x} \tag{9.7}$$

is a linear map between fibers. It is easy to verify that if P_2 is another differential operator, mapping $C^\infty(M, E_1)$ to $C^\infty(M, E_2)$, then

$$\sigma_{P_2P}(x, \xi) = \sigma_{P_2}(x, \xi)\sigma_P(x, \xi). \tag{9.8}$$

If M has a Riemannian metric, and the vector bundles E_j have metrics, then the formal adjoint P^t of a differential operator of order m like (9.6) is a differential operator of order m:

$$P^t : C^\infty(M, E_1) \longrightarrow C^\infty(M, E_0),$$

defined by the condition that

$$(Pu, v) = (u, P^t v) \tag{9.9}$$

if u and v are smooth, compactly supported sections of the bundles E_0 and E_1. If u and v are supported on a coordinate patch $\mathcal{O}$ on M, over which E_j are trivialized, so u and v have components u^σ, v^σ, and if the metrics on E_0 and E_1 are denoted $h_{\sigma\delta}$, $\tilde{h}_{\sigma\delta}$, respectively, while the Riemannian metric is g_{jk}, then we have

$$(Pu, v) = \int_{\mathcal{O}} \tilde{h}_{\sigma\delta}(x)\big(Pu\big)^\sigma \, \overline{v}^\delta \, \sqrt{g(x)}\, dx. \tag{9.10}$$

Substituting (9.1) and integrating by parts produce an expression for P^t, of the form

$$P^t v(x) = \sum_{|\alpha|\le m} p^t_\alpha(x) D^\alpha v(x). \tag{9.11}$$

In particular, one sees that the principal symbol of P^t is given by

$$\sigma_{P^t}(x, \xi) = \sigma_P(x, \xi)^t. \tag{9.12}$$

Compare this with the specific formula (2.22) for the formal adjoint of a real vector field, which has a purely imaginary symbol.

Now suppose M is a compact, smooth manifold with smooth boundary. We want to obtain a generalization of formula (2.24), that is,

$$(Xu, v) - (u, X^t v) = \int_{\partial M} \langle \nu, X\rangle u\overline{v}\, dS, \tag{9.13}$$

to the case where P is a general *first-order* differential operator, acting on sections of a vector bundle as in (9.6). Using a partition of unity, we can suppose that u and v are supported in a coordinate patch $\mathcal{O}$ in M. If the patch is disjoint from ∂M, then of course (9.9) holds. Otherwise, suppose $\mathcal{O}$ is a patch in $\mathbb{R}^n_+$. If the first-order operator P has the form

$$Pu = \sum_{j=1}^n a_j(x)\frac{\partial u}{\partial x_j} + b(x)u, \tag{9.14}$$

then

$$\int_{\mathcal{O}} \langle Pu, v\rangle \sqrt{g}\, dx = \int_{\mathcal{O}} \Big[\sum_{j=1}^n \langle a_j(x)\frac{\partial u}{\partial x_j}, v\rangle + \langle b(x)u, v\rangle\Big]\sqrt{g}\, dx. \tag{9.15}$$

If we apply the fundamental theorem of calculus, the only boundary integral comes from the term involving $\partial u/\partial x_n$. Thus we have
(9.16)

$$\int_{\mathcal{O}} \langle Pu, v\rangle \sqrt{g}\, dx = \int_{\mathcal{O}} \langle u, P^t v\rangle \sqrt{g}\, dx - \int_{\mathbb{R}^{n-1}} \langle a_n(x', 0)u, v\rangle \sqrt{g(x', 0)}\, dx',$$

where $dx' = dx_1 \cdots dx_{n-1}$. If we pick the coordinate patch so that $\partial/\partial x_n$ is the unit inward normal at ∂M, then $\sqrt{g(x', 0)}\, dx'$ is the volume element on ∂M, and we are ready to establish the following Green-Stokes formula:

Proposition 9.1. *If M is a smooth, compact manifold with boundary and P is a first-order differential operator (acting on sections of a vector bundle), then*

$$(Pu, v) - (u, P^t v) = \frac{1}{i} \int_{\partial M} \langle \sigma_P(x, \nu)u, v\rangle \, dS. \tag{9.17}$$

Proof. The formula (9.17), which arose via a choice of local coordinate chart, is invariant and hence valid independent of choices.

As in (9.13), ν denotes the *outward*-pointing unit normal to ∂M; we use the Riemannian metric on M to identify tangent vectors and cotangent vectors.

We will see an important application of (9.17) in the next section, where we consider the Laplace operator on k-forms.

Exercises

1. Consider the divergence operator acting on (complex-valued) vector fields:

$$\operatorname{div} : C^\infty(\Omega, \mathbb{C}^n) \longrightarrow C^\infty(\Omega), \quad \Omega \subset \mathbb{R}^n.$$

Show that its symbol is given by

$$\sigma_{\operatorname{div}}(x, \xi)v = i\langle v, \xi\rangle.$$

2. Consider the gradient operator acting on (complex valued) functions:

$$\operatorname{grad} : C^\infty(\Omega) \longrightarrow C^\infty(\Omega, \mathbb{C}^n), \quad \Omega \subset \mathbb{R}^n.$$

Show that its symbol is

$$\sigma_{\operatorname{grad}}(x, \xi) = i\xi.$$

3. Consider the operator

$$L = \operatorname{grad} \operatorname{div} : C^\infty(\Omega, \mathbb{C}^n) \longrightarrow C^\infty(\Omega, \mathbb{C}^n).$$

Show that its symbol is

$$\sigma_L(x, \xi) = -|\xi|^2 P_\xi,$$

where $P_\xi \in \operatorname{End}(\mathbb{C}^n)$ is the orthogonal projection onto the (complex) linear span of ξ.

4. What is the symbol of the operator

$$P = \mu\Delta + (\lambda + \mu) \text{ grad div},$$

which appears in the equation (1.59) of linear elasticity? What are the eigenvalues of the symbol, for given $\xi \in \mathbb{R}^n$?

5. Generalize Exercises 1–3 to the case of a Riemannian manifold.

6. Let L be a constant-coefficient, second-order, homogeneous, linear differential operator acting on functions on $\mathbb{R}^n$ with values in $\mathbb{C}^k$, of the form

$$Lu = \sum_{|\alpha|=2} A_\alpha \, D^\alpha u, \quad A_\alpha \in \text{End}(\mathbb{C}^k).$$

Let $\xi \in \mathbb{R}^n \setminus 0$. A "plane wave" solution to $u_{tt} - Lu = 0$ is a $\mathbb{C}^k$-valued function $u(t, x)$ of the form

$$u(t, x) = v(t, x \cdot \xi),$$

with $v(t, y)$ a $\mathbb{C}^k$-valued function on $\mathbb{R} \times \mathbb{R}$. Show that the PDE for v becomes

$$v_{tt} - M v_{yy} = 0,$$

with

$$M = -\sigma_L(x, \xi).$$

In case $\sigma_L(x, \xi)$ is negative-definite with eigenvalues $-c_j^2 = -c_j(\xi)^2$, show that the initial-value problem for v can be solved in terms of the formula for the one-dimensional wave equation derived in §1.

7. Consider the equation of linear elasticity from (1.59):

$$m w_{tt} - \mu\Delta w - (\lambda + \mu) \text{ grad div } w = 0.$$

Suppose $\mu > 0$, $2\mu + \lambda > 0$. Fix $\xi \in \mathbb{R}^n \setminus 0$. Using the results of Exercises 4 and 6, analyze plane wave solutions $w(t, x) = v(t, x \cdot \xi)$. Show that if $n \geq 2$, there are *two* propagation speeds. The faster and slower waves are called "p-waves" (pressure waves) and "s-waves" (shear waves), respectively. If $n = 1$, only p-waves arise.

10. The Hodge Laplacian on k-forms

If M is an n-dimensional Riemannian manifold, recall the exterior derivative

$$d : \Lambda^k(M) \longrightarrow \Lambda^{k+1}(M), \tag{10.1}$$

satisfying

$$d^2 = 0. \tag{10.2}$$

The Riemannian metric on M gives rise to an inner product on T_x^* for each $x \in M$, and then to an inner product on $\Lambda^k T_x^*$, via

$$\langle v_1 \wedge \cdots \wedge v_k, w_1 \wedge \cdots \wedge w_k \rangle = \sum_\pi (\text{sgn } \pi) \langle v_1, w_{\pi(1)} \rangle \cdots \langle v_k, w_{\pi(k)} \rangle, \tag{10.3}$$

where π ranges over the set of permutations of $\{1, \dots, k\}$. Equivalently, if $\{e_1, \dots, e_n\}$ is an orthonormal basis of $T_x^* M$, then $\{e_{j_1} \wedge \cdots \wedge e_{j_k} : j_1 < j_2 <$

$\cdots < j_k\}$ is an orthonormal basis of $\Lambda^k T_x^* M$. Consequently, there is an inner product on k-forms (that is, sections of Λ^k) given by

$$(u, v) = \int_M \langle u, v \rangle \, dV(x). \tag{10.4}$$

Thus there is a first-order differential operator

$$\delta : \Lambda^{k+1}(M) \longrightarrow \Lambda^k(M), \tag{10.5}$$

which is the formal adjoint of d, that is, δ is characterized by

$$(du, v) = (u, \delta v), \quad u \in \Lambda^k(M), \ v \in \Lambda^{k+1}(M), \text{ compactly supported.} \tag{10.6}$$

We set $\delta = 0$ on 0-forms. Of course, (10.2) implies

$$\delta^2 = 0. \tag{10.7}$$

There is a useful formula for δ, involving d and the "Hodge star operator," which will be derived in Chapter 5, §8.

The Hodge Laplacian on k-forms,

$$\Delta : \Lambda^k(M) \longrightarrow \Lambda^k(M), \tag{10.8}$$

is defined by

$$-\Delta = (d + \delta)^2 = d\delta + \delta d. \tag{10.9}$$

Consequently,

$$(-\Delta u, v) = (du, dv) + (\delta u, \delta v), \quad \text{for } u, v \in C_0^\infty(M, \Lambda^k). \tag{10.10}$$

Since $\delta = 0$ on $\Lambda^0(M)$, we have $-\Delta = \delta d$ on $\Lambda^0(M)$.

We will obtain an analogue of (10.10) for the case where M is a compact manifold with boundary, so a boundary integral appears. To obtain such a formula, we specialize the general Green-Stokes formula (9.17) to the cases $P = d$ and $P = \delta$. First, we compute the symbols of d and δ. Since, for a k-form u,

$$d(u\, e^{i\lambda\psi}) = i\lambda e^{i\lambda\psi}(d\psi) \wedge u + e^{i\lambda\psi}\, du, \tag{10.11}$$

we see that

$$\frac{1}{i}\sigma_d(x, \xi)u = \xi \wedge u. \tag{10.12}$$

As a special case of (9.12), we have

$$\sigma_\delta(x, \xi) = \sigma_d(x, \xi)^t. \tag{10.13}$$

The adjoint of the map (10.12) from $\Lambda^k T_x^*$ to $\Lambda^{k+1} T_x^*$ is given by the interior product

$$\iota_\xi u = u \rfloor X, \tag{10.14}$$

where $X \in T_x$ is the vector corresponding to $\xi \in T_x^*$ under the isomorphism $T_x \approx T_x^*$ given by the Riemannian metric. Consequently,

$$\frac{1}{i}\,\sigma_\delta(x,\xi)u = -\iota_\xi u. \tag{10.15}$$

Now, the Green-Stokes formula (9.17) implies, for M a compact Riemannian manifold with boundary,

$$\begin{aligned}(du, v) &= (u, \delta v) + \frac{1}{i}\int_{\partial M} \langle \sigma_d(x,\nu)u, v\rangle\, dS \\ &= (u, \delta v) + \int_{\partial M} \langle \nu \wedge u, v\rangle\, dS,\end{aligned} \tag{10.16}$$

and

$$\begin{aligned}(\delta u, v) &= (u, dv) + \frac{1}{i}\int_{\partial M} \langle \sigma_\delta(x,\nu)u, v\rangle\, dS \\ &= (u, dv) - \int_{\partial M} \langle \iota_\nu u, v\rangle\, dS.\end{aligned} \tag{10.17}$$

Recall that ν is the outward-pointing unit normal to ∂M.

Consequently, our generalization of (10.10), and also of (4.8), is

$$\begin{aligned}-(\Delta u, v) &= (du, dv) + (\delta u, \delta v) \\ &\quad + \frac{1}{i}\int_{\partial M} \big[\langle \sigma_d(x,\nu)\,\delta u, v\rangle + \sigma_\delta(x,\nu)\, du, v\rangle\big]\, dS\end{aligned} \tag{10.18}$$

or, equivalently,

$$\begin{aligned}-(\Delta u, v) &= (du, dv) + (\delta u, \delta v) \\ &\quad + \int_{\partial M} \big[\langle \nu \wedge (\delta u), v\rangle - \langle \iota_\nu(du), v\rangle\big]\, dS.\end{aligned} \tag{10.19}$$

Taking adjoints of the symbol maps, we can also write

$$\begin{aligned}-(\Delta u, v) &= (du, dv) + (\delta u, \delta v) \\ &\quad + \int_{\partial M} \big[\langle \delta u, \iota_\nu v\rangle - \langle du, \nu \wedge v\rangle\big]\, dS.\end{aligned} \tag{10.20}$$

Let us note what the symbol of Δ is. By (10.12) and (10.15),

$$-\sigma_\Delta(x,\xi)u = \iota_\xi \xi \wedge u + \xi \wedge \iota_\xi u. \tag{10.21}$$

If we perform the calculation by picking an orthonormal basis for T_x^* of the form $\{e_1, \dots, e_n\}$ with $\xi = |\xi| e_1$, we see that

$$\sigma_\Delta(x,\xi)u = -|\xi|^2 u. \tag{10.22}$$

In other words, in a local coordinate system, we have, for a k-form u,

$$\Delta u = g^{j\ell}(x)\,\partial_j \partial_\ell u + Y_k u, \tag{10.23}$$

where Y_k is a first-order differential operator.

A differential operator $P : C^\infty(M, E_0) \to C^\infty(M, E_1)$ is said to be *elliptic* provided $\sigma_P(x, \xi) : E_{0x} \to E_{1x}$ is invertible for each $x \in M$ and each $\xi \neq 0$. By (10.22), the Laplace operator on k-forms is elliptic.

Of course, the definition $-\Delta = \delta d$ for the Laplace operator on 0-forms coincides with the definition given in §4. In this regard, it is useful to note explicitly the following result about δ on 1-forms. Let X be a vector field and ξ the 1-form corresponding to X under a given metric:

$$g(Y, X) = \langle Y, \xi \rangle. \tag{10.24}$$

Then

$$\delta \xi = -\operatorname{div} X. \tag{10.25}$$

This identity is equivalent to (2.18) and the definition of δ as the formal adjoint of d.

We end this section with some algebraic implications of the symbol formula (10.21)–(10.22) for the Laplace operator. If we define $\wedge_\xi : \Lambda^* T_x^* \to \Lambda^* T_x^*$ by $\wedge_\xi(\omega) = \xi \wedge \omega$, and define ι_ξ as above, by (10.14), then the content of this calculation is

$$\wedge_\xi \iota_\xi + \iota_\xi \wedge_\xi = |\xi|^2. \tag{10.26}$$

As we have mentioned, this can be established by picking $\xi/|\xi|$ to be the first member of an orthonormal basis of T_x^*. Extending the identity (10.26), we have

$$\wedge_\xi \iota_\eta + \iota_\eta \wedge_\xi = \langle \xi, \eta \rangle, \tag{10.27}$$

a result that follows from the formula (13.37) of Chapter 1. Note also the connection with (2.26).

Exercises

1. Show that the adjoint of the exterior product operator $\xi \wedge$ is ι_ξ, as asserted in (10.14).
2. If $\alpha = \sum a_{jk}(x)\, dx_j \wedge dx_k$ and $a_j{}^k = g^{k\ell} a_{j\ell}$, relate $\delta\alpha$ to the divergence $a_j{}^k{}_{;k}$, as defined in (3.29).
3. Using (10.20), write down an expression for
$$(\Delta u, v) - (u, \Delta v)$$
as a boundary integral, when u and v are k-forms.
4. Relate the characterization (10.3) of the inner product on $\Lambda^* T_x^*$ arising from an inner product on T_x^*, to that given in the following section, before (11.24).
5. Let $\omega \in \Lambda^n(M)$, $n = \dim M$, be the volume form of an oriented Riemannian manifold M. Show that $\delta\omega = 0$. (*Hint*: Compare (10.6) with the special case of Stokes' formula $\int_M du = 0$ for $u \in \Lambda^{n-1}(M)$, compactly supported.)

6. Given the result of Exercise 5, show that Stokes' formula $\int_M du = \int_{\partial M} u$, for $u \in \Lambda^{n-1}(M)$, follows from (10.16).
7. If $f \in C^\infty(M)$ and $u \in \Lambda^k(M)$, show that

$$\delta(fu) = f\delta u - \iota_{(df)}u.$$

8. For a vector field u on the Riemannian manifold M, let $\widetilde{u}$ denote the associated 1-form. Show that

$$\delta(\widetilde{u} \wedge \widetilde{v}) = (\operatorname{div} v)\widetilde{u} - (\operatorname{div} u)\widetilde{v} - \widetilde{[u, v]},$$

for $\widetilde{u}, \widetilde{v} \in \Lambda^1(M)$. Reconsider this problem after reading Chapter 5, §8.

11. Maxwell's equations

The equations governing the electromagnetic field are one of the major triumphs of theoretical physics. We list them here, for the electric field E and the magnetic field B, in a vacuum:

$$\operatorname{div} B = 0, \tag{11.1}$$

$$\frac{\partial B}{\partial t} + \operatorname{curl} E = 0, \tag{11.2}$$

$$\operatorname{div} E = 4\pi\rho, \tag{11.3}$$

$$\frac{\partial E}{\partial t} - \operatorname{curl} B = -4\pi J. \tag{11.4}$$

Here, ρ is the charge density and J the electric current. Units are chosen so that the speed of light is 1. Here we are glossing over the distinction between two types of electric field, typically denoted E and D, and two types of magnetic field, typically denoted B and H, and their relation via "dielectric constants." Material on this may be found in texts on electromagnetism, such as [Ja].

Of the four equations above, (11.1) and (11.3) have a relatively elementary character. Equation (11.3), known as Gauss' law, follows in the case of stationary charges from the statement that a charge e at a point $p \in \mathbb{R}^3$ produces an electric field

$$E(x) = e\,\frac{x - p}{|x - p|^3},$$

which is Coulomb's law. Equation (11.1) is the statement that there are no magnetic charges. Both of these laws are well supported by experiments. We note parenthetically that there is reason to believe that at high energies magnetic charges might exist. A theoretical framework for this is provided by a modification of the theory of the electromagnetic field, called the "electroweak theory." But that is a story that we will not try to relate in this book. As one reference, we mention [IZ].

The equations (11.2) and (11.4) are more subtle. Equation (11.2), which implies that a changing magnetic field produces an electric field, is called Faraday's law. One implication of (11.4) is that an electric current produces a magnetic field;

this is exploited in electric motors. The first quantitative expression of this effect written down was

$$\operatorname{curl} B = 4\pi J,$$

which is valid when all quantities involved are independent of time. It breaks down when variation with time is allowed. Indeed, the left side must have vanishing divergence, but in the time-varying case one has, not $\operatorname{div} J = 0$, but rather the following law of conservation of charge:

$$\frac{\partial \rho}{\partial t} + \operatorname{div} J = 0. \tag{11.5}$$

Maxwell produced the modification (11.4), which completed the set of equations for the electromagnetic field.

Careful thought about the implications of Maxwell's equations, together with the experimental fact that two observers moving with respect to each other would measure the speed of light to be the same, led to the development of Einstein's theory of relativity. We will not discuss how this was done. Rather, following J. Wheeler, we will reverse the historical order. We will rewrite equations (11.1)–(11.4) in an invariant fashion, depending only on the Lorentz metric $-dx_0^2 + dx_1^2 + dx_2^2 + dx_3^2$ on Minkowski spacetime $\mathbb{R}^4$ rather than a particular Cartesian product decomposition of $\mathbb{R}^4$ into time $\mathbb{R}$ and space $\mathbb{R}^3$. We can then show that, within the relativistic framework, the subtle equations (11.2) and (11.4) actually follow from the "simple" equations (11.1) and (11.3).

We bring in the 2-form (with $t = x_0$)

$$\mathcal{F} = \sum_1^3 E_j \, dx_j \wedge dt + B_1 \, dx_2 \wedge dx_3 + B_2 \, dx_3 \wedge dx_1 + B_3 \, dx_1 \wedge dx_2. \tag{11.6}$$

In §18 of Chapter 1 it was shown how this form arises naturally in the relativistic expression of how the electromagnetic field acts on a charged particle to make it move. A calculation of the exterior derivative gives

$$d\mathcal{F} = \sum_1^3 \Big(\frac{\partial B}{\partial t} + \operatorname{curl} E \Big)_j (*dx_j) \wedge dt + (\operatorname{div} B) \, dx_1 \wedge dx_2 \wedge dx_3, \tag{11.7}$$

where, for $1 \le j \le 3$, we set

$$*dx_j = dx_k \wedge dx_\ell, \quad (j, k, \ell) \text{ a cyclic permutation of } (1, 2, 3).$$

Consequently, (11.1) and (11.2) together are equivalent to the equation

$$d\mathcal{F} = 0. \tag{11.8}$$

On the other hand, (11.1) alone is equivalent to the following. For fixed T, define $\kappa_T : \mathbb{R}^3 \to \mathbb{R}^4$ by $\kappa_T(x') = (T, x')$. Then (11.1) holds at $t = T$ if and only if

$$\kappa_T^* d\mathcal{F} = 0. \tag{11.9}$$

Now, in the relativistic set-up, any physical law that is valid on all surfaces $t =$ const. in $\mathbb{R}^4$ should be valid on *all* spacelike hyperplanes in $\mathbb{R}^4$. But the following result is easy to establish.

Lemma 11.1. *Let $0 \le k \le 3$, and suppose $\alpha \in \Lambda^k(\mathbb{R}^4)$ has the property that*

$$\kappa^*\alpha = 0, \tag{11.10}$$

for every inclusion $\kappa : S \to \mathbb{R}^4$ of spacelike hyperplanes in $\mathbb{R}^4$. Then $\alpha = 0$.

Applying this to $\alpha = d\mathcal{F}$, we see how (11.1) yields (11.2).

We will be able to rewrite (11.3)–(11.4) using the "adjoint" to d:

$$d^\star : \Lambda^k(\mathbb{R}^4) \longrightarrow \Lambda^{k-1}(\mathbb{R}^4), \tag{11.11}$$

defined like $\delta = d^*$ in §10, but using an inner product coming from the Lorentz metric. Thus, for compactly supported u,

$$L(du, v) = L(u, d^\star v), \tag{11.12}$$

for a $(k-1)$-form u and a k-form v, where the inner product of two k-forms v_j is

$$L(v_1, v_2) = \int \langle v_1, v_2\rangle \, dx_0 \cdots dx_3, \tag{11.13}$$

the integral of the pointwise inner product, characterized as follows.

A form $dx_{j_1} \wedge \cdots \wedge dx_{j_k}$ with distinct j_ν's has square norm $\varepsilon_{j_1} \cdots \varepsilon_{j_k}$, where $\varepsilon_0 = -1,\ \varepsilon_1 = \varepsilon_2 = \varepsilon_3 = 1$. Two such forms are orthogonal unless their sets of indices $\{j_1, \dots, j_k\}$ coincide. A straightforward calculation yields

$$d^\star g_{k\ell}(x)\, dx_k \wedge dx_\ell = -\sum_{i,j} \varepsilon_j \varepsilon(i, j; k, \ell)\, \frac{\partial g_{k\ell}}{\partial x_i}\, dx_j, \tag{11.14}$$

where

$$\varepsilon(i, j; k, \ell) = \langle dx_i \wedge dx_j, dx_k \wedge dx_\ell\rangle \tag{11.15}$$

is characterized above. This is 0 unless $\{i, j\} = \{k, \ell\}$, and we can rewrite (11.14) as

$$d^\star g_{k\ell}(x)\, dx_k \wedge dx_\ell = \varepsilon(k, \ell; k, \ell)\Big[\varepsilon_k \frac{\partial g_{k\ell}}{\partial x_\ell}\, dx_k - \varepsilon_\ell \frac{\partial g_{k\ell}}{\partial x_k}\, dx_\ell\Big]. \tag{11.16}$$

This implies

$$d^\star \sum_1^3 E_j\, dx_j \wedge dx_0 = -(\operatorname{div} E)\, dx_0 - \sum_1^3 \frac{\partial E_j}{\partial t}\, dx_j, \tag{11.17}$$

and

$$d^\star\big(B_1\, dx_2 \wedge dx_3 + B_2\, dx_3 \wedge dx_1 + B_3\, dx_1 \wedge dx_2\big) = \sum_1^3 (\operatorname{curl} B)_j\, dx_j. \tag{11.18}$$

Thus (11.3) and (11.4) together are equivalent to the equation

$$d^\star \mathcal{F} = 4\pi \mathcal{J}^b, \tag{11.19}$$

where

$$\mathcal{J}^b = -\rho\, dt + \sum_1^3 J_k\, dx_k. \tag{11.20}$$

Thus $\mathcal{J}^b$ is the 1-form associated via the Lorentz metric to the vector

$$\mathcal{J} = (\rho, J), \tag{11.21}$$

called the charge-current 4-vector.

In this case, (11.3) alone is equivalent to the identity

$$(d^\star \mathcal{F} - 4\pi \mathcal{J}^b) \rfloor \frac{\partial}{\partial t} = 0. \tag{11.22}$$

Again, in the relativistic set-up, such a physical law ought to be independent of the choice of timelike vector field with which to take the interior product. Thus, if we assume that $\mathcal{F}$ has an invariant significance as a 2-form and also that $\mathcal{J}^b$ has an invariant significance as a 1-form, we are in a position to apply the following.

Lemma 11.2. *If* $1 \le k \le 4$ *and* $\alpha \in \Lambda^k(\mathbb{R}^4)$ *has the property that*

$$\alpha \rfloor V = 0 \tag{11.23}$$

for all timelike vectors V*, then* $\alpha = 0$.

Applying this to $\alpha = d^\star \mathcal{F} - 4\pi \mathcal{J}^b$, we see how (11.3) yields (11.4).

The pair of Maxwell equations (11.8), (11.19) make sense on any Lorentz manifold of dimension 4 and provide the appropriate equations for an electromagnetic field in curved spacetime. To define $d^\star$, one uses the formula (11.12), replacing $dx_0 \cdots dx_3$ by the natural volume element on a general Lorentz manifold M in (11.13).

This construction defines $d^\star$ for Lorentz manifolds of any dimension. The inner product in the integrand in (11.13) can be characterized as follows. To the Lorentz inner product on $V = T_x M$ corresponds an isomorphism $Q : V \to V'$ satisfying $Q' = Q$ (with $V'' = V$). This induces isomorphisms

$$Q_k : \Lambda^k V \to \Lambda^k V' \approx (\Lambda^k V)',$$

with the same symmetry property, yielding inner products on $\Lambda^k V$, $0 \le k \le m = \dim M$. Equivalently, if you pick an "orthonormal" basis $\{v_0, \dots, v_{m-1}\}$ of V, satisfying $\langle v_0, v_0 \rangle = -1$, $\langle v_j, v_j \rangle = 1$ for $1 \le j \le m-1$, then the characterization given after (11.13) is easily extended.

In analogy with (10.9), it is of interest to form the second-order operator

$$-\Box = (d + d^\star)^2 = dd^\star + d^\star d. \tag{11.24}$$

A calculation similar to (10.23) gives

$$\square u = h^{j\ell}(x)\,\partial_j\partial_\ell u + Y_k u, \tag{11.25}$$

for a k-form u, where $(h^{j\ell})$ is formed from the Lorentz metric tensor, as in (7.7), and Y_k is a first-order differential operator. On 0-forms, this operator is exactly (7.7). For Minkowski spacetime $\mathbb{R}^4$, $\square$ is just $-\partial^2/\partial x_0^2 + \sum_1^3 \partial^2/\partial x_j^2$, acting on each component of a k-form.

The equations $d\mathcal{F} = 0$, $d^\star\mathcal{F} = 4\pi\mathcal{J}^b$ imply that $\mathcal{F}$ satisfies the "wave equation"

$$\square\mathcal{F} = -4\pi\, d\mathcal{J}^b. \tag{11.26}$$

The results developed in §8 for scalar hyperbolic operators of the type (8.2) are easily extended to cover the operator $\square$ constructed here, which by (11.25) has scalar principal part.

In particular, finite propagation speed arguments apply to solutions to Maxwell's equations. Existence of solutions, including propagation of electromagnetic waves in regions bounded by perfect conductors, is studied in Chapter 6.

The energy in an electromagnetic field in $\mathbb{R}^4 = \mathbb{R}\times\mathbb{R}^3$ is

$$V(t) = \frac{1}{8\pi}\int_{\mathbb{R}^3}\big[|E(t,x)|^2 + |B(t,x)|^2\big]\,dx. \tag{11.27}$$

If (11.1)–(11.4) hold, then

$$\begin{aligned} 4\pi\,\frac{dV}{dt} &= \Big(\frac{\partial E}{\partial t}, E\Big) + \Big(\frac{\partial B}{\partial t}, B\Big) \\ &= (\text{curl } B, E) - (\text{curl } E, B) - 4\pi(J, E). \end{aligned} \tag{11.28}$$

If $E(t,x)$ and $B(t,x)$ decrease sufficiently rapidly as $|x|\to\infty$, we have

$$(\text{curl } B, E) = (B, \text{curl } E), \tag{11.29}$$

as can be established by integration by parts. Hence

$$\frac{dV}{dt} = -\int_{\mathbb{R}^3} J(t,x)\cdot E(t,x)\,dx. \tag{11.30}$$

In particular, for $J = 0$ we have conservation of $V(t)$.

One can construct a stress-energy tensor $\mathcal{T}$ due to the electromagnetic field, by an argument similar to that of §7. First note that, with $\mathcal{F}$ given by (11.6), we have

$$\langle\mathcal{F},\mathcal{F}\rangle = |B|^2 - |E|^2. \tag{11.31}$$

Equivalently,

$$\operatorname{Tr}\widetilde{\mathcal{F}}^2 = 2(|E|^2 - |B|^2), \tag{11.32}$$

where $\widetilde{\mathcal{F}}$ is the tensor field of type (1, 1) associated to $\mathcal{F}$. Note also that $(\widetilde{\mathcal{F}}^2)^0{}_0 = |E|^2$. Thus a natural construction of $\mathcal{T}$ giving rise to $\mathcal{T}_{00} = (1/8\pi)(|E|^2 + |B|^2)$

is

$$\widetilde{\mathcal{T}} = -\frac{1}{4\pi}\Big(\widetilde{\mathcal{F}}^2 - \frac{1}{4}(\operatorname{Tr}\widetilde{\mathcal{F}}^2)I\Big) = -\frac{1}{4\pi}\Big(\widetilde{\mathcal{F}}^2 + \frac{1}{2}\langle\mathcal{F},\mathcal{F}\rangle I\Big), \tag{11.33}$$

where $\widetilde{\mathcal{T}}$ is the tensor field of type (1, 1) associated with $\mathcal{T}$. In index notation,

$$\mathcal{T}_{ij} = \frac{1}{4\pi}\big(\mathcal{F}_{im}\mathcal{F}_j{}^m - \frac{1}{4}h_{ij}\mathcal{F}^{mn}\mathcal{F}_{mn}\big), \tag{11.34}$$

where (h_{ij}) is the Lorentz metric tensor. In this case, in analogy with (7.10), one obtains

$$\mathcal{T}^{jk}{}_{;k} = -\mathcal{F}^j{}_k\mathcal{J}^k, \tag{11.35}$$

provided the Maxwell equations (11.8) and (11.19) hold. Equivalently, with $\widehat{\mathcal{T}}$ denoting the tensor field of type (2, 0) associated with $\mathcal{T}$,

$$\operatorname{div}\widehat{\mathcal{T}} = -\widetilde{\mathcal{F}}\mathcal{J}. \tag{11.36}$$

If the electromagnetic field $\mathcal{F}$ is defined on a Lorentz 4-manifold which is simply connected, the equation $d\mathcal{F} = 0$ implies the existence of a 1-form $\mathcal{A}$ such that $\mathcal{F} = d\mathcal{A}$. We can define the Lagrangian

$$L = -\frac{1}{8\pi}\langle\mathcal{F},\mathcal{F}\rangle = -\frac{1}{8\pi}\langle d\mathcal{A}, d\mathcal{A}\rangle, \tag{11.37}$$

with inner product as in (11.31). The action integral $I(\mathcal{A}) = \int L\,dV$ satisfies, for a compactly supported 1-form β,

$$\frac{d}{d\tau}I(\mathcal{A}+\tau\beta)\Big|_{\tau=0} = -\frac{1}{4\pi}\int\langle d\beta, d\mathcal{A}\rangle\,dV = -\frac{1}{4\pi}\int\langle\beta, d^\star d\mathcal{A}\rangle\,dV, \tag{11.38}$$

so the stationary condition $\delta\int L\,dV = 0$ is equivalent to $d^\star d\mathcal{A} = 0$, that is, to the rest of Maxwell's equations (11.19), in case $\mathcal{J} = 0$. Thus (11.37) is the appropriate Lagrangian for the electromagnetic field, in order to produce Maxwell's equations in empty space. If the current $\mathcal{J}$ is *given* (subject to the condition $d^\star\mathcal{J} = 0$), and $\mathcal{F} = d\mathcal{A}$, then the equation (11.19) is the stationary condition $\delta\int L\,dV = 0$ for the Lagrangian

$$L = -\frac{1}{8\pi}\langle\mathcal{F},\mathcal{F}\rangle + \langle\mathcal{A},\mathcal{J}\rangle. \tag{11.39}$$

In typical problems the current is not given in advance, but is itself influenced by the electromagnetic force. The nature of the influence involves the masses of the substances that carry charges, whose motion produces the current. Then the Maxwell equations are coupled to other equations, which are often nonlinear. We describe a model for one example.

Suppose we have a diffuse cloud of electrons, in otherwise empty space. We model this as a continuous charged substance, whose motion is described by a 4-velocity vector field u, satisfying $\langle u, u\rangle = -1$, yielding a current $\mathcal{J} = \sigma u$, where $\sigma\,dV$ is the charge density, measured by an observer whose velocity is u. Taking a cue from the Lagrangian (18.20) of Chapter 1, derived to reflect the relativistic Lorenz force law, we use the Lagrangian

$$(11.40)\qquad L = -\frac{1}{8\pi}\langle \mathcal{F}, \mathcal{F}\rangle + \langle \mathcal{A}, \mathcal{J}\rangle + \frac{1}{2}\mu\langle u, u\rangle = L_1 + L_2 + L_3,$$

where $\mu\, dV$ is the mass density, measured by an observer whose velocity is u. We are assuming that only one type of matter is present, so σ is a constant multiple of μ. In more general cases there would be additional terms in the Lagrangian.

We look at $I(\mathcal{A}, u) = I_1 + I_2 + I_3$. The term I_3 is independent of $\mathcal{A}$, and as above we have

$$(11.41)\qquad \frac{\partial}{\partial \tau} I(\mathcal{A} + \tau\beta, u)\big|_{\tau=0} = \int \Big[-\frac{1}{4\pi}\langle \beta, d^\star d\mathcal{A}\rangle + \langle \beta, \mathcal{J}\rangle\Big]\, dV.$$

The stationary condition this yields is again the Maxwell equation (11.19). Next we compute $(\partial/\partial\tau) I\big(\mathcal{A}, u(\tau)\big)\big|_{\tau=0}$, where $u(\tau)$ is a one-parameter family of velocity fields on M, obtained by varying the electron trajectories. There is no variation in I_1, so we need to consider I_2 and I_3.

We first treat the variation of I_3, in a manner parallel to the calculations (11.17)–(11.26) in Chapter 1, leading to the geodesic equations. To do this, we parameterize the electron trajectories by $X : \Omega \times I \to M$, $X(y, s) = x$, $u = \partial_s X$. We suppose the mass density is constant in (y, s)-coordinates, say m, so $m\, dy\, ds = \mu\, dV$. Since $u = \partial/\partial s$ in (y, s)-coordinates, this implies $\mathcal{L}_u(\mu\, dV) = 0$, or

$$(11.42)\qquad \operatorname{div}(\mu u) = 0,$$

where div is computed using the Lorentz metric on M. Our hypothesis amounts to the law of conservation of matter. If we vary this map, using $X(y, s, \tau)$, then

$$(11.43)\qquad \frac{d}{d\tau}\int \frac{1}{2}\mu\langle u, u\rangle\, dV = \int \frac{1}{2} m\mathcal{L}_w\langle u, u\rangle\, dy\, ds = \int m\langle \nabla_w u, u\rangle\, dy\, ds,$$

where $\partial_\tau X = w$. Using $[\partial_s, \partial_\tau] = 0$, convert this last integral to

$$(11.44)\qquad -\int m\langle w, \nabla_u u\rangle\, dy\, ds + m\int \mathcal{L}_u\langle w, u\rangle\, dy\, ds.$$

The last integral here vanishes for a compactly supported perturbation, by the fundamental theorem of calculus, so

$$(11.45)\qquad \frac{d}{d\tau} I_3\big(\mathcal{A}, u(\tau)\big)\big|_{\tau=0} = -\int \langle w, \nabla_u u\rangle m\, dy\, ds = -\int \langle w, \mu\nabla_u u\rangle\, dV.$$

We now treat the variation of I_2, also using (y, s)-coordinates. Since σ is a constant multiple of μ, we have $\sigma\, dV = e\, dy\, ds$ for some constant e, and, parallel to (11.42), we have conservation of electric charge, $\operatorname{div}(\sigma u) = 0$ (i.e., div $\mathcal{J} = 0$), which is equivalent to (11.5) when M is Minkowski space. We have

$$(11.46)\qquad \frac{d}{d\tau}\int \langle \mathcal{A}, \mathcal{J}\rangle\, dV = \int e\mathcal{L}_w\langle \mathcal{A}, u\rangle\, dy\, ds.$$

We use the identity $\mathcal{L}_u \mathcal{A} = d\mathcal{A} \rfloor u + d(\mathcal{A} \rfloor u)$ to write

$$\begin{aligned}\mathcal{L}_w \langle \mathcal{A}, u \rangle &= -(d\mathcal{A})(u, w) + \langle \mathcal{L}_u \mathcal{A}, w \rangle \\ &= -(d\mathcal{A})(u, w) + \mathcal{L}_u \langle \mathcal{A}, w \rangle - \langle \mathcal{A}, \mathcal{L}_u w \rangle.\end{aligned} \tag{11.47}$$

Since $d\mathcal{A} = \mathcal{F}$, $[\partial_s, \partial_\tau] = 0$, and $\mathcal{L}_u \langle \mathcal{A}, w \rangle$ integrates to zero, we have

$$\frac{d}{d\tau} \int \langle \mathcal{A}, \mathcal{J} \rangle \, dV = \int e \langle \tilde{\mathcal{F}} u, w \rangle \, dy \, ds = \int \langle \tilde{\mathcal{F}} \mathcal{J}, w \rangle \, dV. \tag{11.48}$$

Together with (11.45), this gives

$$\frac{\partial}{\partial \tau} I(\mathcal{A}, u(\tau))\big|_{\tau=0} = -\int \langle \mu \nabla_u u - \tilde{\mathcal{F}} \mathcal{J}, w \rangle \, dV. \tag{11.49}$$

Thus the stationary condition for variation of u is

$$\mu \nabla_u u - \tilde{\mathcal{F}} \mathcal{J} = 0 \text{ or, equivalently, } \nabla_u u - \frac{e}{m} \tilde{\mathcal{F}} u = 0, \tag{11.50}$$

which is the Lorentz force law in this context.

It is useful to consider what the stress-energy tensor should be when we have the Lagrangian (11.40). It is reasonable to take it to be the sum of the stress-energy tensor $\mathcal{T}_e$ for the electromagnetic field, given by (11.34), and a stress-energy tensor $\mathcal{T}_m$ associated with the "matter field." If we want $\mathcal{T}_m(Z, Z) dV$ to be the mass-energy density of the electrons observed by one moving with velocity Z, then it is natural to set

$$\widehat{\mathcal{T}}_m = \mu u \otimes u, \tag{11.51}$$

(i.e., $\mathcal{T}_m^{jk} = \mu u^j u^k$). Then the total stress-energy tensor is given by

$$\mathcal{T}^{jk} = \frac{1}{4\pi} \Big(\mathcal{F}^j{}_\ell \mathcal{F}^{k\ell} - \frac{1}{4} h^{jk} \mathcal{F}^{i\ell} \mathcal{F}_{i\ell} \Big) + \mu u^j u^k. \tag{11.52}$$

The divergence of $\widehat{\mathcal{T}}_e$ is given by (11.36), provided the Maxwell equation (11.19) holds. Furthermore, $(\mu u^j u^k)_{;k} = (\mu u^k)_{;k} u^j + \mu u^k u^j{}_{;k}$, so

$$\operatorname{div} \widehat{\mathcal{T}}_m = \operatorname{div}(\mu u) u + \mu \nabla_u u. \tag{11.53}$$

Thus, for $\widehat{\mathcal{T}} = \widehat{\mathcal{T}}_e + \widehat{\mathcal{T}}_m$, we have (granted (11.19))

$$\operatorname{div} \widehat{\mathcal{T}} = \operatorname{div}(\mu u) + \mu \nabla_u u - \tilde{\mathcal{F}} \mathcal{J}. \tag{11.54}$$

We have the conservation law $\operatorname{div} \widehat{\mathcal{T}} = 0$ for a solution to the coupled Maxwell-Lorentz equations. Indeed, the vanishing of the first term on the right side of (11.54) is equivalent to the matter conservation law (11.42), and the vanishing of the sum of the other terms on the right side of (11.54) is equivalent to the Lorentz force law (11.50).

Exercises

1. Demonstrate Lemmas 11.1 and 11.2.

2. Verify the calculations (11.14)–(11.18).
3. Show that the inner product of forms defined after (11.13) depends only on the Lorentz metric on $\mathbb{R}^4$, not on the coordinate representation.
4. Show that div curl = 0 is a special case of $dd = 0$.
5. Show that (11.3)–(11.4) imply the "conservation law" (11.5).
   (*Hint*: Apply $\partial/\partial t$ to (11.3) and div to (11.4); use div curl = 0.)
   Show that (11.5) is equivalent to $d^\star \mathcal{J}^b = 0$.
6. Verify the identity (11.29), for any compactly supported vector fields $E(x)$ and $B(x)$ on $\mathbb{R}^3$.
7. Prove the conservation law (11.36), as a consequence of Maxwell's equations.
8. Show that the identity $d\mathcal{F} = 0$ is equivalent to

$$\mathcal{F}_{jk;\ell} + \mathcal{F}_{k\ell;j} + \mathcal{F}_{\ell j;k} = 0.$$

9. Show that the identity $d^\star \mathcal{F} = 4\pi \mathcal{J}^b$ is equivalent to

$$\mathcal{F}^{jk}{}_{;k} = 4\pi \mathcal{J}^j.$$

10. The equation $d\mathcal{F} = 0$ on $\mathbb{R}^4$ implies $\mathcal{F} = \mathcal{A}$ for some 1-form $\mathcal{A}$ on $\mathbb{R}^4$. $\mathcal{A}$ is not unique, as any 1-form du can be added. Show that $\mathcal{A}$ can be picked to satisfy $d^\star \mathcal{A} = 0$ and that, for such $\mathcal{A}$,

$$\Box \mathcal{A} = -4\pi \mathcal{J}^b.$$

   (*Hint*: Set up a PDE for u. Look for the relevant existence theorem in Chapter 3.)
11. The calculation (11.31) of $\langle \mathcal{F}, \mathcal{F} \rangle$ shows that $|B|^2 - |E|^2$ is Lorentz invariant. Calculate $\mathcal{F} \wedge \mathcal{F}$ and show that $E \cdot B$ is also Lorentz invariant.
12. Think about the fact that the tensor $\widetilde{T}$ given by (11.33) is trace-free, i.e., Tr $\widetilde{T} = 0$. What is the trace of the stress-energy tensor defined by (7.5) or, equivalently (7.11)?
13. As mentioned in Exercise 5 in §18, Chapter 1, a sign change in the Lorentz metric, from one of signature $(-, +, +, +)$ to one of signature $(+, -, -, -)$ (which some people prefer), leads to a sign change in the formula for the 2-form $\mathcal{F}$ (though no change in the tensor field $\widetilde{\mathcal{F}}$ of type (1, 1)). Show that it leads to a sign change in the formula (11.34) for the stress-energy tensor of the electromagnetic field.
    What sign changes arise in the formula (11.40) for the Lagrangian of an electromagnetic field coupled to charged matter?

References

[Ar] V. Arnold, *Mathematical Methods of Classical Mechanics*, Springer-Verlag, New York, 1978.

[Au] T. Aubin, *Nonlinear Analysis on Manifolds. Monge-Ampere Equations*, Springer-Verlag, New York, 1982.

[Ba] H. Bateman, *Partial Differential Equations of Mathematical Physics*, Dover, New York, 1944.

[BGM] M. Berger, P. Gauduchon, and E. Mazet, *Le Spectre d'une Variété Riemannienne*, LNM no.194, Springer-Verlag, New York, 1971.

[BJS] L. Bers, F. John, and M. Schechter, *Partial Differential Equations*, Wiley, New York, 1964.

[Car] C. Caratheodory, *Calculus of Variations and Partial Differential Equations of the First Order*, Holden-Day, San Francisco, 1965.

[ChM] A. Chorin and J. Marsden, *A Mathematical Introduction to Fluid Mechanics*, Springer-Verlag, New York, 1979.

[CK] D. Christodoulu and S. Klainerman, *The Global Nonlinear Stability of the Minkowski Space*, Princeton Univ. Press, Princeton, N. J., 1993.

[CH] R. Courant and D. Hilbert, *Methods of Mathematical Physics II*, J.Wiley, New York, 1966.

[ES] J. Eells and J. Sampson, Harmonic mappings of Riemannian manifolds, *Amer. J. Math.* 86(1964), 109–160.

[FM] A. Fischer and J. Marsden, General relativity, partial differential equations, and dynamical systems, *AMS Proc. Symp. Pure Math.* 23(1973), 309–327.

[Fo1] G. Folland, *Introduction to Partial Differential Equations*, Princeton Univ. Press, Princeton, N. J., 1976.

[Frl] F. G. Friedlander, *Sound Pulses*, Cambridge Univ. Press, Cambridge, 1958.

[Frd] A. Friedman, *Generalized Functions and Partial Differential Equations*, Prentice-Hall, Englewood Cliffs, N. J., 1963.

[Ga] P. Garabedian, *Partial Differential Equations*, Wiley, New York, 1964.

[Go] H. Goldstein, *Classical Mechanics*, Addison-Wesley, New York, 1950.

[Had] J. Hadamard, *Le Probleme de Cauchy et les Equations aux Derivées Partielles Linéaires Hyperboliques*, Hermann, Paris, 1932.

[HE] S. Hawking and G. Ellis, *The Large Scale Structure of Space-Time*, Cambridge Univ. Press, Cambridge, 1973.

[Hel] L. Helms, *Introduction to Potential Theory*, Wiley, New York, 1969.

[HM] T. Hughes and J. Marsden, *A Short Course in Fluid Mechanics*, Publish or Perish Press, Boston, 1976.

[IZ] C. Itzykson and J. Zuber, *Quantum Field Theory*, McGraw-Hill, New York, 1980.

[Ja] J. Jackson, *Classical Electrodynamics*, J. Wiley, New York, 1962.

[Jo] F. John, *Partial Differential Equations*, Springer-Verlag, New York, 1975.

[Jos] J. Jost, *Nonlinear Methods in Riemannian and Kahlerian Geometry*, Birkhauser, Boston, 1988.

[Keg] O. Kellogg, *Foundations of Potential Theory*, Dover, New York, 1954.

[Lax] P. Lax, *The theory of hyperbolic equations*. Stanford Lecture Notes, 1963.

[Lry] J. Leray, *Hyperbolic Differential Equations*, Princeton Univ. Press, Princeton, N. J., 1952.

[MS] H. McKean and I. Singer, Curvature and the eigenvalues of the Laplacian, *J. Diff. Geom.* 1(1967), 43–69.

[Mj] A. Majda, *Compressible Fluid Flow and Systems of Conservation Laws in Several Space Variables*, Appl. Math. Sci. #53, Springer-Verlag, New York, 1984.

[MH] J. Marsden and T. Hughes, *Mathematical Foundations of Elasticity*, Prentice-Hall, Englewood Cliffs, N. J., 1983.

[Miz] S. Mizohata, *The Theory of Partial Differential Equations*, Cambridge Univ. Press, Cambridge, 1973.

[Mor] C. B. Morrey, *Multiple Integrals in the Calculus of Variations*, Springer-Verlag, New York, 1966.

[Os] R. Osserman, *A Survey of Minimal Surfaces*, Van Nostrand, New York, 1969.

[Poo] W. Poor, *Differential Geometric Structures*, McGraw-Hill, New York, 1981.

[Rau] J. Rauch, *Partial Differential Equations*, Springer-Verlag, New York, 1992.

[Ray] Lord Rayleigh, On waves propagated along the plane surface of an elastic solid, *Proc. London Math. Soc.* 17(1885), 4–11.

[RS] M. Reed and B. Simon, *Methods of Mathematical Physics*, Academic Press, New York, Vols. 1,2, 1975; Vols. 3,4, 1978.

[Sm] J. Smoller, *Shock Waves and Reaction-Diffusion Equations*, Springer-Verlag, New York, 1983.

[So] S. Sobolev, *Partial Differential Equations of Mathematical Physics*, Dover, New York, 1964.

[Stk] I. Stakgold, *Boundary Value Problems of Mathematical Physics*, Macmillan, New York, 1968.

[Stb] S. Sternberg, *Lectures on Differential Geometry*, Prentice Hall, Englewood Cliffs, N. J., 1964.

[Str] M. Struwe, *Variational Problems*, Springer-Verlag, New York, 1990.

[VMF] R. von Mises and K. Friedrichs, *Fluid Dynamics*, Springer-Verlag, New York, 1971.

3

Fourier Analysis, Distributions, and Constant-Coefficient Linear PDE

Introduction

Fourier analysis is perhaps the most important single tool in the study of linear partial differential equations. It serves in several ways, the most basic—and historically the first—being to give specific formulas for solutions to various linear PDE with constant coefficients, particularly the three classics, the Laplace, wave, and heat equations:

$$\Delta u = f, \quad \frac{\partial^2 u}{\partial t^2} - \Delta u = f, \quad \frac{\partial u}{\partial t} - \Delta u = f, \tag{0.1}$$

with $\Delta = \partial^2/\partial x_1^2 + \cdots + \partial^2/\partial x_n^2$. The Fourier transform accomplishes this by transforming the operation of $\partial/\partial x_j$ to the algebraic operation of multiplication by $i\xi_j$. Thus the equations (0.1) are transformed to algebraic equations and to ODE with parameters.

Before introducing the Fourier transform of functions on Euclidean space $\mathbb{R}^n$, we discuss the Fourier series associated to functions on the torus $\mathbb{T}^n$ in §1. Methods developed to establish the Fourier inversion formula for Fourier series, in the special case of the circle $S^1 = \mathbb{T}^1$, provide for free a development of the basic results on harmonic functions in the plane, and we give such results in §2, noting that these results specialize further to yield standard basic results in the theory of holomorphic functions of one complex variable, such as power-series expansions and Cauchy's integral formula.

In §3 we define the Fourier transform of functions on $\mathbb{R}^n$ and prove the Fourier inversion formula. The proof shares with the argument for Fourier series in §1 the property of simultaneously yielding explicit solutions to a PDE, this time the heat equation.

It turns out that representations of solutions to such PDE as listed in (0.1) are most naturally done in terms of objects more general than functions, called distributions. We develop the theory of distributions in §4; Fourier analysis works very naturally with the class of distributions known as tempered distributions.

Section 5, in some sense the heart of this chapter, derives explicit solutions to the classical linear PDE (0.1) via Fourier analysis. The use of Fourier analysis and distribution theory to represent solutions to these PDE gives rise to numerous interesting identities, involving both elementary functions and "special functions," such as the gamma function and Bessel functions, and we present some of these identities here, only a smattering from a rich area of classical analysis. Further development of harmonic analysis in Chapter 8 will bring in additional studies of special functions.

Fourier analysis and distribution theory are also useful tools for general investigations of linear PDE, in cases where explicit formulas might not be obtainable. We illustrate a couple of cases of this in the present chapter, discussing the existence and behavior of "parametrices" for elliptic PDE with constant coefficients, and applications to smoothness of solutions to such PDE, in §9, and proving local solvability of general linear PDE with constant coefficients in §10. Fourier analysis and distribution theory will acquire further power in the next chapter as tools for investigations of existence and qualitative properties of solutions to various classes of PDE, with the development of Sobolev spaces.

Sections 11 and 12 deal with the discrete Fourier transform, particularly with Fourier analysis on finite cyclic groups. We study this both as an approximation to Fourier analysis on the torus and Euclidean space, sometimes useful for numerical work, and as a subject with its intrinsic interest, and with implications for number theory. In §12 we give a brief description of "fast" algorithms for computing discrete Fourier transforms.

1. Fourier series

Let f be an integrable function on the torus $\mathbb{T}^n$, naturally isomorphic to $\mathbb{R}^n/\mathbb{Z}^n$ and to the Cartesian product of n copies $S^1 \times \cdots \times S^1$ of the circle. Its Fourier series is by definition a function on $\mathbb{Z}^n$ given by

$$\hat{f}(k) = \frac{1}{(2\pi)^n} \int_{\mathbb{T}^n} f(\theta) e^{-ik\cdot\theta}\, d\theta, \tag{1.1}$$

where $k = (k_1, \dots, k_n)$, $k \cdot \theta = k_1\theta_1 + \cdots + k_n\theta_n$. We use the notation

$$\mathcal{F}f(k) = \hat{f}(k). \tag{1.2}$$

Clearly, we have a continuous linear map

$$\mathcal{F} : L^1(\mathbb{T}^n) \longrightarrow \ell^\infty(\mathbb{Z}^n), \tag{1.3}$$

where $\ell^\infty(\mathbb{Z}^n)$ denotes the space of bounded functions on $\mathbb{Z}^n$, with the sup norm. If $f \in C^\infty(\mathbb{T}^n)$, then we can integrate by parts to get

$$k^\alpha \hat{f}(k) = \frac{1}{(2\pi)^n} \int_{\mathbb{T}^n} (D^\alpha f)(\theta)\, e^{-ik\cdot\theta}\, d\theta, \tag{1.4}$$

where $k^\alpha = k_1^{\alpha_1} \cdots k_n^{\alpha_n}$, and

$$D^\alpha = D_1^{\alpha_1} \cdots D_n^{\alpha_n}, \quad D_j = \frac{1}{i} \frac{\partial}{\partial \theta_j}. \tag{1.5}$$

It follows easily that

$$\mathcal{F} : C^\infty(\mathbb{T}^n) \longrightarrow s(\mathbb{Z}^n), \tag{1.6}$$

where $s(\mathbb{Z}^n)$ consists of functions u on $\mathbb{Z}^n$ which are rapidly decreasing, in the sense that, for each N,

$$p_N(u) = \sup_{k \in \mathbb{Z}^n} \langle k \rangle^N \, |u(k)| < \infty. \tag{1.7}$$

Here, we use the notation

$$\langle k \rangle = \left(1 + |k|^2\right)^{1/2},$$

where $|k|^2 = k_1^2 + \cdots + k_n^2$. If we use the inner product

$$(f, g) = (f, g)_{L^2} = \frac{1}{(2\pi)^n} \int_{\mathbb{T}^n} f(\theta) \overline{g(\theta)} \, d\theta, \tag{1.8}$$

for $f, g \in C^\infty(\mathbb{T}^n)$, or more generally for $f, g \in L^2(\mathbb{T}^n)$, and if on $s(\mathbb{Z}^n)$, or more generally on $\ell^2(\mathbb{Z}^n)$, the space of square summable functions on $\mathbb{Z}^n$, we use the inner product

$$(u, v) = (u, v)_{\ell^2} = \sum_{k \in \mathbb{Z}^n} u(k) \overline{v(k)}, \tag{1.9}$$

we have the formula

$$(\mathcal{F} f, u)_{\ell^2} = (f, \mathcal{F}^* u)_{L^2}, \tag{1.10}$$

valid for $f \in C^\infty(\mathbb{T}^n), \ u \in s(\mathbb{Z}^n)$, where

$$\mathcal{F}^* : s(\mathbb{Z}^n) \longrightarrow C^\infty(\mathbb{T}^n) \tag{1.11}$$

is given by

$$\big(\mathcal{F}^* u\big)(\theta) = \sum_{k \in \mathbb{Z}^n} u(k) \, e^{ik \cdot \theta}. \tag{1.12}$$

Another identity that we will find useful is

$$\frac{1}{(2\pi)^n} \int_{\mathbb{T}^n} e^{ik \cdot \theta} \, e^{-i\ell \cdot \theta} \, d\theta = \delta_{k\ell}, \tag{1.13}$$

where $\delta_{k\ell} = 1$ if $k = \ell$ and $\delta_{k\ell} = 0$ otherwise.

Our main goal here is to establish the Fourier inversion formula

$$f(\theta) = \sum_{k \in \mathbb{Z}^n} \hat{f}(k) \, e^{ik \cdot \theta}, \tag{1.14}$$

the sum on the right in (1.14) converging in the appropriate function space, depending on the nature of f. Let us single out another space of functions on $\mathbb{T}^n$, the *trigonometric polynomials*:

$$\mathcal{TP} = \Big\{\sum_{k\in\mathbb{Z}^n} a(k)e^{ik\cdot\theta} : a(k) = 0 \text{ except for finitely many } k\Big\}. \tag{1.15}$$

Clearly,

$$\mathcal{F} : \mathcal{TP} \longrightarrow c_{00}(\mathbb{Z}^n), \tag{1.16}$$

where $c_{00}(\mathbb{Z}^n)$ consists of functions on $\mathbb{Z}^n$ which vanish except at a finite number of points; this follows from (1.13). The formula (1.12) gives

$$\mathcal{F}^* : c_{00}(\mathbb{Z}^n) \longrightarrow \mathcal{TP}, \tag{1.17}$$

and the formula (1.13) easily yields

$$\mathcal{F}\mathcal{F}^* = I \;\text{ on } c_{00}(\mathbb{Z}^n), \tag{1.18}$$

and even

$$\mathcal{F}\mathcal{F}^* = I \;\text{ on } s(\mathbb{Z}^n). \tag{1.19}$$

By comparison, the inversion formula (1.14) states

$$\mathcal{F}^*\mathcal{F} = I, \tag{1.20}$$

on $C^\infty(\mathbb{T}^n)$, or some other space of functions on $\mathbb{T}^n$, as specified below. Before getting to this, let us note one other implication of (1.13), namely, if

$$f_j(\theta) = \sum_k \varphi_j(k)e^{ik\cdot\theta} \tag{1.21}$$

are elements of $\mathcal{TP}$, or more generally, if $\varphi_j \in s(\mathbb{Z}^n)$, then we have the Parseval identity

$$(f_1, f_2)_{L^2} = \sum_{k\in\mathbb{Z}^n} \varphi_1(k)\overline{\varphi_2(k)}; \tag{1.22}$$

in particular, the Plancherel identity

$$\|f_j\|^2_{L^2} = \sum_{k\in\mathbb{Z}^n} |\varphi_j(k)|^2, \tag{1.23}$$

for $f_j \in \mathcal{TP}$, or more generally for any f_j of the form (1.21) with $\varphi_j \in s(\mathbb{Z}^n)$. In particular, the map $\mathcal{F}^*$ given by (1.12), and satisfying (1.11) and (1.17), has a unique continuous extension to $\ell^2(\mathbb{Z}^n)$, and

$$\mathcal{F}^* : \ell^2(\mathbb{Z}^n) \longrightarrow L^2(\mathbb{T}^n) \tag{1.24}$$

is an isometry of $\ell^2(\mathbb{Z}^n)$ onto its range. Part of the inversion formula will be that the map (1.24) is also surjective.

Let us note that if $f_j \in \mathcal{TP}$, satisfying (1.21), then (1.13) implies $\hat{f}_j(k) = \varphi_j(k)$, so we have directly in this case:

$$\mathcal{F}^*\mathcal{F} = I \text{ on } \mathcal{TP}. \tag{1.25}$$

One approach to more general inversion formulas would be to establish that $\mathcal{TP}$ is dense in various function spaces, on which $\mathcal{F}^*\mathcal{F}$ can be shown to act continuously. For more details on this approach, see the exercises at the end of §1 and §2 in the Functional Analysis appendix. Here, we will take a superficially different approach. We will make use of such basic results from real analysis as the denseness of $C(\mathbb{T}^n)$ in $L^p(\mathbb{T}^n)$, for $1 \le p < \infty$.

Our approach to (1.14) will be to establish the following *Abel summability* result. Consider

$$J_r f(\theta) = \sum_{k\in\mathbb{Z}^n} \hat{f}(k)\, r^{|k|}\, e^{ik\cdot\theta}, \tag{1.26}$$

where $|k| = |k_1| + \cdots + |k_n|$, $r \in [0, 1)$. We will show that

$$J_r f \to f, \quad \text{as } r \nearrow 1, \tag{1.27}$$

in the appropriate spaces. The operator J_r in (1.26) is defined for any $f \in L^1(\mathbb{T}^n)$, if $r < 1$, and we have the formula

$$J_r f(\theta) = (2\pi)^{-n} \int_{\mathbb{T}^n} f(\theta') \sum_{k\in\mathbb{Z}^n} r^{|k|}\, e^{ik\cdot(\theta-\theta')}\, d\theta'. \tag{1.28}$$

The sum over $\mathbb{Z}^n$ inside the integral can be written as

$$\begin{aligned}\sum_{k\in\mathbb{Z}^n} r^{|k|}\, e^{ik\cdot(\theta-\theta')} &= P_n(r, \theta-\theta') \\ &= p(r,\theta_1-\theta_1')\cdots p(r,\theta_n-\theta_n'),\end{aligned} \tag{1.29}$$

where

$$\begin{aligned}p(r,\theta) &= \sum_{k=-\infty}^{\infty} r^{|k|}\, e^{ik\theta} \\ &= 1 + \sum_{k=1}^{\infty} \left(r^k e^{ik\theta} + r^k e^{-ik\theta}\right) \\ &= \frac{1-r^2}{1-2r\cos\theta + r^2}.\end{aligned} \tag{1.30}$$

Then we have the explicit integral formula

$$\begin{aligned}J_r f(\theta) &= (2\pi)^{-n} \int_{\mathbb{T}^n} f(\theta') P_n(r,\theta-\theta')\, d\theta' \\ &= (2\pi)^{-n} \int_{\mathbb{T}^n} f(\theta-\theta') P_n(r,\theta')\, d\theta'.\end{aligned} \tag{1.31}$$

Let us examine $p(r,\theta)$. It is clear that the numerator and denominator on the right side of (1.30) are positive, so $p(r,\theta) > 0$ for each $r \in [0, 1)$, $\theta \in S^1$. Of course, as $r \nearrow 1$, the numerator tends to 0; as $r \nearrow 1$, the denominator tends to a nonzero limit, except at $\theta = 0$. Since it is clear that

$$(1.32) \qquad (2\pi)^{-1} \int_{S^1} p(r,\theta)\, d\theta = (2\pi)^{-1} \int_{-\pi}^{\pi} \sum r^{|k|} e^{ik\theta}\, d\theta = 1,$$

we see that, for r close to 1, $p(r,\theta)$ as a function of θ is highly peaked near $\theta = 0$ and small elsewhere, as in Fig. 1.1.

We are now prepared to prove the following result giving Abel summability (1.27).

Proposition 1.1. *If $f \in C(\mathbb{T}^n)$, then*

$$(1.33) \qquad J_r f \to f \ \text{ uniformly on } \mathbb{T}^n \text{ as } r \nearrow 1.$$

Furthermore, for any $p \in [1,\infty)$, if $f \in L^p(\mathbb{T}^n)$, then

$$(1.34) \qquad J_r f \to f \ \text{ in } L^p(\mathbb{T}^n) \text{ as } r \nearrow 1.$$

The proof of (1.33) is an immediate consequence of (1.31) and the peaked nature of $p(r,\theta)$ near $\theta = 0$ discussed above, together with the observation that, if f is continuous at θ, then it does not vary very much near θ. The convergence in (1.34) is in the L^p-norm, defined by

$$(1.35) \qquad \|g\|_{L^p} = \Big[(2\pi)^{-n} \int_{\mathbb{T}^n} |g(\theta)|^p\, d\theta\Big]^{1/p}.$$

We have the well-known triangle inequality in such a norm:

$$(1.36) \qquad \|g_1 + g_2\|_{L^p} \le \|g_1\|_{L^p} + \|g_2\|_{L^p},$$

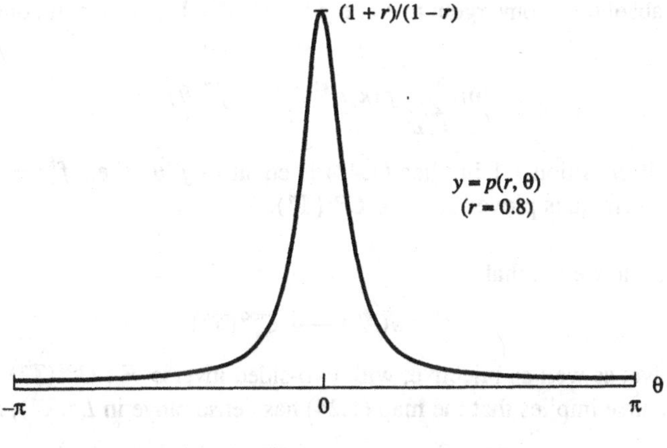

FIGURE 1.1

and this implies, via (1.31) and (1.32),

$$\begin{aligned} \|J_r f\|_{L^p} &= (2\pi)^{-n} \Big\| \int_{\mathbb{T}^n} P_n(r,\theta')\tau_{\theta'} f \, d\theta' \Big\|_{L^p} \\ &\le (2\pi)^{-n} \int P_n(r,\theta') \|\tau_{\theta'} f\|_{L^p} \, d\theta' \\ &= \|f\|_{L^p}, \end{aligned} \tag{1.37}$$

where

$$\tau_{\theta'} f(\theta) = f(\theta - \theta'), \tag{1.38}$$

which implies $\|\tau_{\theta'} f\|_{L^p} = \|f\|_{L^p}$. In other words,

$$\|J_r\|_{\mathcal{L}(L^p)} \le 1, \quad 1 \le p < \infty, \tag{1.39}$$

where we are using the operator norm on L^p:

$$\|T\|_{\mathcal{L}(L^p)} = \sup \{\|Tf\|_{L^p} : \|f\|_{L^p} \le 1\}. \tag{1.40}$$

Using this, we can deduce (1.34) from (1.33), and the denseness of $C(\mathbb{T}^n)$ in each space $L^p(\mathbb{T}^n)$, for $1 \le p < \infty$. Indeed, given $f \in L^p(\mathbb{T}^n)$, and given $\varepsilon > 0$, find $g \in C(\mathbb{T}^n)$ such that $\|f - g\|_{L^p} < \varepsilon$. Note that, generally, $\|g\|_{L^p} \le \|g\|_{\sup}$. Now we have

$$\begin{aligned} \|J_r f - f\|_{L^p} &\le \|J_r(f-g)\|_{L^p} + \|J_r g - g\|_{L^p} + \|g - f\|_{L^p} \\ &< \varepsilon + \|J_r g - g\|_{L^\infty} + \varepsilon, \end{aligned} \tag{1.41}$$

making use of (1.39). By (1.33), the middle term is $< \varepsilon$ if r is close enough to 1, so this proves (1.34).

Corollary 1.2. *If $f \in C^\infty(\mathbb{T}^n)$, then the Fourier inversion formula (1.14) holds.*

Proof. In such a case, as noted, we have $\hat{f} \in s(\mathbb{Z}^n)$, so certainly the right side of (1.14) is absolutely convergent to some $f^\# \in C(\mathbb{T}^n)$. In such a case, one a fortiori has

$$\lim_{r \nearrow 1} \sum_{k \in \mathbb{Z}^n} \hat{f}(k) r^{|k|} e^{ik\cdot\theta} = f^\#(\theta). \tag{1.42}$$

But now Proposition 1.1 implies (1.42) is equal to $f(\theta)$ (i.e., $f^\# = f$), so the inversion formula is proved for $f \in C^\infty(\mathbb{T}^n)$.

As a result, we see that

$$\mathcal{F}^* : s(\mathbb{Z}^n) \longrightarrow C^\infty(\mathbb{T}^n) \tag{1.43}$$

is surjective, as well as injective, with two-sided inverse $\mathcal{F} : C^\infty(\mathbb{T}^n) \to s(\mathbb{Z}^n)$. This of course implies that the map (1.24) has dense range in $L^2(\mathbb{T}^n)$; hence

$$\mathcal{F}^* : \ell^2(\mathbb{Z}^n) \longrightarrow L^2(\mathbb{T}^n) \text{ is unitary.} \tag{1.44}$$

Another way of stating this is

$$\{e^{ik\cdot\theta} : k \in \mathbb{Z}^n\} \text{ is an orthonormal basis of } L^2(\mathbb{T}^n), \tag{1.45}$$

with inner product given by (1.8). Also, the inversion formula

$$\mathcal{F}^*\mathcal{F} = I \quad \text{on } C^\infty(\mathbb{T}^n) \tag{1.46}$$

implies

$$\|\mathcal{F}f\|_{\ell^2} = \|f\|_{L^2}, \tag{1.47}$$

so therefore $\mathcal{F}$ extends by continuity from $C^\infty(\mathbb{T}^n)$ to a map

$$\mathcal{F} : L^2(\mathbb{T}^n) \longrightarrow \ell^2(\mathbb{Z}^n), \ \text{unitary}, \tag{1.48}$$

inverting (1.44). The denseness $C^\infty(\mathbb{T}^n) \subset L^2(\mathbb{T}^n) \subset L^1(\mathbb{T}^n)$ implies that this $\mathcal{F}$ coincides with the restriction to $L^2(\mathbb{T}^n)$ of the map (1.3). Note that the fact that (1.44) and (1.48) are inverses of each other extends the inversion result of Corollary 1.2.

We devote a little space to conditions implying that the Fourier series (1.14) is absolutely convergent, weaker than the hypothesis that $f \in C^\infty(\mathbb{T}^n)$. Note that since $|e^{ik\cdot\theta}| = 1$, the absolute convergence of (1.14) implies uniform convergence. By (1.4), we see that

$$f \in C^\ell(\mathbb{T}^n) \Longrightarrow |\hat{f}(k)| \le C\langle k\rangle^{-\ell}, \tag{1.49}$$

which in turn clearly gives absolute convergence provided

$$\ell \ge n + 1. \tag{1.50}$$

Using Plancherel's identity and Cauchy's inequality, we can do somewhat better:

Proposition 1.3. *If $f \in C^\ell(\mathbb{T}^n)$, then the Fourier series for f is absolutely convergent provided*

$$\ell > \frac{n}{2}. \tag{1.51}$$

Proof. We have

$$\begin{aligned}
\sum_k |\hat{f}(k)| &= \sum_k \langle k\rangle^{-\ell}\langle k\rangle^{\ell}|\hat{f}(k)| \\
&\le \Big[\sum_k \langle k\rangle^{-2\ell}\Big]^{1/2} \cdot \Big[\sum_k \langle k\rangle^{2\ell}|\hat{f}(k)|^2\Big]^{1/2} \\
&\le C\Big[\sum_k \langle k\rangle^{2\ell}|\hat{f}(k)|^2\Big]^{1/2},
\end{aligned} \tag{1.52}$$

as long as (1.51) holds. The square of the right side is dominated by

$$\begin{aligned}
C'\sum_k \sum_{|\gamma|\le\ell} \big|k^\gamma \hat{f}(k)\big|^2 &= C' \sum_{|\gamma|\le\ell} \|D^\gamma f\|_{L^2}^2 \\
&\le C''\|f\|_{C^\ell}^2,
\end{aligned} \tag{1.53}$$

so the proposition is proved.

Sharper results on absolute convergence of Fourier series will be given in Chapter 4. See also some of the exercises below for more on convergence when $n = 1$.

Exercises

1. Given $f, g \in L^1(\mathbb{T}^n)$, show that

$$\hat{f}(k)\hat{g}(k) = \hat{u}(k),$$

with

$$u(\theta) = (2\pi)^{-n} \int_{\mathbb{T}^n} f(\varphi) g(\theta - \varphi)\, d\varphi.$$

2. Given $f, g \in C(\mathbb{T}^n)$, show that

$$\widehat{(fg)}(k) = \sum_m \hat{f}(k - m)\hat{g}(m).$$

3. Using the proof of Proposition 1.3, show that every $f \in \text{Lip}(S^1)$ has an absolutely convergent Fourier series.
4. Show that for any $f \in L^1(\mathbb{T}^n)$, $\hat{f}(k) \to 0$ as $|k| \to \infty$.
(*Hint*: Given $\varepsilon > 0$, pick $f_\varepsilon \in C^\infty(\mathbb{T}^n)$, $\|f - f_\varepsilon\|_{L^1} < \varepsilon$. Compare $\hat{f}_\varepsilon(k)$ and $\hat{f}(k)$.)
This result is known as the Riemann-Lebesgue lemma.
5. For $f \in L^1(S^1)$, set

$$S_N f(\theta) = \sum_{k=-N}^{N} \hat{f}(k) e^{ik\theta}. \tag{1.54}$$

Show that $S_N f(\theta) = (1/2\pi) \int_{-\pi}^{\pi} f(\theta - \varphi) D_N(\varphi)\, d\varphi$, where

$$D_N(\theta) = \sum_{k=-N}^{N} e^{ik\theta} = \frac{\sin(N + \frac{1}{2})\theta}{\sin \frac{1}{2}\theta}. \tag{1.55}$$

(*Hint*: To evaluate the sum, recall how to sum a finite geometrical series.)
$D_N(\theta)$ is called the Dirichlet kernel. See Fig. 1.2.
6. Let $f \in L^1(S^1)$ have the following property of "vanishing" at $\theta = 0$:

$$\frac{f(\theta)}{\sin \frac{1}{2}\theta} = g(\theta) \in L^1(-\pi, \pi).$$

Show that $S_N f(0) \to 0$ as $N \to \infty$.
(*Hint*. Adapt the Riemann-Lebesgue lemma to show that

$$g \in L^1(-\pi, \pi) \Rightarrow \int_{-\pi}^{\pi} g(\theta) \sin(N + \tfrac{1}{2})\theta\, d\theta \to 0 \text{ as } N \to \infty.)$$

7. Deduce that if $f \in L^1(S^1)$ is Lipschitz continuous at θ_0, then $S_N f(\theta_0) \to f(\theta_0)$ as $N \to \infty$. Furthermore, if f is Lipschitz on an open interval $J \subset S^1$, then $S_N f \to f$ uniformly on compact subsets of J.

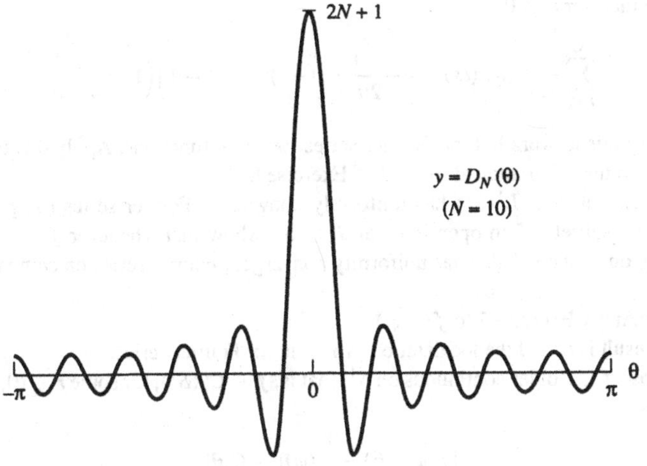

FIGURE 1.2

8. Let $f \in L^\infty(S^1)$ be piecewise Lipschitz, with a finite number of simple jumps. Show that $S_N f(\theta) \to f(\theta)$ at points of continuity. If f has a jump at θ_j, with limiting values $f_\pm(\theta_j)$, show that

$$S_N f(\theta_j) \to \frac{1}{2}[f_+(\theta_j) + f_-(\theta_j)], \tag{1.56}$$

as $N \to \infty$.
(*Hint*: By Exercise 7, it remains only to establish (1.56). Show that this can be reduced to the case $\theta_j = \pi,\ f(\theta) = \theta$, for $-\pi \le \theta < \pi$. Verify that this function has Fourier series

$$2 \sum_{k=1}^{\infty} \frac{(-1)^k}{k} \sin k\theta.)$$

Alternative: Reduce to the case $\theta_j = 0$ and note that $S_N f(0)$ depends only on the even part of f, $(1/2)[f(\theta) + f(-\theta)]$.

9. Work out the Fourier series of the function $f \in \text{Lip}(S^1)$ given by

$$f(\theta) = |\theta|, \quad -\pi \le \theta \le \pi.$$

Examining this at $\theta = 0$, establish that

$$\sum_{k=1}^{\infty} \frac{1}{k^2} = \frac{\pi^2}{6}. \tag{1.57}$$

10. One can obtain Fourier coefficients of functions θ^k and $|\theta|^k$ on $[-\pi, \pi]$ in terms of the Fourier coefficients of

$$q_k(\theta) = \theta^k \quad \text{on } [0, \pi],$$
$$0 \quad \text{on } [-\pi, 0].$$

Show that, for $n \neq 0$,

$$\sum_{k=0}^{\infty} \frac{1}{k!} \hat{q}_k(n)\,(is)^k = -\frac{1}{2\pi i n}\Big[(-1)^n \cdot e^{\pi i s} - 1\Big]\Big(1 - \frac{s}{n}\Big)^{-1},$$

and use this to work out the Fourier series for these functions. Apply this to Exercise 9, and to the calculation at the end of Exercise 8.

11. Assume that $g \in L^1(S^1)$ has uniformly convergent Fourier series ($S_N g \to g$) on compact subsets of an open interval $J \subset S^1$. Show that whenever $f \in L^1(S^1)$ and $f = g$ on J, then f also has uniformly convergent Fourier series on compact subsets of J.
(*Hint*: Apply Exercise 7 to $f - g$.)
This result is called the localization principle for Fourier series.
12. Suppose f is Hölder continuous on S^1, that is, $f \in C^r(S^1)$, for some $r \in (0, 1)$, which means

$$|f(\varphi + \theta) - f(\varphi)| \le C|\theta|^r.$$

Show that f has uniformly convergent Fourier series on S^1.
(*Hint*: We have

$$\left|\frac{f(\theta + \varphi) - f(\varphi)}{\sin \frac{1}{2}\theta}\right| \le C\left|\sin \tfrac{1}{2}\theta\right|^{-(1-r)}.$$

Apply Exercise 6.)
13. If $\omega : [0, \infty) \to [0, \infty)$ is continuous and increasing and $\omega(0) = 0$, we say a function f on S^1 is continuous with modulus of continuity ω provided

$$|f(\varphi + \theta) - f(\varphi)| \le C\omega(|\theta|).$$

Formulate the most general condition you can to establish uniform convergence of the Fourier series of a function with such a modulus of continuity. Note that Exercise 12 deals with the case $\omega(s) = s^r$, $r \in (0, 1)$.
14. Consider the *Cesaro sum* of the Fourier series of f:

$$C_N f(\theta) = \sum_{k=-N}^{N} \Big(1 - \frac{|k|}{N}\Big)\hat{f}(k)e^{ik\theta} = \frac{1}{N}\sum_{\ell=0}^{N-1} S_\ell f(\theta).$$

Show that $C_N f(\theta) = (1/2\pi)\int_{-\pi}^{\pi} f(\theta - \varphi)F_N(\varphi)d\varphi$, where

$$F_N(\theta) = \frac{1}{N}\sum_{\ell=0}^{N-1} D_\ell(\theta) = \frac{1}{N \sin \frac{1}{2}\theta}\sum_{\ell=0}^{N-1} \sin\Big(\ell + \frac{1}{2}\Big)\theta = \frac{1}{N}\left(\frac{\sin \frac{N}{2}\theta}{\sin \frac{1}{2}\theta}\right)^2. \tag{1.58}$$

The function $F_N(\theta)$ is called the Fejer kernel (see Fig. 1.3). Modify the proof of Proposition 1.1 to show that

$$C_N f \to F \text{ in } B, \quad \text{for } f \in B,$$

where B is one of the Banach spaces $C(S^1)$ or $L^p(S^1)$, $1 \le p < \infty$.
(*Hint*: To evaluate the second sum in (1.58), use $\sin(\ell + \frac{1}{2})\theta = \operatorname{Im} e^{i\theta/2}e^{i\ell\theta}$ and sum a finite geometrical series. Also use the identity $2\sin^2 z = 1 - \cos 2z$.)

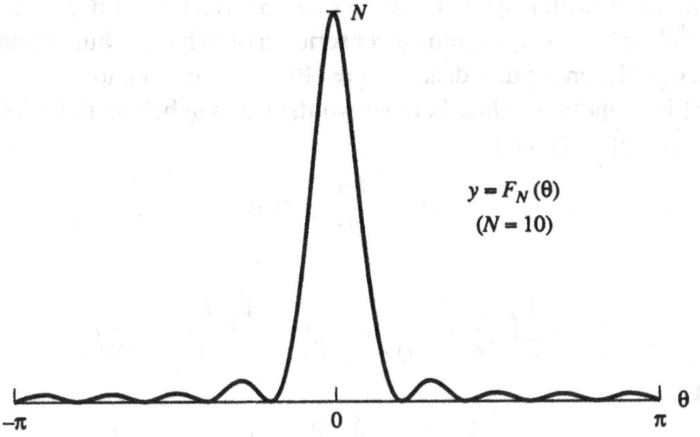

FIGURE 1.3

2. Harmonic functions and holomorphic functions in the plane

The method of proof of the Abel summability (1.26)–(1.27) of Fourier series, specialized to $\mathbb{T}^1 = S^1$, has important implications for the theory of harmonic functions on a domain $\Omega \subset \mathbb{R}^2$, which we will discuss here. In the case of S^1, let us rewrite (1.26),

$$J_r f(\theta) = \sum_{k=-\infty}^{\infty} \hat{f}(k)\, r^{|k|}\, e^{ik\theta}, \tag{2.1}$$

as

$$(\mathrm{PI}\, f)(r, \theta) = \sum_{k=-\infty}^{\infty} \hat{f}(k)\, r^{|k|}\, e^{ik\theta}. \tag{2.2}$$

The function $u(r, \theta) = \mathrm{PI}\, f(r, \theta)$ is called the *Poisson integral* of f. If we use polar coordinates in the complex plane $\mathbb{C} = \mathbb{R}^2$,

$$z = r\, e^{i\theta}, \tag{2.3}$$

then (2.2) becomes

$$\begin{aligned}(\mathrm{PI}\, f)(z) &= \sum_{k=0}^{\infty} \hat{f}(k) z^k + \sum_{k=1}^{\infty} \hat{f}(-k) \bar{z}^k \\ &= (\mathrm{PI}_+ f)(z) + (\mathrm{PI}_- f)(z),\end{aligned} \tag{2.4}$$

defined on the unit disk $|z| < 1$. Note also, from (1.30), that

$$\mathrm{PI}\, f(z) = \frac{1 - |z|^2}{2\pi} \int_{S^1} \frac{f(w)}{|w - z|^2}\, ds(w), \tag{2.5}$$

the integral being with respect to arclength on S^1. Recall that if $f \in L^1(S^1)$, the function $\hat{f}(k)$ is bounded, so both power series in (2.4) have radius of convergence at least 1. Clearly, on the unit disk, $v(z) = (\mathrm{PI}_+ f)(z)$ is holomorphic and $w(z) = (\mathrm{PI}_- f)(z)$ is antiholomorphic. In other words, v and w belong to $C^\infty(\mathcal{D})$, where $\mathcal{D} = \{z \in \mathbb{C} : |z| < 1\}$, and

$$\frac{\partial v}{\partial \bar{z}} = 0, \quad \frac{\partial w}{\partial z} = 0 \text{ on } \mathcal{D}, \tag{2.6}$$

where

$$\frac{\partial}{\partial \bar{z}} = \frac{1}{2}\Big(\frac{\partial}{\partial x} + i\frac{\partial}{\partial y}\Big), \quad \frac{\partial}{\partial z} = \frac{1}{2}\Big(\frac{\partial}{\partial x} - i\frac{\partial}{\partial y}\Big). \tag{2.7}$$

Note that

$$\frac{\partial}{\partial z}\frac{\partial}{\partial \bar{z}} = \frac{\partial}{\partial \bar{z}}\frac{\partial}{\partial z} = \frac{1}{4}\Delta,$$

where Δ is the Laplace operator on $\mathbb{R}^2$, a special case of the Laplace operator introduced in Chapter 2. Since $v, w \in C^\infty(\mathcal{D})$, we have $\Delta v = 0$ and $\Delta w = 0$, and hence

$$\Delta(\mathrm{PI}\, f) = 0. \tag{2.8}$$

In light of the results of §1, we have the following.

Proposition 2.1. *If $f \in C(S^1)$, then*

$$u = PI\, f(z) \in C^\infty(\mathcal{D}) \cap C(\overline{\mathcal{D}}) \tag{2.9}$$

is harmonic, with boundary value f, that is, u solves the Dirichlet problem

$$\Delta u = 0 \textit{ in } \mathcal{D}, \quad u|_{\partial\mathcal{D}} = f. \tag{2.10}$$

One should expect that if f has extra smoothness on S^1, so does PI f on $\overline{\mathcal{D}}$. The following result is crude compared to results established in Chapters 4 and 5, but it will be of some interest.

Proposition 2.2. *For $\ell = 1, 2, 3, \ldots$, we have*

$$PI : C^{\ell+1}(S^1) \longrightarrow C^\ell(\overline{\mathcal{D}}). \tag{2.11}$$

Proof. We begin with the case $\ell = 1$. Since we know from (2.4) that PI $f \in C^\infty(\mathcal{D})$, we need merely check smoothness near $\partial\mathcal{D} = S^1$. Clearly,

$$\frac{\partial}{\partial\theta}\mathrm{PI}\, f = \mathrm{PI}\,\frac{\partial f}{\partial\theta}, \tag{2.12}$$

so if $f \in C^1(S^1)$, then $\partial f/\partial\theta \in C(S^1)$ and we have $(\partial/\partial\theta)\mathrm{PI}\, f$ continuous on $\overline{\mathcal{D}}$. Also, by (2.2),

$$r\frac{\partial}{\partial r}\mathrm{PI}\, f = \mathrm{PI}\,(Nf), \tag{2.13}$$

where Nf is characterized by the Fourier series representation

$$Nf(\theta) = \sum_{k=-\infty}^{\infty} \hat{f}(k)|k|e^{ik\theta}. \tag{2.14}$$

Thus

$$Nf = -i\frac{\partial}{\partial\theta}Hf = -iH\frac{\partial f}{\partial\theta}, \tag{2.15}$$

where H has the Fourier series representation

$$Hg(\theta) = \sum_{k=-\infty}^{\infty} (\operatorname{sgn} k)\hat{g}(k)e^{ik\theta}. \tag{2.16}$$

We claim that, for $\ell \geq 0$,

$$H : C^{\ell+1}(S^1) \longrightarrow C^{\ell}(S^1), \tag{2.17}$$

and hence $N : C^{\ell+2}(S^1) \to C^{\ell}(S^1)$. Given this, for $f \in C^2(S^1)$, the quantity (2.13) is seen to belong to $C(\overline{\mathcal{D}})$, and this finishes the $\ell = 1$ case of (2.11).

In turn, since H commutes with $\partial/\partial\theta$, it suffices to establish the $\ell = 0$ case of (2.17). Now, by Proposition 1.3, the Fourier series for $g \in C^1(S^1)$ is absolutely convergent, giving (2.17).

To prove the general case of (2.11), a short calculation yields

$$\Bigl(r\frac{\partial}{\partial r}\Bigr)^j\Bigl(\frac{\partial}{\partial\theta}\Bigr)^k \mathrm{PI}\, f = \mathrm{PI}\Bigl(\Bigl(\frac{\partial}{\partial\theta}\Bigr)^k N^j f\Bigr). \tag{2.18}$$

Note that $N^j = (-i)^j(\partial/\partial\theta)^j H^{o(j)}$, where $o(j)$ is zero if j is even and one if j is odd, so, for $\ell \geq 0$,

$$N^j : C^{\ell+j+1}(S^1) \longrightarrow C^{\ell}(S^1),$$

the left side being improved to $C^{\ell+j}$ if j is even. Therefore, if $f \in C^{\ell+1}(S^1)$ and $j + k = \ell$, the right side of (2.18) is $PI\ f_{jk}$ with $f_{jk} \in C(S^1)$, which proves (2.11) in general.

The implication $f \in C^{\ell}(S^1) \Rightarrow \mathrm{PI}\, f \in C^{\ell}(\overline{\mathcal{D}})$ does not quite work, as we will see later, essentially because H does not map $C^{\ell}(S^1)$ to itself. It is true that $f \in C^{\ell,\alpha}(S^1) \Rightarrow \mathrm{PI}\, f \in C^{\ell,\alpha}(\overline{\mathcal{D}})$, for $\alpha \in (0, 1)$. This is a special case of Hölder estimates that will be established in §7 of Chapter 13. Similarly there are "sharp" results on regularity of PI f in Sobolev spaces, discussed in Chapter 4, and in much greater generality, in Chapter 5.

It is important to know that PI f provides the *unique* solution to the Dirichlet problem (2.10). We will establish several general uniqueness results, starting with the following.

Proposition 2.3. *Let $\Omega \subset \mathbb{R}^n$ be a bounded region with smooth boundary, say, $\Omega = \mathcal{D}$. Suppose $u, v \in C^2(\overline{\Omega})$, with $u = v = f$ on $\partial\Omega$, and $\Delta u = \Delta v = 0$ in Ω. Then $u = v$ on all of Ω.*

Proof. Set $w = u - v \in C^2(\overline{\Omega})$; $w = 0$ on $\partial\Omega$. We can apply the Green identity (3.15) of Chapter 2, to write

$$(dw, dw) = -(\Delta w, w) + \int_{\partial\Omega} w \frac{\partial \overline{w}}{\partial \nu} \, dS. \tag{2.19}$$

By hypothesis the right side of (2.19) is 0. Thus w is constant on each component of Ω, and the boundary condition forces $w = 0$.

In view of Proposition 2.2, we could apply this to $u = \mathrm{PI}\, f$ if $f \in C^3(S^1)$, but this is not a satisfactory result, and we will do much better below.

A result related to our uniqueness question is the *mean-value property*, a special case of which is the following.

Proposition 2.4. *If $f \in C(S^1)$, $u = PI\, f$, then*

$$u(0) = \frac{1}{2\pi} \int_{-\pi}^{\pi} f(\theta) \, d\theta. \tag{2.20}$$

Proof. It follows from the series (2.2) that $u(0) = \hat{f}(0)$, which gives (2.20).

A more general result is the following.

Proposition 2.5. *If $B_R \subset \mathbb{R}^n$ is the open ball of radius R, centered at the origin, with $\partial B_R = S_R$, of area $A(R)$, then for $u \in C^2(B_R) \cap C(\overline{B}_R)$, $\Delta u = 0$, we have*

$$u(0) = \frac{1}{A(R)} \int_{S_R} u(x) \, dS. \tag{2.21}$$

Proof. We apply Green's identity

$$\int_{\Omega} \left[u \Delta v - v \Delta u \right] dx = \int_{\partial\Omega} \left[u \frac{\partial v}{\partial \nu} - v \frac{\partial u}{\partial \nu} \right] dS, \tag{2.22}$$

to $\Omega = B_r$, $0 < r < R$, $v(x) = |x|^2$, with $\Delta v = 2n$, to get, when $\Delta u = 0$,

$$n \int_{B_r} u(x) \, dx = r \int_{S_r} u(x) \, dS, \tag{2.23}$$

noting that substituting $v = 1$ in (2.22) gives $\int_{\partial\Omega} (\partial u / \partial \nu) \, dS = 0$. If we let $\varphi(r) = \int_{B_r} u(x) \, dx$, this implies $\varphi'(r) = (n/r)\varphi(r)$, and hence $\varphi(r) = K r^n$, i.e., $V(r)^{-1} \int_{B_r} u(x) \, dx$ is constant. Passing to the limit $r \to 0$ gives (2.21).

Corollary 2.6. *For any $\Omega \subset \mathbb{R}^n$ open, any $u \in C^2(\Omega)$ harmonic, any ball B_p, centered at p and contained in Ω, we have*

$$u(p) = \mathrm{Avg}_{\partial B_p} \, u(z). \tag{2.24}$$

We can now prove the following important *maximum principle* for harmonic functions. Much more general versions of this will be given in Chapter 5.

Proposition 2.7. *Let $\Omega \subset \mathbb{R}^n$ be connected and open, and let $u \in C^2(\Omega)$ be harmonic and real-valued. Then u has no interior maximum, or minimum, unless u is constant. In particular, if Ω is bounded and $u \in C(\overline{\Omega})$, then*

$$\sup_{p\in\Omega} u(p) = \sup_{q\in\partial\Omega} u(q). \tag{2.25}$$

Also, even for u complex-valued,

$$\sup_{p\in\Omega} |u(p)| = \sup_{q\in\partial\Omega} |u(q)|. \tag{2.26}$$

Proof. That a non-constant, real harmonic function has no interior extremum is an obvious consequence of (2.24), and the other consequences, (2.25) and (2.26), follow immediately.

Corollary 2.8. *The uniqueness result of Proposition 2.3 holds for any bounded open $\Omega \subset \mathbb{R}^n$, with no smoothness on $\partial\Omega$, and for any harmonic*

$$u, v \in C^2(\Omega) \cap C(\overline{\Omega}). \tag{2.27}$$

Proof. Apply the maximum principle to $u - v$.

Here, we have used the mean-value property to prove the maximum principle. The more general maximum principle established in Chapter 5 will not use the mean-value property; indeed, together with a symmetry argument, it can be made a basis for a proof of the mean-value property. We will leave these considerations until Chapter 5.

With our uniqueness result in hand, we can easily establish the following interior regularity result for harmonic functions.

Proposition 2.9. *Let $\Omega \subset \mathbb{R}^2$ be open, and let $u \in C^2(\Omega)$ be harmonic. Then in fact, $u \in C^\infty(\Omega)$; u is even real analytic on Ω.*

Proof. By translations and dilations, we can reduce to the case $\Omega = \mathcal{D}$, $u \in C^2(\overline{\mathcal{D}})$. The uniqueness result of Corollary 2.8 implies

$$u = \text{PI}\, f, \quad \text{where } f = u|_{S^1}. \tag{2.28}$$

Then the conclusion that u is real analytic on $\mathcal{D}$ follows directly from the power-series expansion (2.4).

Parenthetically, we remark that, by Corollary 2.8, the identity (2.28) holds for any $u \in C^2(\mathcal{D}) \cap C(\overline{\mathcal{D}})$ harmonic in $\mathcal{D}$.

Using the results we have developed, via Fourier series, about harmonic functions, we can quickly draw some basic conclusions about holomorphic functions. If $\Omega \subset \mathbb{C}$ is open, $f : \Omega \to \mathbb{C}$ is by definition holomorphic if and only if $f \in C^1(\Omega)$ and $\partial f/\partial \bar{z} = 0$, where $\partial/\partial \bar{z}$ is given by (2.7). Clearly, if $f \in C^2(\Omega)$ is holomorphic, then it is also harmonic, and so are its real and imaginary parts. Suppose $u \in C^2(\mathcal{D}) \cap C(\overline{\mathcal{D}})$ is holomorphic in $\mathcal{D}$. Then the series representation (2.4) is valid, since (2.28) holds. This series is a sum of two terms:

$$u(z) = \sum_{k=0}^{\infty} \hat{f}(k) z^k + \sum_{k=1}^{\infty} \hat{f}(-k) \bar{z}^k \tag{2.29}$$

$$= u_1(z) + u_2(z),$$

where $\partial u_1/\partial \bar{z} = 0$ and $\partial u_2/\partial z = 0$. But if we are given that $\partial u/\partial \bar{z} = 0$, then also $\partial u_2/\partial \bar{z} = 0$, so $\partial u_2/\partial x = \partial u_2/\partial y = 0$. Thus u_2 is constant, and since $u_2(0) = 0$, this forces $u_2 = 0$. In other words, the holomorphic function $u(z)$ has the power series

$$u(z) = \sum_{k=0}^{\infty} a_k z^k, \quad z \in \mathcal{D}, \tag{2.30}$$

where $a_k = \hat{f}(k),\ k \ge 0$. Note that differentiation of (2.30) gives

$$a_k = \frac{1}{k!} \Big(\frac{\partial}{\partial z}\Big)^k u(0). \tag{2.31}$$

By the usual method of translating and dilating coordinates, we deduce the following.

Proposition 2.10. *If $\Omega \subset \mathbb{C}$ is open, $u \in C^2(\Omega)$ holomorphic, $p \in \Omega$, and $\mathcal{D}_p$ a disk centered at p, $\overline{\mathcal{D}}_p \subset \Omega$, then on $\mathcal{D}_p$, $u(z)$ is given by a convergent power-series expansion*

$$u(z) = \sum_{k=0}^{\infty} b_k (z-p)^k. \tag{2.32}$$

We can relax the C^2-hypothesis to $u \in C^1(\Omega)$. As much stronger and more general results are given in Chapter 5, we omit the details here.

For further use, we record the following result, whose proof is trivial.

Lemma 2.11. *If $u \in C^\infty(\Omega)$ is holomorphic, and*

$$P = \sum a_{jk} D_x^j D_y^k$$

is any constant-coefficient differential operator, then Pu is holomorphic in Ω.

Proof. $(\partial/\partial \bar{z}) Pu = P(\partial u/\partial \bar{z})$.

We can use the power-series representation (2.30)–(2.32) to prove the fundamental result on uniqueness of analytic continuation, which we give below. Here is the first result, of a very general nature.

Proposition 2.12. *Let $\Omega \subset \mathbb{R}^n$ be open and connected, and let u be a real-analytic function on Ω. If $p \in \Omega$ and all derivatives $D^\alpha u(p) = 0$, then $u = 0$ on all of Ω.*

Proof. Let $\mathcal{K} = \{x \in \Omega : D^\alpha u(x) = 0 \text{ for all } \alpha \geq 0\}$. Since $u \in C^\infty(\Omega)$, $\mathcal{K}$ is closed in Ω. However, for each $p \in \mathcal{K}$, since u is given in a neighborhood of p by a power series

$$u(q) = \sum_{\alpha \geq 0} \frac{u^{(\alpha)}(p)}{\alpha!} (q-p)^\alpha,$$

we also see that $\mathcal{K}$ is open in Ω. This proves the proposition.

Our basic corollary for holomorphic functions is the following.

Corollary 2.13. *Let $\Omega \subset \mathbb{C}$ be open and connected, u holomorphic on Ω. Let γ be a line segment contained in Ω. If $u|_\gamma = 0$, then $u = 0$ on Ω.*

Proof. Translating and rotating, we can assume γ is a segment in the real axis, with $0 \in \gamma$. Near 0, $u(z)$ has a power-series expansion of the form (2.30), with a_k given by (2.31). Using Lemma 2.11, we see that

$$\left(\frac{\partial}{\partial z}\right)^k u(0) = \left(\frac{\partial}{\partial x}\right)^k u(0), \tag{2.33}$$

which vanishes for all k. Thus $u = 0$ on a nonempty open set in Ω, and the rest follows by Proposition 2.12.

Actually, a much stronger result is true. If $\Omega \subset \mathbb{C}$ is connected, $p_j \in \Omega$ are distinct, $p_j \to p \in \Omega$, u is holomorphic in Ω, and $u(p_j) = 0$ for each j, then u must vanish identically. In other words, u can have only isolated zeros if it does not vanish identically. Indeed, say $u(p) = 0$. If u is not identically zero, some coefficient in the series (2.32) is nonzero; let b_m be the first such coefficient:

$$\begin{aligned} u(z) &= (z-p)^m \sum_{k=0}^{\infty} b_{m+k} (z-p)^k \\ &= (z-p)^m v(z), \end{aligned} \tag{2.34}$$

where $v(z)$ is holomorphic on $\mathcal{D}_p$ and $v(0) = b_m \neq 0$. Thus, by continuity, $v(\zeta) \neq 0$ for $|\zeta - p| < \varepsilon$ if ε is sufficiently small, which implies $u(\zeta) \neq 0$ if $\zeta \neq p$ but $|\zeta - p| < \varepsilon$.

A typical use of Corollary 2.13 is in computations of integrals. We will see an example of this in the next section.

We end this section by recalling the classical Cauchy integral theorem and integral formula. Throughout, Ω will be a bounded open domain in $\mathbb{C}$ with smooth boundary. Stokes' formula, proved in Chapter 1, §13, states

$$\iint_{\Omega} d\alpha = \int_{\partial\Omega} \alpha, \tag{2.35}$$

for a 1-form α with coefficients in $C^1(\overline{\Omega})$. If $\alpha = p\,dx + q\,dy$, this gives the classical Green's formula

$$\int_{\partial\Omega} p\,dx + q\,dy = \iint_{\Omega} \Big(\frac{\partial q}{\partial x} - \frac{\partial p}{\partial y}\Big)\,dx\,dy. \tag{2.36}$$

If $u(x, y) \in C^1(\overline{\Omega})$ is a complex-valued function, we consequently have

$$\begin{aligned} \int_{\partial\Omega} u\,dz &= \int_{\partial\Omega} (u\,dx + iu\,dy) \\ &= \iint_{\Omega} \Big(i\frac{\partial u}{\partial x} - \frac{\partial u}{\partial y}\Big)\,dx\,dy \\ &= 2i \iint_{\Omega} \frac{\partial u}{\partial \overline{z}}\,dx\,dy. \end{aligned} \tag{2.37}$$

In the special case when u is holomorphic in Ω, we have the Cauchy integral theorem:

Theorem 2.14. *If $\Omega \subset \mathbb{C}$ is bounded with smooth boundary and $u \in C^1(\overline{\Omega})$ is holomorphic, then*

$$\int_{\partial\Omega} u(z)\,dz = 0. \tag{2.38}$$

Using various limiting arguments, one can relax the hypotheses on smoothness of $\partial\Omega$ and of u near $\partial\Omega$; we won't go into this here. Next we prove Cauchy's integral formula.

Proposition 2.15. *With Ω as above, $u \in C^2(\overline{\Omega})$ holomorphic in Ω, we have*

$$u(\zeta) = \frac{1}{2\pi i} \int_{\partial\Omega} \frac{u(z)}{\zeta - z}\,dz, \quad \textit{for } \zeta \in \Omega. \tag{2.39}$$

Proof. Write

$$\begin{aligned} u(z)(\zeta - z)^{-1} &= (\zeta - z)^{-1}\big[u(z) - u(\zeta)\big] + u(\zeta)(\zeta - z)^{-1} \\ &= v(z) + u(\zeta)(\zeta - z)^{-1}. \end{aligned} \tag{2.40}$$

By the series expansion for $u(z)$ about ζ, we see that $v(z)$ is holomorphic near ζ; clearly, it is holomorphic on the rest of Ω, and it belongs to $C^1(\overline{\Omega})$, so $\int_{\partial\Omega} v(z)\,dz = 0$. Thus, to prove (2.39), it suffices to show that

$$\int_{\partial\Omega} (\zeta - z)^{-1}\,dz = 2\pi i, \quad \text{for } \zeta \in \Omega. \tag{2.41}$$

Indeed, if ε is small enough that $B(\zeta, \varepsilon) = \{z \in \mathbb{C} : |z - \zeta| \le \varepsilon\}$ is contained in Ω, then Cauchy's theorem implies that the left side of (2.41) is equal to

$$\int_{\partial B(\zeta,\varepsilon)} (\zeta - z)^{-1}\,dz \tag{2.42}$$

since $(\zeta - z)^{-1}$ is holomorphic in z for $z \neq \zeta$. Making a change of variable, we see that (2.42) is equal to

$$\int_0^{2\pi} \left(\varepsilon\, e^{i\theta}\right)^{-1} i\varepsilon\, e^{i\theta}\,d\theta = 2\pi i,$$

so the proof is complete.

As stated before, the C^2-hypothesis can be relaxed to C^1.

A function $u(z)$ of the form $u(z) = v(z)/(z-a)^k$, $k \in \mathbb{Z}^+$, where v is holomorphic on a neighborhood $\mathcal{O}$ of a, is said to have a pole of order k at $z = a$ if $v(a) \neq 0$. In such a case, a variant of the preceding calculations yields

$$\frac{1}{2\pi i}\int_\gamma u(z)\,dz = \frac{1}{(k-1)!}v^{(k-1)}(a),$$

the coefficient of $(z-a)^{k-1}$ in the power series of $v(z)$ about $z = a$, if γ is a smooth, simple, closed curve about a such that v is holomorphic on a neighborhood of the closed region bounded by γ. This quantity is called the *residue* of $u(z)$ at $z = a$.

One can often evaluate integrals by evaluating residues. We give a simple illustration here; others are given in (2.48), (3.32), (A.14), and (A.15). Here we evaluate

$$\int_{-\infty}^{\infty} \frac{dx}{1+x^2} = \lim_{R\to\infty} \int_{\gamma_R} \frac{dz}{1+z^2}, \tag{2.43}$$

where γ_R is the closed curve, going from $-R$ to R along the real axis, then from R to $-R$ counterclockwise on the circle of radius R centered at 0, that is, $\gamma_R = \partial\mathcal{O}_R$, where $\mathcal{O}_R = \{z : \operatorname{Re} z > 0, |z| < R\}$. There is just one pole of $(1+z^2)^{-1}$ in $\mathcal{O}_R$, located at $z = i$. Since $(1+z^2)^{-1} = (z+i)^{-1}(z-i)^{-1}$, we see that the residue of $(1+z^2)^{-1}$ at $z = i$ is $1/2i$, so

$$\int_{-\infty}^{\infty} \frac{dx}{1+x^2} = \pi.$$

Exercises

1. Suppose u satisfies the following Neumann boundary problem in the disk $\mathcal{D}$:

$$\Delta u = 0 \text{ in } \mathcal{D}, \quad \frac{\partial u}{\partial r} = g \text{ on } S^1. \tag{2.44}$$

If $u = \mathrm{PI}(f)$, show that f and g must be related by $\hat{g}(k) = |k|\hat{f}(k)$, for all $k \in \mathbb{Z}$, that is,

$$g = Nf, \tag{2.45}$$

with N defined by (2.13).

2. Define the function k_N by

$$k_N(\theta) = \sum_{k\neq 0} |k|^{-1} e^{ik\theta}.$$

Show that $k_N \in L^2(S^1) \subset L^1(S^1)$. Also show that, provided $g \in L^2(S^1)$ and

$$\int_{S^1} g(\theta)\, d\theta = 0, \tag{2.46}$$

a solution to (2.45) is given by

$$f(\theta) = (2\pi)^{-1} \int_{S^1} k_N(\theta - \varphi) g(\varphi)\, d\varphi = Tg(\theta). \tag{2.47}$$

3. If T is defined by (2.47), show that $T : L^p(S^1) \to L^p(S^1)$ for $p \in [1, \infty)$ and $T : C^\ell(S^1) \to C^\ell(S^1)$ for $\ell = 0, 1, 2, \dots$.
4. Given $g \in C^1(S^1)$, show that (2.44) has a solution $u \in C^1(\overline{\mathcal{D}})$ if and only if (2.46) holds. If $g \in C^\ell(S^1)$, show that (2.44) has a solution $u \in C^\ell(\overline{\mathcal{D}})$.
Note: Regularity results of a more precise nature are given in Chapter 4, §4, in Exercise 1. See also Chapter 5, §7, for more general results.
5. Let $\Omega \subset \mathbb{R}^2$ be a smooth, bounded, connected region. Show that if $w \in C^2(\overline{\Omega})$, $\Delta w = 0$ on Ω, and $\partial w/\partial \nu = 0$ on $\partial\Omega$, then w is constant. (*Hint*: Use (2.19).)
Note: One can weaken the C^2-hypothesis to $w \in C^1(\overline{\Omega})$; compare Proposition 2.2 of Chapter 5. For another type of relaxation, see Chapter 4, §4, Exercise 3.
6. Show that a C^1-function $f : \Omega \to \mathbb{C}$ is holomorphic if and only if, at each $z \in \Omega$, $Df(z)$, a priori a real linear map on $\mathbb{R}^2$, is in fact complex linear on $\mathbb{C}$.
Note: This exercise has already been given in Chapter 1, §1.
7. Let f be a holomorphic function on $\Omega \subset \mathbb{C}$, with $f : \Omega \to \mathcal{O}$, and let u be harmonic on $\mathcal{O}$. Show that $v = u \circ f$ is harmonic on Ω. (*Hint*: For a short proof, write u locally as a sum of a holomorphic and an anti-holomorphic function.)
8. Let $g(z) = \sum_1^\infty a_k z^k$, and form the harmonic function $u = 2\,\mathrm{Re}\, g = g + \overline{g}$. Show that, under appropriate hypotheses on (a_k), $g|_{S^1} = P_+(u|_{S^1})$, where P_+ is given by (2.18).
9. Find a holomorphic function on $\mathcal{D}$ that is *unbounded* but whose real part has a continuous extension to $\overline{\mathcal{D}}$.
Reconsider this problem after reading §6 of Chapter 5.
10. Hence show that P_+ does not map $C(S^1)$ to itself, nor does it map any $C^\ell(S^1)$ to itself, for any integer $\ell \geq 0$.
11. Hence find $f \in C^1(S^1)$ such that $\mathrm{PI}\, f \notin C^1(\overline{\mathcal{D}})$.

12. Use the method of residues to calculate

$$\int_{-\infty}^{\infty} \frac{e^{i\xi x}}{1+x^2}\, dx, \quad \xi \ge 0. \tag{2.48}$$

(*Hint*: Write this as $\lim_{R\to\infty} \int_{\gamma_R} e^{i\xi z}/(1+z^2)\, dz$, with γ_R as in (2.43), given $\xi \ge 0$. Then find the residue of $e^{i\xi z}/(1+z^2)$ at $z = i$.)

3. The Fourier transform

The Fourier transform is defined by

$$\mathcal{F}f(\xi) = \hat{f}(\xi) = (2\pi)^{-n/2} \int f(x) e^{-ix\cdot\xi}\, dx \tag{3.1}$$

when $f \in L^1(\mathbb{R}^n)$. It is clear that

$$\mathcal{F} : L^1(\mathbb{R}^n) \longrightarrow L^\infty(\mathbb{R}^n). \tag{3.2}$$

This is analogous to (1.3). The analogue for $C^\infty(\mathbb{T}^n)$, and simultaneously for $s(\mathbb{Z}^n)$, of §1, in this case is the Schwartz space of rapidly decreasing functions:

$$\mathcal{S}(\mathbb{R}^n) = \big\{u \in C^\infty(\mathbb{R}^n) : x^\beta D^\alpha u \in L^\infty(\mathbb{R}^n) \text{ for all } \alpha, \beta \ge 0\big\}, \tag{3.3}$$

where $x^\beta = x_1^{\beta_1} \cdots x_n^{\beta_n}$, $D^\alpha = D_1^{\alpha_1} \cdots D_n^{\alpha_n}$, with $D_j = -i\partial/\partial x_j$. It is easy to verify that

$$\mathcal{F} : \mathcal{S}(\mathbb{R}^n) \longrightarrow \mathcal{S}(\mathbb{R}^n) \tag{3.4}$$

and

$$\xi^\alpha D_\xi^\beta \mathcal{F}f(\xi) = (-1)^{|\beta|} \mathcal{F}(D^\alpha x^\beta f)(\xi). \tag{3.5}$$

We define $\mathcal{F}^*$ by

$$\mathcal{F}^* f(\xi) = \check{f}(\xi) = (2\pi)^{-n/2} \int f(x) e^{ix\cdot\xi}\, dx, \tag{3.6}$$

which differs from (3.1) only in the sign of the exponent. It is clear that $\mathcal{F}^*$ satisfies the mapping properties (3.2), (3.4), and

$$(\mathcal{F}u, v) = (u, \mathcal{F}^* v), \tag{3.7}$$

for $u, v \in \mathcal{S}(\mathbb{R}^n)$, where (u, v) denotes the usual L^2-inner product, $(u, v) = \int u(x)\, \overline{v(x)}\, dx$.

As in the theory of Fourier series, the first major result is the Fourier inversion formula. The following is our first version.

Proposition 3.1. *We have the inversion formula*

$$\mathcal{F}^*\mathcal{F} = \mathcal{F}\mathcal{F}^* = I \text{ on } \mathcal{S}(\mathbb{R}^n). \tag{3.8}$$

As in the proof of the inversion formula for Fourier series, via Proposition 1.1, in the present proof we will sneak up on the inversion formula by throwing in a convergence factor that will allow interchange of orders of integration (in the proof of Proposition 1.1, the orders of an integral and an infinite series were interchanged). Also, as we will see in §5, this method will have serendipitous applications to PDE. So, let us write, for $f \in \mathcal{S}(\mathbb{R}^n)$,

$$\begin{aligned} \mathcal{F}^*\mathcal{F}f(x) &= (2\pi)^{-n} \int \Big[\int f(y)e^{-iy\cdot\xi}\,dy\Big]e^{ix\cdot\xi}\,d\xi \\ &= (2\pi)^{-n} \lim_{\varepsilon\searrow 0} \iint f(y)\,e^{-\varepsilon|\xi|^2}\,e^{i(x-y)\cdot\xi}\,dy\,d\xi. \end{aligned} \tag{3.9}$$

We can interchange the order of integration on the right for any $\varepsilon > 0$, to obtain

$$\mathcal{F}^*\mathcal{F}f(x) = \lim_{\varepsilon\searrow 0} \int f(y)p(\varepsilon, x-y)\,dy, \tag{3.10}$$

where

$$p(\varepsilon, x) = (2\pi)^{-n} \int e^{-\varepsilon|\xi|^2+ix\cdot\xi}\,d\xi. \tag{3.11}$$

Note that

$$p(\varepsilon, x) = \varepsilon^{-n/2}\,q(\varepsilon^{-1/2}x), \tag{3.12}$$

where $q(x) = p(1, x)$. In a moment we will show that

$$p(\varepsilon, x) = (4\pi\varepsilon)^{-n/2}\,e^{-|x|^2/4\varepsilon}. \tag{3.13}$$

The derivation of this identity will also show that

$$\int_{\mathbb{R}^n} q(x)\,dx = 1. \tag{3.14}$$

From this, it follows as in the proof of Proposition 1.1 that

$$\lim_{\varepsilon\searrow 0} \int f(y)p(\varepsilon, x-y)\,dy = f(x), \tag{3.15}$$

for any $f \in \mathcal{S}(\mathbb{R}^n)$, even for f bounded and continuous, so we have proved $\mathcal{F}^*\mathcal{F} = I$ on $\mathcal{S}(\mathbb{R}^n)$; the proof that $\mathcal{F}\mathcal{F}^* = I$ on $\mathcal{S}(\mathbb{R}^n)$ is identical.

It remains to verify (3.13). We observe that $p(\varepsilon, x)$, defined by (3.11), is an entire holomorphic function of $x \in \mathbb{C}^n$, for any $\varepsilon > 0$. It is convenient to verify that

$$p(\varepsilon, ix) = (4\pi\varepsilon)^{-n/2}\,e^{|x|^2/4\varepsilon}, \quad x \in \mathbb{R}^n, \tag{3.16}$$

from which (3.13) follows by analytic continuation. Now

$$
\begin{aligned}
p(\varepsilon, ix) &= (2\pi)^{-n} \int e^{-x\cdot\xi-\varepsilon|\xi|^2}\, d\xi \\
&= (2\pi)^{-n} e^{|x|^2/4\varepsilon} \int e^{-|x/2\sqrt{\varepsilon}+\sqrt{\varepsilon}\xi|^2}\, d\xi \qquad (3.17) \\
&= (2\pi)^{-n} \varepsilon^{-n/2} e^{|x|^2/4\varepsilon} \int_{\mathbb{R}^n} e^{-|\xi|^2}\, d\xi.
\end{aligned}
$$

To prove (3.16), it remains to show that

$$\int_{\mathbb{R}^n} e^{-|\xi|^2}\, d\xi = \pi^{n/2}. \tag{3.18}$$

Indeed, if

$$A = \int_{-\infty}^{\infty} e^{-\xi^2}\, d\xi, \tag{3.19}$$

then the left side of (3.18) is equal to A^n. But for $n = 2$ we can use polar coordinates:

$$A^2 = \int_{\mathbb{R}^2} e^{-|\xi|^2}\, d\xi = \int_0^{2\pi} \int_0^{\infty} e^{-r^2} r\, dr\, d\theta = \pi. \tag{3.20}$$

This completes the proof of the identity (3.16) and hence of (3.13).

In light of (3.7) and the Fourier inversion formula (3.8), we see that, for $u, v \in \mathcal{S}(\mathbb{R}^n)$,

$$(\mathcal{F}u, \mathcal{F}v) = (u, v) = (\mathcal{F}^* u, \mathcal{F}^* v). \tag{3.21}$$

Thus $\mathcal{F}$ and $\mathcal{F}^*$ extend uniquely from $\mathcal{S}(\mathbb{R}^n)$ to isometries on $L^2(\mathbb{R}^n)$ and are inverses to each other. Thus we have the Plancherel theorem:

Proposition 3.2. *The Fourier transform*

$$\mathcal{F} : L^2(\mathbb{R}^n) \longrightarrow L^2(\mathbb{R}^n) \tag{3.22}$$

is unitary, with inverse $\mathcal{F}^*$.

The inversion formulas of Propositions 3.1 and 3.2 do not provide for the inversion of $\mathcal{F}$ in (3.2). We will obtain this as a byproduct of the Fourier inversion formula for tempered distributions, in the next section.

We make a remark about the computation of the Fourier integral (3.11), done above via analytic continuation. The following derivation does not make any direct use of complex analysis. It suffices to handle the case $\varepsilon = 1/2$, that is, to show

$$\hat{G}(\xi) = e^{-|\xi|^2/2} \text{ if } G(x) = e^{-|x|^2/2}, \text{ on } \mathbb{R}^n. \tag{3.23}$$

We have interchanged the roles of x and ξ compared to those in (3.11) and (3.13). It suffices to get (3.23) in the case $n = 1$, by the obvious multiplicativity. Now the Gaussian function $G(x) = e^{-x^2/2}$ satisfies the differential equation

$$\left(\frac{d}{dx} + x\right)G(x) = 0. \tag{3.24}$$

By the intertwining property (3.5), it follows that $(d/d\xi + \xi)\hat{G}(\xi) = 0$, and uniqueness of solutions to this ODE yields $\hat{G}(\xi) = Ce^{-\xi^2/2}$. The constant C is evaluated via the identity (3.20); $C = 1$; and we are done.

As for the necessity of computing the Fourier integral (3.11) to prove the Fourier inversion formula, let us note the following. For *any* $g \in \mathcal{S}(\mathbb{R}^n)$ with $g(0) = 1$, ($g(\xi) = e^{-|\xi|^2}$ being an example), we have (replacing ε by δ^2), just as in (3.9),

$$\begin{aligned}\mathcal{F}^*\mathcal{F}f(x) &= (2\pi)^{-n} \lim_{\delta \searrow 0} \iint f(y)g(\delta\xi)e^{i(x-y)\cdot\xi}\, dy\, d\xi \\ &= \lim_{\delta\searrow 0} \int f(y)h_\delta(x-y)\, dy,\end{aligned} \tag{3.25}$$

where

$$\begin{aligned}h_\delta(x) &= (2\pi)^{-n} \int g(\delta\xi)e^{ix\cdot\xi}\, d\xi \\ &= (2\pi)^{-n/2}\delta^{-n}\tilde{g}(\delta^{-1}x).\end{aligned} \tag{3.26}$$

By the peaked nature of h_δ as $\delta \to 0$, we see that the limit in (3.25) is equal to

$$C\, f(x), \tag{3.27}$$

where

$$C = \int h_1(x)\, dx = (2\pi)^{-n/2} \int \tilde{g}(x)\, dx. \tag{3.28}$$

The argument (3.25)–(3.27) shows that C is independent of the choice of $g \in \mathcal{S}(\mathbb{R}^n)$, and we need only find a single example g such that $\tilde{g}(x)$ can be evaluated explicitly and then the integral on the right in (3.28) can be evaluated explicitly. In most natural examples one picks g to be even, so $\tilde{g} = \hat{g}$.

We remark that one does not need to have $g \in \mathcal{S}(\mathbb{R}^n)$ in the argument above; it suffices to have $g \in L^1(\mathbb{R}^n)$, bounded and continuous, and such that $\hat{g} \in L^1(\mathbb{R}^n)$. An example, in the case $n = 1$, is

$$g(\xi) = e^{-|\xi|}. \tag{3.29}$$

In this case, elementary calculations give

$$\hat{g}(x) = \left(\frac{2}{\pi}\right)^{1/2} \frac{1}{x^2+1}; \tag{3.30}$$

compare (5.21). In this case, (3.28) can be evaluated in terms of the arctangent. Another example, in the case $n = 1$, is

$$g(\xi) = \begin{cases} 1 - |\xi| & \text{if } |\xi| \le 1, \\ 0 & \text{if } |\xi| \ge 1. \end{cases} \tag{3.31}$$

In this case,

$$\hat{g}(x) = (2\pi)^{-1/2}\left(\frac{\sin \frac{1}{2}x}{\frac{1}{2}x}\right)^2, \tag{3.32}$$

and (3.28) can be evaluated by the method of residues. The calculation of (3.32) can be achieved by evaluating

$$\int_0^1 (1 - \xi) \cos x\xi \, d\xi$$

via an integration by parts, though there is a more painless way, mentioned below.

We now make some comments on the relation between the Fourier transform and convolutions. The convolution $u * v$ of two functions on $\mathbb{R}^n$ is defined by

$$\begin{aligned} u * v(x) &= \int u(y)v(x - y)\, dy \\ &= \int u(x - y)v(y)\, dy. \end{aligned} \tag{3.33}$$

Note that $u * v = v * u$. If $u, v \in \mathcal{S}(\mathbb{R}^n)$, so is $u * v$. Also

$$\|u * v\|_{L^p(\mathbb{R}^n)} \le \|u\|_{L^1} \|v\|_{L^p},$$

so the convolution has a unique, continuous extension to a bilinear map

$$L^1(\mathbb{R}^n) \times L^p(\mathbb{R}^n) \longrightarrow L^p(\mathbb{R}^n), \tag{3.34}$$

for $1 \le p < \infty$; one can directly perceive that this also works for $p = \infty$. Note that the right side of (3.10), for any $\varepsilon > 0$, is an example of a convolution. Computing the Fourier transform of (3.33) leads immediately to the formula

$$\mathcal{F}(u * v)(\xi) = (2\pi)^{n/2}\hat{u}(\xi)\hat{v}(\xi). \tag{3.35}$$

We also note that if

$$P = \sum_{|\alpha| \le k} a_\alpha D^\alpha \tag{3.36}$$

is a constant-coefficient differential operator, we have

$$P(u * v) = (Pu) * v = u * (Pv) \tag{3.37}$$

if $u, v \in \mathcal{S}(\mathbb{R}^n)$. This also generalizes; if $u \in \mathcal{S}(\mathbb{R}^n)$, $v \in L^p(\mathbb{R}^n)$, the first identity continues to hold; as we will see in the next section, so does the second identity, once we are able to interpret what it means.

We mention the following simple application of (3.35), to a short calculation of (3.32). With g given by (3.31), we have $g = g_1 * g_1$, where

$$g_1(\xi) = 1 \text{ for } \xi \in \left[-\frac{1}{2}, \frac{1}{2}\right], \quad 0 \text{ otherwise.} \tag{3.38}$$

Thus

$$\hat{g}_1(x) = (2\pi)^{-1/2} \int_{-1/2}^{1/2} e^{-ix\xi}\, d\xi = (2\pi)^{-1/2} \frac{\sin \frac{1}{2}x}{\frac{1}{2}x}, \tag{3.39}$$

and then (3.32) follows immediately from (3.35).

Exercises

1. Show that $\mathcal{F} : L^1(\mathbb{R}^n) \to C_0(\mathbb{R}^n)$, where $C_0(\mathbb{R}^n)$ denotes the space of functions v, continuous on $\mathbb{R}^n$, such that $v(\xi) \to 0$ as $|\xi| \to \infty$.
(*Hint*: Use the denseness of $\mathcal{S}(\mathbb{R}^n)$ in $L^1(\mathbb{R}^n)$.)
This result is the Riemann-Lebesgue lemma for the Fourier transform.
2. Show that the Fourier transforms (3.1) and (3.22) coincide on $L^1(\mathbb{R}^n) \cap L^2(\mathbb{R}^n)$.
3. For $f \in L^1(\mathbb{R})$, set $S_R f(x) = (2\pi)^{-1/2} \int_{-R}^{R} \hat{f}(\xi) e^{ix\xi}\, d\xi$. Show that

$$S_R f(x) = D_R * f(x) = \int_{-\infty}^{\infty} D_R(x-y) f(y)\, dy,$$

where

$$D_R(x) = (2\pi)^{-1} \int_{-R}^{R} e^{ix\xi}\, d\xi = \frac{\sin Rx}{\pi x}.$$

Compare Exercise 5 of §1.
4. Show that $f \in L^2(\mathbb{R}) \Rightarrow S_R f \to f$ in L^2-norm as $R \to \infty$.
5. Show that there exist $f \in L^1(\mathbb{R})$ such that $S_R f \notin L^1(\mathbb{R})$ for any $R \in (0, \infty)$.
(*Hint*: Note that $D_R \notin L^1(\mathbb{R})$.)
6. For $f \in L^1(\mathbb{R})$, set

$$C_R f(x) = (2\pi)^{-1/2} \int_{-R}^{R} \left(1 - \frac{|\xi|}{R}\right) \hat{f}(\xi) e^{ix\xi}\, d\xi.$$

Show that $C_R f(x) = E_R * f(x)$, where

$$E_R(x) = (2\pi)^{-1} \int_{-R}^{R} \left(1 - \frac{|\xi|}{R}\right) e^{ix\xi}\, d\xi = \frac{2}{\pi R} \left[\frac{\sin \frac{1}{2}Rx}{x}\right]^2.$$

Note that $E_R \in L^1(\mathbb{R})$. Show that, for $1 \le p < \infty$,

$$f \in L^p(\mathbb{R}) \Longrightarrow C_R f \to f \text{ in } L^p\text{-norm, as } R \to \infty.$$

We say the Fourier transform of f is Cesaro-summable if $C_R f \to f$ as $R \to \infty$.

In Exercises 7–13, suppose $f \in \mathcal{S}(\mathbb{R})$ has the following properties: $f \ge 0$, $\int_{-\infty}^{\infty} f(x)\, dx = 1$, and $\int_{-\infty}^{\infty} x f(x)\, dx = 0$. Set $F(\xi) = (2\pi)^{1/2} \hat{f}(\xi)$. The point of the exercises is to obtain a version of the *central limit theorem*.

7. Show that $F(0) = 1$, $F'(0) = 0$, $F''(0) = -2a < 0$. Also, $\xi \ne 0 \Rightarrow |F(\xi)| < 1$.

8. Set $F_n(\xi) = F(\xi/\sqrt{n})^n$. Relate $(2\pi)^{-1/2}\tilde{F}_n(x)$ to the convolution of n copies of f.
9. Show that there exist $A > 0$ and $G \in C^\infty([-A, A])$ such that $f(\xi) = e^{-a\xi^2}G(\xi)$ for $|\xi| \le A$, and $G(0) = 1,\ G'(0) = G''(0) = 0$. Hence

$$F_n(\xi) = e^{-a\xi^2}G(n^{-1/2}\xi)^n, \quad \text{for } |\xi| \le A\sqrt{n}.$$

10. Show that $\left|G(\xi/\sqrt{n})^n - 1\right| \le Cn^{-\alpha}$ if $|\xi| \le n^{(1/2-\alpha)/3}$, for n large.
    Fix $\alpha \in (0, \frac{1}{2})$, and set $\gamma = (1/2 - \alpha)/3 \in (0, \frac{1}{6})$.
11. Show that, for $|\xi| \ge n^\gamma$, $\left|F(\xi/\sqrt{n})\right| \le 1 - \frac{1}{2}an^{-(1-2\gamma)}$, for n large, so $|F_n(\xi)| \le e^{-an^{2\gamma}/4} = \delta_n$. Deduce that

$$\int\limits_{|\xi|\ge n^\gamma} |F_n(\xi)|\, d\xi \le C\, \delta_n^{(n-1)/n}\sqrt{n} \to 0, \quad \text{as } n \to \infty.$$

12. From Exercises 9–11, deduce that $F_n \to e^{-a\xi^2}$ in $L^1(\mathbb{R})$ as $n \to \infty$.
13. Deduce now that $(2\pi)^{-1/2}\tilde{F}_n \to (4\pi a)^{-1/2}e^{-x^2/4a}$ in both $C_0(\mathbb{R})$ and $L^1(\mathbb{R})$, as $n \to \infty$. Relate this to the central limit theorem of probability theory. Weaken the hypotheses on f as much as you can.
    (*Hint*: In passing from the C_0-result to the L^1-result, positivity of $\tilde{F}_n$ will be useful.)
14. With $p_\varepsilon(x) = (4\pi\varepsilon)^{-1/2}e^{-x^2/4\varepsilon}$, as in (3.13) for $n = 1$, show that, for any $u(x)$, continuous and compactly supported on $\mathbb{R}$, $p_\varepsilon * u \to u$ uniformly as $\varepsilon \to 0$. Show that for each $\varepsilon > 0$, $p_\varepsilon * u(x)$ is the restriction to $\mathbb{R}$ of an entire holomorphic function of $x \in \mathbb{C}$.
15. Using Exercise 14, prove the *Weierstrass approximation theorem*:
    Any $f \in C([a, b])$ is a uniform limit of polynomials.
    (*Hint*: Extend f to u as above, approximate u by $p_\varepsilon * u$, and expand this in a power series.)
16. Suppose $f \in \mathcal{S}(\mathbb{R}^n)$ is supported in $B_R = \{x \in \mathbb{R}^n : |x| \le R\}$. Show that $\hat{f}(\xi)$ is holomorphic in $\xi \in \mathbb{C}^n$ and satisfies

$$|\hat{f}(\xi + i\eta)| \le C_N\langle\xi\rangle^{-N}e^{R|\eta|}, \quad \xi, \eta \in \mathbb{R}^n. \tag{3.40}$$

17. Conversely, suppose $g(\xi) = \hat{f}(\xi) \in \mathcal{S}(\mathbb{R}^n)$ has a holomorphic extension to $\mathbb{C}^n$ satisfying (3.40). Show that f is supported in $|x| \le R$.
    (*Hint*: With $\omega = x/|x|$, $r \ge 0$, write

$$f(x) = (2\pi)^{-n/2}\int\limits_{\mathbb{R}^n} \hat{f}(\xi + ir\omega)e^{ix\cdot\xi - rx\cdot\omega}\, d\xi, \tag{3.41}$$

with

$$|\hat{f}(\xi + ir\omega)\, e^{ix\cdot\xi - rx\cdot\omega}| \le C_N\langle\xi\rangle^{-N}\, e^{r(R-|x|)},$$

and let $r \to +\infty$.)
    This is a basic case of the *Paley-Wiener theorem*.
18. Given $f \in L^1(\mathbb{R}^n)$, show that f is supported in B_R if and only if $\hat{f}(\xi)$ is holomorphic in $\mathbb{C}^n$, satisfying

$$|\hat{f}(\xi + i\eta)| \le C\, e^{R|\eta|}, \quad \xi, \eta \in \mathbb{R}^n.$$

Reconsider this problem after reading §4.

19. Show that

$$f(x) = \frac{1}{\cosh x} \Longrightarrow \hat{f}(\xi) = \frac{\sqrt{\pi/2}}{\cosh \frac{\pi}{2}\xi}.$$

(*Hint*: See (A.13)–(A.15).)

4. Distributions and tempered distributions

L. Schwartz's theory of distributions has proved to be not only a wonderful tool in partial differential equations, but also a device that lends clarity to many aspects of Fourier analysis. We sketch the basic concepts of distribution theory here, making use of such basic concepts as Fréchet spaces and weak topologies, which are treated in Appendix A, Functional Analysis.

We begin with the concept of a tempered distribution. This is a continuous linear functional

$$w : \mathcal{S}(\mathbb{R}^n) \longrightarrow \mathbb{C}, \tag{4.1}$$

where $\mathcal{S}(\mathbb{R}^n)$ is the Schwartz space defined in §3. The space $\mathcal{S}(\mathbb{R}^n)$ has a topology, determined by the seminorms

$$p_k(u) = \sum_{|\alpha| \le k} \sup_{x \in \mathbb{R}^n} \langle x \rangle^k |D^\alpha u(x)|. \tag{4.2}$$

The distance function

$$d(u, v) = \sum_{k=0}^{\infty} 2^{-k} \frac{p_k(u - v)}{1 + p_k(u - v)} \tag{4.3}$$

makes $\mathcal{S}(\mathbb{R}^n)$ a complete metric space; with such a topology it is a Fréchet space. For a linear map w as in (4.1) to be continuous, it is necessary and sufficient that, for some k, C,

$$|w(u)| \le C\, p_k(u), \quad \text{for all } u \in \mathcal{S}(\mathbb{R}^n). \tag{4.4}$$

The action of w is often written as follows:

$$w(u) = \langle u, w \rangle. \tag{4.5}$$

The set of all continuous linear functionals on $\mathcal{S}(\mathbb{R}^n)$ is denoted

$$\mathcal{S}'(\mathbb{R}^n) \tag{4.6}$$

and is called the space of tempered distributions.

The space $\mathcal{S}'(\mathbb{R}^n)$ has a topology, called the weak* topology, or sometimes simply the weak topology, in terms of which a directed family w_γ converges to w weakly in $\mathcal{S}'(\mathbb{R}^n)$ if and only if, for each $u \in \mathcal{S}(\mathbb{R}^n)$, $\langle u, w_\gamma \rangle \to \langle u, w \rangle$. One can also consider the strong topology on $\mathcal{S}'(\mathbb{R}^n)$, the topology of uniform convergence on bounded subsets of $\mathcal{S}(\mathbb{R}^n)$, but we will not consider this explicitly. For more on the topology of $\mathcal{S}$ and $\mathcal{S}'$, see [H], [Sch], and [Yo]. We now consider examples of tempered distributions.

There is a natural injection

$$L^p(\mathbb{R}^n) \hookrightarrow \mathcal{S}'(\mathbb{R}^n), \tag{4.7}$$

for any $p \in [1, \infty]$, given by

$$\langle u, f\rangle = \int u(x) f(x)\, dx, \quad u \in \mathcal{S}(\mathbb{R}^n),\ f \in L^p(\mathbb{R}^n). \tag{4.8}$$

Similarly any finite measure on $\mathbb{R}^n$ gives an element of $\mathcal{S}'(\mathbb{R}^n)$. The basic example is the Dirac "delta function" δ, defined by

$$\langle u, \delta\rangle = u(0). \tag{4.9}$$

Also, each differential operator $D_j = -i\partial/\partial x_j$ acts on $\mathcal{S}'(\mathbb{R}^n)$, by the definition

$$\langle u, D_j w\rangle = -\langle D_j u, w\rangle, \quad u \in \mathcal{S},\ w \in \mathcal{S}'. \tag{4.10}$$

Iterating, we see that each $D^\alpha = D_1^{\alpha_1} \cdots D_n^{\alpha_n}$ acts on $\mathcal{S}'$:

$$D^\alpha : \mathcal{S}'(\mathbb{R}^n) \longrightarrow \mathcal{S}'(\mathbb{R}^n), \tag{4.11}$$

and we have

$$\langle u, D^\alpha w\rangle = (-1)^{|\alpha|}\langle D^\alpha u, w\rangle \tag{4.12}$$

for $u \in \mathcal{S}$, $w \in \mathcal{S}'$. Similarly,

$$\langle u, fw\rangle = \langle fu, w\rangle$$

defines fw for $w \in \mathcal{S}'$, provided that f and each of its derivatives is polynomially bounded.

To illustrate, consider on $\mathbb{R}$ the Heaviside function

$$\begin{aligned} H(x) &= 1 \ \text{ if } x \geq 0, \\ &\ 0 \ \text{ if } x < 0. \end{aligned} \tag{4.13}$$

Then $H \in L^\infty(\mathbb{R}) \subset \mathcal{S}'(\mathbb{R})$, and the definition (4.10) gives

$$\frac{d}{dx} H = \delta \tag{4.14}$$

as a consequence of the fundamental theorem of calculus. The derivative of δ is characterized by

$$\langle u, \delta'\rangle = -u'(0) \tag{4.15}$$

in this case.

The Fourier transform $\mathcal{F} : \mathcal{S}(\mathbb{R}^n) \to \mathcal{S}(\mathbb{R}^n)$, studied in §3, extends to $\mathcal{S}'$ by the formula

$$\langle u, \mathcal{F}w\rangle = \langle \mathcal{F}u, w\rangle; \tag{4.16}$$

we can also set

$$\langle u, \mathcal{F}^* w\rangle = \langle \mathcal{F}^* u, w\rangle \tag{4.17}$$

to get

$$\mathcal{F},\ \mathcal{F}^* : \mathcal{S}'(\mathbb{R}^n) \longrightarrow \mathcal{S}'(\mathbb{R}^n). \tag{4.18}$$

The maps (4.18) are continuous when $\mathcal{S}'(\mathbb{R}^n)$ is given the weak* topology, as follows easily from the definitions.

The Fourier inversion formula of Proposition 3.1 yields:

Proposition 4.1. *We have*

$$\mathcal{F}^*\mathcal{F} = \mathcal{F}\mathcal{F}^* = I \ \text{ on } \mathcal{S}'(\mathbb{R}^n). \tag{4.19}$$

Proof. Using (4.16) and (4.17), if $u \in \mathcal{S},\ w \in \mathcal{S}'$,

$$\langle u, \mathcal{F}^*\mathcal{F}w\rangle = \langle \mathcal{F}^*u, \mathcal{F}w\rangle = \langle \mathcal{F}\mathcal{F}^*u, w\rangle = \langle u, w\rangle,$$

and a similar analysis works for $\mathcal{F}\mathcal{F}^*w$.

As an example of a Fourier transform of a tempered distribution, the definition gives directly

$$\mathcal{F}\delta = (2\pi)^{-n/2}; \tag{4.20}$$

the Fourier transform of the delta function is a constant function. One has the same result for $\mathcal{F}^*\delta$. By the Fourier inversion formula,

$$\mathcal{F}1 = (2\pi)^{n/2}\delta. \tag{4.21}$$

Next, let us consider on the line $\mathbb{R}$, for any $\varepsilon > 0$,

$$\begin{aligned} H_\varepsilon(x) &= e^{-\varepsilon x} && \text{for } x \geq 0, \\ &\ 0 && \text{for } x < 0. \end{aligned} \tag{4.22}$$

We have, by elementary calculation,

$$\hat{H}_\varepsilon(\xi) = (2\pi)^{-1/2} \int_0^\infty e^{-\varepsilon x - i\xi x}\, dx = (2\pi)^{-1/2}(\varepsilon + i\xi)^{-1}, \tag{4.23}$$

for each $\varepsilon > 0$. Now it is clear that

$$H_\varepsilon \rightarrow H \ \text{ as } \varepsilon \searrow 0, \ \text{ in } \mathcal{S}'(\mathbb{R}), \tag{4.24}$$

in the weak* topology. It follows that

$$\hat{H}(\xi) = (2\pi)^{-1/2} \lim_{\varepsilon \searrow 0} (\varepsilon + i\xi)^{-1} \ \text{ in } \mathcal{S}'(\mathbb{R}). \tag{4.25}$$

In particular, the limit on the right *exists* in $\mathcal{S}'(\mathbb{R})$. Changing the sign of x in (4.22)–(4.24) and noting that $H(-x) = 1 - H(x)$, we also have

$$(2\pi)^{1/2}\delta - \hat{H} = (2\pi)^{-1/2} \lim_{\varepsilon \searrow 0} (\varepsilon - i\xi)^{-1} \ \text{ in } \mathcal{S}'(\mathbb{R}). \tag{4.26}$$

Let us set

$$(\xi \pm i0)^{-1} = \lim_{\varepsilon \searrow 0} (\xi \pm i\varepsilon)^{-1}. \tag{4.27}$$

Then (4.25) and (4.26) give

$$\hat{H} = -i(2\pi)^{-1/2}(\xi - i0)^{-1} \tag{4.28}$$

and

$$(\xi + i0)^{-1} - (\xi - i0)^{-1} = -2\pi i \delta. \tag{4.29}$$

The last identity is often called the Plemelj jump relation. Also, subtracting (4.25) from (4.26) gives

$$(\xi + i0)^{-1} + (\xi - i0)^{-1} = (2\pi)^{1/2} i \, \widehat{\operatorname{sgn}}(\xi), \tag{4.30}$$

where

$$\operatorname{sgn}(x) = \begin{matrix} 1 & \text{if } x \geq 0, \\ -1 & \text{if } x < 0. \end{matrix} \tag{4.31}$$

It is also an easy exercise to show that

$$(\xi + i0)^{-1} + (\xi - i0)^{-1} = 2\, PV\Big(\frac{1}{\xi}\Big), \tag{4.32}$$

where the "principal value" distribution

$$PV\Big(\frac{1}{x}\Big) \in \mathcal{S}'(\mathbb{R}) \tag{4.33}$$

is defined by

$$\begin{aligned} \Big\langle u, PV\Big(\frac{1}{x}\Big)\Big\rangle &= \lim_{h \searrow 0} \int_{\mathbb{R}\setminus(-h,h)} \frac{u(x)}{x}\, dx \\ &= \lim_{h \searrow 0} \int_h^\infty \Big[\frac{u(x)}{x} - \frac{u(-x)}{x}\Big]\, dx \\ &= \int_0^\infty \frac{u(x) - u(-x)}{x}\, dx. \end{aligned} \tag{4.34}$$

Note that if we replace the left side of (4.29) by

$$\frac{1}{\xi + i\varepsilon} - \frac{1}{\xi - i\varepsilon} = -\frac{2i\varepsilon}{\xi^2 + \varepsilon^2} = -2i\, \frac{1}{\varepsilon} \cdot \frac{1}{(\xi/\varepsilon)^2 + 1},$$

the conclusion (4.29) is a special case of the following obvious result.

Proposition 4.2. *If $f \in L^1(\mathbb{R}^n)$, $\int f(x)dx = C_0$, then, as $\varepsilon \to 0$,*

$$\varepsilon^{-n} f(\varepsilon^{-1}x) \to C_0 \delta \ \text{ in } \mathcal{S}'(\mathbb{R}^n). \tag{4.35}$$

That δ is the limit of a sequence of elements of $\mathcal{S}(\mathbb{R}^n)$ is a special case of the fact that $\mathcal{S}(\mathbb{R}^n)$ is dense in $\mathcal{S}'(\mathbb{R}^n)$, which will be established shortly.

Given $w \in \mathcal{S}'(\mathbb{R}^n)$, if $\Omega \subset \mathbb{R}^n$ is open, one says w vanishes on Ω provided $\langle u, w\rangle = 0$ for all $u \in C_0^\infty(\Omega)$. By a partition of unity argument, it follows that if

w vanishes on Ω_j, then it vanishes on their union; w is said to be *supported* on a closed set $K \subset \mathbb{R}^n$ if w vanishes on $\mathbb{R}^n \setminus K$. The smallest closed set K on which w is supported exists; it is denoted supp w. Note that if w vanishes on all of $\mathbb{R}^n$, then $w = 0$, since $C_0^\infty(\mathbb{R}^n)$ is dense in $\mathcal{S}(\mathbb{R}^n)$. (See the first part of the proof of Proposition 4.4 below.) If $w \in \mathcal{S}'(\mathbb{R}^n)$ is supported on a compact set $K \subset \mathbb{R}^n$, we say w has compact support. The space of compactly supported distributions on $\mathbb{R}^n$ is denoted by

$$\mathcal{E}'(\mathbb{R}^n). \tag{4.36}$$

If $w \in \mathcal{S}'(\mathbb{R}^n)$ is supported on a compact set $K \subset \mathbb{R}^n$, then w can be extended to a continuous linear functional

$$w : C^\infty(\mathbb{R}^n) \longrightarrow \mathbb{C} \tag{4.37}$$

by setting

$$\langle u, w \rangle = \langle \chi u, w \rangle, \quad u \in C^\infty(\mathbb{R}^n), \tag{4.38}$$

for any $\chi \in C_0^\infty(\mathbb{R}^n)$ such that $\chi = 1$ on a neighborhood of K. The space $C^\infty(\mathbb{R}^n)$ is also a Fréchet space, with topology defined by the seminorms

$$p_{R,k}(u) = \sup_{|x| \le R} \sum_{|\alpha| \le k} |D^\alpha u(x)|. \tag{4.39}$$

To say a linear map (4.37) is continuous is to say there exist R, k, and C such that

$$|w(u)| \le C\, p_{R,k}(u), \quad \text{for all } u \in C^\infty(\mathbb{R}^n). \tag{4.40}$$

Such a linear functional restricts to $\mathcal{S}(\mathbb{R}^n) \subset C^\infty(\mathbb{R}^n)$, so it defines an element of $\mathcal{S}'(\mathbb{R}^n)$, and from (4.40) it follows that such an element must be supported in the compact set $\{x \in \mathbb{R}^n : |x| \le R\}$. Thus the space (4.36) is precisely the dual space of $C^\infty(\mathbb{R}^n)$.

Fourier transforms of compactly supported distributions have some special properties.

Proposition 4.3. *If $w \in \mathcal{E}'(\mathbb{R}^n)$, then $\hat{w} \in C^\infty(\mathbb{R}^n)$ and, with $e_\xi(x) = e^{-ix\cdot\xi}$,*

$$\hat{w}(\xi) = (2\pi)^{-n/2} \langle e_\xi, w \rangle, \tag{4.41}$$

for all $\xi \in \mathbb{R}^n$. Furthermore, $\hat{w}$ extends to an entire holomorphic function of $\xi \in \mathbb{C}^n$.

Proof. For any $u \in \mathcal{S}(\mathbb{R}^n)$, $\langle u, \hat{w} \rangle = \langle \hat{u}, w \rangle$. Now we can write

$$\hat{u}(x) = (2\pi)^{-n/2} \int u(\xi)\, e_\xi(x)\, d\xi, \tag{4.42}$$

the integral converging in the Fréchet space topology of $C^\infty(\mathbb{R}^n)$, and the continuity of w acting on $C^\infty(\mathbb{R}^n)$ implies

$$\langle \hat{u}, w \rangle = (2\pi)^{-n/2} \int u(\xi) \langle e_\xi, w \rangle\, d\xi,$$

which gives (4.41). The right side of (4.41) is clearly holomorphic in $\xi \in \mathbb{C}^n$.

We next obtain the promised denseness of $\mathcal{S}(\mathbb{R}^n)$ in $\mathcal{S}'(\mathbb{R}^n)$.

Proposition 4.4. *$C_0^\infty(\mathbb{R}^n)$ is dense in $\mathcal{S}'(\mathbb{R}^n)$, with its weak* topology.*

Proof. Pick $\varphi \in C_0^\infty(\mathbb{R}^n)$, $\varphi(0) = 1$. It is easy to see that, given $w \in \mathcal{S}'(\mathbb{R}^n)$, φw is well defined, by $\langle u, \varphi w\rangle = \langle \varphi u, w\rangle$. Also, if $\varphi_j(x) = \varphi(x/j)$, then for $u \in \mathcal{S}(\mathbb{R}^n)$,

$$\varphi_j u \to u \ \text{ in } \mathcal{S}(\mathbb{R}^n), \tag{4.43}$$

which gives sequential denseness of $C_0^\infty(\mathbb{R}^n)$ in $\mathcal{S}(\mathbb{R}^n)$. Hence $\varphi_j w \to w$ in $\mathcal{S}'(\mathbb{R}^n)$ as $j \to \infty$. Since $\mathcal{F}$ and $\mathcal{F}^*$ are continuous on $\mathcal{S}'(\mathbb{R}^n)$, we have $\mathcal{F}^*(\varphi_j \hat{w}) \to w$ as $j \to \infty$. Now for each j, $w_{jk} = \varphi_k(\mathcal{F}^* \varphi_j \hat{w}) \to \mathcal{F}^*(\varphi_j \hat{w})$ as $k \to \infty$. But by Proposition 4.3, $\mathcal{F}^*(\varphi_j \hat{w})$ is smooth, so $w_{jk} \in C_0^\infty(\mathbb{R}^n)$, and the result follows.

One useful result is the following classification of distributions supported at a single point.

Proposition 4.5. *If $w \in \mathcal{S}'(\mathbb{R}^n)$ is supported by $\{0\}$, then there exist k and complex numbers a_α such that*

$$w = \sum_{|\alpha| \le k} a_\alpha D^\alpha \delta. \tag{4.44}$$

Proof. We can suppose w satisfies the estimate (4.4). Thus w extends to $\mathcal{B}_k$, the closure of $\mathcal{S}(\mathbb{R}^n)$ in the space of C^k-functions on $\mathbb{R}^n$ for which the norm p_k is finite. By hypothesis, w annihilates the linear space $\mathcal{E}_0$ of elements of $C_0^\infty(\mathbb{R}^n)$ vanishing on a neighborhood of 0; thus w annihilates the closure of $\mathcal{E}_0$ in $\mathcal{B}_k$; call this closure $\mathcal{E}_k$. It is not hard to prove that

$$\mathcal{E}_k = \{u \in \mathcal{B}_k : D^\alpha u(0) = 0 \text{ for } |\alpha| \le k\}. \tag{4.45}$$

See Exercise 7 below for some hints. Now, for general $u \in \mathcal{B}_k$, write

$$u(x) = \chi \Big[\sum_{|\alpha| \le k} \frac{u^{(\alpha)}(0)}{\alpha!} x^\alpha \Big] + u^b(x), \tag{4.46}$$

where $\chi \in C_0^\infty(\mathbb{R}^n)$, $\chi(x) = 1$ for $|x| < 1$, and $u^b \in \mathcal{E}_k$. Applying w to both sides, we have an expression of the form (4.44), with $a_\alpha = (-1)^{|\alpha|} \langle \chi x^\alpha, w\rangle / \alpha!$.

As an application of Proposition 4.5, we establish the following result, which is an extension of the classical Liouville theorem for harmonic functions.

Proposition 4.6. *Suppose $u \in \mathcal{S}'(\mathbb{R}^n)$ satisfies*

$$\Delta u = 0 \ \text{ in } \mathbb{R}^n. \tag{4.47}$$

Then u is a polynomial in $(x_1, \ldots, x_n)$.

Proof. As in §3, the identity

$$\Delta u = f \in \mathcal{S}'(\mathbb{R}^n) \tag{4.48}$$

is equivalent to

$$-|\xi|^2 \hat{u} = \hat{f} \text{ in } \mathcal{S}'(\mathbb{R}^n). \tag{4.49}$$

In particular, (4.47) for $u \in \mathcal{S}'(\mathbb{R}^n)$ implies

$$|\xi|^2 \hat{u} = 0 \text{ in } \mathcal{S}'(\mathbb{R}^n). \tag{4.50}$$

This of course implies

$$\operatorname{supp} \hat{u} \subset \{0\}. \tag{4.51}$$

By Proposition 4.5, $\hat{u}$ must have the form (4.44), and the result follows.

It is clear that any nonconstant polynomial blows up, so we have:

Corollary 4.7. *If u is harmonic on* $\mathbb{R}^n$ *and bounded, then u is constant.*

This is the classical version of the Liouville theorem. We remind the reader of one of its uses. If $p(z)$ is a polynomial on $\mathbb{C}$, and if it has no zeros, then $q(z) = 1/p(z)$ is holomorphic (hence harmonic) on all of $\mathbb{C}$; clearly, $|q(z)| \to 0$ as $|z| \to \infty$ if $\deg p \geq 1$. Corollary 4.7 yields the obvious contradiction that $q(z)$ would have to be constant. This proves the fundamental theorem of algebra: Any nonconstant polynomial $p(z)$ must have a complex root.

Consider the function

$$\Psi(z) = \frac{1}{z}. \tag{4.52}$$

This is holomorphic on $\mathbb{C} \setminus \{0\}$. It is integrable near 0 and bounded outside a neighborhood of 0, and hence it defines an element of $\mathcal{S}'(\mathbb{R}^2)$. $(\partial/\partial\bar{z})\Psi \in \mathcal{S}'(\mathbb{R}^2)$ is supported at $\{0\}$. In fact, we claim:

Proposition 4.8. *We have*

$$\frac{\partial}{\partial \bar{z}} \frac{1}{z} = \pi\delta. \tag{4.53}$$

Proof. Let $u \in C_0^\infty(\mathbb{R}^2)$. We have

$$\begin{aligned} \left\langle u, \frac{\partial}{\partial \bar{z}} \frac{1}{z} \right\rangle &= -\iint_{\mathbb{R}^2} \frac{\partial u}{\partial \bar{z}} \frac{1}{z} \, dx \, dy \\ &= -\lim_{\varepsilon \to 0} \iint_{\mathbb{R}^2 \setminus B_\varepsilon} \frac{\partial (z^{-1} u)}{\partial \bar{z}} \, dx \, dy, \end{aligned} \tag{4.54}$$

where $B_\varepsilon = \{(x, y) \in \mathbb{R}^2 : x^2 + y^2 < \varepsilon^2\}$. By Green's formula, in the form (2.37), the right side is equal to

$$\frac{1}{2i} \int_{\partial B_\varepsilon} \frac{u}{z}\, dz, \tag{4.55}$$

which is clearly equal in the limit $\varepsilon \to 0$ to $\pi u(0)$. This proves the proposition.

We say $(\pi z)^{-1}$ is a *fundamental solution* of $(\partial/\partial \bar{z})$. We will say more about the use of fundamental solutions later. Let us look at the task of producing a fundamental solution for the Laplace operator Δ on $\mathbb{R}^n$. In view of the rotational invariance of Δ, we are led to look for a function of $r = |x|$, for $x \neq 0$. The form

$$\Delta = \frac{\partial^2}{\partial r^2} + \frac{n-1}{r}\frac{\partial}{\partial r} + \frac{1}{r^2}\Delta_S, \tag{4.56}$$

for Δ in polar coordinates, where Δ_S is the Laplace operator on the unit sphere S^{n-1}, shows that, for $n \geq 3$,

$$|x|^{2-n} \tag{4.57}$$

is harmonic for $x \neq 0$. As it is locally integrable near 0, it defines an element of $\mathcal{S}'(\mathbb{R}^n)$. We have the following result.

Proposition 4.9. *If $n \geq 3$,*

$$\Delta\big(|x|^{2-n}\big) = C_n \delta \ \text{ on } \mathbb{R}^n, \tag{4.58}$$

with $C_n = -(n-2)\cdot \mathrm{Area}(S^{n-1})$. Also,

$$\Delta\big(\log|x|\big) = C_2 \delta \ \text{ on } \mathbb{R}^2, \tag{4.59}$$

with $C_2 = 2\pi$.

Proof. This will use Green's formula, in the form

$$\int_\Omega \big(\Delta u \cdot v - u \cdot \Delta v\big)\, dx = \int_{\partial\Omega} \Big[v \frac{\partial u}{\partial \nu} - u \frac{\partial v}{\partial \nu}\Big]\, dS. \tag{4.60}$$

Let $u \in C_0^\infty(\mathbb{R}^n)$ be arbitrary. Let $v = |x|^{2-n}$, and let $\Omega = \Omega_\varepsilon = \mathbb{R}^n \setminus B_\varepsilon$, where $B_\varepsilon = \{x \in \mathbb{R}^n : |x| < \varepsilon\}$. We have

$$\begin{aligned} \langle \Delta u, |x|^{2-n}\rangle &= \lim_{\varepsilon \to 0} \int_{\Omega_\varepsilon} \Delta u \cdot |x|^{2-n}\, dx \\ &= \lim_{\varepsilon \to 0} \int_{\Omega_\varepsilon} \big[\Delta u \cdot |x|^{2-n} - u \cdot \Delta |x|^{2-n}\big]\, dx \end{aligned} \tag{4.61}$$

since $\Delta|x|^{2-n} = 0$ for $x \neq 0$. Applying (4.60), we have this equal to

$$-\lim_{\varepsilon \to 0} \int_{\partial B_\varepsilon} \Big[\varepsilon^{2-n} \frac{\partial u}{\partial r} - (2-n)\varepsilon^{1-n} u\Big]\, dS. \tag{4.62}$$

Since the area of ∂B_ε is $\varepsilon^{n-1} \cdot \text{Area } S^{n-1}$, this limit is seen to be

$$-(n-2)u(0) \cdot \text{Area } S^{n-1}, \tag{4.63}$$

which proves (4.58). The proof of (4.59) is similar.

Calculations yielding expressions for the area of S^{n-1} will be given in the appendix to this chapter.

Note that the equation

$$\Delta\Phi = \delta \text{ on } \mathbb{R}^n, \tag{4.64}$$

with $\Phi \in \mathcal{S}'(\mathbb{R}^n)$, is equivalent to

$$-|\xi|^2\hat{\Phi} = (2\pi)^{-n/2}. \tag{4.65}$$

If $n \geq 3$, $|\xi|^{-2} \in L^1_{\text{loc}}(\mathbb{R}^n)$ and one solution to (4.65) is

$$\hat{\Phi}(\xi) = -(2\pi)^{-n/2}|\xi|^{-2} \in \mathcal{S}'(\mathbb{R}^n), \tag{4.66}$$

in such a case. We can relate this directly to (4.58) as follows. The orthogonal group $\mathrm{O}(n)$ acts on $\mathcal{S}(\mathbb{R}^n)$, by

$$\pi(g)u(x) = u(g^{-1}x), \quad x \in \mathbb{R}^n,\ g \in \mathrm{O}(n), \tag{4.67}$$

and this extends to an action on $\mathcal{S}'(\mathbb{R}^n)$, via $\langle u, \pi(g)v\rangle = \langle \pi(g^{-1})u, v\rangle$. This action commutes with $\mathcal{F}$. Thus the Fourier transform of an element like $|x|^{2-n}$ which is invariant under the $\mathrm{O}(n)$-action will also be $\mathrm{O}(n)$-invariant. There is also the dilation action on $\mathcal{S}(\mathbb{R}^n)$,

$$D(s)u(x) = u(sx), \quad s > 0,\ x \in \mathbb{R}^n, \tag{4.68}$$

which extends to $\mathcal{S}'(\mathbb{R}^n)$, via $\langle u, D(t)v\rangle = t^{-n}\langle D(1/t)u, v\rangle$. We have

$$D(s)\mathcal{F} = s^{-n}\mathcal{F}D(s^{-1}). \tag{4.69}$$

The element $|x|^{2-n} \in \mathcal{S}'(\mathbb{R}^n)$ is homogeneous of degree $2-n$, that is,

$$D(s)(|x|^{2-n}) = s^{2-n}|x|^{2-n}, \tag{4.70}$$

so $\mathcal{F}(|x|^{2-n})$ will be homogeneous of degree -2. This establishes that $\Phi(x) = C_n|x|^{2-n}$ satisfies (4.66), up to a constant factor. Note that $\delta \in \mathcal{S}'(\mathbb{R}^n)$ is homogeneous of degree $-n$. Since the Laplace operator Δ decreases the order of homogeneity of a distribution by two units, these considerations of orthogonal invariance and homogeneity directly suggest a constant times $|x|^{2-n}$ as a suitable candidate for a fundamental solution for the Laplace operator on $\mathbb{R}^n$.

We mention some extensions of the convolution

$$u * v(x) = \int u(y)v(x-y)\, dy, \tag{4.71}$$

which gives a bilinear map

$$\mathcal{S}(\mathbb{R}^n) \times \mathcal{S}(\mathbb{R}^n) \to \mathcal{S}(\mathbb{R}^n). \tag{4.72}$$

Note that if $u, v, w \in \mathcal{S}(\mathbb{R}^n)$,

$$\langle u * v, w\rangle = \langle u, v^\# * w\rangle, \tag{4.73}$$

where $v^\#(x) = v(-x)$, so the convolution extends in a straightforward way to

$$\mathcal{S}(\mathbb{R}^n) \times \mathcal{S}'(\mathbb{R}^n) \to \mathcal{S}'(\mathbb{R}^n), \tag{4.74}$$

with $\mathcal{S}(\mathbb{R}^n) \times \mathcal{E}'(\mathbb{R}^n) \to \mathcal{S}(\mathbb{R}^n)$, and hence

$$\mathcal{E}'(\mathbb{R}^n) \times \mathcal{S}'(\mathbb{R}^n) \to \mathcal{S}'(\mathbb{R}^n). \tag{4.75}$$

In either case, the identity

$$\mathcal{F}(u * v) = (2\pi)^{n/2}\hat{u}(\xi)\hat{v}(\xi) \tag{4.76}$$

continues to hold. For more on this, see Exercises 11–13 below. If P is any constant-coefficient differential operator, then

$$P(u * v) = (Pu) * v = u * Pv \tag{4.77}$$

in cases (4.74) and (4.75). For example, if $\Phi \in \mathcal{S}'(\mathbb{R}^n)$ and

$$P\Phi = \delta \tag{4.78}$$

(we say Φ is a fundamental solution of P), then a solution to

$$Pu = f, \tag{4.79}$$

for any given $f \in \mathcal{E}'(\mathbb{R}^n)$, is provided by

$$u = f * \Phi. \tag{4.80}$$

An object more general than a tempered distribution is a *distribution*. In general, a distribution on $\mathbb{R}^n$ is a continuous linear map

$$w : C_0^\infty(\mathbb{R}^n) \longrightarrow \mathbb{C}. \tag{4.81}$$

Here, continuity can be characterized as follows. For each $\varphi \in C_0^\infty(\mathbb{R}^n)$, the identity $\langle u, \varphi w\rangle = \langle \varphi u, w\rangle$ makes φw a linear functional on $C^\infty(\mathbb{R}^n)$. We require that each such linear functional be continuous, in the sense specified in (4.40). For further discussion, including a direct discussion of the natural "inductive limit" topology on $C_0^\infty(\mathbb{R}^n)$, see [RS] and [Sch]. The space of all distributions on $\mathbb{R}^n$ is denoted by

$$\mathcal{D}'(\mathbb{R}^n). \tag{4.82}$$

More generally, if M is any smooth, paracompact manifold, the space of continuous linear functionals on $C^\infty(M)$ is denoted by $\mathcal{E}'(M)$ and the space of continuous linear functionals on $C_0^\infty(M)$ is denoted by $\mathcal{D}'(M)$. Of course, if M is compact, $\mathcal{E}'(M) = \mathcal{D}'(M)$.

The case $M = \mathbb{T}^n$ is of interest, with respect to Fourier series. Given $w \in \mathcal{D}'(\mathbb{T}^n)$, we can define

$$\mathcal{F}w(k) = \hat{w}(k) = (2\pi)^{-n}\langle e_k, w\rangle, \tag{4.83}$$

where

$$e_k(\theta) = e^{-ik\cdot\theta} \in C^\infty(\mathbb{T}^n). \tag{4.84}$$

Since w must satisfy an estimate of the form

$$|\langle u, w\rangle| \le C\|u\|_{C^\ell(\mathbb{T}^n)}, \tag{4.85}$$

it is clear that

$$\mathcal{F} : \mathcal{D}'(\mathbb{T}^n) \longrightarrow s'(\mathbb{Z}^n), \tag{4.86}$$

where

$$s'(\mathbb{Z}^n) = \{a : \mathbb{Z}^n \to \mathbb{C} : |a(k)| \le C\langle k\rangle^N \text{ for some } C, N\} \tag{4.87}$$

consists of polynomially bounded functions on $\mathbb{Z}^n$. Note that $s'(\mathbb{Z}^n)$ is the dual space to $s(\mathbb{Z}^n)$, defined in §1, and the map $\mathcal{F}$ in (4.86) is the adjoint of the map $\mathcal{F}^* : s(\mathbb{Z}^n) \longrightarrow C^\infty(\mathbb{T}^n)$ given by (1.11) and (1.12). Here we use the Hermitian inner product $(u, w) = \langle u, \overline{w}\rangle = \overline{\langle \overline{u}, w\rangle}$. The map $\mathcal{F} : C^\infty(\mathbb{T}^n) \to s(\mathbb{Z}^n)$ given by (1.1) and (1.6) also has an adjoint

$$\mathcal{F}^* : s'(\mathbb{Z}^n) \to \mathcal{D}'(\mathbb{T}^n), \tag{4.88}$$

extending the map (1.11)–(1.12), which, we recall, is

$$(\mathcal{F}^* a)(\theta) = \sum_{k\in\mathbb{Z}^n} a(k)e^{ik\cdot\theta}. \tag{4.89}$$

The Fourier inversion formulas

$$\mathcal{F}^*\mathcal{F} = I \text{ on } C^\infty(\mathbb{T}^n), \quad \mathcal{F}\mathcal{F}^* = I \text{ on } s(\mathbb{Z}^n), \tag{4.90}$$

extend by duality (or by continuity, and denseness of $C^\infty(\mathbb{T}^n)$ in $\mathcal{D}'(\mathbb{T}^n)$ and of $s(\mathbb{Z}^n)$ in $s'(\mathbb{Z}^n)$) to

$$\mathcal{F}^*\mathcal{F} = I \text{ on } \mathcal{D}'(\mathbb{T}^n), \quad \mathcal{F}\mathcal{F}^* = I \text{ on } s'(\mathbb{Z}^n); \tag{4.91}$$

consequently, the map (4.86) is an isomorphism.

Exercises

1. Define $M_f u$ by $\langle v, M_f u\rangle = \langle fv, u\rangle$, for $v \in \mathcal{S}(\mathbb{R}^n)$, $u \in \mathcal{S}'(\mathbb{R}^n)$. $M_f u$ is also denoted by fu. Show that $M_f : \mathcal{S}'(\mathbb{R}^n) \to \mathcal{S}'(\mathbb{R}^n)$ continuously, provided $f \in C^\infty(\mathbb{R}^n)$ and each derivative is polynomially bounded, that is, $|D^\alpha f(x)| \le C_\alpha\langle x\rangle^{N(\alpha)}$.
2. Show that the identity $\xi^\alpha D_\xi^\beta \mathcal{F}f(\xi) = (-1)^{|\beta|}\mathcal{F}(D^\alpha x^\beta f)(\xi)$ from §3 continues to hold for $f \in \mathcal{S}'(\mathbb{R}^n)$.
3. Calculate $\mathcal{F}$ of x^α and of $D^\alpha\delta$.
4. Verify the identity (4.32) involving $PV(1/\xi)$.
5. Give a proof of Proposition 4.4, that $\mathcal{S}(\mathbb{R}^n)$ is dense in $\mathcal{S}'(\mathbb{R}^n)$, using convolutions in place of the Fourier transform.
6. Show the denseness of $\mathcal{S}(\mathbb{R}^n)$ in $\mathcal{S}'(\mathbb{R}^n)$ follows from the Hahn-Banach theorem. See Appendix A for a discussion of the Hahn-Banach theorem. On the other hand, sharpen the proof of Proposition 4.4, to obtain *sequential* denseness.

7. Prove the identity (4.45), used in the proof of Proposition 4.5.
(*Hint*: Fix $\eta \in C^\infty(\mathbb{R}^n)$ so that $\eta(x) = 0$ for $|x| \le 1/2$, 1 for $|x| \ge 1$. Show that $\eta(Rx)u \to u$ in $\mathcal{B}_k$, as $R \to \infty$, for any $u \in \mathcal{E}_k$. This, plus a couple of further approximations, yields (4.45).)
8. Let $f(x, y) = 1/z \in \mathcal{S}'(\mathbb{R}^2)$. Using (4.53), compute $\mathcal{F}f \in \mathcal{S}'(\mathbb{R}^2)$. Using Proposition 4.9, compute the Fourier transform of $\log |x|$ on $\mathbb{R}^2$, and of $|x|^{2-n}$ on $\mathbb{R}^n$, $n \ge 3$. Reconsider this problem after reading §8.
9. Let $u \in \mathcal{E}'(\mathbb{R}^n)$, and suppose $\langle p, u \rangle = 0$ for every *polynomial* p on $\mathbb{R}^n$. Show that $u = 0$. (*Hint*: Show that $D^\alpha \hat{u}(0) = 0$ for all α; but $\hat{u}$ is analytic.)
Show that this result implies the Weierstrass approximation theorem, discussed in Exercise 15 of §3.
10. Let $f \in C^\infty(\mathbb{R}^n)$ be real-valued, $\Sigma = \{x : f(x) = 0\}$. Define $\delta(f(x)) \in \mathcal{D}'(\mathbb{R}^n)$ to be

$$\delta(f(x)) = \lim_{\varepsilon \searrow 0} \delta_\varepsilon(f(x)), \tag{4.92}$$

where $\delta_\varepsilon(x) = 1/\varepsilon$ for $|t| < \varepsilon/2$, 0 otherwise, provided this limit exists, with respect to the weak* topology on $\mathcal{D}'(\mathbb{R}^n)$. Show that if $\nabla f \ne 0$ on Σ, the limit does exist and that, for $u \in C_0^\infty(\mathbb{R}^n)$,

$$\big\langle u, \delta(f(x))\big\rangle = \int_\Sigma u(y) |\nabla f(y)|^{-1}\, dS(y),$$

where dS is the $(n-1)$-dimensional measure on Σ. Consider cases where the limit in (4.92) exists though ∇f vanishes on a variety in Σ.
11. Using an argument like that in the proof of Proposition 4.5, show that if $u \in \mathcal{S}'(\mathbb{R}^n)$ has support in a closed ball B, then, for some C, k,

$$|\langle f, u\rangle| \le C \sup_{x \in B, |\alpha| \le k} |D^\alpha f(x)|.$$

(*Hint*: Establish the following analogue of (4.45). If $\mathcal{E}_B$ is the linear space of elements of $C_0^\infty(\mathbb{R}^n)$ vanishing on a neighborhood of B, and $\mathcal{E}_k$ is the closure of $\mathcal{E}_B$ in $\mathcal{B}_k$, then

$$\mathcal{E}_k = \{u \in \mathcal{B}_k : u = 0 \text{ on } B\}.$$

Then show that if $u : \mathcal{B}_k \to \mathbb{C}$ is continuous and supp $u \subset B$,

$$\langle f, u \rangle = \langle E\rho(f), u\rangle,$$

where $\rho(f) = f\big|_B$ and $E : C^k(B) \to \mathcal{B}_k$ is an extension operator. For help in constructing E, look ahead to §4 of Chapter 4.)
12. If $u \in \mathcal{E}'(\mathbb{R}^n)$, show that $\hat{u} \in C^\infty(\mathbb{R}^n)$ satisfies an estimate

$$|\hat{u}(\xi)| \le C\langle \xi \rangle^m, \quad \xi \in \mathbb{R}^n, \tag{4.93}$$

for some $m \in \mathbb{R}$. More generally, show that if a distribution u has support in $B_R = \{x \in \mathbb{R}^n : |x| \le R\}$, then

$$|\hat{u}(\xi + i\eta)| \le C\langle \xi + i\eta\rangle^m\, e^{R|\eta|}, \quad \xi, \eta \in \mathbb{R}^n. \tag{4.94}$$

(*Hint*. Use (4.39)–(4.41). For (4.94), use the result of Exercise 11.)
13. Given the formula (4.76) for $\mathcal{F}(u * v)$ when $u \in \mathcal{S}(\mathbb{R}^n)$, $v \in \mathcal{S}'(\mathbb{R}^n)$, show that if $u \in \mathcal{S}(\mathbb{R}^n)$ and $v \in \mathcal{E}'(\mathbb{R}^n)$, then $\mathcal{F}(u * v) \in \mathcal{S}(\mathbb{R}^n)$, hence $u * v \in \mathcal{S}(\mathbb{R}^n)$, as asserted above (4.75). (*Hint*: Use (4.93).)
14. Show that the convolution product extends to

$$\mathcal{E}'(\mathbb{R}^n) \times \mathcal{D}'(\mathbb{R}^n) \longrightarrow \mathcal{D}'(\mathbb{R}^n), \quad \mathcal{E}'(\mathbb{R}^n) \times \mathcal{E}'(\mathbb{R}^n) \to \mathcal{E}'(\mathbb{R}^n).$$

15. Given $u \in \mathcal{E}'(\mathbb{R}^n)$, show that there exist $k \in \mathbb{Z}^+$, $f \in L^2(\mathbb{R}^n)$ such that $u = (1-\Delta)^k f$. (*Hint*: Obtain $\langle\xi\rangle^{-2k+m} \in L^2(\mathbb{R}^n)$.)
Show that there exist *compactly supported* $f_\alpha \in L^2(\mathbb{R}^n)$ such that $u = \sum_{|\alpha|\le k} D^\alpha f_\alpha$.
16. Assume that $u \in \mathcal{S}'(\mathbb{R}^n)$ and $\hat{u}$ is holomorphic in $\mathbb{C}^n$ and satisfies (4.94). Show that u is supported in the ball B_R. This is the distributional version of the Paley-Wiener theorem.
(*Hint*: Pick $\varphi \in C_0^\infty(\mathbb{R}^n)$, supported in B_1, $\int \varphi\, dx = 1$, let $\varphi_\varepsilon(x) = \varepsilon^{-n}\varphi(x/\varepsilon)$, and consider $u_\varepsilon = \varphi_\varepsilon * u \in \mathcal{S}(\mathbb{R}^n)$. Apply Exercises 17 and 18 of §3.)

5. The classical evolution equations

In this section we analyze solutions to the classical heat equation on $\mathbb{R}^+ \times \mathbb{R}^n$, to the Laplace equation on $\mathbb{R}^{n+1}_+$, and to the wave equation on $\mathbb{R} \times \mathbb{R}^n$. We begin with the heat equation for $u = u(t,x)$,

$$\frac{\partial u}{\partial t} - \Delta u = 0, \tag{5.1}$$

where Δ is the Laplace operator on $\mathbb{R}^n$,

$$\Delta u = \frac{\partial^2 u}{\partial x_1^2} + \cdots + \frac{\partial^2 u}{\partial x_n^2}. \tag{5.2}$$

We pose an initial condition

$$u(0,x) = f(x). \tag{5.3}$$

We suppose that $f \in \mathcal{S}'(\mathbb{R}^n)$, and we look for a solution $u \in C^\infty(\overline{\mathbb{R}}^+, \mathcal{S}'(\mathbb{R}^n))$, via Fourier analysis. Taking the Fourier transform of u with respect to x, we obtain the ODE with parameters

$$\frac{\partial \hat{u}}{\partial t} = -|\xi|^2 \hat{u}(t,\xi), \tag{5.4}$$

with initial condition

$$\hat{u}(0,\xi) = \hat{f}(\xi). \tag{5.5}$$

The unique solution to (5.4)–(5.5) is

$$\hat{u}(t,\xi) = e^{-t|\xi|^2} \hat{f}(\xi). \tag{5.6}$$

Set

$$G(t,\xi) = e^{-t|\xi|^2}. \tag{5.7}$$

Note that, for each $t > 0$, $G(t,\cdot) \in \mathcal{S}(\mathbb{R}^n)$. By (4.76), we have

$$u(t,x) = (2\pi)^{-n/2}\hat{G}(t,\cdot) * f(x). \tag{5.8}$$

The computation of the Fourier transform of such a Gaussian function was made in §3. From (3.18), we deduce that

$$u(t,x) = p(t,\cdot) * f(x), \tag{5.9}$$

where

$$p(t,x) = (4\pi t)^{-n/2} e^{-|x|^2/4t}, \tag{5.10}$$

for $t > 0$. The function $p(t,x)$ is called the fundamental solution to the heat equation. It satisfies

$$\begin{aligned} &(\partial/\partial t - \Delta)p = 0, \quad \text{for } t > 0, \\ &\lim_{t \searrow 0} p(t,x) = \delta(x) \ \text{ in } \mathcal{S}'(\mathbb{R}^n). \end{aligned} \tag{5.11}$$

We record what Fourier analysis has yielded for the heat equation.

Proposition 5.1. *The heat equation (5.1)–(5.3), with $f \in \mathcal{S}'(\mathbb{R}^n)$, has a unique solution $u \in C^\infty(\overline{\mathbb{R}}^+, \mathcal{S}'(\mathbb{R}^n))$, given by (5.9)–(5.10). The solution is C^∞ on $(0,\infty) \times \mathbb{R}^n$. If $f \in \mathcal{S}(\mathbb{R}^n)$, then $u \in C^\infty(\overline{\mathbb{R}}^+, \mathcal{S}(\mathbb{R}^n))$.*

Note carefully that uniqueness of the solution is asserted only within the class $C^\infty(\overline{\mathbb{R}}^+, \mathcal{S}'(\mathbb{R}^n))$; this entails bounds on the solution considered, near infinity. If one removes such growth restrictions, uniqueness fails. There exist nontrivial solutions to

$$\Big(\frac{\partial}{\partial t} - \Delta\Big) v = 0, \quad \text{for } t > 0, \quad v(0,x) = 0, \tag{5.12}$$

outlined in Exercise 2 at the end of this section. For fixed $t > 0$, such solutions blow up too fast to belong to $\mathcal{S}'(\mathbb{R}^n)$.

In view of the boundedness and continuity properties of (5.7), we have the following:

Corollary 5.2. *Suppose $f \in L^2(\mathbb{R}^n)$. Then the solution u to (5.1)–(5.3) of Proposition 5.1 belongs to $C(\overline{\mathbb{R}}^+, L^2(\mathbb{R}^n))$.*

We cannot say that $u \in C^\infty(\overline{\mathbb{R}}^+, L^2(\mathbb{R}^n))$, or even that $\partial u/\partial t$ belongs to $C(\overline{\mathbb{R}}^+, L^2(\mathbb{R}^n))$, in such a case, without further restrictions on f. The appropriate behavior of $\partial^j u/\partial t^j$ in such a case is best described in terms of *Sobolev spaces*, which will be discussed in Chapter 4.

Next we look at the following boundary problem for functions harmonic in an upper half space:

$$\Big(\frac{\partial^2}{\partial y^2} + \Delta\Big) u(y,x) = 0, \quad \text{for } y > 0, \ x \in \mathbb{R}^n, \tag{5.13}$$

$$u(0,x) = f(x). \tag{5.14}$$

Here, Δ is given by (5.2). In view of such simple examples as

$$u(y,x) = y, \tag{5.15}$$

which satisfy (5.13) and (5.14) with $f = 0$, we will need to make appropriate restrictions on u in order to obtain uniqueness. As before, we suppose that $f \in$

$\mathcal{S}'(\mathbb{R}^n)$ and look for $u \in C^\infty(\overline{\mathbb{R}}^+, \mathcal{S}'(\mathbb{R}^n))$. Fourier transforming with respect to x gives the second-order ODE, with parameters,

$$\left(\frac{d^2}{dy^2} - |\xi|^2\right)\hat{u}(y,\xi) = 0, \tag{5.16}$$

$$\hat{u}(0,\xi) = \hat{f}(\xi). \tag{5.17}$$

The general solution to (5.16)–(5.17), for any fixed $\xi \neq 0$, is

$$\hat{u}(y,\xi) = c_0(\xi)e^{y|\xi|} + c_1(\xi)e^{-y|\xi|}, \tag{5.18}$$

with $c_0(\xi) + c_1(\xi) = \hat{f}(\xi)$. Let us restrict attention to f such that $\hat{f}(\xi)$ is continuous, and look for $\hat{u}(y,\xi)$ continuous in (y,ξ); as (5.15) illustrates, things can be more complicated if $\hat{u}(y,\cdot)$ is a singular distribution near $\xi = 0$. In view of the blow-up of $e^{y|\xi|}$ as $y \nearrow \infty$, it is natural to require that $c_0(\xi) = 0$, so

$$\hat{u}(y,\xi) = e^{-y|\xi|}\hat{f}(\xi). \tag{5.19}$$

In partial analogy with Proposition 5.1, we have obtained the following result.

Proposition 5.3. *Let $f \in \mathcal{S}'(\mathbb{R}^n)$, and suppose $\hat{f}$ is continuous. Then there is a unique solution $u(y,x)$ of (5.13) and (5.14), belonging to the space $C^\infty(\overline{\mathbb{R}}^+, \mathcal{S}'(\mathbb{R}^n))$ and satisfying the condition that $\hat{u}(y,\xi)$ is continuous on $\overline{\mathbb{R}}^+ \times \mathbb{R}^n$, and furthermore that $u(y,\cdot)$ is a bounded function of y taking values in $\mathcal{S}'(\mathbb{R}^n)$. It is given by (5.19).*

Note that $\hat{f}(\xi)$ is continuous provided f is a finite measure. It is also continuous if $f \in \mathcal{E}'(\mathbb{R}^n)$.

We want to find the "fundamental solution" $P(y,x)$ for (5.13)–(5.14), corresponding to $f = \delta$. In other words,

$$\hat{P}(y,\xi) = (2\pi)^{-n/2}e^{-y|\xi|}. \tag{5.20}$$

This computation is elementary in the case $n = 1$. We have

$$\begin{aligned} P(y,x) &= (2\pi)^{-1}\int_{-\infty}^{\infty} e^{-y|\xi|+ix\xi}\,d\xi \\ &= (2\pi)^{-1}\left[\int_0^\infty e^{-y\xi+ix\xi}\,d\xi + \int_0^\infty e^{-y\xi-ix\xi}\,d\xi\right] \\ &= (2\pi)^{-1}\left[(y-ix)^{-1} + (y+ix)^{-1}\right] \\ &= \frac{1}{\pi}\,\frac{y}{y^2+x^2}. \end{aligned} \tag{5.21}$$

For $n > 1$, a direct calculation of such a Fourier transform is not so elementary. One way to perform the computation is to use the following *subordination identity*:

$$e^{-yA} = \frac{y}{2\pi^{1/2}}\int_0^\infty e^{-y^2/4t}\,e^{-tA^2}\,t^{-3/2}\,dt, \quad A > 0,\ y > 0. \tag{5.22}$$

We will give a proof of (5.22) shortly. First we will show how it leads to a computation of the Fourier transform of (5.20). We let $A = |\xi|$, and we use our prior computation of the Fourier transform of $e^{-|\xi|^2}$. Thus, for any $n \geq 1$,

$$(5.23)\quad \begin{aligned} P(y,x) &= (2\pi)^{-n} \int e^{-y|\xi|+ix\cdot\xi}\, d\xi \\ &= (2\pi)^{-n} \frac{y}{2\pi^{1/2}} \int_0^\infty e^{-y^2/4t} \left[\int e^{-t|\xi|^2+ix\cdot\xi}\, d\xi\right] t^{-3/2}\, dt, \end{aligned}$$

and substituting in the calculation (5.7)–(5.10), we have

$$(5.24)\quad \begin{aligned} P(y,x) &= (4\pi)^{-(n+1)/2} y \int_0^\infty e^{-y^2/4t} e^{-|x|^2/4t} t^{-(n+3)/2}\, dt \\ &= c_n \frac{y}{(y^2+|x|^2)^{(n+1)/2}}, \end{aligned}$$

where the last integral is evaluated using the substitution $s = 1/t$. The constant c_n is given by

$$(5.25)\quad c_n = \pi^{-(n+1)/2}\,\Gamma\Big(\frac{n+1}{2}\Big),$$

where $\Gamma(z) = \int_0^\infty e^{-s} s^{z-1} ds$ is Euler's gamma function, which is discussed in the appendix to this chapter. Note that $c_1 = 1/\pi$, so the calculation (5.24) agrees with (5.21), in the case $n = 1$.

The observation that the calculations (5.21) and (5.24) coincide for $n = 1$ can be used to provide a simple proof of the subordination identity (5.22). Indeed, with $|\xi|$ substituted for A in (5.22), we know that the operation of Fourier multiplication by the left side coincides with the operation of Fourier multiplication by the right side, so the two functions of $|\xi|$ ($\xi \in \mathbb{R}$) must coincide.

There are other proofs of (5.22). In Appendix A we note the equivalence of such an identity and a classical identity involving Euler's gamma function. While the proof of (5.22) given above is complete, it leaves one with an unsatisfied feeling, since the right side of the formula (5.22) seems to have been pulled out of a hat. We want to introduce a setting where such a formula arises naturally, a setting involving the use of operator notations, as follows. For a decent function f defined on $[0,\infty)$, define $f(\sqrt{-\Delta})$ on $\mathcal{S}(\mathbb{R}^n)$, or on $L^2(\mathbb{R}^n)$, or even on $\mathcal{S}'(\mathbb{R}^n)$, when it makes sense, by

$$(5.26)\quad \big(f(\sqrt{-\Delta})u\big)^\wedge(\xi) = f(|\xi|)\hat{u}(\xi).$$

Thus, the content of (5.10) is

$$(5.27)\quad e^{t\Delta}\delta(x) = (4\pi t)^{-n/2} e^{-|x|^2/4t},$$

for $t > 0$. The formula (5.24) is a formula for $e^{-y\sqrt{-\Delta}}\delta(x)$, $x \in \mathbb{R}^n$, and the formula (5.21) is the special case of this for $n = 1$.

We will approach the subordination identity via the formula

$$(5.28)\quad (\lambda^2 - \Delta)^{-1} = \int_0^\infty e^{-(\lambda^2-\Delta)t}\, dt,$$

for the *resolvent* $(\lambda^2 - \Delta)^{-1}$ of the Laplace operator Δ. This identity follows via the Fourier transform, as in (5.26), from

$$(\lambda^2 + |\xi|^2)^{-1} = \int_0^\infty e^{-(\lambda^2+|\xi|^2)t}\, dt. \tag{5.29}$$

In order to derive the subordination identity, we will apply both sides of (5.28) to δ; we will do this in the special case of $\Delta = d^2/dx^2$ acting on $\mathcal{S}'(\mathbb{R}^1)$. For the special case $\xi \in \mathbb{R}^1$, we have the Fourier integral formula

$$\int_{-\infty}^\infty (\lambda^2 + \xi^2)^{-1} e^{ix\xi}\, d\xi = \frac{\pi}{\lambda} e^{-\lambda|x|} \quad (\lambda > 0), \tag{5.30}$$

a fact that can be established either by residue calculus or by applying the Fourier inversion formula to the computation (5.21). Thus applying both sides of (5.28) to $\delta \in \mathcal{S}'(\mathbb{R})$ and using $e^{t\Delta}\delta = (4\pi t)^{-1/2} e^{-x^2/4t}$ in this case, we have

$$\lambda^{-1} e^{-\lambda|x|} = \frac{1}{\sqrt{\pi}} \int_0^\infty e^{-x^2/4t} e^{-\lambda^2 t} t^{-1/2}\, dt. \tag{5.31}$$

Making the change of variables $y = |x|$, $A = \lambda$, and taking the y-derivative of the resulting identity give (5.22). Also note that taking the A-derivative of (5.22) gives (5.31). The identity (5.31) is also called the subordination identity. One can see that it arises very naturally in this context, from (5.28). We will return to the calculation of $(\lambda^2 - \Delta)^{-1}\delta(x)$ in case $n > 1$, later in this section.

We next consider the wave equation on $\mathbb{R} \times \mathbb{R}^n$:

$$\frac{\partial^2 u}{\partial t^2} - \Delta u = 0, \tag{5.32}$$

with initial data

$$u(0, x) = f(x), \quad u_t(0, x) = g(x). \tag{5.33}$$

As before, we suppose that f and g belong to $\mathcal{S}'(\mathbb{R}^n)$ and look for a solution u in $C^\infty(\mathbb{R}, \mathcal{S}'(\mathbb{R}^n))$. Taking the Fourier transform with respect to x again yields an ODE with parameters:

$$\frac{d^2\hat{u}}{dt^2} + |\xi|^2 \hat{u} = 0, \tag{5.34}$$

$$\hat{u}(0, \xi) = \hat{f}(\xi), \quad \hat{u}_t(0, \xi) = \hat{g}(\xi). \tag{5.35}$$

The general solution to (5.28) (for $\xi \neq 0$) is of the form

$$\hat{u}(t, \xi) = c_2(\xi) \sin t|\xi| + c_1(\xi) \cos t|\xi|,$$

which it is convenient to write as

$$\hat{u}(t, \xi) = c_0(\xi)|\xi|^{-1} \sin t|\xi| + c_1(\xi) \cos t|\xi|,$$

since the right side is well defined for any $c_0, c_1 \in \mathcal{S}'(\mathbb{R}^n)$. The initial conditions (5.35) imply $c_1 = \hat{f}$, $c_0 = \hat{g}$, so the solution to (5.34)–(5.35) is

$$\hat{u}(t, \xi) = \hat{g}(\xi)|\xi|^{-1} \sin t|\xi| + \hat{f}(\xi) \cos t|\xi|. \tag{5.36}$$

This is clearly the unique solution in $C^\infty(\mathbb{R}, \mathcal{S}(\mathbb{R}^n))$, if $f, g \in \mathcal{S}(\mathbb{R}^n)$. The uniqueness in $C^\infty(\mathbb{R}, \mathcal{S}'(\mathbb{R}^n))$ for general $f, g \in \mathcal{S}'(\mathbb{R}^n)$ will be proved shortly.

If $f = 0,\ g = \delta$, the solution given by (5.36) is called the fundamental solution, or the "Riemann function." Of course, it is actually a distribution. It is characterized by

$$\hat{R}(t, \xi) = (2\pi)^{-n/2}|\xi|^{-1} \sin t|\xi|. \tag{5.37}$$

We want a direct formula for $R(t, x)$. We will be able to deduce this formula from the formula (5.24) for the Fourier transform of $e^{-y|\xi|}$, via analytic continuation. To bring in the factor $|\xi|^{-1}$, integrate (5.24) with respect to y. Thus, if

$$\hat{F}(y, \xi) = (2\pi)^{-n/2}|\xi|^{-1}e^{-y|\xi|}, \tag{5.38}$$

which belongs to $\mathcal{S}'(\mathbb{R}^n)$ for each $y \geq 0$ if $n \geq 2$, we have

$$F(y, x) = c_n'(y^2 + |x|^2)^{-(n-1)/2} \tag{5.39}$$

with

$$c_n' = \frac{c_n}{n-1} = \frac{1}{2}\pi^{-(n+1)/2}\Gamma\Big(\frac{n-1}{2}\Big). \tag{5.40}$$

This has been verified for real $y > 0$. But (5.38) is holomorphic in y, with values in $\mathcal{S}'(\mathbb{R}^n)$, for all y such that $\operatorname{Re} y > 0$. Also, it is continuous in the right half-plane $\{y \in \mathbb{C} : \operatorname{Re} y \geq 0\}$. In view of the continuity of the Fourier transform on $\mathcal{S}'(\mathbb{R}^n)$, we deduce that if

$$\hat{\Phi}(t, \xi) = (2\pi)^{-n/2}|\xi|^{-1}e^{it|\xi|}, \tag{5.41}$$

$t \in \mathbb{R}$, then

$$\Phi(t, x) = \lim_{\varepsilon \searrow 0} c_n'\big(|x|^2 - (t - i\varepsilon)^2\big)^{-(n-1)/2}, \tag{5.42}$$

the limit existing in $\mathcal{S}'(\mathbb{R}^n)$ for each $t \in \mathbb{R}$, since $\hat{\Phi}(t, \cdot) = \lim_{\varepsilon \searrow 0} \hat{\Phi}(t - i\varepsilon, \cdot)$ in $\mathcal{S}'(\mathbb{R}^n)$. Consequently, for the Riemann function, we have

$$R(t, x) = \lim_{\varepsilon \searrow 0} c_n' \operatorname{Im}\big(|x|^2 - (t - i\varepsilon)^2\big)^{-(n-1)/2}. \tag{5.43}$$

Note that if $|x| > |t|$, $\lim_{\varepsilon \searrow 0}\big(|x|^2 - (t - i\varepsilon)^2\big)^{-(n-1)/2} = (|x|^2 - t^2)^{-(n-1)/2}$ is real, so

$$R(t, x) = 0, \quad \text{for } |x| > |t|. \tag{5.44}$$

This is a reflection of the *finite propagation speed*, which was discussed in Chapter 2. Note also that if n is odd, then $(n-1)/2$ is an integer, so (5.43) vanishes also for $|x| < |t|$. In other words,

$$n \geq 3 \text{ odd} \Rightarrow \operatorname{supp} R(t, \cdot) \subset \{x \in \mathbb{R}^n : |x| = |t|\}. \tag{5.45}$$

This is the *strict Huygens principle*. Of course, it does not hold when n is even. When $n = 2$, the computation of the limit in (5.43) is elementary. We have

$$R(t,x) = \begin{cases} c_2'(t^2 - |x|^2)^{-1/2} \cdot \operatorname{sgn}(t), & \text{for } |x| < |t|, \\ 0, & \text{for } |x| > |t|, \end{cases} \tag{5.46}$$

for $n = 2$. For $n = 3$, the Plemelj jump relation (4.29) yields

$$R(t,x) = (4\pi t)^{-1} \delta(|x| - |t|). \tag{5.47}$$

The discussion above has to be modified for $n = 1$, since (5.41) is not locally integrable near $\xi = 0$ in that case. This simple case ($n = 1$) was treated in §1 of Chapter 2; see (1.24)–(1.28) in that chapter.

The solution to (5.32)–(5.33) given by (5.36) can be expressed as

$$u(t,\cdot) = R(t,\cdot) * g + \frac{\partial}{\partial t} R(t,\cdot) * f. \tag{5.48}$$

We record our result on solutions to the wave equation.

Proposition 5.4. *Given $f, g \in \mathcal{S}'(\mathbb{R}^n)$, there is a unique solution $u \in C^\infty(\mathbb{R}, \mathcal{S}'(\mathbb{R}^n))$ to the initial-value problem (5.32)–(5.33). It is given by (5.48).*

Proof. The only point remaining to be established is the uniqueness. Suppose $u \in C^\infty(\mathbb{R}, \mathcal{S}'(\mathbb{R}^n))$ solves (5.32)–(5.33) with $f = g = 0$. Then

$$v(t,\cdot) = u(t,\cdot) * \varphi$$

solves the same equation, for any $\varphi \in C_0^\infty(\mathbb{R}^n)$; $\varphi = \varphi(x)$. We have $v \in C^\infty(\mathbb{R} \times \mathbb{R}^n)$. Thus the energy estimates of Chapter 2 are applicable to v, and we have $v = 0$ everywhere. Taking a sequence $\varphi_j \in C_0^\infty(\mathbb{R}^n)$ approaching δ, we have $v_j \to u$; since each $v_j = 0$, it follows that $u = 0$, and the proof is complete.

We note that the argument above yields uniqueness for u in the class $C^\infty(\mathbb{R}, \mathcal{D}'(\mathbb{R}^n))$. We also remark that any $u \in \mathcal{D}'(\mathbb{R} \times \mathbb{R}^n)$ solving (5.32) actually belongs to $C^\infty(\mathbb{R}, \mathcal{D}'(\mathbb{R}^n))$, and that (5.48) gives the unique solution to (5.32)–(5.33) for any $f, g \in \mathcal{D}'(\mathbb{R}^n)$. Justification of these statements is left as an exercise.

Returning to the operator notation (5.26), we have

$$R(t,x) = (-\Delta)^{-1/2} \sin t\sqrt{-\Delta}\, \delta(x) \tag{5.49}$$

and

$$\frac{\partial}{\partial t} R(t,x) = \cos t\sqrt{-\Delta}\, \delta(x). \tag{5.50}$$

We also denote (5.49) by $R(t)$ and (5.50) by $R'(t)$.

Having introduced in (5.28) the notion of synthesizing some operators from other operators, we want to mention the particular desirability of synthesizing functions of the Laplace operator from the fundamental solution of the wave equation. If $f(s)$ is an even function, the Fourier inversion formula implies

$$f(\sqrt{-\Delta}) = (2\pi)^{-1/2} \int_{-\infty}^{\infty} \hat{f}(t) \cos t\sqrt{-\Delta}\, dt. \tag{5.51}$$

Note that

$$\cos t\sqrt{-\Delta}\, u = R'(t) * u, \tag{5.52}$$

where $R(t)$ is the Riemann function constructed above. We have the following rather general calculation of $f(\sqrt{-\Delta})\delta$, using the formula (5.37) for the Riemann function on $\mathbb{R}^n$.

Proposition 5.5. *Let $f \in \mathcal{S}(\mathbb{R})$ be even. Then, on $\mathbb{R}^n$, we have*

$$f(\sqrt{-\Delta})\delta(x) = \frac{1}{\sqrt{2\pi}}\Big[-\frac{1}{2\pi r}\frac{\partial}{\partial r}\Big]^k \hat{f}(r) \tag{5.53}$$

if $n = 2k+1$ is odd, and

$$f(\sqrt{-\Delta})\delta(x) = \frac{1}{\sqrt{\pi}} \int_r^{\infty} \Big\{\Big[-\frac{1}{2\pi s}\frac{\partial}{\partial s}\Big]^k \hat{f}(s)\Big\} \frac{s}{\sqrt{s^2 - r^2}}\, ds \tag{5.54}$$

if $n = 2k$ is even. Here, $r = |x|$.

Proof. When $n = 3$, we have

$$\begin{aligned} f(\sqrt{-\Delta})\delta &= -\frac{2}{\sqrt{2\pi}} \int_0^{\infty} \hat{f}'(t)(4\pi t)^{-1}\delta(r-t)\, dt \\ &= \frac{1}{\sqrt{2\pi}}\Big[-\frac{1}{2\pi r}\frac{\partial}{\partial r}\Big]\hat{f}(r), \end{aligned} \tag{5.55}$$

giving (5.53) in this case. When $n = 2$, we have, from (5.46),

$$f(\sqrt{-\Delta})\delta = -\frac{2c_2'}{\sqrt{2\pi}} \int_r^{\infty} \hat{f}'(t)(t^2 - r^2)^{-1/2}\, dt, \tag{5.56}$$

giving (5.54) in this case, once one checks that $c_2' = (1/2)\pi^{-3/2}\Gamma(1/2) = 1/2\pi$. To pass to the general case, we note that if $R_n(t,r)$ denotes the formula (5.43) for the Riemann function, in view of the evaluation of c_n' in (5.40), we have the formal relation

$$R_{n+2}(t,r) = \Big[-\frac{1}{2\pi r}\frac{\partial}{\partial r}\Big] R_n(t,r), \tag{5.57}$$

so (5.53) and (5.54) follow by induction.

It is clear that Proposition 5.5 holds for a more general class of even functions f than those in $\mathcal{S}(\mathbb{R})$, by simple limiting arguments. For example, the function $f(s) = (\lambda^2+s^2)^{-1}$, $\lambda > 0$, giving the resolvent of Δ, can be treated. We leave the formulation of general results on classes of f which can be treated as an exercise.

In the case of using Proposition 5.5 to treat the resolvent of Δ, we have the following formula. With $f(s) = (\lambda^2 + s^2)^{-1}$, from (5.30) we have

$$\hat{f}(t) = \left(\frac{\pi}{2}\right)^{1/2} \lambda^{-1} e^{-\lambda|t|}, \tag{5.58}$$

to plug into (5.53)–(5.54). For example, for $n = 3$, we have

$$(\lambda^2 - \Delta)^{-1}\delta = (4\pi|x|)^{-1} e^{-\lambda|x|} \quad \text{on } \mathbb{R}^3. \tag{5.59}$$

Note that computing $(\lambda^2 - \Delta)^{-1}\delta$ by evaluating the right side of (5.28) gives in this case

$$(\lambda^2 - \Delta)^{-1}\delta = \int_0^\infty e^{-|x|^2/4t} e^{-\lambda^2 t} (4\pi t)^{-3/2}\, dt; \tag{5.60}$$

comparing (5.59) and (5.60) again reveals the subordination identity, in the original form (5.22).

The fact that the answer comes out "in closed form" for n odd is a consequence of the strict Huygens principle. For n even, one tends not to get elementary functions. Note that, for $n = 2$, the formula (5.54) gives

$$(\lambda^2 - \Delta)^{-1}\delta = c \int_{|x|}^\infty e^{-\lambda s} (s^2 - r^2)^{-1/2}\, ds \quad \text{on } \mathbb{R}^2; \tag{5.61}$$

the formula (5.28) gives

$$(\lambda^2 - \Delta)^{-1}\delta = \frac{1}{4\pi} \int_0^\infty e^{-|x|^2/4t - t}\, t^{-1}\, dt \quad \text{on } \mathbb{R}^2. \tag{5.62}$$

Both of these integrals can be expressed in terms of the modified Bessel function K_0; we say a little more about this in the next section; see (6.46)–(6.54).

In general, the use of results on the wave equation together with (5.51) provides a tool of tremendous power and flexibility in the analysis of numerous functions of the Laplace operator. We will see more of this in Chapter 8.

Exercises

1. A function $g \in C^\infty(\mathbb{R}^n)$ is said to be in the Gevrey class $G^\sigma(\mathbb{R}^n)$ provided, for each compact $K \subset \mathbb{R}^n$, there exist C and R, such that

$$|D^\alpha g(x)| \le C R^k k^{\sigma k}, \quad |\alpha| = k,\ x \in K. \tag{5.63}$$

   The class $G^1(\mathbb{R}^n)$ is equal to the space of real-analytic functions on $\mathbb{R}^n$. If $\sigma > 1$, show that there exist *compactly supported* elements of $G^\sigma(\mathbb{R}^n)$, not identically zero.
   Remark: This is part of the Denjoy-Carleman theorem; see [Ru].
2. Consider the following "sideways heat equation" for $u = u(t,x)$ on $\mathbb{R} \times \mathbb{R}$:

$$u_t = u_{xx}, \quad u(t,0) = g(t),\ u_x(t,0) = 0.$$

Show that if $g \in G^\sigma(\mathbb{R})$ for some $\sigma \in (1,2)$, then a solution on all of $\mathbb{R} \times \mathbb{R}$ is given by the convergent series

$$u(t,x) = \sum_{k=0}^{\infty} \frac{x^{2k}}{(2k)!} g^{(k)}(t), \tag{5.64}$$

which is the power series for the "formal" object $\left(\cos x\sqrt{-\partial/\partial t}\right)g$. Using Exercise 1, find nontrivial solutions to $u_t = u_{xx}$ which are supported in a strip $a \le t \le b$. This construction is due to J. Rauch.
(*Hint*: To prove convergence, use Stirling's formula:

$$n! \sim (2\pi n)^{1/2} \cdot e^{-n} \cdot n^n \quad \text{as } n \to \infty.) \tag{5.65}$$

3. Given the formula (5.37) for $\hat{R}(t,\xi)$, that is,

$$\hat{R}(t,\xi) = (2\pi)^{-n/2}\, |\xi|^{-1} \sin t|\xi|, \quad \xi \in \mathbb{R}^n,$$

show that the fact that $R(t,\cdot) \in \mathcal{S}'(\mathbb{R}^n)$ is supported in $B_{|t|} = \{x \in \mathbb{R}^n : |x| \le |t|\}$ follows from the Paley-Wiener theorem for distributions, given in Exercise 16 of §4.
4. Making use of (5.43), show that when $n = 2k+1$ is odd,

$$\frac{\sin t\sqrt{-\Delta}}{\sqrt{-\Delta}} f(x) = C_k \left(\frac{1}{t}\frac{\partial}{\partial t}\right)^{k-1} \left(t^{2k-1}\overline{f}(x,t)\right), \tag{5.66}$$

where $\overline{f}(x,t)$ is the mean value of f on $\{y \in \mathbb{R}^n : |x-y| = |t|\}$, and $C_k = 1/(2k-1)!!$, where $(2k-1)!! = 3 \cdot 5 \cdots (2k-1)$. Use $f(x) = 1$ to check the value of C_k.

6. Radial distributions, polar coordinates, and Bessel functions

The rotational invariance of the Laplace operator on $\mathbb{R}^n$ directly suggests the use of polar coordinates; one has

$$\Delta = \frac{\partial^2}{\partial r^2} + \frac{n-1}{r}\frac{\partial}{\partial r} + \frac{1}{r^2}\Delta_S, \tag{6.1}$$

where Δ_S is the Laplace operator on the unit sphere S^{n-1}. This formula has been used in (4.56) and follows from the formula given in Chapter 2 for the Laplace operator in a general coordinate system; see (4.4) in that chapter.

Related is the fact that in treating the equations of §5 via Fourier analysis, one computes the Fourier transforms of various radial functions, such Fourier transforms also being radial functions (or rotationally invariant tempered distributions). Bessel functions arise naturally in either approach, and we will develop a little of the theory of Bessel functions here. More results on Bessel functions will appear in Chapter 8, which discusses spectral theory, and in Chapter 9, which treats scattering theory. One can find a great deal more material on this subject in the treatise [Wat].

We begin by considering the Fourier transform of a radial function, $F(x) = f(r)$, $r = |x|$. We have

$$\hat{F}(\xi) = (2\pi)^{-n/2} \int_0^\infty f(r)\psi_n(r|\xi|)r^{n-1}\,dr, \tag{6.2}$$

where

$$\psi_n(|\xi|) = \Psi_n(\xi) \tag{6.3}$$

with

$$\Psi_n(\xi) = \int_{S^{n-1}} e^{i\xi\cdot\omega}\,dS(\omega). \tag{6.4}$$

In other words, with A_{n-2} the volume of S^{n-2},

$$\psi_n(r) = A_{n-2}\int_{-1}^{1} e^{irs}(1-s^2)^{(n-3)/2}\,ds. \tag{6.5}$$

From (A.4) we have $A_{n-2} = 2\pi^{(n-1)/2}\big/\Gamma\big((n-1)/2\big)$. It is common to write

$$\psi_n(r) = (2\pi)^{n/2} r^{1-n/2} J_{n/2-1}(r), \tag{6.6}$$

where, for general ν satisfying $\operatorname{Re}\nu > -1/2$, the Bessel function $J_\nu(z)$ is defined to be

$$J_\nu(z) = \Big[\Gamma\Big(\frac{1}{2}\Big)\Gamma\Big(\nu+\frac{1}{2}\Big)\Big]^{-1}\Big(\frac{z}{2}\Big)^\nu \int_{-1}^{1}(1-t^2)^{\nu-1/2}e^{izt}\,dt. \tag{6.7}$$

For example, $\psi_2(r) = 2\pi J_0(r)$. Since $(1-t^2)^{\nu-1/2}$ is even in t, one can replace e^{izt} by $\cos zt$ in this formula. Now, (6.2) becomes

$$\hat{F}(\xi) = |\xi|^{1-n/2}\int_0^\infty f(r)J_{n/2-1}(r|\xi|)r^{n/2}\,dr. \tag{6.8}$$

We want to consider the ODE, known as Bessel's equation, solved by $J_\nu(r)$. First we consider the case $\nu = n/2-1$. Since Ψ_n is the Fourier transform of a measure supported on the unit sphere, we have that

$$(\Delta+1)\Psi_n = 0. \tag{6.9}$$

Using the polar coordinate expression (6.1) for Δ, we have

$$\Big(\frac{d^2}{dr^2} + \frac{n-1}{r}\frac{d}{dr} + 1\Big)\psi_n(r) = 0. \tag{6.10}$$

Substituting (6.6) yields Bessel's equation

$$\Big[\frac{d^2}{dr^2} + \frac{1}{r}\frac{d}{dr} + \Big(1-\frac{\nu^2}{r^2}\Big)\Big]J_\nu(r) = 0 \tag{6.11}$$

in case $\nu = n/2-1$. We want to verify this for all ν, from the integral formula (6.7). This is an exercise, but we will present the details to one approach, which yields further interesting identities for Bessel functions. For notational simplicity, let us set

$$c(\nu) = \left[\Gamma\left(\frac{1}{2}\right)\Gamma\left(\nu + \frac{1}{2}\right)\right]^{-1} \cdot 2^{-\nu}.$$

Differentiating (6.7) with respect to z yields

$$\frac{d}{dz} J_\nu(z) = \left(\frac{\nu c(\nu)}{z}\right) z^\nu \int_{-1}^{1} e^{izt}(1-t^2)^{\nu-1/2}\,dt \tag{6.12}$$
$$+\, ic(\nu) z^\nu \int_{-1}^{1} e^{izt} t (1-t^2)^{\nu-1/2}\,dt.$$

The first term on the right is equal to $(\nu/z)J_\nu(z)$. The second is equal to

$$\frac{c(\nu)}{z} z^\nu \int_{-1}^{1} \left(\frac{d}{dt} e^{izt}\right) t(1-t^2)^{\nu-1/2}\,dt$$
$$= -\frac{c(\nu)}{z} z^\nu \int_{-1}^{1} e^{izt}\left[(1-t^2)^{\nu-1/2} - (2\nu-1)t^2(1-t^2)^{\nu-1-1/2}\right] dt$$
$$= -\frac{c(\nu)}{z} z^\nu \int_{-1}^{1} e^{izt}\left[2\nu(1-t^2)^{\nu-1/2} - (2\nu-1)(1-t^2)^{\nu-1-1/2}\right] dt$$
$$= -\frac{2\nu}{z} J_\nu(z) + \frac{(2\nu-1)c(\nu)}{c(\nu-1)} J_{\nu-1}(z).$$

Since $c(\nu)/c(\nu-1) = 1/(2\nu-1)$, we have the formula

$$\frac{d}{dz} J_\nu(z) = -\frac{\nu}{z} J_\nu(z) + J_{\nu-1}(z), \tag{6.13}$$

or

$$\left(\frac{d}{dz} + \frac{\nu}{z}\right) J_\nu(z) = J_{\nu-1}(z). \tag{6.14}$$

As we have stated, the formula (6.7) for $J_\nu(z)$ is convergent for Re $\nu > -1/2$. The formula (6.14) provides an analytic continuation for all complex ν. In fact, one can see directly that the integral in (6.7) is meromorphic in ν, with simple poles at $\nu + 1/2 = -1, -2, \ldots$. The factor $\Gamma(\nu + 1/2)^{-1}$ cancels these poles. This serves to explain the desirability of throwing in this factor in the definition (6.7) of $J_\nu(z)$. Of course, the factor $\Gamma(1/2)^{-1}$ is more arbitrary.

Next, we note that

$$J_{\nu+1}(z) = c(\nu+1) z^{\nu+1} \int_{-1}^{1} e^{izt}(1-t^2)^{\nu+1-1/2}\,dt$$
$$= -ic(\nu+1) z^\nu \int_{-1}^{1} \left(\frac{d}{dt} e^{izt}\right)(1-t^2)^{\nu+1-1/2}\,dt$$
$$= -ic(\nu+1)(2\nu+1) z^\nu \int_{-1}^{1} e^{izt} t(1-t^2)^{\nu-1/2}\,dt,$$

and since $c(\nu+1) = c(\nu)/(2\nu+1)$, this is equal to the negative of the second term on the right in (6.12). Hence we have

$$\left(\frac{d}{dz} - \frac{\nu}{z}\right) J_\nu(z) = -J_{\nu+1}(z), \tag{6.15}$$

complementing (6.14). Putting together (6.14) and (6.15), we have

$$\left(\frac{d}{dz} - \frac{\nu-1}{z}\right)\left(\frac{d}{dz} + \frac{\nu}{z}\right) J_\nu(z) = -J_\nu(z), \tag{6.16}$$

which is equivalent to Bessel's equation, (6.11). Note that adding and subtracting (6.14) and (6.15) produce the identities

$$\begin{aligned} 2J_\nu'(z) &= J_{\nu-1}(z) - J_{\nu+1}(z), \\ \frac{2\nu}{z} J_\nu(z) &= J_{\nu-1}(z) + J_{\nu+1}(z). \end{aligned} \tag{6.17}$$

Note that, by analytic continuation, $J_{-\nu}(z)$ is also a solution to Bessel's equation. This equation, for each ν, has a two-dimensional solution space. We will examine when $J_\nu(z)$ and $J_{-\nu}(z)$ are linearly independent. First, we will obtain a power-series expansion for $J_\nu(z)$. This is done by replacing e^{itz} by its power-series expansion in (6.7). To simplify the expression for the coefficients, one uses identities for the beta function and the gamma function established in Appendix A. From (6.7), we have

$$J_\nu(z) = \left[\Gamma\left(\frac{1}{2}\right)\Gamma\left(\nu+\frac{1}{2}\right)\right]^{-1} \left(\frac{z}{2}\right)^\nu \sum_{k=0}^\infty \frac{1}{(2k)!} \int_{-1}^1 (izt)^{2k}(1-t^2)^{\nu-1/2}\, dt. \tag{6.18}$$

The identity (A.24) implies

$$\int_{-1}^1 t^{2k}(1-t^2)^{\nu-1/2}\, dt = \frac{\Gamma(k+\frac{1}{2})\Gamma(\nu+\frac{1}{2})}{\Gamma(k+\nu+1)},$$

so

$$J_\nu(z) = \frac{(z/2)^\nu}{\Gamma(\frac{1}{2})\Gamma(\nu+\frac{1}{2})} \sum \frac{1}{(2k)!}(iz)^{2k} \frac{\Gamma(k+\frac{1}{2})\Gamma(\nu+\frac{1}{2})}{\Gamma(k+\nu+1)}.$$

Setting $(2k)! = \Gamma(2k+1)$, and using the duplication formula (A.22), which implies

$$\frac{\Gamma(k+\frac{1}{2})}{\Gamma(\frac{1}{2})\Gamma(2k+1)} = \frac{2^{-2k}}{\Gamma(k+1)},$$

we obtain the formula

$$J_\nu(z) = \left(\frac{z}{2}\right)^\nu \sum_{k=0}^\infty \frac{(-1)^k}{k!\Gamma(k+\nu+1)} \left(\frac{z}{2}\right)^{2k}. \tag{6.19}$$

This follows from (6.18) if Re $\nu > -1/2$, and then for general ν by analytic continuation. In particular, we note the leading behavior as $z \to 0$,

$$J_\nu(z) = \frac{z^\nu}{2^\nu \Gamma(\nu+1)} + O(z^{\nu+1}) \tag{6.20}$$

and

$$J'_\nu(z) = \frac{z^{\nu-1}}{2^\nu \Gamma(\nu)} + O(z^\nu). \tag{6.21}$$

The leading coefficients are nonzero as long as ν is not a negative integer (or 0, for (6.21)).

From the expression (6.19) it is clear that $J_\nu(z)$ and $J_{-\nu}(z)$ are linearly independent provided ν is not an integer. On the other hand, comparison of power series shows

$$J_{-n}(z) = (-1)^n J_n(z), \quad n = 0, 1, 2, \dots . \tag{6.22}$$

We want to construct a basis of solutions to Bessel's equation, uniformly good for all ν. This construction can be motivated by a calculation of the *Wronskian*.

Generally, for a pair of solutions u_1 and u_2 to a second-order ODE

$$a(z)u'' + b(z)u' + c(z)u = 0,$$

u_1 and u_2 are linearly independent if and only if their Wronskian

$$W(z) = W(u_1, u_2)(z) = u_1 u_2' - u_2 u_1' \tag{6.23}$$

is nonvanishing. Note that the Wronskian satisfies the first-order ODE

$$W'(z) = -\frac{b}{a} W(z). \tag{6.24}$$

In the case of Bessel's equation, (6.11), this becomes

$$W'(z) = -\frac{W(z)}{z}, \tag{6.25}$$

so

$$W(z) = \frac{K}{z}, \tag{6.26}$$

for some K (independent of z, but perhaps depending on ν). If $u_1 = J_\nu$, $u_2 = J_{-\nu}$, we can compute K by considering the limiting behavior of $W(z)$ as $z \to 0$. From (6.20) and (6.21), we get

$$W(J_\nu, J_{-\nu})(z) = -\left[\frac{1}{\Gamma(\nu)\Gamma(1-\nu)} - \frac{1}{\Gamma(\nu+1)\Gamma(-\nu)}\right]\frac{1}{z} = -2\frac{\sin \pi\nu}{\pi z}, \tag{6.27}$$

making use of the identity (A.10). This recaptures the observation that J_ν and $J_{-\nu}$ are linearly independent, and consequently a basis of solutions to (6.11), if and only if ν is not an integer.

To construct a basis of solutions uniformly good for all ν, it is natural to set

$$Y_\nu(z) = \frac{J_\nu(z)\cos\pi\nu - J_{-\nu}(z)}{\sin\pi\nu} \tag{6.28}$$

when ν is not an integer, and define

$$Y_n(z) = \lim_{\nu\to n} Y_\nu(z) = \frac{1}{\pi}\Big[\frac{\partial J_\nu(z)}{\partial\nu} - (-1)^n\frac{\partial J_{-\nu}(z)}{\partial\nu}\Big]\Big|_{\nu=n}. \tag{6.29}$$

We have

$$W(J_\nu, Y_\nu)(z) = \frac{2}{\pi z}, \tag{6.30}$$

for all ν. Another important pair of solutions to Bessel's equation is the pair of Hankel functions

$$H_\nu^{(1)}(z) = J_\nu(z) + iY_\nu(z), \quad H_\nu^{(2)}(z) = J_\nu(z) - iY_\nu(z). \tag{6.31}$$

For $H_\nu^{(1)}$, there is the integral formula

$$H_\nu^{(1)}(z) = \frac{2e^{-\pi i\nu}}{i\sqrt{\pi}\,\Gamma(\nu+\frac{1}{2})}\Big(\frac{z}{2}\Big)^\nu \int_1^\infty e^{izt}(t^2-1)^{\nu-1/2}\,dt, \tag{6.32}$$

for $\operatorname{Re}\nu > -1/2$, $\operatorname{Im} z > 0$. Another formula, valid for $\operatorname{Re}\nu > \frac{1}{2}$ and $\operatorname{Re} z > 0$, is

$$H_\nu^{(1)}(z) = \Big(\frac{2}{\pi z}\Big)^{1/2}\frac{e^{i(z-\pi\nu/2-\pi/4)}}{\Gamma(\nu+\frac{1}{2})}\int_0^\infty e^{-s}s^{\nu-1/2}\Big(1-\frac{s}{2iz}\Big)^{\nu-1/2}\,ds. \tag{6.33}$$

To prove these identities, one can show as above that each of the right sides of (6.32) and (6.33) satisfies the same recursion formulas as $J_\nu(z)$ and hence solves the Bessel equation; thus it is a linear combination of $J_\nu(z)$ and $Y_\nu(z)$. The coefficients can be found by examining the limiting behavior as $z \to 0$, to establish the asserted identity. Hankel functions are important in scattering theory; see Chapter 9.

It is worth pointing out that the Bessel functions $J_{k+1/2}(z)$, etc., for k an integer, are elementary functions, particularly since they arise in analysis on odd-dimensional Euclidean space. For $\nu = k + 1/2$, the integrand in (6.7) involves $(1-t^2)^k$, so the integral can be evaluated explicitly. We have, in particular,

$$J_{1/2}(z) = \Big(\frac{2}{\pi z}\Big)^{1/2}\sin z. \tag{6.34}$$

Then (6.14) gives

$$J_{-1/2}(z) = \Big(\frac{2}{\pi z}\Big)^{1/2}\cos z, \tag{6.35}$$

which by (6.28) is equal to $-Y_{1/2}(z)$. Applying (6.16) and (6.14) repeatedly gives

$$J_{k+1/2}(z) = (-1)^k\Big\{\prod_{j=1}^k\Big(\frac{d}{dz} - \frac{j-\frac{1}{2}}{z}\Big)\Big\}\frac{\sin z}{\sqrt{2\pi z}} \tag{6.36}$$

and the same sort of formula for $J_{-k-1/2}(z)$, with the $(-1)^k$ removed, and $\sin z$ replaced by $\cos z$. Similarly,

$$H^{(1)}_{1/2}(z) = -i\Big(\frac{2}{\pi z}\Big)^{1/2} e^{iz}, \tag{6.37}$$

with a formula for $H^{(1)}_{k+1/2}(z)$ similar to (6.36).

We now make contact between the formulas (6.2)–(6.6) and some of the formulas of §5, particularly from Proposition 5.5. Note that if $F(x) = f(|x|)$, then

$$\hat{F}(x) = (2\pi)^{n/2} f(\sqrt{-\Delta})\delta. \tag{6.38}$$

Thus, as in (5.51), we have

$$\hat{F}(x) = (2\pi)^{n/2-1} \iint f(r) e^{itr} R_n'(t,x)\, dt\, dr, \tag{6.39}$$

where $R_n'(t,x) = (\partial/\partial t)R_n(t,x)$, and $R_n(t,x)$ is the Riemann function given by (5.43). Comparison with (6.2) gives

$$\psi_n(r|x|) = (2\pi)^{n-1} r^{1-n} \int_{-\infty}^{\infty} e^{itr} R_n'(t,x)\, dt$$

or, equivalently,

$$J_{n/2-1}(r|x|) = 2(2\pi)^{n/2-1}\Big(\frac{|x|}{r}\Big)^{n/2-1} \int_{-\infty}^{\infty} (\sin tr) R_n(t,x)\, dt. \tag{6.40}$$

Note that using $R_3(t,x) = (4\pi t)^{-1}\delta(|t| - |x|)$ gives again the formula (6.34) for $J_{1/2}(r)$. Note also that the recursive formula

$$R_{n+2}(t,s) = -\frac{1}{2\pi s}\frac{\partial}{\partial s} R_n(t,s)$$

used in the proof of Proposition 5.5, when substituted into (6.40), gives rise to the formula (6.15), in the case $\nu = n/2 - 1$.

Instead of synthesizing functions of Δ via the formula (5.51), we could use

$$g(-\Delta) = (2\pi)^{-1/2} \int \hat{g}(t) e^{-it\Delta}\, dt, \tag{6.41}$$

where the operator $e^{-it\Delta}$ is obtained from the solution operator $e^{t\Delta}$ to the heat equation by analytic continuation:

$$e^{-it\Delta}\delta(x) = (-4\pi i t)^{-n/2} e^{|x|^2/4it}, \quad t \neq 0. \tag{6.42}$$

If $f(r) = g(r^2)$, with g real-valued and even, we get

$$\hat{g}(t) = \frac{4}{\sqrt{2\pi}} \int_0^{\infty} f(r)\big(\cos r^2 t\big) r\, dr \tag{6.43}$$

and hence

$$g(-\Delta)\delta = \frac{2}{\pi} \int_{-\infty}^{\infty}\int_0^{\infty} (-4\pi i t)^{-n/2}\big(\cos r^2 t\big) e^{|x|^2/4it} f(r) r\, dr\, dt. \tag{6.44}$$

Comparison of this with (6.2) gives

$$\psi_n(r|x|) = c_n r^{2-n} \int_{-\infty}^{\infty} (\cos r^2 t) e^{|x|^2/4it} t^{-n/2}\, dt, \tag{6.45}$$

where, for n odd, we take $t^{-n/2} = \lim_{\varepsilon \searrow 0} (t - i\varepsilon)^{-n/2}$. Note that (6.45) is an improper integral near $t = 0$.

We will not look in detail at implications of (6.41)–(6.44), which are generally not as incisive as those of Proposition 5.5, but we will briefly make a connection with the idea, used in §5, of synthesizing operators from the heat semigroup. Recall particularly the formula for the resolvent:

$$(\lambda^2 - \Delta)^{-1} = \int_0^{\infty} e^{-(\lambda^2-\Delta)t}\, dt. \tag{6.46}$$

Generalizing (5.28)–(5.31), we have, for $\lambda > 0$,

$$(\lambda^2 - \Delta)^{-1}\delta = \int_0^{\infty} e^{-|x|^2/4t - \lambda^2 t} (4\pi t)^{-n/2}\, dt. \tag{6.47}$$

A superficial resemblance with (6.45) suggests that this function is related to Bessel functions. This is consistent with the fact that the resolvent kernel $R_\lambda(x) = (\lambda^2 - \Delta)^{-1}\delta = R^b_{n,\lambda}(|x|)$ satisfies the ODE

$$\left[\frac{d^2}{dr^2} + \frac{n-1}{r} - \lambda^2\right] R^b_{n,\lambda}(r) = 0, \quad r > 0, \tag{6.48}$$

as a consequence of the formula (6.1) for the Laplace operator in polar coordinates; this is similar to (6.10), with 1 replaced by $-\lambda^2$. In fact, there is the following result. From (6.47),

$$(\lambda^2 - \Delta)^{-1}\delta = (2\pi)^{-n/2} \left(\frac{|x|}{\lambda}\right)^{1-n/2} K_{n/2-1}(\lambda|x|), \quad x \in \mathbb{R}^n \setminus 0, \tag{6.49}$$

where $K_\nu(r)$ is defined by

$$K_\nu(r) = \frac{1}{2}\left(\frac{r}{2}\right)^{\nu} \int_0^{\infty} e^{-r^2/4t - t}\, t^{-1-\nu}\, dt. \tag{6.50}$$

Simple manipulations of (6.50) produce the following analogues of (6.14)–(6.16):

$$\begin{gathered} \left(\frac{d}{dr} - \frac{\nu}{r}\right) K_\nu(r) = -K_{\nu+1}(r), \\ K_{\nu+1}(r) - K_{\nu-1}(r) = \frac{2\nu}{r} K_\nu(r), \end{gathered} \tag{6.51}$$

so we have the ODE

$$\left[\frac{d^2}{dr^2} + \frac{1}{r}\frac{d}{dr} - \left(1 + \frac{\nu^2}{r^2}\right)\right] K_\nu(r) = 0, \quad r > 0, \tag{6.52}$$

which differs from Bessel's equation (6.11), only in one sign. The ODE (6.52) is solved by $J_\nu(ir)$ and by $Y_\nu(ir)$, so $K_\nu(r)$ must be a linear combination of these functions. In fact,

$$K_\nu(r) = \frac{1}{2}\pi i e^{\pi i\nu/2} H_\nu^{(1)}(ir), \quad r > 0. \tag{6.53}$$

A proof of (6.53) can be found in [Leb], Chapter 5; see also Exercise 4 below. When $\nu = 1/2$, (6.53) follows from (6.33) and (6.34) together with the identity

$$K_{1/2}(r) = \left(\frac{\pi}{2r}\right)^{1/2} e^{-r}, \tag{6.54}$$

which in turn, given (6.50), follows from the subordination identity. Then the recursion relation (6.51) and analogues for the Bessel functions imply (6.53) for ν of the form $\nu = k + 1/2$, when k is a positive integer.

We mention that it is customary to take a second, linearly independent, solution to (6.52) to be

$$I_\nu(r) = e^{-\pi i\nu/2} J_\nu(ir), \quad r > 0. \tag{6.55}$$

The functions $K_\nu(r)$ and $I_\nu(r)$ are called modified Bessel functions; also $K_\nu(r)$ is sometimes called MacDonald's function.

Exercises

1. Using the integral formula (6.7), show that, for fixed $\nu > -1/2$, as $z \to +\infty$,

$$J_\nu(z) = \left(\frac{2}{\pi z}\right)^{1/2} \cos\left(z - \frac{\nu\pi}{2} - \frac{\pi}{4}\right) + O(z^{-3/2}). \tag{6.56}$$

(*Hint*: The endpoint contributions from the integral give exponentials times Fourier transforms of functions with simple singularities at the origin.)
Reconsider this problem after reading §§7 and 8.

Similarly, using (6.33), show that, for fixed $\nu > -1/2$, as $z \to +\infty$,

$$H_\nu^{(1)}(z) = \left(\frac{2}{\pi z}\right)^{1/2} e^{i(z-\nu\pi/2-\pi/4)} + O(z^{-3/2}). \tag{6.57}$$

2. Using the integral formula (6.50), show that, for fixed ν, as $r \to +\infty$,

$$K_\nu(r) = \left(\frac{\pi}{2r}\right)^{1/2} e^{-r}\left[1 + O(r^{-1})\right].$$

(*Hint*: Use the Laplace asymptotic method, such as applied to the gamma function in the appendix to this chapter; compare (A.34)–(A.39). To implement this, rewrite (6.50) as

$$K_\nu(r) = 2^{-1-\nu} r^{-1} \int_0^\infty e^{-r(s+1/4s)} s^{-1-\nu}\, ds.$$

Note that $\varphi(s) = s + 1/4s$ has its minimum at $s = 1/2$.)

3. Using the definition (6.55) for $I_\nu(r)$, and plugging $z = ir$ into the integral formula (6.7) for $J_\nu(z)$, show that, for fixed $\nu > -1/2$, as $r \to +\infty$,

$$I_\nu(r) = \left(\frac{1}{2\pi r}\right)^{1/2} e^r\left[1 + O(r^{-1})\right].$$

4. Show that, for $r > 0$,

$$K_\nu(r) = \frac{1}{\Gamma(\frac{1}{2})\Gamma(\nu+\frac{1}{2})}\left(\frac{r}{2}\right)^\nu \int_1^\infty e^{-rt}(t^2-1)^{\nu-\frac{1}{2}}\,dt,$$

by showing that the function on the right solves the modified Bessel equation (6.52) and has the same asymptotic behavior as $r \to +\infty$ as $K_\nu(r)$ does, according to Exercise 2. Hence establish the identity (6.53).

5. For $y, a > 0$, $\xi \in \mathbb{R}^n$, consider

$$F(\xi) = e^{-y(|\xi|^2+a^2)^{1/2}}.$$

Applying the subordination identity (5.22) to $A^2 = |\xi|^2 + a^2$, and taking Fourier transforms of both sides of the resulting identity, show that

$$\begin{aligned}\hat{F}(x) &= c_n y \int_0^\infty e^{-(y^2+|x|^2)/4t} e^{-ta^2} t^{-(n+3)/2}\,dt \\ &= c_n'\, y\, a^\nu\, r^{-\nu}\, K_\nu(ar),\end{aligned}$$

with $\nu = (n+1)/2$, $r^2 = |x|^2 + y^2$.

6. Using analytic continuation involving both y and a, find an expression for the fundamental solution to

$$u_{tt} + 2au_t - \Delta u = 0,$$

for $u = u(t,x)$, $t \in \mathbb{R}$, $x \in \mathbb{R}^n$, where a is a real number. Be explicit in the case $n = 2$, using the elementary character of $K_{3/2}(z)$.

7. Show that, under the change of variable $u(r) = r^\alpha f(cr)$, Bessel's equation (6.11), $u''(r) + (1/r)u'(r) + (1 - \nu^2/r^2)u(r) = 0$, is transformed to

$$f''(r) + \frac{A}{r}f'(r) + \left(\mu^2 - \frac{\lambda^2}{r^2}\right)f(r) = 0, \tag{6.58}$$

with

$$A = 2\alpha + 1, \quad \mu = c^{-1}, \quad \lambda^2 = \nu^2 - \alpha^2.$$

8. Suppose in particular that $v(x) = f(r)w(\theta)$, $\theta \in \mathcal{O} \subset S^{n-1}$, and $\Delta_S w = -\lambda^2 w$. Show that the equation $\Delta v = -\mu^2 v$ is equivalent to (6.58), with $A = n-1$, so $\alpha = n/2 - 1$. Thus f is a linear combination of $r^{1-n/2}J_\nu(\mu r)$ and $r^{1-n/2}H_\nu^{(1)}(\mu r)$, with $\nu = \left[\lambda^2 + (n-2)^2/4\right]^{1/2}$.

9. Show that, complementary to (6.20), we have, for $\nu > 0$,

$$H_\nu^{(1)}(z) \sim -i\frac{\Gamma(\nu)}{\pi}\left(\frac{2}{z}\right)^\nu, \quad z \searrow 0.$$

7. The method of images and Poisson's summation formula

We discuss here techniques for solving such problems as the Dirichlet problem for the heat equation on a rectangular solid in $\mathbb{R}^n$, defined by

$$\Omega = \{x \in \mathbb{R}^n : 0 \le x_j \le a_j,\ 1 \le j \le n\}. \tag{7.1}$$

That is, we want to solve

$$\frac{\partial u}{\partial t} - \Delta u = 0, \tag{7.2}$$

for $u = u(t, x)$, $t \geq 0$, $x \in \Omega$, subject to the boundary condition

$$u\big|_{\mathbb{R}^+ \times \partial\Omega} = 0 \tag{7.2}$$

and the initial condition

$$u(0, x) = f(x), \quad x \in \Omega. \tag{7.4}$$

There are two ways of doing this. One involves using Fourier series on the torus $\mathbb{R}^n/2\Gamma$, where Γ is the lattice in $\mathbb{R}^n$ generated by $a_j e_j$ (e_j being the standard basis of $\mathbb{R}^n$). The other is to use the solution on $\mathbb{R}^+ \times \mathbb{R}^n$ constructed in §5 together with the method of images, described below. Comparing these methods provides interesting analytical identities.

The method of images works as follows. Let $u^\#$ solve the heat equation

$$\frac{\partial u^\#}{\partial t} - \Delta u^\# = 0 \ \text{ on } \mathbb{R}^+ \times \mathbb{R}^n, \tag{7.5}$$

with initial data

$$u^\#(0, x) = f^\#(x), \tag{7.6}$$

where $f^\# = f$ on Ω and $f^\#$ is *odd* with respect to reflections across the walls of all the translates of Ω by elements of Γ. The set of such translates is a set of rectangles tiling $\mathbb{R}^n$, and $f^\#$ is uniquely determined by this prescription. Since reflections are isometries, it follows that, for each $t > 0$, $u^\#(t, \cdot)$ is odd with respect to such reflections; since $u^\#$ is smooth for $t > 0$, it must therefore vanish on all these walls. The restriction of $u^\#(t, x)$ to $\mathbb{R}^+ \times \Omega$ is hence the desired solution to (7.2)–(7.4).

The same sort of technique works for the wave equation on $\mathbb{R} \times \Omega$,

$$\frac{\partial^2 u}{\partial t^2} - \Delta u = 0 \ \text{ on } \mathbb{R} \times \Omega, \tag{7.7}$$

with Dirichlet boundary condition

$$u(t, x) = 0, \ \text{ for } x \in \partial\Omega, \tag{7.8}$$

and initial condition

$$u(0, x) = f(x), \quad u_t(0, x) = g(x) \ \text{ on } \Omega. \tag{7.9}$$

One takes odd extensions $f^\#, g^\#$, as above.

One can apply the method of images to regions other than rectangular solids. It applies when Ω is a half-space, for example; in that case, only one reflection, across the hyperplane $\partial\Omega$, is involved. Similarly, one can treat slabs, bounded by parallel hyperplanes, quadrants, and so on. One can also treat different boundary conditions. If one extends f above to be *even* with respect to these reflections, one obtains solutions with *Neumann* boundary condition satisfied on $\partial\Omega$.

Another type of boundary condition to impose is a periodic boundary condition:

$$u(t,x)=u(t,x+\gamma)\ \text{ if } x,x+\gamma\in\partial\Omega,\ \gamma\in\Gamma. \tag{7.10}$$

The solution to (7.1), (7.4), (7.10) is obtained as follows. Let $f^0(x)=f(x)$ for $x\in\Omega$, let $f^0(x)=0$ for $x\notin\Omega$, set

$$f^b(x)=\sum_{\gamma\in\Gamma}f^0(x+\gamma), \tag{7.11}$$

and let $u^b(t,x)$ be the solution to

$$\frac{\partial u^b}{\partial t}-\Delta u^b=0\ \text{ on } \mathbb{R}^+\times\mathbb{R}^n,\quad u^b(0,x)=f^b(x). \tag{7.12}$$

Note that if $u^0(t,x)$ is defined by

$$\frac{\partial u^0}{\partial t}-\Delta u^0=0\ \text{ on } \mathbb{R}^+\times\mathbb{R}^n,\quad u^0(0,x)=f^0(x), \tag{7.13}$$

then

$$u^b(t,x)=\sum_{\gamma\in\Gamma}u^0(t,x+\gamma). \tag{7.14}$$

In this case, $u(t,x)$ is the restriction of $u^b(t,x)$ to $\mathbb{R}^+\times\Omega$.

Let us specialize to the case $\Gamma=(2\pi\mathbb{Z})^n$, $f=\delta$. We have the fundamental solution, satisfying periodic boundary conditions, given by

$$H(t,x)=(4\pi t)^{-n/2}\sum_{k\in\mathbb{Z}^n}e^{-|x+2\pi k|^2/4t}. \tag{7.15}$$

On the other hand, identifying $\mathbb{R}^n/(2\pi\mathbb{Z})^n$ with $\mathbb{T}^n$, we obtain via Fourier series

$$H(t,x)=(2\pi)^{-n}\sum_{\ell\in\mathbb{Z}^n}e^{-t|\ell|^2+i\ell\cdot x}. \tag{7.16}$$

Comparing these formulas gives the following important case of Poisson's summation formula:

$$\sum_{k\in\mathbb{Z}^n}e^{-|x+2\pi k|^2/4t}=\left(\frac{t}{\pi}\right)^{n/2}\sum_{\ell\in\mathbb{Z}^n}e^{i\ell\cdot x-|\ell|^2t}. \tag{7.17}$$

We now show how the special case of this for $n=1$ implies the famous functional equation for the Riemann zeta function. With $n=1$, $x=0$, and $t=\pi/\tau$, (7.17) yields the identity

$$\sum_{n=-\infty}^{\infty}e^{-n^2\pi\tau}=\left(\frac{1}{\tau}\right)^{1/2}\sum_{n=-\infty}^{\infty}e^{-\pi n^2/\tau}. \tag{7.18}$$

In other words, with $g_1(\tau)$ denoting the left side of (7.18), we have $g_1(\tau)=\tau^{-1/2}g_1(1/\tau)$. This is a transformation formula of Jacobi. It follows that if

$$g(t)=\sum_{n=1}^{\infty}e^{-n^2\pi t}, \tag{7.19}$$

then

$$g(t) = -\frac{1}{2} + \frac{1}{2}t^{-1/2} + t^{-1/2}g(t^{-1}). \tag{7.20}$$

Now (7.19) is related to the *Riemann zeta function*

$$\zeta(s) = \sum_{n=1}^{\infty} n^{-s} \quad (\text{Re } s > 1) \tag{7.21}$$

via the *Mellin transform*, discussed briefly in Appendix A, at the end of this chapter. Indeed, we have

$$\begin{aligned} \int_0^\infty g(t)t^{s-1}\,dt &= \int_0^\infty \sum_{n=1}^{\infty} e^{-n^2\pi t}\, t^{s-1}\,dt \\ &= \sum_{n=1}^{\infty} n^{-2s}\pi^{-s} \int_0^\infty e^{-t}t^{s-1}\,dt \\ &= \zeta(2s)\,\pi^{-s}\,\Gamma(s). \end{aligned} \tag{7.22}$$

Consequently, for Re $s > 1$,

$$\begin{aligned} \Gamma\Big(\frac{s}{2}\Big)\pi^{-s/2}\zeta(s) &= \int_0^\infty g(t)t^{s/2-1}\,dt \\ &= \int_0^1 g(t)t^{s/2-1}\,dt + \int_1^\infty g(t)t^{s/2-1}\,dt. \end{aligned} \tag{7.23}$$

Now, into the integral over $[0, 1]$, substitute the right side of (7.20) for $g(t)$, to obtain

$$\begin{aligned} \Gamma\Big(\frac{s}{2}\Big)\pi^{-s/2}\zeta(s) &= \int_0^1 \Big(-\frac{1}{2} + \frac{1}{2}t^{-1/2}\Big)t^{s/2-1}\,dt \\ &\quad + \int_0^1 g(t^{-1})t^{s/2-3/2}\,dt + \int_1^\infty g(t)t^{s/2-1}\,dt. \end{aligned} \tag{7.24}$$

We evaluate the first integral on the right, and replace t by $1/t$ in the second integral, to obtain, for Re $s > 1$,

$$\Gamma\Big(\frac{s}{2}\Big)\pi^{-s/2}\zeta(s) = \frac{1}{s-1} - \frac{1}{s} + \int_1^\infty \big[t^{s/2} + t^{(1-s)/2}\big]g(t)t^{-1}\,dt. \tag{7.25}$$

Note that $g(t) \le Ce^{-\pi t}$ for $t \in [1, \infty)$, so the integral on the right is an entire analytic function of s. Since $1/\Gamma(s/2)$ is entire, with simple zeros at $s = 0, -2, -4, \dots$, as shown in Appendix A at the end of this chapter, this implies that $\zeta(s)$ is continued as a meromorphic function on $\mathbb{C}$, with one simple pole, at $s = 1$. The punch line is this: The right side of (7.25) is *invariant* under replacing s by $1 - s$. Thus we have Riemann's functional equation

$$\Gamma\Big(\frac{s}{2}\Big)\pi^{-s/2}\zeta(s) = \Gamma\Big(\frac{1-s}{2}\Big)\pi^{-(1-s)/2}\zeta(1-s). \tag{7.26}$$

The functional equation is often written in an alternative form, obtained by multiplying both sides by $\Gamma\big((1+s)/2\big)$, and using the identities

$$\Gamma\Big(\frac{1-s}{2}\Big)\Gamma\Big(\frac{1+s}{2}\Big)=\frac{\pi}{\sin\frac{1}{2}\pi(1-s)},$$
$$\Gamma\Big(\frac{s}{2}\Big)\Gamma\Big(\frac{1+s}{2}\Big)=2^{-s+1}\pi^{1/2}\Gamma(s), \tag{7.27}$$

which follow from (A.10) and (A.22). We obtain

$$\zeta(1-s)=2^{1-s}\pi^{-s}\Big(\cos\frac{\pi s}{2}\Big)\Gamma(s)\zeta(s). \tag{7.28}$$

Exercises

1. Apply the method of images to find the solution to the heat equation on a half-line:
$$\frac{\partial u}{\partial t}=u_{xx},\quad t\ge 0,\ x\ge 0,$$
$$u(0,x)=f(x),\quad u(t,0)=0.$$
2. Similarly, treat the wave equation on a half-space:
$$u_{tt}-\Delta u=0,\quad t\in\mathbb{R},\ x_1>0,$$
$$u(0,x)=f(x),\quad u_t(0,x)=g(x),\quad u(t,0,x')=0,$$
where $x=(x_1,x')\in\mathbb{R}^n$.
3. Given $u\in\mathcal{S}(\mathbb{R}^n)$, define $f\in C^\infty(\mathbb{T}^n)$ by $f(x)=\sum_{\nu\in\mathbb{Z}^n}u(x+2\pi\nu)$. Show that, for $\ell\in\mathbb{Z}^n$, we have $\hat{f}(\ell)=(2\pi)^{-n/2}\hat{u}(\ell)$ and hence
$$\sum_k u(x+2\pi k)=(2\pi)^{-n/2}\sum_\ell \hat{u}(\ell)e^{i\ell\cdot x}. \tag{7.29}$$
Show that this generalizes the identity (7.17).
4. Let (a_ℓ) be polynomially bounded, and consider $v=\sum_{\ell\in\mathbb{Z}^n}a_\ell e^{i\ell\cdot x}$, pictured as a $2\pi\mathbb{Z}^n$-periodic (tempered) distribution on $\mathbb{R}^n$ rather than as a distribution on $\mathbb{T}^n$; $v\in\mathcal{S}'(\mathbb{R}^n)$. Show that
$$\hat{v}=(2\pi)^{n/2}\sum_{\ell\in\mathbb{Z}^n}a_\ell\,\delta_\ell\in\mathcal{S}'(\mathbb{R}^n). \tag{7.30}$$
Relate this to the result in Exercise 3.
5. Show that $\zeta(s)$ satisfies the identity
$$\zeta(s)=\prod_p\big(1-p^{-s}\big)^{-1},\quad \operatorname{Re}s>1,$$
the product taken over all the primes. This is known as the Euler product formula.

8. Homogeneous distributions and principal value distributions

Recall from §4 that the fundamental solution of the Laplace operator Δ on $\mathbb{R}^n$ is $c_n|x|^{2-n}$ (if $n \geq 3$), which is homogeneous. It is useful to consider homogeneous distributions in general. The notion of homogeneity is determined by the action of the group of dilations,

$$D(t)f(x) = f(tx), \quad t > 0. \tag{8.1}$$

Note that $D(t) : \mathcal{S}(\mathbb{R}^n) \to \mathcal{S}(\mathbb{R}^n)$. Also, if $f, g \in \mathcal{S}(\mathbb{R}^n)$, a change of variable gives

$$\int g(x)\, D(t)f(x)\, dx = t^{-n} \int f(x)\, D(t^{-1})g(x)\, dx. \tag{8.2}$$

Thus we can define

$$D(t) : \mathcal{S}'(\mathbb{R}^n) \longrightarrow \mathcal{S}'(\mathbb{R}^n) \tag{8.3}$$

by

$$\langle f, D(t)u \rangle = t^{-n} \langle D(t^{-1})f, u \rangle, \tag{8.4}$$

for $f \in \mathcal{S}(\mathbb{R}^n)$, $u \in \mathcal{S}'(\mathbb{R}^n)$. We say that $u \in \mathcal{S}'(\mathbb{R}^n)$ is homogeneous of degree m if

$$D(t)u = t^m\, u, \quad \text{for all } t > 0. \tag{8.5}$$

Here, m can be any complex number. Let us denote the space of elements of $\mathcal{S}'(\mathbb{R}^n)$ which are homogeneous of degree m by $\mathcal{H}_m(\mathbb{R}^n)$. It is easy to see that if $\mathcal{F}$ is the Fourier transform, then

$$\mathcal{F}D(t) = t^{-n} D(t^{-1})\mathcal{F}, \tag{8.6}$$

so

$$\mathcal{F} : \mathcal{H}_m(\mathbb{R}^n) \longrightarrow \mathcal{H}_{-m-n}(\mathbb{R}^n). \tag{8.7}$$

Before we delve any further into $\mathcal{H}_m(\mathbb{R}^n)$, we should aver that one's real interest is in elements of $\mathcal{H}_m(\mathbb{R}^n)$ which are *smooth* outside the origin, so we consider

$$\mathcal{H}^{\#}_m(\mathbb{R}^n) = \{u \in \mathcal{H}_m(\mathbb{R}^n) : u \in C^\infty(\mathbb{R}^n \setminus 0)\}. \tag{8.8}$$

It is easy to see that

$$u \in \mathcal{H}^{\#}_m(\mathbb{R}^n) \Rightarrow D^\alpha u \in \mathcal{H}^{\#}_{m-|\alpha|}(\mathbb{R}^n) \text{ and } x^\alpha u \in \mathcal{H}^{\#}_{m+|\alpha|}(\mathbb{R}^n). \tag{8.9}$$

We claim (8.7) can be strengthened as follows.

Proposition 8.1. *We have*

$$\mathcal{F} : \mathcal{H}^{\#}_m \longrightarrow \mathcal{H}^{\#}_{-m-n}(\mathbb{R}^n). \tag{8.10}$$

The only point left to prove is that if $u \in \mathcal{H}^{\#}_m(\mathbb{R}^n)$, then $\hat{u}$ is smooth on $\mathbb{R}^n \setminus 0$. Taking $\varphi \in C_0^\infty(\mathbb{R}^n)$, $\varphi(x) = 1$ for $|x| \le 1$, we can write $u = \varphi u + (1-\varphi)u = u_1 + u_2$ with $u_1 \in \mathcal{E}'(\mathbb{R}^n)$ and $u_2 \in C^\infty(\mathbb{R}^n)$, homogeneous for $|x|$ large. We know that $\hat{u}_1 \in C^\infty(\mathbb{R}^n)$, so it suffices to show that $\hat{u}_2 \in C^\infty(\mathbb{R}^n \setminus 0)$. This is a special case of the following important result.

For $m \in \mathbb{R}$, we define the class $S_1^m(\mathbb{R}^n)$ of C^∞-functions by

$$p \in S_1^m(\mathbb{R}^n) \Longleftrightarrow |D_x^\alpha p(x)| \le C_\alpha \langle x\rangle^{m-|\alpha|}, \quad \text{for all } \alpha \ge 0. \tag{8.11}$$

Clearly, $S_1^m(\mathbb{R}^n) \subset \mathcal{S}'(\mathbb{R}^n)$. It is also clear that $u_2 \in S_1^{\operatorname{Re} m}(\mathbb{R}^n)$, so the proof of Proposition 8.1 is finished once we establish the following.

Proposition 8.2. *If $p \in S_1^m(\mathbb{R}^n)$, then $\hat{p} \in C^\infty(\mathbb{R}^n \setminus 0)$. Also, if $\varphi \in C_0^\infty(\mathbb{R}^n)$, and $\varphi(x) = 1$ for $|x| < a$ $(a > 0)$, then $(1-\varphi)\hat{p} \in \mathcal{S}(\mathbb{R}^n)$.*

Proof. We will show that if β is large, then $x^\beta \hat{p}$ is bounded and continuous, and so are lots of derivatives, which will suffice. Clearly,

$$\mathcal{F} : S_1^\mu(\mathbb{R}^n) \longrightarrow L^\infty(\mathbb{R}^n) \cap C(\mathbb{R}^n), \quad \text{for } \mu < -n.$$

Now, given $p \in S_1^m(\mathbb{R}^n)$, then $D^\beta p \in S_1^{m-|\beta|}(\mathbb{R}^n)$, so

$$x^\beta \hat{p} = \mathcal{F}(D^\beta p) \in L^\infty \cap C, \quad \text{for } |\beta| > m+n,$$

and more generally $x^\alpha D^\beta p \in S_1^{m-|\beta|+|\alpha|}(\mathbb{R}^n)$, so

$$D^\alpha(x^\beta \hat{p}) = \mathcal{F}(x^\alpha D^\beta p) \in L^\infty \cap C, \quad \text{for } |\beta| > m+n+|\alpha|. \tag{8.12}$$

This proves Proposition 8.2.

Generally, there is going to be a singularity at the origin for an element of $\mathcal{H}^{\#}_m(\mathbb{R}^n)$. In fact, there is the following result, whose proof we leave as an exercise.

Proposition 8.3. *If there is a nonzero $u \in \mathcal{H}^{\#}_m(\mathbb{R}^n) \cap C^\infty(\mathbb{R}^n)$, then m is a nonnegative integer and u is a homogeneous polynomial.*

Let us consider other examples of homogeneous distributions. It is easy to see from the definition (8.4) of the action of $D(t)$ that

$$\delta \in \mathcal{H}^{\#}_{-n}(\mathbb{R}^n). \tag{8.13}$$

Of course, δ is *zero* on $\mathbb{R}^n \setminus 0$! Since $\mathcal{F}\delta = (2\pi)^{-n/2} \in \mathcal{H}^{\#}_0(\mathbb{R}^n)$, this result is consistent with Proposition 8.1. For more examples, choose any

$$w \in C^\infty(S^{n-1}), \tag{8.14}$$

and consider, for any $m \in \mathbb{C}$,

$$u_m(x) = |x|^m \, w(|x|^{-1}x), \quad x \in \mathbb{R}^n \setminus 0. \tag{8.15}$$

If $\operatorname{Re} m > -n$, then $u_m \in L^1_{\mathrm{loc}}(\mathbb{R}^n)$, so it defines in a natural manner an element of $\mathcal{S}'(\mathbb{R}^n)$, which belongs to $\mathcal{H}^\#_m(\mathbb{R}^n)$. Thus

$$D^\alpha u_m \in \mathcal{H}^\#_{m-|\alpha|}(\mathbb{R}^n) \quad (\operatorname{Re} m > -n). \tag{8.16}$$

If $\operatorname{Re} m \le -n$, then $u_m \notin L^1_{\mathrm{loc}}(\mathbb{R}^n)$. In the borderline case $m = -n$, it is significant that there is a natural identification of u_m with an element of $\mathcal{S}'(\mathbb{R}^n)$, under the further condition that

$$\int_{S^{n-1}} w(x)\, dS = 0. \tag{8.17}$$

The element of $\mathcal{S}'(\mathbb{R}^n)$ is called a *principal value* distribution and is denoted $PV\ u_m$. We establish this as follows. Pick any radial $\varphi \in \mathcal{S}(\mathbb{R}^n)$ such that $\varphi(0) = 1$, such as $\varphi(x) = e^{-|x|^2}$. Then, for any $v \in \mathcal{S}(\mathbb{R}^n)$, with u_{-n} as in (8.15), $u_{-n}(x)\big[v(x) - v(0)\varphi(x)\big]$ belongs to $L^1(\mathbb{R}^n)$, so we can define

$$\langle v, PV\ u_{-n}\rangle = \int_{\mathbb{R}^n} u_{-n}(x)\big[v(x) - v(0)\varphi(x)\big]\, dx. \tag{8.18}$$

Note that (8.17) is precisely what is required to guarantee that the right side of (8.18) is independent of the choice of φ (satisfying the conditions given above). Thus we can write, for any $t > 0$,

$$\begin{aligned} \langle D(t)v, PV\ u_{-n}\rangle &= \int_{\mathbb{R}^n} u_{-n}(x)\big[v(tx) - v(0)\varphi(tx)\big]\, dx \\ &= t^{-n} \int_{\mathbb{R}^n} u_{-n}(x/t)\big[v(x) - v(0)\varphi(x)\big]\, dx \\ &= \langle v, PV\ u_{-n}\rangle. \end{aligned} \tag{8.19}$$

In light of (8.4), this implies

$$PV\ u_{-n} \in \mathcal{H}^\#_{-n}(\mathbb{R}^n), \tag{8.20}$$

provided (8.17) holds. By Proposition 8.1, we have

$$\mathcal{F}(PV\ u_{-n}) \in \mathcal{H}^\#_0(\mathbb{R}^n). \tag{8.21}$$

In particular, this Fourier transform is bounded. Consequently, the convolution operator

$$Tv = (PV\ u_{-n}) * v, \tag{8.22}$$

a priori taking $\mathcal{S}(\mathbb{R}^n)$ to $\mathcal{S}'(\mathbb{R}^n)$, has the property that

$$T : L^2(\mathbb{R}^n) \longrightarrow L^2(\mathbb{R}^n). \tag{8.23}$$

Continuity properties of such a convolution operator on $L^p(\mathbb{R}^n)$, for $1 < p < \infty$, will be demonstrated in Chapter 13.

The special one-dimensional case of a principal value distribution has been discussed in §4. In analogy with (4.34), we have the following.

Proposition 8.4. *Under the hypothesis (8.17), we have, for $v \in \mathcal{S}(\mathbb{R}^n)$,*

$$\langle v, PV\, u_{-n}\rangle = \lim_{\varepsilon\to 0} \int\limits_{\mathbb{R}^n\setminus B_\varepsilon} v(x)\, u_{-n}(x)\, dx, \tag{8.24}$$

where $B_\varepsilon = \{x \in \mathbb{R}^n : |x| < \varepsilon\}$.

Proof. Since $u_{-n}(x)\big[v(x) - v(0)\varphi(x)\big] \in L^1(\mathbb{R}^n)$, via (8.18), we have

$$\langle v, PV\, u_{-n}\rangle = \lim_{\varepsilon\to 0} \int\limits_{\mathbb{R}^n\setminus B_\varepsilon} u_{-n}(x)\big[v(x) - v(0)\varphi(x)\big]\, dx,$$

so (8.24) follows from the observation that if (8.17) holds, then

$$\int\limits_{\mathbb{R}^n\setminus B_\varepsilon} u_{-n}(x)\varphi(x)\, dx = 0, \quad \text{for all } \varepsilon > 0,$$

for any radial $\varphi \in \mathcal{S}(\mathbb{R}^n)$.

In general, if $u(x)$ has the form (8.15) with $m = -n$, then u is a sum of a term to which (8.17) applies and a constant times r^{-n}. Now one can still define a distribution in $\mathcal{S}'(\mathbb{R}^n)$, equal to r^{-n} on $\mathbb{R}^n \setminus 0$, by the prescription

$$\langle v, E_\varphi r^{-n}\rangle = \int\limits_{\mathbb{R}^n} r^{-n}\big[v(x) - v(0)\varphi(x)\big]\, dx, \tag{8.25}$$

for any given radial $\varphi \in \mathcal{S}(\mathbb{R}^n)$ satisfying $\varphi(0) = 1$. This time, $E_\varphi r^{-n} \in \mathcal{S}'(\mathbb{R}^n)$ depends on the choice of φ. One has

$$E_\varphi r^{-n} - E_\psi r^{-n} = \Big(\int \big[\psi(x) - \varphi(x)\big] r^{-n} dx\Big)\delta. \tag{8.26}$$

Also, $E_\varphi r^{-n}$ is not homogeneous. Instead, one has

$$D(t)(E_\varphi r^{-n}) = t^{-n} E_{D(t)\varphi} r^{-n}, \tag{8.27}$$

and by (8.26) this yields, after a brief calculation,

$$D(t)(E_\varphi r^{-n}) = t^{-n} E_\varphi r^{-n} + A_{n-1} t^{-n} (\log t)\delta, \tag{8.28}$$

where $A_{n-1} = \text{vol}(S^{n-1})$. This implies for the Fourier transform of $E_\varphi r^{-n}$ that

$$D(t)\mathcal{F}(E_\varphi r^{-n}) = \mathcal{F}(E_\varphi r^{-n}) + (2\pi)^{-n/2} A_{n-1} \log t, \tag{8.29}$$

which, in view of rotational invariance, implies that

$$\mathcal{F}(E_\varphi r^{-n})(\xi) = (2\pi)^{-n/2} A_{n-1} \log |\xi| + B, \tag{8.30}$$

where B is a constant, depending in an affine manner on φ. A "canonical" choice of $E_\varphi r^{-n}$ would be one for which $B = 0$; such a distribution $E_\varphi r^{-n} \in \mathcal{S}'(\mathbb{R}^n)$ is denoted $PF\, r^{-n}$ (for "finite part"); we have

$$\mathcal{F}(PF\, r^{-n})(\xi) = (2\pi)^{-n/2} A_{n-1} \log|\xi|. \tag{8.31}$$

Note that this is consistent with (4.59) when $n = 2$.

It turns out that r^{-m}, which is holomorphic in $\{m \in \mathbb{C} : \operatorname{Re} m > -n\}$, with values in $\mathcal{S}'(\mathbb{R}^n)$, has a meromorphic continuation. This can be perceived as follows. First note that if $-n < \operatorname{Re} m < 0$, then both r^m and r^{-n-m} belong to $L^1_{\text{loc}}(\mathbb{R}^n)$, so from Proposition 8.1 and rotational invariance we deduce that

$$\mathcal{F}(r^m) = c(m)\, r^{-m-n}, \tag{8.32}$$

for a certain factor $c(m)$, which we want to work out. We claim that

$$\mathcal{F}(r^m) = 2^{m+n/2} \frac{\Gamma(\frac{1}{2}(m+n))}{\Gamma(-\frac{1}{2}m)}\, r^{-m-n}, \tag{8.33}$$

for $-n < \operatorname{Re} m < 0$. This can be deduced from (8.32) and Parseval's identity, which gives

$$\langle u, r^m \rangle = c(m) \langle \hat{u}, r^{-m-n} \rangle. \tag{8.34}$$

If we plug in $u(x) = e^{-|x|^2/2} = \hat{u}(x)$, both sides of (8.34) can be evaluated by integrating in polar coordinates. The left side is

$$\begin{aligned} A_{n-1} \int_0^\infty r^{m+n-1} e^{-r^2/2}\, dr &= 2^{(m+n-1)/2} A_{n-1} \int_0^\infty s^{(m+n)/2-1}\, e^{-s}\, ds \\ &= 2^{(m+n-1)/2} \Gamma\big(\tfrac{1}{2}(m+n)\big) A_{n-1}, \end{aligned} \tag{8.35}$$

and the right side of (8.34) is similarly evaluated, giving (8.33).

Now the left side of (8.33) extends to be holomorphic in $\operatorname{Re} m > -n$, with values in $\mathcal{S}'(\mathbb{R}^n)$, while the right side extends to be meromorphic in $\operatorname{Re} m < 0$, with poles at $m = -n, -n-2, -n-4, \dots$, due to the factor $\Gamma((m+n)/2)$. Thus we have the desired meromorphic continuation. With r^m so defined,

$$r^m \in \mathcal{H}^\#_m(\mathbb{R}^n), \quad m \neq -n, -n-2, -n-4, \dots; \tag{8.36}$$

indeed, $D(t)r^m - t^m r^m$ is a meromorphic function of m which vanishes on a nonempty open set. As we have seen, $PF\, r^{-n}$ can be defined by a "renormalization," though it does not belong to $\mathcal{H}^\#_{-n}(\mathbb{R}^n)$.

Let us now consider the possibility of extending u_m, of the form (8.15), to an element of $\mathcal{S}'(\mathbb{R}^n)$, in case

$$m = -n - j, \quad j = 1, 2, 3, \dots. \tag{8.37}$$

In analogy with (8.18) and (8.25), we can define $E_{j,\varphi} u_m$ in this case by

$$\langle v, E_{j,\varphi} u_m \rangle = \int u_m(x) \Big[v(x) - \sum_{|\alpha| \le j} \frac{v^{(\alpha)}(0)}{\alpha!} x^\alpha \varphi(x) \Big]\, dx, \tag{8.38}$$

provided $\varphi \in \mathcal{S}(\mathbb{R}^n)$ is a radial function such that $\varphi(0) = 1$ and $1 - \varphi$ vanishes to order at least j at 0; for example, we could require $\varphi(x) = 1$ for $|x| \le c$. The dependence on φ is given by

$$E_{j,\varphi}u_m - E_{j,\psi}u_m = \sum_{|\alpha|\le j} \beta_\alpha(\varphi - \psi)\, \delta^{(\alpha)}, \tag{8.39}$$

where

$$\beta_\alpha(\varphi - \psi) = -\frac{1}{\alpha!}\int x^\alpha [\varphi(x) - \psi(x)] u_m(x)\, dx. \tag{8.40}$$

In analogy with (8.27), we have

$$D(t)E_{j,\varphi}u_m = t^m E_{j,D(t)\varphi}u_m, \tag{8.41}$$

and hence, given (8.37), by a calculation similar to that establishing (8.28),

$$\begin{aligned} D(t)(E_{j,\varphi}u_m) = t^m E_{j,\varphi}u_m &+ t^m \sum_{|\alpha|<j} \gamma_\alpha \big(t^{|\alpha|-j} - 1\big)\delta^{(\alpha)} \\ &+ t^m \log t \sum_{|\alpha|=j} \gamma_\alpha\, \delta^{(\alpha)}, \end{aligned} \tag{8.42}$$

for certain constants γ_α, which depend in an affine fashion on φ. Consequently, if we set

$$\widetilde{E}u_m = E_{j,\varphi}u_m - \sum_{|\alpha|<j} \gamma_\alpha \delta^{(\alpha)}, \tag{8.43}$$

we have another element of $\mathcal{S}'(\mathbb{R}^n)$ which agrees with u_m on $\mathbb{R}^n \setminus 0$, and

$$D(t)(\widetilde{E}u_m) = t^m \widetilde{E}u_m + t^m (\log t) \sum_{|\alpha|=j} \gamma_\alpha\, \delta^{(\alpha)}. \tag{8.44}$$

It follows for the Fourier transform $\mathcal{F}(\widetilde{E}u_m)$ that

$$D(t)\mathcal{F}(\widetilde{E}u_m) = t^j \mathcal{F}(\widetilde{E}u_m) + t^j (\log t) \sum_{|\alpha|=j} \gamma'_\alpha\, \xi^\alpha.$$

Consequently, if $\mathcal{F}(\widetilde{E}u_m) = \omega(\xi)$ for $|\xi| = 1$, we have

$$\mathcal{F}(\widetilde{E}u_m)(t\xi) = t^j \omega(\xi) + (\log t) \sum_{|\alpha|=j} \gamma'_\alpha (t\xi)^\alpha, \quad \text{for } |\xi| = 1,$$

and hence

$$\mathcal{F}(\widetilde{E}u_m)(\xi) = w_j(\xi) + p_j(\xi) \log |\xi|, \tag{8.45}$$

where

$$w_j \in \mathcal{H}^{\#}_j(\mathbb{R}^n) \text{ and } p_j \text{ is a homogeneous polynomial, of degree } j. \tag{8.46}$$

We leave it as an exercise to the reader to construct a similar extension of u_m to an element of $\mathcal{S}'(\mathbb{R}^n)$, when $\operatorname{Re} m \le -n$ and m is not an integer. In such a case one can produce an element of $\mathcal{H}^{\#}_m(\mathbb{R}^n)$; log terms do not arise.

Exercises

1. More generally than $S_1^m(\mathbb{R}^n)$, for $0 < \rho \le 1$, define $S_\rho^m(\mathbb{R}^n)$ by
$$p \in S_\rho^m(\mathbb{R}^n) \Longleftrightarrow |D_x^\alpha p(x)| \le C_\alpha \langle x\rangle^{m-\rho|\alpha|}, \quad \text{for all } \alpha \ge 0.$$
Show that $\hat{p} \in C^\infty(\mathbb{R}^n \setminus 0)$ in this case, as in Proposition 8.2.
2. Define $p \in C^\infty(\mathbb{R}^n \setminus 0)$ by $p(\xi) = (i\xi_1 + |\xi'|^2)^{-1}$, $\xi = (\xi_1, \xi_2, \dots, \xi_n) = (\xi_1, \xi')$. Show that $p(\xi)$ agrees outside any neighborhood of the origin with a member of $S_{1/2}^{-1}(\mathbb{R}^n)$.
3. Prove Proposition 8.3.
4. If $-n < \operatorname{Re} m < 0$ and u_m is of the form (8.15), then u_m and $\hat{u}_m$ belong to $L^1_{\text{loc}}(\mathbb{R}^n)$, with $u_m \in \mathcal{H}_m^\#$, $\hat{u}_m \in \mathcal{H}_{-n-m}^\#$. Hence
$$\hat{u}(x) = |x|^{-n-m} W_m(|x|^{-1}x), \quad W_m \in C^\infty(S^{n-1}).$$
Study the transformation $w \mapsto W_m$. Use this to produce a meromorphic continuation of u_m.
5. Study the residue of the meromorphic distribution-valued function r^z at $z = -n$, and relate this to the failure of $PF\, r^{-n}$ to be homogeneous.
6. In case $n = 1$ and $m = -s$, the formula (8.33) says
$$\mathcal{F}(r^{-s}) = 2^{1/2-s} \frac{\Gamma(\frac{1-s}{2})}{\Gamma(\frac{s}{2})} r^{s-1}, \quad \text{for } 0 < \operatorname{Re} s < 1,$$
while Riemann's functional equation (7.26) can be written
$$\frac{\zeta(s)}{\zeta(1-s)} = \pi^{s-1/2} \frac{\Gamma(\frac{1-s}{2})}{\Gamma(\frac{s}{2})}.$$
Is this a coincidence? (See [Pat], Chapter 2.) Note that these formulas yield
$$\frac{(2\pi)^{s/2}}{\zeta(s)} \mathcal{F}(r^{-s}) = \frac{(2\pi)^{(1-s)/2}}{\zeta(1-s)} r^{s-1}, \quad 0 < \operatorname{Re} s < 1.$$

9. Elliptic operators

A partial differential operator $P(D)$ of order m,

$$P(D) = \sum_{|\alpha| \le m} a_\alpha D^\alpha, \tag{9.1}$$

is said to be *elliptic* provided

$$|P(\xi)| \ge C|\xi|^m, \quad \text{for } |\xi| \text{ large}. \tag{9.2}$$

Here $P(\xi) = \sum a_\alpha \xi^\alpha$. The paradigm example is the Laplace operator $\Delta = P(D)$, with $P(\xi) = -|\xi|^2$, which is elliptic of order 2. In this section we consider some important properties of solutions to

$$P(D)u = f \tag{9.3}$$

when $P(D)$ is elliptic.

The hypothesis (9.2) implies the following. If (9.2) holds for $|\xi| \ge C_1$, and if $\varphi \in C_0^\infty(\mathbb{R}^n)$ is equal to 1 for $|\xi| \le C_1$, then

$$q(\xi) = \bigl(1 - \varphi(\xi)\bigr)\, P(\xi)^{-1} \in S_1^{-m}(\mathbb{R}^n), \tag{9.4}$$

where $S_1^m(\mathbb{R}^n)$ is the space defined by (8.11); we call it a space of "symbols." Now consider

$$E = (2\pi)^{-n/2}\check{q} \in \mathcal{S}'(\mathbb{R}^n). \tag{9.5}$$

By Proposition 8.2, we know that E is smooth on $\mathbb{R}^n \setminus 0$ and rapidly decreasing as $|x| \to \infty$. If we set

$$v = P(D)E, \tag{9.6}$$

then

$$\hat{v}(\xi) = (2\pi)^{-n/2}\bigl(1 - \varphi(\xi)\bigr). \tag{9.7}$$

In other words,

$$P(D)E = \delta + w, \tag{9.8}$$

with

$$w = -(2\pi)^{-n/2}\check{\varphi} \in \mathcal{S}(\mathbb{R}^n). \tag{9.9}$$

We say E is a *parametrix* for $P(D)$. It is almost as useful as a fundamental solution, for some qualitative purposes. For example, it enables us to say a great deal about the *singular support* of a solution u to (9.3), given $f \in \mathcal{D}'(\mathbb{R}^n)$. The singular support of a general distribution $u \in \mathcal{D}'(\mathbb{R}^n)$ is defined as follows. Let $\Omega \subset \mathbb{R}^n$ be open. We say u is smooth on Ω if there exists $v \in C^\infty(\Omega)$ such that $u = v$ on Ω. The smallest set K for which u is smooth on $\mathbb{R}^n \setminus K$ is the singular support of u, denoted

$$\text{sing supp}\, u.$$

For example, sing supp $\delta = \{0\}$; also sing supp $|x|^{2-n} = \{0\}$, if $n \ne 2$. Now, suppose that $u \in \mathcal{E}'(\mathbb{R}^n)$ and (9.3) holds. Then

$$E * f = E * P(D)u = (P(D)E) * u = u + w * u \tag{9.10}$$

and, of course,

$$w * u \in C^\infty(\mathbb{R}^n).$$

On the other hand, it is easy to see that, for any $f \in \mathcal{E}'(\mathbb{R}^n)$,

$$\text{sing supp}\, E * f \subset \text{sing supp}\, f, \tag{9.11}$$

provided sing supp $E \subset \{0\}$. More generally, for any $f_1, f_2 \in \mathcal{E}'(\mathbb{R}^n)$, if sing supp $f_j \subset K_j$, then

$$\text{sing supp}\, f_1 * f_2 \subset K_1 + K_2, \tag{9.12}$$

a result we leave as an exercise.

Noting that we can multiply distributions by cut-offs $\chi \in C_0^\infty(\mathbb{R}^n)$, equal to 1 on an arbitrarily large set, we deduce the following result, known as *elliptic regularity*.

Proposition 9.1. *For any $u \in \mathcal{D}'(\mathbb{R}^n)$, if (9.3) holds with $P(D)$ elliptic, then*

$$\text{sing supp}\, u = \text{sing supp}\, f. \tag{9.13}$$

Finally, we want to make a detailed analysis of the behavior of the singularity at the origin of the parametrix E for an elliptic operator $P(D)$. Since E is given by (9.4) and (9.5), with $P(\xi) = \sum_{|\alpha|\le m} a_\alpha \xi^\alpha$ a polynomial, it follows that, for $|\xi|$ large,

$$q(\xi) \sim \sum_{j\ge 0} q_j(\xi), \tag{9.14}$$

where each $q_j \in C^\infty(\mathbb{R}^n)$, and, for $|\xi| \ge C,\ q_j(\xi)$ is homogeneous in ξ of degree $-m-j$. The meaning of (9.14) is that, for any N,

$$q(\xi) - \sum_{j=0}^{N-1} q_j(\xi) = r_N(\xi) \in S_1^{-m-N}(\mathbb{R}^n). \tag{9.15}$$

Consequently,

$$E \sim (2\pi)^{-n/2} \sum_{j\ge 0} \tilde{q}_j \tag{9.16}$$

in the sense that, for any K, one can take N large enough that

$$E - (2\pi)^{-n/2} \sum_{j=0}^{N-1} \tilde{q}_j = (2\pi)^{-n/2} \tilde{r}_N \in C^K(\mathbb{R}^n). \tag{9.17}$$

Now, we can replace each q_j by $q_j^b \in C^\infty(\mathbb{R}^n \setminus 0)$, equal to q_j for $|\xi|$ large, and homogeneous of degree $-m-j$ on $\mathbb{R}^n \setminus 0$, and replace each q_j^b by $q_j^\# \in \mathcal{S}'(\mathbb{R}^n)$, equal to q_j^b on $\mathbb{R}^n \setminus 0$, such that $q_j^\# \in \mathcal{H}^\#_{-m-j}$ if $m+j < n$, or in any event satisfying the counterpart of (8.44)–(8.46). Note that, for each $j,\ q_j - q_j^\# \in \mathcal{E}'(\mathbb{R}^n)$, so the Fourier transform of the difference belongs to $C^\infty(\mathbb{R}^n)$. We have established the following.

Proposition 9.2. *A parametrix E for an elliptic operator $P(D)$ of order m satisfies the condition that $E \in C^\infty(\mathbb{R}^n \setminus 0)$, and the singularity is given by*

$$E \sim \sum_{\ell\ge 0} (E_\ell + p_\ell(x) \log|x|), \tag{9.18}$$

where

$$E_\ell \in \mathcal{H}^\#_{m-n+\ell}(\mathbb{R}^n) \tag{9.19}$$

and $p_\ell(x)$ is a polynomial homogeneous of degree $m-n+\ell$; these log coefficients appear only for $\ell \ge n-m$.

More generally, this result holds for $E = (2\pi)^{-n/2}\tilde{q}$ whenever $q \in S^m_1$ has an expansion of the form (9.14), for any $m \in \mathbb{R}$, and log terms do not arise if m is not an integer.

Exercises

1. Using Exercises 1 and 2 of §8, establish an analogue of the regularity result in Proposition 9.1 when $P(D)$ is the (nonelliptic) "heat operator":
$$P(D) = \frac{\partial}{\partial x_1} - \left(\frac{\partial^2}{\partial x_2^2} + \cdots + \frac{\partial^2}{\partial x_n^2}\right).$$
2. Give a detailed proof of (9.12), in order to deduce (9.11).
(*Hint*: Use
$$f \in \mathcal{E}'(\mathbb{R}^n),\ g \in C^\infty(\mathbb{R}^n) \Longrightarrow f * g \in C^\infty(\mathbb{R}^n).$$
Break up f_1 and f_2 into pieces. For nonsmooth pieces, establish and use
$$\operatorname{supp} \varphi_j \subset \widetilde{K}_j \Longrightarrow \operatorname{supp} \varphi_1 * \varphi_2 \subset \widetilde{K}_1 + \widetilde{K}_2.)$$

10. Local solvability of constant-coefficient PDE

In the previous sections we have mainly used Fourier analysis as a tool to provide explicit solutions to the classical linear PDEs. Here we use Fourier series to prove an existence theorem for solutions to a general constant-coefficient linear PDE

$$P(D)u = f. \tag{10.1}$$

We show that, given any $f \in \mathcal{D}'(\mathbb{R}^n)$, and any $R < \infty$, there exists $u \in \mathcal{D}'(\mathbb{R}^n)$ solving (10.1) on the ball $|x| < R$. This result was originally established by Malgrange and Ehrenpreis. If $f \in C^\infty(\mathbb{R}^n)$, we produce $u \in C^\infty(\mathbb{R}^n)$. We do not produce a global solution, and other references, particularly [H] and [Tre], contain much more information on solutions to (10.1) than is presented here. Our method, due to Dadok and Taylor [DT], does have the advantage of being fairly straightforward and short.

For any $\alpha \in \mathbb{R}^n$, solving (10.1) on $B_R = \{x \in \mathbb{R}^n : |x| < R\}$ is equivalent to solving

$$P(D+\alpha)v = g, \tag{10.2}$$

where $v = e^{-i\alpha\cdot x}u$ and $g = e^{-i\alpha\cdot x}f$. To solve (10.2) on B_R, we can cut off g to be supported on $B_{3R/2}$ and work on $\mathbb{R}^n/2R\mathbb{Z}^n$. Without loss of generality (altering $P(D)$), we can rescale and suppose $R = \pi$, so $\mathbb{R}^n/2R\mathbb{Z}^n = \mathbb{T}^n$. The following result then implies solvability on B_R.

Proposition 10.1. *For almost every* $\alpha \in A = \{(\alpha_1, \dots, \alpha_n) : 0 \le \alpha_\nu \le 1\}$,

$$P(D+\alpha) : \mathcal{D}'(\mathbb{T}^n) \longrightarrow \mathcal{D}'(\mathbb{T}^n) \tag{10.3}$$

is an isomorphism, as is $P(D+\alpha) : C^\infty(\mathbb{T}^n) \to C^\infty(\mathbb{T}^n)$.

In view of the characterizations of Fourier series of elements of $\mathcal{D}'(\mathbb{T}^n)$ and of $C^\infty(\mathbb{T}^n)$, it suffices to establish the following.

Proposition 10.2. *Let* $P(\xi)$ *be a polynomial of order* m *on* $\mathbb{R}^n$. *For almost all* $\alpha \in A$, *there are constants* C, N *such that*

$$|P(k+\alpha)^{-1}| \le C\langle k\rangle^N, \quad \text{for all } k \in \mathbb{Z}^n. \tag{10.4}$$

We will prove this using the following elementary fact about the behavior of a polynomial near its zero set.

Lemma 10.3. *Let* $P(\xi)$ *be a polynomial of order* m *on* $\mathbb{R}^n$, *not identically zero. Then there exists* $\delta > 0$ *such that*

$$|P(\xi)|^{-\delta} \in L^1_{\text{loc}}(\mathbb{R}^n). \tag{10.5}$$

Before proving Lemma 10.3, we show how it yields (10.4). First, we claim that, for any polynomial of order m on $\mathbb{R}^n$, not identically zero, there exist $\delta > 0$ and M such that

$$\int |P(\xi)|^{-\delta}\langle\xi\rangle^{-M}\, d\xi < \infty. \tag{10.6}$$

Indeed, Lemma 10.3 guarantees

$$\int\limits_{|\xi|\le 1} |P(\xi)|^{-\delta}\, d\xi < \infty,$$

while, for M sufficiently large,

$$\int\limits_{|\xi|\ge 1} |P(\xi)|^{-\delta}|\xi|^{-M}\, d\xi \le C \int\limits_{|\xi|\le 1} |P(|\xi|^{-2}\xi)|^{-\delta}\, d\xi,$$

and Lemma 10.3 also implies that, for $\delta > 0$ small enough,

$$|P(|\xi|^{-2}\xi)|^{-\delta} \in L^1_{\text{loc}}.$$

Now, using (10.6), note that

$$\int\limits_A \sum_{k\in\mathbb{Z}^n} |P(k+\alpha)|^{-\delta}\langle k\rangle^{-M}\, d\alpha \le C \int\limits_{\mathbb{R}^n} |P(\xi)|^{-\delta}\langle\xi\rangle^{-M}\, d\xi < \infty. \tag{10.7}$$

Thus, for almost all $\alpha \in A$,

$$\sum_{k\in\mathbb{Z}^n} |P(k+\alpha)|^{-\delta}\langle k\rangle^{-M} < \infty, \tag{10.8}$$

which immediately gives (10.4).

We now prove the lemma for any $\delta < 1/m$. We must prove that $|P(\xi)|^{-\delta}$ is integrable on any bounded subset of $\mathbb{R}^n$. Rotating coordinates, we can suppose that $P(\xi_1, 0)$ is a polynomial of order exactly m:

$$P(\xi_1, 0) = a_m \xi_1^m + \cdots + a_0, \quad a_m \neq 0. \tag{10.9}$$

It follows that, with $\xi' = (\xi_2, \dots, \xi_n)$,

$$P(\xi) = a_m \xi_1^m + \sum_{\ell=0}^{m-1} a_\ell(\xi')\xi_1^\ell, \tag{10.10}$$

where $a_\ell(\xi')$ is a polynomial on $\mathbb{R}^{n-1}$ of order $\leq m - \ell$. Consequently, we have

$$P(\xi) = a_m \prod_{j=1}^{m} \big(\xi_1 - \lambda_j(\xi')\big). \tag{10.11}$$

Hence it is clear that, for any $C_1 < \infty$, there is a $C_2 < \infty$ such that if $\delta < 1/m$,

$$\int_{-C_1}^{C_1} |P(\xi)|^{-\delta} \, d\xi_1 \leq C_2, \quad \text{for } |\xi'| \leq C_1. \tag{10.12}$$

This completes the proof.

Exercises

1. Consider the following boundary problem on $[0, A] \times \mathbb{T}^n$:

$$u_{tt} - \Delta u = 0,$$
$$u(0, x) = f_1(x), \quad u(A, x) = f_2(x),$$

where $f_j \in C^\infty(\mathbb{T}^n)$. Show that, for almost all $A \in \mathbb{R}^+$, this has a unique solution $u \in C^\infty([0, A] \times \mathbb{T}^n)$, for all $f_j \in C^\infty(\mathbb{T}^n)$. Show that, for a dense set of A, this solvability fails.

11. The discrete Fourier transform

When doing numerical work involving Fourier series, it is convenient to discretize, and replace S^1, pictured as the group of complex numbers of modulus 1, by the group Γ_n generated by $\omega = e^{2\pi i/n}$. One can also approximate $\mathbb{T}^d$ by $(\Gamma_n)^d$, a product of d copies of Γ_n. We will restrict attention to the case $d = 1$ here; results for general d are obtained similarly.

The cyclic group Γ_n is isomorphic to the group $\mathbb{Z}_n = \mathbb{Z}/(n)$, but we will observe a distinction between these two groups; an element of Γ_n is a certain complex number of modulus 1, and an element of $\mathbb{Z}_n$ is an equivalence class of integers. For n large, we think of Γ_n as an approximation to S^1 and $\mathbb{Z}_n$ as an approximation to $\mathbb{Z}$. We note the natural dual pairing $\Gamma_n \times \mathbb{Z}_n \to \mathbb{C}$ given by $(\omega^j, \ell) \mapsto \omega^{j\ell}$, which is well defined since $\omega^{jn} = 1$.

Now, given a function $f : \Gamma_n \to \mathbb{C}$, its discrete Fourier transform $f^\# = \Phi_n f$, mapping $\mathbb{Z}_n$ to $\mathbb{C}$, is defined by

$$f^\#(\ell) = \frac{1}{n} \sum_{\omega^j \in \Gamma_n} f(\omega^j)\omega^{-j\ell}. \tag{11.1}$$

Similarly, given a function $g : \mathbb{Z}_n \to \mathbb{C}$, its "inverse Fourier transform" $g^b : \Gamma_n \to \mathbb{C}$ is defined by

$$g^b(\omega^j) = \sum_{\ell \in \mathbb{Z}_n} g(\ell)\omega^{j\ell}. \tag{11.2}$$

The following is the Fourier inversion formula in this context.

Proposition 11.1. *The map*

$$\Phi_n : L^2(\Gamma_n) \longrightarrow L^2(\mathbb{Z}_n) \tag{11.3}$$

is a unitary isomorphism, with inverse defined by (11.2), so

$$f(\omega^j) = \sum_{\ell \in \mathbb{Z}_n} f^\#(\ell)\omega^{j\ell}. \tag{11.4}$$

Here the space $L^2(\mathbb{Z}_n)$ is defined by counting measure and $L^2(\Gamma_n)$ by $1/n$ times counting measure, that is,

$$(u, v)_{L^2(\Gamma_n)} = \frac{1}{n} \sum_{\omega^j \in \Gamma_n} u(\omega^j)\overline{v(\omega^j)}. \tag{11.5}$$

Note that if we define functions e_j on Γ_n by

$$e_j(\omega^k) = \omega^{jk}, \tag{11.6}$$

then Proposition 11.1 is equivalent to:

Proposition 11.2. *The functions e_j, $1 \le j \le n$, form an orthonormal basis of $L^2(\Gamma_n)$.*

Proof. Since $L^2(\Gamma_n)$ has dimension n, we need only check that the e_js are mutually orthogonal. Note that

$$(e_k, e_\ell) = \frac{1}{n} \sum_{\omega^j \in \Gamma_n} \omega^{mj}, \quad m = k - \ell.$$

Denote the sum by S_m. If we multiply by ω^m, we have a sum of the same set of powers of ω, so $S_m = \omega^m S_m$. Thus $S_m = 0$ whenever $\omega^m \neq 1$, which completes the proof. Alternatively, the series is easily summed as a finite geometrical series.

Note that the functions e_j in (11.6) are the restrictions to Γ_n of $e^{ij\theta}$ (i.e., values at $\theta = 2\pi k/n$). These restrictions depend only on the residue class of j mod n, which leads to the following simple but fundamental connection between Fourier series on S^1 and on Γ_n.

Proposition 11.3. *If $f \in C(S^1)$ has absolutely convergent Fourier series, then*

$$f^{\#}(\ell) = \sum_{j=-\infty}^{\infty} \hat{f}(\ell + jn). \tag{11.7}$$

We will use (11.7) as a tool to see how well a function on S^1 is approximated by discretization, involving restriction to Γ_n. Precisely, we consider the operators

$$R_n : C(S^1) \longrightarrow L^2(\Gamma_n), \quad E_n : L^2(\Gamma_n) \longrightarrow C^\infty(S^1) \tag{11.8}$$

given by

$$(R_n f)(\omega^j) = f\Big(\frac{2\pi j}{n}\Big), \tag{11.9}$$

for $f = f(\theta)$, $0 \le \theta \le 2\pi$, and

$$E_n\Big(\sum_{\ell \in \mathbb{Z}_n} g(\ell)\omega^{j\ell}\Big) = \sum_{\ell=-\nu}^{\nu-1} g(\ell)e^{i\ell\theta}, \ n = 2\nu. \tag{11.10}$$

We assume $n = 2\nu$ is even; one can also treat $n = 2\nu - 1$, changing the upper limit in the last sum from $\nu - 1$ to ν. Clearly, $R_n E_n$ is the identity operator on $L^2(\Gamma_n)$. The question of interest to us is: How close is $E_n R_n f$ to f, a function on S^1? The answer depends on smoothness properties of f and is expressed in terms involving (typically) negative powers of n.

We compare $E_n R_n$ and the partial summing operator

$$P_n f = \sum_{\ell=-\nu}^{\nu-1} \hat{f}(\ell)e^{i\ell\theta} \tag{11.11}$$

for Fourier series. Note that

$$E_n R_n f(\theta) = \sum_{\ell=-\nu}^{\nu-1} f^{\#}(\ell)e^{i\ell\theta}. \tag{11.12}$$

Consequently,

$$E_n R_n f = P_n f + Q_n f, \tag{11.13}$$

with

$$Q_n f(\theta) = \sum_{\ell=-\nu}^{\nu-1} \big[f^{\#}(\ell) - \hat{f}(\ell)\big]e^{i\ell\theta}. \tag{11.14}$$

By (11.7), we have, for $-\nu \le \ell \le \nu - 1$,

$$f^{\#}(\ell) - \hat{f}(\ell) = \sum_{j \in \mathbb{Z}\setminus 0} \hat{f}(\ell + jn). \tag{11.15}$$

Consequently, the sup norm of $Q_n f$ is bounded by

$$\sum_{\ell=-\nu}^{\nu-1} \left| f^{\#}(\ell) - \hat{f}(\ell) \right| \le \sum_{|k|\ge\nu} |\hat{f}(k)|. \tag{11.16}$$

The right side also dominates the sup norm of $f - P_n f$, proving:

Proposition 11.4. *If $f \in C(S^1)$ has absolutely convergent Fourier series, then*

$$\|f - E_n R_n f\|_{L^\infty} \le 2 \sum_{|k|\ge\nu} |\hat{f}(k)|. \tag{11.17}$$

The estimates of various norms of $f - P_n f$ is an exercise in Fourier analysis on S^1. There are many estimates involving Sobolev spaces; see Chapter 4. Here we note a simple estimate, for $m \ge 1$:

$$\begin{aligned} \|f - P_n f\|_{C^\ell(S^1)} &\le \sum_{|k|\ge\nu} |k|^\ell |\hat{f}(k)| \\ &\le C_{m\ell} \|f\|_{C^{\ell+m+1}(S^1)} \cdot n^{-m}, \end{aligned} \tag{11.18}$$

the last inequality following from (1.49). As the reader can verify, use of the proof of Proposition 1.3 can lead to a sharper estimate. As for an estimate of the contribution of Q_n to the discretization error, from (11.14)–(11.16) we easily obtain

$$\begin{aligned} \|Q_n f\|_{C^\ell(S^1)} &\le \left(\frac{n}{2}\right)^\ell \sum_{|k|\ge\nu} |\hat{f}(k)| \\ &\le C_{\ell m} \|f\|_{C^{\ell+m+1}(S^1)} \cdot n^{-m}. \end{aligned} \tag{11.19}$$

We reiterate that sharper estimates are possible.

Recall that solutions to a number of evolution equations are given by Fourier multipliers on $L^2(S^1)$, of the form

$$F(D)u(\theta) = \sum_{\ell=-\infty}^{\infty} F(\ell)\hat{u}(\ell)e^{i\ell\theta}. \tag{11.20}$$

We want to compare such an operator with its discretized version on $L^2(\Gamma_n)$:

$$F(D_n)\Big[\sum_{\ell\in\mathbb{Z}_n} g(\ell)\omega^{j\ell}\Big] = \sum_{\ell=-\nu}^{\nu-1} F(\ell)g(\ell)\omega^{j\ell}. \tag{11.21}$$

In fact, a simple calculation yields

$$E_n F(D_n) R_n u(\theta) = \sum_{\ell=-\nu}^{\nu-1} F(\ell)u^{\#}(\ell)e^{i\ell\theta} \tag{11.22}$$

and hence

$$E_n F(D_n) R_n u = P_n F(D) u + \Psi_n u, \tag{11.23}$$

where

$$\Psi_n u(\theta) = \sum_{\ell=-\nu}^{\nu-1} F(\ell)\Big[\sum_{j\in\mathbb{Z}\setminus 0} \hat{u}(\ell + jn)\Big] e^{i\ell\theta}. \tag{11.24}$$

This implies the estimate

$$\|\Psi_n u\|_{L^\infty} \le \Big[\sup_{|\ell|\le\nu} |F(\ell)|\Big] \sum_{|k|\ge\nu} |\hat{u}(k)|. \tag{11.25}$$

Also, as in (11.18), we have, for $m \ge 1$,

$$\|\Psi_n u\|_{C^\ell(S^1)} \le C_{\ell m}\Big[\sup_{|\ell|\le\nu} |F(\ell)|\Big]\, \|u\|_{C^{\ell+m+1}(S^1)} \cdot n^{-m}. \tag{11.26}$$

The significance of these statements is that, for u smooth and n large, the discretized $F(D_n)$ provides a very accurate approximation to $F(D)$. This is of practical importance for a number of numerical problems.

Note the distinction between D_n and the centered difference operator Δ_n, defined by

$$(\Delta_n f)(\omega^j) = \frac{n}{4\pi i}\big[f(\omega^{j+1}) - f(\omega^{j-1})\big].$$

We have, in place of (11.21),

$$F(\Delta_n)\Big[\sum_{\ell\in\mathbb{Z}_n} g(\ell)\omega^{j\ell}\Big] = \sum_{\ell=-\nu}^{\nu-1} F\Big(\frac{n}{2\pi}\sin\Big(\frac{2\pi\ell}{n}\Big)\Big) g(\ell)\omega^{j\ell}, \tag{11.27}$$

so, for $g^b \in L^2(\Gamma_n)$ given by (11.2),
(11.28)

$$F(\Delta_n)g^b(\omega^j) - F(D_n)g^b(\omega^j) = \sum_{\ell=-\nu}^{\nu-1}\Big[F\Big(\frac{n}{2\pi}\sin\Big(\frac{2\pi\ell}{n}\Big)\Big) - F(\ell)\Big] g(\ell)\omega^{j\ell}.$$

This identity leads to a variety of estimates, of which the following is a simple example. If $|F'(\lambda)| \le K$ for $-\nu \le \lambda \le \nu$, then

$$\|F(\Delta_n)u - F(D_n)u\|_{L^\infty} \le \frac{2}{3}\cdot\pi^2 K\Big[\sum_{\ell=-\nu}^{\nu-1} |\ell|^3\, |u^\#(\ell)|\Big]\cdot n^{-2}, \tag{11.29}$$

since, for $-\pi \le x \le \pi$, $|\sin x - x| \le (1/6)|x|^3$. The basic content of this is that $F(\Delta_n)$ furnishes a *second-order*-accurate approximation to $F(D)$ (as $n \to \infty$). This is an improvement over the first-order accuracy one would get by using a one-sided difference operator, such as

$$(\Delta_n^+ f)(\omega^j) = \frac{n}{2\pi i}\big[f(\omega^{j+1}) - f(\omega^j)\big],$$

but not as good as the "infinite-order accuracy" one gets for $F(D_n)$ as a consequence of (11.23)–(11.26).

Similar to the case of functions on S^1, we have, for $u \in L^2(\Gamma_n)$,

$$F(D_n)u(\omega^j) = (k_F * u)(\omega^j) = \frac{1}{n}\sum_{\ell\in\mathbb{Z}_n} k_F(\omega^{j-\ell})u(\omega^\ell), \tag{11.30}$$

where

$$k_F(\omega^j) = \sum_{\ell=-\nu}^{\nu-1} F(\ell)\omega^{j\ell}. \tag{11.31}$$

For example, with $F(\lambda) = e^{-y|\lambda|}$, we get the discrete version of the Poisson kernel:

$$k_F(\omega^j) = p_y(\omega^j) = \sum_{\ell=-\nu}^{\nu-1} e^{-y|\ell|}\omega^{j\ell}, \tag{11.32}$$

which we can write as a sum of two finite geometrical series to get

$$p_y(\omega^j) = \frac{1-r^2-2r^{\nu+1}(-1)^j\cos(2\pi j/n)}{1+r^2-2r\cos(2\pi j/n)} + r^\nu\omega^{-j\nu}, \tag{11.33}$$

with $r = e^{-y}$ and, as usual, $\omega = e^{2\pi i/n}$, $n = 2\nu$. Compare with (1.30). The reader can produce a similar formula for n odd.

As in the case of S^1, the sum (11.31) for the (discretized) heat kernel, with $F(\ell) = e^{-t\ell^2}$, cannot generally be simplified to an expression whose size is independent of n. However, when t is an imaginary integer, such an evaluation can be performed. Such expressions are called *Gauss sums*, and their evaluation is regarded as one of the pearls of early nineteenth-century mathematics. We present one such result here.

Proposition 11.5. *For any $n \geq 1$, even or odd,*

$$\sum_{k=0}^{n-1} e^{2\pi ik^2/n}e^{2\pi i\ell k/n} = \frac{1}{2}(1+i)e^{-\pi i\ell^2/2n}\left[1+(-1)^\ell i^{-n}\right]n^{1/2}. \tag{11.34}$$

Proof. The sum on the left is $n\cdot f^\#(-\ell)$, where $f \in C(S^1)$ is given by

$$f(\theta) = e^{in\theta^2/2\pi}, \quad 0 \leq \theta \leq 2\pi.$$

Note that f is Lipschitz on S^1, with a simple jump in its derivative, so $\hat{f}(k) = O(|k|^{-2})$. Hence Proposition 11.3 applies, and (11.7) yields

$$f^\#(-\ell) = \sum_{j=-\infty}^{\infty}\int_0^1 e^{2\pi in[y^2+(j+\ell/n)y]}\,dy. \tag{11.35}$$

To evaluate this, we use the "Gaussian integral" (convergent though not absolutely convergent):

$$\int_{-\infty}^{\infty} e^{2\pi iny^2}\,dy = n^{-1/2}\gamma, \quad \gamma = \frac{1}{2}(1+i), \tag{11.36}$$

obtained from (3.20) by a change of variable and analytic continuation, as in (6.42). We will break up the real line as a countable union of intervals $\bigcup_k [k+a, k+a+1]$, in two different ways, and then evaluate (11.35). Note that

$$\int_{k+a}^{k+a+1} e^{2\pi i n y^2}\, dy = \int_0^1 e^{2\pi i n[y^2+2(k+a)y]}\, dy \cdot e^{2\pi i n(k+a)^2}. \tag{11.37}$$

If we pick $a = \ell/2n$, then $2(k+a) = 2k + \ell/n$, and as k runs over $\mathbb{Z}$, we get those integrands in (11.35) for which j is even. If we pick $a = -1/2+\ell/2n$, then $2(k+a) = 2k - 1 + \ell/n$. Furthermore, we have

$$e^{2\pi i n(k+a)^2} = e^{\pi i \ell^2/2n} \text{ and } e^{\pi i(\ell-n)^2/2n}, \tag{11.38}$$

respectively, for these two choices of a. Thus the sum in (11.35) is equal to $n^{-1/2}\gamma$ times $e^{-\pi i\ell^2/2n} + e^{-\pi i(\ell-n)^2/2n}$, which gives the desired formula, (11.34).

The basic case of this sum is the $\ell = 0$ case:

$$\sum_{k=0}^{n-1} e^{2\pi i k^2/n} = \frac{1}{2}(1+i)(1+i^{-n})n^{1/2} = \sigma_n \cdot n^{1/2}, \tag{11.39}$$

where σ_n is periodic of period 4 in n, with

$$\sigma_0 = 1+i, \quad \sigma_1 = 1, \quad \sigma_2 = 0, \quad \sigma_3 = i. \tag{11.40}$$

This result, particularly when $n = p$ is a prime, is used as a tool to obtain fascinating number-theoretical results. For more on this, see the exercises and the references [Hua], [Land], and [Rad].

Exercises

1. Generalize the Gauss sum identity (11.34) to

$$\sum_{k=0}^{n-1} e^{2\pi i k^2 m/n} e^{2\pi i \ell k/n} = \frac{1+i}{2}\left(\frac{n}{m}\right)^{1/2} e^{-\pi i \ell^2/2mn} \sum_{\nu=0}^{2m-1} e^{-\pi i n\nu^2/2m} e^{-\pi i \nu\ell/m}. \tag{11.41}$$

(*Hint*: The left side is $n \cdot f^\#(-\ell)$, with

$$f(\theta) = e^{inm\theta^2/2\pi}, \quad 0 \le \theta \le 2\pi.$$

For this, one has a formula like (11.35):

$$f^\#(-\ell) = \sum_{j=-\infty}^{\infty} \int_0^1 e^{2\pi i n m[y^2+(1/m)(j+\ell/n)y]}\, dy.$$

Write $j = 2m\mu + \nu$, so

$$\sum_{j=-\infty}^{\infty} a_j = \sum_{\nu=0}^{2m-1} \sum_{j=\nu \bmod 2m} a_j.$$

For fixed ν, the sum becomes a multiple of the Gaussian integral (11.36), with n replaced by nm, and the formula (11.41) arises.)

Note the $\ell = 0$ case of this:

$$\sum_{k=0}^{n-1} e^{2\pi i k^2 m/n} = \frac{1+i}{2}\left(\frac{n}{m}\right)^{1/2} \sum_{\nu=0}^{2m-1} e^{-\pi i n \nu^2/2m}.$$

2. Let Δ be d^2/dx^2 on $S^1 = \mathbb{R}/(2\pi\mathbb{Z})$. Using Fourier series, show that, for $t = 2\pi m/n$, where m and n are positive integers, $e^{-it\Delta}\delta(x) = H(t, x)$ has the form

$$H\left(2\pi \frac{m}{n}, x\right) = \frac{1}{n} \sum_{\ell=0}^{n-1} G(m, n, \ell)\, \delta_{2\pi\ell/n}(x), \tag{11.42}$$

where $G(m, n, \ell)$ is given by the left side of (11.41). On the other hand, applying $e^{-it\Delta}$, acting on $\mathcal{S}'(\mathbb{R})$, to $\sum_\nu \delta(x - 2\pi\nu)$, show that (11.42) holds, with $G(m, n, \ell)$ given by the right side of (11.41). Hence deduce another proof of this Gauss sum identity.

3. Let $\#(\ell, n)$ denote the number of solutions $k \in \mathbb{Z}_n$ to

$$\ell = k^2 \pmod{n}.$$

Show that, with $\omega = e^{2\pi i/n}$,

$$\sum_{k=0}^{n-1} \omega^{jk^2} = \sum_{\ell=0}^{n-1} \#(\ell, n)\omega^{j\ell}.$$

4. Show that, more generally,

$$\left(\sum_{k=0}^{n-1} \omega^{jk^2}\right)^{\nu} = \sum_{\ell=0}^{n-1} \#(\ell, n; \nu)\, \omega^{j\ell},$$

where $\#(\ell, n; \nu)$ denotes the number of solutions $(k_1, \dots, k_\nu) \in (\mathbb{Z}_n)^\nu$ to

$$\ell = k_1^2 + \cdots + k_\nu^2 \pmod{n}.$$

5. Let p be a prime. The *Legendre symbol* $(\ell|p)$ is defined to be $+1$ if $\ell = k^2 \bmod p$ for some k and $\ell \neq 0$, 0 if $\ell = 0$, and -1 otherwise. If p is an odd prime, $\#(\ell, p) = (\ell|p) + 1$. The Legendre symbol has the useful *multiplicative property*: $(\ell_1\ell_2|p) = (\ell_1|p)(\ell_2|p)$. Check this. Show that, with $\omega = e^{2\pi i/p}$, if p is an odd prime,

$$\sum_{k=0}^{p-1} \omega^{k^2} = \sum_{\ell=0}^{p-1} (\ell|p)\omega^{\ell},$$

and, more generally,

$$\sum_{k=0}^{p-1} \omega^{jk^2} = \sum_{\ell=0}^{p-1} (\ell|p)\omega^{j\ell} + p\delta_{j0},$$

where $\delta_{j0} = 1$ if $j = 0 \pmod{p}$, 0 otherwise. (*Hint*: Use Exercise 3.)

6. Denoting $\sum_{k=0}^{p-1} \omega^{k^2}$ by G_p, p an odd prime, show that

$$\sum_{k=0}^{p-1} \omega^{jk^2} = (j|p) \cdot G_p + p \cdot \delta_{j0}.$$

(*Hint*: If $1 \le j \le p-1$, use $\sum_{\ell=0}^{p-1} (\ell|p)\omega^{\ell} = \sum_{\ell=0}^{p-1} (j\ell|p)\omega^{j\ell}$ and $(j\ell|p) = (j|p)(\ell|p)$.)

Denote by $S(m,n)$ the Gauss sum

$$S(m,n) = \sum_{k=0}^{n-1} e^{2\pi i k^2 m/n}.$$

Then the content of Exercise 6 is that $S(j,p) = (j|p)S(1,p)$, for $1 \le j \le p-1$, when p is an odd prime.

7. Assume p and q are distinct odd primes. Show that

$$S(1,pq) = S(q,p)S(p,q).$$

(*Hint*: To resum $\sum_{k=0}^{pq-1} e^{2\pi i k^2/pq}$, use the fact that, as μ runs over $\{0,1,\dots,p-1\}$ and ν runs over $\{0,1,\dots,q-1\}$, then $k = \mu q + \nu p$ runs once over $\mathbb{Z}$ mod pq.)

8. From Exercises 6 and 7, it follows that when p and q are distinct odd primes,

$$(p|q)(q|p) = \frac{S(1,pq)}{S(1,p)S(1,q)}.$$

Use the evaluation (11.39) of $S(1,n)$ to deduce the *quadratic reciprocity law*:

$$(p|q)(q|p) = (-1)^{(p-1)(q-1)/4}.$$

This law, together with the complementary results

$$(-1|p) = (-1)^{(p-1)/2}, \quad (2|p) = (-1)^{(p^2-1)/8},$$

allows for an effective computation of $(\ell|p)$, as one application, but the significance of quadratic reciprocity goes beyond this. It and other implications of Gauss sums are absolutely fundamental in number theory. For material on this, see [Hua], [Land], and [Rad].

12. The fast Fourier transform

In the last section we discussed some properties of the discrete Fourier transform

$$f^{\#}(\ell) = \frac{1}{n} \sum_{\omega^j \in \Gamma_n} f(\omega^j)\omega^{-j\ell}, \tag{12.1}$$

where $\ell \in \mathbb{Z}_n = \mathbb{Z}/(n)$ and Γ_n is the multiplicative group of unit complex numbers generated by $\omega = e^{2\pi i/n}$. We now turn to a discussion of the efficient numerical computation of the discrete Fourier transform. Note that, for any fixed ℓ, computing the right side of (12.1) involves $n-1$ additions and n multiplications of complex numbers, plus n integer products $j\ell = m$ and looking up ω^m and $f(\omega^j)$. If the computations for varying ℓ are done independently, the total effort to compute $f^{\#}$ involves n^2 multiplications and $n(n-1)$ additions of complex numbers, plus some further chores. The *fast Fourier transform* (denoted FFT) is a method for computing $f^{\#}$ in $Cn(\log n)$ steps, in case n is a power of 2.

The possibility of doing this arises from observing redundancies in the calculation of the Fourier coefficients $f^{\#}(\ell)$. Let us illustrate this in the case of Γ_4. We can write

$$(12.2)\qquad \begin{aligned} 4f^{\#}(0) &= \left[f(1)+f(i^2)\right]+\left[f(i)+f(i^3)\right],\\ 4f^{\#}(2) &= \left[f(1)+f(i^2)\right]-\left[f(i)+f(i^3)\right], \end{aligned}$$

and

$$(12.3)\qquad \begin{aligned} 4f^{\#}(1) &= \left[f(1)-f(i^2)\right]-i\left[f(i)-f(i^3)\right],\\ 4f^{\#}(3) &= \left[f(1)-f(i^2)\right]+i\left[f(i)-f(i^3)\right]. \end{aligned}$$

Note that each term in square brackets appears twice. Note also that (12.2) gives the Fourier coefficients of a function on Γ_2; namely, if

$$(12.4)\qquad {}^0f(1) = f(1)+f(-1),\quad {}^0f(-1) = f(i)+f(i^3),$$

then

$$(12.5)\qquad 2f^{\#}(2\ell) = {}^0f^{\#}(\ell),\quad \text{for } \ell = 0 \text{ or } 1.$$

Similarly, if we set

$$(12.6)\qquad {}^1f(1) = f(1)-f(-1),\quad {}^1f(-1) = -i\left[f(i)-f(i^3)\right],$$

then

$$(12.7)\qquad 2f^{\#}(2\ell+1) = {}^1f^{\#}(\ell),\quad \text{for } \ell = 0 \text{ or } 1.$$

This phenomenon is a special case of a more general result that leads to a fast inductive procedure for evaluating the Fourier transform $f^{\#}$.

Suppose $n = 2^k$; let us use the notation $G_k = \Gamma_n$. Note that G_{k-1} is a subgroup of G_k. Furthermore, there is a homomorphism of G_k onto G_{k-1}, given by $\omega^j \mapsto \omega^{2j}$. Given $f : G_k \to \mathbb{C}$, define the following functions ${}^0f$ and ${}^1f$ on G_{k-1}, with $\omega_1 = \omega^2$, generating G_{k-1}:

$$(12.8)\qquad {}^0f(\omega_1^j) = f(\omega^j)+f(\omega^{j+n/2}),$$

$$(12.9)\qquad {}^1f(\omega_1^j) = \overline{\omega}^j\left[f(\omega^j)-f(\omega^{j+n/2})\right].$$

Note that the factor $\overline{\omega}^j$ in (12.9) makes ${}^1f(\omega_1^j)$ well defined for $j \in \mathbb{Z}_{n/2}$, that is, the right side of (12.9) is unchanged if j is replaced by $j + n/2$. Then ${}^0f^{\#}$ and ${}^1f^{\#}$, the discrete Fourier transforms of the functions ${}^0f$ and ${}^1f$, respectively, are functions on $\mathbb{Z}_{n/2} = \mathbb{Z}/(2^{k-1})$.

Proposition 12.1. *We have the following identities relating the Fourier transforms of ${}^0f$, ${}^1f$, and f:*

$$(12.10)\qquad 2f^{\#}(2\ell) = {}^0f^{\#}(\ell)$$

and

$$(12.11)\qquad 2f^{\#}(2\ell+1) = {}^1f^{\#}(\ell),$$

for $\ell \in \{0, 1, \dots, n/2-1\}$.

Proof. Recall that we set $\omega_1 = \omega^2$. Since $\omega^n = 1$ and $\omega_1^{n/2} = 1$, we have

$$(12.12)\qquad \begin{aligned} nf^{\#}(2\ell) &= \sum_{\omega^j \in G_k} f(\omega^j)\overline{\omega}^{2j\ell} \\ &= \sum_{\omega_1^j = \omega^{2j} \in G_{k-1}} \big[f(\omega^j) + f(\omega^{j+n/2})\big]\overline{\omega}_1^{j\ell}, \end{aligned}$$

proving (12.10), and, since $\omega^{n/2} = -1$,

$$(12.13)\qquad \begin{aligned} nf^{\#}(2\ell+1) &= \sum_{\omega^j \in G_k} f(\omega^j)\overline{\omega}^j\, \overline{\omega}^{2j\ell} \\ &= \sum_{\omega_1^j = \omega^{2j} \in G_{k-1}} \overline{\omega}^j\big[f(\omega^j) - f(\omega^{j+n/2})\big]\overline{\omega}_1^{j\ell}, \end{aligned}$$

proving (12.11).

Thus the problem of computing $f^{\#}$, given $f \in L^2(G_k)$, is transformed, after $n/2$ multiplications and n additions of complex numbers in (12.8) and (12.9), to the problem of computing the Fourier transforms of *two* functions on G_{k-1}. After $n/4$ new multiplications and $n/2$ new additions for each of these functions ${}^0f$ and ${}^1f$, that is, after an additional total of $n/2$ new multiplications and n additions, this is reduced to the problem of computing *four* Fourier transforms of functions on G_{k-2}. After k iterations, we obtain 2^k functions on $G_0 = \{1\}$, which precisely give the Fourier coefficients of f. Doing this hence takes $kn = (\log_2 n)n$ additions and $kn/2 = (\log_2 n)n/2$ multplications of complex numbers, plus a comparable number of integer operations and fetching from memory values of given or previously computed functions.

To describe an explicit implementation of Proposition 12.1 for a computation of $f^{\#}$, let us identify an element $\ell \in \mathbb{Z}_n$ ($n = 2^k$) with a k-tuple $L = (L_{k-1}, \dots, L_1, L_0)$ of elements of $\{0, 1\}$ giving the binary expansion of the integer in $\{0, \dots, n-1\}$ representing ℓ (i.e., $L_0 + L_1 \cdot 2 + \cdots + L_{k-1} \cdot 2^{k-1} = \ell$ mod n). To be a little fussy, we use the notation

$$(12.14)\qquad f^{\#}(\ell) = f^{\#\#}(L).$$

Then the formulas (12.10) and (12.11) state that

$$(12.15)\qquad 2f^{\#\#}(L_{k-1}, \dots, L_1, 0) = {}^0f^{\#\#}(L_{k-1}, \dots, L_1)$$

and

$$(12.16)\qquad 2f^{\#\#}(L_{k-1}, \dots, L_1, 1) = {}^1f^{\#\#}(L_{k-1}, \dots, L_1).$$

The inductive procedure described above gives, from ${}^0f$ and ${}^1f$ defined on G_{k-1}, the functions

$$(12.17)\qquad {}^{00}f = {}^0({}^0f),\quad {}^{10}f = {}^1({}^0f),\quad {}^{01}f = {}^0({}^1f),\quad {}^{11}f = {}^1({}^1f)$$

defined on G_{k-2}, and so forth, and we see from (12.15) and (12.16) that

$$f^{\#}(\ell) = \frac{1}{n}\, {}^{L}f, \tag{12.18}$$

where ${}^{L}f = {}^{L}f(1)$ is defined on $G_0 = \{1\}$. From (12.8) and (12.9) we have the following inductive formula for ${}^{L_{m+1}L_m\cdots L_1}f$ on G_{k-m-1}:

$$\begin{aligned} {}^{0L_m\cdots L_1}f(\omega_{m+1}^j) &= {}^{L_m\cdots L_1}f(\omega_m^j) + {}^{L_m\cdots L_1}f(\omega_m^{j+2^{k-m-1}}), \\ {}^{1L_m\cdots L_1}f(\omega_{m+1}^j) &= \overline{\omega}_m^j\big[{}^{L_m\cdots L_1}f(\omega_m^j) - {}^{L_m\cdots L_1}f(\omega_m^{j+2^{k-m-1}})\big], \end{aligned} \tag{12.19}$$

where ω_m is the generator of G_{k-m}, defined by $\omega_0 = \omega = e^{2\pi i/n}$ $(n = 2^k)$, $\omega_{m+1} = \omega_m^2$, that is, $\omega_m = \omega^{2^m}$.

When doing computations, particularly in a higher-level language, it may be easier to work with integers ℓ than with m-tuples $(L_1, \dots, L_1)$. Therefore, let us set

$${}^{L_m\cdots L_1}f(\omega_m^j) = F_m(2^m \cdot j + \ell), \tag{12.20}$$

where

$$\ell = L_1 + L_2 \cdot 2 + \cdots + L_m \cdot 2^{m-1} \in \{0, 1, \dots, 2^m - 1\}$$

and

$$j \in \{0, 1, \dots, 2^{k-m} - 1\}.$$

Note that this precisely defines F_m on $\{0, 1, \dots, 2^k - 1\}$. For $m = 0$, we have

$$F_0(j) = f(\omega^j), \quad 0 \le j \le 2^k - 1. \tag{12.21}$$

The iterative formulas (12.19) give

$$\begin{aligned} F_{m+1}(2^{m+1}j + \ell) &= F_m(2^m j + \ell) + F_m(2^m j + 2^{k-1} + \ell), \\ F_{m+1}(2^m + 2^{m+1}j + \ell) &= \overline{\omega}_m^j\big[F_m(2^m j + \ell) - F_m(2^m j + 2^{k-1} + \ell)\big], \end{aligned} \tag{12.22}$$

for $0 \le j \le 2^{k-m-1} - 1$, $0 \le \ell \le 2^m - 1$. It is easy to write a computer program to implement such an iteration. The formula (12.18) for the Fourier transform of f becomes

$$f^{\#}(\ell) = n^{-1} F_k(\ell), \quad 0 \le \ell \le 2^k - 1. \tag{12.23}$$

While (12.21)–(12.23) provide an easily implementable FFT algorithm, it is not necessarily the best. One drawback is the following. In passing from F_m to F_{m+1} via (12.22), you need two different arrays of n complex numbers. A variant of (12.19), where ${}^{L_{m+1}L_m\cdots L_1}f$ is replaced by $f^{L_1\cdots L_m L_{m+1}}$, leads to an iterative procedure where a transformation of the type (12.19) is performed "in place," and only one such array needs to be used. If memory is expensive and one needs to make the best use of it, this savings can be important. At the end of such an iteration, one needs to perform a "bit reversal" to produce $f^{\#}$. Details, including sample programs, can be found in [PFTV].

On any given computer, a number of factors would influence the choice of the best FFT algorithm. These include such things as relative speed of memory access and floating-point performance, efficiency of computing trigonometric functions (e.g., whether this is implemented in hardware), degree of accuracy required, and other factors. Also, special features, such as computing the Fourier transform of a real-valued function or of a function whose Fourier transform is known to be real-valued, would affect specific computer programs designed for maximum efficiency. Working out how best to implement FFTs on various computers presents many interesting problems.

Exercises

1. Write a computer program to implement the FFT via (12.21)–(12.23). Try to make it run as fast as possible.
2. Using the FFT, write a computer program to solve numerically the initial-value problem for the heat equation $\partial u/\partial t - u_{xx} = 0$ on $\mathbb{R}^+ \times S^1$.
3. Consider multidimensional generalizations of the discrete Fourier transform, and in particular the FFT. What size three-dimensional FFT could be handled by a computer with 4 megabytes of RAM? With 256 MB?
4. Generalize the FFT algorithm to a cyclic group Γ_n with $n = 3^k$. Also, generalize to the case $n = p_1 \cdots p_k$ where p_j are "small" primes.

A. The mighty Gaussian and the sublime gamma function

The Gaussian function $e^{-|x|^2}$ on $\mathbb{R}^n$ is an object whose study yields many wonderful identities. We will use the identity

$$\int_{\mathbb{R}^n} e^{-|x|^2}\,dx = \pi^{n/2}, \tag{A.1}$$

which was established in (3.18), to compute the area A_{n-1} of the unit sphere S^{n-1} in $\mathbb{R}^n$. This computation will bring in Euler's gamma function, and other results will flow from this. Switching to polar coordinates for the right side of (A.1), we have

$$\begin{aligned}\pi^{n/2} &= A_{n-1}\int_0^\infty e^{-r^2}\,r^{n-1}\,dr\\ &= \frac{1}{2}A_{n-1}\int_0^\infty e^{-t}\,t^{n/2-1}\,dt\\ &= \frac{1}{2}A_{n-1}\Gamma\Big(\frac{n}{2}\Big),\end{aligned} \tag{A.2}$$

where the gamma function is defined by

$$\Gamma(z) = \int_0^\infty e^{-t}\,t^{z-1}\,dt, \tag{A.3}$$

for Re $z > 0$. Thus we have the formula

$$A_{n-1} = \frac{2\pi^{n/2}}{\Gamma(\frac{1}{2}n)}. \tag{A.4}$$

To be satisfied with this, we need an explicit evaluation of $\Gamma(n/2)$. This can be obtained from $\Gamma(1/2)$ and $\Gamma(1)$ via the following identity:

$$\begin{aligned} \Gamma(z+1) &= \int_0^\infty e^{-t}\, t^z\, dt \\ &= -\int_0^\infty \frac{d}{dt}(e^{-t})\, t^z\, dt \\ &= z\Gamma(z), \end{aligned} \tag{A.5}$$

for Re $z > 0$, where we used integration by parts. The definition (A.3) clearly gives

$$\Gamma(1) = 1. \tag{A.6}$$

Thus, for any integer $k \geq 1$,

$$\Gamma(k) = (k-1)\Gamma(k-1) = \cdots = (k-1)!\,. \tag{A.7}$$

Note that, for $n = 2$, we have $A_1 = 2\pi/\Gamma(1)$, so (A.6) agrees with the fact that the circumference of the unit circle is 2π (which, of course, figured into the proof of (3.18), via (3.20)). In case $n = 1$, we have $A_0 = 2$, which by (A.4) is equal to $2\pi^{1/2}/\Gamma(1/2)$, so

$$\Gamma\Big(\frac{1}{2}\Big) = \pi^{1/2}. \tag{A.8}$$

Again using (A.5), we see that, when $k \geq 1$ is an integer,

$$\begin{aligned} \Gamma\Big(k+\frac{1}{2}\Big) &= \Big(k-\frac{1}{2}\Big)\Gamma\Big(k-\frac{1}{2}\Big) = \cdots \\ &= \Big(k-\frac{1}{2}\Big)\Big(k-\frac{3}{2}\Big)\cdots\Big(\frac{1}{2}\Big)\Gamma\Big(\frac{1}{2}\Big) \\ &= \pi^{1/2}\Big(k-\frac{1}{2}\Big)\Big(k-\frac{3}{2}\Big)\cdots\Big(\frac{1}{2}\Big). \end{aligned} \tag{A.9}$$

In particular, $\Gamma(3/2) = (1/2)\Gamma(1/2) = \pi^{1/2}/2$, so $A_2 = 2\pi^{3/2}/(\pi^{1/2}/2) = 4\pi$, which agrees with the well known formula for the area of the unit sphere in $\mathbb{R}^3$.

Note that while $\Gamma(z)$ defined by (A.3) is a priori holomorphic for Re z positive, the equation (A.5) shows that $\Gamma(z)$ has a meromorphic extension to the entire complex plane, with simple poles at $z = 0, -1, -2, \dots$. It turns out that $\Gamma(z)$ has no zeros, so $1/\Gamma(z)$ is an entire analytic function. This is a consequence of the identity

$$\Gamma(z)\Gamma(1-z) = \frac{\pi}{\sin \pi z}, \tag{A.10}$$

which we now establish. From (A.4) we have (for $0 < \operatorname{Re} z < 1$)

$$\begin{aligned}\Gamma(z)\Gamma(1-z) &= \int_0^\infty\int_0^\infty e^{-(s+t)}s^{-z}t^{z-1}\,ds\,dt\\ &= \int_0^\infty\int_0^\infty e^{-u}v^{z-1}(1+v)^{-1}\,du\,dv \qquad \text{(A.11)}\\ &= \int_0^\infty (1+v)^{-1}v^{z-1}\,dv,\end{aligned}$$

where we have used the change of variables $u = s+t,\ v = t/s$. With $v = e^x$, the last integral is

$$\int_{-\infty}^\infty (1+e^x)^{-1}\,e^{xz}\,dx, \tag{A.12}$$

which is holomorphic for $0 < \operatorname{Re} z < 1$, and we want to show that it is equal to the right side of (A.10) on this strip. It suffices to prove identity on the line $z = 1/2 + i\xi,\ \xi \in \mathbb{R}$; then (A.12) is equal to the Fourier integral

$$\int_{-\infty}^\infty \left(2\cosh\frac{x}{2}\right)^{-1} e^{ix\xi}\,dx. \tag{A.13}$$

To evaluate this, shift the contour of integration from the real line to the line Im $x = -2\pi$. There is a pole of the integrand at $x = -\pi i$, and we have (A.13) equal to

$$-\int_{-\infty}^\infty \left(2\cosh\frac{x}{2}\right)^{-1} e^{2\pi\xi}e^{ix\xi}\,dx - \text{Residue}\cdot(2\pi i). \tag{A.14}$$

Consequently, (A.13) is equal to

$$-2\pi i\,\frac{\text{Residue}}{1+e^{2\pi\xi}} = \frac{\pi}{\cosh\pi\xi}, \tag{A.15}$$

and since $\pi/\sin\pi(1/2+i\xi) = \pi/\cosh\pi\xi$, the demonstration of (A.10) is complete.

The integral (A.3) and also the last integral in (A.11) are special cases of the *Mellin transform*:

$$\mathcal{M}f(z) = \int_0^\infty f(t)t^{z-1}\,dt. \tag{A.16}$$

If we evaluate this on the imaginary axis:

$$\mathcal{M}^\# f(s) = \int_0^\infty f(t)t^{is-1}\,dt, \tag{A.17}$$

given appropriate growth restrictions on f, this is related to the Fourier transform by a change of variable:

$$\mathcal{M}^\# f(s) = \int_{-\infty}^\infty f(e^x)e^{isx}\,dx. \tag{A.18}$$

The Fourier inversion formula and Plancherel formula imply

(A.19) $$f(r) = (2\pi)^{-1} \int_{-\infty}^{\infty} (\mathcal{M}^\# f)(s) r^{-is}\, ds$$

and

(A.20) $$\int_{-\infty}^{\infty} |\mathcal{M}^\# f(s)|^2\, ds = (2\pi) \int_0^{\infty} |f(r)|^2 r^{-1}\, dr.$$

In some cases, as seen above, one evaluates $\mathcal{M}f(z)$ on a vertical line other than the imaginary axis, which introduces only a slight wrinkle.

An important identity for the gamma function follows from taking the Mellin transform with respect to y of both sides of the subordination identity

(A.21) $$e^{-yA} = \frac{1}{2} y\pi^{-1/2} \int_0^{\infty} e^{-y^2/4t} e^{-tA^2} t^{-3/2}\, dt$$

(if $y > 0,\ A > 0$), established in §5; see (5.22). The Mellin transform of the left side is clearly $\Gamma(z)A^{-z}$. The Mellin transform of the right side is a double integral, which is readily converted to a product of two integrals, each defining gamma functions. After a few changes of variables, there results the identity

(A.22) $$\pi^{1/2}\Gamma(2z) = 2^{2z-1}\Gamma(z)\Gamma\Bigl(z + \frac{1}{2}\Bigr),$$

known as the *duplication formula* for the gamma function. In view of the uniqueness of Mellin transforms, following from (A.18) and (A.19), the identity (A.22) conversely implies (A.21). In fact, (A.22) was obtained first (by Legendre) and this argument produces one of the standard proofs of the subordination identity (A.21).

There is one further identity, which, together with (A.5), (A.10), and (A.22), completes the list of the basic elementary identities for the gamma function. Namely, if the beta function is defined by

(A.23) $$B(x, y) = \int_0^1 s^{x-1}(1-s)^{y-1}\, ds = \int_0^1 (1+u)^{-x-y} u^{x-1}\, du$$

(with $u = s/(1-s)$), then

(A.24) $$B(x, y) = \frac{\Gamma(x)\Gamma(y)}{\Gamma(x+y)}.$$

To prove this, note that since

(A.25) $$\Gamma(z)p^{-z} = \int_0^{\infty} e^{-pt} t^{z-1}\, dt,$$

we have

$$(1+u)^{-x-y} = \frac{1}{\Gamma(x+y)} \int_0^{\infty} e^{-(1+u)t} t^{x+y-1}\, dt,$$

so

$$\begin{aligned} B(x,y) &= \frac{1}{\Gamma(x+y)}\int_0^\infty e^{-t}t^{x+y-1}\int_0^\infty e^{-ut}u^{x-1}\,du\,dt \\ &= \frac{\Gamma(x)}{\Gamma(x+y)}\int_0^\infty e^{-t}t^{y-1}\,dt \\ &= \frac{\Gamma(x)\Gamma(y)}{\Gamma(x+y)}, \end{aligned}$$

as asserted.

The four basic identities proved above are the workhorses for most applications involving gamma functions, but fundamental insight is provided by the identities (A.27) and (A.31) below. First, since $0 \le e^{-t} - (1-t/n)^n \le e^{-t}\cdot t^2/n$ for $0 \le t \le n$, we have, for $\operatorname{Re} z > 0$,

$$\begin{aligned} \Gamma(z) &= \int_0^\infty e^{-t}t^{z-1}\,dt \\ &= \lim_{n\to\infty}\int_0^n \Big(1-\frac{t}{n}\Big)^n t^{z-1}\,dt \qquad \text{(A.26)} \\ &= \lim_{n\to\infty} n^z\int_0^1 (1-s)^n s^{z-1}\,ds. \end{aligned}$$

Repeatedly integrating by parts gives

$$\Gamma(z) = \lim_{n\to\infty} n^z \frac{n(n-1)\cdots 1}{z(z+1)\cdots(z+n-1)}\int_0^1 s^{z+n-1}\,ds,$$

which yields the following result of Euler:

$$\text{(A.27)} \qquad \Gamma(z) = \lim_{n\to\infty} n^z \frac{1\cdot 2\cdots n}{z(z+1)\cdots(z+n)}.$$

Using the identity (A.5), analytically continuing $\Gamma(z)$, we have (A.27) for all z, other than $0, -1, -2, \dots$. We can rewrite (A.27) as

$$\text{(A.28)} \qquad \Gamma(z) = \lim_{n\to\infty} n^z\, z^{-1}(1+z)^{-1}\Big(1+\frac{z}{2}\Big)^{-1}\cdots\Big(1+\frac{z}{n}\Big)^{-1}.$$

If we denote by γ Euler's constant:

$$\text{(A.29)} \qquad \gamma = \lim_{n\to\infty}\Big(1+\frac{1}{2}+\cdots+\frac{1}{n}-\log n\Big),$$

then (A.28) is equivalent to

$$\text{(A.30)} \quad \Gamma(z) = \lim_{n\to\infty} e^{-\gamma z}e^{z(1+1/2+\cdots+1/n)}z^{-1}(1+z)^{-1}\Big(1+\frac{z}{2}\Big)^{-1}\cdots\Big(1+\frac{z}{n}\Big)^{-1},$$

that is, to the Euler product expansion

$$\text{(A.31)} \qquad \frac{1}{\Gamma(z)} = z\, e^{\gamma z}\prod_{n=1}^\infty\Big(1+\frac{z}{n}\Big)e^{-z/n}.$$

It follows that the entire analytic function $1/\Gamma(z)\Gamma(-z)$ has the product expansion

$$\frac{1}{\Gamma(z)\Gamma(-z)} = -z^2 \prod_{n=1}^{\infty}\left(1 - \frac{z^2}{n^2}\right). \tag{A.32}$$

Since $\Gamma(1-z) = -z\Gamma(-z)$, by virtue of (A.10) this last identity is equivalent to the Euler product expansion

$$\sin \pi z = \pi z \prod_{n=1}^{\infty}\left(1 - \frac{z^2}{n^2}\right). \tag{A.33}$$

It is quite easy to deduce the formula (A.5) from the Euler product expansion (A.31). Also, to deduce the duplication formula (A.22) from the Euler product formula is a fairly straightforward exercise.

Finally, we derive Stirling's formula, for the asymptotic behavior of $\Gamma(z)$ as $z \to +\infty$. The approach uses the Laplace asymptotic method, which has many other applications. We begin by setting $t = sz$ and then $s = e^y$ in the integral formula (A.3), obtaining

$$\begin{aligned}\Gamma(z) &= z^z \int_0^\infty e^{-z(s-\log s)} s^{-1}\, ds \\ &= z^z \int_{-\infty}^{\infty} e^{-z(e^y - y)}\, dy.\end{aligned} \tag{A.34}$$

The last integral is of the form

$$\int_{-\infty}^{\infty} e^{-z\varphi(y)}\, dy, \tag{A.35}$$

where $\varphi(y) = e^y - y$ has a nondegenerate minimum at $y = 0$; $\varphi(0) = 1$, $\varphi'(0) = 0$, $\varphi''(0) = 1$. If we write $1 = A(y) + B(y)$, $A \in C_0^\infty((-2, 2))$, $A(y) = 1$ for $|y| \leq 1$, then the integral (A.35) is readily seen to be

$$\int_{-\infty}^{\infty} A(y) e^{-z\varphi(y)}\, dy + O(e^{-(1+1/e)z}). \tag{A.36}$$

We can make a smooth change of variable $x = \xi(y)$ such that $\xi(y) = y + O(y^2)$, $\varphi(y) = 1 + x^2/2$, and the integral in (A.36) becomes

$$e^{-z} \int_{-\infty}^{\infty} A_1(x) e^{-zx^2/2}\, dx, \tag{A.37}$$

where $A_1 \in C_0^\infty(\mathbb{R})$, $A_1(0) = 1$, and it is easy to see that, as $z \to +\infty$,

$$\int_{-\infty}^{\infty} A_1(x) e^{-zx^2/2}\, dx \sim \left(\frac{2\pi}{z}\right)^{1/2}\left[1 + \frac{a_1}{z} + \cdots\right]. \tag{A.38}$$

In fact, if $z = 1/2t$, then (A.38) is equal to $(4\pi t)^{1/2} u(t, 0)$, where $u(t, x)$ solves the heat equation, $u_t - u_{xx} = 0$, $u(0, x) = A_1(x)$. Returning to (A.34), we have Stirling's formula:

$$\Gamma(z) = \left(\frac{2\pi}{z}\right)^{1/2} z^z\, e^{-z}\, [1 + O(z^{-1})]. \tag{A.39}$$

Since $n! = \Gamma(n+1)$, we have in particular that

$$n! = (2\pi n)^{1/2}\, n^n\, e^{-n}\big[1 + O(n^{-1})\big] \tag{A.40}$$

as $n \to \infty$.

Regarding this approach to the Laplace asymptotic method, compare the derivation of the stationary phase method in Appendix B of Chapter 6.

References

[BJS] L. Bers, F. John, and M. Schechter, *Partial Differential Equations*, Wiley, New York, 1964.

[Ch] R. Churchill, *Fourier Series and Boundary Value Problems*, McGraw Hill, New York, 1963.

[DT] J. Dadok and M. Taylor, Local solvability of constant coefficient linear PDE as a small divisor problem, *Proc. AMS* 82(1981), 58–60.

[Do] W. Donoghue, *Distributions and Fourier Transforms*, Academic Press, New York, 1969.

[Er] A. Erdelyi, *Asymptotic Expansions*, Dover, New York, 1965.

[Er2] A. Erdelyi et al., *Higher Transcendental Functions*, Vols. I–III (Bateman Manuscript Project), McGraw-Hill, New York, 1953.

[Fol] G. Folland, *Introduction to Partial Differential Equations*, Princeton Univ. Press, Princeton, N. J., 1976.

[Frl] F. G. Friedlander, *Sound Pulses*, Cambridge Univ. Press, Cambridge, 1958.

[Frd] A. Friedman, *Generalized Functions and Partial Differential Equations*, Prentice-Hall, Englewood Cliffs, N. J., 1963.

[GS] I. M. Gelfand and I. Shilov, *Generalized Functions*, Vols. 1–3, Moscow, 1958.

[Go] R. Goldberg, *Fourier Transforms*, Cambridge Univ. Press, Cambridge, 1962.

[HP] E. Hille and R. Phillips, *Functional Analysis and Semi-groups*, Colloq. Publ. AMS, 1957.

[Hob] E. Hobson, *The Theory of Spherical Harmonics*, Chelsea, New York, 1965.

[H] L. Hörmander, *The Analysis of Linear Partial Differential Equations*, Vols. 1–2, Springer-Verlag, New York, 1983.

[Hua] L. K. Hua, *Introduction to Number Theory*, Springer-Verlag, New York, 1982.

[Jo] F. John, *Partial Differential Equations*, Springer-Verlag, New York, 1975.

[Kat] Y. Katznelson, *An Introduction to Harmonic Analysis*, Dover, New York, 1976.

[Keg] O. Kellogg, *Foundations of Potential Theory*, Dover, New York, 1954.

[Kor] T. Körner, *Fourier Analysis*, Cambridge Univ. Press, Cambridge, 1988.

[Land] E. Landau, *Elementary Number Theory*, Chelsea, New York, 1958.

[Leb] N. Lebedev, *Special Functions and Their Applications*, Dover, New York, 1972.

[Lry] J. Leray, *Hyperbolic Differential Equations*, Princeton Univ. Press, Princeton, N. J., 1952.

[Miz] S. Mizohata, *The Theory of Partial Differential Equations*, Cambridge Univ. Press, Cambridge, 1973.

[Ol] F. Olver, *Asymptotics and Special Functions*, Academic Press, New York, 1974.

[Pat] S. Patterson, *An Introduction to the Theory of the Riemann Zeta-Function*, Cambridge Univ. Press, Cambridge, 1988.

[PT] M. Pinsky and M. Taylor, Pointwise Fourier inversion: a wave equation approach, IMA Preprint #1369, 1995.

[PFTV] W. Press, B. Flannery, S. Teukolsky, and W. Vetterling, *Numerical Recipes; the Art of Scientific Computing*, Cambridge Univ. Press, Cambridge, 1986.

[Rad] H. Rademacher, *Lectures on Elementary Number Theory*, Chelsea, New York, 1958.

[RS] M. Reed and B. Simon, *Methods of Mathematical Physics*, Academic Press, New York, Vols. 1,2, 1975; Vols. 3,4, 1978.

[RN] F. Riesz and B. S. Nagy, *Functional Analysis*, Ungar, New York, 1955.

[Ru] W. Rudin, *Real and Complex Analysis*, McGraw-Hill, New York, 1976.

[Sch] L. Schwartz, *Théorie des Distributions*, Hermann, Paris, 1950.

[Stk] I. Stakgold, *Boundary Value Problems of Mathematical Physics*, Macmillan, New York, 1968.

[St] E. Stein, *Singular Integrals and Differentiability Properties of Functions*, Princeton Univ. Press, Princeton, N. J., 1970.

[SW] E. Stein and G. Weiss, *Introduction to Fourier Analysis on Euclidean Space*, Princeton Univ. Press, Princeton, N. J., 1971.

[Tay] M. Taylor, *Noncommutative Harmonic Analysis*, Math. Surveys and Monographs, No. 22, AMS, 1986.

[Tre] F. Treves, *Linear Partial Differential Equations with Constant Coefficients*, Gordon and Breach, New York, 1966.

[Wat] G. Watson, *A Treatise on the Theory of Bessel Functions*, Cambridge Univ. Press, Cambridge, 1944.

[WW] E. Whittaker and G. Watson, *A Course of Modern Analysis*, Cambridge Univ. Press, Cambridge, 1927.

[Wie] N. Wiener, Tauberian theorems, *Ann. of Math.* 33(1932), 1–100.

[Yo] K. Yosida, *Functional Analysis*, Springer-Verlag, New York, 1965.

[Zy] A. Zygmund, *Trigonometrical Series*, Cambridge Univ. Press, Cambridge, 1969.

4

Sobolev Spaces

Introduction

In this chapter we develop the elements of the theory of Sobolev spaces, a tool that, together with methods of functional analysis, provides for numerous successful attacks on the questions of existence and smoothness of solutions to many of the basic partial differential equations. For a positive integer k, the Sobolev space $H^k(\mathbb{R}^n)$ is the space of functions in $L^2(\mathbb{R}^n)$ such that, for $|\alpha| \leq k$, $D^\alpha u$, regarded a priori as a distribution, belongs to $L^2(\mathbb{R}^n)$. This space can be characterized in terms of the Fourier transform, and such a characterization leads to a notion of $H^s(\mathbb{R}^n)$ for all $s \in \mathbb{R}$. For $s < 0$, $H^s(\mathbb{R}^n)$ is a space of distributions. There is an invariance under coordinate transformations, permitting an invariant notion of $H^s(M)$ whenever M is a compact manifold. We also define and study $H^s(\Omega)$ when Ω is a compact manifold with boundary.

The tools from Sobolev space theory discussed in this chapter are of great use in the study of linear PDE; this will be illustrated in the following chapter. Chapter 13 will develop further results in Sobolev space theory, which will be seen to be of use in the study of nonlinear PDE.

1. Sobolev spaces on $\mathbb{R}^n$

When $k \geq 0$ is an integer, the Sobolev space $H^k(\mathbb{R}^n)$ is defined as follows:

$$H^k(\mathbb{R}^n) = \{u \in L^2(\mathbb{R}^n) : D^\alpha u \in L^2(\mathbb{R}^n) \text{ for } |\alpha| \leq k\}, \tag{1.1}$$

where $D^\alpha u$ is interpreted a priori as a tempered distribution. Results from Chapter 3 on Fourier analysis show that, for such k, if $u \in L^2(\mathbb{R}^n)$, then

$$u \in H^k(\mathbb{R}^n) \Longleftrightarrow \langle\xi\rangle^k \hat{u} \in L^2(\mathbb{R}^n). \tag{1.2}$$

Recall that

$$\langle\xi\rangle = \left(1 + |\xi|^2\right)^{1/2}. \tag{1.3}$$

We can produce a definition of the Sobolev space $H^s(\mathbb{R}^n)$ for general $s \in \mathbb{R}$, parallel to (1.2), namely

$$H^s(\mathbb{R}^n) = \{u \in \mathcal{S}'(\mathbb{R}^n) : \langle\xi\rangle^s \hat{u} \in L^2(\mathbb{R}^n)\}. \tag{1.4}$$

We can define the operator Λ^s on $\mathcal{S}'(\mathbb{R}^n)$ by

$$\Lambda^s u = \mathcal{F}^{-1}\big(\langle\xi\rangle^s \hat{u}\big). \tag{1.5}$$

Then (1.4) is equivalent to

$$H^s(\mathbb{R}^n) = \{u \in \mathcal{S}'(\mathbb{R}^n) : \Lambda^s u \in L^2(\mathbb{R}^n)\}, \tag{1.6}$$

or $H^s(\mathbb{R}^n) = \Lambda^{-s} L^2(\mathbb{R}^n)$. Each space $H^s(\mathbb{R}^n)$ is a Hilbert space, with inner product

$$\big(u, v\big)_{H^s(\mathbb{R}^n)} = \big(\Lambda^s u, \Lambda^s v\big)_{L^2(\mathbb{R}^n)}. \tag{1.7}$$

We note that the dual of $H^s(\mathbb{R}^n)$ is $H^{-s}(\mathbb{R}^n)$.

Clearly, we have

$$D_j : H^s(\mathbb{R}^n) \longrightarrow H^{s-1}(\mathbb{R}^n), \tag{1.8}$$

and hence

$$D^\alpha : H^s(\mathbb{R}^n) \longrightarrow H^{s-|\alpha|}(\mathbb{R}^n). \tag{1.9}$$

Furthermore, it is easy to see that, given $u \in H^s(\mathbb{R}^n)$,

$$u \in H^{s+1}(\mathbb{R}^n) \Longleftrightarrow D_j u \in H^s(\mathbb{R}^n), \quad \forall\, j. \tag{1.10}$$

We can relate difference quotients to derivatives of elements of Sobolev spaces. Define τ_y, for $y \in \mathbb{R}^n$, by

$$\tau_y u(x) = u(x + y). \tag{1.11}$$

By duality this extends to $\mathcal{S}'(\mathbb{R}^n)$:

$$\langle \tau_{-y} u, v\rangle = \langle u, \tau_y v\rangle.$$

Note that

$$\tau_y v = \mathcal{F}^{-1}\big(e^{iy\cdot\xi}\, \hat{v}\big), \tag{1.12}$$

so it is clear that $\tau_y : H^s(\mathbb{R}^n) \to H^s(\mathbb{R}^n)$ is norm-preserving for each $s \in \mathbb{R}$, $y \in \mathbb{R}^n$. Also, for each $u \in H^s(\mathbb{R}^n)$, $\tau_y u$ is a continuous fuction of y with values in $H^s(\mathbb{R}^n)$. The following result is of frequent use, as we will see in the next chapter.

Proposition 1.1. *Let $(e_1, \dots, e_n)$ be the standard basis of $\mathbb{R}^n$; let $u \in H^s(\mathbb{R}^n)$. Then*

$$\sigma^{-1}(\tau_{\sigma e_j} u - u) \text{ is bounded in } H^s(\mathbb{R}^n),$$

for $\sigma \in (0, 1]$, if and only if $D_j u \in H^s(\mathbb{R}^n)$.

Proof. We have $\sigma^{-1}(\tau_{\sigma e_j} u - u) \to i D_j u$ in $H^{s-1}(\mathbb{R}^n)$ as $\sigma \to 0$ if $u \in H^s(\mathbb{R}^n)$. The hypothesis of boundedness implies that there is a sequence $\sigma_\nu \to 0$ such that

$\sigma_\nu^{-1}(\tau_{\sigma_\nu e_j}u - u)$ converges weakly to an element of $H^s(\mathbb{R}^n)$; call it w. Since the natural inclusion $H^s(\mathbb{R}^n) \hookrightarrow H^{s-1}(\mathbb{R}^n)$ is easily seen to be continuous, it follows that $w = iD_j u$. Since $w \in H^s(\mathbb{R}^n)$, this gives the desired conclusion.

Corollary 1.2. *Given $u \in H^s(\mathbb{R}^n)$, then u belongs to $H^{s+1}(\mathbb{R}^n)$ if and only if $\tau_y u$ is a Lipschitz-continuous function of y with values in $H^s(\mathbb{R}^n)$.*

Proof. This follows easily, given the observation (1.10).

We now show that elements of $H^s(\mathbb{R}^n)$ are smooth in the classical sense for sufficiently large positive s. This is a Sobolev imbedding theorem.

Proposition 1.3. *If $s > n/2$, then each $u \in H^s(\mathbb{R}^n)$ is bounded and continuous.*

Proof. By the Fourier inversion formula, it suffices to prove that $\hat{u}(\xi)$ belongs to $L^1(\mathbb{R}^n)$. Indeed, using Cauchy's inequality, we get

$$\int |\hat{u}(\xi)|\, d\xi \le \Big(\int |\hat{u}(\xi)|^2 \langle\xi\rangle^{2s}\, d\xi\Big)^{1/2} \cdot \Big(\int \langle\xi\rangle^{-2s}\, d\xi\Big)^{1/2}. \tag{1.13}$$

Since the last integral on the right is finite precisely for $s > n/2$, this completes the proof.

Corollary 1.4. *If $s > n/2 + k$, then $H^s(\mathbb{R}^n) \subset C^k(\mathbb{R}^n)$.*

If $s = n/2 + \alpha$, $0 < \alpha < 1$, we can establish Hölder continuity. For $\alpha \in (0, 1)$, we say

$$u \in C^\alpha(\mathbb{R}^n) \Longleftrightarrow u \text{ bounded and } |u(x+y) - u(x)| \le C|y|^\alpha. \tag{1.14}$$

An alternative notation is $\text{Lip}^\alpha(\mathbb{R}^n)$; then the definition above is effective for $\alpha \in (0, 1]$.

Proposition 1.5. *If $s = n/2 + \alpha$, $0 < \alpha < 1$, then $H^s(\mathbb{R}^n) \subset C^\alpha(\mathbb{R}^n)$.*

Proof. For $u \in H^s(\mathbb{R}^n)$, use the Fourier inversion formula to write

$$\begin{aligned} |u(x+y) - u(x)| &= (2\pi)^{-n/2}\Big|\int \hat{u}(\xi) e^{ix\cdot\xi}(e^{iy\cdot\xi} - 1)\, d\xi\Big| \\ &\le C\Big(\int |\hat{u}(\xi)|^2 \langle\xi\rangle^{n+2\alpha}\, d\xi\Big)^{1/2} \cdot \Big(\int |e^{iy\cdot\xi} - 1|^2 \langle\xi\rangle^{-n-2\alpha}\, d\xi\Big)^{1/2}. \end{aligned} \tag{1.15}$$

Now, if $|y| \le 1/2$, write

$$\begin{aligned} &\int |e^{iy\cdot\xi} - 1|^2 \langle\xi\rangle^{-n-2\alpha}\, d\xi \\ &\le C \int_{|\xi| \le \frac{1}{|y|}} |y|^2 |\xi|^2 \langle\xi\rangle^{-n-2\alpha}\, d\xi + 4 \int_{|\xi| \ge \frac{1}{|y|}} \langle\xi\rangle^{-n-2\alpha}\, d\xi. \end{aligned} \tag{1.16}$$

If we use polar coordinates, the right side is readily dominated by

$$C|y|^2 + C|y|^2 \frac{|y|^{2\alpha-2} - 1}{2\alpha - 2} + C|y|^{2\alpha}, \tag{1.17}$$

provided $0 < \alpha < 1$. This implies that, for $|y| \le 1/2$,

$$|u(x+y) - u(x)| \le C_\alpha |y|^\alpha, \tag{1.18}$$

given $u \in H^s(\mathbb{R}^n)$, $s = n/2 + \alpha$, and the proof is complete.

We remark that if one took $\alpha = 1$, the middle term in (1.17) would be modified to $C|y|^2 \log(1/|y|)$, so when $u \in H^{n/2+1}(\mathbb{R}^n)$, one gets the estimate

$$|u(x+y) - u(x)| \le C|y| \Big(\log \frac{1}{|y|}\Big)^{1/2}.$$

Elements of $H^{n/2+1}(\mathbb{R}^n)$ need not be Lipschitz, and elements of $H^{\frac{1}{2}n}(\mathbb{R}^n)$ need not be bounded.

We indicate an example of the last phenomenon. Let us define u by

$$\hat{u}(\xi) = \frac{\langle \xi \rangle^{-n}}{1 + \log\langle \xi \rangle}. \tag{1.19}$$

It is easy to show that $u \in H^{n/2}(\mathbb{R}^n)$. But $\hat{u} \notin L^1(\mathbb{R}^n)$. Now one can show that if $\hat{u} \in L^1_{\text{loc}}(\mathbb{R}^n)$ is positive and belongs to $\mathcal{S}'(\mathbb{R}^n)$, but does not belong to $L^1(\mathbb{R}^n)$, then $u \notin L^\infty(\mathbb{R}^n)$; and this is what happens in the case of (1.19). For more on this, see Exercises 2 and 3 below.

A result dual to Proposition 1.3 is

$$\delta \in H^{-n/2-\varepsilon}(\mathbb{R}^n), \quad \text{for all } \varepsilon > 0, \tag{1.20}$$

which follows directly from the definition (1.4) together with the fact that $\mathcal{F}\delta = (2\pi)^{-n/2}$, by the same sort of estimate on $\int \langle \xi \rangle^{-2s} d\xi$ used to prove Proposition 1.3. Consequently,

$$D^\alpha \delta \in H^{-n/2-|\alpha|-\varepsilon}(\mathbb{R}^n), \quad \text{for all } \varepsilon > 0. \tag{1.21}$$

Next we consider the trace map τ, defined initially from $\mathcal{S}(\mathbb{R}^n)$ to $\mathcal{S}(\mathbb{R}^{n-1})$ by $\tau u = f$, where $f(x') = u(0, x')$ if $x = (x_1, \dots, x_n)$, $x' = (x_2, \dots, x_n)$.

Proposition 1.6. *The map τ extends uniquely to a continuous linear map*

$$\tau : H^s(\mathbb{R}^n) \longrightarrow H^{s-1/2}(\mathbb{R}^{n-1}), \quad \textit{for } s > \frac{1}{2}. \tag{1.22}$$

Proof. If $f = \tau u$, we have

$$\hat{f}(\xi') = \int \hat{u}(\xi)\, d\xi_1. \tag{1.23}$$

Thus

$$|\hat{f}(\xi')|^2 \le \left(\int |\hat{u}(\xi)|^2 \langle\xi\rangle^{2s} d\xi_1\right) \cdot \left(\int \langle\xi\rangle^{-2s} d\xi_1\right),$$

where the last integral is finite if $s > 1/2$. In such a case, we have

$$\tag{1.24} \begin{aligned} \int \langle\xi\rangle^{-2s}\, d\xi_1 &= \int \left(1 + |\xi'|^2 + \xi_1^2\right)^{-s} d\xi_1 \\ &= C\left(1 + |\xi'|^2\right)^{-s+1/2} = C\langle\xi'\rangle^{-2(s-1/2)}. \end{aligned}$$

Thus

$$\tag{1.25} \langle\xi'\rangle^{2(s-1/2)} |\hat{f}(\xi')|^2 \le C \int |\hat{u}(\xi)|^2 \langle\xi\rangle^{2s}\, d\xi_1,$$

and integrating with respect to ξ' gives

$$\tag{1.26} \|f\|^2_{H^{s-1/2}(\mathbb{R}^{n-1})} \le C\|u\|^2_{H^s(\mathbb{R}^n)}.$$

Proposition 1.6 has a converse:

Proposition 1.7. *The map (1.22) is surjective, for each $s > 1/2$.*

Proof. If $g \in H^{s-1/2}(\mathbb{R}^{n-1})$, we can let

$$\tag{1.27} \hat{u}(\xi) = \hat{g}(\xi')\, \frac{\langle\xi'\rangle^{2(s-1/2)}}{\langle\xi\rangle^{2s}}.$$

It is easy to verify that this defines an element $u \in H^s(\mathbb{R}^n)$ and $u(0, x') = cg(x')$ for a nonzero constant c, using (1.24) and (1.23); this provides the proof.

In the next section we will develop a tool that establishes the continuity of a number of natural transformations on $H^s(\mathbb{R}^n)$, as an automatic consequence of the (often more easily checked) continuity for integer s. This will be useful for the study of Sobolev spaces on compact manifolds, in §§3 and 4.

Exercises

1. Show that $\mathcal{S}(\mathbb{R}^n)$ is dense in $H^s(\mathbb{R}^n)$ for each s.
2. Assume $v \in \mathcal{S}'(\mathbb{R}^n) \cap L^1_{\text{loc}}(\mathbb{R}^n)$ and $v(\xi) \ge 0$. Show that if $\hat{v} \in L^\infty(\mathbb{R}^n)$, then $v \in L^1(\mathbb{R}^n)$ and
$$(2\pi)^{n/2}\|\hat{v}\|_{L^\infty} = \|v\|_{L^1}.$$
(*Hint*: Consider $v_k(\xi) = \chi(\xi/k)v(\xi)$, with $\chi \in C_0^\infty(\mathbb{R}^n)$, $\chi(0) = 1$.)
3. Verify that (1.19) defines $u \in H^{n/2}(\mathbb{R}^n)$, $u \notin L^\infty(\mathbb{R}^n)$.
4. Show that the pairing
$$\langle u, v\rangle = \int \hat{u}(\xi)\bar{\hat{v}}(\xi)\, d\xi = \int \hat{u}(\xi)\langle\xi\rangle^s \bar{\hat{v}}(\xi)\langle\xi\rangle^{-s}\, d\xi$$

gives an isomorphism of $H^{-s}(\mathbb{R}^n)$ and the space $H^s(\mathbb{R}^n)'$, dual to $H^s(\mathbb{R}^n)$.

5. Show that the trace map (1.22) satisfies the estimate

$$\|\tau u\|^2_{L^2(\mathbb{R}^{n-1})} \le C\|u\|_{L^2} \cdot \|\nabla u\|_{L^2},$$

given $u \in H^1(\mathbb{R}^n)$, where on the right L^2 means $L^2(\mathbb{R}^n)$.

6. Show that $H^k(\mathbb{R}^n)$ is an algebra for $k > n/2$, that is,

$$u, v \in H^k(\mathbb{R}^n) \Longrightarrow uv \in H^k(\mathbb{R}^n).$$

Reconsider this problem after doing Exercise 5 in §2.

7. Let $f : \mathbb{R} \to \mathbb{R}$ be C^∞. Show that $u \mapsto f(u)$ defines a continuous map $F : H^k(\mathbb{R}^n) \to H^k(\mathbb{R}^n)$, for $k > n/2$. Show that F is a C^1-map, with $DF(u)v = f'(u)v$. Show that F is a C^∞-map.
8. Show that a continuous map $F : H^{k+m}(\mathbb{R}^n) \to H^k(\mathbb{R}^n)$ is defined by $F(u) = f(D^m u)$, where $D^m u = \{D^\alpha u : |\alpha| \le m\}$, assuming f is smooth in its arguments, and $k > n/2$. Show that F is C^1, and compute $DF(u)$. Show F is a C^∞-map from $H^{k+m}(\mathbb{R}^n)$ to $H^k(\mathbb{R}^n)$.
9. Suppose $P(D)$ is an elliptic differential operator of order m, as in Chapter 3. If $\sigma < s + m$, show that

$$u \in H^\sigma(\mathbb{R}^n),\ P(D)u = f \in H^s(\mathbb{R}^n) \Longrightarrow u \in H^{s+m}(\mathbb{R}^n).$$

(*Hint*: Estimate $\langle\xi\rangle^{s+m}\hat{u}$ in terms of $\langle\xi\rangle^\sigma \hat{u}$ and $\langle\xi\rangle^s P(\xi)\hat{u}$.)

10. Given $0 < s < 1$ and $u \in L^2(\mathbb{R}^n)$, show that

$$u \in H^s(\mathbb{R}^n) \Longleftrightarrow \int_0^\infty t^{-(2s+1)} \|\tau_{te_j} u - u\|^2_{L^2}\, dt < \infty, \quad 1 \le j \le n, \tag{1.28}$$

where τ_y is as in (1.12).
(*Hint*: Show that the right side of (1.28) is equal to

$$\int_{\mathbb{R}^n} \psi_s(\xi_j) |\hat{u}(\xi)|^2\, d\xi, \tag{1.29}$$

where, for $0 < s < 1$,

$$\psi_s(\xi_j) = 2\int_0^\infty t^{-(2s+1)} \big(1 - \cos t\xi_j\big)\, dt = C_s|\xi_j|^{2s}.) \tag{1.30}$$

11. The fact that $u \in H^s(\mathbb{R}^n)$ implies that $\sigma^{-1}(\tau_{\sigma e_j} u - u) \to iD_j u$ in $H^{s-1}(\mathbb{R}^n)$ was used in the proof of Proposition 1.1. Give a detailed proof of this. Use it to provide details for a proof of Corollary 1.4.
12. Establish the following, as another approach to justifying Corollary 1.4.

Lemma. *If $u \in C(\mathbb{R}^n)$ and $D_j u \in C(\mathbb{R}^n)$ for each j ($D_j u$ regarded a priori as a distribution), then $u \in C^1(\mathbb{R}^n)$.*

(*Hint*: Consider $\varphi_\varepsilon * u$ for $\varphi_\varepsilon(x) = \varepsilon^{-n}\varphi(x/\varepsilon)$, $\varphi \in C_0^\infty(\mathbb{R}^n)$, $\int \varphi\, dx = 1$, and let $\varepsilon \to 0$.)

2. The complex interpolation method

It is easy to see from the product rule that if M_φ is defined by

$$M_\varphi u = \varphi(x)u(x), \tag{2.1}$$

then, for any integer $k \geq 0$,

$$M_\varphi : H^k(\mathbb{R}^n) \longrightarrow H^k(\mathbb{R}^n), \tag{2.2}$$

provided φ is C^∞ and

$$D^\alpha \varphi \in L^\infty(\mathbb{R}^n), \quad \text{for all } \alpha. \tag{2.3}$$

By duality, (2.2) also holds for negative integers. We claim it holds when k is replaced by any real s, but it is not so simple to deduce this directly from the definition (1.4) of $H^s(\mathbb{R}^n)$. Similarly, suppose

$$\chi : \mathbb{R}^n \longrightarrow \mathbb{R}^n \tag{2.4}$$

is a diffeomorphism, which is linear outside some compact set, and define χ^* on functions by

$$\chi^* u(x) = u(\chi(x)). \tag{2.5}$$

The chain rule easily gives

$$\chi^* : H^k(\mathbb{R}^n) \longrightarrow H^k(\mathbb{R}^n), \tag{2.6}$$

for any integer $k \geq 0$. Since the adjoint of χ^* is ψ^* composed with the operation of multiplication by $|\det D\psi(x)|$, where $\psi = \chi^{-1}$, we see that (2.6) also holds for negative integers k. Again, it is not so straightforward to deduce (2.6) when k is replaced by any real number s. A convenient tool for proving appropriate generalizations of (2.2) and (2.6) is provided by the complex interpolation method, introduced by A. P. Calderon, which we now discuss.

Let E and F be Banach spaces. We suppose that F is included in E, and the inclusion $F \hookrightarrow E$ is continuous. If Ω is the vertical strip in the complex plane,

$$\Omega = \{z \in \mathbb{C} : 0 < \operatorname{Re} z < 1\}, \tag{2.7}$$

we define

$$\begin{aligned} \mathcal{H}_{E,F}(\Omega) = \{&u(z) \text{ bounded and continuous on } \overline{\Omega} \text{ with values in } E; \\ &\text{holomorphic on } \Omega : \|u(1+iy)\|_F \text{ is bounded, for } y \in \mathbb{R}\}. \end{aligned} \tag{2.8}$$

We define the *interpolation* spaces $[E, F]_\theta$ by

$$[E, F]_\theta = \{u(\theta) : u \in \mathcal{H}_{E,F}(\Omega)\}, \quad \theta \in [0, 1]. \tag{2.9}$$

We give $[E, F]_\theta$ the Banach space topology, making it isomorphic to the quotient

$$\mathcal{H}_{E,F}(\Omega)/\{u : u(\theta) = 0\}. \tag{2.10}$$

We will also use the convention

$$[F, E]_\theta = [E, F]_{1-\theta}. \tag{2.11}$$

The following result is of basic importance.

Proposition 2.1. *Let E, F be as above; suppose $\widetilde{E}$, $\widetilde{F}$ are Banach spaces with $\widetilde{F}$ continuously injected in $\widetilde{E}$. Suppose $T : E \to \widetilde{E}$ is a continuous linear map, and suppose $T : F \to \widetilde{F}$. Then, for all $\theta \in [0, 1]$,*

$$T : [E, F]_\theta \to [\widetilde{E}, \widetilde{F}]_\theta. \tag{2.12}$$

Proof. Given $v \in [E, F]_\theta$, let $u \in \mathcal{H}_{E,F}(\Omega)$, $u(\theta) = v$. It follows that $Tu(z) \in \mathcal{H}_{\widetilde{E},\widetilde{F}}(\Omega)$, so $Tv = Tu(\theta) \in [\widetilde{E}, \widetilde{F}]_\theta$, as asserted.

We next identify $[H, \mathcal{D}(A)]_\theta$ when H is a Hilbert space and $\mathcal{D}(A)$ is the domain of a positive, self-adjoint operator on H. By the spectral theorem, this means the following. There is a unitary map $U : H \to L^2(X, \mu)$ such that $B = UAU^{-1}$ is a multiplication operator on $L^2(X, \mu)$:

$$Bu(x) = M_b u(x) = b(x)u(x). \tag{2.13}$$

Then $\mathcal{D}(A) = U^{-1}\mathcal{D}(B)$, where

$$\mathcal{D}(B) = \{u \in L^2(X, \mu) : bu \in L^2(X, \mu)\}.$$

We will assume $b(x) \geq 1$, though perhaps b is unbounded. (Of course, if b is bounded, then $\mathcal{D}(B) = L^2(X, \mu)$ and $\mathcal{D}(A) = H$.) This is equivalent to assuming $(Au, u) \geq \|u\|^2$. In such a case, we define A^θ to be $U^{-1}B^\theta U$, where $B^\theta u(x) = b(x)^\theta u(x)$, if $\theta \geq 0$, and $\mathcal{D}(A^\theta) = U^{-1}\mathcal{D}(B^\theta)$, where $\mathcal{D}(B^\theta) = \{u \in L^2(X, \mu) : b^\theta u \in L^2(X, \mu)\}$. We will give a proof of the spectral theorem in Chapter 8. In this chapter we will apply this notion only to operators A for which such a representation is explicitly implemented by a Fourier transform. Our characterization of interpolation spaces $[H, \mathcal{D}(A)]_\theta$ is given as follows.

Proposition 2.2. *For $\theta \in [0, 1]$,*

$$[H, \mathcal{D}(A)]_\theta = \mathcal{D}(A^\theta). \tag{2.14}$$

Proof. First suppose $v \in \mathcal{D}(A^\theta)$. We want to write $v = u(\theta)$, for some $u \in \mathcal{H}_{H,\mathcal{D}(A)}(\Omega)$. Let

$$u(z) = A^{-z+\theta} v.$$

Then $u(\theta) = v$, u is bounded with values in H, and furthermore $u(1 + iy) = A^{-1}A^{-iy}(A^\theta v)$ is bounded in $\mathcal{D}(A)$.

Conversely, suppose $u(z) \in \mathcal{H}_{H,\mathcal{D}(A)}(\Omega)$. We need to prove that $u(\theta) \in \mathcal{D}(A^\theta)$. Let $\varepsilon > 0$, and note that, by the maximum principle,

$$\begin{aligned} &\|A^z(I + i\varepsilon A)^{-1}u(z)\|_H \\ &\leq \sup_{y\in\mathbb{R}} \max \Big\{ \|(I + i\varepsilon A)^{-1}A^{iy}u(y)\|_H, \\ &\qquad \|A^{1+iy}(I + i\varepsilon A)^{-1}u(1 + iy)\|_H \Big\} \leq C, \end{aligned} \tag{2.15}$$

with C independent of ε. This implies $u(\theta) \in \mathcal{D}(A^\theta)$, as desired.

Now the definition of the Sobolev spaces $H^s(\mathbb{R}^n)$ given in §1 makes it clear that, for $s \geq 0$, $H^s(\mathbb{R}^n) = \mathcal{D}(\Lambda^s)$, where Λ^s is the self-adjoint operator on $L^2(\mathbb{R}^n)$ defined by

$$\Lambda^s = \mathcal{F} M_{\langle\xi\rangle^s} \mathcal{F}^{-1}, \tag{2.16}$$

where $\mathcal{F}$ is the Fourier transform. Thus it follows that, for $k \geq 0$,

$$[L^2(\mathbb{R}^n), H^k(\mathbb{R}^n)]_\theta = H^{k\theta}(\mathbb{R}^n), \quad \theta \in [0, 1]. \tag{2.17}$$

In fact, the same sort of reasoning applies more generally. For any $\sigma, s \in \mathbb{R}$,

$$[H^\sigma(\mathbb{R}^n), H^s(\mathbb{R}^n)]_\theta = H^{\theta s+(1-\theta)\sigma}(\mathbb{R}^n), \quad \theta \in [0, 1]. \tag{2.18}$$

Consequently Proposition 2.1 is applicable to (2.4) and (2.6), to give

$$M_\varphi : H^s(\mathbb{R}^n) \longrightarrow H^s(\mathbb{R}^n) \tag{2.19}$$

and

$$\chi^* : H^s(\mathbb{R}^n) \longrightarrow H^s(\mathbb{R}^n), \tag{2.20}$$

for all $s \in \mathbb{R}$.

It is often convenient to have a definition of $[E, F]_\theta$ when neither Banach space E nor F is contained in the other. Suppose they are both continuously injected into a locally convex topological vector space V. Then $G = \{e + f : e \in E, f \in F\}$ has a natural structure of a Banach space, with norm

$$\|a\|_G = \inf\{\|e\|_E + \|f\|_F : a = e + f \text{ in } V,\ e \in E,\ f \in F\}.$$

In fact, G is naturally isomorphic to the quotient $(E \oplus F)/L$ of the Banach space $E \oplus F$, with the product norm, by the closed linear subspace $L = \{(e, -e) : e \in E \cap F \subset V\}$. Generalizing (2.8), we set

(2.21)

$$\mathcal{H}_{E,F}(\Omega) = \{u(z) \text{ bounded and continuous in } \overline{\Omega} \text{ with values in } G;\ \text{holo-}$$
$$\text{morphic in } \Omega : \|u(iy)\|_E \text{ and } \|u(1+iy)\|_F \text{ bounded, } y \in \mathbb{R}\},$$

where Ω is the vertical strip (2.7). Then we define the interpolation space $[E, F]_\theta$ by (2.9), as before. In this context, the identity (2.11) is a (simple) proposition rather than a definition.

Typical cases where it is of interest to apply such a construction include $E = L^{p_1}(X, \mu)$, $F = L^{p_2}(X, \mu)$. If (X, μ) is a measure space that is neither finite nor atomic (e.g., $\mathbb{R}^n$ with Lebesgue measure), typically neither of these L^p-spaces is contained in the other. We have the following useful result.

Proposition 2.3. *For* $0 < \theta < 1$,

$$[L^{p_1}(X, \mu), L^{p_2}(X, \mu)]_\theta = L^q(X, \mu), \tag{2.22}$$

where p_1, p_2, and q are related by

$$\frac{1}{q} = \frac{1-\theta}{p_1} + \frac{\theta}{p_2}. \tag{2.23}$$

Proof. Given $f \in L^q$, one can define

$$u(z) = |f(x)|^{c(\theta - z)} f(x), \tag{2.24}$$

by convention zero when $f(x) = 0$, with c chosen so that u belongs to $\mathcal{H}_{L^{p_1}, L^{p_2}}$, which gives $L^q \subset [L^{p_1}, L^{p_2}]_\theta$.

Conversely, suppose that one is given $f \in [L^{p_1}, L^{p_2}]_\theta$; say $f = u(\theta)$ with $u \in \mathcal{H}_{L^{p_1}, L^{p_2}}(\Omega)$. For $g \in L^{q'}$, you can define $v(z) = |g(x)|^{b(\theta - z)} g(x)$ with b chosen so that $v \in \mathcal{H}_{L^{p_1'}, L^{p_2'}}(\Omega)$. Then the maximum principle implies

$$|\langle f, g\rangle| \le \sup_{y \in \mathbb{R}} \max \Big\{ \big|\langle u(iy), v(iy)\rangle\big|, \big|\langle u(1+iy), v(1+iy)\rangle\big| \Big\}, \tag{2.25}$$

for any simple function g. This implies

$$\begin{aligned} \Big| \int_X f(x) g(x)\, d\mu(x) \Big| &\le C \Big\| |g|^{b\theta + 1} \Big\|_{L^{p_1'}}^{1-\theta} \cdot \Big\| |g|^{b(\theta - 1) + 1} \Big\|_{L^{p_2'}}^{\theta} \\ &= C\, \|g\|_{L^{q'}}, \end{aligned} \tag{2.26}$$

which implies $f \in L^q$.

We record a couple of consequences of this last result, together with Proposition 2.1. Recall that the Fourier transform has the following mapping properties:

$$\mathcal{F} : L^1(\mathbb{R}^n) \longrightarrow L^\infty(\mathbb{R}^n); \quad \mathcal{F} : L^2(\mathbb{R}^n) \longrightarrow L^2(\mathbb{R}^n).$$

Thus interpolation yields

$$\mathcal{F} : L^p(\mathbb{R}^n) \longrightarrow L^{p'}(\mathbb{R}^n), \quad \text{for } p \in [1, 2], \tag{2.27}$$

where p' is defined by $1/p + 1/p' = 1$. Also, for the convolution product $f * g$, we clearly have

$$L^p * L^1 \subset L^p; \quad L^p * L^{p'} \subset L^\infty.$$

Fixing $f \in L^p$ and interpolating between L^1 and $L^{p'}$ give

$$L^p * L^q \subset L^r, \quad \text{for } q \in [1, p'], \quad \frac{1}{r} = \frac{1}{p} + \frac{1}{q} - 1. \tag{2.28}$$

We return to Hilbert spaces, and an interpolation result that is more general than Proposition 2.2, in that it involves $\mathcal{D}(A)$ for not necessarily self-adjoint A.

Proposition 2.4. *Let P^t be a uniformly bounded, strongly continuous semigroup on a Hilbert space H_0, whose generator A has domain $\mathcal{D}(A) = H_1$. Let $f \in H_0$, $0 < \theta < 1$. Then the following are equivalent:*

$$f \in [H_0, H_1]_\theta; \tag{2.29}$$

for some u,

$$f = u(0), \quad t^{1/2-\theta} u \in L^2(\mathbb{R}^+, H_1), \quad t^{1/2-\theta} \frac{du}{dt} \in L^2(\mathbb{R}^+, H_0); \tag{2.30}$$

$$\int_0^\infty t^{-(2\theta+1)} \|P^t f - f\|_{H_0}^2 \, dt < \infty. \tag{2.31}$$

Proof. First suppose (2.30) holds; then $u'(t) - Au(t) = g(t)$ satisfies $t^{1/2-\theta} g \in L^2(\mathbb{R}^+, H_0)$. Now, $u(t) = P^t f + \int_0^t P^{t-s} g(s)\, ds$, by Duhamel's principle, so

$$P^t f - f = \big(u(t) - f\big) - \int_0^t P^{t-s} g(s)\, ds, \tag{2.32}$$

and hence

$$\|t^{-1}(P^t f - f)\|_{H_0} \le \frac{1}{t} \int_0^t \|u'(s)\|_{H_0}\, ds + \frac{C}{t} \int_0^t \|g(s)\|_{H_0}\, ds. \tag{2.33}$$

This implies (2.31), via the elementary inequality (see Exercise 4 below)

$$\begin{gathered} \|\Phi h\|_{L^2(\mathbb{R}^+, t^\beta dt)} \le K \|h\|_{L^2(\mathbb{R}^+, t^\beta dt)}, \quad \beta < 1, \\ \Phi h(t) = \frac{1}{t} \int_0^t h(s)\, ds, \end{gathered} \tag{2.34}$$

where we set $\beta = 1 - 2\theta$ and take $h(t) = \|u'(t)\|_{H_0}$ or $h(t) = \|g(t)\|_{H_0}$.

Next we show that (2.31) $\Rightarrow$ (2.30). If f satisfies (2.31), set

$$u(t) = \frac{\varphi(t)}{t} \int_0^t P^s f\, ds, \tag{2.35}$$

where $\varphi \in C_0^\infty(\mathbb{R})$ and $\varphi(0) = 1$. Then $u(0) = f$. We need to show that

$$t^{1/2-\theta} Au \in L^2(\mathbb{R}^+, H_0) \;\text{ and }\; t^{1/2-\theta} u' \in L^2(\mathbb{R}^+, H_0). \tag{2.36}$$

Now, $t^{1/2-\theta} Au = \varphi(t) t^{-1/2-\theta}(P^t f - f)$, so the first part of (2.36) follows directly from (2.31). The second part of (2.36) will be proved once we show that $t^{1/2-\theta} v' \in L^2(\mathbb{R}^+, H_0)$, where

$$v(t) = \frac{1}{t} \int_0^t P^s f\, ds. \tag{2.37}$$

Now

$$v'(t) = \frac{1}{t}(P^t f - f) - \frac{1}{t^2} \int_0^t (P^s f - f)\, ds, \tag{2.38}$$

and since the first term on the right has been controlled, it suffices to show that

$$w(t) = t^{1/2-\theta-2} \int_0^t (P^s f - f)\, ds \in L^2(\mathbb{R}^+, H_0). \tag{2.39}$$

Indeed, since $s \le t$ in the integrand,

$$\|w(t)\|_{H_0} \le \frac{t^{\frac{1}{2}-\theta}}{t} \int_0^t h(s)\,ds, \tag{2.40}$$
$$h(t) = t^{-1}\|P^t f - f\|_{H_0} \in L^2(\mathbb{R}^+, t^{1-2\theta}dt),$$

so (2.39) follows from (2.34).

We now tackle the equivalence (2.29) $\Leftrightarrow$ (2.31). Since we have (2.30) $\Leftrightarrow$ (2.31) and (2.30) is independent of the choice of P^t, it suffices to show that (2.29) $\Leftrightarrow$ (2.31) for a *single* choice of P^t such that $\mathcal{D}(A) = H_1$. Now, we can pick a positive self-adjoint operator B such that $\mathcal{D}(B) = H_1$ (see Exercise 2 below), and take $A = iB$, so $P^t = e^{itB}$ is a unitary group. In such a case, the spectral decomposition yields the identity

$$\|B^\theta f\|_{H_0}^2 = C_\theta \int_0^\infty t^{-(2\theta+1)}\|e^{itB}f - f\|_{H_0}^2\,dt; \tag{2.41}$$

compare (1.28)–(1.30); and the proof is easily completed.

Exercises

1. Show that the class of interpolation spaces $[E, F]_\theta$ defined in (2.9) and (2.15) is unchanged if one replaces various norm bounds $\|u(x+iy)\|$ by bounds on $e^{-K|y|}\|u(x+iy)\|$.

   In Exercises 2 and 3, let $H_0 = E$ and $H_1 = F$ be two *Hilbert spaces* satisfying the hypotheses of Proposition 2.1. Assume H_1 is dense in H_0.

2. Show that there is a positive self adjoint operator A on H_0 such that $\mathcal{D}(A) = H_1$. (*Hint*: Use the Friedrichs method.)
3. Let $H_\theta = [H_0, H_1]_\theta$, $0 < \theta < 1$. Show that if $0 \le r < s \le 1$, then
$$[H_r, H_s]_\theta = H_{(1-\theta)r+\theta s}, \quad 0 < \theta < 1.$$
   Relate this to (2.18).
4. Prove the estimate (2.34). (*Hint*: Make the change of variable $e^{(\beta-1)\tau/2}h(e^\tau) = \tilde{h}(\tau)$, and convert Φ into a convolution operator on $L^2(\mathbb{R})$.)
5. Show that, for $0 \le s < n/2$,
$$H^s(\mathbb{R}^n) \subset L^p(\mathbb{R}^n), \quad \forall\, p \in \Big[2, \frac{2n}{n-2s}\Big). \tag{2.42}$$
   (*Hint*: Use interpolation.)
   Use (2.42) to estimate $(D^\alpha u)(D^\beta v)$, given $u, v \in H^k(\mathbb{R}^n)$, $k > n/2$, $|\alpha| + |\beta| \le k$. Sharper and more general results will be obtained in Chapter 13.

3. Sobolev spaces on compact manifolds

Let M be a compact manifold. If $u \in \mathcal{D}'(M)$, we say $u \in H^s(M)$ provided that, on any coordinate patch $U \subset M$, any $\psi \in C_0^\infty(U)$, the element $\psi u \in \mathcal{E}'(U)$

belongs to $H^s(U)$, if U is identified with its image in $\mathbb{R}^n$. By the invariance under coordinate changes derived in §2, it suffices to work with any single coordinate cover of M. If $s = k$, a nonnegative integer, then $H^k(M)$ is equal to the set of $u \in L^2(M)$ such that, for any ℓ smooth vector fields $X_1, \dots, X_\ell$ on M, $\ell \leq k$, $X_1 \cdots X_\ell u \in L^2(M)$. Parallel to (2.17), we have the following result.

Proposition 3.1. *For $k \geq 0$ an integer, $\theta \in [0, 1]$,*

$$[L^2(M), H^k(M)]_\theta = H^{k\theta}(M). \tag{3.1}$$

More generally, for any $\sigma, s \in \mathbb{R}$,

$$[H^\sigma(M), H^s(M)]_\theta = H^{\theta s + (1-\theta)\sigma}(M). \tag{3.2}$$

Proof. These results follow directly from (2.17) and (2.18), with the aid of a partition of unity on M subordinate to a coordinate cover. We leave the details as an exercise.

Similarly, the duality of $H^s(\mathbb{R}^n)$ and $H^{-s}(\mathbb{R}^n)$ can easily be used to establish:

Proposition 3.2. *If M is a compact Riemannian manifold, $s \in \mathbb{R}$, there is a natural isomorphism*

$$H^s(M)^* \approx H^{-s}(M). \tag{3.3}$$

Furthermore, Propositions 1.3–1.5 easily yield:

Proposition 3.3. *If M is a smooth compact manifold of dimension n, and $u \in H^s(M)$, then*

$$u \in C(M) \text{ provided } s > \frac{n}{2}. \tag{3.4}$$

$$u \in C^k(M) \text{ provided } s > \frac{n}{2} + k, \tag{3.5}$$

$$u \in C^\alpha(M) \text{ provided } s = \frac{n}{2} + \alpha, \ \alpha \in (0, 1). \tag{3.6}$$

In the case $M = \mathbb{T}^n$, the torus, we know from results on Fourier series given in Chapter 3 that, for $k \geq 0$ an integer,

$$u \in H^k(\mathbb{T}^n) \Longleftrightarrow \sum_{m \in \mathbb{Z}^n} |\hat{u}(m)|^2 \langle m \rangle^{2k} < \infty. \tag{3.7}$$

By duality, this also holds for k a negative integer. Now interpolation, via Proposition 2.2, implies that, for any $s \in \mathbb{R}$,

$$u \in H^s(\mathbb{T}^n) \Longleftrightarrow \sum_{m \in \mathbb{Z}^n} |\hat{u}(m)|^2 \langle m \rangle^{2s} < \infty. \tag{3.8}$$

Alternatively, if we define Λ^s on $\mathcal{D}'(\mathbb{T}^n)$ by

$$(3.9) \qquad \Lambda^s u = \sum_{m \in \mathbb{Z}^n} \langle m \rangle^s \, \hat{u}(m) \, e^{im \cdot \theta},$$

then, for $s \in \mathbb{R}$,

$$(3.10) \qquad H^s(\mathbb{T}^n) = \Lambda^{-s} L^2(\mathbb{T}^n).$$

Thus, for any $s, \sigma \in \mathbb{R}$,

$$(3.11) \qquad \Lambda^s : H^\sigma(\mathbb{T}^n) \longrightarrow H^{\sigma - s}(\mathbb{T}^n)$$

is an isomorphism.

It is clear from (3.9) that, for any $\sigma > 0$,

$$\Lambda^{-\sigma} : H^s(\mathbb{T}^n) \longrightarrow H^s(\mathbb{T}^n)$$

is a norm limit of finite rank operators, hence compact. Consequently, if j denotes the natural injection, we have, for any $s \in \mathbb{R}$,

$$(3.12) \qquad j : H^{s+\sigma}(\mathbb{T}^n) \longrightarrow H^s(\mathbb{T}^n) \text{ compact}, \quad \forall \sigma > 0.$$

This is a special case of the following result.

Proposition 3.4. *For any compact M, $s \in \mathbb{R}$,*

$$(3.13) \qquad j : H^{s+\sigma}(M) \longrightarrow H^s(M) \text{ is compact}, \quad \forall \sigma > 0.$$

Proof. This follows easily from (3.12), by using a partition of unity to break up an element of $H^{s+\sigma}(M)$ and transfer it to a finite set of elements of $H^{s+\sigma}(\mathbb{T}^n)$, if $n = \dim M$.

This result is a special case of a theorem of Rellich, which also deals with manifolds with boundary, and will be treated in the next section. Rellich's theorem will play a fundamental role in Chapter 5.

We next mention the following observation, an immediate consequence of (3.8) and Cauchy's inequality, which provides a refinement of Proposition 1.3 of Chapter 3.

Proposition 3.5. *If $u \in H^s(\mathbb{T}^n)$, then the Fourier series of u is absolutely convergent, provided $s > n/2$.*

Exercises

1. Fill in the details in the proofs of Propositions 3.1–3.4.
2. Show that $C^\infty(M)$ is dense in each $H^s(M)$, when M is a compact manifold.
3. Consider the projection P defined by

$$Pf(\theta) = \sum_{n=0}^{\infty} \hat{f}(n) e^{in\theta}.$$

Show that $P : H^s(S^1) \to H^s(S^1)$, for all $s \in \mathbb{R}$.

4. Let $a \in C^\infty(S^1)$, and define M_a by $M_a f(\theta) = a(\theta) f(\theta)$. Thus $M_a : H^s(S^1) \to H^s(S^1)$. Consider the commutator $[P, M_a] = PM_a - M_a P$. Show that

$$[P, M_a]f = \sum_{k \geq 0, m > 0} \hat{a}(k+m)\hat{f}(-m)e^{ik\theta} - \sum_{k > 0, m \geq 0} \hat{a}(-k-m)\hat{f}(m)e^{-ik\theta},$$

and deduce that, for all $s \in \mathbb{R}$,

$$[P, M_a] : H^s(S^1) \longrightarrow C^\infty(S^1).$$

(*Hint*: The Fourier coefficients $(\hat{a}(n))$ form a rapidly decreasing sequence.)

5. Let $a_j, b_j \in C^\infty(S^1)$, and consider $T_j = M_{a_j} P + M_{b_j}(I - P)$. Show that

$$T_1 T_2 = M_{a_1 a_2} P + M_{b_1 b_2}(I - P) + R,$$

where, for each $s \in \mathbb{R}$, $R : H^s(S^1) \to C^\infty(S^1)$.

6. Suppose $a, b \in C^\infty(S^1)$ are both nowhere vanishing. Let

$$T = M_a P + M_b(I - P), \quad S = M_{a^{-1}} P + M_{b^{-1}}(I - P).$$

Show that $ST = I + R_1$ and $TS = I + R_2$, where $R_j : H^s(S^1) \to C^\infty(S^1)$, for all $s \in \mathbb{R}$. Deduce that, for each $s \in \mathbb{R}$,

$$T : H^s(S^1) \longrightarrow H^s(S^1) \text{ is Fredholm.}$$

Remark: The theory of Fredholm operators is discussed in §7 of Appendix A, Functional Analysis.

7. Let $e_j(\theta) = e^{ij\theta}$. Describe explicitly the kernel and range of

$$T_{jk} = M_{e_j} P + M_{e_k}(I - P).$$

Hence compute the *index* of T_{jk}. Using this, if a and b are nowhere-vanishing, complex-valued smooth functions on S^1, compute the index of $T_a = M_a P + M_b(I - P)$, in terms of the winding numbers of a and b. (*Hint*: If a and b are homotopic to e_j and e_k, respectively, as maps from S^1 to $\mathbb{C} \setminus 0$, then T and T_{jk} have the same index.)

4. Sobolev spaces on bounded domains

Let $\overline{\Omega}$ be a smooth, compact manifold with boundary $\partial\Omega$ and interior Ω. Our goal is to describe Sobolev spaces $H^s(\Omega)$. In preparation for this, we will consider Sobolev spaces $H^s(\mathbb{R}^n_+)$, where $\mathbb{R}^n_+$ is the half-space

$$\mathbb{R}^n_+ = \{x \in \mathbb{R}^n : x_1 > 0\},$$

with closure $\overline{\mathbb{R}^n_+}$. For $k \geq 0$ an integer, we want

$$H^k(\mathbb{R}^n_+) = \{u \in L^2(\mathbb{R}^n_+) : D^\alpha u \in L^2(\mathbb{R}^n_+) \text{ for } |\alpha| \leq k\}. \tag{4.1}$$

Here, $D^\alpha u$ is regarded a priori as a distribution on the interior $\mathbb{R}^n_+$. The space $H^k(\mathbb{R}^n)$ defined above has a natural Hilbert space structure. It is not hard to show that the space $\mathcal{S}(\overline{\mathbb{R}^n_+})$ of restrictions to $\mathbb{R}^n_+$ of elements of $\mathcal{S}(\mathbb{R}^n)$ is dense in $H^k(\mathbb{R}^n_+)$, from the fact that, if $\tau_s u(x) = u(x_1 + s, x_2, \ldots, x_n)$, then $\tau_s u \to u$ in

$H^k(\mathbb{R}^n_+)$ as $s \searrow 0$, if $u \in H^k(\mathbb{R}^n_+)$. Now, we claim that each $u \in H^k(\mathbb{R}^n_+)$ is the restriction to $\mathbb{R}^n_+$ of an element of $H^k(\mathbb{R}^n)$. To see this, fix an integer N, and let

$$Eu(x) = u(x), \quad \text{for } x_1 \geq 0,$$

(4.2)
$$\sum_{j=1}^{N} a_j u(-jx_1, x'), \quad \text{for } x_1 < 0,$$

defined a priori for $u \in \mathcal{S}(\overline{\mathbb{R}^n_+})$. We have the following.

Lemma 4.1. *One can pick $\{a_1, \dots, a_N\}$ such that the map E has a unique continuous extension to*

(4.3)
$$E : H^k(\mathbb{R}^n_+) \longrightarrow H^k(\mathbb{R}^n), \quad \text{for } k \leq N-1.$$

Proof. Given $u \in \mathcal{S}(\mathbb{R}^n)$, we get an H^k-estimate on Eu provided all the derivatives of Eu of order $\leq N-1$ match up at $x_1 = 0$, that is, provided

(4.4)
$$\sum_{j=1}^{N} (-j)^\ell a_j = 1, \quad \text{for } \ell = 0, 1, \dots, N-1.$$

The system (4.4) is a linear system of N equations for the N quantities a_j; its determinant is a Vandermonde determinant that is seen to be nonzero, so appropriate a_j can be found.

Corollary 4.2. *The restriction map*

(4.5)
$$\rho : H^k(\mathbb{R}^n) \longrightarrow H^k(\mathbb{R}^n_+)$$

is surjective.

Indeed, this follows from

(4.6)
$$\rho E = I \text{ on } H^k(\mathbb{R}^n_+).$$

Suppose $s \geq 0$. We can define $H^s(\mathbb{R}^n_+)$ by interpolation:

(4.7)
$$H^s(\mathbb{R}^n_+) = [L^2(\mathbb{R}^n_+), H^k(\mathbb{R}^n_+)]_\theta, \quad k \geq s, s = \theta k.$$

We can show that (4.7) is independent of the choice of an integer $k \geq s$. Indeed, interpolation from (4.3) gives

(4.8)
$$E : H^s(\mathbb{R}^n_+) \longrightarrow H^s(\mathbb{R}^n);$$

interpolation of (4.5) gives

(4.9)
$$\rho : H^s(\mathbb{R}^n) \longrightarrow H^s(\mathbb{R}^n_+);$$

and we have

(4.10)
$$\rho E = I \text{ on } H^s(\mathbb{R}^n_+).$$

This gives

$$H^s(\mathbb{R}^n_+) \approx H^s(\mathbb{R}^n)/\{u \in H^s(\mathbb{R}^n) : u\big|_{\mathbb{R}^n_+} = 0\}, \tag{4.11}$$

for $s \geq 0$, a characterization that is manifestly independent of the choice of $k \geq s$ in (4.7).

Now let $\overline{\Omega}$ be a smooth, compact manifold with smooth boundary. We can suppose that $\overline{\Omega}$ is imbedded as a submanifold of a compact (boundaryless) manifold M of the same dimension. If $\overline{\Omega} \subset \mathbb{R}^n$, $n = \dim \Omega$, you can arrange this by putting $\overline{\Omega}$ in a large box and identifying opposite sides to get $\overline{\Omega} \subset \mathbb{T}^n$. In the general case, one can construct the "double" of $\overline{\Omega}$, as follows. Using a vector field X on $\partial\Omega$ that points into Ω at each point, that is, X is nowhere vanishing on $\partial\Omega$ and in fact nowhere tangent to $\partial\Omega$, we can extend X to a vector field on a neighborhood of $\partial\Omega$ in $\overline{\Omega}$, and using its integral curves construct a neighborhood of $\partial\Omega$ in $\overline{\Omega}$ diffeomorphic to $[0, 1) \times \partial\Omega$, a so-called "collar neighborhood" of $\partial\Omega$. Using this, one can glue together two copies of $\overline{\Omega}$ along $\partial\Omega$ in such a fashion as to produce a smooth, compact M as desired.

If $k \geq 0$ is an integer, we define $H^k(\Omega)$ to consist of all $u \in L^2(\Omega)$ such that $Pu \in L^2(\Omega)$ for all differential operators P of order $\leq k$ with coefficients in $C^\infty(\overline{\Omega})$. We use Ω to denote $\overline{\Omega} \setminus \partial\Omega$. Similar to the case of $\mathbb{R}^n_+$, one shows that $C^\infty(\overline{\Omega})$ is dense in $H^k(\Omega)$. By covering a neighborhood of $\partial\Omega \subset M$ with coordinate patches and locally using the extension operator E from above, we get, for each finite N, an extension operator

$$E : H^k(\Omega) \longrightarrow H^k(M), \quad 0 \leq k \leq N - 1. \tag{4.12}$$

If, for real $s \geq 0$, we define $H^s(\Omega)$ by

$$H^s(\Omega) = [L^2(\Omega), H^k(\Omega)]_\theta, \quad k \geq s, \ s = \theta k, \tag{4.13}$$

we see that

$$E : H^s(\Omega) \longrightarrow H^s(M), \tag{4.14}$$

so the restriction $\rho : H^s(M) \to H^s(\Omega)$ is onto, and

$$H^s(\Omega) \approx H^s(M)/\{u \in H^s(M) : u\big|_\Omega = 0\}, \tag{4.15}$$

which shows that (4.13) is independent of the choice of $k \geq s$.

The characterization (4.15) can be used to define $H^s(\Omega)$ when s is a negative real number. In that case, one wants to show that the space $H^s(\Omega)$ so defined is independent of the inclusion $\Omega \subset M$. We will take care of this point in the next section.

The existence of the extension map (4.14) allows us to draw the following immediate consequence from Proposition 3.3.

Proposition 4.3. *If* $\dim \Omega = n$ *and* $u \in H^s(\Omega)$, *then*

$$u \in C(\overline{\Omega}) \text{ provided } s > \frac{n}{2};$$
$$u \in C^k(\overline{\Omega}) \text{ provided } s > \frac{n}{2} + k;$$
$$u \in C^\alpha(\overline{\Omega}) \text{ provided } s = \frac{n}{2} + \alpha,\ \alpha \in (0, 1).$$

We now extend Proposition 3.4, obtaining the full version of Rellich's theorem.

Proposition 4.4. *For any* $s \geq 0$, $\sigma > 0$, *the natural inclusion*

$$j : H^{s+\sigma}(\Omega) \longrightarrow H^s(\Omega) \text{ is compact.} \tag{4.16}$$

Proof. Using E and ρ, we can factor the map (4.16) through the map (3.9):

$$\begin{array}{ccc} H^{s+\sigma}(\Omega) & \xrightarrow{\ j\ } & H^s(\Omega) \\ {\scriptstyle E}\big\downarrow & & \big\uparrow{\scriptstyle \rho} \\ H^{s+\sigma}(M) & \xrightarrow{\ j\ } & H^s(M) \end{array}$$

which immediately gives (4.16) as a consequence of Proposition 3.4.

The boundary $\partial\Omega$ of Ω is a smooth, compact manifold, on which Sobolev spaces have been defined. By using local coordinate systems flattening out $\partial\Omega$, together with the extension map (4.14) and the trace theorem, Proposition 1.6, we have the following result on the trace map:

$$\tau u = u\big|_{\partial\Omega}. \tag{4.17}$$

Proposition 4.5. *For* $s > 1/2$, τ *extends uniquely to a continuous map*

$$\tau : H^s(\Omega) \longrightarrow H^{s-1/2}(\partial\Omega). \tag{4.18}$$

We close this section with a consideration of mapping properties on Sobolev spaces of the Poisson integral considered in §2 of Chapter 3:

$$\mathrm{PI} : C(S^1) \longrightarrow C(\overline{\mathcal{D}}), \tag{4.19}$$

where

$$\mathcal{D} = \{(x, y) \in \mathbb{R}^2 : x^2 + y^2 < 1\}, \tag{4.20}$$

given explicitly by

$$\mathrm{PI}\, f(z) = \sum_{k=0}^{\infty} \hat{f}(k) z^k + \sum_{k=1}^{\infty} \hat{f}(-k) \overline{z}^k, \tag{4.21}$$

as in (2.4) of Chapter 3, and satisfying the property that

$$u = \mathrm{PI}\, f \Longrightarrow \Delta u = 0 \text{ in } \mathcal{D} \text{ and } u\big|_{S^1} = f. \tag{4.22}$$

The following result can be compared with Proposition 2.2 in Chapter 3.

Proposition 4.6. *The Poisson integral gives a continuous map*

$$PI : H^s(S^1) \longrightarrow H^{s+1/2}(\mathcal{D}), \quad \text{for } s \geq -\frac{1}{2}. \tag{4.23}$$

Proof. It suffices to prove this for $s = k - 1/2$, $k = 0, 1, 2, \dots$; this result for general $s \geq -1/2$ will then follow by interpolation. Recall that to say $f \in H^{k-1/2}(S^1)$ means

$$\sum_{n=-\infty}^{\infty} |\hat{f}(n)|^2 \langle k \rangle^{2k-1} < \infty. \tag{4.24}$$

Now the functions $\{r^{|n|}e^{in\theta} : n \in \mathbb{Z}\}$ are mutually orthogonal in $L^2(\mathcal{D})$, and

$$\iint_{\mathcal{D}} |r^{|n|}e^{in\theta}|^2 \, dx\, dy = 2\pi \int_0^1 r^{2|n|} r\, dr = \frac{\pi}{|n| + 1}. \tag{4.25}$$

In particular, $f \in H^{-1/2}(S^1)$ implies

$$\sum_{n=-\infty}^{\infty} |\hat{f}(n)|^2 \langle n \rangle^{-1} < \infty,$$

which implies $\mathrm{PI}\, f \in L^2(\mathcal{D})$, by (4.25).

Next, if $f \in H^{k-1/2}(S^1)$, then $(\partial/\partial\theta)^\nu f \in H^{-1/2}(S^1)$, for $0 \leq \nu \leq k$, so $(\partial/\partial\theta)^\nu \mathrm{PI}\, f = \mathrm{PI}(\partial/\partial\theta)^\nu f \in L^2(\mathcal{D})$. We need to show that

$$\Big(r\frac{\partial}{\partial r}\Big)^\mu \Big(\frac{\partial}{\partial\theta}\Big)^\nu \mathrm{PI}\, f \in L^2(\mathcal{D}),$$

for $0 \leq \mu + \nu \leq k$. Indeed, set

$$Nf = \sum_{n=-\infty}^{\infty} |n| \hat{f}(n) e^{in\theta}. \tag{4.26}$$

It follows from Plancherel's theorem that $(\partial/\partial\theta)^\nu N^\mu f \in H^{-1/2}(S^1)$, for $0 \leq \mu + \nu \leq k$, if $f \in H^{k-1/2}(S^1)$, while, as in (2.18) of Chapter 2, we have

$$\Big(r\frac{\partial}{\partial r}\Big)^\mu \Big(\frac{\partial}{\partial\theta}\Big)^\nu \mathrm{PI}\, f = \mathrm{PI}\Big(\frac{\partial}{\partial\theta}\Big)^\nu N^\mu f, \tag{4.27}$$

which hence belongs to $L^2(\mathcal{D})$. Since $\mathrm{PI}\, f$ is smooth in a neighborhood of the origin $r = 0$, this finishes the proof.

The Poisson integral taking functions on the sphere S^{n-1} to harmonic functions on the ball in $\mathbb{R}^n$, and more generally the map taking functions on the boundary of

$\partial\Omega$ of a compact Riemannian manifold $\overline{\Omega}$ (with boundary), to harmonic functions on Ω, will be studied in Chapter 5.

Exercises

1. Let $\mathcal{D}$ be the unit disk in $\mathbb{R}^2$, with boundary $\partial\mathcal{D} = S^1$. Consider the solution to the Neumann problem

$$\Delta u = 0 \text{ on } \mathcal{D}, \quad \frac{\partial u}{\partial r} = g \text{ on } S^1, \tag{4.28}$$

studied in Chapter 3, §2, Exercises 1–4. Show that, for $s \geq 1/2$,

$$g \in H^s(S^1) \Longrightarrow u \in H^{s+3/2}(\mathcal{D}). \tag{4.29}$$

(*Hint*: Write $u = \text{PI}\, f$, with $Nf = g$, where N is given by (4.26).)

2. Let $\overline{\Omega}$ be a smooth, compact manifold with boundary. Show that the following versions of the divergence theorem and Green's formula hold:

$$\int_{\Omega} \big[(\operatorname{div} X)uv + (Xu)v + u(Xv)\big]\, dV = \int_{\partial\Omega} \langle X, \nu\rangle uv\, dS, \tag{4.30}$$

when, among X, u, and v, one is smooth and two belong to $H^1(\Omega)$. Also show that

$$-(u, \Delta v)_{L^2(\Omega)} = (du, dv)_{L^2(\Omega)} - \int_{\partial\Omega} u \frac{\partial \overline{v}}{\partial \nu}\, dS, \tag{4.31}$$

for $u \in H^1(\Omega)$, $v \in H^2(\Omega)$. (*Hint*: Approximate.)

3. Show that if $u \in H^2(\Omega)$ satisfies $\Delta u = 0$ on Ω and $\partial u/\partial \nu = 0$ on $\partial\Omega$, then u must be constant, if Ω is connected. (*Hint*: Use (4.31) with $v = u$.)

Exercises 4–9 deal with the "oblique derivative problem" for the Laplace operator on the disk $\mathcal{D} \subset \mathbb{R}^2$. The oblique derivative problem on higher-dimensional regions is discussed in exercises in §12 of Chapter 5.

4. Consider the oblique derivative problem

$$\Delta u = 0 \text{ on } \mathcal{D}, \quad a\frac{\partial u}{\partial r} + b\frac{\partial u}{\partial \theta} + cu = g \text{ on } S^1, \tag{4.32}$$

where $a, b, c \in C^\infty(S^1)$ are given. If $u = \text{PI}\, f$, show that u is a solution if and only if $Qf = g$, where

$$Q = M_a N + M_b \frac{\partial}{\partial \theta} + M_c : H^{s+1}(S^1) \longrightarrow H^s(S^1). \tag{4.33}$$

5. Recall $\Lambda : H^{s+1}(S^1) \to H^s(S^1)$, defined by

$$\Lambda f(\theta) = \sum \langle k\rangle \hat{f}(k) e^{ik\theta}, \tag{4.34}$$

as in (3.9). Show that Λ is an isomorphism and that

$$\Lambda - N : H^s(S^1) \longrightarrow H^s(S^1). \tag{4.35}$$

6. With Q as in (4.33), show that $Q = T\Lambda$ with

$$T = M_{a+ib}P + M_{a-ib}(I - P) + R : H^s(S^1) \longrightarrow H^s(S^1), \tag{4.36}$$

where

$$R : H^s(S^1) \longrightarrow H^{s+1}(S^1).$$

Here P is as in Exercise 3 of §3. (*Hint*: Note that $\partial/\partial\theta = iPN - i(I-P)N$.)

7. Deduce that the operator Q in (4.33) is Fredholm provided $a+ib$ and $a-ib$ are nowhere vanishing on S^1. In particular, if a and b are real-valued, Q is Fredholm provided a and b have no common zeros on S^1. (*Hint*: Recall Exercises 4–6 of §3.)
8. Let $\mathcal{H} = \{u \in C^2(\mathcal{D}) : \Delta u = 0 \text{ in } \mathcal{D}\}$. Take $s > 0$. Using the commutative diagram

$$\begin{array}{ccc} H^{s+1}(S^1) & \xrightarrow{PI} & H^{s+\frac{3}{2}}(\mathcal{D}) \cap \mathcal{H} \\ {\scriptstyle Q}\downarrow & & \downarrow{\scriptstyle B} \\ H^s(S^1) & \xrightarrow{I} & H^s(S^1) \end{array} \tag{4.37}$$

where Q is as in (4.33) and

$$Bu = a\frac{\partial u}{\partial r} + b\frac{\partial u}{\partial \theta} + cu\big|_{S^1}, \tag{4.38}$$

deduce that B is Fredholm provided $a, b \in C^\infty(S^1)$ are real-valued and have no common zeros on S^1. In such a case, compute the index of B. (*Hint*: Recall Exercise 7 from §3. Also note that the two horizontal arrows in (4.37) are isomorphisms.)
9. Let B be as above; assume $a, b, c \in C^\infty(S^1)$ are all real-valued. Also assume that a is nowhere vanishing on S^1. If $c/a \geq 0$ on S^1, show that Ker B consists at most of constant functions. (*Hint*: See Zaremba's principle, in §2 of Chapter 5.)
If, in addition, c is not identically zero, show that Ker $B = 0$. Using Exercise 8, show that B has index zero in this case. Draw conclusions about the solvability of the oblique derivative problem (4.32).
10. Prove that $C^\infty(\overline{\Omega})$ is dense in $H^s(\Omega)$ for all $s \geq 0$.
(*Hint*: With E as in (4.14), approximate Eu by elements of $C^\infty(M)$.)
11. Consider the Vandermonde determinant

$$\Delta_{n+1}(x_0, \ldots, x_n) = \begin{vmatrix} 1 & 1 & \cdots & 1 \\ x_0 & x_1 & \cdots & x_n \\ \vdots & \vdots & \ddots & \vdots \\ x_0^n & x_1^n & \cdots & x_n^n \end{vmatrix}.$$

Show that $\Delta_{n+1}(x_0, \ldots, x_{n-1}, t)$ is a polynomial of degree n in t, with roots $x_0, \ldots, x_{n-1}$, hence equal to $K(t-x_0)\cdots(t-x_{n-1})$; the coefficient K of t^n is equal to $\Delta_n(x_0, \ldots, x_{n-1})$. Deduce by induction that

$$\Delta_{n+1}(x_0, \ldots, x_n) = \prod_{0 \leq j < k \leq n} (x_k - x_j).$$

12. Given $0 < s < 1$ and $f \in L^2(\mathbb{R}^+)$, show that

$$f \in H^s(\mathbb{R}^+) \Longleftrightarrow \int_0^\infty t^{-(2s+1)} \|\tau_t f - f\|_{L^2(\mathbb{R}^+)}^2 \, dt < \infty, \tag{4.39}$$

where $\tau_t f(x) = f(x+t)$. (*Hint*: Use Proposition 2.4, with $P^t f(x) = f(x+t)$, whose infinitesimal generator is d/dx, with domain $H^1(\mathbb{R}^+)$. Note that "$\Rightarrow$" also follows from (4.14) plus (1.28).)

More generally, given $0 < s < 1$ and $f \in L^2(\mathbb{R}^n_+)$, show that

$$f \in H^s(\mathbb{R}^n_+) \Longleftrightarrow \int_0^\infty t^{-(2s+1)} \|\tau_{te_j} f - f\|^2_{L^2(\mathbb{R}^n_+)} \, dt < \infty, \quad 1 \le j \le n, \tag{4.40}$$

where τ_y is as in (1.12).

5. The Sobolev spaces $H_0^s(\Omega)$

Let $\overline{\Omega}$ be a smooth, compact manifold with boundary; we denote the interior by Ω, as before. As before, we can suppose $\overline{\Omega}$ is contained in a compact, smooth manifold M, with $\partial\Omega$ a smooth hypersurface. For $s \ge 0$, we define $H_0^s(\Omega)$ to consist of the closure of $C_0^\infty(\Omega)$ in $H^s(\Omega)$. For $s = k$ a nonnegative integer, it is not hard to show that

$$H_0^k(\Omega) = \{u \in H^k(M) : \operatorname{supp} u \subset \overline{\Omega}\}. \tag{5.1}$$

This is because a norm giving the topology of $H^k(\Omega)$ can be taken to be the square root of

$$\sum_{j=1}^K \|P_j u\|^2_{L^2(\Omega)}, \tag{5.2}$$

for a certain finite number of differential operators P_j of order $\le k$, which implies that the closure of $C_0^\infty(\Omega)$ in $H^k(\Omega)$ can be identified with its closure in $H^k(M)$. Since the topology of $H^s(M)$ for $s \notin \mathbb{Z}^+$ is not defined in such a localizable fashion, such an argument does not work for general real s. For a general closed set B in M, set

$$H_B^s(M) = \{u \in H^s(M) : \operatorname{supp} u \subset B\}. \tag{5.3}$$

It has been proved in [Fu] that, for $s \ge 0$,

$$H_0^s(\Omega) \approx H_{\overline{\Omega}}^s(M) \ \text{ if } s + \frac{1}{2} \notin \mathbb{Z}. \tag{5.4}$$

See the exercises below for some related results.

Recall our characterization of the space $H^s(\Omega)$ given in (4.15), which we rewrite as

$$H^s(\Omega) \approx H^s(M)/H_K^s(\Omega), \quad K = \overline{M \setminus \Omega}. \tag{5.5}$$

This characterization makes sense for any $s \in \mathbb{R}$, not just for $s \ge 0$, and we use it as a definition of $H^s(\Omega)$ for $s < 0$. For $k \in \mathbb{Z}^+$, we can redefine $H^{-k}(\Omega)$ in a fashion intrinsic to $\overline{\Omega}$, making use of the following functional analytic argument.

In general, if E is a Banach space, with dual E^*, and F a closed linear subspace of E, we have a natural isomorphism of dual spaces:

$$F^* \approx E^*/F^\perp, \tag{5.6}$$

where

$$F^\perp = \{u \in E^* : \langle v, u \rangle = 0 \text{ for all } v \in F\}. \tag{5.7}$$

If $E = H^k(M)$, we take $F = H_0^k(\Omega)$, which, as discussed above, we can regard as the closure of $C_0^\infty(\Omega)$ in $H^k(M) = E$. Then it is clear that $F^\perp = H_K^{-k}(M)$, with $K = \overline{M \setminus \Omega}$, so we have proved:

Proposition 5.1. *For Ω open in M with smooth boundary, $k \geq 0$ an integer, we have a natural isomorphism*

$$H_0^k(\Omega)^* \approx H^{-k}(\Omega). \tag{5.8}$$

Let P be a differential operator of order $2k$, with smooth coefficients on $\overline{\Omega}$. Suppose

$$P = \sum_{j=1}^{L} A_j B_j, \tag{5.9}$$

where A_j and B_j are differential operators of order k, with coefficients smooth on $\overline{\Omega}$. Then we have a well-defined continuous linear map

$$P : H_0^k(\Omega) \longrightarrow H^{-k}(\Omega), \tag{5.10}$$

and, if A_j^t denotes the formal adjoint of A_j on $\overline{\Omega}$, endowed with a smooth Riemannian metric, then, for $u, v \in H_0^k(\Omega)$, we have

$$\langle u, Pv \rangle = \sum_{j=1}^{L} (A_j^t u, B_j v)_{L^2(\Omega)}, \tag{5.11}$$

the dual pairing on the left side being that of (5.8). In fact, the formula (5.5) gives

$$P : H^s(\Omega) \longrightarrow H^{s-2k}(\Omega) \tag{5.12}$$

for all real s, and in particular

$$P : H^k(\Omega) \longrightarrow H^{-k}(\Omega), \tag{5.13}$$

and the identity (5.11) holds for $v \in H^k(\Omega)$, provided $u \in H_0^k(\Omega)$. In Chapter 5 we will study in detail properties of the map (5.10) when P is the Laplace operator (so $k = 1$).

The following is an elementary but useful result.

Proposition 5.2. *Suppose $\overline{\Omega}$ is a smooth, connected, compact manifold with boundary, endowed with a Riemannian metric. Suppose $\partial\Omega \neq \emptyset$. Then there exists a constant $C = C(\Omega) < \infty$ such that*

$$\|u\|_{L^2(\Omega)}^2 \leq C \|du\|_{L^2(\Omega)}^2, \quad \textit{for } u \in H_0^1(\Omega). \tag{5.14}$$

It suffices to establish (5.14) for $u \in C^\infty(\overline{\Omega})$. Given $u\big|_{\partial\Omega} = 0$, one can write

$$u(x) = -\int_{\gamma(x)} du, \tag{5.15}$$

for any $x \in \Omega$, where $\gamma(x)$ is some path from x to $\partial\Omega$. Upon making a reasonable choice of $\gamma(x)$, obtaining (5.14) is an exercise, which we leave to the reader.

Finding a sharp value of C such that (5.14) holds is a challenging problem, for which a number of interesting results have been obtained. As will follow from results in Chapter 5, this is equivalent to the problem of estimating the smallest eigenvalue of $-\Delta$ on Ω, with Dirichlet boundary conditions.

Below, there is a sequence of exercises, one of whose implications is that

$$[L^2(\Omega), H_0^1(\Omega)]_s = H_0^s(\Omega) = H^s(\Omega), \quad 0 < s < \frac{1}{2}. \tag{5.16}$$

Here we will establish a result that is useful for the proof.

Proposition 5.3. *Let $\Omega \subset \mathbb{R}^n$ be a bounded region with smooth boundary. If $0 \le s < 1/2$, and $Tu = \chi_{\overline{\Omega}} u$, then*

$$T : H^s(\mathbb{R}^n) \longrightarrow H^s(\mathbb{R}^n). \tag{5.17}$$

Proof. It is easy to reduce this to the case $\Omega = \mathbb{R}^n_+$, and then to the case $n = 1$, which we will treat here. Also, the case $s = 0$ is trivial, so we take $0 < s < 1/2$. By (1.28), it suffices to estimate

$$\int_0^\infty t^{-(2s+1)} \|\tau_t \widetilde{u} - \widetilde{u}\|^2_{L^2(\mathbb{R})}\, dt, \tag{5.18}$$

where $\widetilde{u}(x) = Tu(x)$, so, for $t > 0$,

$$\tau_t\widetilde{u}(x) - \widetilde{u}(x) = \begin{cases} u(t+x) - u(x), & x > 0 \\ u(t+x), & -t < x < 0 \\ 0, & x < -t \end{cases} \tag{5.19}$$

Hence (5.18) is

$$\le \int_0^\infty t^{-(2s+1)} \|\tau_t u - u\|^2_{L^2(\mathbb{R})}\, dt + \int_0^\infty t^{-(2s+1)} \int_{-t}^0 |u(t+x)|^2\, dx\, dt. \tag{5.20}$$

The first term in (5.20) is finite for $u \in H^s(\mathbb{R})$, $0 < s < 1$, by (1.28). The last term in (5.20) is equal to

$$\begin{aligned} \int_0^\infty \int_0^t t^{-(2s+1)} |u(t-x)|^2\, dx\, dt &= \int_0^\infty \int_0^t t^{-(2s+1)} |u(x)|^2\, dx\, dt \\ &= C_s \int_0^\infty |x|^{-2s} |u(x)|^2\, dx. \end{aligned} \tag{5.21}$$

The next lemma implies that this is finite for $u \in H^s(\mathbb{R})$, $0 < s < 1/2$.

Lemma 5.4. *If $0 < s < 1/2$, then*

$$u \in H^s(\mathbb{R}^n) \Longrightarrow |x_1|^{-s} u \in L^2(\mathbb{R}^n). \tag{5.22}$$

Proof. The general case is easily deduced from the case $n = 1$, which we establish here. Also, it suffices to show that, for $0 < s < 1/2$,

$$u \in H^s(\mathbb{R}) \Longrightarrow x^{-s}\widetilde{u} \in L^2(\mathbb{R}^+), \tag{5.23}$$

where $\widetilde{u} = u\big|_{\mathbb{R}^+}$. Now, for $x > 0$, $u \in C_0^\infty(\mathbb{R})$, set

$$v(x) = \frac{1}{x}\int_0^x \big[u(x) - u(y)\big]\,dy, \quad w(x) = \int_x^\infty \frac{v(y)}{y}\,dy. \tag{5.24}$$

We claim that

$$u(x) = v(x) - w(x), \quad x > 0. \tag{5.25}$$

In fact, if $u \in C_0^\infty(\mathbb{R})$, then $v(x) \to 0$ and $w(x) \to 0$ as $x \to +\infty$, and one verifies easily that $u'(x) = v'(x) - w'(x)$. Thus it suffices to show that, for $0 < s < 1/2$,

$$\|x^{-s}v\|_{L^2(\mathbb{R}^+)} \le C\|u\|_{H^s(\mathbb{R})}, \quad \|x^{-s}w\|_{L^2(\mathbb{R}^+)} \le C\|u\|_{H^s(\mathbb{R})}, \tag{5.26}$$

for $u \in C_0^\infty(\mathbb{R})$.

To verify the first estimate in (5.26), we will use the simple fact that $|v(x)|^2 \le (1/x)\int_0^x |u(x) - u(y)|^2\,dy$. Hence

$$\begin{aligned}\int_0^\infty x^{-s}|v(x)|^2\,dx &\le \int_0^\infty \int_0^x x^{-(2s+1)}\big|u(x) - u(y)\big|^2\,dy\,dx\\ &= \int_0^\infty \int_0^\infty (y+t)^{-(2s+1)}\big|u(y+t) - u(y)\big|^2\,dt\,dy\\ &\le \int_0^\infty y^{-(2s+1)}\|\tau_t u - u\|^2_{L^2(\mathbb{R}^+)}\,dy.\end{aligned} \tag{5.27}$$

Since the $L^2(\mathbb{R}^+)$-norm is less than the $L^2(\mathbb{R})$-norm, it follows from (1.28) that the last integral in (5.27) is dominated by $C\|u\|^2_{H^s(\mathbb{R})}$, for $0 < s < 1$.

Thus, to prove the rest of (5.26), it suffices to show that

$$\|x^{-s}w\|_{L^2(\mathbb{R}^+)} \le C\|x^{-s}v\|_{L^2(\mathbb{R}^+)}, \quad 0 < s < \frac{1}{2}, \tag{5.28}$$

or equivalently, that $\|w\|_{L^2(\mathbb{R}^+, x^{-2s}dx)} \le C\|v\|_{L^2(\mathbb{R}^+, x^{-2s}dx)}$. In turn, this follows from the estimate (2.34), with $\beta = 2s$, since we have $w = \Phi^* v$, where Φ^* acting on $L^2(\mathbb{R}^+, x^{-\beta}dx)$ is the adjoint of Φ in (2.34). This completes the proof of the lemma, hence of Proposition 5.3.

Corollary 5.5. *If $Sv(x) = v(x)$ for $x \in \Omega$, and $Sv(x) = 0$ for $x \in \mathbb{R}^n \setminus \Omega$, then*

$$S : H^s(\Omega) \longrightarrow H^s(\mathbb{R}^n), \quad 0 \le s < \frac{1}{2}. \tag{5.29}$$

Proof. Apply Proposition 5.3 to $u = Ev$, where $E : H^s(\Omega) \to H^s(\Omega)$ is any extension operator that works for $0 \le s \le 1$.

Exercises

1. Give the a detailed proof of (5.1).
2. With $\tau u = u\big|_{\partial\Omega}$, as in (4.17), prove that

$$H_0^1(\Omega) = \{u \in H^1(\Omega) : \tau u = 0\}. \tag{5.30}$$

(*Hint*: Given $u \in H^1(\Omega)$ and $\tau u = 0$, define $\tilde{u} = u(x)$ for $x \in \Omega$, $\tilde{u}(x) = 0$ for $x \in M \setminus \Omega$. Use (4.30) to show that $\tilde{u} \in H^1(M)$.)

3. Let $u \in H^k(\Omega)$. Prove that $u \in H_0^k(\Omega)$ if and only if $\tau(Pu) = 0$ for all differential operators P (with smooth coefficients) of order $\le k-1$ on M.
4. Give a detailed proof of Proposition 5.2 along the lines suggested, involving (5.15).
5. Give an alternative proof of Proposition 5.2, making use of the compactness of the inclusion $H^1(\Omega) \hookrightarrow L^2(\Omega)$. (*Hint*: If (5.14) is false, take $u_j \in H_0^1(\Omega)$ such that $\|du_j\|_{L^2} \to 0$, $\|u_j\|_{L^2} = 1$. The compactness yields a subsequence $u_j \to v$ in $H^1(\Omega)$. Hence $\|v\|_{L^2} = 1$ while $\|dv\|_{L^2} = 0$.)
6. Suppose $\Omega \subset \mathbb{R}^n$ lies between two parallel hyperplanes, $x_1 = A$ and $X_1 = B$. Show that the estimate (5.14) holds with $C = (B-A)^2/\pi^2$.
Reconsider this problem after reading §1 of Chapter 5.
7. Show that $C^\infty(\overline{\Omega})$ is dense in $H^{-s}(\Omega)$, for $s \ge 0$. Compare Exercise 10 of §4.
8. Give a detailed proof that (5.11) is true for $u \in H_0^k(\Omega)$, $v \in H^k(\Omega)$.
(*Hint*: Approximate u by $u_j \in C_0^\infty(\Omega)$ and v by $v_j \in C^\infty(\overline{\Omega})$.)
9. Show that if P^t is the formal adjoint of P, then $\langle u, Pv\rangle = \langle P^t u, v\rangle$ for $u, v \in H_0^k(\Omega)$.

In the following problems, let Ω be an open subset of a compact manifold M, with smooth boundary $\partial\Omega$ and closure $\overline{\Omega}$. Let $\mathcal{O} = M \setminus \overline{\Omega}$.

10. Define $Z : L^2(\Omega) \to L^2(M)$ by $Zu(x) = u(x)$ for $x \in \Omega$, 0 for $x \in \mathcal{O}$. Show that

$$Z : H_0^k(\Omega) \longrightarrow H^k_{\overline{\Omega}}(M), \quad k = 0, 1, 2, \dots \tag{5.31}$$

and that Z is an isomorphism in these cases. Deduce that

$$Z : [L^2(\Omega), H_0^k(\Omega)]_\theta \longrightarrow H^{k\theta}_{\overline{\Omega}}(M), \quad 0 < \theta < 1,\ k \in \mathbb{Z}^+. \tag{5.32}$$

11. For fixed but large N, let $E : H^s(\mathcal{O}) \to H^s(M)$ be an extension operator, similar to (4.14), for $0 \le s \le N$. Define $Tu = u - ERu$, where $Ru = u\big|_{\mathcal{O}}$. Show that

$$T : H^s(M) \longrightarrow H^s_{\overline{\Omega}}(M), \quad 0 \le s \le N. \tag{5.33}$$

Note that $Tu = u$ for $u \in H^s_{\overline{\Omega}}(M)$.

12. Set $T^b u = Tu\big|_{\Omega}$, so $T^b : H^k(M) \to H_0^k(\Omega)$, for $0 \le k \le N$, and hence

$$T^b : H^{k\theta}(M) \longrightarrow [L^2(\Omega), H_0^k(\Omega)]_\theta.$$

Show that

$$T^b j Z = id. \text{ on } [L^2(\Omega), H_0^k(\Omega)]_\theta,$$

where $j : H^s_{\overline{\Omega}}(M) \hookrightarrow H^s(M)$ is the natural inclusion. Deduce that (5.15) is an isomorphism. Conclude that

$$[L^2(\Omega), H_0^k(\Omega)]_\theta \approx [H^0_{\overline{\Omega}}(M), H^k_{\overline{\Omega}}(M)]_\theta = H^{k\theta}_{\overline{\Omega}}(M), \quad 0 < \theta < 1. \tag{5.34}$$

13. Show that $H^s_{\overline{\Omega}}(M)$ is equal to the closure of $C_0^\infty(\Omega)$ in $H^s(M)$. (This can fail when $\partial\Omega$ is not smooth.) Conclude that there is a natural injective map

$$\kappa : H^s_{\overline{\Omega}}(M) \longrightarrow H_0^s(\Omega), \quad s \ge 0.$$

(*Hint*: Recall that $H_0^s(\Omega)$ is the closure of $C_0^\infty(\Omega)$ in $H^s(\Omega) \approx H^s(M)/H^s_{\overline{\mathcal{O}}}(M)$.)

14. If Z is defined as in Exercise 10, use Corollary 5.5 to show that

$$Z : H_0^s(\Omega) \longrightarrow H^s(M), \quad 0 \le s < \frac{1}{2}. \tag{5.35}$$

15. If $v \in C^\infty(\overline{\Omega})$, and $w = v$ on Ω, 0 on $\mathcal{O}$, show that $w \in H^s(M)$, for all $s \in [0, 1/2)$. If $v = 1$, show that $w \notin H^{1/2}(M)$.
16. Show that

$$H_0^s(\Omega) = H^s(\Omega), \quad \text{for } 0 \le s \le \frac{1}{2}. \tag{5.36}$$

(*Hint*: To show that $C_0^\infty(\Omega)$ is dense in $H^s(\Omega)$, show that $\{u \in C^\infty(M) : u = 0 \text{ near } \partial\Omega\}$ is dense in $H^s(M)$, for $0 \le s \le 1/2$.)

17. Using the results of Exercises 10–16, show that, for $k \in \mathbb{Z}^+$,

$$[L^2(\Omega), H_0^k(\Omega)]_\theta = H_0^s(\Omega) = H^s(\Omega) \ \text{ if } \ s = k\theta \in [0, \tfrac{1}{2}). \tag{5.37}$$

See [LM], pp. 60–62, for a demonstration that, for $s > 0$,

$$Z : H_0^s(\Omega) \longrightarrow H^s(M) \Longleftrightarrow s - \frac{1}{2} \notin \mathbb{Z},$$

which, by Exercise 12, implies (5.4) and also, for $k \in \mathbb{Z}^+$,

$$[L^2(\Omega), H_0^k(\Omega)]_\theta = H_0^s(\Omega) \ \text{ if } \ s = k\theta \notin \mathbb{Z} + \frac{1}{2}.$$

6. The Schwartz kernel theorem

Let M and N be compact manifolds. Suppose

$$T : C^\infty(M) \longrightarrow \mathcal{D}'(N) \tag{6.1}$$

is a linear map that is continuous. We give $C^\infty(M)$ its usual Fréchet space topology and $\mathcal{D}'(N)$ its weak* topology. Consequently, we have a bilinear map

$$B : C^\infty(M) \times C^\infty(N) \longrightarrow \mathbb{C}, \tag{6.2}$$

separately continuous in each factor, given by

$$B(u, v) = \langle v, Tu \rangle, \quad u \in C^\infty(M), \ v \in C^\infty(N). \tag{6.3}$$

For such u, v, define

$$u \otimes v \in C^\infty(M \times N) \tag{6.4}$$

by

$$(u \otimes v)(x, y) = u(x)v(y), \quad x \in M, \ y \in N. \tag{6.5}$$

We aim to prove the following result, known as the Schwartz kernel theorem.

Theorem 6.1. *Given B as in (6.2), there exists a distribution*

$$\kappa \in \mathcal{D}'(M \times N) \tag{6.6}$$

such that

$$B(u, v) = \langle u \otimes v, \kappa \rangle, \tag{6.7}$$

for all $u \in C^\infty(M)$, $v \in C^\infty(N)$.

We note that the right side of (6.7) defines a bilinear map (6.2) that is continuous in each factor, so Theorem 6.1 establishes an isomorphism between $\mathcal{D}'(M \times N)$ and the space of maps of the form (6.2), or equivalently the space of continuous linear maps (6.1).

The first step in the proof is to elevate the hypothesis of separate continuity to an apparently stronger condition. Generally speaking, let E and F be Fréchet spaces, and let

$$\beta : E \times F \longrightarrow \mathbb{C} \tag{6.8}$$

be a separately continuous bilinear map. Suppose the topology of E is defined by seminorms $p_1 \le p_2 \le p_3 \le \cdots$ and that of F by seminorms $q_1 \le q_2 \le q_3 \le \cdots$. We have the following result.

Proposition 6.2. *If* β *in (6.8) is separately continuous, then there exist seminorms* p_K *and* q_L *and a constant* C' *such that*

$$|\beta(u, v)| \le C' p_K(u) q_L(v), \quad u \in E, \ v \in F. \tag{6.9}$$

Proof. This will follow from the Baire category theorem, in analogy with the proof of the uniform boundedness theorem. Let $S_{C,j} \subset E$ consist of $u \in E$ such that

$$|\beta(u, v)| \le C q_j(v), \quad \text{for all } v \in F. \tag{6.10}$$

The hypothesis that β is continuous in v for each u implies

$$\bigcup_{C,j} S_{C,j} = E. \tag{6.11}$$

The hypothesis that β is continuous in u implies that each $S_{C,j}$ is closed. The Baire category theorem implies that some $S_{C,L}$ has nonempty interior. Hence $S_{1/2,L} = (2C)^{-1} S_{C,L}$ has nonempty interior. Since $S_{c,L} = -S_{c,L}$ and $S_{1/2,L} + S_{1/2,L} = S_{1,L}$, it follows that $S_{1,L}$ is a neighborhood of 0 in E. Picking K so large that, for some C_1, the set of $u \in E$ with $p_K(u) \le C_1$ is contained in this neighborhood, we have (6.9) with $C' = C/C_1$. This proves the proposition.

Returning to the bilinear map B of (6.2), we use Sobolev norms to define the topology of $C^\infty(M)$ and of $C^\infty(N)$:

$$p_j(u) = \|u\|_{H^j(M)}, \quad q_j(v) = \|v\|_{H^j(N)}. \tag{6.12}$$

In the case of $M = \mathbb{T}^m$, we can take

$$p_j(u) = \Big(\sum_{|\alpha| \le j} \|D^\alpha u\|^2_{L^2(\mathbb{T}^m)} \Big)^{1/2}, \tag{6.13}$$

and similarly for $p_j(v)$ if $N = \mathbb{T}^n$. Proposition 6.2 implies that there are C, K, L such that

$$|B(u, v)| \le C\|u\|_{H^K(M)} \|v\|_{H^L(N)}. \tag{6.14}$$

Recalling that the dual of $H^L(N)$ is $H^{-L}(N)$, we have the following result.

Proposition 6.3. *Let B be as in Theorem 6.1. Then for some K, L, there is a continuous linear map*

$$T : H^K(M) \longrightarrow H^{-L}(N) \tag{6.15}$$

such that

$$B(u, v) = \langle v, Tu \rangle, \quad \text{for } u \in C^\infty(M),\ v \in C^\infty(N). \tag{6.16}$$

Thus, if a continuous linear map of the form (6.1) is given, it has a continuous linear extension of the form (6.15).

In the next few steps of the proof of Theorem 6.1, it will be convenient to work with the case $M = \mathbb{T}^m$, $N = \mathbb{T}^n$. Once Theorem 6.1 is established in this case, it can readily be extended to the general case.

Recall from (3.7) the isomorphisms

$$\Lambda^s : H^\sigma(\mathbb{T}^m) \longrightarrow H^{\sigma-s}(\mathbb{T}^m), \tag{6.17}$$

for all real s, σ, where $\Lambda^2 = I - \Delta$. It follows from (6.15) that

$$T_{jk} = (I - \Delta)^{-j} T (I - \Delta)^{-k} : L^2(\mathbb{T}^m) \longrightarrow H^s(\mathbb{T}^n) \tag{6.18}$$

as long as $k \ge K/2$ and $j \ge L/2 + s$. Note that

$$T = (I - \Delta)^j T_{jk} (I - \Delta)^k. \tag{6.19}$$

The next step in our analysis will exploit the fact that if j is picked sufficiently large in (6.18), then T_{jk} is a *Hilbert-Schmidt* operator from $L^2(\mathbb{T}^m)$ to $L^2(\mathbb{T}^n)$.

We recall here the notion of a Hilbert-Schmidt operator, which is discussed in detail in §6 of Appendix A. Let H_1 and H_2 be two separable infinite dimensional Hilbert spaces, with orthonormal bases $\{u_j\}$ and $\{v_j\}$, respectively. Then $A : H_1 \to H_2$ is Hilbert-Schmidt if and only if

$$\sum_j \|Au_j\|^2 = \sum_{j,k} |a_{jk}|^2 < \infty, \tag{6.20}$$

where $a_{jk} = (Au_j, v_k)$. The quantity on the left is denoted $\|A\|^2_{HS}$. It is not hard to show that this property is independent of choices of orthonormal bases. Also, if there are bounded operators $V_1 : X_1 \to H_1$ and $V_2 : H_2 \to X_2$ between Hilbert spaces, we have

$$\|V_2 A V_1\|_{HS} \le \|V_2\| \cdot \|A\|_{HS} \cdot \|V_1\|, \tag{6.21}$$

where of course $\|V_j\|$ are operator norms. If V_j are both unitary, there is identity in (6.21). For short, we call a Hilbert-Schmidt operator an "HS operator."

From the definition, and using the exponential functions for Fourier series as an orthonormal basis, it easily follows that

$$\Lambda^{-s} \text{ is HS on } L^2(\mathbb{T}^n) \Longleftrightarrow s > \frac{n}{2}. \tag{6.22}$$

Consequently, we can say of the operator T_{jk} given by (6.18) that

$$T_{jk} : L^2(\mathbb{T}^m) \longrightarrow L^2(\mathbb{T}^n) \text{ is HS if } 2k \ge K \text{ and } 2j > L + n. \tag{6.23}$$

Our next tool, which we call the Hilbert-Schmidt kernel theorem, is proved in §6 of Appendix A.

Theorem 6.4. *Given a Hilbert-Schmidt operator*

$$T_1 : L^2(X_1, \mu_1) \longrightarrow L^2(X_2, \mu_2),$$

there exists $K \in L^2(X_1 \times X_2, \mu_1 \times \mu_2)$ such that

$$(T_1 u, v)_{L^2} = \iint K(x_1, x_2) u(x_1) \overline{v(x_2)}\, d\mu_1(x_1)\, d\mu_2(x_2). \tag{6.24}$$

To proceed with the proof of the Schwartz kernel theorem, we can now establish the following.

Proposition 6.5. *The conclusion of Theorem 6.1 holds when $M = \mathbb{T}^m$ and $N = \mathbb{T}^n$.*

Proof. By Theorem 6.4, there exists $K \in L^2(\mathbb{T}^m \times \mathbb{T}^n)$ such that

$$\langle v, T_{jk} u\rangle = \iint K(x, y) u(x) v(y)\, dx\, dy, \tag{6.25}$$

for $u \in C^\infty(\mathbb{T}^m)$, $v \in C^\infty(\mathbb{T}^n)$, provided T_{jk}, given by (6.18), satisfies (6.23). In view of (6.19), this implies

$$\begin{aligned} \langle v, Tu\rangle &= \langle (I - \Delta)^j v, T_{jk} (I - \Delta)^k u\rangle \\ &= \iint K(x, y)\, (I - \Delta_y)^j v(y)\, (I - \Delta_x)^k u(x)\, dx\, dy, \end{aligned} \tag{6.26}$$

so (6.7) holds with

$$\kappa = (I - \Delta_x)^k (I - \Delta_y)^j K(x, y) \in \mathcal{D}'(\mathbb{T}^m \times \mathbb{T}^n). \tag{6.27}$$

Now Theorem 6.1 for general compact M and N can be proved by writing

$$B(u, v) = \sum_{j,k} B(\varphi_j u, \psi_k v), \tag{6.28}$$

for partitions of unity $\{\varphi_j\}$, $\{\psi_k\}$ subordinate to coordinate covers of M and N, and transferring the problem to the case of tori.

Exercises

1. Extend Theorem 6.1 to treat the case of
$$B : C_0^\infty(M) \times C_0^\infty(N) \longrightarrow \mathbb{C},$$
when M and N are smooth, paracompact manifolds. State carefully an appropriate continuity hypothesis on B.
2. What is the Schwartz kernel of the identity map $I : C^\infty(\mathbb{T}^n) \to C^\infty(\mathbb{T}^n)$?

References

[Ad] R. Adams, *Sobolev Spaces*, Academic Press, New York, 1975.

[Ag] S. Agmon, *Lectures on Elliptic Boundary Problems*, Van Nostrand, New York, 1964.

[Au] T. Aubin, *Nonlinear Analysis on Manifolds. Monge-Ampere Equations*, Springer-Verlag, New York, 1982.

[BL] J. Bergh and J. Löfstrom, *Interpolation spaces, an Introduction*, Springer-Verlag, New York, 1976.

[BJS] L. Bers, F. John, and M. Schechter, *Partial Differential Equations*, Wiley, New York, 1964.

[Ca] A. P. Calderon, Intermediate spaces and interpolation, the complex method, *Studia Math.* 24(1964), 113–190.

[Do] W. Donoghue, *Distributions and Fourier Transforms*, Academic Press, New York, 1969.

[Fo1] G. Folland, *Introduction to Partial Differential Equations*, Princeton Univ. Press, Princeton, N. J., 1976.

[Frd] A. Friedman, *Generalized Functions and Partial Differential Equations*, Prentice-Hall, Englewood Cliffs, N. J., 1963.

[Fu] D. Fujiwara, Concrete characterizations of the domains of fractional powers of some elliptic differential operators of the second order, *Proc. Japan Acad.* 43(1967), 82–86.

[Ho1] L. Hörmander, *The Analysis of Linear Partial Differential Operators*, Vol. 1, Springer-Verlag, New York, 1983.

[L1] P. Lax, The theory of hyperbolic equations. Stanford Lecture Notes, 1963.

[LM] J. Lions and E. Magenes, *Non-homogeneous Boundary Problems and Applications* I, II, Springer-Verlag, New York, 1972.

[Miz] S. Mizohata, *The Theory of Partial Differential Equations*, Cambridge Univ. Press, Cambridge, 1973.

[Mor] C. B. Morrey, *Multiple Integrals in the Calculus of Variations*, Springer-Verlag, New York, 1966.
[Sch] L. Schwartz, *Théorie des Distributions*, Hermann, Paris, 1950.
[So] S. Sobolev, On a theorem of functional analysis, *Mat. Sb.* 4(1938), 471–497; *AMS Transl.* 34(1963), 39–68.
[So2] S. Sobolev, *Partial Differential Equations of Mathematical Physics*, Dover, New York, 1964.
[S1] E. Stein, *Singular Integrals and Differentiability Properties of Functions*, Princeton Univ. Press, Princeton, N. J., 1970.
[Tre1] F. Treves, *Basic Linear Partial Differential Equations*, Academic Press, New York, 1975.
[Tri] H. Triebel, *Theory of Function Spaces*, Birkhauser, Boston, 1983.
[Yo] K. Yosida, *Functional Analysis*, Springer-Verlag, New York, 1965.

5

Linear Elliptic Equations

Introduction

The first major topic of this chapter is the Dirichlet problem for the Laplace operator on a compact domain with boundary:

$$\Delta u = 0 \text{ on } \Omega, \quad u\big|_{\partial\Omega} = f. \tag{0.1}$$

We also consider the nonhomogeneous problem $\Delta u = g$ and allow for lower-order terms. As in Chapter 2, Δ is the Laplace operator determined by a Riemannian metric. In §1 we establish some basic results on existence and regularity of solutions, using the theory of Sobolev spaces. In §2 we establish maximum principles, which are useful for uniqueness theorems and for treating (0.1) for f continuous, among other things.

For general Ω, one does not expect to write down an explicit integral formula for solutions to (0.1), but when Ω is the unit ball in $\mathbb{R}^n$ this is possible. The resulting formula, called the Poisson integral formula, is derived in §3, generalizing the formula for the disk in $\mathbb{R}^2$ derived in §2 of Chapter 3.

One of the most famous classical applications of the solvability of (0.1) is to a proof of the Riemann mapping theorem. We prove this theorem for bounded, simply connected domains, with *smooth* boundary, in §4. To prove the Riemann mapping theorem for general simply connected planar domains, it is necessary to extend the existence theory of §1 to compact domains whose boundaries are not smooth. We provide results on this in §5, not giving an exhaustive treatment but going far enough to accomplish the goal of proving the Riemann mapping theorem in general in §6. The analysis in §5 makes strong use of the maximum principle established in §2. Further results on irregular boundaries will be established in Chapter 11, via the use of Brownian motion.

Sections 7 through 9 include material on other boundary conditions. Section 7 looks at the Neumann boundary condition

$$\Delta u = g \text{ on } \Omega, \quad \frac{\partial u}{\partial \nu} = f \text{ on } \partial\Omega. \tag{0.2}$$

It is shown that the methods of §1 extend to treat this when Ω has smooth boundary. Unlike the case of the Dirichlet problem, we do not discuss the Neumann boundary condition on domains with nonsmooth boundary, though much has been done on this; we refer to [Gri], [DK], [Wil] and works cited therein. In §8 we consider the Laplace operator on k-forms, and derive the Hodge theorem. When Ω has a boundary, there arise natural boundary conditions, which we treat in §9. The Hodge decomposition is extended to the case of manifolds with boundary. These results have topological significance, providing useful tools in deRham cohomology. We develop some of these topological consequences, particularly in exercise sets following §§8 and 9. The results of these sections also have physical significance, as will be seen in the analysis of Maxwell's equations for the electromagnetic field, in Chapter 6. Further use of this material will be made in Chapter 17, on fluid mechanics.

In §10 there is a brief return to the Dirichlet problem for the Laplace operator, in order to prove the existence of isothermal coordinates on any two-dimensional Riemannian manifold. We treat this topic so late in the chapter only to have the luxury of exploiting the Hodge star operator, introduced in §8.

In §11 we discuss general elliptic boundary problems. The method of freezing coefficients, introduced in §9, plays a major role here in producing Sobolev space estimates for variable-coefficient equations out of estimates for constant-coefficient equations (and flat boundaries). The latter estimates can be obtained via Fourier analysis. We analyze which boundary-value problems lead to estimates and regularity of the sort obtained in earlier sections for the Dirichlet and Neumann problems. These are called regular elliptic boundary problems. Further study of regular boundary problems is made in §12. We mention that Hölder space estimates for solutions to regular elliptic boundary problems will be obtained in §8 of Chapter 13.

At the end of this chapter are two appendices. One studies spaces of functions and generalized functions on a compact manifold with boundary, arising from a self-adjoint elliptic boundary problem. This material will be useful for the discussion of fundamental solutions to parabolic and hyperbolic equations in the next chapter. The second appendix, on the Mayer-Vietoris sequence, complements some results on deRham cohomology obtained in §§8 and 9. We illustrate the use of this sequence with several applications to topology, including a proof of a variant of the Jordan-Brouwer separation theorem, in the smooth case.

1. Existence and regularity of solutions to the Dirichlet problem

Let $\overline{\Omega}$ be a smooth, compact Riemannian manifold with boundary, Ω the interior of $\overline{\Omega}$. Let Δ denote the Laplace operator. We have

$$(-\Delta u, u) = \|du\|^2_{L^2(\Omega)}, \quad \text{for } u \in C_0^\infty(\Omega). \tag{1.1}$$

We will asume here that each connected component of Ω has nonempty boundary. Thus we have the estimate

$$\|u\|^2_{L^2(\Omega)} \le C\|du\|^2_{L^2(\Omega)}, \quad u \in C_0^\infty(\Omega), \tag{1.2}$$

by Proposition 5.2 of Chapter 4. Hence

$$\|du\|^2_{L^2(\Omega)} \approx \|u\|^2_{H^1(\Omega)}, \quad \text{for } u \in H_0^1(\Omega). \tag{1.3}$$

Recall from §5 of Chapter 4 that

$$\Delta : H_0^1(\Omega) \longrightarrow H^{-1}(\Omega) \tag{1.4}$$

is well defined. It follows that (1.1) continues to hold for $u \in H_0^1(\Omega)$. Consequently, (1.3) implies

$$(-\Delta u, u) \ge C\|u\|^2_{H^1(\Omega)} \quad \text{if } u \in H_0^1(\Omega). \tag{1.5}$$

Furthermore, we have

$$\|\Delta u\|_{H^{-1}(\Omega)} \ge C\|u\|_{H^1(\Omega)} \quad \text{if } u \in H_0^1(\Omega). \tag{1.6}$$

We can now obtain our first existence theorem.

Proposition 1.1. *In (1.4), Δ is one-to-one and onto.*

Proof. Clearly, (1.6) implies Δ is injective, with closed range. If it is not surjective, there must be an element of $\big(H^{-1}(\Omega)\big)^* = H_0^1(\Omega)$ that is orthogonal to the range, that is, an element $u_0 \in H_0^1(\Omega)$ that satisfies

$$(-\Delta u, u_0) = 0, \quad \text{for all } u \in H_0^1(\Omega).$$

Setting $u = u_0$, we deduce from (1.5) that $u_0 = 0$, so the proposition is proved.

Thus there is a uniquely determined inverse

$$T : H^{-1}(\Omega) \longrightarrow H_0^1(\Omega). \tag{1.7}$$

Note that if $\varphi = \Delta u$, $\psi = \Delta v$, with $u, v \in H_0^1(\Omega)$, then

$$\begin{aligned}(T\varphi, \psi) &= (T\Delta u, \Delta v) = (u, \Delta v) \\ &= -(du, dv) = (\Delta u, v) \\ &= (\varphi, T\psi),\end{aligned} \tag{1.8}$$

where we have used the fact that (1.1) extends to

$$(-\Delta u, v) = (du, dv)_{L^2}, \quad \text{for } u, v \in H_0^1(\Omega). \tag{1.9}$$

If we consider the restriction of T to $L^2(\Omega)$, we have

$$T = T^*. \tag{1.10}$$

Since $T : L^2(\Omega) \to H_0^1(\Omega)$, we have by Rellich's theorem that T is compact on $L^2(\Omega)$. We record this useful fact:

Proposition 1.2. *The inverse T to Δ in (1.4) is a compact (negative) self adjoint operator on $L^2(\Omega)$.*

Hence there is an orthonormal basis $\{u_j\}$ of $L^2(\Omega)$ consisting of eigenfunctions of T:

$$Tu_j = -\mu_j u_j, \quad \mu_j \searrow 0. \tag{1.11}$$

In view of (1.7), we have

$$u_j \in H_0^1(\Omega), \quad \text{for each } j. \tag{1.12}$$

Furthermore, it is clear that

$$\Delta u_j = -\lambda_j u_j, \quad \lambda_j = \frac{1}{\mu_j} \nearrow +\infty. \tag{1.13}$$

We next investigate higher-order regularity of solutions to $\Delta u = f$, and more generally to

$$Lu = f, \quad u \in H_0^1(\Omega). \tag{1.14}$$

We consider operators L of the form

$$Lu = -\Delta u + Xu, \tag{1.15}$$

where X is a first-order differential operator, with smooth coefficients on $\overline{\Omega}$.

Theorem 1.3. *Given $f \in H^{k-1}(\Omega)$, for $k = 0, 1, 2, \dots$, a solution $u \in H_0^1(\Omega)$ to (1.14) belongs to $H^{k+1}(\Omega)$, and we have the estimate*

$$\|u\|^2_{H^{k+1}} \le C\|Lu\|^2_{H^{k-1}} + C\|u\|^2_{H^k}, \tag{1.16}$$

for all $u \in H^{k+1}(\Omega) \cap H_0^1(\Omega)$.

Proof. First we establish the estimate (1.16) for $k = 0$. By (1.5), together with the estimate

$$|(Xu, u)| \le C\|u\|_{H^1}\|u\|_{L^2} \le \frac{C}{2}\Big[\varepsilon\|u\|^2_{H^1} + \frac{1}{\varepsilon}\|u\|^2_{L^2}\Big],$$

we have

$$\text{Re}\,(Lu, u) \ge C\|u\|^2_{H^1} - C'\|u\|^2_{L^2}, \tag{1.17}$$

for $u \in H_0^1(\Omega)$. Hence

$$\|u\|^2_{H^1} \le C\,\text{Re}\,(Lu, u) + C'\|u\|^2_{L^2}. \tag{1.18}$$

Cauchy's inequality gives

$$\begin{aligned} \text{Re}\,(Lu, u) &\le C\|Lu\|_{H^{-1}}\|u\|_{H^1} \\ &\le C\varepsilon\|u\|^2_{H^1} + \frac{C}{\varepsilon}\|Lu\|^2_{H^{-1}}, \end{aligned} \tag{1.19}$$

and taking ε small enough, we can absorb the $\|u\|^2_{H^1}$-term into the left side of (1.18), obtaining

$$\|u\|^2_{H^1} \leq C\|Lu\|^2_{H^{-1}} + C\|u\|^2_{L^2}, \quad u \in H^1_0(\Omega). \tag{1.20}$$

We now proceed to prove Theorem 1.3 by induction on k. Given that

$$u \in H^1_0(\Omega),\ Lu = f \in H^{k-1}(\Omega) \Longrightarrow u \in H^{k+1}(\Omega)$$

and that (1.16) is true, suppose now that

$$u \in H^1_0(\Omega), \quad Lu \in H^k(\Omega). \tag{1.21}$$

So, we know that $u \in H^{k+1}(\Omega)$, and we want to establish that $u \in H^{k+2}(\Omega)$ and also show that u satisfies the estimate (1.16), with k replaced by $k+1$.

First, note that, for any $\chi \in C^\infty(\overline{\Omega})$,

$$L(\chi u) = \chi(Lu) + [L, \chi]u, \tag{1.22}$$

and since the commutator $[L, \chi]$ is a *first-order* differential operator, the hypothesis (1.21), together with the observation that $u \in H^{k+1}(\Omega)$, gives $L(\chi u) \in H^k(\Omega)$, so our analysis of u on $\overline{\Omega}$ can be *localized*.

So suppose u, belonging to $H^{k+1}(\Omega)$ and satisfying (1.21), is supported on a coordinate neighborhood $\mathcal{O}$, either one with no boundary or one in which $\partial\Omega$ is given by $\{x_n = 0\}$. In either case, we now apply (1.16), with u replaced by

$$D_{j,h}u(x) = \frac{1}{h}\big[\tau_{j,h}u(x) - u(x)\big] = \frac{1}{h}\big[u(x + he_j) - u(x)\big], \tag{1.23}$$

where $e_1, \dots, e_n$ are the standard coordinate vectors in $\mathbb{R}^n$. In case $\mathcal{O}$ has no boundary, we can take $1 \leq j \leq n$; otherwise $1 \leq j \leq n-1$. By (1.16), we have

$$\begin{aligned} \|D_{j,h}u\|^2_{H^{k+1}} &\leq C\|LD_{j,h}u\|^2_{H^{k-1}} + C\|u\|^2_{H^{k+1}} \\ &\leq C\|D_{j,h}Lu\|^2_{H^{k-1}} + C\|[L, D_{j,h}]u\|^2_{H^{k-1}} + C\|u\|^2_{H^{k+1}}. \end{aligned} \tag{1.24}$$

As in (1.22), we have a commutator to estimate. This time, there is the following result.

Lemma 1.4. *As $h \searrow 0$, $[L, D_{j,h}]$ is a bounded family of operators of order two:*

$$\|[L, D_{j,h}]u\|_{H^{k-1}} \leq C\|u\|_{H^{k+1}}, \quad k \geq 0, \tag{1.25}$$

given $u \in H^1_0(\Omega) \cap H^{k+1}(\Omega)$, supported in $\mathcal{O}$.

Proof. The estimate (1.25) follows directly from

$$\|[M_\varphi, D_{j,h}]v\|_{H^k} \leq C\|v\|_{H^k}, \quad k \geq -1, \varphi \in C^\infty(\overline{\Omega}), \tag{1.26}$$

which in turn is easy to demonstrate, as

$$[M_\varphi, D_{j,h}]v = -M_{(D_{j,h}\varphi)} \circ \tau_{j,h}v.$$

Using (1.25), we can deduce from (1.24) that

$$\|D_{j,h}u\|^2_{H^{k+1}} \le C\|Lu\|^2_{H^k} + C\|u\|^2_{H^{k+1}}, \tag{1.27}$$

and passing to the limit $h \searrow 0$ gives

$$D_j u \in H^{k+1}(\Omega). \tag{1.28}$$

If $\mathcal{O}$ has no boundary, then (1.28) is valid for $1 \le j \le n$, and we have $u \in H^{k+2}(\Omega)$. Otherwise, we have (1.28) for $1 \le j \le n-1$, and it remains to establish

$$D_n u \in H^{k+1}(\Omega). \tag{1.29}$$

Recall that $k \ge 0$. Thus we need to know

$$D_j D_n u \in H^k(\Omega), \quad 1 \le j \le n. \tag{1.30}$$

But $D_j D_n u = D_n D_j u \in H^k(\Omega)$ if $1 \le j \le n-1$, since we have (1.28) for $1 \le j \le n-1$. It remains only to establish

$$D_n^2 u \in H^k(\Omega). \tag{1.31}$$

To see this, write

$$g^{nn}(x) D_n^2 u = Lu - \sum_{(j,k)\neq(n,n)} g^{jk}(x) D_j D_k u - Yu, \tag{1.32}$$

where Y is a first-order differential operator. All the terms on the right side of (1.32) have been shown to be in $H^k(\Omega)$. This establishes (1.31) and completes the proof of Theorem 1.3.

From Theorem 1.3, we can draw an immediate corollary about the eigenfunctions u_j of Δ, satisfying (1.11)–(1.13). We have

$$L_j u_j = (-\Delta - \lambda_j) u_j = 0,$$

which gives the following.

Corollary 1.5. *The eigenfunctions u_j of Δ belong to $C^\infty(\overline{\Omega})$.*

We note that the localization argument from (1.22) gives the following local regularity result.

Proposition 1.6. *Let $\mathcal{O} \subset\subset \Omega$. Say $u \in H^1(\Omega)$ and $Lu = f \in H^{k-1}(\Omega)$, $k \ge 0$. Then $u \in H^{k+1}(\mathcal{O})$. Thus if $f \in C^\infty(\Omega)$, then $u \in C^\infty(\mathcal{O})$ for all $\mathcal{O} \subset\subset \Omega$, so $u \in C^\infty(\Omega)$. Furthermore, if $\Omega = M$ is a compact manifold without boundary, then, for $k \ge 0$,*

$$u \in H^1(M),\ Lu = f \in H^{k-1}(M) \Longrightarrow u \in H^{k+1}(M). \tag{1.33}$$

We also remark that the first order operator X in (1.15) could have matrix coefficients. The regularity result being localizable, we could suppose L operates on sections of a vector bundle, as long as the principal part of L has scalar coefficients. For example, Proposition 1.6 holds when L is the Laplace operator on p-forms. We will pursue this further in §8.

We now turn to a consideration of the following boundary problem for u:

$$\Delta u = 0 \text{ on } \Omega, \quad u\big|_{\partial\Omega} = f, \tag{1.34}$$

where

$$f \in C^\infty(\partial\Omega) \tag{1.35}$$

is given. Let $F \in C^\infty(\overline{\Omega})$ be constructed so that $F\big|_{\partial\Omega} = f$. Then (1.34) is equivalent to

$$u = F + v, \tag{1.36}$$

where

$$\Delta v = g = -\Delta F, \quad v\big|_{\partial\Omega} = 0. \tag{1.37}$$

Since $g \in C^\infty(\overline{\Omega})$, we see that

$$v = Tg \in H_0^1(\Omega) \tag{1.38}$$

satisfies (1.37), and by virtue of Theorem 1.3, $v \in C^\infty(\overline{\Omega})$. Thus, for any $f \in C^\infty(\partial\Omega)$, we have a unique $u \in C^\infty(\overline{\Omega})$ solving (1.34), assuming each connected component of Ω has nonempty boundary. We denote the solution to (1.34) by

$$u = \text{PI}\, f. \tag{1.39}$$

In analogy with Proposition 4.6 of Chapter 4, we have

Proposition 1.7. *The map (1.39) has a unique continuous extension*

$$PI : H^s(\partial\Omega) \longrightarrow H^{s+1/2}(\Omega), \quad s \geq \frac{1}{2}. \tag{1.40}$$

Proof. It suffices to prove this for $s = k + 1/2, k = 0, 1, 2, \dots$, by interpolation. Given $f \in H^{k+1/2}(\partial\Omega)$, there exists $F \in H^{k+1}(\Omega)$ such that $F\big|_{\partial\Omega} = f$, by Proposition 1.7 of Chapter 4. Then PI $f = F + v$, where v is defined by

$$\Delta v = -\Delta F \in H^{k-1}(\Omega), \quad v \in H_0^1(\Omega).$$

The regularity result of Theorem 1.3 gives $v \in H^{k+1}(\Omega)$, which establishes (1.40) for $s = k + 1/2$.

We note that Proposition 4.6 of Chapter 4 was established for a slightly greater range of s than in (1.40), namely for $s \geq -1/2$. With further effort, both results can be extended to all $s \in \mathbb{R}$. Compare Proposition 11.14.

Amalgamating the equations $\Delta u = f, u\big|_{\partial\Omega} = 0$ and $\Delta u = 0, u\big|_{\partial\Omega} = g$, we can solve the nonhomogeneous Dirichlet boundary problem

$$\Delta u = f, \quad u\big|_{\partial\Omega} = g. \tag{1.41}$$

Given $g \in H^{k+1/2}(\partial\Omega)$ and $f \in H^{k-1}(\Omega)$, $k = 0, 1, 2, \dots$, there exists a unique solution $u \in H^{k+1}(\Omega)$. Generalizing (1.16), we have the estimate

$$\|u\|^2_{H^{k+1}(\Omega)} \le C\|\Delta u\|^2_{H^{k-1}(\Omega)} + C\|u\|^2_{H^{k+1/2}(\partial\Omega)} + C\|u\|^2_{H^k(\Omega)}, \tag{1.42}$$

for all $u \in H^{k+1}(\Omega)$.

Next, we briefly consider existence of solutions to the more general equation

$$Lu = f, \quad u \in H_0^1(\Omega), \tag{1.43}$$

where, as in (1.14), $L = -\Delta + X$, X being a first-order differential operator on $\overline{\Omega}$. With T denoting the inverse of Δ, as in (1.7), we look for a solution of the form $u = Tv$, for some $v \in H^{-1}(\Omega)$. The equation (1.43) becomes

$$(I - XT)v = -f, \quad v \in H^{-1}(\Omega). \tag{1.44}$$

Note that

$$XT : H^{-1}(\Omega) \longrightarrow L^2(\Omega). \tag{1.45}$$

By Rellich's theorem, XT is a compact operator on $H^{-1}(\Omega)$. Thus the Fredholm alternative applies to the map $I - XT : H^{-1}(\Omega) \to H^{-1}(\Omega)$; this map is surjective if and only if it is injective. Note that v is in the kernel of this map if and only if $u = Tv \in H_0^1(\Omega)$ is annihilated by $-\Delta + X$. We have established the following.

Proposition 1.8. *Given a first-order differential operator X on $\overline{\Omega}$, the map*

$$-\Delta + X : H_0^1(\Omega) \longrightarrow H^{-1}(\Omega) \tag{1.46}$$

is Fredholm of index zero; hence it is surjective if and only if it is injective.

Of course, given a solution to (1.43), the regularity results of Theorem 1.3 apply. In particular, any element of the kernel of $-\Delta + X$ belongs to $C^\infty(\overline{\Omega})$. We will see in the next section that the map (1.46) is injective when X is a real vector field, so one has solvability in that case.

To close this section, we mention a few situations other than the Dirichlet problem on a connected manifold with nonempty, smooth boundary. For example, given a smooth, compact M without boundary, we can consider an open subset Ω, making no smoothness assumptions on $\partial\Omega$. Then one can still define $H_0^1(\Omega)$ as the completion of $C_0^\infty(\Omega)$ with respect to either of the equivalent norms

$$\left(\|du\|^2_{L^2(\Omega)} + \|u\|^2_{L^2(\Omega)}\right)^{1/2} \quad \text{or} \quad \|du\|_{L^2(\Omega)}.$$

The estimate (1.4) continues to hold. One can no longer identify $H_0^1(\Omega)^*$ with $H^{-1}(\Omega)$, but we still have

$$\Delta : H_0^1(\Omega) \longrightarrow H_0^1(\Omega)^*, \tag{1.47}$$

and the proof of Proposition 1.1 extends to show that the map (1.47) is bijective. We have a natural injection $L^2 \hookrightarrow H_0^1(\Omega)^*$, and the inverse operator

$$T : H_0^1(\Omega)^* \longrightarrow H_0^1(\Omega) \tag{1.48}$$

to Δ in (1.47) still restricts to a compact, self-adjoint operator on $L^2(\Omega)$. The global regularity result of Theorem 1.3 does not extend, although of course, by (1.22), one has such a regularity result on the interior. We will take a further look at the Dirichlet problem on domains with nonsmooth boundaries in §5.

Another variation is the case where Ω is compact, without boundary. Then the map $\Delta : H^1(\Omega) \to H^{-1}(\Omega)$ is not injective, since $1 \in \text{Ker}\,\Delta$. But we have

$$\big((-\Delta + 1)u, u\big) = \|du\|^2_{L^2(\Omega)} + \|u\|^2_{L^2(\Omega)}, \tag{1.49}$$

which gives

$$-\Delta + 1 : H^1(\Omega) \longrightarrow H^{-1}(\Omega) \text{ bijective.} \tag{1.50}$$

Its inverse,

$$T_1 : H^{-1}(\Omega) \longrightarrow H^1(\Omega), \tag{1.51}$$

is again seen to define a compact, self-adjoint operator on $L^2(\Omega)$, so we again have an orthonormal basis $\{u_j\}$ of $L^2(\Omega)$ satisfying $T_1 u_j = -\mu_j u_j$, with $\mu_j \searrow 0$ and $\mu_0 = 1$. Hence

$$\Delta u_j = -\lambda_j u_j, \quad \lambda_j = \frac{1}{\mu_j} - 1 \nearrow \infty. \tag{1.52}$$

Of course, $\lambda_0 = 0$, with corresponding $u_0 = \text{const}$. By (1.22), the regularity result of Theorem 1.3 extends to this case; we have $T_1 : H^{k-1}(\Omega) \to H^{k+1}(\Omega)$, for $k = 0, 1, 2, \ldots$, giving a two-sided inverse of the operator $-\Delta + 1 : H^{k+1}(\Omega) \to H^{k-1}(\Omega)$. By interpolation, we have

$$T_1 : H^s(\Omega) \longrightarrow H^{s+2}(\Omega), \tag{1.53}$$

for real $s \geq -1$, giving a two-sided inverse of

$$-\Delta + 1 : H^{s+2}(\Omega) \longrightarrow H^s(\Omega). \tag{1.54}$$

Since $T_1 = T_1^*$, by duality (1.53) holds for all real s.

Returning to the case of Ω with nonempty smooth boundary, we remark that boundary problems other than the Dirichlet problem arise naturally, such as the Neumann problem. We discuss some of these other boundary problems later in this chapter.

Exercises

1. Prove the following local boundary regularity result. If $u \in H_0^1(\Omega)$ and $Lu = f$, as in Theorem 1.3, and if $f\big|_{\mathcal{O}} \in H^k(\mathcal{O})$, for some open $\mathcal{O} \subset \Omega$, with $\overline{\mathcal{O}} \cap \partial\Omega$ perhaps nonempty, then $u \in H^{k+2}(\mathcal{O}')$ for any open $\mathcal{O}' \subset \mathcal{O}$ such that $\overline{\mathcal{O}'} \subset \mathcal{O} \cup \partial\Omega$.
(*Hint*: Recall the observation about (1.22).)
2. Let T be the operator inverting Δ, as in Proposition 1.2. Show that the largest eigenvalue μ_0 of $-T$ satisfies

$$\mu_0 = \sup\big\{(-Tu, u) : u \in L^2(\Omega);\ \|u\|_{L^2} = 1\big\},$$

and this supremum is achieved for $u = u_0$; in fact any v for which this supremum is achieved satisfies $Tv = -\mu_0 v$. Deduce that

$$\lambda_0 = \inf \left\{ \|du\|^2_{L^2(\Omega)} : u \in H^1_0(\Omega),\ \|u\|_{L^2} = 1 \right\}, \tag{1.55}$$

and furthermore, for any $v \in H^1_0(\Omega)$ for which this infimum is achieved, v is a λ_0-eigenfunction of $-\Delta$.

3. Suppose Ω is an open region in $\mathbb{R}^n$, lying between two hyperplanes $x_1 = A$ and $x_1 = B$. If λ_0 is the smallest eigenvalue of $-\Delta$, as in (1.55), show that

$$\lambda_0 \geq \frac{\pi^2}{(B-A)^2}.$$

(*Hint*: First consider the case $n = 1$.)

4. Show that the argument preceding Proposition 1.2 has the following generalization. Let $\mathcal{H}^0$ be a Hilbert space, $\mathcal{H}^1$ a dense linear subspace, with a Hilbert space structure, continuously injected in $\mathcal{H}^0$. Denote by $\mathcal{H}^{-1}$ the conjugate dual of $\mathcal{H}^1$, so there are continuous inclusions

$$\mathcal{H}^1 \subset \mathcal{H}^0 \subset \mathcal{H}^{-1}.$$

Suppose $L : \mathcal{H}^1 \to \mathcal{H}^{-1}$ is continuous, bijective, and Hermitian symmetric. Let T denote the restriction of $L^{-1} : \mathcal{H}^{-1} \to \mathcal{H}^1$ to $\mathcal{H}^0$:

$$T : \mathcal{H}^0 \longrightarrow \mathcal{H}^0.$$

Show that T is a bounded self-adjoint operator on $\mathcal{H}^0$. Relate this to the Friedrichs extension method, discussed in §8 of Appendix A.

5. Extend Proposition 1.1 to the case where Δ is the Laplace operator on $\overline{\Omega}$, endowed with a *continuous* metric tensor.
6. Show that Theorem 1.3 holds in the case $k = 1$, provided g_{ij} are *Lipschitz* on $\overline{\Omega}$.
7. Show that if Ω is a bounded open set in $\mathbb{R}^n$ with a $C^{1,1}$-boundary, one can smooth out the boundary, transforming the Laplace operator to an operator to which Problem 6 applies. Why doesn't this work if Ω merely has a Lipschitz boundary?
8. Consider $Lu = \Delta u - V(x)u$, that is, L of the form (1.15) with $Xu = V(x)u$. Show that (1.45), and hence Proposition 1.8, hold, provided

$$V \in L^n(\Omega), \quad n \geq 3,$$

where $n = \dim \Omega$, given that

$$H^1(\Omega) \subset L^{2n/(n-2)}(\Omega), \text{ for } n \geq 3,$$

a result that will be established in §1 of Chapter 13. Try to show that Proposition 1.8 holds under the even weaker hypothesis

$$V \in L^q(\Omega), \quad q > \frac{n}{2}.$$

2. The weak and strong maximum principles

In this section, we take M to be a smooth, compact Riemannian manifold without boundary, and Ω to be a connected open subset of M, with nonempty boundary.

We will derive several results related to the maximum principle for second-order differential operators of the form

$$L = \Delta + X, \tag{2.1}$$

where X is a real vector field on M. In local coordinates, L has the form

$$L = g^{jk}(x)\partial_j\partial_k + b^j(x)\partial_j, \tag{2.2}$$

with $\big(g^{jk}(x)\big)$ the metric on cotangent vectors, a positive-definite matrix, and $b^j(x)$ smooth and real-valued. We begin with the following, a weak maximum principle.

Proposition 2.1. *Suppose Ω is an open bounded domain in $\mathbb{R}^n$ and L is given by (2.2), with coefficients smooth on a neighborhood of $\overline{\Omega}$. If $u \in C(\overline{\Omega}) \cap C^2(\Omega)$ and*

$$Lu \geq 0 \text{ on } \Omega, \tag{2.3}$$

then

$$\sup_{x\in\Omega} u(x) = \sup_{y\in\partial\Omega} u(y). \tag{2.4}$$

Furthermore, if

$$Lu = 0 \text{ on } \Omega, \tag{2.5}$$

then also

$$\sup_{x\in\Omega} |u(x)| = \sup_{y\in\partial\Omega} |u(y)|. \tag{2.6}$$

Proof. First note that if $Lu > 0$ on Ω, an interior maximum is impossible, since $\partial_j u(x) = 0$ and $\big(\partial_j\partial_k u(x)\big)$ is negative semidefinite at any interior maximum. So we certainly have (2.4) in that case. To show that (2.3)$\Rightarrow$(2.4), note that if $\Omega \subset\subset \mathbb{R}^n$,

$$L(e^{\gamma x_1}) = \big(\gamma^2 g^{11}(x) + \gamma b^1(x)\big)e^{\gamma x_1} > 0,$$

for $\gamma > 0$ large enough. Fix γ so large that $L(e^{\gamma x_1}) > 0$. Then, for any $\varepsilon > 0$, $L(u + \varepsilon e^{\gamma x_1}) > 0$, so we have

$$\sup_{x\in\Omega}\ u(x) + \varepsilon e^{\gamma x_1} = \sup_{y\in\partial\Omega}\ u(y) + \varepsilon e^{\gamma y_1},$$

for each $\varepsilon > 0$. Passing to the limit $\varepsilon \searrow 0$ yields (2.4). If (2.5) holds, then (2.4) also holds with u replaced by $-u$, which gives (2.6).

In the following proposition, we will not need to suppose $\Omega \subset \mathbb{R}^n$; we resume the hypotheses on Ω made at the beginning of this section. Proposition 2.3 will contain the extension of Proposition 2.1 to this general class of domains; it will also be sharper than Proposition 2.1 in other respects.

The following result, sometimes called *Zaremba's principle*, has many important uses, including providing a tool to establish the strong maximum principle.

Proposition 2.2. *In addition to the hypotheses above, suppose $\partial\Omega$ is smooth and $u \in C^1(\overline{\Omega}) \cap C^2(\Omega)$. If $Lu \geq 0$ and if $y \in \partial\Omega$ is a point such that*

$$u(y) > u(x), \quad \text{for all } x \in \Omega, \tag{2.7}$$

then, if ν denotes the inward-pointing normal to $\partial\Omega$,

$$\frac{\partial u}{\partial \nu}(y) < 0. \tag{2.8}$$

Proof. Pick a coordinate system centered at y. In this coordinate system, put a small ball $\mathcal{O}$ in Ω, whose boundary is tangent to $\partial\Omega$ at y, as illustrated in Fig. 2.1.

Let p denote the center of $\mathcal{O}$; let R denote the radius of $\mathcal{O}$ (in this coordinate system; forget about the Riemannian metric on Ω). Let

$$r(x)^2 = |x - p|^2, \tag{2.9}$$

for $x \in \mathcal{O}$. A short calculation gives

$$\begin{aligned} L\big(e^{-\alpha r^2} - e^{-\alpha R^2}\big) &= L\big(e^{-\alpha r^2}\big) \\ &= e^{-\alpha r^2}\Big[4\alpha^2 g^{jk}(x_j - p_j)(x_k - p_k) - 2\alpha\big(g^j{}_j + b^j(x_j - p_j)\big)\Big]. \end{aligned} \tag{2.10}$$

What can be deduced from this calculation is that if $\rho \in (0, R)$ is fixed, then, for $\alpha > 0$ sufficiently large,

$$w = e^{-\alpha r^2} - e^{-\alpha R^2} \tag{2.11}$$

implies

$$Lw > 0 \text{ on the shell } \mathcal{A}, \tag{2.12}$$

defined by

$$\mathcal{A} = \{x \in \mathcal{O} : r(x) > \rho\} \tag{2.13}$$

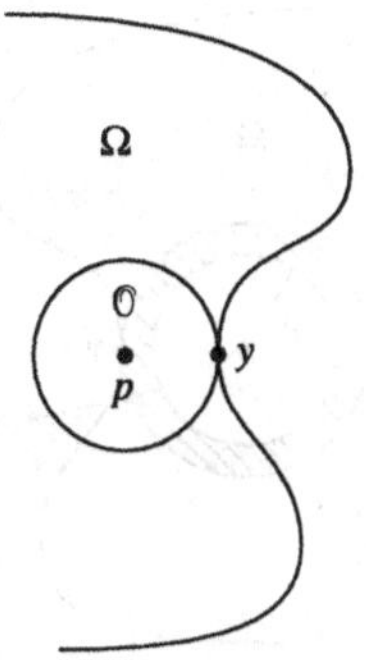

FIGURE 2.1

(see Fig. 2.2). Consequently, for any $\varepsilon > 0$, if w is given by (2.11), then $Lu \geq 0$ implies

$$L(u + \varepsilon w) > 0 \text{ on } \mathcal{A}, \tag{2.14}$$

so, by Proposition 2.1, we have

$$\sup_{\mathcal{A}} (u + \varepsilon w) = \sup_{\partial \mathcal{A}} (u + \varepsilon w). \tag{2.15}$$

Note that $w = 0$ on $\partial \mathcal{O} = \{r(x) = R\}$. Since, by the hypothesis (2.7),

$$\sup_{\{r(x)=\rho\}} u(x) < u(y), \tag{2.16}$$

we see that, for $\varepsilon > 0$ sufficiently small, the right side of (2.15) is equal to $u(y)$. Fix ε, sufficiently small. Then (2.15) yields

$$u(x) + \varepsilon w(x) \leq u(y), \quad \text{for all } x \in \mathcal{A}, \tag{2.17}$$

and hence

$$\frac{u(y) - u(x)}{|y - x|} \geq \varepsilon \frac{w(x)}{|y - x|} = \varepsilon \frac{w(x) - w(y)}{|y - x|}, \tag{2.18}$$

since $w(y) = 0$. Now the formula (2.11) for w implies

$$\frac{\partial w}{\partial \nu}(y) > 0, \tag{2.19}$$

so letting $x \to y$ along the normal to $\partial\Omega$ at y gives (2.8), as a consequence of (2.18).

We can now elevate Proposition 2.1 to the strong maximum principle. In this result, we do not need any smoothness on $\partial\Omega$.

Proposition 2.3. *If $u \in C(\overline{\Omega}) \cap C^2(\Omega)$ and $Lu \geq 0$, then either u is constant, or*

$$u(x) < \sup_{z \in \partial\Omega} u(z), \quad \text{for all } x \in \Omega. \tag{2.20}$$

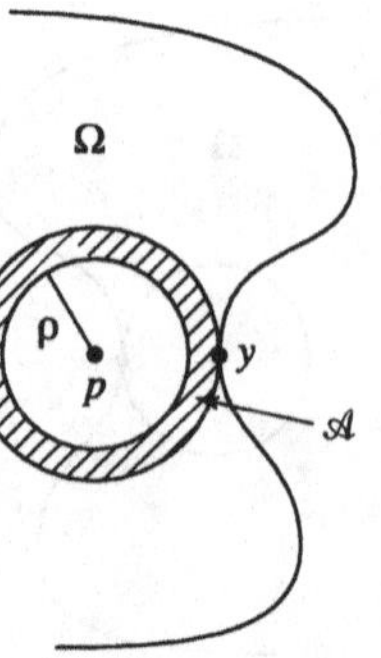

FIGURE 2.2

Proof. First we prove the weaker estimate (2.4) in this case. Indeed, if this estimate fails, u must have an interior maximum, say on a nonempty compact set $K \subset \Omega$. If we put a ball $\mathcal{D}$ in $\Omega \setminus K$, touching K at a point y, then Zaremba's principle (2.8), applied to functions on $\mathcal{D}$, contradicts the fact that we must have $du = 0$ at y. This shows that K is empty.

Now, if u is not constant, we see that $\mathcal{O}$, the set of points $x \in \Omega$, where $u(x) < \sup_{z\in\partial\Omega} u(z)$, is nonempty and open in Ω. If $\mathcal{O}$ is not all of Ω, pick p_0 in the boundary of $\mathcal{O}$ in Ω, as illustrated in Fig. 2.3. Then pick $q_0 \in \mathcal{O}$ closer to p_0 than to $\partial\Omega$, and let $\mathcal{D}$ be the largest ball, centered at q_0, lying in $\mathcal{O}$. Then $\partial\mathcal{D}$ intersects $\Omega \setminus \mathcal{O}$ at (at least) one point; call it y.

Since $y \notin \mathcal{O}$, we must have $u(y) = \sup\limits_{z\in\partial\Omega} u(z) = \sup\limits_{x\in\Omega} u(x)$. This implies both that

$$du(y) = 0 \tag{2.21}$$

and that

$$u(y) > u(x), \quad \text{for all } x \in \mathcal{D}. \tag{2.22}$$

Again, this contradicts Zaremba's principle for $u \in C^1(\overline{\mathcal{D}}) \cap C^2(\mathcal{D})$ satisfying $Lu \geq 0$ in $\mathcal{D}$. Proposition 2.3 is proved.

In case $\overline{\Omega}$ is a smooth, connected, compact manifold with nonempty smooth boundary, recall from §1 that we have a map

$$\mathrm{PI} : C^\infty(\partial\Omega) \longrightarrow C^\infty(\overline{\Omega}) \tag{2.23}$$

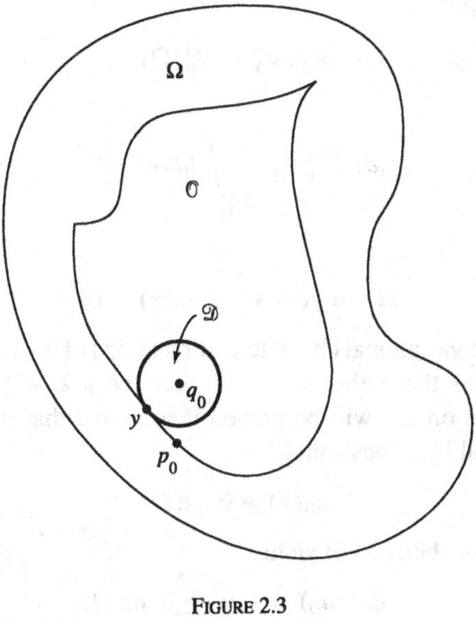

FIGURE 2.3

with the property that, for $f \in C^\infty(\partial\Omega)$, PI $f = u$ is the unique element of $C^\infty(\overline{\Omega})$ satisfying

$$\Delta u = 0 \text{ on } \Omega, \quad u\big|_{\partial\Omega} = f. \tag{2.24}$$

It follows from Proposition 2.3 that

$$\sup_{\overline{\Omega}} |u| = \sup_{\partial\Omega} |f|. \tag{2.25}$$

Consequently, the map (2.23) has a unique continuous extension to

$$\mathrm{PI} : C(\partial\Omega) \longrightarrow C(\overline{\Omega}). \tag{2.26}$$

We will discuss the situation where $\partial\Omega$ is not smooth in §5.

Using the strong maximum principle, we draw a conclusion about the fundamental eigenspace of Δ. Let λ_0 be the smallest eigenvalue of $-\Delta$, as in (1.13); $\lambda_0 > 0$. Assume $\overline{\Omega}$ is a connected, compact manifold with nonempty smooth boundary.

Proposition 2.4. *If $u_0 \in H_0^1(\Omega)$ is an eigenfunction for $-\Delta$ corresponding to λ_0, that is,*

$$\Delta u_0 = -\lambda_0 u_0, \tag{2.27}$$

then u_0 is nowhere vanishing on the interior of Ω.

Proof. We have $u_0 \in C^\infty(\overline{\Omega})$. Define u_0^+ and u_0^-, respectively, by

$$\begin{aligned} u_0^+(x) &= \max\,(u_0(x), 0), \\ u_0^-(x) &= \min\,(u_0(x), 0). \end{aligned}$$

It is easy to see that

$$u_0^+, u_0^- \in H_0^1(\Omega) \tag{2.28}$$

and

$$\|du_0^\pm\|_{L^2(\Omega)}^2 = \int_{\Omega^\pm} |du_0|^2 \, dV, \tag{2.29}$$

where

$$\Omega^\pm = \{x \in \Omega : \pm u_0(x) > 0\}.$$

Next we invoke the variational characterization (1.55) of λ_0 and associated eigenfunctions. It follows that either u_0^+ or u_0^- must be a λ_0-eigenfunction of $-\Delta$. Therefore, Proposition 2.4 will be proved if we show that its conclusion holds under the additional hypothesis that

$$u_0(x) \geq 0 \text{ on } \Omega. \tag{2.30}$$

Indeed, if this holds, then (2.27) yields

$$\Delta(-u_0) = \lambda_0 u_0 \geq 0 \text{ on } \Omega. \tag{2.31}$$

Thus Proposition 2.3 applies to $-u_0$, so, since $u_0\big|_{\partial\Omega} = 0$,

$$-u_0(x) < 0, \quad \text{for all } x \in \Omega. \tag{2.32}$$

This finishes the proof of Proposition 2.4.

Corollary 2.5. *If λ_0 is the smallest eigenvalue of $-\Delta$ for Ω, with Dirichlet boundary conditions, as in Proposition 2.4, then the corresponding λ_0-eigenspace is one-dimensional.*

Proof. If there were a λ_0-eigenvector u_1 orthogonal to u_0, then u_1 would have to change sign in Ω, contradicting Proposition 2.4.

The following result, involving a zero-order term, is often useful. With L as in (2.2), let

$$\mathcal{L}u = Lu - c(x)u. \tag{2.33}$$

We assume $c \in C(\overline{\Omega})$, with $\Omega \subset \mathbb{R}^n$, bounded.

Proposition 2.6. *Supose $c(x) \geq 0$ in (2.33). For $u, v \in C^2(\Omega) \cap C(\overline{\Omega})$,*

$$\mathcal{L}u \leq \mathcal{L}v \text{ on } \Omega, \ u \geq v \text{ on } \partial\Omega \Longrightarrow u \geq v \text{ on } \Omega. \tag{2.34}$$

Proof. By linearity, it suffices to show that

$$\mathcal{L}v \geq 0 \text{ on } \Omega, \ v \leq 0 \text{ on } \partial\Omega \Longrightarrow v \leq 0 \text{ on } \Omega.$$

If we let $\mathcal{O} = \{x \in \Omega : v(x) > 0\}$, then $Lv = cv \geq 0$ on $\mathcal{O}$, and $v = 0$ on $\partial\mathcal{O}$. But Proposition 2.1 implies $\sup_{\mathcal{O}} v = \sup_{\partial\mathcal{O}} v$. This is impossible if $\mathcal{O} \neq \emptyset$.

Corollary 2.7. *If $c(x) \geq 0$ and $\mathcal{L}u = 0$, then, with $\alpha = \sup_{\partial\Omega} u$, we have*

$$\alpha \geq 0 \Rightarrow \sup_{\Omega} u = \alpha, \quad \text{and} \quad \alpha < 0 \Rightarrow \sup_{\Omega} u < 0. \tag{2.35}$$

Proof. The first implication follows from (2.34), with (u, v) replaced by (α, u), since $\alpha \geq 0 \Rightarrow \mathcal{L}\alpha \leq 0$. For the second implication, let $\mathcal{O} = \{x \in \Omega : u(x) > 0\}$. If $\mathcal{O} \neq \emptyset$, we must have $\overline{\mathcal{O}} \subset \Omega$ and $u = 0$ on $\partial\mathcal{O}$. But now the first implication of (2.35) applies to $u\big|_{\mathcal{O}}$, so we have a contradiction.

In case $L = \Delta$, there is the following useful strengthening of Proposition 2.6.

Proposition 2.8. *Assume $c \in C(\overline{\Omega})$ and that $\mathcal{L} = \Delta - c$ is negative-definite with the Dirichlet boundary condition, that is,*

$$-\|du\|_{L^2}^2 - (cu, u) < 0, \quad \textit{for nonzero } u \in H_0^1(\Omega). \tag{2.36}$$

Then, for $v \in H^1(\Omega)$,

$$(\Delta - c)v \geq 0 \text{ on } \Omega, \ v \leq 0 \text{ on } \partial\Omega \Longrightarrow v \leq 0 \text{ on } \Omega. \tag{2.37}$$

Proof. Let $v_+ = \max(v, 0)$. Then the hypotheses in (2.37) imply that $v_+ \in H_0^1(\Omega)$ and

$$-(dv, dv_+) - (cv, v_+) \geq 0.$$

Since $(dv, dv_+) = (dv_+, dv_+)$ in this case, it follows that $-(dv_+, dv_+) - (cv_+, v_+) \geq 0$. By (2.36) this implies $v_+ = 0$, proving the proposition.

Further results involving zero-order terms are given in the exercises.

Exercises

1. If $\mathcal{L}u$ is as in (2.33) with $c(x) \geq 0$ and $u \in C^2(\Omega) \cap C(\overline{\Omega})$ satisfies $\mathcal{L}u = 0$, show that $\|u\|_{L^\infty(\Omega)} = \|u\|_{L^\infty(\partial\Omega)}$. (*Hint*: Supplement (2.35) with the following for $\beta = \inf_{\partial\Omega} u$:
$$\beta \leq 0 \Rightarrow \inf_\Omega u = \beta, \quad \beta > 0 \Rightarrow \inf_\Omega u > 0.)$$

2. Show that if $u, v \in C^2(\Omega) \cap C(\overline{\Omega})$, $f \in C^1(\mathbb{R})$, $f'(t) \geq 0$, and
$$-Lu + f(u) \leq -Lv + f(v) \text{ on } \Omega, \quad u \leq v \text{ on } \partial\Omega,$$
then $u \leq v$ on Ω. (*Hint*: Let $w = u - v$. Then $-Lw + c(x)w \leq 0$, with
$$c(x) = \frac{f(u(x)) - f(v(x))}{u(x) - v(x)}.)$$

3. Suppose $u \in C^2(\Omega) \cap C(\overline{\Omega})$ satisfies
$$Lu = f, \quad u\big|_{\partial\Omega} = g. \tag{2.38}$$
Suppose $V \in C^2(\Omega) \cap C(\overline{\Omega})$ satisfies
$$LV \geq 1, \quad V\big|_{\partial\Omega} \leq 0. \tag{2.39}$$
(Note that $V < 0$ in Ω.) Show that, for $x \in \Omega$,
$$\begin{aligned} u(x) &\geq (\sup f_+)V(x) + (\inf g), \\ u(x) &\leq (\inf f_-)V(x) + (\sup g). \end{aligned} \tag{2.40}$$
(*Hint*: Compare u respectively with $v = (\sup f_+)V + (\inf g)$ and with $v = (\inf f_-)V + (\sup g)$. In the first case, show that $Lu \leq Lv$ on Ω and $u \geq v$ on $\partial\Omega$.)

   In case Ω is a bounded region in $\mathbb{R}^n$ and Δ is the flat Laplacian, apply this with
$$V(x) = \frac{1}{2n}\big(|x - x_0|^2 - R^2\big), \quad R^2 = \sup_{x\in\Omega} |x - x_0|^2.$$

4. Extend estimates of Exercise 3 to the case
$$\big[L - c(x)\big]u = f, \quad u\big|_{\partial\Omega} = g, \tag{2.41}$$
under the hypothesis $c(x) \geq 0$. Show that if V satisfies (2.39), then (2.40) holds, with $\inf g$ replaced by $\inf g_-$ and $\sup g$ replaced by $\sup g_+$.

In Exercises 5 and 6, we outline an approach to estimates for a solution v to
$$\big[\Delta - c(x)\big]v = f, \quad v\big|_{\partial\Omega} = g, \tag{2.42}$$

where, rather than $c(x) \geq 0$, we assume that $\Delta - c(x)$ is negative-definite, with the Dirichlet boundary condition, as in Proposition 2.8. For example, we might have

$$c(x) \geq \mu > -\lambda_0, \tag{2.43}$$

λ_0 being the smallest eigenvalue of $-\Delta$ on $\overline{\Omega}$, with the Dirichlet boundary condition.

5. Set $v = Fu$ with $F \in C^2(\overline{\Omega})$, $F \geq 1$ on $\overline{\Omega}$. Show that (2.42) is equivalent to

$$Lu - c_1(x)u = \frac{f}{F}, \quad u\big|_{\partial\Omega} = \frac{g}{F},$$

where

$$Lu = \Delta u + 2F^{-1}\langle \nabla F, \nabla u\rangle, \text{ and } c_1(x) = -F^{-1}\big[\Delta F - c(x)F\big].$$

6. Estimates derived in Exercise 4 will apply to u in Exercise 5, provided

$$\Delta F - c(x)F \leq 0 \text{ on } \Omega, \quad F \geq 1 \text{ on } \overline{\Omega}. \tag{2.44}$$

By Proposition 2.8, this holds for $F = 1 + F_1$, provided

$$\Delta F_1 - c(x)F_1 \leq c_-(x) \text{ on } \Omega, \quad F_1 \geq 0 \text{ on } \partial\Omega.$$

Given (2.43), with $\mu < 0$, this holds provided

$$(\Delta - \mu)F_1 \leq \mu \text{ on } \Omega, \quad F_1 \geq 0 \text{ on } \partial\Omega.$$

Using these results, provide estimates for solutions to (2.42), under the hypothesis (2.43).

3. The Dirichlet problem on the ball in $\mathbb{R}^n$

If $\mathcal{B} = \{x \in \mathbb{R}^n : |x| < 1\}$ is the unit ball in $\mathbb{R}^n$, with boundary $\partial\mathcal{B} = S^{n-1}$, the unit sphere, we know there is a unique map

$$\mathrm{PI} : C(S^{n-1}) \longrightarrow C(\overline{\mathcal{B}}) \cap C^\infty(\mathcal{B}) \tag{3.1}$$

satisfying

$$u = \mathrm{PI}\, f \Longrightarrow \Delta u = 0 \text{ on } \mathcal{B}, \quad u\big|_{S^{n-1}} = f. \tag{3.2}$$

We also know that

$$\mathrm{PI} : H^s(S^{n-1}) \longrightarrow H^{s+1/2}(\mathcal{B}), \quad \text{for } s \geq \frac{1}{2}, \tag{3.3}$$

and in particular

$$\mathrm{PI} : C^\infty(S^{n-1}) \longrightarrow C^\infty(\overline{\mathcal{B}}). \tag{3.4}$$

Our goal here is to produce an explicit integral formula for this solution operator. Before deriving this explicit formula, we record the classical mean-value property, which has been proved once, in §2 of Chapter 3.

Proposition 3.1. *For $f \in C(S^{n-1})$, $u = PI\, f$ satisfies*

$$u(0) = Avg_{S^{n-1}}\, f = \frac{1}{A_{n-1}} \int\limits_{S^{n-1}} f(\omega)\, dS(\omega), \tag{3.5}$$

where A_{n-1} *is the area of* S^{n-1}; $A_{n-1} = 2\pi^{n/2}/\Gamma(n/2)$.

In view of its fundamental nature, we give two more proofs of this result, one based on the rotational symmetry of the Laplace operator, the other based on Green's formula.

For the first proof, let

$$v(x) = \mathrm{Avg}_{SO(n)}\, u(g \cdot x) = \int_{SO(n)} u(g \cdot x)\, dg \tag{3.6}$$

be the average of the set of rotates of $u(x)$. Then, since Δ is rotationally invariant, we have

$$\Delta v = 0 \ \text{ on } \mathcal{B}. \tag{3.7}$$

Now, clearly,

$$v\big|_{S^{n-1}} = \mathrm{Avg}_{S^{n-1}}\, f = C \tag{3.8}$$

and

$$v(0) = u(0). \tag{3.9}$$

But a solution to (3.7)–(3.8) is

$$v_0(x) = C, \tag{3.10}$$

and by the maximum principle this solution must be unique. Thus the conclusion (3.5) follows from (3.9) and (3.10).

As was already noted in §2 of Chapter 3, we could also obtain uniqueness by applying Green's formula

$$(dw, dw) = -(\Delta w, w) + \int_{\partial\mathcal{B}} w\, \frac{\partial \overline{w}}{\partial \nu}\, dS \tag{3.11}$$

to $w = v - v_0$, at least if we know $w \in C^2(\overline{\mathcal{B}})$, which in this case would follow from $u \in C^2(\overline{\mathcal{B}})$. To pass to general $u \in C(\overline{\mathcal{B}})$, harmonic in $\mathcal{B}$, we can replace $u(x)$ by $u_\rho(x) = u(\rho x)$ for $\rho < 1$, which belongs to $C^\infty(\overline{\mathcal{B}})$ since we know $u \in C^\infty(\mathcal{B})$. Then passing to the limit $\rho \nearrow 1$ yields another variation on the proof of Proposition 3.1 (which is not counted as the second proof).

Our second proof uses Green's formula:

$$(\Delta u, v)_{L^2(\mathcal{B})} = (u, \Delta v)_{L^2(\mathcal{B})} + \int_{\partial\mathcal{B}} \left(u \frac{\partial v}{\partial \nu} - v \frac{\partial u}{\partial \nu} \right) dS. \tag{3.12}$$

We will use this in a slightly different context than before, via the following result, which is an exercise in distribution theory.

Lemma 3.2. *The formula (3.12) is valid provided* $u \in C^\infty(\overline{\mathcal{B}})$ *and* v *is a distribution on* $\mathcal{B}$, *equal near* $\partial\mathcal{B}$ *to a function in* $C^\infty(\overline{\mathcal{B}})$.

Thus, we apply (3.12) to $u = \mathrm{PI}\, f$, assumed to be in $C^\infty(\overline{\mathcal{B}})$, and to

$$v(x) = 1 - |x|^{2-n}; \quad v(x) = \log|x| \ \text{ if } n = 2. \tag{3.13}$$

As shown in Proposition 4.9 of Chapter 3, we have

$$\Delta v = C_n \delta, \quad C_n = (n-2)A_{n-1}, \ C_2 = 2\pi. \tag{3.14}$$

Since $v = 0$ on $\partial\mathcal{B}$, while $\partial v/\partial r = n-2$ on $\partial\mathcal{B}$, (3.12) yields

$$(n-2)A_{n-1}u(0) = (n-2) \int_{S^{n-1}} u(x)\, dS(x), \tag{3.15}$$

with an obvious modification for $n = 2$. We can go from $u \in C^\infty(\overline{\mathcal{B}})$ to $u \in C(\overline{\mathcal{B}})$ by the limiting argument described above. This completes the second proof. See the exercises for yet another proof.

Of course, one could use the mean-value property, established via the second proof, to derive the maximum principle for harmonic functions on open regions in $\mathbb{R}^n$, as was done in Chapter 3, §2. The advantage of the method of §2 of this chapter is its much more general applicability.

We now tackle our main goal of this section, which is to obtain an explicit integral formula for the map (3.1). First we recall analogous computations performed in Chapter 3. As shown in (5.21) of Chapter 3,

$$\mathrm{PI} : \mathcal{S}(\mathbb{R}) \longrightarrow C^\infty(\mathbb{R}^2_+)$$

is given by

$$u(y,x) = \frac{y}{\pi} \int_{-\infty}^{\infty} \frac{f(x')}{y^2 + (x-x')^2}\, dx'. \tag{3.16}$$

Formula (5.24) of that chapter shows that, more generally,

$$\mathrm{PI} : \mathcal{S}(\mathbb{R}^{n-1}) \longrightarrow C^\infty(\mathbb{R}^n_+)$$

is given by

$$u(y,x) = c_{n-1} y \int_{\mathbb{R}^{n-1}} \frac{f(x')}{\left[y^2 + |x-x'|^2\right]^{n/2}}\, dx'. \tag{3.17}$$

Also, formula (2.5) of Chapter 3 shows that

$$\mathrm{PI} : C^\infty(S^1) \longrightarrow C^\infty(\overline{\mathcal{D}})$$

is given by

$$u(x) = \frac{1-|x|^2}{2\pi} \int_{S^1} \frac{f(x')}{|x-x'|^2}\, dS(x'). \tag{3.18}$$

In order to define PI on $C^\infty(S^{n-1})$, there are systematic methods, involving conformal transformations, which are used in many texts, such as [Hel] and [Keg]. The method we will use here is the method of the "inspired guess," based on extrapolation from (3.16)–(3.18). Note that (3.16) and (3.17) differ only in the constant factor in front and in the exponent on $(y^2 + |x - x'|^2)$ in the integrand. The denominator in the integrand in (3.17) is the nth power of the distance from $(0, x')$ to (y, x) in $\mathbb{R}^n$. This makes it very tempting to try to generalize (3.18) to

$$u(x) = c_n'(1 - |x|^2) \int_{S^{n-1}} \frac{f(x')}{|x - x'|^n} \, dS(x'), \tag{3.19}$$

for $u = \text{PI}\, f$, $f \in C^\infty(S^{n-1})$. We have only to show that this works. First we show that u is harmonic in $\mathcal{B}$. This is a consequence of the following.

Lemma 3.3. *For a given $x' \in S^{n-1}$ (i.e., $|x'| = 1$), set*

$$v(x) = (1 - |x|^2)|x - x'|^{-n}. \tag{3.20}$$

Then v is harmonic on $\mathbb{R}^n \setminus \{x'\}$.

One can apply Δ to (3.20) in a straightforward manner, but the formulas can get very bulky if produced naively, so we give a clean route to the calculations. It suffices to show that $w(x) = v(x + x')$ is harmonic on $\mathbb{R}^n \setminus 0$. Since $1 - |x + x'|^2 = -(2x \cdot x' + |x|^2)$ provided $|x'| = 1$, we have

$$-w(x) = 2(x' \cdot x)|x|^{-n} + |x|^{2-n}. \tag{3.21}$$

That $|x|^{2-n}$ is harmonic on $\mathbb{R}^n \setminus 0$ we already know, as a consequence of the formula for Δ in polar coordinates, which yields

$$g(x) = \varphi(r) \Longrightarrow \Delta g = \varphi''(r) + \frac{n-1}{r}\varphi'(r). \tag{3.22}$$

Now, applying $\partial/\partial x_j$ to a harmonic function on an open set in $\mathbb{R}^n$ gives another, so the following are harmonic on $\mathbb{R}^n \setminus 0$:

$$w_j(x) = \frac{\partial}{\partial x_j} |x|^{2-n} = (2 - n)x_j|x|^{-n}. \tag{3.23}$$

For $n = 2$, we take

$$\frac{\partial}{\partial x_j} \log |x| = x_j|x|^{-2}. \tag{3.24}$$

Thus the first term on the right side of (3.21) is a linear combination of these functions, so the lemma is established.

To justify (3.19), it remains to show that if u is given by this formula, and c_n' is chosen correctly, then $u = f$ on S^{n-1}. Note that if we write $x = r\omega$, $\omega \in S^{n-1}$, then (3.19) gives

$$u(r\omega) = \int_{S^{n-1}} p(r, \omega, \omega') f(\omega') \, dS(\omega'), \tag{3.25}$$

where

$$p(r, \omega, \omega') = c_n'(1 - r^2)|r\omega - \omega'|^{-n}. \tag{3.26}$$

It is clear that

$$p(r,\omega,\omega') \to 0 \ \text{ as } r \nearrow 1 \ \text{ if } \omega \neq \omega'. \tag{3.27}$$

We claim that

$$\int_{S^{n-1}} p(r,\omega,\omega')\, dS(\omega') = c''_n, \tag{3.28}$$

a constant independent of r. By rotational invariance, this integral is clearly independent of ω. Thus we could integrate with respect to ω. But Lemma 3.3 implies that

$$p(r,x,\omega') = c'_n\big(1 - r^2|x|^2\big)|rx - \omega'|^{-n} \tag{3.29}$$

is *harmonic* in x, for $|x| < 1/r$, so the mean-value theorem gives

$$\frac{1}{A_{n-1}} \int_{S^{n-1}} p(r,\omega,\omega')\, dS(\omega) = c'_n, \tag{3.30}$$

for all $r < 1$, $\omega' \in S^{n-1}$. This implies (3.28), with $c''_n = c'_n A_{n-1}$. Thus, in view of (3.27), $p(r,\omega,\omega')$ is highly peaked near $\omega = \omega'$ as $r \nearrow 1$, so the limit of (3.25) as $r \nearrow 1$ is equal to $c'_n A_{n-1} f(\omega)$, for any $f \in C(S^{n-1})$. This justifies the formula (3.19) and fixes the constant:

$$c'_n = \frac{1}{A_{n-1}}. \tag{3.31}$$

We summarize:

Proposition 3.4. *The map (3.1) is given by the Poisson integral formula*

$$u(x) = \frac{1 - |x|^2}{A_{n-1}} \int_{S^{n-1}} \frac{f(x')}{|x - x'|^n}\, dS(x'). \tag{3.32}$$

Exercises

1. If $\Omega \subset\subset \mathbb{R}^n$ is smooth, and $u \in C^2(\overline{\Omega})$ is harmonic, show that

$$\int_{\partial\Omega} \frac{\partial u}{\partial \nu}\, dS = 0.$$

(*Hint*: Set $v = 1$ in Green's formula (3.12), with $\mathcal{B}$ replaced by Ω.)

2. Derive the mean-value property as follows. For u harmonic on a neighborhood $B_R = \{x \in \mathbb{R}^n : |x| < R\}$ of 0, if $0 < r < R$, it follows from Exercise 1 that

$$\frac{d}{dr} \int_{S^{n-1}} u(r\omega)\, dS(\omega) = \int_{S^{n-1}} \frac{\partial}{\partial r} u(r\omega)\, dS(\omega) = 0,$$

so $\mathrm{Avg}_{\partial B_r} u$ is constant for $0 < r \le R$.

3. Modify the approach to Exercise 2 to show that, if u is *subharmonic* (i.e., $\Delta u \geq 0$ on B_R), then

$$u(0) \leq \operatorname{Avg}_{\partial B_R} u.$$

4. If $u = \mathrm{PI}\, f$ as in (3.1)–(3.2), show that, for any $\omega \in S^{n-1}$,

$$\omega \cdot \nabla u(0) = \frac{n-1}{A_{n-1}} \int\limits_{S^{n-1}} (\omega \cdot \theta) f(\theta)\, dS(\theta). \tag{3.33}$$

Deduce that if u is harmonic on $\Omega \subset \mathbb{R}^n$ and $p \in B_r(p) \subset \Omega$, then

$$\begin{aligned} \omega \cdot \nabla u(p) &= \frac{n-1}{A_{n-1}} \frac{1}{r^{n+1}} \int\limits_{\partial B_r(p)} \omega \cdot (y-p) u(y)\, dS(y) \\ &= \frac{n-1}{r^2} \operatorname{Avg}_{\partial B_r(p)} \big\{\omega \cdot (y-p) u(y)\big\}. \end{aligned} \tag{3.34}$$

5. Deduce from Proposition 3.4 that a harmonic function on a domain $\Omega \subset \mathbb{R}^n$ is real analytic on Ω.

4. The Riemann mapping theorem (smooth boundary)

Let Ω be a bounded domain in $\mathbb{C}$, with smooth boundary. Assume Ω is connected and simply connected. In particular, this implies that $\partial\Omega$ is connected, so diffeomorphic to the circle S^1. Let p be a point in Ω. We aim to construct a holomorphic function Φ on Ω such that $\Phi(p) = 0$ and $\Phi : \overline{\Omega} \to \overline{\mathcal{D}}$ is a diffeomorphism, where $\mathcal{D} = \{z \in \mathbb{C} : |z| < 1\}$ is the unit disk. This will be done via solving a Dirichlet problem for the Laplace operator on Ω.

Note that the function $\log|z-p|$ is harmonic on $\mathbb{C} \setminus p$. Let $G_0(x, y)$ be the solution to the Dirichlet problem

$$\Delta G_0 = 0 \text{ in } \Omega, \quad G_0\big|_{\partial\Omega} = -\log|z-p|\big|_{\partial\Omega}. \tag{4.1}$$

As we know, there is a unique such $G_0 \in C^\infty(\overline{\Omega})$. Then

$$G(x, y) = \log|z-p| + G_0(x, y) \tag{4.2}$$

is harmonic on $\Omega \setminus \{p\}$ and vanishes on $\partial\Omega$. This is a *Green function*.

We next construct $H_0 \in C^\infty(\overline{\Omega})$, the *harmonic conjugate* of G_0. It is given by

$$H_0(z) = \int_p^z \Big[-\frac{\partial G_0}{\partial y}\, dx + \frac{\partial G_0}{\partial x}\, dy\Big], \tag{4.3}$$

the integral being along any path from p to z in Ω. Green's theorem, and the harmonicity of G_0, imply that the integral is independent of the choice of path. Making appropriate choices of path, we readily see that

$$\frac{\partial H_0}{\partial x} = -\frac{\partial G_0}{\partial y}, \quad \frac{\partial H_0}{\partial y} = \frac{\partial G_0}{\partial x}, \tag{4.4}$$

so $G_0 + iH_0$ is holomorphic. Now

$$H(x, y) = \operatorname{Im} \log(z-p) + H_0(x, y) \tag{4.5}$$

is multivalued, but

$$\Phi(z) = e^{G+iH} = (z-p)e^{G_0+iH_0} \tag{4.6}$$

is a single-valued holomorphic function on Ω, with $\Phi(p) = 0$. Note that $z \in \partial\Omega \Rightarrow G = \operatorname{Re}(G+iH) = 0$, so

$$\Phi : \partial\Omega \longrightarrow S^1 \tag{4.7}$$

and hence, by the maximum modulus principle,

$$\Phi : \Omega \longrightarrow \mathcal{D}. \tag{4.8}$$

The Riemann mapping theorem (for this class of domains) asserts the following.

Theorem 4.1. *Φ is a holomorphic diffeomorphism of $\overline{\Omega}$ onto $\overline{\mathcal{D}}$.*

Proof. We must show that $\Phi : \overline{\Omega} \to \overline{\mathcal{D}}$ is one-to-one and onto, with nowhere-vanishing derivative. This will be easy once we establish that

$$\varphi = \Phi\big|_{\partial\Omega} : \partial\Omega \longrightarrow S^1 \tag{4.9}$$

has nowhere-vanishing derivative. Note that since $G\big|_{\partial\Omega} = 0$, $\varphi = e^{iH}\big|_{\partial\Omega}$. In view of the Cauchy-Riemann equations yielding holomorphy of $G + iH$, to say that the tangential derivative of H on $\partial\Omega$ is nowhere zero is equivalent to saying

$$\frac{\partial G}{\partial \nu}(z) \neq 0, \quad \text{for all } z \in \partial\Omega. \tag{4.10}$$

On the other hand, since $G(z) \to -\infty$ as $z \to p$, $G(z)$ is maximal on $\partial\Omega$, and so Zaremba's principle implies (4.10). Thus (4.9) is a local diffeomorphism, and hence a covering map. To finish off the argument, we make use of the following result, known as the *argument principle.*

Proposition 4.2. *Let $\Phi \in C^1(\overline{\Omega})$ be holomorphic inside Ω, a bounded region in $\mathbb{C}$ with smooth boundary, $\partial\Omega = \gamma$. Take $q \in \mathbb{C}$, not in the image of γ under Φ. Then the number of points p_j in Ω, counting multiplicity, for which $\Phi(p_j) = q$ is equal to the winding number of the curve $\Phi(\gamma)$ about q.*

Here, if $\Phi(p_j) = q$, we say p_j has multiplicity 1 if $\Phi'(p_j) \neq 0$, and we say it has multiplicity $k+1$ if $\Phi'(p_j) = \cdots = \Phi^{(k)}(p_j) = 0$ but $\Phi^{(k+1)}(p_j) \neq 0$. A proof of this elementary result can be found in most complex analysis texts, e.g., [Ahl] or [Hil]. A substantial generalization of this result can be found in Exercises 19–22 in the set of exercises on cohomology, after §9 of this chapter.

Now, to finish off the proof of Theorem 4.1, we show that the map (4.9) has winding number 1, by appealing to the argument principle, with $q = 0$. We see from (4.6) that p is the unique zero of Φ, a simple zero. Hence the map (4.9) is a diffeomorphism. Thus, again by the argument principle, any $q \in \mathcal{D}$ is equal to $\Phi(w)$ for precisely one $w \in \Omega$. This implies $\Phi'(w) \neq 0$ for all $w \in \Omega$, and the proof of Theorem 4.1 is complete.

Two smooth, bounded domains in $\mathbb{C}$ that are homeomorphic may not be holomorphically equivalent if they are not simply connected. We discuss the analogue of the Riemann mapping theorem in the next simplest case, when $\overline{\Omega}$ is a smooth, bounded domain in $\mathbb{C}$ whose boundary has *two* connected components, say γ_0 and γ_1. Assume γ_0 is the "outer" boundary component, touching the unbounded component of $\mathbb{C} \setminus \overline{\Omega}$.

In such a case, let $u(x, y)$ be the solution to the Dirichlet problem

$$\Delta u = 0 \text{ on } \Omega, \quad u\big|_{\gamma_0} = 0, \; u\big|_{\gamma_1} = -1. \tag{4.11}$$

Given $c > 0$, consider $G = cu$, which will play a role analogous to the function G in (4.2). Consider the 1-form

$$\beta = -\frac{\partial G}{\partial y}\, dx + \frac{\partial G}{\partial x}\, dy, \tag{4.12}$$

which is closed by the harmonicity of G. If γ_0 is oriented in the clockwise direction, we have $\int_{\gamma_0} \beta = cA$, for

$$A = -\int\limits_{\gamma_0} \frac{\partial u}{\partial \nu}\, ds > 0.$$

Hence there is a unique value of $c \in (0, \infty)$ for which $\int_{\gamma_0} \beta = 2\pi$. In that case we can write $\beta = dH$, where H, a harmonic conjugate of G, is a smooth, real-valued "function" on Ω, well defined mod 2π. Hence

$$\Psi(z) = e^{G+iH} \tag{4.13}$$

is a single-valued holomorphic function on Ω. It maps γ_0 to the circle $|z| = 1$ and it maps γ_1 to the circle $|z| = e^{-c}$. Using Zaremba's principle as in the proof of Theorem 4.1, we see that Ψ maps γ_0 to $S^1 = \{z : |z| = 1\}$ and γ_1 to $\{z : |z| = e^{-c}\}$, locally diffeomorphically. The fact that $\int_{\gamma_0} \beta = 2\pi$ implies that γ_0 is mapped diffeomorphically onto S^1. From there, an application of the argument principle yields:

Theorem 4.3. *If $\overline{\Omega}$ is a smooth, bounded domain in $\mathbb{C}$ with two boundary components, and Ψ is constructed by (4.11)–(4.13), then Ψ is a holomorphic diffeomorphism of $\overline{\Omega}$ onto the annular region*

$$\mathfrak{A}_\rho = \{z \in \mathbb{C} : \rho \le |z| \le 1\}, \quad \rho = e^{-c}. \tag{4.14}$$

It is easy to show that if $0 < \rho < \sigma < 1$, then $\mathfrak{A}_\rho$ and $\mathfrak{A}_\sigma$ are not holomorphically equivalent. If there were a holomorphic diffeomorphism $F : \mathfrak{A}_\rho \to \mathfrak{A}_\sigma$, then, using an inversion if necessary, we could assume F maps $|z| = 1$ to itself and that it maps $|z| = \rho$ to $|z| = \sigma$. Then, applying the Schwartz reflection principle an infinite sequence of times, we can extend F to a holomorphic diffeomorphism of $\overline{\mathcal{D}} = \{|z| = 1\}$ onto itself, preserving the origin. Then we must have $F(z) = az$, $|a| = 1$ (see Exercise 4 below), which would imply $\rho = \sigma$.

Exercises

1. Let Ω be the unit disk in $\mathbb{C}$, $f \in C^\infty(S^1)$, real-valued, with mean zero. Let $u = PI\, f$. Show that, for $g \in C^\infty(S^1)$ real-valued, $v = PI\, g$ is the harmonic conjugate of u (satisfying $v(0) = 0$) if and only if

$$g = -iHf,$$

where H is the operator (2.16) of Chapter 3, that is,

$$Hf(\theta) = \sum_{k=-\infty}^{\infty} (\operatorname{sgn} k)\, \hat{f}(k)\, e^{ik\theta}.$$

2. Let Ω be a bounded, simply connected domain in $\mathbb{C}$, and suppose $F : \overline{\Omega} \to \overline{\mathcal{D}}$ is holomorphic, taking $\partial\Omega$ to $\partial\mathcal{D}$. Suppose $F(p) = 0$, $p \in \Omega$, $F'(p) \neq 0$, and F has no other zeros. Show that $F(z) = (z-p)e^{f(z)}$ with $f : \Omega \to \mathbb{C}$ holomorphic, and $e^{\operatorname{Re} f(z)} = |z-p|^{-1}$ on $\partial\Omega$. Use this to motivate the constructions used in this section to prove the Riemann mapping theorem.
3. Given $a, b \in \mathbb{C}$, $|a|^2 - |b|^2 = 1$, set

$$A = \begin{pmatrix} a & b \\ \overline{b} & \overline{a} \end{pmatrix}.$$

We say $A \in SU(1,1)$. Define the map

$$F_A(z) = \frac{az + b}{\overline{b}z + \overline{a}}.$$

Show that each such F_A maps $\mathcal{D}$ one-to-one and onto itself. Show that $F_{AB}(z) = F_A(F_B(z))$. Show that, for any $q \in \mathcal{D}$, there exists $A \in SU(1,1)$ such that $F_A(q) = 0$.
4. Suppose $F : \mathcal{D} \to \mathcal{D}$ is a holomorphic diffeomorphism such that $F(0) = 0$. Show that $F(z) = az$, for some $a \in \mathbb{C}$, $|a| = 1$. (*Hint*: Consider the behavior of $F(z)/z$ and of $z/F(z)$.)
5. Deduce that *every* holomorphic diffeomorphism $F : \mathcal{D} \to \mathcal{D}$ is of the form F_A of Exercise 3. (*Hint*: First construct F_{A_j} such that $F_{A_2} \circ F \circ F_{A_1}(0) = 0$.)
6. Given $p \in \Omega$, simply connected, etc., show that there is a *unique* holomorphic diffeomorphism $\Phi : \Omega \to \mathcal{D}$ such that $\Phi(p) = 0$ and $\Phi'(p) > 0$.

5. The Dirichlet problem on a domain with a rough boundary

Let Ω be an open connected subset of the interior of M, a smooth, compact, connected Riemannian manifold with nonempty boundary. The boundary of Ω can be quite wild. We want to formulate and study the Dirichlet problem for the Laplace operator on Ω.

Let us start with a function $\varphi \in C^\infty(M)$ given; let $\psi = \varphi\big|_{\partial\Omega}$. We want to find $u \in C^\infty(\Omega)$ such that

$$\Delta u = 0 \text{ in } \Omega, \text{ and } u\big|_{\partial\Omega} = \psi, \text{ in some sense.} \tag{5.1}$$

In the best cases, we will have $u \in C(\overline{\Omega})$, but not always.

A particular example of the problem we're interested in solving is the following, when Ω is a bounded open subset of $\mathbb{R}^n$. Let $E_p(x)$ be (a multiple of) the fundamental solution to the Laplace equation on $\mathbb{R}^n$, with pole at p, given by

$$\begin{aligned} E_p(x) &= \log|x-p| \qquad (n=2), \\ &\quad -|x-p|^{2-n} \quad (n \geq 3). \end{aligned} \tag{5.2}$$

We want to construct a function G_p, harmonic on $\Omega \setminus \{p\}$, with the same type of singularity as E_p at p, so

$$\Delta G_p = c_n \delta_p; \quad G_p\big|_{\partial\Omega} = 0, \text{ in some sense.}$$

Recall from §4 that such a function was constructed for $\Omega \subset\subset \mathbb{R}^2$ with smooth boundary, as a tool to prove the Riemann mapping theorem for smooth, simply connected domains. One motivating force pushing our analysis here will be to generalize Theorem 4.1 to an arbitrary bounded, simply connected domain Ω in $\mathbb{C}$, with no smoothness assumptions whatsoever on $\partial\Omega$. To relate G_p to (5.1), note that if we write

$$G_p = E_p + F, \tag{5.3}$$

then $u = F$ solves (5.1), with $\psi = -E_p\big|_{\partial\Omega}$. We then have $\psi = \varphi\big|_{\partial\Omega}$, where $\psi = -\chi E_p$, with $\chi \in C^\infty(\mathbb{R}^n)$ equal to zero on a neighborhood of p, 1 on a neighborhood of $\partial\Omega$.

A construction of the solution to (5.1) is given by

$$u = v + \varphi, \tag{5.4}$$

where v is defined by

$$v \in H_0^1(\Omega), \quad \Delta v = -\Delta\varphi = \Phi. \tag{5.5}$$

Even for $\partial\Omega$ not smooth, the estimate (1.5) still holds, so such an element v is uniquely determined. We proceed to give a more precise sense to the assertion that $u\big|_{\partial\Omega} = \psi$.

We will analyze the behavior of the solution u defined by (5.4)–(5.5) by the following limiting process. Pick a sequence of connected domains Ω_j with smooth boundary such that

$$\Omega_j \subset\subset \Omega_{j+1}, \quad \bigcup_j \Omega_j = \Omega. \tag{5.6}$$

Then as shown in §1, we have $u_j \in C^\infty(\overline{\Omega}_j)$ such that

$$\Delta u_j = 0 \text{ on } \Omega_j, \quad u_j\big|_{\partial\Omega_j} = \psi_j = \varphi\big|_{\partial\Omega_j}. \tag{5.7}$$

Parallel to (5.4)–(5.5), we have

$$u_j = v_j + \varphi_j, \tag{5.8}$$

where

$$\varphi_j = \varphi\big|_{\Omega_j}, \tag{5.9}$$

and $v_j \in C^\infty(\overline{\Omega}_j)$ is uniquely determined by

$$(5.10) \qquad v_j \in H_0^1(\Omega_j), \quad \Delta v_j = \Phi_j = \Phi\big|_{\Omega_j},$$

with $\Phi \in C^\infty(M)$ defined as in (5.5). Extending each $v_j \in C^\infty(\overline{\Omega}_j)$ to be zero in $\Omega \setminus \Omega_j$, we can regard each v_j as an element of $H_0^1(\Omega)$. We then have the following.

Lemma 5.1. *The set $\{v_j\}$ is bounded in $H_0^1(\Omega)$.*

Proof. We have

$$\|v_j\|^2_{H^1(\Omega)} = \|v_j\|^2_{H^1(\Omega_j)} = \|dv_j\|^2_{L^2(\Omega_j)} + \|v_j\|^2_{L^2(\Omega_j)}.$$

By (5.10),

$$(5.11) \qquad \|dv_j\|^2_{L^2} = -(\Delta v_j, v_j) = -(\Phi, v_j) \le \|\Phi\|_{L^2}\|v_j\|_{L^2}.$$

Now there is a constant K such that

$$(5.12) \qquad \|u\|_{L^2} \le K\|du\|_{L^2}, \quad \text{for all } u \in H_0^1(\Omega),$$

indeed, for all $u \in H_0^1(M)$. Inserting this estimate into (5.11) and cancelling a factor of $\|dv_j\|_{L^2}$, we have

$$(5.13) \qquad \|dv_j\|_{L^2} \le K\|\Phi\|_{L^2}.$$

Appealing again to (5.12), we have a bound on the $H^1(\Omega)$-norm of v_j.

Since any closed ball in the Hilbert space $H_0^1(\Omega)$ is compact and metrizable in the weak topology, any subsequence of $\{v_j\}$ in turn has a weakly convergent subsequence. Any limit must satisfy (5.5). Since the solution to (5.5) is unique, we have

$$(5.14) \qquad v_j \longrightarrow v \text{ weakly in } H_0^1(\Omega).$$

Since $\Delta v_j = \Phi$ on Ω_j, we deduce from interior regularity estimates the following.

Lemma 5.2. *Let $\mathcal{O} \subset\subset \Omega$; say $\mathcal{O} \subset\subset \Omega_J$. Then*

$$\{v_j\big|_{\mathcal{O}} : j \ge J\} \textit{ is bounded in } C^\infty(\mathcal{O}).$$

It follows that

$$(5.15) \qquad v_j \longrightarrow v \ \text{ in } C^\infty(\mathcal{O}).$$

Thus, by (5.4) and (5.8),

$$(5.16) \qquad u_j \longrightarrow u \text{ in } C^\infty(\mathcal{O}), \quad \text{for any } \mathcal{O} \subset\subset \Omega.$$

We can use this to obtain the following version of the strong maximum principle for u.

Proposition 5.3. *The function u defined by (5.4)–(5.5) satisfies*

$$\inf_{\partial\Omega} \psi(y) < u(x) < \sup_{\partial\Omega} \psi(y), \quad x \in \Omega, \tag{5.17}$$

unless u is constant.

Proof. For $u_j \in C^\infty(\overline{\Omega}_j)$, the strong maximum principle established in §2 implies

$$\inf_{\partial\Omega_j} \varphi(y) < u_j(x) < \sup_{\partial\Omega_j} \varphi(y), \quad \text{for } x \in \Omega_J,\ j \geq J \tag{5.18}$$

(unless u_j is constant). It follows from (5.16) that

$$\inf_{\partial\Omega} \varphi(y) \leq u(x) \leq \sup_{\partial\Omega} \varphi(y), \tag{5.19}$$

for all $x \in \Omega_J$, for all J, that is, (5.19) holds for all $x \in \Omega$. Since the strong maximum principle holds for $u\big|_{\Omega_j}$, we see that, unless $u\big|_{\Omega_j}$ is constant,

$$\inf_{\partial\Omega_j} u(y) < u(x) < \sup_{\partial\Omega_j} u(y), \quad \text{for } x \in \Omega_J,\ j \geq J,$$

so the estimate (5.17) follows.

One obvious consequence of Proposition 5.3, or even of (5.19), is that u is uniquely determined by $\psi = \varphi\big|_{\partial\Omega}$, independent of the extension φ to M. We hence have a map

$$\mathrm{PI} : \mathcal{E}(\partial\Omega) \longrightarrow L^\infty(\Omega) \cap C^\infty(\Omega), \tag{5.20}$$

where $\mathcal{E}(\partial\Omega)$ denotes the space of restrictions of elements of $C^\infty(M)$ to $\partial\Omega$. In (5.20), PI $\psi = u$ for $\psi = \varphi\big|_{\partial\Omega}$, with u given by (5.4)–(5.5). This map preserves the sup norm. It can be shown, by a mollification procedure, or by the Stone-Weierstrass theorem, that $\mathcal{E}(\partial\Omega)$ is dense in $C(\partial\Omega)$, so there is a unique continuous extension

$$\mathrm{PI} : C(\partial\Omega) \longrightarrow L^\infty(\Omega) \cap C^\infty(\Omega). \tag{5.21}$$

If $\psi \in C(\partial\Omega)$, $u = \mathrm{PI}\ \psi$ satisfies $\Delta u = 0$ in Ω. It is clear that Proposition 5.3 continues to hold for $u = \mathrm{PI}\ \psi$, given $\psi \in C(\partial\Omega)$.

We now examine conditions for PI $\psi = u$ to be continuous at a given boundary point $z_0 \in \partial\Omega$, involving the use of *barriers*. By definition, a function $w \in C^2(\Omega)$ is a barrier at z_0 for Ω provided $\Delta w \leq 0$ in Ω, $w(x) \to 0$ as $x \to z_0$, and, for any neighborhood U of z_0 in M, there is a $\delta > 0$ such that

$$w(x) > \delta, \quad \text{for } x \in \Omega \setminus U. \tag{5.22}$$

There are more general concepts of barriers, and we will use some of them later on, though for clarity we will give them different names, like "weak barriers," and so on. A point $z_0 \in \partial\Omega$ is called a *regular point* provided the conclusion of the following proposition holds.

Proposition 5.4. *If there is a barrier at $z_0 \in \partial\Omega$, then, for $\psi \in C(\partial\Omega)$, $u = PI\,\psi$,*

$$\lim_{x \to z_0} u(x) = \psi(z_0). \tag{5.23}$$

Proof. By a simple limiting argument, it suffices to prove the result for $\psi \in \mathcal{E}(\partial\Omega)$; suppose $\psi = \varphi\big|_{\partial\Omega}$, $\varphi \in C^\infty(M)$. Fix $\varepsilon > 0$. Then there exists $k > 0$ such that, for each j,

$$-\varepsilon - kw + \varphi(z_0) \le u_j(x) \le \varphi(z_0) + \varepsilon + kw \tag{5.24}$$

on $\partial\Omega_j$. This is arranged by picking k so large that $|\varphi(y) - \varphi(z_0)| \le \varepsilon + kw$ on $\partial\Omega_j$ for all j, so (5.24) holds on $\partial\Omega_j$. By the maximum principle, (5.24) must hold on Ω_j if w satisfies $\Delta w \le 0$. Letting $j \to \infty$, by (5.16), we have

$$-\varepsilon - kw(x) + \varphi(z_0) \le u(x) \le \varphi(z_0) + \varepsilon + kw(x), \quad x \in \Omega. \tag{5.25}$$

Since $w(x) \to 0$ as $x \to z_0$, this implies

$$\varphi(z_0) - \varepsilon \le \liminf_{x \to z_0} u(x) \le \limsup_{x \to z_0} u(x) \le \varphi(z_0) + \varepsilon, \tag{5.26}$$

for all $\varepsilon > 0$, which proves the proposition.

It turns out to be easier to construct an object that we call a *weak barrier*, defined as follows. A function $w \in C^2(\Omega)$ is a weak barrier at $z_0 \in \partial\Omega$ provided $\Delta w \le 0$ in Ω, $w(x) > 0$ for $x \in \Omega$, and $w(x) \to 0$ as $x \to z_0$.

We give a couple of examples of barriers and weak barriers, for planar domains.

Proposition 5.5. *Let $\Omega \subset\subset \mathbb{R}^2 = \mathbb{C}$, $z_0 \in \partial\Omega$. Suppose there is a simple curve γ, lying in $\mathbb{C} \setminus \Omega$, connecting z_0 to ∞. Then there is a barrier at z_0, so z_0 is a regular point.*

Proof. Cut $\mathbb{C}$ along γ; for any $K > 0$, $\log\big[(z - z_0)/K\big]$ can be defined as a single-valued holomorphic function in $\mathbb{C} \setminus \gamma$. For $K > \operatorname{diam} \Omega$, the harmonic function

$$V = -\operatorname{Re} \frac{1}{\log\left(\frac{z - z_0}{K}\right)} \tag{5.27}$$

is easily verified to be a barrier, so z_0 is a regular point.

We note that if $(z - z_0)/K = re^{i\theta}$, with θ continuous on $\mathbb{C} \setminus \gamma$, then

$$V(z) = -\frac{\log r}{(\log r)^2 + \theta^2}. \tag{5.28}$$

A larger class of planar domains is treated by the following result.

Proposition 5.6. *If $\Omega \subset \mathbb{C}$ is any bounded, simply connected domain, $z_0 \in \partial\Omega$, then we can define a single-valued branch of (5.27) on Ω, which will be a weak barrier function.*

Proof. This is clear. Note that the conclusion also holds if Ω is *contained* in a simply connected region Ω', with $z_0 \in \partial\Omega'$.

We remark that there exist domains satisfying the hypotheses of Proposition 5.6, for which V, given by (5.27), is not a genuine barrier, in the sense of the first definition. We indicate one example in Fig. 5.1. The region Ω illustrated there winds infinitely often around the circle that is its inner boundary, and z_0 lies on this circle. Below we will show that whenever a weak barrier exists, then a genuine barrier exists. Indeed, somewhat more will be demonstrated, in Proposition 5.12.

First, we show how to use the concept of weak barrier directly to examine the continuity at the boundary of Green functions (5.3). Let G_{pj} be such Green functions defined on the domain Ω_j, with smooth boundary, so

$$G_{pj} = E_p + F_j, \tag{5.29}$$

where $F_j \in C^\infty(\overline{\Omega_j})$ satisfies $\Delta F_j = 0$, $F_j\big|_{\partial\Omega_j} = -E_p\big|_{\partial\Omega_j}$. Thus

$$G_p - G_{pj} = F - F_j \quad \text{on } \Omega_j, \tag{5.30}$$

and hence, by (5.15),

$$G_p - G_{pj} \longrightarrow 0, \quad \text{in } C^\infty(\mathcal{O}), \quad \text{for } \mathcal{O} \subset\subset \Omega. \tag{5.31}$$

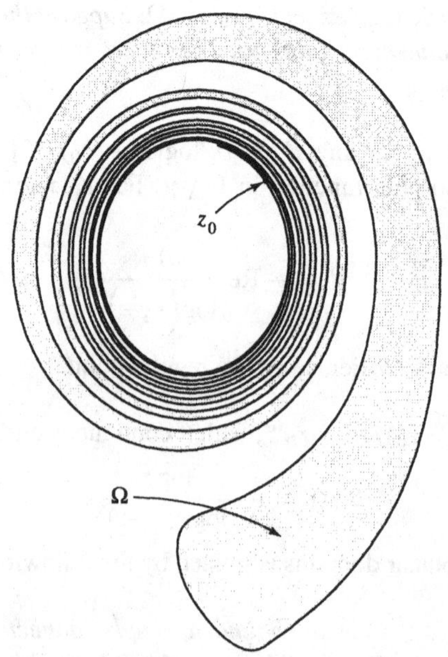

FIGURE 5.1

Since $G_{p\ell}(x) \to -\infty$ as $x \to p$ and $G_{p\ell}(x) = 0$ for $x \in \partial\Omega_\ell$, then, by the maximum principle,

$$G_{p\ell}(x) < 0, \quad \text{for } x \in \Omega_\ell \setminus \{p\},$$

and hence for $x \in \Omega_j \setminus \{p\}$, for all $\ell \geq j$, so we certainly have $G_p(x) \leq 0$ on $\Omega \setminus \{p\}$. Applying the strong maximum principle on Ω_j, $j \to \infty$, we can strengthen this to

$$G_p(x) < 0 \text{ on } \Omega \setminus \{p\}. \tag{5.32}$$

Now we show directly that weak barriers yield continuity of the Green function G_p at boundary points.

Proposition 5.7. *Let $z_0 \in \partial\Omega$. Suppose there exists a function $V \in C^2(\Omega)$ that is a weak barrier at z_0. Then $G_p(x) \to 0$ as $x \to z_0$.*

Proof. Fix a compact set $K \subset \Omega_1$, containing a neighborhood of p. Then there exists a k such that, for all j,

$$-kV < G_{pj} \text{ on } \partial(\Omega_j \setminus K),$$

since we know $\{G_{pj}\}$ is uniformly bounded on ∂K, and $G_{pj} = 0$ on $\partial\Omega_j$. Then, by the maximum principle,

$$-kV < G_{pj} \text{ on } \Omega_j \setminus K.$$

By (5.31) and (5.32), we have

$$-kV \leq G_p < 0 \text{ on } \Omega \setminus K, \tag{5.33}$$

which yields the proposition.

Propositions 5.6 and 5.7 will suffice for our treatment of the Riemann mapping theorem for general, simply connected domains in the next section, but we will proceed with some further results.

First, we consider local versions of barriers. A function $w \in C^2(\Omega)$ is a *local barrier* (resp., *weak local barrier*) at $z_0 \in \partial\Omega$ provided there is a neighborhood U of z_0 in M such that $w\big|_{\Omega\cap U}$ is a barrier (resp., weak barrier) at z_0 for $\Omega \cap U$.

The motivation for studying this concept is that local barriers and weak local barriers are frequently easier to construct than their global counterparts. However, when the local objects exist, their global counterparts do, too. This is easy to prove for (genuine) barriers.

Proposition 5.8. *If w is a local barrier at z_0, then there exists a barrier for Ω at z_0, equal to w in some neighborhood of z_0.*

Proof. Let $f : \mathbb{R} \to \mathbb{R}$ be a C^∞-function, with $f(0) = 0$, $f'(0) > 0$. A simple calculation shows

$$\Delta f(u) = f'(u)\Delta u + f''(u)|du|^2, \tag{5.34}$$

where $|du|^2 = g^{jk}(x)\partial_j u\ \partial_k u$. Thus, if $\Delta w \le 0$ on $\Omega \cap U$, we have $\Delta f(w) \le 0$ on $\Omega \cap U$ provided

$$f'(u) \ge 0, \quad f''(u) \le 0. \tag{5.35}$$

Take f to be such a function, with the additional property of being identically 1 for $u \ge \delta$, so the graph of f is as depicted in Fig. 5.2.

If w satisfies the barrier condition on $\Omega \cap U$ and in particular $w(x) > \delta$ outside $U_1 \subset\subset U$, define $w_1(x)$ by

$$w_1(x) = \begin{array}{ll} f(w(x)), & \text{for } x \in U_1 \cap \Omega, \\ 1, & \text{for } x \in \Omega \setminus U_1. \end{array} \tag{5.36}$$

Then w_1 is a barrier for Ω at z_0.

The argument above does not work if w is a weak local barrier. We now tackle the problem of dealing with weak barriers.

Proposition 5.9. *If w is a weak local barrier for Ω at $z_0 \in \partial\Omega$, then there exists a local barrier for Ω at z_0.*

Proof. Start with the function

$$h(x) = \sum_j (x_j - z_{0,j})^2 \tag{5.37}$$

in a local coordinate patch about z_0, chosen to be normal at z_0. Thus $h(z_0) = 0$ and this is the strict minimum of h. While Δ may not be the flat Laplacian Δ_0 in these coordinates, their coefficients do coincide at z_0. Thus, if U_1 is a sufficiently small neighborhood of z_0 in M,

$$\Delta h(x) \ge C > 0 \ \text{ in } U_1. \tag{5.38}$$

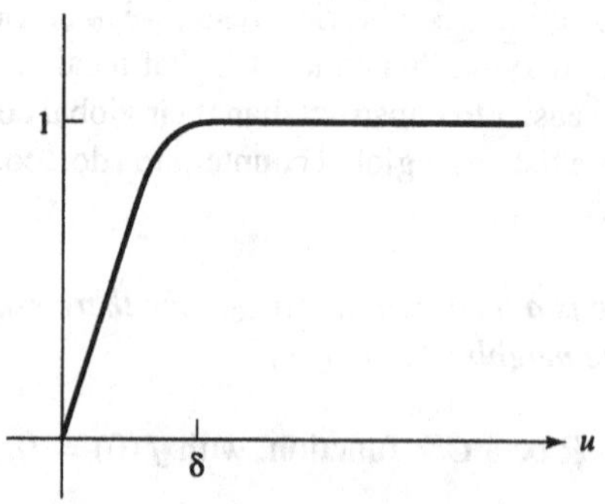

FIGURE 5.2

Also pick U_1 sufficiently small that w is a weak barrier for $\Omega \cap U_1$ at z_0. Then define w_1 to be the Poisson integral of $h\big|_{\partial(\Omega\cap U_1)}$, where

$$\mathrm{PI} : C\big(\partial(\Omega \cap U_1)\big) \longrightarrow L^\infty(\Omega \cap U_1).$$

We claim that w_1 is a barrier for $\Omega \cap U_1$, hence is the desired local barrier. Clearly $\Delta w_1 = 0$ on $\Omega \cap U_1$. Next we claim that

$$w_1(x) \geq h(x), \quad \text{for } x \in \Omega \cap U_1. \tag{5.39}$$

Indeed, we can write

$$w_1(x) = \lim_{j\to\infty} u_j(x), \tag{5.40}$$

where u_j are harmonic on $\mathcal{O}_j \nearrow \Omega \cap U_1$, with $\partial\mathcal{O}_j$ smooth and

$$u_j\big|_{\partial\mathcal{O}_j} = h(x),$$

and we can apply the maximum principle, Proposition 2.1, to $h - u_j$. This proves (5.39). To prove Proposition 5.9, it remains to establish that

$$\lim_{x\to z_0} w_1(x) = h(z_0) = 0. \tag{5.41}$$

This takes some effort, so we will take a break and advertise this formally.

Lemma 5.10. *The function w_1 constructed above satisfies (5.41) in $\Omega \cap U_1$.*

Proof. For convenience, we relabel $\Omega \cap U_1$, calling it Ω; also denote $\mathcal{O}_j$ above by Ω_j. Recall that we are working in an exponential coordinate system centered at z_0; in particular, $g_{jk}(z_0) = \delta_{jk}$. Let B_ρ be the ball of radius ρ in $\mathbb{R}^n$ centered at the origin (identified with z_0), as illustrated in Fig. 5.3. Pick $\rho > 0$ small. We can suppose that $\partial B_\rho \cap \Omega \neq \emptyset$. Let F be a compact subset of $\partial B_\rho \cap \Omega$ such that the $(n-1)$-dimensional measure of $\partial B_\rho \cap \Omega \setminus F$ is less than $\rho/2$ times the measure of ∂B_ρ. Here we are using measure determined by the flat metric on $\mathbb{R}^n$. Assume that $F \subset \partial B_\rho$ has a smooth $(n-2)$-dimensional boundary.

Let f be the product of the characteristic function of $\partial B_\rho \cap \Omega$ with a non-negative C^∞ function, ≤ 1, equal to 1 on $\partial B_\rho \cap \Omega \setminus F$, such that f has mean value $< \rho$ on ∂B_ρ, and let $p(x)$ be the Poisson integral of f on B_ρ, with respect to the flat Laplacian $\Delta_0 = \partial_1^2 + \cdots + \partial_n^2$; $\Delta_0 p = 0$. Even though f is not continuous everywhere on ∂B_ρ, study of the Poisson integral formula (3.32) shows that

$$p(x) \to f(y) \;\text{ as } x \to y \in \partial B_\rho \cap \Omega. \tag{5.42}$$

In fact, $p(x) \to f(y)$ as $x \to y$, for all $y \in \partial B_\rho \setminus \partial\Omega$. Note that, by the mean-value property and the maximum principle,

$$p(z_0) = \rho_0 \in (0, \rho); \quad 0 < p(x) < 1 \text{ on } B_\rho. \tag{5.43}$$

We want to alter $p(x)$, producing a function $q(x)$ such that $\Delta q \leq 0$ on B_ρ, where Δ is the Laplace operator of interest to us:

$$\Delta q = \Delta_0 q + L_2 q + L_1 q, \tag{5.44}$$

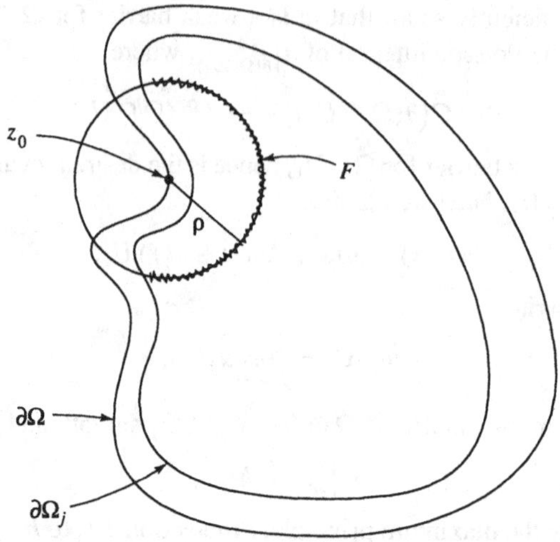

FIGURE 5.3

where L_2 is a second-order differential operator whose coefficients vanish to second order at z_0 and L_1 is a first-order differential operator whose coefficients vanish at z_0. Set

$$q(x) = p(x) + \alpha(\rho^2 - h(x)), \tag{5.45}$$

with $h(x)$ given by (5.37), so

$$\begin{aligned} \Delta_0 q &= -2n\alpha \text{ on } B_\rho, \quad q = p \text{ on } \partial B_\rho, \\ q(x) &> 0 \text{ on } B_\rho, \quad q(z_0) = p(z_0) + \alpha\rho^2. \end{aligned} \tag{5.46}$$

Fix α to be some positive constant. Then, by (5.44), if $\rho > 0$ is sufficiently small, we have

$$\Delta q \le -\alpha < 0 \text{ on } B_\rho. \tag{5.47}$$

Now let

$$M = \sup_\Omega h(x) = \sup_\Omega w_1(x), \tag{5.48}$$

let

$$k = \inf_F w(x) > 0, \tag{5.49}$$

where w is the weak (local) barrier hypothesized in Proposition 5.9, and consider

$$s(x) = w_1(x) - \rho - \frac{M}{k} w(x) - Mq(x), \quad x \in \mathcal{O}_\rho, \tag{5.50}$$

where $\mathcal{O}_\rho = \Omega \cap B_\rho$. We know that

$$s(x) = \lim_{j\to\infty} s_j(x), \tag{5.51}$$

where

$$s_j(x) = u_j(x) - \rho - \frac{M}{k}\, w(x) - Mq(x) \ \text{ on } \overline{\Omega_j \cap B_\rho}, \tag{5.52}$$

u_j being given by (5.40). Now $s_j(x)$ is continuous on $\overline{\Omega_j \cap B_\rho}$, and $\Delta s_j \geq 0$ on the interior, by (5.47). Also, using (5.42) and (5.46), and noting that, on $\partial\Omega_j \cap B_\rho$, $u_j = h \leq \rho^2 < \rho$ for ρ small, we see that

$$s_j(x) \leq 0 \tag{5.53}$$

on $\partial(\Omega_j \cap B_\rho) = (\partial\Omega_j \cap B_\rho) \cup (\overline{\Omega}_j \cap \partial B_\rho)$. By the maximum principle, (5.53) holds on $\Omega_j \cap B_\rho$. Passing to the limit gives $s(x) \leq 0$ on $\Omega \cap B_\rho$, hence

$$w_1(x) \leq \rho + \frac{M}{k}\, w(x) + Mq(x), \quad \text{for } x \in \Omega \cap B_\rho. \tag{5.54}$$

This implies

$$\limsup_{x \to z_0, x \in B_\rho \cap \Omega} w_1(x) \leq \rho + M(\rho + \alpha\rho^2), \tag{5.55}$$

since $w(x) \to 0$ as $x \to z_0$, and hence

$$\limsup_{x \to z_0} w_1(x) \leq 0. \tag{5.56}$$

Together with (5.39), this gives (5.41). The proof that w_1 is a barrier is complete.

Combining Propositions 5.4, 5.8, and 5.9, we have the following conclusion, essentially due to G. Bouligand.

Proposition 5.11. *Given $z_0 \in \partial\Omega$, the following are equivalent:*

(5.57) *there is a weak local barrier at z_0;*

(5.58) *there is a barrier at z_0;*

(5.59) *z_0 is a regular point.*

We also record the following consequence of localizability of the concept of a regular point.

Corollary 5.12. *Suppose Ω and Ω' are open subsets of M, with a common boundary point z_0. Suppose there is a neighborhood U of z_0 such that*

$$\Omega' \cap U \supset \Omega \cap U. \tag{5.60}$$

If z_0 is regular for Ω', then z_0 is regular for Ω.

Proof. A barrier at z_0 for Ω' gives a local barrier at z_0 for Ω.

As an application, we can localize the result on simply connected planar domains given in Proposition 5.6, to obtain the following result of H. Lebesgue.

Proposition 5.13. *Let $\Omega \subset \mathbb{R}^2$ be an open, bounded, connected region, and let $z_0 \in \partial\Omega$. Suppose the connected component of $\partial\Omega$ containing z_0 consists of more than one point (i.e., is a "continuum"). Then z_0 is a regular point.*

Proof. Let Γ be the connected component of $\partial\Omega$ containing z_0. Let Ω' be the connected component of $\mathbb{C} \setminus \Gamma$ containing Ω. Thus $\partial\Omega' = \Gamma$. If Ω' is bounded, then $\partial\Omega'$ connected implies that the planar domain Ω' is simply connected, so, by Proposition 5.6, z_0 is regular for Ω'. Hence, by Corollary 5.12, z_0 is regular for Ω in this case.

On the other hand, if Ω' is not bounded, pick $z_1 \in \Gamma$, at a maximal distance from z_0 ($z_1 \neq z_0$, under the hypothesis of the proposition). Let ρ denote the ray from z_1 to infinity, directly away from z_0. Then let Ω'' be $\Omega' \setminus \rho$, intersected with some disk D of large radius centered at z_1, so $\partial\Omega'' = \Gamma \cup (\rho \cap \overline{D}) \cup \partial D$. Thus Ω'' is simply connected. Since Ω'' coincides with Ω' on a neighborhood of z_0, we again have z_0 regular for Ω, and the proof is complete.

It is not hard to show that an isolated boundary point of Ω is always irregular, when $\dim \Omega \geq 2$. This can be obtained as a consequence of the following simple result.

Proposition 5.14. *The space $\mathcal{F}$ of functions in $C_0^\infty(\mathbb{R}^n)$ vanishing in a neighborhood of a given point p is dense in $H^1(\mathbb{R}^n)$ if $n \geq 2$.*

Proof. The annihilator of $\mathcal{F}$ is the space of elements of $H^{-1}(\mathbb{R}^n)$ supported at p. But any distribution supported at p is a linear combination of derivatives of the delta function δ_p, and none of these belong to $H^{-1}(\mathbb{R}^n)$, except for 0.

More generally, we say a compact set K in the interior of M is negligible if it supports no nonzero elements of $H^{-1}(M)$. For example, a smooth submanifold of codimension ≥ 2 is negligible.

Proposition 5.15. *Suppose a boundary point $z_0 \in \partial\Omega$ has a neighborhood U in M whose intersection with $\partial\Omega$ is negligible. Then z_0 is an irregular boundary point.*

Proof. Let $\Omega' = \Omega \cup U$. Thus z_0 is an interior point of Ω'. The hypothesis implies

$$H_0^1(\Omega') = H_0^1(\Omega).$$

Now, pick $f \in \mathcal{E}(\partial\Omega)$ such that $f = 0$ near z_0, $f \geq 0$ on all of $\partial\Omega$, and $f > 0$ somewhere. We claim that if we consider

$$u = \mathrm{PI}\, f; \quad \mathrm{PI} : C(\partial\Omega) \longrightarrow C^\infty(\Omega),$$

then $u(x)$ does not tend to 0 as $x \to z_0$. Indeed, u is simply the restriction to Ω of $u' = \mathrm{PI}\, f'$, where

$$\mathrm{PI} : C(\partial\Omega') \longrightarrow C^\infty(\Omega')$$

and f' is the restriction of f to $\partial\Omega'$. The strong maximum principle implies that $u' > 0$ everywhere in Ω', including at z_0, so the proof is done.

N. Wiener obtained a precise characterization of regular and irregular points in terms of "capacity," a certain countably subadditive set function defined on Borel sets. A boundary point $z_0 \in \partial\Omega$ is irregular provided the capacity of $\partial\Omega$ intersected with a small ball centered at z_0 decreases fast enough. The negligible sets defined above are precisely the compact sets of capacity zero. This characterization has a natural probabilistic analysis, using the theory of Brownian motion. In Chapter 11 we will discuss Brownian motion and present such a proof of Wiener's theorem.

We will derive one more sufficient condition for $z_0 \in \partial\Omega$ to be a regular point, due to S. Zaremba.

Proposition 5.16. *Let Ω be a bounded, open, connected subset of $\mathbb{R}^n$, with its flat metric. Suppose $z_0 \in \partial\Omega$ and there exists a cone C with vertex at z_0 such that, for some ball $\mathcal{B}$ centered at z_0,*

$$\mathcal{B} \cap C \setminus \{z_0\} \subset \mathbb{R}^n \setminus \overline{\Omega}. \tag{5.61}$$

Then z_0 is a regular point for Ω.

Proof. By Corollary 5.12, it suffices to show that z_0 is a regular point for $\mathcal{B} \setminus C$, where $\mathcal{B}$ is some ball centered at z_0. We can translate coordinates so that z_0 is the origin. We will construct a weak barrier for $\mathcal{B} \setminus C$ at $z_0 = 0$ of the form

$$v(x) = r^\alpha \varphi_0(\omega), \quad x = r\omega, \tag{5.62}$$

where $\varphi_0(\omega)$ is an eigenfunction of the Laplace operator Δ_S on the region $\mathcal{O} = S^{n-1} \setminus C$, an open subset of the sphere with nonempty smooth boundary:

$$\varphi_0 \in H_0^1(\mathcal{O}), \quad \Delta_S \varphi_0 = -\mu\varphi_0, \tag{5.63}$$

and $\mu > 0$ is the smallest eigenvalue of $-\Delta_S$ on $\mathcal{O}$. The formula for the Laplace operator in polar coordinates

$$\Delta v = \frac{\partial^2 v}{\partial r^2} + \frac{n-1}{r}\frac{\partial v}{\partial r} + \frac{1}{r^2}\Delta_S v \tag{5.64}$$

shows that (5.62) defines a function harmonic in $\Omega_1 = \mathcal{B} \setminus C$, continuous on $\overline{\Omega}_1$, and vanishing at z_0, if we take

$$\alpha = \nu - \frac{n-2}{2}, \quad \nu = \left[\mu + \frac{(n-2)^2}{4}\right]^{1/2}. \tag{5.65}$$

Note that $\alpha > 0$. Furthermore, as shown in Proposition 2.4, since μ is minimal, the eigenfunction φ_0 is nowhere vanishing on the interior of $\mathcal{O}$; thus it can be taken to be positive there. This makes v a weak barrier and completes the proof. Note that if the cone is shrunk, such construction produces a (genuine) local barrier for Ω at z_0.

We remark that, by considering $v_1 = f(v)$ with $f' > 0$, $f'' < 0$, $f(0) = 0$, we can construct a weak barrier v_1 satisfying $\Delta_0 v_1 \le -C < 0$ on $\mathcal{B} \setminus \mathcal{C}$, where $\Delta_0 = \partial_1^2 + \cdots + \partial_n^2$ is the flat Laplacian considered in Proposition 5.16. Such v_1 would also be a weak barrier with the flat Laplacian Δ_0 replaced by a small perturbation, of the form (5.44). In this way we can obtain an analogue of Proposition 5.16 for the general class of subdomains of a Riemannian manifold with boundary, which we have been dealing with in this section. Details are left as an exercise.

Exercises

1. Suppose Ω is a bounded region in $\mathbb{R}^n$. Suppose there is a point $p \in \mathbb{R}^n \setminus \overline{\Omega}$ such that z_0 is the point in $\partial\Omega$ *closest* to p. Show that

$$w(x) = -|x-p|^{2-n} \quad (\log|x-p| \text{ if } n=2) \tag{5.66}$$

is a barrier at z_0. Note that such p exists provided there exists a sphere in $\mathbb{R}^n \setminus \Omega$, touching $\partial\Omega$ precisely at z_0. If this happens, we say Ω satisfies the exterior sphere condition at z_0. Show that this condition holds for every C^2-boundary, but not for every C^1-boundary. Show it holds whenever Ω is convex.

2. Denote by $C^k(\partial\Omega)$ the space of restrictions to $\partial\Omega$ of elements of $C^k(\mathbb{R}^n)$. Let $f \in C^2(\mathbb{R}^n)$, and assume that Ω satisfies the exterior sphere condition. Given $z_0 \in \partial\Omega$, show that there exists a barrier of the form (5.66) and a $K < \infty$, such that, for all $z \in \partial\Omega$,

$$-K\big[w(z) - w(z_0)\big] \le f(z) - f(z_0) - (z - z_0)\cdot\nabla f(z_0) \le K\big[w(z) - w(z_0)\big],$$

with strict inequality except at $z_0 \in \partial\Omega$. Deduce from the maximum principle that such an inequality holds inside Ω, for $u = \mathrm{PI}(f)$. When can you deduce that

$$\mathrm{PI} : C^2(\partial\Omega) \longrightarrow \mathrm{Lip}(\overline{\Omega})?$$

(*Hint*: Look for uniform estimates on Ω_j.)
When can you replace $\mathrm{Lip}(\overline{\Omega})$ by $C^1(\overline{\Omega})$?

3. Replace barriers (5.66) by barriers of the form (5.62), and obtain boundary regularity results for more general domains Ω and less regular f, such as

$$\mathrm{PI} : \mathrm{Lip}(\partial\Omega) \longrightarrow C^\alpha(\overline{\Omega}),$$

for an appropriate class of domains $\Omega \subset \mathbb{R}^n$, with $0 < \alpha < 1$.
For a systematic treatment of Hölder estimates, see Chapter 6 of [GT], and references given there.

4. For the Green functions G_{pj} on $\Omega_j \nearrow \Omega$, approaching G_p as in (5.31), show that

$$G_{pj} \searrow G_p.$$

5. For the approximating solutions $v_j \in H_0^1(\Omega_j)$ in (5.9)–(5.14), show that (5.14) can be strengthened to

$$v_j \longrightarrow v \text{ in the } H^1(\Omega)\text{-}\textit{norm}.$$

(*Hint*: Show that $\|dv_j\|_{L^2}^2 \to \|dv\|_{L^2}^2$.)

6. Show that if $\Omega \subset \mathbb{R}^n$ is open and bounded (with smooth boundary) and $\Delta u = f$ on Ω, $u\big|_{\partial\Omega} = 0$, then

$$\sum_{j,k} \int_\Omega \left|\partial_j \partial_k u(x)\right|^2 dx = \int_\Omega |\Delta u(x)|^2 \, dx + (n-1) \int_{\partial\Omega} \left|\frac{\partial u}{\partial \nu}\right|^2 H(x) \, dS(x), \tag{5.67}$$

where $H(x)$ is the mean curvature of $\partial\Omega$ (with respect to the outward-pointing normal). This is known as Kadlec's formula.

7. Using Exercise 6, deduce that, for Ω *convex*, but with no other regularity assumed,

$$u \in H_0^1(\Omega), \ \Delta u = f \in L^2(\Omega) \Longrightarrow u \in H^2(\Omega).$$

(*Hint*: Look for uniform estimates on Ω_j. Each mean curvature $H_j(x)$ is ≤ 0.) Compare results in [Gri].

6. The Riemann mapping theorem (rough boundary)

Let Ω be a bounded open domain in $\mathbb{R}^2 = \mathbb{C}$ which is connected and simply connected. We aim to construct a one-to-one holomorphic map

$$\Phi : \Omega \longrightarrow \mathcal{D} \tag{6.1}$$

of Ω onto the unit disk $\mathcal{D}$. The construction of Φ will be similar to that given in §4 for domains with smooth boundary, but the proof that (6.1) is one-to-one and onto will be slightly different from the smooth case, and of course the conclusion will be weaker.

With $p \in \Omega$ given, we take the Green function $G = G_p$ constructed in §5. Thus $\Delta G = c\delta_p$ $(c > 0)$, $G(z) < 0$ on $\Omega \setminus \{p\}$. By Propositions 5.6 and 5.11, every point of $\partial\Omega$ is regular, so $\lim_{z \to z_0} G(z) = 0$ for each $z_0 \in \partial\Omega$ (i.e., G is continuous on $\overline{\Omega} \setminus \{p\}$). We can write, for $z \in \Omega$,

$$G(z) = \log|z - p| + G_0(z), \tag{6.2}$$

with $G_0 \in C(\overline{\Omega}) \cap C^\infty(\Omega)$, $\Delta G_0 = 0$, and we can construct the harmonic conjugate of G_0,

$$H_0 \in C^\infty(\Omega), \tag{6.3}$$

as

$$H_0(z) = \int_p^z \left[-\frac{\partial G_0}{\partial y}\, dx + \frac{\partial G_0}{\partial x}\, dy\right], \tag{6.4}$$

the integral being along any path from p to z in Ω, as before. As opposed to the case where $\partial\Omega$ is smooth, in general we cannot guarantee that $H_0 \in C(\overline{\Omega})$. In particular, there is no guarantee that Φ extends continuously to $\overline{\Omega}$ unless some restrictions are placed on $\partial\Omega$. As before, Φ is defined by

$$\Phi(z) = e^{G + iH} = (z - p)e^{G_0 + iH_0}, \tag{6.5}$$

where $H(z) = \operatorname{Im} \log(z - p) + H_0$. We aim to prove the following Riemann mapping theorem.

Theorem 6.1. *If Ω is a bounded, simply connected domain, then the map $\Phi : \Omega \to \mathcal{D}$ given by (6.5) is one-to-one and onto.*

Proof. Since G is continuous on $\overline{\Omega} \setminus \{p\}$, we see that

$$|\Phi| : \overline{\Omega} \longrightarrow [0, 1] \tag{6.6}$$

is continuous, hence uniformly continuous; it takes $\partial\Omega$ to $\{1\}$. Fix $\varepsilon > 0$. If $\gamma_\varepsilon \subset \Omega$ is a simple closed curve, enclosing p, which stays sufficiently close to $\partial\Omega$, then

$$\sigma_\varepsilon = \Phi(\gamma_\varepsilon) \subset \mathcal{D} \setminus \overline{\mathcal{D}}_{1-\varepsilon}, \tag{6.7}$$

where $\mathcal{D}_\rho = \{z \in \mathbb{C} : |z| < \rho\}$. By the argument principle, for any $c \in \mathcal{D}_{1-\varepsilon}$, the degree of σ_ε about c is equal to the number (counting multiplicities) of points $q_j \in \Omega_\varepsilon$ (the region enclosed by γ_ε) such that $\Phi(q_j) = c$. This winding number is independent of $c \in \mathcal{D}_{1-\varepsilon}$. But for $c = 0$, we see from (6.5) that p is the only zero of Φ, a simple zero, so the winding number is one. Thus, for all $c \in \mathcal{D}_{1-\varepsilon}$, there is a unique $q \in \Omega_\varepsilon$ such that $\Phi(q) = c$. Letting $\varepsilon \to 0$, we have the theorem.

As noted in Exercise 6 of §4, such a map Φ is essentially unique. It is called the Riemann mapping function.

The Riemann mapping function Φ does not always extend to be a homeomorphism of $\overline{\Omega}$ onto $\overline{\mathcal{D}}$; clearly a necessary condition for this is that $\partial\Omega$ be homeomorphic to S^1, that is, that it be Jordan curve. In fact, C. Caratheodory proved that this condition is also sufficient. A proof can be found in [Ts]. Here we establish a simpler result.

Proposition 6.2. *Assume $\Omega \subset \mathbb{R}^2$ is a simply connected region whose boundary $\partial\Omega$ is a finite union of smooth curves. Then the Riemann mapping function Φ extends to a homeomorphism $\Phi : \overline{\Omega} \to \overline{\mathcal{D}}$.*

Proof. Local elliptic regularity implies G_0 and hence H_0 and Φ extend smoothly to the smooth part of $\partial\Omega$. Also, an application of Zaremba's principle as in §4 shows that the smooth parts of $\partial\Omega$ are mapped diffeomorphically onto open intervals in $S^1 = \partial\mathcal{D}$. Let J_1 and J_2 be smooth curves in $\partial\Omega$, meeting at p, as illustrated in Fig. 6.1, and denote by I_ν the images in S^1, $I_\nu = \Phi(J_\nu)$. It will suffice to show that I_1 and I_2 meet, that is, the endpoints q_1 and q_2 pictured in Fig. 6.1 coincide.

Let γ_r be the intersection $\Omega \cap \{z : |z - p| = r\}$, and let $\ell(r)$ be the length of $\Phi(\gamma_r) = \sigma_r$. Clearly, $|q_1 - q_2| \le \ell(r)$ for all (small) $r > 0$, so we would like to show that $\ell(r)$ is small for (some) small r.

We have $\ell(r) = \int_{\gamma_r} |\Phi'(z)|\,|dz|$, and Cauchy's inequality implies

$$\frac{\ell(r)^2}{r} \le 2\pi \int_{\gamma_r} |\Phi'(z)|^2\, ds. \tag{6.8}$$

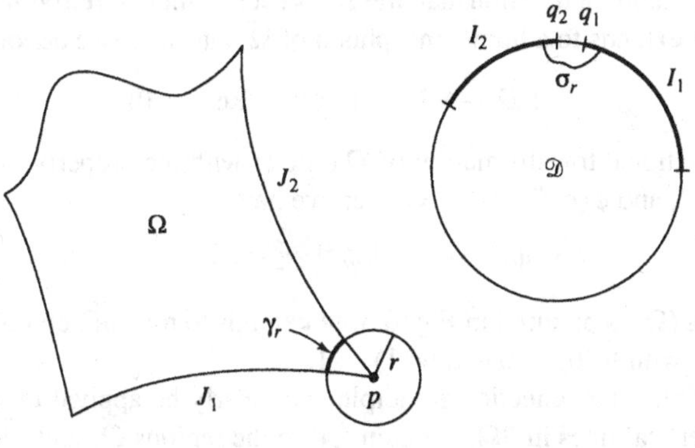

FIGURE 6.1

If $\ell(r) \geq \delta$ for $\varepsilon \leq r \leq R$, then integrating over $r \in [\varepsilon, R]$ implies

$$\delta^2 \log \frac{R}{\varepsilon} \leq 2\pi \iint\limits_{\Omega(\varepsilon, R)} |\Phi'(z)|^2 \, dx \, dy = 2\pi \cdot \text{Area}\, \Phi\big(\Omega(\varepsilon, R)\big) \leq 2\pi^2, \tag{6.9}$$

where $\Omega(\varepsilon, R) = \Omega \cap \{z : \varepsilon \leq |z - p| \leq R\}$. Since $\log(1/\varepsilon) \to \infty$ as $\varepsilon \searrow 0$, this implies that, for any $\delta > 0$, there exists arbitrarily small $r > 0$ such that $\ell(r) < \delta$. Hence $|q_1 - q_2| < \delta$, so $q_1 = q_2$, as needed to complete the proof.

We next discuss a particularly important case of the Riemann mapping function of a domain Ω whose boundary is not smooth. Namely, Ω is a subdomain of the unit disk $\mathcal{D}$, whose boundary consists of three circles, intersecting $\partial\mathcal{D}$ at right angles, at the points $\{1, e^{2\pi i/3}, e^{-2\pi i/3}\}$ (see Fig. 6.2). Denote by

$$\Psi : \Omega \longrightarrow \mathcal{D} \tag{6.10}$$

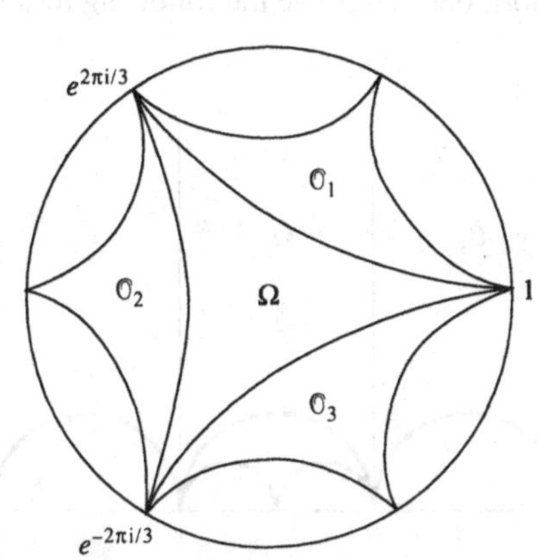

FIGURE 6.2

the Riemann mapping function that preserves each of these three points. By Proposition 6.2, Ψ extends to a homeomorphism of $\overline{\Omega}$ onto $\overline{\mathcal{D}}$. If we denote by

$$\varphi : \mathcal{D} \longrightarrow \mathcal{U} = \{z \in \mathbb{C} : \operatorname{Re} z > 0\}$$

the linear fractional transformation of $\mathcal{D}$ onto $\mathcal{U}$ with the property that $\varphi(1) = 0$, $\varphi(e^{2\pi i/3}) = 1$, and $\varphi(e^{-2\pi i/3}) = \infty$, then we have

$$\widetilde{\Psi} = \varphi \circ \Psi \circ \varphi^{-1} : \widetilde{\Omega} \longrightarrow \mathcal{U}, \tag{6.11}$$

where $\widetilde{\Omega} = \varphi(\Omega)$ is pictured in Fig. 6.3. $\widetilde{\Psi}$ extends to map $\partial\widetilde{\Omega}$ continuously onto the real axis, with $\widetilde{\Psi}(0) = 0$ and $\widetilde{\Psi}(1) = 1$.

Now the Schwartz reflection principle can clearly be applied to $\widetilde{\Psi}$, reflecting across the vertical lines in $\partial\widetilde{\Omega}$, to extend $\widetilde{\Psi}$ to the regions $\widetilde{\mathcal{O}}_2$ and $\widetilde{\mathcal{O}}_3$ in Fig. 6.3. A variant extends $\widetilde{\Psi}$ to $\widetilde{\mathcal{O}}_1$. Note that this extension maps the closure in $\mathcal{U}$ of $\widetilde{\Omega} \cup \widetilde{\mathcal{O}}_1 \cup \widetilde{\mathcal{O}}_2 \cup \widetilde{\mathcal{O}}_3$ onto $\mathbb{C} \setminus \{0, 1\}$. Now we can iterate this reflection process indefinitely, obtaining

$$\widetilde{\Psi} : \mathcal{U} \longrightarrow \mathbb{C} \setminus \{0, 1\}, \tag{6.12}$$

which is a holomorphic *covering map*. Composing on the right with φ gives the holomorphic covering map

$$\sigma = \widetilde{\Psi} \circ \varphi : \mathcal{D} \longrightarrow \mathbb{C} \setminus \{0, 1\}. \tag{6.13}$$

The existence of such a covering map is very significant. One simple application is to the following result of Picard.

Proposition 6.3. *If $u : \mathbb{C} \to \mathbb{C} \setminus \{0, 1\}$ is holomorphic, then it is constant.*

Proof. Using σ, we lift u to a holomorphic function $v : \mathbb{C} \to \mathcal{D}$, such that $u = \sigma \circ v$. But Liouville's theorem implies that v is constant.

With some more effort, one can prove the following result of Montel.

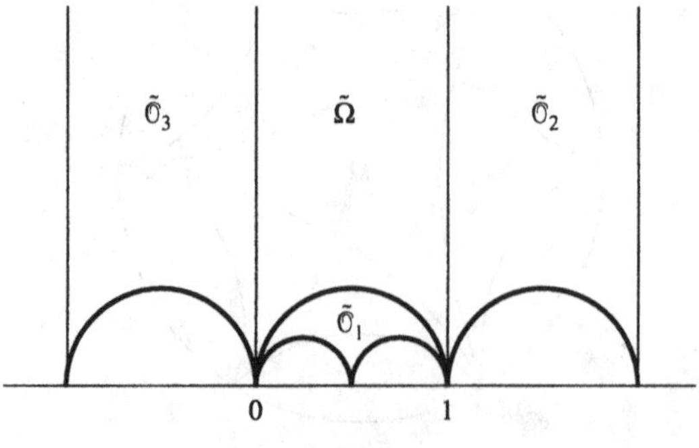

FIGURE 6.3

Proposition 6.4. *If $\mathcal{F}$ is a family of holomorphic maps $u_\alpha : \mathcal{D} \to S^2 = \mathbb{C} \cup \{\infty\}$ with range in $S^2 \setminus \{0, 1, \infty\}$, then $\mathcal{F}$ is equicontinuous.*

We leave the proof to the reader, with the comment that the trick is to make a careful choice of lifts $v_\alpha : \mathcal{D} \to \mathcal{D}$.

Exercises

1. With how little regularity of $\partial\Omega$ can you show that $G_0 \in C^1(\overline{\Omega})$? With how little regularity can you show that $H_0 \in C^1(\overline{\Omega})$? When can you show that $\Phi : \overline{\Omega} \to \overline{\mathcal{D}}$ is a C^1-diffeomorphism?
2. Extend Proposition 6.2 to the case where $\partial\Omega$ is assumed only to be a Jordan curve.
3. Let Ω be the following (unbounded) region in $\mathbb{C}$:
$$\Omega = \{z = x + iy : 0 < x < 1,\ 0 < y < x^{-1}\}.$$
Consider a Riemann mapping function $\Phi : \Omega \to \mathcal{D}$, with inverse $\Phi^{-1} : \mathcal{D} \to \Omega$. Show that $\operatorname{Re} \Phi^{-1}$ is *continuous* on $\overline{\mathcal{D}}$, while $\operatorname{Im} \Phi^{-1}$ is *unbounded.*
4. Let Ω be a simply connected, *unbounded* region in $\mathbb{C}$, with nonempty complement. Show that, given $z_0 \in \mathbb{C} \setminus \Omega$, the function $(z - z_0)^{1/2}$ can be defined on Ω as a one-to-one holomorphic map of Ω onto a domain $\mathcal{O} \subset \mathbb{C}$ whose complement has nonempty interior.
(*Hint*: If $w \in \mathcal{O}$, then $-w \notin \mathcal{O}$.)
Using this, extend the Riemann mapping theorem to all such Ω.
(*Hint*: Use an inversion to map $\mathcal{O}$ to a bounded region.)
5. State and prove an analogue of Theorem 4.3 for bounded $\Omega \subset \mathbb{C}$ of the form $\Omega = \mathcal{O}_1 \setminus \overline{\mathcal{O}}_2$, where $\mathcal{O}_j$ and $\overline{\mathcal{O}}_2$ are simply connected, and $\overline{\mathcal{O}}_2 \subset \mathcal{O}_1$, in case $\partial\Omega$ is rough.
6. In the proof of Proposition 6.2, it was stated that it sufficed to show that I_1 and I_2 must meet. Why can't they overlap?

7. The Neumann boundary problem

Let $\overline{\Omega}$ be a connected, compact manifold with nonempty smooth boundary, as in §1. We want to study the existence and regularity of solutions to the Neumann problem

$$\Delta u = f \text{ on } \Omega, \quad \frac{\partial u}{\partial \nu} = 0 \text{ on } \partial\Omega. \tag{7.1}$$

Recall that, by Green's formula, if u and v are smooth on $\overline{\Omega}$,

$$(-\Delta u, v) = (du, dv) - \int_{\partial\Omega} \overline{v}\, \frac{\partial u}{\partial \nu}\, dS. \tag{7.2}$$

By continuity, this identity holds for $u \in H^2(\Omega)$, $v \in H^1(\Omega)$. The boundary integral vanishes if $\partial u / \partial \nu = 0$ on $\partial\Omega$, so we are motivated to consider the operator

$$\mathcal{L}_N : H^1(\Omega) \longrightarrow H^1(\Omega)^* \tag{7.3}$$

defined by

$$(\mathcal{L}_N u, v) = (du, dv), \quad u, v \in H^1(\Omega). \tag{7.4}$$

The operator $\mathcal{L}_N$ is not injective, since it annihilates constants, but

$$\big((\mathcal{L}_N + 1)u, u\big) = \|du\|_{L^2}^2 + \|u\|_{L^2}^2, \tag{7.5}$$

so we have

Proposition 7.1. *The map*

$$\mathcal{L}_N + 1 : H^1(\Omega) \longrightarrow H^1(\Omega)^*$$

is one-to-one and onto.

As in §1, it is clear that the inverse map

$$T_N : H^1(\Omega)^* \longrightarrow H^1(\Omega) \tag{7.6}$$

restricts to a compact, self-adjoint operator on $L^2(\Omega)$, so there is an orthonormal basis u_j of $L^2(\Omega)$ consisting of eigenfunctions of T_N:

$$T_N u_j = \mu_j u_j, \quad \mu_j \searrow 0, \ u_j \in H^1(\Omega). \tag{7.7}$$

It follows that

$$\mathcal{L}_N u_j = \lambda u_j, \quad \lambda_j = \frac{1}{\mu_j} - 1 \nearrow \infty. \tag{7.8}$$

Note that since (7.4) is equal to $(-\Delta u, v)$ for $u \in H^1(\Omega)$, $v \in C_0^\infty(\Omega)$, we have

$$\iota(\mathcal{L}_N u) = -\Delta u \ \text{ in } \mathcal{D}'(\Omega), \tag{7.9}$$

for $u \in H^1(\Omega)$, where $\iota : H^1(\Omega)^* \to \mathcal{D}'(\Omega)$ is the adjoint of the inclusion $C_0^\infty(\Omega) \hookrightarrow H^1(\Omega)$, but ι is not injective. Nevertheless, (7.8) implies that, in the distributional sense, the eigenvectors u_j satisfy

$$\Delta u_j = -\lambda_j u_j \ \text{ on } \Omega. \tag{7.10}$$

We will establish regularity theorems that imply that each u_j belongs to $C^\infty(\overline{\Omega})$ and satisfies the Neumann boundary condition. The proof of such regularity results is just slightly more elaborate than the proof of Theorem 1.3. We divide it into two parts.

Proposition 7.2. *Given $f \in L^2(\Omega)$, $u = T_N f$ satisfies*

$$u \in H^2(\Omega), \quad \frac{\partial u}{\partial \nu}\bigg|_{\partial\Omega} = 0, \tag{7.11}$$

and

$$(-\Delta + 1)u = f. \tag{7.12}$$

Furthermore, we have the estimate

$$\|u\|_{H^2}^2 \le C\|\Delta u\|_{L^2}^2 + C\|u\|_{H^1}^2, \tag{7.13}$$

for all u satisfying (7.11).

Proof. First we establish the estimate

$$\|u\|_{H^1}^2 \le C\|\mathcal{L}_N u\|_{H^{1*}}^2 + C\|u\|_{L^2}^2. \tag{7.14}$$

Indeed, by (7.4) and Cauchy's inequality, we have

$$\begin{aligned} \|u\|_{H^1}^2 &= (\mathcal{L}_N u, u) + \|u\|_{L^2}^2 \\ &\le \|\mathcal{L}_N u\|_{H^{1*}} \cdot \|u\|_{H^1} + \|u\|_{L^2}^2 \\ &\le \frac{1}{2}\|u\|_{H^1}^2 + \frac{1}{2}\|\mathcal{L}_N u\|_{H^{1*}}^2 + \|u\|_{L^2}^2, \end{aligned} \tag{7.15}$$

which readily gives (7.14). To proceed, we localize to coordinate patches, as in the proof of Theorem 1.3. Suppose $\chi \in C^\infty(\overline{\Omega})$ is supported in a coordinate patch, and either $\chi \in C_0^\infty(\Omega)$ or $\partial\chi/\partial\nu = 0$ on $\partial\Omega$. We need to analyze the commutator $[\mathcal{L}_N, M_\chi]$, where $M_\chi\, u = \chi u$. Note that M_χ acts continuously on $H^1(\Omega)$ and on $H^1(\Omega)^*$. For $u, v \in H^1(\Omega)$,

$$(\mathcal{L}_N M_\chi u, v) = (d(\chi u), dv) = ((d\chi)u, dv) + (\chi du, dv), \tag{7.16}$$

while

$$(M_\chi \mathcal{L}_N u, v) = (\mathcal{L}_N u, \chi v) = (du, (d\chi)v) + (du, \chi dv), \tag{7.17}$$

so

$$([\mathcal{L}_N, M_\chi]u, v) = ((d\chi)u, dv) - (\langle d\chi, du\rangle, v). \tag{7.18}$$

We can integrate the first term on the right by parts, using formula (9.17) of Chapter 2, extended to $u, v \in H^1(\Omega)$. The boundary integral is

$$\int_{\partial\Omega} \nu \cdot (d\chi)u\overline{v}\, dS = 0,$$

by the hypothesis $\partial\chi/\partial\nu = 0$ on $\partial\Omega$, so we have

$$[\mathcal{L}_N, M_\chi]u = d^*((d\chi)u) - \langle d\chi, du\rangle = -(\Delta\chi)u - 2\langle d\chi, du\rangle, \tag{7.19}$$

for $u \in H^1(\Omega)$, in view of the identity

$$d^*(u\alpha) = ud^*\alpha - \langle du, \alpha\rangle \tag{7.20}$$

when u is a scalar function and α a 1-form. (Compare formula (2.19) of Chapter 2, and also Exercise 7 in §10 of Chapter 2.) In particular,

$$[\mathcal{L}_N, M_\chi] : H^1(\Omega) \longrightarrow L^2(\Omega). \tag{7.21}$$

Consequently, it suffices to prove Proposition 7.2 for u supported in a coordinate patch.

We proceed by applying the estimate (7.14) to $D_{j,h}u$, as in the proof of Theorem 1.3, where $1 \le j \le n-1$ if u is supported in a coordinate patch with boundary.

Of course, interior regularity results proved in §1 apply here. We have

$$\begin{aligned}(7.22)\qquad \|D_{j,h}u\|_{H^1}^2 &\le C\|\mathcal{L}_N D_{j,h}u\|_{H^{1*}}^2 + C\|D_{j,h}u\|_{L^2}^2 \\ &\le C\|D_{j,h}\mathcal{L}_N u\|_{H^{1*}}^2 + C\|[\mathcal{L}_N, D_{j,h}]u\|_{H^{1*}}^2 + C\|D_{j,h}u\|_{L^2}^2,\end{aligned}$$

where $D_{j,h}$ is defined on H^{1*} in a natural fashion by duality. We need to estimate $[\mathcal{L}_N, D_{j,h}]u$. We have

$$(7.23)\qquad (\mathcal{L}_N D_{j,h}u, v) = (dD_{j,h}u, dv) = (D_{j,h}^{(1)}du, dv),$$

where, if translation $x \mapsto x + he_j$ is denoted $\tau_{j,h}$, we define $D_{j,h}^{(1)}$ on 1-forms as

$$(7.24)\qquad D_{j,h}^{(1)}\varphi = h^{-1}(\tau_{j,h}^*\varphi - \varphi).$$

In order to analyze $(D_{j,h}\mathcal{L}_N u, v)$, we simplify the calculation by requiring that the coordinate map of a piece of Ω to a part of $\mathbb{R}^n_+$ preserve volume elements, which is easily arranged. Then the adjoint of $D_{j,h}$ is $D_{j,-h}$, so

$$(7.25)\qquad (D_{j,h}\mathcal{L}_N u, v) = (\mathcal{L}_N u, D_{j,-h}v) = (du, D_{j,-h}^{(1)}\,dv),$$

and hence

$$(7.26)\qquad ([\mathcal{L}_N, D_{j,h}]u, v) = ([D_{j,-h}^{(1)*} - D_{j,h}^{(1)}]\,du, dv).$$

We have a uniform bound on the right side of (7.26):

Lemma 7.3. *If β is a 1-form on Ω,*

$$\|[D_{j,-h}^{(1)*} - D_{j,h}^{(1)}]\beta\|_{L^2} \le C\|\beta\|_{L^2}.$$

Proof. This is similar to Lemma 1.4, and we leave it as an exercise.

We note that, as $h \to 0$, $D_{j,h}^{(1)}\varphi$ tends to the Lie derivative $\mathcal{L}_{\partial_j}\varphi$, for a 1-form φ. Thus the uniform estimate is related to the fact that the difference between $\mathcal{L}_{\partial_j}$ and $-\mathcal{L}_{\partial_j}^*$ is a zero-order operator. Compare with formula (3.43) of Chapter 2.

Applying the lemma to (7.26), we have

$$(7.27)\qquad \|[\mathcal{L}_N, D_{j,h}]u\|_{H^{1*}} \le C\|du\|_{L^2}.$$

Hence, (7.22) yields

$$(7.28)\qquad \|D_{j,h}u\|_{H^1}^2 \le C\|D_{j,h}\mathcal{L}_N u\|_{H^{1*}}^2 + C\|u\|_{H^1}^2.$$

Given $\mathcal{L}_N u = f_1 \in L^2(\Omega)$, we have $(D_{j,h}f_1, v) = (f_1, D_{j,-h}v)$ for $v \in H^1(\Omega)$, and hence

$$(7.29)\qquad \|D_{j,h}f_1\|_{H^{1*}}^2 \le C\|f_1\|_{L^2}^2,$$

so we get

$$(7.30)\qquad \|D_{j,h}u\|_{H^1}^2 \le C\|f_1\|_{L^2}^2 + C\|u\|_{H^1}^2,$$

provided $\mathcal{L}_N u = f_1 \in L^2$. Letting $h \to 0$, we get $D_j u \in H^1(\Omega)$, and an accompanying norm estimate.

As in §1, the rest of the proof that $u \in H^2(\Omega)$ comes down to showing $D_n^2 u \in L^2(\Omega)$. But by (7.9) we have $\Delta u = -f_1$ in the distributional sense, and

$$g^{nn}(x) D_n^2 u = -f_1 - \sum_{(j,k)\neq(n,n)} g^{jk}(x) D_j D_k u - \sum b^j(x) D_j u, \tag{7.31}$$

so the proof that $u \in H^2(\Omega)$ is complete.

It remains to show that u satisfies the Neumann boundary condition. However, for $u = T_N f \in H^2(\Omega)$, $v \in H^1(\Omega)$, the identity (7.2) holds, so

$$\begin{aligned}(f, v) &= (\mathcal{L}_N u, v) + (u, v) = (du, dv) + (u, v) \\ &= ((-\Delta + 1)u, v) + \int_{\partial\Omega} \overline{v} \, \frac{\partial u}{\partial \nu} \, dS \\ &= (f, v) + \int_{\partial\Omega} \overline{v} \, \frac{\partial u}{\partial \nu} \, dS.\end{aligned} \tag{7.32}$$

Since this holds for all $v \in H^1(\Omega)$, this forces $\partial u/\partial\nu$ to vanish on $\partial\Omega$. The proof of Proposition 7.2 is complete.

To complete the parallel with Theorem 1.3, we have the following.

Proposition 7.4. *For $k = 1, 2, 3, \ldots$, given $f_1 \in H^k(\Omega)$, a function $u \in H^{k+1}(\Omega)$ satisfying*

$$\Delta u = f_1 \text{ on } \Omega, \quad \frac{\partial u}{\partial \nu} = 0 \text{ on } \partial\Omega \tag{7.33}$$

belongs to $H^{k+2}(\Omega)$, and we have an estimate

$$\|u\|^2_{H^{k+2}} \leq C\|\Delta u\|^2_{H^k} + C\|u\|^2_{H^{k+1}}, \tag{7.34}$$

for all $u \in H^{k+2}(\Omega)$ such that $\partial u/\partial\nu = 0$ on $\partial\Omega$.

Proof. We proceed from Proposition 7.2 inductively, using cut-offs χ and difference operators $D_{j,h}$, as in the proof of Theorem 1.3. We need to require $\partial\chi/\partial\nu = 0$ on $\partial\Omega$, so χu satisfies the Neumann boundary condition. In order for $D_{j,h} u$ to satisfy the Neumann boundary condition, this time pick coordinate charts so that the normal ν to $\partial\Omega$ is mapped to $\partial/\partial x_n$. Then the proof works out just as in Theorem 1.3.

One can also analyze nonhomogeneous boundary problems, such as

$$(-\Delta + 1)u = f \text{ in } \Omega, \quad \frac{\partial u}{\partial \nu} = g \text{ on } \partial\Omega. \tag{7.35}$$

Given $g \in H^{k+1/2}(\partial\Omega)$, $k = 0, 1, 2, \dots$, you can pick $h \in H^{k+2}(\Omega)$ such that $\partial h/\partial\nu = g$ on $\partial\Omega$, and then write $u = v + h$, where v solves

$$(7.36) \qquad (-\Delta + 1)v = f + (\Delta - 1)h \text{ in } \Omega, \quad \frac{\partial v}{\partial \nu} = 0 \text{ on } \partial\Omega.$$

Then $v \in H^{k+2}(\Omega)$ if $f \in H^k(\Omega)$, so also $u \in H^{k+2}(\Omega)$, and one has the estimate

$$(7.37) \qquad \|u\|^2_{H^{k+2}(\Omega)} \leq C\|\Delta u\|^2_{H^k(\Omega)} + C\left\|\frac{\partial u}{\partial \nu}\right\|^2_{H^{k+1/2}(\partial\Omega)} + C\|u\|^2_{H^{k+1}(\Omega)},$$

valid for all $u \in H^{k+2}$. Let us formally record this as the following generalization of Proposition 7.4.

Proposition 7.5. *For $k = 0, 1, 2, \dots$, given $f \in H^k(\Omega)$, $g \in H^{k+1/2}(\partial\Omega)$, there is a unique solution $u \in H^{k+2}(\Omega)$ to (7.35), and the estimate (7.37) holds.*

We note that to prove this result, one could bypass Proposition 7.4 and proceed as follows. For $k = 0$, the construction (7.36) gets the result as a consequence of Proposition 7.2. Then you can proceed by induction on k, using cut-offs and difference operators as in the proof of Theorem 1.3. The (slight) advantage of doing this is that one does not need to preserve the homogeneous boundary condition, so there is no need to arrange $\partial\chi/\partial\nu = 0$ on $\partial\Omega$ or use coordinate charts mapping ν to $\partial/\partial x_n$. In the case of more elaborate boundary conditions, such as considered in §9, the flexibility gained by this sort of strategy will be of greater importance.

Returning to the original Neumann boundary problem (7.1), we see that the fact that 0 is an eigenvalue in (7.8), with eigenspace consisting of constants, implies

Proposition 7.6. *Given $f \in L^2(\Omega)$, the boundary problem (7.1) has a solution $u \in H^2(\Omega)$ if and only if*

$$(7.38) \qquad \int_\Omega f(x)\, dV(x) = 0.$$

Provided this condition holds, the solution u is unique up to an additive constant and belongs to $H^{k+2}(\Omega)$ if $f \in H^k(\Omega)$, $k \geq 0$.

We have an extension of this for the nonhomogeneous boundary problem

$$(7.39) \qquad \Delta u = f \text{ on } \Omega, \quad \frac{\partial u}{\partial \nu} = g \text{ on } \partial\Omega.$$

Note that if we set $v = 1$ in (7.2), we get

$$(7.40) \qquad \int_\Omega \Delta u(x)\, dV(x) = \int_{\partial\Omega} \frac{\partial u}{\partial \nu}\, dS.$$

Thus a necessary condition for (7.39) to have a solution is

$$(7.41) \qquad \int_\Omega f(x)\, dV(x) = \int_{\partial\Omega} g(x)\, dS.$$

This condition is also sufficient.

Proposition 7.7. *If $k \geq 0$, $f \in H^k(\Omega)$, and $g \in H^{k+\frac{1}{2}}(\partial\Omega)$, then (7.39) has a solution $u \in H^{k+2}(\Omega)$ if and only if (7.41) holds.*

Proof. Define the linear operator

$$T : H^{k+2}(\Omega) \longrightarrow H^k(\Omega) \oplus H^{k+1/2}(\partial\Omega), \tag{7.42}$$

$$\mathcal{T}u = \left(\Delta u, \frac{\partial u}{\partial \nu}\right). \tag{7.43}$$

The estimate (7.37) implies that $\mathcal{T}$ has closed range, by Proposition 6.7 in Appendix A. We know that the kernel of $\mathcal{T}$ consists of constants. The identity (7.41) implies that

$$(-1, 1) \in C^\infty(\overline{\Omega}) \oplus C^\infty(\partial\Omega) \tag{7.44}$$

is orthogonal to the range $\mathcal{R}(\mathcal{T})$. It remains to show that this is all of the orthogonal complement of $\mathcal{R}(\mathcal{T})$, which follows if we show that $\mathcal{T}$ in (7.42) is Fredholm of *index zero*. Now $\mathcal{T}$ differs from

$$\mathcal{T}^\# : H^{k+2}(\Omega) \longrightarrow H^k(\Omega) \oplus H^{k+1/2}(\partial\Omega), \quad \mathcal{T}^\# u = \left((\Delta - 1)u, \frac{\partial u}{\partial \nu}\right) \tag{7.45}$$

by the operator $\mathcal{K}u = (-u, 0)$, which is *compact*, by Rellich's theorem. Proposition 7.5 implies that $\mathcal{T}^\#$ is an isomorphism, and by Corollary 7.5 of Appendix A, this implies that $\mathcal{T}$ is Fredholm of index zero. This completes the proof of Proposition 7.7.

Exercises

1. Given two Riemannian manifolds M and N with boundary, $p \in \partial M$, and $q \in \partial N$, show that there exists a diffeomorphism Φ from a neighborhood of p to a neighborhood of q, $\Phi(p) = q$, which preserves volumes. (*Hint*: Set up a first order PDE for one component of Φ.)
   Use this to justify (7.25). Show that Φ can also be arranged to preserve unit normals to the boundaries.
2. Give a detailed proof of Lemma 7.3.
3. If $\overline{\Omega}$ is a compact Riemannian manifold with boundary, show that the *Robin* boundary condition

$$\frac{\partial u}{\partial \nu} = a(x)u(x), \quad \text{for } x \in \partial\Omega,$$

   given $a \in C^\infty(\partial\Omega)$, has the regularity properties established in this section for the Neumann condition (which is the $a = 0$ case). (*Hint*: Make use of (7.37).)
   If a is real-valued, show that Δ is self-adjoint on $L^2(\Omega)$, with domain

$$\mathcal{D}(\Delta) = \{u \in H^2(\Omega) : \partial_\nu u = a(x)u \text{ on } \partial\Omega\}.$$

   Reconsider this problem when reading §12.

4. Let $\Omega \subset \mathbb{R}^n$ be bounded, but do not assume $\partial\Omega$ is smooth. Note that the map T_N in (7.6) is well defined. Assume there exist smoothly bounded $\Omega_j \nearrow \Omega$ satisfying the following hypotheses:

 (i) There exist extension maps $E_j : H^1(\Omega_j) \to H^1(\Omega)$ of uniformly bounded norm.
 (ii) The inclusion $H^1(\Omega) \hookrightarrow L^2(\Omega)$ is compact.
 (iii) $\text{Meas}(\Omega \setminus \Omega_j) \to 0$.

 Then show that if $f \in L^2(\Omega)$, $f_j = f\big|_{\Omega_j}$, we have

$$T_{Nj} f_j \longrightarrow T_N f \ \text{ in } \ L^2(\Omega),$$

 where T_{Nj} is as in (7.6), with Ω replaced by Ω_j.
 More information on this type of problem is given in [RT].

8. The Hodge decomposition and harmonic forms

Let M be a compact Riemannian manifold, without boundary. Recall from Chapter 2 the Hodge Laplacian on k-forms,

$$\Delta : C^\infty(M, \Lambda^k) \longrightarrow C^\infty(M, \Lambda^k), \tag{8.1}$$

defined by

$$-\Delta = (d + \delta)^2 = d\delta + \delta d, \tag{8.2}$$

where d is the exterior derivative operator and δ its formal adjoint, satisfying

$$(du, v) = (u, \delta v), \tag{8.3}$$

for a smooth k-form u and $(k+1)$-form v; $\delta = 0$ on 0-forms. The local coordinate expression

$$\Delta u = g^{j\ell}(x)\, \partial_j \partial_\ell u + Y_j u, \tag{8.4}$$

where Y_k are first-order differential operators, derived in (10.23) of Chapter 2, indicates that the Hodge Laplacian on k-forms is amenable to an analysis similar to that for the Laplace operator on functions in §1. Note that, for smooth k-forms, by (8.3),

$$-(\Delta u, v) = (du, dv) + (\delta u, \delta v). \tag{8.5}$$

Now we have Δ operating on Sobolev spaces; in particular,

$$\Delta : H^1(M, \Lambda^k) \longrightarrow H^{-1}(M, \Lambda^k), \tag{8.6}$$

and (8.5) holds for $u, v \in H^1(M, \Lambda^k)$. We want to study invertibility of the operator $-\Delta + C_1$, where C_1 is a convenient positive constant, and to produce consequences of this. Our first result is the following analogue of the estimates (1.5), (1.49), and (7.15).

Proposition 8.1. *There exist positive constants C_0 and C_1 such that*

$$-(\Delta u, u) \geq C_0 \|u\|_{H^1}^2 - C_1 \|u\|_{L^2}^2 \tag{8.7}$$

for a k-form $u \in H^1$.

Proof. Cover M with coordinate patches U_j, and pick $\varphi_j \in C_0^\infty(U_j)$ such that $\sum \varphi_j^2 = 1$, so

$$-(\Delta u, u) = -\sum_j (\Delta(\varphi_j^2 u), u) \tag{8.8}$$
$$= -\sum_j (\Delta(\varphi_j u), \varphi_j u) + (Yu, u),$$

where Y is a first-order differential operator, $Y = \sum [\Delta, \varphi_j]$. The local coordinate formula (8.4) and integration by parts yield

$$-(\Delta(\varphi_j u), \varphi_j u) \geq C\|\varphi_j u\|_{H^1}^2 - C'\|\varphi_j u\|_{L^2}^2, \tag{8.9}$$

and summing gives

$$-(\Delta u, u) \geq C_2\|u\|_{H^1}^2 - C_3\|u\|_{L^2}^2 - C_4\|Yu\|_{L^2}\|u\|_{L^2}. \tag{8.10}$$

We can dominate the last term in (8.10) by $\varepsilon\|u\|_{H^1}^2 + (C/\varepsilon)\|u\|_{L^2}^2$, and absorb $\varepsilon\|u\|_{H^1}^2$ into the first term on the right side of (8.10), to prove (8.7).

From here, a number of results follow just as in §§1 and 7. We have the estimate $\|(-\Delta + C_1)u\|_{H^{-1}} \geq C_0\|u\|_{H^1}$, and hence

$$-\Delta + C_1 : H^1(M, \Lambda^k) \longrightarrow H^{-1}(M, \Lambda^k) \tag{8.11}$$

is injective with closed range. The annihilator of the range, in $H^{-1*} = H^1$, belongs to the kernel of $-\Delta + C_1$, and so is zero, so the map (8.11) is bijective. We have a two-sided inverse

$$T : H^{-1}(M, \Lambda^k) \longrightarrow H^1(M, \Lambda^k). \tag{8.12}$$

As in (1.8), $T = T^*$, and by Rellich's theorem T is a compact self adjoint operator on $L^2(M, \Lambda^k)$. The identity (8.5) implies

$$0 < (Tu, u) \leq C_1^{-1}\|u\|_{L^2}^2, \tag{8.13}$$

for nonzero u. The space $L^2(M, \Lambda^k)$ has an orthonormal basis $u_j^{(k)}$ consisting of eigenfunctions of T:

$$Tu_j^{(k)} = \mu_j^{(k)} u_j^{(k)}; \quad u_j^{(k)} \in H^1(M, \Lambda^k). \tag{8.14}$$

By (8.13), we have

$$0 < \mu_j^{(k)} \leq C_1^{-1}. \tag{8.15}$$

For each k, we can order the $u_j^{(k)}$ so that $\mu_j^{(k)} \searrow 0$, as $j \nearrow \infty$. It follows that

$$-\Delta u_j^{(k)} = \lambda_j^{(k)} u_j^{(k)}, \tag{8.16}$$

with

$$\lambda_j^{(k)} = \frac{1}{\mu_j^{(k)}} - C_1, \tag{8.17}$$

so

$$\lambda_j^{(k)} \geq 0, \quad \lambda_j^{(k)} \nearrow \infty \text{ as } j \to \infty. \tag{8.18}$$

The local regularity results proved in Theorem 1.3 apply to Δ, by (8.4), and since M has no boundary, we conclude that

$$u_j^{(k)} \in C^\infty(M, \Lambda^k). \tag{8.19}$$

In particular, the 0-eigenspace of Δ on k-forms is finite-dimensional and consists of smooth k-forms. These are called *harmonic forms*. We denote this 0-eigenspace by $\mathcal{H}_k$. By (8.5), we see that

$$u \in \mathcal{H}_k \Longleftrightarrow u \in C^\infty(M, \Lambda^k), \ du = 0, \text{ and } \delta u = 0 \text{ on } M. \tag{8.20}$$

Denote by P_k the orthogonal projection of $L^2(M, \Lambda^k)$ onto $\mathcal{H}_k$. We also define a continuous linear map

$$G : L^2(M, \Lambda^k) \longrightarrow L^2(M, \Lambda^k) \tag{8.21}$$

by

$$Gu_j^{(k)} = \begin{cases} 0 & \text{if } \lambda_j^{(k)} = 0, \\ \dfrac{1}{\lambda_j^{(k)}} u_j^{(k)} & \text{if } \lambda_j^{(k)} > 0. \end{cases} \tag{8.22}$$

Hence $-\Delta G u_j^{(k)} = (I - P_k) u_j^{(k)}$. Since $\Delta : L^2(M, \Lambda^k) \to H^{-2}(M, \Lambda^k)$ continuously, it follows that

$$-\Delta G u = (I - P_k) u, \quad \text{for } u \in L^2(M, \Lambda^k). \tag{8.23}$$

Now the local regularity implies

$$G : L^2(M, \Lambda^k) \longrightarrow H^2(M, \Lambda^k), \tag{8.24}$$

and more generally

$$G : H^j(M, \Lambda^k) \longrightarrow H^{j+2}(M, \Lambda^k), \tag{8.25}$$

for $j \geq 0$. Using (8.2), we write (8.23) in the following form, known as the *Hodge decomposition*.

Proposition 8.2. *Given $u \in H^j(M, \Lambda^k)$, we have*

$$u = d\delta G u + \delta d G u + P_k u. \tag{8.26}$$

The three terms on the right are mutually orthogonal in $L^2(M, \Lambda^k)$.

Proof. Only the orthogonality remains to be established. But if $u \in H^1(M, \Lambda^{k-1})$ and $v \in H^1(M, \Lambda^{k+1})$, then

$$(du, \delta v) = (d^2 u, v) = 0, \tag{8.27}$$

and if $w \in \mathcal{H}_k$, so $dw = \delta w = 0$, we have

$$(du, w) = (u, \delta w) = 0 \ \text{ and } \ (\delta v, w) = (v, dw) = 0, \tag{8.28}$$

so the orthogonality is established.

A smooth k-form u is said to be *exact* if $u = dv$ for some smooth $(k-1)$-form v, and *closed* if $du = 0$. Since $d^2 = 0$, every exact form is closed:

$$\mathcal{E}^k(M) \subset \mathcal{C}^k(M), \tag{8.29}$$

where $\mathcal{E}^k(M)$ and $\mathcal{C}^k(M)$ respectively denote the spaces of exact and closed k-forms. Similarly, a k-form u is said to be *co-exact* if $u = \delta v$ for some smooth $(k+1)$-form v, and *co-closed* if $\delta u = 0$, and since $\delta^2 = 0$ we have

$$\mathcal{CE}^k(M) \subset \mathcal{CC}^k(M), \tag{8.30}$$

with obvious notation. The *deRham cohomology* groups are defined as quotient spaces:

$$\mathcal{H}^k(M) = \mathcal{C}^k(M)/\mathcal{E}^k(M). \tag{8.31}$$

The following is one of the most important consequences of the Hodge decomposition (8.26).

Proposition 8.3. *If M is a compact Riemannian manifold, there is a natural isomorphism*

$$\mathcal{H}^k(M) \approx \mathcal{H}_k. \tag{8.32}$$

Proof. Since every harmonic form is closed, there is an injection

$$j : \mathcal{H}_k \hookrightarrow \mathcal{C}^k(M), \tag{8.33}$$

which hence gives rise to a natural map

$$J : \mathcal{H}_k \longrightarrow \mathcal{H}^k(M), \tag{8.34}$$

by passing to the quotient (8.31). It remains to show that J is bijective. The orthogonality (8.28) shows that

$$(\text{Image } j) \cap \mathcal{E}^k(M) = 0,$$

so J is injective. Also (8.28) shows that if $u \in \mathcal{C}^k(M)$, then $\delta d G u = 0$ in (8.26), so $u = d\delta G u + P_k u$, or $u = P_k u \bmod \mathcal{E}^k(M)$. Hence J is surjective, and the proof is complete.

Clearly, the space $\mathcal{H}^k(M)$ is independent of the Riemannian metric chosen for M. Thus the *dimension* of the space $\mathcal{H}_k$ of harmonic k-forms is independent of the metric. Indeed, since the isomorphism (8.32) is natural, we can say the following. Given two Riemannian metrics g and g' for M, with associated spaces $\mathcal{H}_k$ and $\mathcal{H}'_k$ of harmonic k-forms, there is a natural isomorphism $\mathcal{H}_k \approx \mathcal{H}'_k$. Otherwise said, each $u \in \mathcal{H}_k$ is cohomologous to a unique $u' \in \mathcal{H}'_k$.

An important theorem of deRham states that $\mathcal{H}^k(M)$, defined by (8.31), is isomorphic to a certain singular cohomology group. A variant is an isomorphism of $\mathcal{H}^k(M, \mathbb{R})$ with a certain Cech cohomology group. We refer to [SiT] and [BoT] for material on this.

We now introduce the *Hodge star operator*

$$* : C^\infty(M, \Lambda^k) \longrightarrow C^\infty(M, \Lambda^{m-k}) \quad (m = \dim M), \tag{8.35}$$

in fact, a bundle map

$$* : \Lambda^k T_x^* \longrightarrow \Lambda^{m-k} T_x^*,$$

which will be seen to relate δ to d. For (8.35) to be defined, we need to assume M is an *oriented* Riemannian manifold, so there is a distinguished volume form

$$\omega \in C^\infty(M, \Lambda^m). \tag{8.36}$$

Then the star operator (8.35) is uniquely specified by the relation

$$u \wedge *v = \langle u, v \rangle \omega, \tag{8.37}$$

where $\langle u, v \rangle$ is the inner product on $\Lambda^k T_x^*$, which was defined by (10.3) of Chapter 2. In particular, it follows that $*1 = \omega$. Furthermore, if $\{e_1, \dots, e_m\}$ is an oriented, orthonormal basis of $T_x^* M$, we have

$$*(e_{j_1} \wedge \cdots \wedge e_{j_k}) = (\operatorname{sgn} \pi)\, e_{\ell_1} \wedge \cdots \wedge e_{\ell_{m-k}}, \tag{8.38}$$

where $\{j_1, \dots, j_k, \ell_1, \dots, \ell_{m-k}\} = \{1, \dots, m\}$, and π is the permutation mapping the one ordered set to the other. It follows that

$$** = (-1)^{k(m-k)} \text{ on } \Lambda^k(M), \tag{8.39}$$

where, for short, we are denoting $C^\infty(M, \Lambda^k)$ by $\Lambda^k(M)$. We denote (8.39) by $\overline{w}$, and also set

$$w = (-1)^k \text{ on } \Lambda^k(M), \tag{8.40}$$

so

$$d(u \wedge v) = du \wedge v + w(u) \wedge dv. \tag{8.41}$$

It follows that if $u \in \Lambda^{k-1}(M)$, $v \in \Lambda^k(M)$, then $w(u) \wedge d * v = -u \wedge d * w(v)$, so

$$\begin{aligned} d(u \wedge *v) &= du \wedge *v - u \wedge d * w(v) \\ &= du \wedge *v - u \wedge *\overline{w} * d * w(v), \end{aligned} \tag{8.42}$$

since $*\overline{w}* = \text{id.}$, by (8.39). Integrating over M, since $\partial M = \emptyset$, we have, by Stokes' formula, $\int_M d(u \wedge *v) = 0$ and hence

$$(8.43) \quad (du, v) = \int_M du \wedge *v = \int_M u \wedge *\overline{w} * d * w(v) = (u, \overline{w} * d * w(v)).$$

In other words,

$$(8.44) \qquad \delta = \overline{w} * d * w = (-1)^{k(m-k)-m+k-1} * d * \quad \text{on } \Lambda^k(M).$$

Thus, by the characterization (8.20) of harmonic k-forms, we have

$$(8.45) \qquad * : \mathcal{H}_k \longrightarrow \mathcal{H}_{m-k},$$

and, by (8.39), this map is an isomorphism. In view of Proposition 8.3, we have the following special case of *Poincaré duality*.

Corollary 8.4. *If M is a compact, oriented Riemannian manifold, there is an isomorphism of deRham cohomology groups*

$$(8.46) \qquad \mathcal{H}^k(M) \approx \mathcal{H}^{m-k}(M).$$

As a further application of the Hodge decomposition, we prove the following result on the deRham cohomology groups of a Cartesian product $M \times N$ of two compact manifolds, a special case of the Kunneth formula.

Proposition 8.5. *If M and N are compact manifolds, of dimension m and n respectively, then, for $0 \le k \le m + n$,*

$$(8.47) \qquad \mathcal{H}^k(M \times N) \approx \bigoplus_{i+j=k} \Big[\mathcal{H}^i(M) \otimes \mathcal{H}^j(N)\Big].$$

Proof. Endow M and N with Riemannian metrics, and give $M \times N$ the product metric. If $\{u_\mu^{(i)}\}$ is an orthonormal basis of $L^2(M, \Lambda^i)$ and $\{v_\nu^{(j)}\}$ is an orthonormal basis of $L^2(N, \Lambda^j)$, each consisting of eigenfunctions of the Hodge Laplace operator, then the collection $\{u_\mu^{(i)} \wedge v_\nu^{(j)} : i + j = k\}$ is an orthonormal basis of $L^2(M \times N, \Lambda^k)$, consisting of eigenfunctions of the Hodge Laplacian, and since all these Laplace operators are negative-semidefinite, we have the isomorphism

$$(8.48) \qquad \mathcal{H}_k(M \times N) \approx \bigoplus_{i+j=k} \Big[\mathcal{H}_i(M) \otimes \mathcal{H}_j(N)\Big],$$

where $\mathcal{H}_i(M)$ denotes the space of harmonic i-forms on M, etc., and by (8.32) this proves the proposition.

We define the ith *Betti number* of M to be

$$(8.49) \qquad b_i(M) = \dim \mathcal{H}^i(M).$$

Thus, (8.47) implies the identity

$$b_k(M \times N) = \sum_{i+j=k} b_i(M) b_j(N). \tag{8.50}$$

This identity has an application to the *Euler characteristic* of a product. The Euler characteristic of M is defined by

$$\chi(M) = \sum_{i=0}^{m} (-1)^i \, b_i(M), \tag{8.51}$$

where $m = \dim M$. From (8.50) follows directly the product formula

$$\chi(M \times N) = \chi(M)\chi(N). \tag{8.52}$$

Exercises

1. Let $\alpha \in \Lambda^1(M^n)$, $\beta \in \Lambda^k(M^n)$. Show that

$$*(\iota_\alpha \beta) = \pm \alpha \wedge *\beta. \tag{8.53}$$

   Find the sign. (*Hint*: Start with the identity $\sigma \wedge \alpha \wedge *\beta = \langle \sigma \wedge \alpha, \beta \rangle \omega$, given $\sigma \in \Lambda^{k-1}(M)$.)
   Alternative: Show $*\delta = \pm d*$, which implies (8.53) by passing to symbols.
2. Show that if X is a smooth vector field on M, and $\beta \in \Lambda^k(M^n)$, then

$$\nabla_X(*\beta) = *(\nabla_X \beta).$$

3. Show that if $F : M \to M$ is an isometry that preserves orientation, then $F^*(*\beta) = *(F^*\beta)$.
4. If $f : M \to N$ is a smooth map between compact manifolds, show that the pull-back $f^* : \Lambda^k(N) \to \Lambda^k(M)$ induces a homomorphism $f^* : \mathcal{H}^k(N) \to \mathcal{H}^k(M)$. If f_t, $0 \le t \le 1$, is a smooth family of such maps, show that $f_0^* = f_1^*$ on $\mathcal{H}^k(N)$.
   (*Hint*: For the latter, recall formulas (13.60)–(13.64) of Chapter 1.)
5. If M is compact, connected, and oriented, and $\dim M = n$, show that

$$\mathcal{H}^0(M) \approx \mathcal{H}^n(M) \approx \mathbb{R}.$$

   Relate this to Proposition 19.5 of Chapter 1.

In Exercises 6–8, let G be a compact, connected Lie group, endowed with a bi-invariant Riemannian metric. For each $g \in G$, there are left and right translations $L_g(h) = gh$, $R_g(h) = hg$. Let $\mathcal{B}_k$ denote the space of bi-invariant k-forms on G,

$$\mathcal{B}_k = \{\beta \in \Lambda^k(G) : R_g^*\beta = \beta = L_g^*\beta \text{ for all } g \in G\}. \tag{8.54}$$

6. Show that every harmonic k-form on G belongs to $\mathcal{B}_k$. (*Hint*: If $\beta \in \mathcal{H}_k$, show $R_g^*\beta$ and $L_g^*\beta$ are both harmonic and cohomologous to β.)
7. Show that every $\beta \in \mathcal{B}_k$ is closed (i.e., $d\beta = 0$). Also, show that $* : \mathcal{B}_k \to \mathcal{B}_{n-k}$ ($n = \dim G$). Hence conclude

$$\mathcal{B}_k = \mathcal{H}_k.$$

(*Hint*: To show that $d\beta = 0$, note that if $\iota : G \to G$ is $\iota(g) = g^{-1}$, then $\iota^*\beta \in \mathcal{B}_k$ and $\iota^*\beta(e) = (-1)^k\beta(e)$. Since also $d\beta \in \mathcal{B}_{k+1}$, deduce that $\iota^* d\beta$ equals both $(-1)^k d\beta$ and $(-1)^{k+1}d\beta$.)

8. With G as above, show that $\mathcal{B}_1$ is linearly isomorphic to the *center* $\mathcal{Z}$ of the Lie algebra $\mathfrak{g}$ of G. Conclude that if $\mathfrak{g}$ has trivial center, then $\mathcal{H}^1(G) = 0$.

Exercises 9–10 look at $\mathcal{H}^k(S^n)$.

9. Let β be any harmonic k-form on S^n. Show that $g^*\beta = \beta$, where g is any element of $SO(n+1)$, the group of rotations on $\mathbb{R}^{n+1}$, acting as a group of isometries of S^n. (*Hint*: Compare the argument used in Exercise 6.)
10. Consider the point $p = (0, \dots, 0, 1) \in S^n$. The group $SO(n)$, acting on $\mathbb{R}^n \subset \mathbb{R}^{n+1}$, fixes p. Show that $\mathcal{H}^k(S^n)$ is isomorphic to (a linear subspace of)

$$V_k = \{\beta \in \Lambda^k\mathbb{R}^n : g^*\beta = \beta \text{ for all } g \in SO(n)\}. \tag{8.55}$$

Show that $V_k = 0$ if $0 < k < n$. Deduce that

$$\mathcal{H}^k(S^n) = 0 \quad \text{if } 0 < k < n. \tag{8.56}$$

(*Hint*: Given $\beta \in \Lambda^k\mathbb{R}^n$, $1 \le j, \ell \le n$, average $g^*\beta$ over g in the group of rotations in the $x_j - x_\ell$ plane.)
Note: By Exercise 5, if $n \ge 1$,

$$\mathcal{H}^k(S^n) = \mathbb{R} \quad \text{if } k = 0 \text{ or } n. \tag{8.57}$$

Recall the elementary proof of this, for $k = n$, in Proposition 19.5 of Chapter 1.
11. Suppose M is compact, connected, but not orientable, $\dim M = n$. Show that $\mathcal{H}^n(M) = 0$. (*Hint*: Let $\tilde{M}$ be an orientable double cover, with natural involution ι. A harmonic n-form on M would lift to a harmonic form on $\tilde{M}$, invariant under ι^*; but ι reverses orientation.)

Exercises on the div-curl lemma

This problem set will derive a result known as the "div-curl lemma" of Murat-Tartar [Tar], an ingredient in the method of "compensated compactness." The approach here follows [RRT]; a related approach is used in [Kic]. Further results are given in Chapter 13, §§6, 11, and 12, and there are applications in Chapters 14 and 16.

1. Let $\alpha_{j\nu} \in \Lambda^{\ell_j} M$ be, for each j, a sequence of forms such that

$$\alpha_{j\nu} \longrightarrow \alpha_j \text{ weakly in } H^1 \text{ as } \nu \to \infty. \tag{8.58}$$

Show that

$$d\alpha_{1\nu} \wedge d\alpha_{2\nu} \longrightarrow d\alpha_1 \wedge d\alpha_2 \text{ weakly in } \mathcal{D}' \text{ as } \nu \to \infty.$$

(*Hint*: Write $d\alpha_{1\nu} \wedge d\alpha_{2\nu} = d(\alpha_{1\nu} \wedge d\alpha_{2\nu})$; note that $\alpha_{1\nu} \to \alpha_1$ strongly in L^2.)
2. Let $\sigma_{j\nu} \in \Lambda^{\ell_j} M$ be, for each j, a sequence of forms such that

$$\sigma_{j\nu} \longrightarrow \sigma_j \text{ weakly in } L^2 \text{ as } \nu \to \infty. \tag{8.59}$$

Suppose furthermore that

$$d\sigma_{j\nu} \text{ is compact in } H^{-1}. \tag{8.60}$$

Show that you can write $\sigma_{j\nu} = d\alpha_{j\nu} + \beta_{j\nu}$ where $\alpha_{j\nu}$ satisfies (8.58) and $\{\beta_{j\nu}\}$ is compact in L^2. (*Hint*: Use the Hodge decomposition

$$\sigma = d\delta G\sigma + \delta dG\sigma + P\sigma = d\alpha + \beta.$$

Note that $d\sigma = d\beta$, $\delta\beta = 0$. Then set $\alpha_{j\nu} = \delta G\sigma_{j\nu}$.)

3. Under the hypotheses on $\sigma_{j\nu}$ in Exercise 2, show that

$$\sigma_{1\nu} \wedge \sigma_{2\nu} \longrightarrow \sigma_1 \wedge \sigma_2 \text{ weakly in } \mathcal{D}' \text{ as } \nu \to \infty.$$

Show that this can fail in examples where (8.60) is violated.
(*Hint*: Write

$$\sigma_{1\nu} \wedge \sigma_{2\nu} = d(\alpha_{1\nu} \wedge d\alpha_{2\nu}) + d\alpha_{1\nu} \wedge \beta_{2\nu} + \beta_{1\nu} \wedge d\alpha_{2\nu} + \beta_{1\nu} \wedge \beta_{2\nu}.)$$

4. Let $\dim M = 3$, and let X_ν and Y_ν be two sequences of vector fields such that

(i) $$X_\nu \to X, \quad Y_\nu \to Y \text{ weakly in } L^2,$$

(ii) $$\text{div } X_\nu \text{ and curl } Y_\nu \text{ are compact in } H^{-1}.$$

Show that $X_\nu \cdot Y_\nu \to X \cdot Y$ weakly in $\mathcal{D}'$. Show that the conclusion can fail in cases where (ii) is violated. (*Hint*: Produce equivalent 1-forms, and use the Hodge star operator to deduce this as a special case of Exercise 3.)

Auxiliary exercises on the Hodge star operator

In most of the exercises to follow, adopt the following notational convention. For a vector field u on an oriented Riemannian manifold, let $\tilde{u}$ denote the associated 1-form.

1. Show that

$$f = \text{div } u \Longleftrightarrow f = *d * \tilde{u}.$$

If $M = \mathbb{R}^3$, show that

$$v = \text{curl } u \Longleftrightarrow \tilde{v} = *d\tilde{u}.$$

2. If u and v are vector fields on $\mathbb{R}^3$, show that

$$w = u \times v \Longleftrightarrow \tilde{w} = *(\tilde{u} \wedge \tilde{v}).$$

Show that, for $\tilde{u}, \tilde{v} \in \Lambda^1(M^n)$, $*(\tilde{u} \wedge \tilde{v}) = (*\tilde{u}) \rfloor v$.
If $u \times v$ is defined by this formula for vector fields on an oriented Riemannian 3-fold, show that $u \times v$ is orthogonal to u and v.

3. Show that the identity

$$\text{div } (u \times v) = v \cdot \text{curl } u - u \cdot \text{curl } v, \tag{8.61}$$

for u and v vector fields on $\mathbb{R}^3$, is a special case of

$$*d(\tilde{u} \wedge \tilde{v}) = \langle *d\tilde{u}, \tilde{v} \rangle - \langle \tilde{u}, *d\tilde{v} \rangle, \quad \tilde{u}, \tilde{v} \in \Lambda^1(M^3).$$

Deduce this from $d(\tilde{u} \wedge \tilde{v}) = (d\tilde{u}) \wedge \tilde{v} - \tilde{u} \wedge d\tilde{v}$.

In Exercises 4–6, we produce a generalization of the identity

$$\begin{aligned} \text{curl } (u \times v) &= v \cdot \nabla u - u \cdot \nabla v + (\text{div } v)u - (\text{div } u)v \\ &= [v, u] + (\text{div } v)u - (\text{div u})v, \end{aligned} \tag{8.62}$$

valid for u and v vector fields on $\mathbb{R}^3$.

4. For $\tilde{u}, \tilde{v} \in \Lambda^1(M^n)$, use Exercise 2 to show that
$$d * (\tilde{u} \wedge \tilde{v}) = -(d * \tilde{u}) \rfloor v + \mathcal{L}_v(*\tilde{u}).$$
5. If $\omega \in \Lambda^n(M^n)$ is the volume form, show that $*(\omega \rfloor v) = \tilde{v}$. Deduce that
$$*\big[d(*\tilde{u}) \rfloor v\big] = (\operatorname{div} u)\tilde{v}.$$
6. Applying $\mathcal{L}_v$ to $(*\tilde{u}) \wedge \tilde{w} = \langle u, \tilde{w}\rangle \omega$, show that
$$*\mathcal{L}_v(*\tilde{u}) = \widetilde{[v, u]} + (\operatorname{div} v)\tilde{u},$$
and hence
$$*d * (\tilde{u} \wedge \tilde{v}) = \widetilde{[v, u]} + (\operatorname{div} v)\tilde{u} - (\operatorname{div} u)\tilde{v},$$
generalizing (8.62).

In Exercises 7–10, we produce a generalization of the identity

$$\operatorname{grad}(u \cdot v) = u \cdot \nabla v + v \cdot \nabla u + u \times \operatorname{curl} v + v \times \operatorname{curl} u, \tag{8.63}$$

valid for u and v vector fields on $\mathbb{R}^3$. Only Exercise 10 makes contact with the Hodge star operator.

7. Noting that, for $\tilde{u}, \tilde{v} \in \Lambda^1(M^n)$, $d(\tilde{u} \rfloor v) = \mathcal{L}_v\tilde{u} - (d\tilde{u}) \rfloor v$, show that
$$2d(\tilde{u} \rfloor v) = \mathcal{L}_v\tilde{u} + \mathcal{L}_u\tilde{v} - (d\tilde{u}) \rfloor v - (d\tilde{v}) \rfloor u.$$
8. Show that
$$\mathcal{L}_v\tilde{u} = \widetilde{[v, u]} + (\mathcal{L}_v g)(\cdot, u),$$
where g is the metric tensor, and where $h(\cdot, u) = w$ means $h(X, u) = g(X, w) = \langle X, w\rangle$. Hence
$$\mathcal{L}_v\tilde{u} + \mathcal{L}_u\tilde{v} = (\mathcal{L}_v g)(\cdot, u) + (\mathcal{L}_u g)(\cdot, v).$$
9. Show that
$$(\mathcal{L}_v g)(\cdot, u) + (\mathcal{L}_u g)(\cdot, v) = d(\tilde{u} \rfloor v) + \nabla_u \tilde{v} + \nabla_v \tilde{u}.$$
10. Deduce that
$$d\langle u, v\rangle = \nabla_u \tilde{v} + \nabla_v \tilde{u} - (d\tilde{u}) \rfloor v - (d\tilde{v}) \rfloor u.$$
To relate this to (8.63), show using Exercises 1 and 2 that, for vector fields on $\mathbb{R}^3$,
$$w = v \times \operatorname{curl} u \Longleftrightarrow \tilde{w} = -(d\tilde{u}) \rfloor v.$$
11. If $u, v \in \Lambda^k(M^n)$ and $w \in \Lambda^{n-k}(M^n)$, show that
$$(w, *v) = (-1)^{k(n-k)}(*w, v)$$
and
$$(*u, *v) = (u, v).$$
12. Show that $*d = (-1)^{k+1}\delta *$ on $\Lambda^k(M)$.
13. Verify carefully that $\Delta * = * \Delta$. In particular, on $\Lambda^k(M^n)$,
$$*\Delta = \Delta * = (\pm 1)\big[(\pm 1) d * d + (\pm 1) * d * d *\big].$$

Find the signs.

9. Natural boundary problems for the Hodge Laplacian

Let $\overline{M}$ be a compact Riemannian manifold with boundary, dim $M = m$. We have the Hodge Laplace operator

$$\Delta : C^\infty(\overline{M}, \Lambda^k) \longrightarrow C^\infty(\overline{M}, \Lambda^k).$$

As shown in §10 of Chapter 2, we have a generalization of Green's formula, expressing $-(\Delta u, v)$ as $(du, dv) + (\delta u, \delta v)$ plus a boundary integral. Two forms of this, equivalent to formula (10.18) of Chapter 2, are

$$\begin{aligned} -(\Delta u, v) &= (du, dv) + (\delta u, \delta v) \\ &\quad + \frac{1}{i} \int\limits_{\partial M} \big[\langle \sigma_d(x, \nu)\delta u, v\rangle + \langle du, \sigma_d(x, \nu)v\rangle\big]\, dS \end{aligned} \tag{9.1}$$

and

$$\begin{aligned} -(\Delta u, v) &= (du, dv) + (\delta u, \delta v) \\ &\quad + \frac{1}{i} \int\limits_{\partial M} \big[\langle \delta u, \sigma_\delta(x, \nu)v\rangle + \langle \sigma_\delta(x, \nu)du, v\rangle\big]\, dS. \end{aligned} \tag{9.2}$$

Recall from (10.12)–(10.14) of Chapter 2 that

$$\frac{1}{i}\sigma_d(x, \nu)u = \nu \wedge u, \quad \frac{1}{i}\sigma_\delta(x, \nu)u = -\iota_\nu u. \tag{9.3}$$

We have studied the Dirichlet and Neumann boundary problems for Δ on 0-forms in previous sections. Here we will see that, for each $k \in \{0, \dots, m\}$, there is a pair of boundary conditions generalizing these. To begin, suppose M is half of a compact Riemannian manifold without boundary N, having an *isometric* involution $\tau : N \to N$, fixing ∂M and switching M and $N \setminus M$. For short, we will say N is the *isometric double* of M. Note that elements of $C^\infty(N)$ that are *odd* with respect to τ vanish on ∂M, hence satisfy the Dirichlet boundary condition, while elements *even* with respect to τ have vanishing normal derivatives on ∂M, hence satisfy the Neumann boundary condition. Now, if $u \in \Lambda^k(N)$, then the hypothesis $\tau^* u = -u$ (which implies $\tau^* du = -du$ and $\tau^* \delta u = -\delta u$) implies

$$\sigma_d(x, \nu)u = 0 \text{ and } \sigma_d(x, \nu)\delta u = 0 \text{ on } \partial M, \tag{9.4}$$

while the hypothesis $\tau^* u = u$ (hence $\tau^* du = du$ and $\tau^* \delta u = \delta u$) implies

$$\sigma_\delta(x, \nu)u = 0 \text{ and } \sigma_\delta(x, \nu)du = 0 \text{ on } \partial M. \tag{9.5}$$

We call the boundary conditions (9.4) and (9.5) *relative* boundary conditions and *absolute* boundary conditions, respectively. Thus, specialized to 0-forms, relative boundary conditions are Dirichlet boundary conditions, and absolute boundary conditions are Neumann boundary conditions.

It is easy to see that

$$\nu \wedge u\big|_{\partial M} = 0 \Longleftrightarrow j^* u = 0, \text{ where } j : \partial M \hookrightarrow \overline{M}. \tag{9.6}$$

Thus the relative boundary conditions (9.4) can be rewritten as

$$j^* u = 0, \quad j^*(\delta u) = 0. \tag{9.7}$$

Using (9.3), we can rewrite the absolute boundary conditions (9.5) as

$$u \rfloor \nu = 0 \text{ and } (du) \rfloor \nu = 0 \text{ on } \partial M. \tag{9.8}$$

Also, from Exercise 1 of §8, it follows that

$$\begin{aligned} \sigma_d(x,\nu)(*u) &= \pm * \sigma_\delta(x,\nu) u, \\ \sigma_d(x,\nu)\delta * u &= \pm * \sigma_\delta(x,\nu)\, du. \end{aligned} \tag{9.9}$$

Thus the Hodge star operator interchanges absolute and relative boundary conditions. In particular, the absolute boundary conditions are also equivalent to

$$j^*(*u) = 0, \quad j^*(\delta * u) = 0. \tag{9.10}$$

Note that if u and v satisfy relative boundary conditions, then the boundary integral in (9.1) vanishes. Similarly, if u and v satisfy absolute boundary conditions, then the boundary integral in (9.2) vanishes.

We define the following closed subspaces of Sobolev spaces of k-forms:

$$\begin{aligned} H^1_R(M,\Lambda^k) &= \{u \in H^1(M,\Lambda^k) : \sigma_d(x,\nu)u\big|_{\partial M} = 0\}, \\ H^1_A(M,\Lambda^k) &= \{u \in H^1(M,\Lambda^k) : \sigma_\delta(x,\nu)u\big|_{\partial M} = 0\}, \\ H^2_R(M,\Lambda^k) &= \{u \in H^2(M,\Lambda^k) : (9.4) \text{ holds}\}, \\ H^2_A(M,\Lambda^k) &= \{u \in H^2(M,\Lambda^k) : (9.5) \text{ holds}\}. \end{aligned} \tag{9.11}$$

We have the following simple result, whose proof is left as an exercise.

Lemma 9.1. *Suppose M has an isometric double N, as above. Given $u \in \Lambda^k(M)$, set*

$$\mathcal{O}u = u \text{ on } M, \ -\tau^* u \text{ on } N \setminus M, \quad \mathcal{E}u = u \text{ on } M, \ \tau^* u \text{ on } N \setminus M. \tag{9.12}$$

Then, for $j = 1, 2$,

$$\mathcal{O} : H^j_R(M,\Lambda^k) \to H^j(N,\Lambda^k), \quad \mathcal{E} : H^j_A(M,\Lambda^k) \to H^j(N,\Lambda^k). \tag{9.13}$$

Now the estimates for Δ on k-forms on N established in §8 consequently imply the following.

Lemma 9.2. *If M has an isometric double N, then we have an estimate*

$$\|u\|^2_{H^1(M)} \le C\|du\|^2_{L^2(M)} + C\|\delta u\|^2_{L^2(M)} + C\|u\|^2_{L^2(M)}, \tag{9.14}$$

both for all $u \in H^1_R(M, \Lambda^k)$ *and for all* $u \in H^1_A(M, \Lambda^k)$. *Furthermore, with* $b = R$ *or* A, *if*

$$u \in H^1_b(M, \Lambda^k) \text{ and } (du, dv) + (\delta u, \delta v) \le C\|v\|_{L^2(M)},$$

for all $v \in H^1_b(M, \Lambda^k)$, *then* $u \in H^2_b(M, \Lambda^k)$.

It is convenient to rewrite the estimate (9.14) as the following pair of estimates:

$$\begin{aligned} \|u\|^2_{H^1(M)} &\le C\|du\|^2_{L^2(M)} + C\|\delta u\|^2_{L^2(M)} \\ &\quad + C\|\sigma_d(x, \nu)u\|^2_{H^{1/2}(\partial M)} + C\|u\|^2_{L^2(M)} \end{aligned} \tag{9.15}$$

and

$$\begin{aligned} \|u\|^2_{H^1(M)} &\le C\|du\|^2_{L^2(M)} + C\|\delta u\|^2_{L^2(M)} \\ &\quad + C\|\sigma_\delta(x, \nu)u\|^2_{H^{1/2}(\partial M)} + C\|u\|^2_{L^2(M)}, \end{aligned} \tag{9.16}$$

both valid for all $u \in H^1(M, \Lambda^k)$.

So far, the estimates (9.15) and (9.16) have been shown to hold when M has an isometric double. Now any compact manifold $\overline{M}$ with smooth boundary has a double N, a smooth manifold without boundary, together with a smooth involution τ fixing ∂M. Also, N possesses Riemannian metrics invariant under τ. However, if M is endowed with some Riemannian metric, it may not extend to a smooth invariant Riemannian metric on N. For example, a necessary condition for such a metric to exist on N would be that ∂M is totally geodesic in $\overline{M}$. Our next task will be to show that the estimates (9.15) and (9.16) hold in general.

To begin, if $\chi \in C^\infty(M)$ is a cut-off, since the commutators $[d, \chi]$ and $[\delta, \chi]$ are bounded on $L^2(M)$, we see that it suffices to prove the following.

Lemma 9.3. *For any* $p \in \overline{M}$, *there is a neighborhood* $\mathcal{O}$ *of* p *in* $\overline{M}$ *such that the estimates (9.15) and (9.16) hold for* u *supported in* $\mathcal{O}$.

Of course, for interior points p, such estimates follow from the analysis of §8, so we need only consider $p \in \partial M$.

For $p \in \partial M$, choose a coordinate mapping of a neighborhood $\mathcal{O}_1$ of p in $\overline{M}$ to a neighborhood U_1 of 0 in $\overline{\mathbb{R}^n_+}$, such that the induced Riemannian metric g_{jk} is equal to δ_{jk} at 0. In addition to the induced metric g on U_1 (which gives rise to $\delta = \pm * d *$), we have the *flat* metric g^0 on U_1, $g^0_{jk} = \delta_{jk}$, and associated operator δ^0. The differential operators δ and δ^0 are first-order differential operators whose principal symbols agree at the origin 0. Of course, the exterior derivative operator d is independent of the metric; $d = d^0$. We also note that the unit normal ν to ∂M with respect to the metric g is equal to the normal $\nu^0 = dx_n$ with respect to the flat metric, at the origin, so the 0-order operators $\sigma_d(x, \nu)$ and $\sigma_{d^0}(x, \nu^0)$ agree at 0, and so do the 0-order operators $\sigma_\delta(x, \nu)$ and $\sigma_{\delta^0}(x, \nu^0)$.

Now the reflection argument described above shows that if we have $u \in H^1(U_1, \Lambda^k)$, vanishing on the upper boundary, then

$$(9.17)\quad \|u\|^2_{H^1(U_1)} \le C\|d^0u\|^2_{L^2(U_1)} + C\|\delta^0 u\|^2_{L^2(U_1)} + C\|B^0u\|^2_{H^{1/2}(\Gamma)} + C\|u\|^2_{L^2(M)},$$

where Γ is $\mathbb{R}^{n-1} = \partial\mathbb{R}^n_+$, compactified into a torus by putting $\overline{U}_1 \cap \mathbb{R}^{n-1}$ in a big box and identifying opposite sides. Also, B^0 in (9.17) is either $\sigma_{d^0}(x, \nu^0)$ or $\sigma_{\delta^0}(x, \nu^0)$. On the other hand, if in addition the support of u is in a sufficiently small neighborhood of 0, we have

$$(9.18)\quad \|\delta u - \delta^0 u\|^2_{L^2(U_1)} \le \varepsilon\|u\|^2_{H^1(U_1)} + c(\varepsilon)\|u\|^2_{L^2(U_1)}$$

and

$$(9.19)\quad \|Bu - B^0u\|^2_{H^{1/2}(\Gamma)} \le \varepsilon\|u\|^2_{H^{1/2}(\Gamma)} \le C_0\varepsilon\|u\|^2_{H^1(U_1)},$$

where B is either $\sigma_d(x, \nu)$ or $\sigma_\delta(x, \nu)$, depending on the choice of B^0. Consequently, for u with sufficiently small support, the estimates (9.15) and (9.16) follow from (9.17). This proves Lemma 9.3 and consequently, in view of the observation on cut-offs, we have the following.

Proposition 9.4. *If M is a compact Riemannian manifold with smooth boundary, then the estimates (9.15) and (9.16) hold for all $u \in H^1(M, \Lambda^k)$. Hence the estimate (9.14) holds both for all $u \in H^1_R(M, \Lambda^k)$ and for all $u \in H^1_A(M, \Lambda^k)$.*

In analogy with our treatment of the Neumann boundary condition in §7, we define an operator

$$(9.20)\quad \mathcal{L}_R : H^1_R(M, \Lambda^k) \longrightarrow H^1_R(M, \Lambda^k)^*$$

by

$$(9.21)\quad (\mathcal{L}_R u, v) = (du, dv) + (\delta u, \delta v), \quad u, v \in H^1_R(M, \Lambda^k),$$

and we also define

$$(9.22)\quad \mathcal{L}_A : H^1_A(M, \Lambda^k) \longrightarrow H^1_A(M, \Lambda^k)^*$$

by

$$(9.23)\quad (\mathcal{L}_A u, v) = (du, dv) + (\delta u, \delta v), \quad u, v \in H^1_A(M, \Lambda^k).$$

The estimates (9.15) and (9.16) show that, with $b = R$ or A, and some $C_0 > 0$,

$$(9.24)\quad \big((\mathcal{L}_b + C_0)u, u\big) \ge C\|u\|^2_{H^1(M)}, \quad u \in H^1_b(M, \Lambda^k),$$

which as before leads to the following.

Proposition 9.5. *For $b = R$ or A, the maps*

$$(9.25)\quad \mathcal{L}_b + C_0 : H^1_b(M, \Lambda^k) \longrightarrow H^1_b(M, \Lambda^k)^*$$

are one-to-one and onto.

The maps

$$T_b : H_b^1(M, \Lambda^k)^* \longrightarrow H_b^1(M, \Lambda^k) \tag{9.26}$$

giving two-sided inverses of (9.25) are compact, self-adjoint operators on $L^2(M, \Lambda^k)$, so we have orthonormal bases $\{u_j^{(k)}\}$ and $\{v_j^{(k)}\}$ of $L^2(M, \Lambda^k)$ satisfying

$$T_R u_j^{(k)} = \mu_j^{(k)} u_j^{(k)}, \quad u_j^{(k)} \in H_R^1(M, \Lambda^k), \tag{9.27}$$

and

$$T_A v_j^{(k)} = \nu_j^{(k)} v_j^{(k)}, \quad v_j^{(k)} \in H_A^1(M, \Lambda^k). \tag{9.28}$$

Since clearly $((\mathcal{L}_b + 1)u, u) \geq \|u\|_{L^2}^2$, we can take $C_0 = 1$ in (9.25). Then the eigenvalues of T_R and T_A all have magnitude ≤ 1, and we can order them so that, for each k, $\mu_j^{(k)}$ and $\nu_j^{(k)} \searrow 0$ as $j \to \infty$. It follows that, for each k,

$$\mathcal{L}_R u_j^{(k)} = \rho_j^{(k)} u_j^{(k)}, \quad \rho_j^{(k)} = \frac{1}{\mu_j^{(k)}} - 1 \nearrow \infty, \tag{9.29}$$

and

$$\mathcal{L}_A v_j^{(k)} = \alpha_j^{(k)} v_j^{(k)}, \quad \alpha_j^{(k)} = \frac{1}{\nu_j^{(k)}} - 1 \nearrow \infty. \tag{9.30}$$

Here, $\rho_j^{(k)} \geq 0$ and $\alpha_j^{(k)} \geq 0$, and only finitely many of these quantities are equal to zero.

We can produce higher-order regularity results by the same techniques as used for the Dirichlet and Neumann problems. In analogy with Proposition 7.2, we have

Proposition 9.6. *With $b = R$ or A, given $f \in L^2(M, \Lambda^k)$, $u = T_b f$ satisfies*

$$u \in H_b^2(M, \Lambda^k), \tag{9.31}$$

and there is the estimate

$$\begin{aligned} \|u\|_{H^2(M)}^2 \leq C\|\Delta u\|_{L^2(M)}^2 &+ C\|B_b^{(0)} u\|_{H^{3/2}(\partial M)}^2 \\ &+ C\|B_b^{(1)} u\|_{H^{1/2}(\partial M)}^2 + C\|u\|_{H^1(M)}^2, \end{aligned} \tag{9.32}$$

for all $u \in H^2(M, \Lambda^k)$, where

$$\begin{aligned} B_R^{(0)} u &= \sigma_d(x, \nu)u, \quad & B_A^{(0)} u &= \sigma_\delta(x, \nu)u, \\ B_R^{(1)} u &= \sigma_d(x, \nu)\delta u, \quad & B_A^{(1)} u &= \sigma_\delta(x, \nu)du. \end{aligned} \tag{9.33}$$

This can be proved in the same way as Proposition 7.2. We give details on why the boundary conditions hold in (9.31), which are slightly more involved than before. We claim that, given $u \in H_b^1(M, \Lambda^k)$, with $\Delta u \in L^2(M, \Lambda^k)$, then the boundary term in (9.1)–(9.2) vanishes for all $v \in H_b^1(M, \Lambda^k)$ if and *only if* all the appropriate boundary data for u vanish; for example, $\sigma_d(x, \nu)\delta u = 0$ on ∂M, in case $b = R$. We need to establish the "only if" part. Take the case $b = R$. Pick

$\sigma \in C^\infty(\overline{M}, \mathrm{Hom}(\Lambda^{k-1}, \Lambda^k))$ such that $\sigma(x) = \sigma_d(x, \nu)$, for $x \in \partial M$. Then, for any $w \in \Lambda^{k-1}(\overline{M})$, we have $v = \sigma w \in H^1_R(M, \Lambda^k)$, and hence, for any $u \in H^1_R(M, \Lambda^k)$, the boundary term in (9.1) is equal to

$$\begin{aligned}\beta(u, v) &= \frac{1}{i} \int_{\partial M} \langle \sigma_d(x, \nu)\, \delta u, \sigma_d(x, \nu) w \rangle \, dS \\ &= \frac{1}{i} \int_{\partial M} \langle \sigma_d(x, \nu)^* \sigma_d(x, \nu)\, (\delta u), w \rangle \, dS.\end{aligned}$$

This vanishes for all $w \in \Lambda^{k-1}(\overline{M})$ if and only if $\sigma_d(x, \nu)^* \sigma_d(x, \nu)\, (\delta u) = 0$ on ∂M, which in turn occurs if and only if $\sigma_d(x, \nu)\, (\delta u) = 0$ on ∂M. Thus, obtaining $u \in H^2(M, \Lambda^k)$ by the methods used in Proposition 7.2, we have (9.31), in case $b = R$. The case $b = A$ is similar.

Next, the same arguments proving Proposition 7.3 and Theorem 1.3 establish the following.

Proposition 9.7. *Given $f_1 \in H^j(M, \Lambda^k)$, $j = 1, 2, 3, \dots$, a k-form $u \in H^{j+1}(M, \Lambda^k)$ satisfying*

$$\Delta u = f_1 \text{ on } M \tag{9.34}$$

and either of the boundary conditions (9.4) or (9.5), belongs to $H^{j+2}(M, \Lambda^k)$. Furthermore, we have estimates

$$\begin{aligned}\|u\|^2_{H^{j+2}(M)} \le C\|\Delta u\|^2_{H^j(M)} &+ C\|B_b^{(0)} u\|^2_{H^{j+3/2}(\partial M)} \\ &+ C\|B_b^{(1)} u\|^2_{H^{j+1/2}(\partial M)} + C\|u\|^2_{H^{j+1}(M)},\end{aligned} \tag{9.35}$$

for all $u \in H^{j+2}(N, \Lambda^k)$, where $B_b^{(\ell)}$ are given by (9.33).

One corollary of this is that the eigenfunctions $u_j^{(k)}$ and $v_j^{(k)}$ are in $C^\infty(\overline{M}, \Lambda^k)$ and satisfy the boundary conditions (9.4) and (9.5), respectively. The 0-eigenspaces of $\mathcal{L}_R$ and $\mathcal{L}_A$ are finite-dimensional spaces in $C^\infty(\overline{M}, \Lambda^k)$; denote them by $\mathcal{H}_k^R$ and $\mathcal{H}_k^A$, respectively. We see that, for $b = R$ or A,

$$\begin{aligned}u \in \mathcal{H}_k^b \Longleftrightarrow u \in C^\infty(\overline{M}, \Lambda^k),\ &B_b^{(0)} u = 0 \text{ on } \partial M, \\ &\text{and } du = \delta u = 0 \text{ on } M.\end{aligned} \tag{9.36}$$

Again, $B_b^{(0)}$ are given by (9.33). Equivalently,

$$B_R^{(0)} u = \nu \wedge u, \quad B_A^{(0)} u = u \rfloor \nu. \tag{9.37}$$

Also recall that we can replace $\nu \wedge u$ by $j^* u$. To state the result slightly differently,

$$u \in \mathcal{H}_k^b \Longleftrightarrow u \in H_b^1(M, \Lambda^k) \text{ and } du = \delta u = 0. \tag{9.38}$$

We call $\mathcal{H}_k^R$ and $\mathcal{H}_k^A$ the spaces of harmonic k-forms, satisfying relative and absolute boundary conditions, respectively.

Denote by P_h^R and P_h^A the orthogonal projections of $L^2(M, \Lambda^k)$ onto $\mathcal{H}_k^R$ and $\mathcal{H}_k^A$. Parallel to (8.22) and (8.23), we have continuous linear maps

$$G^b : L^2(M, \Lambda^k) \longrightarrow H_b^2(M, \Lambda^k), \quad b = R \text{ or } A, \tag{9.39}$$

such that G^b annihlates $\mathcal{H}_k^b$ and inverts $-\Delta$ on the orthogonal complement of $\mathcal{H}_k^b$:

$$-\Delta G^b u = (I - P_h^b)u, \quad \text{for } u \in L^2(M, \Lambda^k), \tag{9.40}$$

and furthermore, for $j \geq 0$,

$$G^b : H^j(M, \Lambda^k) \longrightarrow H^{j+2}(M, \Lambda^k). \tag{9.41}$$

The identity (9.40) then produces the following two Hodge decompositions for a compact Riemannian manifold with boundary.

Proposition 9.8. *Given $u \in H^j(M, \Lambda^k)$, $j \geq 0$, we have*

$$u = d\delta G^R u + \delta d G^R u + P_h^R u = P_d^R u + P_\delta^R u + P_h^R u \tag{9.42}$$

and

$$u = d\delta G^A u + \delta d G^A u + P_h^A u = P_d^A u + P_\delta^A u + P_h^A u. \tag{9.43}$$

In both cases, the three terms on the right side are mutually orthogonal in $L^2(M, \Lambda^k)$.

Proof. It remains only to check orthogonality, which requires a slightly longer argument than that used in Proposition 8.2. By continuity, it suffices to check the orthogonality for $u \in C^\infty(\overline{M}, \Lambda^k)$. We will use the identity

$$(du, v) = (u, \delta v) + \gamma(u, v), \tag{9.44}$$

for $u \in \Lambda^{j-1}(\overline{M})$ and $v \in \Lambda^j(\overline{M})$, with

$$\gamma(u, v) = \frac{1}{i} \int_{\partial M} \langle \sigma_d(x, \nu)u, v \rangle \, dS = \frac{1}{i} \int_{\partial M} \langle u, \sigma_\delta(x, \nu)v \rangle \, dS. \tag{9.45}$$

Note that $\gamma(u, v) = 0$ if either $u \in H_R^1(M, \Lambda^{j-1})$ or $v \in H_A^1(M, \Lambda^j)$. In particular, we see that

$$\begin{aligned} u \in H_R^1(M, \Lambda^{k-1}) &\Longrightarrow du \perp \ker \delta \cap H^1(M, \Lambda^k), \\ v \in H_A^1(M, \Lambda^k) &\Longrightarrow \delta v \perp \ker d \cap H^1(M, \Lambda^{k-1}). \end{aligned} \tag{9.46}$$

From the definitions, we have

$$\begin{aligned} \delta : H_R^2(M, \Lambda^j) &\longrightarrow H_R^1(M, \Lambda^{j-1}), \\ d : H_A^2(M, \Lambda^j) &\longrightarrow H_A^1(M, \Lambda^{j+1}), \end{aligned} \tag{9.47}$$

so

$$\begin{aligned} d\delta H_R^2(M, \Lambda^k) &\perp \ker \delta \cap H^1(M, \Lambda^k), \\ \delta d H_A^2(M, \Lambda^k) &\perp \ker d \cap H^1(M, \Lambda^k). \end{aligned} \tag{9.48}$$

Now (9.48) implies for the ranges:

$$\mathcal{R}(P_d^R) \perp \mathcal{R}(P_\delta^R) + \mathcal{R}(P_h^R), \quad \mathcal{R}(P_\delta^A) \perp \mathcal{R}(P_d^A) + \mathcal{R}(P_h^A). \tag{9.49}$$

Furthermore, if $u \in \mathcal{H}_k^R$ and $v = dG^R w$, then $\gamma(u, v) = 0$, so $(u, \delta v) = (du, v) = 0$. Similarly, if $v \in \mathcal{H}_k^A$ and $u = \delta G^A w$, then $\gamma(u, v) = 0$, so $(du, v) = (u, \delta v) = 0$. Thus

$$\mathcal{R}(P_\delta^R) \perp \mathcal{R}(P_h^R), \quad \mathcal{R}(P_d^A) \perp \mathcal{R}(P_h^A). \tag{9.50}$$

The proposition is proved.

We can produce an analogue of Proposition 8.3, relating the spaces $\mathcal{H}_k^b$ to cohomology groups. We first look at the case $b = R$. Set

$$C_r^\infty(\overline{M}, \Lambda^k) = \{u \in C^\infty(\overline{M}, \Lambda^k) : j^* u = 0\}. \tag{9.51}$$

Since $d \circ j^* = j^* \circ d$, it is clear that

$$d : C_r^\infty(\overline{M}, \Lambda^k) \longrightarrow C_r^\infty(\overline{M}, \Lambda^{k+1}). \tag{9.52}$$

Our spaces of "closed" and "exact" forms are

$$\begin{aligned} \mathcal{C}_R^k(\overline{M}) &= \{u \in C_r^\infty(\overline{M}, \Lambda^k) : du = 0\}, \\ \mathcal{E}_R^k(\overline{M}) &= d\, C_r^\infty(\overline{M}, \Lambda^{k-1}). \end{aligned} \tag{9.53}$$

We set

$$\mathcal{H}^k(\overline{M}, \partial M) = \mathcal{C}_R^k(\overline{M}) / \mathcal{E}_R^k(\overline{M}). \tag{9.54}$$

Proposition 9.9. *If $\overline{M}$ is a compact Riemannian manifold with boundary, there is a natural isomorphism*

$$\mathcal{H}^k(\overline{M}, \partial M) \approx \mathcal{H}_k^R. \tag{9.55}$$

Proof. By (9.36) we have an injection

$$j : \mathcal{H}_k^R \longrightarrow \mathcal{C}_R^k(\overline{M}),$$

which yields a map

$$J : \mathcal{H}_k^R \longrightarrow \mathcal{H}^k(\overline{M}, \partial M),$$

by composing with (9.54). The orthogonality of the terms in (9.42) implies (Image $j) \cap \mathcal{E}_R^k(\overline{M}) = 0$, so J is injective. Furthermore, if $u \in \mathcal{C}_R^k(\overline{M})$, then u is orthogonal to δv for any $v \in C^\infty(\overline{M}, \Lambda^{k+1})$, so the term $\delta(dG^R u)$ in (9.42) vanishes, and hence J is surjective. This proves the proposition.

As in §8, it is clear that $\mathcal{H}^k(\overline{M}, \partial M)$ is independent of a metric on M. Thus the dimension of $\mathcal{H}_k^R$ is independent of such a metric.

Associated to absolute boundary conditions is the family of spaces

$$C_a^\infty(\overline{M}, \Lambda^k) = \{u \in C^\infty(\overline{M}, \Lambda^k) : \iota_\nu u = \iota_\nu (du) = 0\}, \tag{9.56}$$

replacing (9.51); we have

$$d : C_a^\infty(\overline{M}, \Lambda^k) \longrightarrow C_a^\infty(\overline{M}, \Lambda^{k+1}), \tag{9.57}$$

and, with $\mathcal{C}_A^k(\overline{M})$ the kernel of d in (9.57) and $\mathcal{E}_A^{k+1}(\overline{M})$ its image, we can form quotients. The following result is parallel to Proposition 9.9.

Proposition 9.10. *There is a natural isomorphism*

$$\mathcal{H}_k^A \approx \mathcal{C}_A^k(\overline{M})/\mathcal{E}_A^k(\overline{M}). \tag{9.58}$$

Proof. This is exactly parallel to the proof of Proposition 9.9.

We have refrained from denoting the right side of (9.58) by $\mathcal{H}^k(\overline{M})$, since the deRham cohomology of $\overline{M}$ has the standard definition

$$\mathcal{H}^k(\overline{M}) = \mathcal{C}^k(\overline{M})/\mathcal{E}^k(\overline{M}), \tag{9.59}$$

where $\mathcal{C}^k(\overline{M})$ is the kernel and $\mathcal{E}^{k+1}(\overline{M})$ the image of d in

$$d : C^\infty(\overline{M}, \Lambda^k) \longrightarrow C^\infty(\overline{M}, \Lambda^{k+1}). \tag{9.60}$$

Note that *no* boundary conditions are imposed here. We now establish that (9.58) is isomorphic to $\mathcal{H}^k(\overline{M})$.

Proposition 9.11. *The quotient spaces $\mathcal{C}_A^k(\overline{M})/\mathcal{E}_A^k(\overline{M})$ and $\mathcal{H}^k(\overline{M})$ are naturally isomorphic. Hence*

$$\mathcal{H}_k^A \approx \mathcal{H}^k(\overline{M}). \tag{9.61}$$

Proof. It is clear that there is a natural map

$$\kappa : \mathcal{C}_A^k(\overline{M})/\mathcal{E}_A^k(\overline{M}) \longrightarrow \mathcal{H}^k(\overline{M}),$$

since $\mathcal{C}_A^k(\overline{M}) \subset \mathcal{C}^k(\overline{M})$ and $\mathcal{E}_A^k(\overline{M}) \subset \mathcal{E}^k(\overline{M})$. To show that κ is surjective, let $\alpha \in C^\infty(\overline{M}, \Lambda^k)$ be closed; we want $\tilde{\alpha} \in \mathcal{C}_a^k(\overline{M})$ such that $\alpha - \tilde{\alpha} = d\beta$ for some $\beta \in C^\infty(\overline{M}, \Lambda^{k-1})$.

To arrange this, we use a 1-parameter family of maps

$$\varphi_t : \overline{M} \longrightarrow \overline{M}, \quad 0 \le t \le 1, \tag{9.62}$$

such that φ_0 is the identity map, and as $t \to 1$, φ_t retracts a collar neighborhood $\mathcal{O}$ of ∂M onto ∂M, along geodesics normal to ∂M. Set $\tilde{\alpha} = \varphi_1^* \alpha$. It is easy to see that $\tilde{\alpha} \in \mathcal{C}_a^k(\overline{M})$. Furthermore, $\alpha - \tilde{\alpha} = d\beta$ with

$$\beta = -\int_0^1 \varphi_t^*\big(\alpha \rfloor X(t)\big)\, dt \in C^\infty(\overline{M}, \Lambda^{k-1}), \tag{9.63}$$

where $X(t) = (d/dt)\varphi_t$. Compare the proof of the Poincaré lemma, Theorem 13.2 of Chapter 1, and formulas (13.61)–(13.64) of that chapter. It follows that κ is surjective.

Consequently, we have a natural surjective homomorphism

$$\tilde{\kappa} : \mathcal{H}_k^A \longrightarrow \mathcal{H}^k(\overline{M}). \tag{9.64}$$

It remains to prove that $\tilde{\kappa}$ is injective. But if $\alpha \in \mathcal{H}_k^A$ and $\alpha = d\beta$, $\beta \in C^\infty(\overline{M}, \Lambda^{k-1})$, then the identity (9.44) with $du = d\beta$, $v = \alpha$ implies $(\alpha, \alpha) = 0$, hence $\alpha = 0$. This completes the proof.

One can give a proof of (9.61) without using such a homotopy argument, in fact without using $C_A^k(\overline{M})/\mathcal{E}_A^k(\overline{M})$ at all. See Exercise 5 in the set of exercises on cohomology after this section. Such an argument will be useful in Chapter 12. On the other hand, homotopy arguments similar to that used above are also useful, and will arise in a number of problems in this set of exercises.

We can now establish the following Poincaré duality theorem, whose proof is immediate, since by (9.9) the Hodge star operator interchanges absolute and relative boundary conditions.

Proposition 9.12. *If $\overline{M}$ is an oriented, compact Riemannian manifold with boundary, then*

$$* : \mathcal{H}_k^R \longrightarrow \mathcal{H}_{m-k}^A \tag{9.65}$$

is an isomorphism, where $m = \dim M$. Consequently,

$$\mathcal{H}^k(\overline{M}, \partial M) \approx \mathcal{H}^{m-k}(\overline{M}). \tag{9.66}$$

We end this section with a brief description of a sequence of maps on cohomology, associated to a compact manifold $\overline{M}$ with boundary. The sequence takes the form

$$\cdots \to \mathcal{H}^{k-1}(\partial M) \xrightarrow{\delta} \mathcal{H}^k(\overline{M}, \partial M) \xrightarrow{\pi} \mathcal{H}^k(\overline{M}) \xrightarrow{\iota} \mathcal{H}^k(\partial M) \to \cdots . \tag{9.67}$$

These maps are defined as follows. The inclusion

$$C_r^\infty(\overline{M}, \Lambda^k) \hookrightarrow C^\infty(\overline{M}, \Lambda^k),$$

yielding $C_R^k(\overline{M}) \subset C^k(\overline{M})$ and $\mathcal{E}_R^k(\overline{M}) \subset \mathcal{E}^k(\overline{M})$, gives rise to π in a natural fashion. The map ι comes from the pull-back

$$j^* : C^\infty(\overline{M}, \Lambda^k) \longrightarrow C^\infty(\partial M, \Lambda^k),$$

which induces a map on cohomology since $j^*d = dj^*$. Note that j^* annihilates $C_r^\infty(\overline{M}, \Lambda^k)$, so $\iota \circ \pi = 0$.

The "coboundary map" δ is defined on the class $[\alpha] \in \mathcal{H}^{k-1}(\partial M, \mathbb{R})$ of a closed form $\alpha \in \Lambda^{k-1}(\partial M)$ by choosing a form $\beta \in C^\infty(\overline{M}, \Lambda^{k-1})$ such that $j^*\beta = \alpha$ and taking the class $[d\beta]$ of $d\beta \in C_R^k(\overline{M})$. Note that $d\beta$ might not belong to $\mathcal{E}_R^k(\overline{M})$ if $j^*\beta$ is not exact. If another $\tilde{\beta}$ is picked such that $j^*\tilde{\beta} = \alpha + d\gamma$, then $d(\beta - \tilde{\beta})$ does belong to $\mathcal{E}_R^k(\overline{M})$, so δ is well defined:

$$\delta[\alpha] = [d\beta], \quad \text{with } j^*\beta = \alpha.$$

Note that if $[\alpha] = \iota[\tilde{\beta}]$, via $\alpha = j^*\tilde{\beta}$ with $\tilde{\beta} \in C^k(\overline{M})$, then $d\tilde{\beta} = 0$, so $\delta \circ \iota = 0$. Also, since $d\beta \in \mathcal{E}^k(\overline{M})$, $\pi \circ \delta = 0$.

In fact, the sequence (9.67) is *exact*, that is, the image of each map is equal to the kernel of the map that follows. This "long exact sequence" in cohomology is a useful computational tool. Exactness will be sketched in some of the following exercises on cohomology.

Another important exact sequence, the Mayer-Vietoris sequence, is discussed in Appendix B at the end of this chapter.

Exercises

1. Let u be a 1-form on M with associated vector field U. Show that the relative boundary conditions (9.4) are equivalent to

$$U \perp \partial M \text{ and } \operatorname{div} U = 0 \text{ on } \partial M.$$

If $\dim M = 3$, show that the absolute boundary conditions (9.5) are equivalent to

$$U \parallel \partial M \text{ and } \operatorname{curl} U \perp \partial M.$$

Treat the case $\dim M = 2$.

2. Let $b = R$ or A. Consider the unbounded operator D_b on $\mathcal{H} = \bigoplus_k L^2(M, \Lambda^k)$:

$$D_b = d + \delta, \quad \mathcal{D}(D_b) = \bigoplus_k H_b^1(M, \Lambda^k).$$

Here $\mathcal{D}(D_b)$ denotes the domain of D_b. Show that D_b is self adjoint, that $\mathcal{D}(D_b^2) = \bigoplus_k H_b^2(M, \Lambda^k)$, and that $D_b^2 = -\Delta$ on this domain. Show that

$$u = (d + \delta)G^b(d + \delta)u + P_k^b u, \quad \text{for } u \in H_b^1(M, \Lambda^k).$$

Reconsider this problem after reading §§11 and 12. For a discussion of unbounded operators defined on dense domains, see §8 in Appendix A.

3. Show that $d, \delta, d\delta$, and δd all map $\mathcal{D}(D_b^{j+1})$ to $\mathcal{D}(D_b^j)$, for $j \geq 0$.
4. Form the orthogonal projections $P_d^b = d\delta G^b$, $P_\delta^b = \delta d G^b$. With $b = R$ or A, show that the four operators

$$G^b, P_h^b, P_d^b, \text{ and } P_\delta^b$$

all commute. Deduce that one can arrange the eigenfunctions $u_j^{(k)}$, forming an orthonormal basis of $L^2(M, \Lambda^k)$, such that each one appears in exactly one term in the Hodge decomposition (9.42), and that the same can be done with the eigenfunctions $v_j^{(k)}$, relative to the decomposition (9.43).

5. If M is oriented, and $*$ the Hodge star operator, show that

$$T_A * = * T_R,$$

where T_A and T_R are as in (9.26). Show that

$$P_h^A * = * P_h^R \text{ and } G^A * = * G^R.$$

Also, with P_d^b and P_δ^b the projections defined above, show that

$$P_d^A * = * P_\delta^R \text{ and } P_\delta^A * = * P_d^R.$$

Exercises on cohomology

1. Let $\overline{M}$ be a compact, connected manifold with nonempty boundary, and double N. Endow N with a Riemannian metric invariant under the involution τ. Show that

$$\mathcal{H}^k(\overline{M}, \partial M) \approx \{u \in \mathcal{H}_k(N) : \tau^* u = -u\}. \tag{9.68}$$

Deduce that if M is also orientable,

$$\mathcal{H}^n(\overline{M}, \partial M) = \mathbb{R}, \quad n = \dim M. \tag{9.69}$$

2. If $\overline{M}$ is connected, show directly that

$$\mathcal{H}^0(\overline{M}) = \mathbb{R}.$$

By Poincaré duality, this again implies (9.69), when M is orientable.

3. Show that if M is connected and $\partial M \neq \emptyset$,

$$\mathcal{H}^0(\overline{M}, \partial M) = 0.$$

Deduce that if M is also orientable, $n = \dim M$, then

$$\mathcal{H}^n(\overline{M}) = 0.$$

Give a proof of this that also works in the nonorientable case.

4. Show directly, using the proof of the Poincaré lemma, Theorem 13.2 of Chapter 1, that

$$\mathcal{H}^k(\overline{B^n}) = 0, \quad 1 \le k \le n, \tag{9.70}$$

where $\overline{B^n}$ is the closed unit ball in $\mathbb{R}^n$, with boundary S^{n-1}. Deduce that

$$\mathcal{H}^k(\overline{B^n}, S^{n-1}) = 0, \quad 0 \le k < n, \qquad \mathbb{R}, \quad k = n. \tag{9.71}$$

5. Use (9.48) to show directly from Proposition 9.8 (not using Proposition 9.11) that, if $\alpha \in C^\infty(\overline{M}, \Lambda^k)$ is closed, then $\alpha = d\beta + P_h^A \alpha$ for some $\beta \in C^\infty(\overline{M}, \Lambda^{k-1})$, in fact, for $\beta = \delta G^A \alpha$. Hence conclude that

$$\mathcal{H}_k^A \approx \mathcal{H}^k(\overline{M})$$

without using the homotopy argument of Proposition 9.11.

Let M be a smooth manifold without boundary. The cohomology with *compact supports* $\mathcal{H}_c^k(M)$ is defined via

$$d : C_0^\infty(M, \Lambda^k) \longrightarrow C_0^\infty(M, \Lambda^{k+1}), \tag{9.72}$$

as

$$\mathcal{H}_c^k(M) = \mathcal{C}_c^k(M)/\mathcal{E}_c^k(M),$$

where the kernel of d in (9.72) is $\mathcal{C}_c^k(M)$ and its image is $\mathcal{E}_c^{k+1}(M)$.

In Exercises 6 and 7, we assume M is the interior of a compact manifold with boundary $\overline{M}$.

6. Via $C_0^\infty(M, \Lambda^k) \hookrightarrow C_r^\infty(\overline{M}, \Lambda^k)$, we have a well-defined homomorphism

$$\rho : \mathcal{H}_c^k(M) \longrightarrow \mathcal{H}^k(\overline{M}, \partial M).$$

Show that ρ is injective. (*Hint*: Let $\varphi_t : \overline{M} \to \overline{M}$ be as in (9.62); also, given $K \subset\subset M$, arrange that each φ_t is the identity on K. If $\alpha \in C_c^k(M)$ has support in K and $\alpha = d\beta$, with $\beta \in C_r^\infty(\overline{M}, \Lambda^{k-1})$, show that $\tilde{\beta} = \varphi_1^*\beta$ has compact support and $d\tilde{\beta} = \alpha$.)

7. Show that ρ is *surjective*, and hence

$$\mathcal{H}_c^k(M) \approx \mathcal{H}^k(\overline{M}, \partial M). \tag{9.73}$$

(*Hint*: If $\alpha \in C_R^k(\overline{M})$, set $\tilde{\alpha} = \varphi_1^*\alpha$ and parallel the argument using (9.63), in the proof of Proposition 9.11.)

8. If M is connected and oriented, and $\dim M = n$, show that

$$\mathcal{H}_c^n(M) = \mathbb{R},$$

even if M cannot be compactified to a manifold with smooth boundary.
(*Hint*: If $\alpha \in C_0^\infty(M, \Lambda^n)$ and $\int_M \alpha = 0$, fit the support of α in the interior Y of a compact, smooth manifold with boundary $\overline{Y} \subset M$. Then apply arguments outlined above.)

9. Let X be a compact, connected manifold; given $p \in X$, let $M = X \setminus \{p\}$. Then $C_0^\infty(M, \Lambda^k) \hookrightarrow C^\infty(X, \Lambda^k)$ induces a homomorphism

$$\gamma : \mathcal{H}_c^k(M) \longrightarrow \mathcal{H}^k(X).$$

Show that γ is an *isomorphism*, for $0 < k \le \dim X$. (*Hint*: Construct a family of maps $\psi_t : X \to X$, with properties like φ_t used in Exercises 6 and 7, this time collapsing a neighborhood $\mathcal{O}$ of p onto p as $t \to 1$. Establish the injectivity and surjectivity of γ by arguments similar to those used in Exercises 6 and 7, noting that the analogue of the argument in Exercise 7 fails in this case when $k = 0$.)

10. Using Exercise 9, deduce that

$$\mathcal{H}^k(S^n) \approx \mathcal{H}_c^k(\mathbb{R}^n), \quad 0 < k \le n. \tag{9.74}$$

In light of Exercises 4 and 7, show that this leads to

$$\mathcal{H}^k(S^n) = \begin{cases} 0 & \text{if } 0 < k < n, \\ \mathbb{R} & \text{if } k = 0 \text{ or } n, \end{cases} \tag{9.75}$$

provided $n \ge 1$, giving therefore a demonstration of (8.56)–(8.57) different from that suggested in Exercise 9 of §8.

Exercises 11–13 establish the exactness of the sequence (9.67).

11. Show that $\ker \iota \subset \operatorname{im} \pi$. (*Hint*: Given $u \in C^k(\overline{M})$, $j^*u = dv$, pick $w \in \Lambda^{k-1}(\overline{M})$ such that $j^*w = v$, to get $u - dw \in C_r^\infty(\overline{M}, \Lambda^k)$, closed.)
12. Show that $\ker \delta \subset \operatorname{im} \iota$. (*Hint*: Given $\alpha \in C^k(\partial M)$, if $\alpha = j^*\beta$ with $[d\beta] = 0$ in $\mathcal{H}^{k+1}(M, \partial M)$, that is, $d\beta = d\tilde{\beta}$, $\tilde{\beta} \in C_r^\infty(\overline{M}, \Lambda^k)$, show that $[\alpha] = \iota[\beta - \tilde{\beta}]$.)
13. Show that $\ker \pi \subset \operatorname{im} \delta$. (*Hint*: Given $u \in C_R^k(\overline{M})$, if $u = dv$, $v \in \Lambda^{k-1}(\overline{M})$, show that $[u] = \delta[v]$.)
14. Applying (9.67) to $\overline{M} = \overline{B^{n+1}}$, the closed unit ball in $\mathbb{R}^{n+1}$, yields

$$\mathcal{H}^k(\overline{B^{n+1}}) \xrightarrow{\iota} \mathcal{H}^k(S^n) \xrightarrow{\delta} \mathcal{H}^{k+1}(\overline{B^{n+1}}, S^n) \xrightarrow{\pi} \mathcal{H}^{k+1}(\overline{B^{n+1}}). \tag{9.76}$$

Deduce that

$$\mathcal{H}^k(S^n) \approx \mathcal{H}^{k+1}(\overline{B^{n+1}}, S^n), \text{ for } k \ge 1,$$

since by (9.70) the endpoints of (9.76) vanish for $k \ge 1$. Then, by (9.71), there follows a third demonstration of the computation (9.75) of $\mathcal{H}^k(S^n)$.

15. Using Exercise 3, show that if M is connected and $\partial M \neq \emptyset$, the long exact sequence (9.67) begins with

$$0 \to \mathcal{H}^0(\overline{M}) \xrightarrow{\iota} \mathcal{H}^0(\partial M) \xrightarrow{\delta} \mathcal{H}^1(\overline{M}, \partial M) \to \cdots$$

and ends with

$$\cdots \to \mathcal{H}^{n-1}(\overline{M}) \xrightarrow{\iota} \mathcal{H}^{n-1}(\partial M) \xrightarrow{\delta} \mathcal{H}^n(\overline{M}, \partial M) \to 0.$$

16. Define the relative Euler characteristic

$$\chi(\overline{M}, \partial M) = \sum_{k\geq 0}(-1)^k \dim \mathcal{H}^k(\overline{M}, \partial M).$$

Define $\chi(\overline{M})$ and $\chi(\partial M)$ as in (8.51). Show that

$$\chi(\overline{M}) = \chi(\overline{M}, \partial M) + \chi(\partial M).$$

(*Hint*: Show that, for any exact sequence of the form

$$0 \to V_1 \to \cdots \to V_N \to 0,$$

with V_k finite-dimensional vector spaces over $\mathbb{R}$, $\sum(-1)^k \dim V_k = 0$.)

17. Using Poincaré duality show that if $\overline{M}$ is orientable, $n = \dim M$,

$$\chi(\overline{M}) = (-1)^n \chi(\overline{M}, \partial M).$$

Deduce that if n is odd and $\overline{M}$ orientable, $\chi(\partial M) = 2\chi(\overline{M})$.

18. If N is the double of $\overline{M}$, show that

$$\dim \mathcal{H}^k(N) = \dim \mathcal{H}^k(\overline{M}) + \dim \mathcal{H}^k(\overline{M}, \partial M).$$

Deduce that if M is orientable and $\dim M$ is even, then $\chi(N) = 2\chi(\overline{M})$.

In Exercises 19–21, let $\overline{\Omega}_j$ be compact, oriented manifolds of dimension n, with boundary. Assume that $\partial\Omega_j \neq \emptyset$ and that Ω_2 is connected. Let $F : \overline{\Omega}_1 \to \overline{\Omega}_2$ be a smooth map with the property that $f = F\big|_{\partial\Omega_1} : \partial\Omega_1 \to \partial\Omega_2$. Recall that we have defined Deg f in §19 of Chapter 1, when $\partial\Omega_2$ is connected.

19. Let $\sigma \in \Lambda^n(\overline{\Omega}_2)$ satisfy $\int_{\Omega_2} \sigma = 1$. Show that $\int_{\Omega_1} F^*\sigma$ is independent of the choice of such σ, using $\mathcal{H}^n(\overline{\Omega}_j, \partial\Omega_j) = \mathbb{R}$. Compare Lemma 19.6 of Chapter 1. Define

$$\text{Deg } F = \int_{\Omega_1} F^*\sigma.$$

20. Produce a formula for Deg F, similar to (19.16) of Chapter 1, making use of $F^{-1}(y_0)$, with $y_0 \in \Omega_2$.
21. Prove that Deg F = Deg f, assuming $\partial\Omega_2$ is connected.
(*Hint*: Pick $\omega \in \Lambda^{n-1}(\partial\Omega_2)$ such that $\int_{\partial\Omega_2} \omega = 1$, pick $\tilde{\omega} \in \Lambda^{n-1}(\overline{\Omega}_2)$ such that $j^*\tilde{\omega} = \omega$, and let $\sigma = d\tilde{\omega}$. Formulate an extension of this result to cases where $\partial\Omega_2$ has several connected components.)
22. Using the results of Exercises 19–21, establish the "argument principle," used in the proof of the Riemann mapping theorem in §4. (*Hint*: A holomorphic map is always orientation preserving.)

In Exercise 23, we assume that $\overline{M}$ is a compact manifold with boundary, with interior M. Define $\mathcal{H}^k(M)$ via the deRham complex, $d : \Lambda^k(M) \to \Lambda^{k+1}(M)$. It is desired to establish the isomorphism of this with $\mathcal{H}^k(\overline{M})$.

23. Let C be a small collar neighborhood of ∂M, so $\overline{M}_1 = \overline{M} \setminus C$ is diffeomorphic to $\overline{M}$. With $j : \overline{M}_1 \hookrightarrow M$, show that the pull-back $j^* : \Lambda^k(M) \to \Lambda^k(\overline{M}_1)$ induces an isomorphism of cohomology:

$$\mathcal{H}^k(M) \approx \mathcal{H}^k(\overline{M}_1).$$

(*Hint*: For part of the argument, it is useful to consider a smooth family

$$\varphi_t : \overline{M} \longrightarrow \overline{M}_t, \quad 0 \le t \le 1$$

of diffeomorphisms of $\overline{M}$ onto manifolds $\overline{M}_t$, with $\overline{M}_0 = \overline{M}$ and $\varphi_0 = id$. If $\beta \in \Lambda^k(M)$ and $d\beta = 0$, and if $\beta_1 = \varphi_1^* j^* \beta$, then

$$\beta = \beta_1 - d\Big(\int_0^1 \varphi_t^* \beta \rfloor X(t)\, dt\Big),$$

where $X(t)(x) = (d/dt)\varphi_t(x)$. Contrast this with the proof of Proposition 9.11.)

Exercises on spaces of gradient and divergence-free vector fields

In this problem set, we will work with the spaces

$$V_\sigma = \{v \in C^\infty(\overline{M}, \Lambda^1) : \delta v = 0 \text{ on } M, \iota_\nu v = 0 \text{ on } \partial M\} \subset H^1_A(M, \Lambda^1) \tag{9.77}$$

and

$$\mathcal{G} = \{dp : p \in H^1(M)\}. \tag{9.78}$$

We assume that $\overline{M}$ is a compact Riemannian manifold with boundary. These are spaces of 1-forms rather than vector fields, but recall that under the correspondence induced by the Riemannian metric, $\delta v \leftrightarrow \operatorname{div} V$ and $dp \leftrightarrow \operatorname{grad} p$.

1. Show that $V_\sigma \perp \mathcal{G}$.
2. Suppose $v \in L^2(M, \Lambda^1)$ is orthogonal to $\mathcal{G}$. Show that $\delta v = 0$ on M, that $\iota_\nu v$ exists on ∂M, and that $\iota_\nu v = 0$ on ∂M, as the identity

$$(v, dp)_{L^2} = (\delta v, p)_{L^2} + \int_{\partial M} \langle \iota_\nu v, p \rangle \, dS$$

is valid under these hypotheses. Conclude that $\mathcal{G}^\perp \subset \widetilde{V}_\sigma$, where

$$\widetilde{V}_\sigma = \{v \in L^2(M, \Lambda^1) : \delta v = 0 \text{ on } M, \iota_\nu v = 0 \text{ on } \partial M\}.$$

Show that actually $\mathcal{G}^\perp = \widetilde{V}_\sigma$. (*Hint*: The space $\{dp : p \in C^\infty(\overline{M})\}$ is dense in $\mathcal{G}$.)
3. Show that $\iota_\nu du\big|_{\partial M} = 0 \Rightarrow \iota_\nu \delta du\big|_{\partial M} = 0$. (*Hint*: Use (9.6) and (9.9).)
4. Show that $v \in L^2(M, \Lambda^1)$ is orthogonal to $\mathcal{G}$ if and only if its Hodge decomposition (9.43) takes the form

$$v = \delta d G^A v + P_1^A v.$$

(*Hint*: Show that $\delta d H^2_A(M, \Lambda^1) \perp \mathcal{G}$. To see this, use either (9.48) or Exercises 2–3.)
5. Deduce that

$$\mathcal{G}^\perp = \widetilde{V}_\sigma = \delta d H^2_A(M, \Lambda^1) \oplus \mathcal{H}^A_1 = \overline{V}_\sigma, \tag{9.79}$$

where $\overline{V}_\sigma$ denotes the closure of V_σ in $L^2(M, \Lambda^1)$, and that the decomposition

$$L^2(M, \Lambda^1) = \mathcal{G} \oplus \overline{V}_\sigma \tag{9.80}$$

is implemented by the Hodge decomposition (9.43), for $k = 1$.
(*Hint*: $H_A^2(M, \Lambda^1)$ has a dense subspace of smooth forms on $\overline{M}$.)

6. Deduce that if $u \in H^j(M, \Lambda^1)$, then its L^2-orthogonal projections onto $\mathcal{G}$ and onto $\overline{V}_\sigma$ belong to $H^j(M, \Lambda^1)$, $j \geq 0$.
7. From Exercise 4, it follows that $dH^1(M) = \mathcal{G} = d\delta H_A^2(M, \Lambda^1)$. Establish that in fact

$$H^1(M) = \delta H_A^2(M, \Lambda^1) + \mathbb{R},$$

via the Hodge decomposition for 0-forms,

$$L^2(M) = \delta d H_A^2(M) \oplus \mathcal{H}_0^A;\ \mathcal{H}_0^A = \mathbb{R}$$

(provided M is connected), where $H_A^2(M) = H_A^2(M, \Lambda^0)$ is given by the Neumann boundary condition. We have $u = \delta d G^A u + P_0^A u$, where

$$G^A : H^j(M) \xrightarrow{\approx} \left\{ v \in H_A^{j+2}(M) : \int_M v \, dV = 0 \right\}$$

comes from solving the Neumann problem.

10. Isothermal coordinates and conformal structures on surfaces

Let M be an oriented manifold of dimension 2, endowed with a Riemannian metric g. We aim to apply some results on the Dirichlet problem to prove the following result.

Proposition 10.1. *There exists a covering U_j of M and coordinate maps*

$$\varphi_j : U_j \longrightarrow \mathcal{O}_j \subset \mathbb{R}^2 \tag{10.1}$$

which are conformal (and orientation preserving).

By definition, a map $\varphi : U \to \mathcal{O}$ between two manifolds, with Riemannian metrics g and g_0, is *conformal* provided

$$\varphi^* g_0 = \lambda g, \tag{10.2}$$

for some positive $\lambda \in C^\infty(U)$. In (10.1), $\mathcal{O}_j$ is of course given the flat metric $dx^2 + dy^2$. Coordinates (10.1) that are conformal are also called "isothermal coordinates." It is clear that the composition of conformal maps is conformal, so if Proposition 10.1 holds, then the transition maps

$$\psi_{jk} = \varphi_j \circ \varphi_k^{-1} : \mathcal{O}_{jk} \longrightarrow \mathcal{O}_{kj} \tag{10.3}$$

are conformal, where $\mathcal{O}_{jk} = \varphi_k(U_j \cap U_k)$. This is particularly significant, in view of the following fact:

Proposition 10.2. *An orientation-preserving conformal map*

$$\psi : \mathcal{O} \longrightarrow \mathcal{O}' \tag{10.4}$$

between two open domains in $\mathbb{R}^2 = \mathbb{C}$ *is a holomorphic map.*

One way to see this is with the aid of the Hodge star operator $*$, introduced in §8, which maps $\Lambda^1(M)$ to $\Lambda^1(M)$ if $\dim M = 2$. Note that, for $M = \mathbb{R}^2$, with its standard orientation and flat metric,

$$*dx = dy, \quad *dy = -dx. \tag{10.5}$$

Since the action of a map (10.4) on 1-forms is given by

$$\begin{aligned} \psi^* dx &= \frac{\partial f}{\partial x}\, dx + \frac{\partial f}{\partial y}\, dy = df, \\ \psi^* dy &= \frac{\partial g}{\partial x}\, dx + \frac{\partial g}{\partial y}\, dy = dg, \end{aligned} \tag{10.6}$$

if $\psi(x, y) = (f, g)$, then the Cauchy-Riemann equations

$$\frac{\partial f}{\partial x} = \frac{\partial g}{\partial y}, \quad \frac{\partial g}{\partial x} = -\frac{\partial f}{\partial y} \quad (\text{i.e., } *df = dg) \tag{10.7}$$

are readily seen to be equivalent to the commutativity relation

$$* \circ (\psi^*) = (\psi^*) \circ * \text{ on 1-forms.} \tag{10.8}$$

Thus Proposition 10.2 is a consequence of the following:

Proposition 10.3. *If M is oriented and of dimension 2, then the Hodge star operator $* : T_p^*M \to T_p^*M$ is conformally invariant.*

In fact, in this case, $*$ can be simply characterized as counterclockwise rotation by 90°, as can be seen by picking a coordinate system centered at $p \in M$ such that $g_{jk} = \delta_{jk}$ at p and using (10.5). This characterization of $*$ is clearly conformally invariant.

Thus Proposition 10.1 implies that an oriented, two-dimensional Riemannian manifold has an associated complex structure. A manifold of (real) dimension two with a complex structure is called a Riemann surface.

To begin the proof of Proposition 10.1, we note that it suffices to show that, for any $p \in M$, there exists a neighborhood U of p and a coordinate map

$$\psi = (f, g) : U \to \mathcal{O} \subset \mathbb{R}^2, \tag{10.9}$$

which is conformal. If $df(p)$ and $dg(p)$ are linearly independent, the map (f, g) will be a coordinate map on some neighborhood of p, and (f, g) will be conformal provided

$$*df = dg. \tag{10.10}$$

Note that if $df(p) \neq 0$, then $df(p)$ and $dg(p)$ are linearly independent. Suppose $f \in C^\infty(U)$ is given. Then, by the Poincaré lemma, if U is diffeomorphic to a disk, there will exist a $g \in C^\infty(U)$ satisfying (10.10) precisely when

$$d * df = 0. \tag{10.11}$$

Now, as we saw in §8, the Laplace operator on $C^\infty(M)$ is given by

$$\Delta f = -\delta df = - * d * df, \tag{10.12}$$

when $\dim M = 2$, so (10.11) is simply the statement that f is a harmonic function on U. Thus Proposition 10.1 will be proved once we establish the following.

Proposition 10.4. *There is a neighborhood U of p and a function $f \in C^\infty(U)$ such that $\Delta f = 0$ on U and $df(p) \neq 0$.*

Proof. In a coordinate system $x = (x_1, x_2)$, we have

$$\begin{aligned}\Delta f(x) &= g(x)^{-1/2}\, \partial_j \big(g^{jk}(x) g(x)^{1/2}\, \partial_k f\big)\\ &= g^{jk}(x)\, \partial_j \partial_k f + b^k(x)\, \partial_k f.\end{aligned}$$

Pick some coordinate system centered at p, identifying the unit disk $\mathcal{D} \subset \mathbb{R}^2$ with some neighborhood U_1 of p. Now dilate the variables by a factor ε, to map the small neighborhood U_ε of p (the image of the disk $\mathcal{D}_\varepsilon$ of radius ε in the original coordinate system) onto the unit disk $\mathcal{D}$. In this dilated coordinate system, we have

$$\Delta f(x) = g^{jk}(\varepsilon x)\, \partial_j \partial_k f + \varepsilon b^k(\varepsilon x)\, \partial_k f. \tag{10.13}$$

Now we define $f = f_\varepsilon$ to be the harmonic function on U_ε equal to x_1/ε on ∂U_ε (in the original coordinate system), hence to x_1 on $\partial \mathcal{D}$ in the dilated coordinate system. We need only show that, for $\varepsilon > 0$ sufficiently small, we can guarantee that $df_\varepsilon(p) \neq 0$.

To see this note that, in the dilated coordinate system, we can write

$$f_\varepsilon = x_1 - \varepsilon v_\varepsilon \ \text{ on } \mathcal{D}, \tag{10.14}$$

where v_ε is defined by

$$\Delta_\varepsilon v_\varepsilon = b^1(\varepsilon x) \ \text{ on } \mathcal{D}, \quad v_\varepsilon\big|_{\partial\mathcal{D}} = 0, \tag{10.15}$$

and Δ_ε is given by (10.13). Now the regularity estimates of Theorem 1.3 hold uniformly in $\varepsilon \in (0, 1]$ in this case, so we have uniform estimates in $H^k(\mathcal{D})$ on v_ε as $\varepsilon \to 0$ for each k, and consequently uniform estimates on v_ε in $C^1(\overline{\mathcal{D}})$ as $\varepsilon \to 0$. This shows that $df_\varepsilon(p) \neq 0$ for ε small and completes the proof.

Exercises

1. Suppose M is an n-dimensional, oriented manifold, with metric tensor g. Let $g' = e^u g$ be a new metric tensor. Use these two metrics to define Hodge star operators $*$ and $*'$,

respectively. Show that

$$*' = e^{(\frac{n}{2}-j)u} * \quad \text{on } \Lambda^j(M).$$

In particular, if $n = 2k$ is even, $*' = *$ on $\Lambda^k(M)$.

2. Express $\delta' u$ in terms of δu and other operators, when $u \in \Lambda^j(M)$, where δ' is the analogue of δ when g is replaced by g'. Do the same for $d\delta' u$, $\delta' du$, and $\Delta' u$.
3. Show that if $n = 2$, $u \in \Lambda^0(M)$, then $\Delta u = 0$ if and only if $\Delta' u = 0$.
4. If M is *compact* and $n = 2k$, show that $u \in \Lambda^k(M)$ is a harmonic form for g if and only if it is a harmonic form for g'.

11. General elliptic boundary problems

An elliptic differential operator of order m on a manifold M is an operator that in local coordinates has the form

$$P(x, D)u = \sum_{|\alpha| \le m} a_\alpha(x) D^\alpha u, \tag{11.1}$$

and whose principal symbol

$$P_m(x, \xi) = \sum_{|\alpha| = m} a_\alpha(x) \xi^\alpha \tag{11.2}$$

is invertible for nonzero $\xi \in \mathbb{R}^n$ ($n =$ dim M). Here $P(x, D)$ could be a $k \times k$ system, or it could map sections of a vector bundle E_0 to sections of E_1. We will assume the coefficients of $P(x, D)$ are smooth. If M is the interior of $\overline{M}$, a compact, smooth manifold with boundary, we require the coefficients to be smooth on $\overline{M}$, and we also want $P_m(x, \xi)$ to be invertible for $x \in \overline{M}$, $\xi \neq 0$. If $\partial M \neq \emptyset$, there will be various boundary conditions to study.

First we study interior regularity for solutions to $P(x, D)u = f$. In Chapter 3 we treated this for constant-coefficient elliptic operators $P(D)$. We will exploit the technique (which was used in the proof of Lemma 9.3) of freezing the coefficients of $P(x, D)$ and using estimates on the resulting constant-coefficient operators. Our interior regularity result is the following.

Theorem 11.1. *If $P(x, D)$ is elliptic of order m and $u \in \mathcal{D}'(M)$, $P(x, D)u = f \in H^s(M)$, then $u \in H^{s+m}_{\text{loc}}(M)$, and, for each $U \subset\subset V \subset\subset M, \sigma < s + m$, there is an estimate*

$$\|u\|_{H^{s+m}(U)} \le C\|P(x, D)u\|_{H^s(V)} + C\|u\|_{H^\sigma(V)}. \tag{11.3}$$

For the proof, we can assume $u\big|_V$ belongs to some Sobolev space, say $u \in H^\tau(V)$. We will first establish the estimate

$$\|u\|_{H^\tau(U)} \le C\|P(x, D)u\|_{H^{\tau-m}(V)} + C\|u\|_{H^{\tau-1}(V)}. \tag{11.4}$$

Once this is done, we will establish $u \in H^{\tau+1}(V)$ if $\tau - m + 1 \le s$, with an analogous estimate, following in outline the program used in §1.

If we pick $\chi \in C_0^\infty(V)$, then

$$P(x, D)(\chi u) = \chi(x) P(x, D)u + Q(x, D)u, \tag{11.5}$$

where $Q(x, D) = [P(x, D), \chi]$ is a differential operator of order $m - 1$, so

$$\|Q(x, D)u\|_{H^{\tau-m}(V)} \leq C\|u\|_{H^{\tau-1}(V)}.$$

Hence, just as in the chain of reasoning involving (1.22), we can localize the task of proving (11.4). We can suppose $u \in H^\tau(V)$ is compactly supported and that V is an open set in $\mathbb{R}^n$, and establish (11.4) in that case.

The next step will be to apply cut-offs with very small support, to effect the freezing of coefficients. Let $\Lambda = \Lambda_\varepsilon$ be the lattice $\varepsilon\mathbb{Z}^n = \{\varepsilon j : j \in \mathbb{Z}^n\}$. Take a partition of unity on $\mathbb{R}^n$ of the form $1 = \sum_{j \in \mathbb{Z}^n} \chi_j(x)$, with $\chi_j(x) = \chi_0(x - j)$, $\chi_0(x) \in C_0^\infty(\mathbb{R}^n)$ supported in $-1 < x_\nu < 1$. Then define a partition of unity

$$1 = \sum_{\lambda \in \Lambda} \chi_\lambda(x), \tag{11.6}$$

with $\chi_\lambda(x) = \chi_0\big((x - \lambda)/\varepsilon\big)$, when $\Lambda = \Lambda_\varepsilon$. We will suppress the dependence on ε in the notation, though of course this dependence is very important.

Now, for each $\lambda \in \Lambda$, set

$$P_\lambda(D) = P(\lambda, D). \tag{11.7}$$

This is the constant-coefficient elliptic operator obtained by freezing the coefficients of $P(x, D)$ at $x = \lambda$. If $P(x, D)u = f$, then

$$\chi_\lambda(x) P_\lambda(D)u = \chi_\lambda f - R_\lambda(x, D)u, \tag{11.8}$$

where

$$\begin{aligned} R_\lambda(x, D) &= \chi_\lambda(x)\big[P(x, D) - P_\lambda(D)\big] \\ &= \chi_\lambda(x) \sum_{|\alpha| \leq m} \big[a_\alpha(x) - a_\alpha(\lambda)\big] D^\alpha. \end{aligned} \tag{11.9}$$

Therefore

$$P_\lambda(D)(\chi_\lambda u) = \chi_\lambda f - R_\lambda(x, D)u - Q_\lambda(x, D)u, \tag{11.10}$$

where

$$Q_\lambda(x, D) = [\chi_\lambda, P_\lambda(D)] \text{ has order } m - 1. \tag{11.11}$$

Now the functions $P_\lambda(\xi)$ are all bounded away from zero on a set

$$\{\xi \in \mathbb{R}^n : |\xi| \geq K\}. \tag{11.12}$$

Thus, taking a cut-off $\varphi(\xi) \in C_0^\infty(\mathbb{R}^n)$, equal to 1 on $|\xi| \leq K$, we can set

$$E_\lambda(\xi) = \big(1 - \varphi(\xi)\big) P_\lambda(\xi)^{-1}. \tag{11.13}$$

Then, as seen in (9.4) of Chapter 3,

$$E_\lambda(\xi) \in S_1^{-m}(\mathbb{R}^n), \tag{11.14}$$

which is to say, there are estimates

$$|D^\alpha E_\lambda(\xi)| \le C_\alpha \langle \xi \rangle^{-m-|\alpha|}. \tag{11.15}$$

We have

$$E_\lambda(D) P_\lambda(D) = I + \rho_\lambda(D), \tag{11.16}$$

with $\rho_\lambda(\xi) \in C_0^\infty(\mathbb{R}^n)$. Furthermore, E_λ and ρ_λ are bounded in their respective spaces, for all $\lambda \in \Lambda_\varepsilon$, independently of $\varepsilon \in (0, 1]$. Applying $E_\lambda(D)$ to each side of (11.10), and summing over λ, we have

$$u = F_\Lambda - \sum_{\lambda \in \Lambda} \Big\{ E_\lambda(D) R_\lambda(x, D) u + E_\lambda(D) Q_\lambda(x, D) u + \rho_\lambda(D)(\chi_\lambda u) \Big\}, \tag{11.17}$$

where

$$F_\Lambda = \sum_{\lambda \in \Lambda} E_\lambda(D)(\chi_\lambda f). \tag{11.18}$$

To prove (11.4), we need (11.15) only for $\alpha = 0$, which implies

$$E_\lambda(D) : H^\sigma(\mathbb{R}^n) \longrightarrow H^{\sigma+m}(\mathbb{R}^n), \tag{11.19}$$

for all $\sigma \in \mathbb{R}$, with norm bound independent of λ, ε, and σ. Since at this point u has compact support in $V \subset \mathbb{R}^n$, all our Sobolev norms in the estimates (11.20)–(11.22) below can be taken to be $H^\gamma(\mathbb{R}^n)$-norms, for various γ. Looking at the first term on the right side of (11.17), we see that

$$\|F_\Lambda\|_{H^\tau} \le C(\varepsilon, \tau) \|f\|_{H^{\tau-m}}. \tag{11.20}$$

In view of (11.11) and the compact support of $\rho_\lambda(\xi)$,

$$\sum_\lambda \big\| E_\lambda(D) Q_\lambda(x, D) u + \rho_\lambda(D)(\chi_\lambda u) \big\|_{H^\tau} \le C(\varepsilon, \tau) \|u\|_{H^{\tau-1}}, \tag{11.21}$$

and by (11.9), we have

$$\Big\| \sum_\lambda E_\lambda(D) R_\lambda(x, D) u \Big\|_{H^\tau} \le C(\tau) \varepsilon \|u\|_{H^\tau} + C(\varepsilon, \tau) \|u\|_{H^{\tau-1}}, \tag{11.22}$$

where $C(\tau)$ is independent of ε. Thus, when we estimate the H^τ-norm of (11.17), the term $C(\tau)\varepsilon\|u\|_{H^\tau}$ can be absorbed into the left side, for $\varepsilon > 0$ sufficiently small. We obtain then the estimate (11.4).

Passing from (11.4) to $H^{\tau+1}$-regularity of u, given $f \in H^{\tau+1-m}$, involves an argument similar to (1.23)–(1.28). Recall we have $u \in H^\tau(\mathbb{R}^n)$, compactly supported. With the difference operators $D_{j,h}$ defined by (1.23), we apply (11.4) to $D_{j,h} u$, obtaining

$$\|D_{j,h} u\|_{H^\tau}^2 \le C \|P(x, D) D_{j,h} u\|_{H^{\tau-m}}^2 + C \|D_{j,h} u\|_{H^{\tau-1}}^2. \tag{11.23}$$

As in (1.24), we replace $P(x, D) D_{j,h}$ by $D_{j,h} P(x, D)$ plus the commutator $[P(x, D), D_{j,h}]$, and use

$$[D_{j,h}, a_\alpha(x) D^\alpha] u = -(D_{j,h} a_\alpha) D^\alpha \tau_{j,h} u, \tag{11.24}$$

where $\tau_{j,h}u(x) = u(x + he_j)$, as in (1.23). Hence

$$\|[D_{j,h}, P(x, D)]u\|^2_{H^{\tau-m}} \le C\|u\|^2_{H^\tau}. \tag{11.25}$$

Thus (11.23) yields

$$\|D_{j,h}u\|^2_{H^\tau} \le C\|u\|^2_{H^\tau} + C\|f\|^2_{H^{\tau-m+1}}, \tag{11.26}$$

and hence taking $h \to 0$ gives $u \in H^{\tau+1}$ and

$$\|u\|^2_{H^{\tau+1}} \le C\|P(x, D)u\|^2_{H^{\tau-m+1}} + C\|f\|^2_{H^\tau}. \tag{11.27}$$

With this advance over (11.4), we have a proof of Theorem 11.1, by a straightforward iteration.

We turn now to boundary conditions. In addition to having the elliptic operator P on $\overline{M}$, we suppose there are differential operators B_j, $j = 1, \dots, \ell$, of order $m_j \le m - 1$, defined on a neighborhood of ∂M, and we consider

$$Pu = f \text{ on } M, \quad B_j u = g_j \text{ on } \partial M, \ 1 \le j \le \ell. \tag{11.28}$$

When $\overline{M}$ is a compact, smooth manifold with boundary, we seek estimates of the form

$$\begin{aligned}\|u\|^2_{H^{m+k}(M)} &\le C\|Pu\|^2_{H^k(M)} \\ &\quad + C\sum_j \|B_j u\|^2_{H^{m+k-m_j-1/2}(\partial M)} + C\|u\|^2_{H^{m+k-1}(M)}\end{aligned} \tag{11.29}$$

and corresponding regularity theorems. Such estimates are called *coercive* estimates.

Applying a cut-off as in (11.5), we see that it suffices to establish the estimate (11.29) for u supported near ∂M, indeed, for u supported in a boundary coordinate patch.

We now introduce the hypothesis of regularity upon freezing coefficients. Given $q \in \partial M$, pick a coordinate neighborhood $\mathcal{O}$ of q, mapped diffeomorphically onto a compact subset $\mathcal{O}'$ of $\overline{\mathbb{R}^n_+} = \{x \in \mathbb{R}^n : x_n \ge 0\}$. The operators P and B_j are transformed to operators on functions on $\mathcal{O}'$. Now freeze their coefficients at q, obtaining constant-coefficient operators $P_q(D)$ and $B_{qj}(D)$. The hypothesis of regularity upon freezing coefficients is that there are estimates

$$\begin{aligned}\|u\|^2_{H^{m+k}} &\le C\|P_q(D)u\|^2_{H^k} \\ &\quad + C\sum_j |B_{jq}(D)u|^2_{H^{m+k-m_j-1/2}} + C\|u\|^2_{H^{m+k-1}},\end{aligned} \tag{11.30}$$

valid for smooth u with bounded support in $\overline{\mathbb{R}^n_+}$, with constants C uniform in $q \in \partial M$. Here, $\|u\|_{H^k} = \|u\|_{H^k(\overline{\mathbb{R}^n_+})}$ and $|v|_{H^s} = \|v\|_{H^s(\mathbb{R}^{n-1})}$. The following result reduces the study of (11.29) to the constant-coefficient case.

Proposition 11.2. *Suppose P is elliptic on $\overline{M}$ and the boundary problem (11.28) satisfies the hypothesis of regularity upon freezing coefficients. Given*

$u \in H^m(M), k \in \mathbb{Z}^+$, if $Pu \in H^k(M)$ and $B_j u \in H^{m+k-m_j-1/2}(\partial M)$, then $u \in H^{m+k}(M)$, and the estimate (11.29) holds.

To prove this, let $\Lambda = \Lambda_\varepsilon$ be the lattice in $\mathbb{R}^n$ used before, except we restrict attention to $\lambda \in \overline{\mathbb{R}^n_+}$. We use the partition of unity (11.6). For $P_\lambda(D)(\chi_\lambda u)$, we still have (11.10), and similarly, if $\lambda \in \mathbb{R}^{n-1} \subset \overline{\mathbb{R}^n_+}$,

$$B_{j\lambda}(D)(\chi_\lambda u) = \chi_\lambda g_j - R_{j\lambda}(x, D)u - Q_{j\lambda}(x, D)u, \tag{11.31}$$

where

$$R_{j\lambda}(x, D) = \chi_\lambda(x)\big[Q_j(x, D) - Q_{j\lambda}(D)\big] \tag{11.32}$$

and

$$Q_{j\lambda}(x, D) = [\chi_\lambda, Q_{j\lambda}(D)] \text{ has order } m_j - 1. \tag{11.33}$$

Thus, granted the hypothesis of Proposition 1.2, for each $\lambda \in \mathbb{R}^{n-1} \cap \Lambda_\varepsilon$, we have an estimate

$$\begin{aligned} &\|\chi_\lambda u\|_{H^{m+k}} \\ &\le C\Big[\|\chi_\lambda f\|_{H^k} + \|R_\lambda(x, D)u\|_{H^k} + \|Q_\lambda(x, D)u\|_{H^k} + \|\chi_\lambda u\|_{H^{m+k-1}}\Big] \\ &\quad + C\sum_j \Big[|\chi_\lambda g_j|_{H^{\mu(j,k)}} + |R_{j\lambda}(x, D)u|_{H^{\mu(j,k)}} + |Q_{j\lambda}(x, D)u|_{H^{\mu(j,k)}}\Big], \end{aligned} \tag{11.34}$$

where, in the last three terms, $\mu(j, k) = m + k - m_j - 1/2$. If $\lambda \in \Lambda \setminus \mathbb{R}^{n-1}$, we can estimate $\|\chi_\lambda u\|_{H^{m+k}}$ by the sum of the first three terms on the right. Summing over λ, we get an estimate for $\|u\|_{H^{m+k}}$. Note that

$$\sum_\lambda \|R_\lambda(x, D)u\|_{H^k} \le C\varepsilon\|u\|_{H^{m+k}} + C(\varepsilon)\|u\|_{H^{m+k-1}}, \tag{11.35}$$

as in (11.22). Using the trace theorem, we can also estimate the quantity $\sum_{j,\lambda} |R_{j\lambda}(x, D)u|_{H^{\mu(j,k)}}$ by the right side of (11.35), and we can also use the trace theorem to estimate $\sum_{j,\lambda} |Q_{j\lambda}(x, D)u|_{H^{\mu(j,k)}}$. We can absorb the $C\varepsilon\|u\|_{H^{m+k}}$ into the left side, obtaining the estimate (11.29), given $u \in H^{m+k}$.

To obtain the associated regularity theorem, we use the difference quotients $D_{j,h}$, $1 \le j \le n-1$, as in (1.23)–(1.32). Given $u \in H^{m+k}$ while $f \in H^{k+1}$, $g_j \in H^{m+k-m_j+1/2}$, if we apply (11.29) to $D_{j,h}u$ (localized to have support in a coordinate patch) and use (11.24) together with the analogous result for $[D_{j,h}, B_i(x, D)]$, just as in (11.26) and (11.27), we get $D_j u \in H^{m+k}$, for $1 \le j \le n-1$, and

$$\|D_j u\|^2_{H^{m+k}} \le C\|Pu\|^2_{H^{k+1}} + C\sum |B_i u|_{H^{m+k-m_i+\frac{1}{2}}} + C\|u\|^2_{H^{m+k}}, \tag{11.36}$$

for $1 \le j \le n$. From here, as in (1.29)–(1.32), we proceed as follows. We need to know that $D_n u \in H^{m+k}$, that is,

$$D^\alpha D_n u \in H^{k+1}, \quad |\alpha| \le m - 1. \tag{11.37}$$

Now if $D^\alpha D_n \neq D_n^m$, we can write $D^\alpha D_n u = D_j D^\beta u$, with $|\beta| \le m-1$, $1 \le j \le n-1$, and conclude from (11.36) that this belongs to H^{k+1}, with an appropriate bound. Finally, to estimate $D_n^m u$, we use the PDE $Pu = f$ to write

$$D_n^m u = a(x) f - \sum_{|\alpha| \le m} b_\alpha(x) D^\alpha u, \tag{11.38}$$

where $D^\alpha \neq D_n^m$ in the last sum, and then estimate the H^{k+1}-norm of the right side of (11.38) by $\|aPu\|_{H^{k+1}}$ plus the right side of (11.36). This completes the proof of Proposition 11.2.

We now turn to the problem of establishing an estimate of the form (11.30), for constant-coefficient operators, that is,

$$\begin{aligned} \|u\|^2_{H^{m+k}} &\le C\|P(D)u\|^2_{H^k} \\ &\quad + C\sum_j |B_j(D)u|^2_{H^{m+k-m_j-1/2}} + C\|u\|^2_{H^{m+k-1}}. \end{aligned} \tag{11.39}$$

We will take $u \in \mathcal{S}(\overline{\mathbb{R}^n_+})$, that is, u will be the restriction to $\overline{\mathbb{R}^n_+}$ of an element of $\mathcal{S}(\mathbb{R}^n)$. Also, we will assume that u vanishes for $x_n \ge 1$. It is convenient to relabel the coordinate variables; set $x = (x_1, \dots, x_{n-1})$, $y = x_n$. We write $P(D)$ in the form

$$P(D) = \frac{\partial^m}{\partial y^m} + \sum_{j=0}^{m-1} A_j(D_x) \frac{\partial^j}{\partial y^j}, \quad \text{order } A_j(D_x) = m - j. \tag{11.40}$$

We convert $P(D)u = f$ to a first-order system; set $v = (v_1, \dots, v_m)^t$, with

$$v_1 = \Lambda^{m-1} u, \dots, v_j = \partial_y^{j-1} \Lambda^{m-j} u, \dots, v_m = \partial_y^{m-1} u. \tag{11.41}$$

Then $P(D)u = f$ becomes the system

$$\frac{\partial v}{\partial y} = K(D_x) v + F, \tag{11.42}$$

where $F = (0, \dots, 0, f)^t$ and

$$K = \begin{pmatrix} 0 & \Lambda & & & \\ & 0 & \Lambda & & \\ & & \ddots & \ddots & \\ & & & & \Lambda \\ C_0 & C_1 & C_2 & \dots & C_{m-1} \end{pmatrix}, \tag{11.43}$$

where

$$C_j = -A_j(D_x) \Lambda^{1-(m-j)}. \tag{11.44}$$

As in Chapter 4, we define $\Lambda : H^s \to H^{s-1}$ by

$$(\Lambda u)\widehat{\ }(\xi) = \langle \xi \rangle \hat{u}(\xi). \tag{11.45}$$

Note that the matrix entries of K are not differential operators, though they are well-behaved Fourier multipliers:

$$K : H^s(\mathbb{R}^{n-1}) \longrightarrow H^{s-1}(\mathbb{R}^{n-1}). \tag{11.46}$$

In fact, $K \in S_1^1(\mathbb{R}^{n-1})$, that is, estimates of the form (11.15) hold for $D^\alpha K(\xi)$, with $-m$ replaced by 1. This fact will be explored further in Chapter 7. Now let us note that $K(\xi) = K_1(\xi) + K_0(\xi)$, where K_0 is bounded and $K_1(\xi)$ is homogeneous of degree 1; $K_1(\xi)$ has the form (11.43) with Λ replaced by $|\xi|$ and A_j replaced by the principal symbol $A_j^0(\xi)$, homogeneous of degree $m - j$.

Lemma 11.3. *The operator $P(D)$ is elliptic if and only if, for all nonzero $\xi \in \mathbb{R}^{n-1}$, $K_1(\xi)$ has no purely imaginary eigenvalues.*

Proof. Indeed, $\det(i\eta - K_1(\xi)) = P_m(\xi, \eta)$ is the principal symbol of the operator (11.40) if $P(D)$ is scalar. If $P(D)$ is a $k \times k$ system, the equivalence of $P_m(\xi, \eta)$ having a nonzero eigenvector in $\mathbb{C}^k$ and of $i\eta - K_1(\xi)$ having a nonzero eigenvector in $\mathbb{C}^{km}$ follows by the same reasoning as the reduction of $P(D)u = f$ to (11.42).

We also rewrite the boundary conditions $B_j(D)u = g_j$ at $x_n = 0$. Let

$$B_j = B_j\Big(D_x, \frac{\partial}{\partial y}\Big) = \sum_{k \le m_j} b_{jk}(D_x) \frac{\partial^k}{\partial y^k}. \tag{11.47}$$

Then we have the boundary conditions

$$\sum_{k \le m_j} b_{jk}(D_x) \Lambda^{m_j - k} v_k(0) = \Lambda^{m - m_j - 1} g_j = h_j, \quad 1 \le j \le \ell, \tag{11.48}$$

which we write as

$$B(D_x)v(0) = h, \tag{11.49}$$

with $B(\xi) \in S_1^0(\mathbb{R}^{n-1})$.

The estimate (11.39) translates to the estimate

$$\|v\|^2_{H^{k+1}} \le C\|Lv\|^2_{H^k} + C|Bv(0)|^2_{H^{k+1/2}} + C\|v\|^2_{H^k}, \tag{11.50}$$

where

$$Lv = \Big[\frac{\partial}{\partial y} - K(D_x)\Big]v, \tag{11.51}$$

and we assume $v \in \mathcal{S}(\overline{\mathbb{R}^n_+})$, with $v(y) = v(y, \cdot) = 0$ for $y \ge 1$.

We want to decouple the equation (11.42) into a forward evolution equation and a backward evolution equation. Let $\gamma = \gamma(\xi)$ be a curve in the right half-plane of $\mathbb{C}$, encircling all the eigenvalues of $K_1(\xi)$ with positive real part, and set

$$E_0(\xi) = \frac{1}{2\pi i} \int_\gamma \big(\zeta - K_1(\xi)\big)^{-1} d\zeta. \tag{11.52}$$

Then $E_0(\xi)$ is smooth on $\mathbb{R}^{n-1} \setminus 0$, homogeneous of degree zero, and, for each ξ, it is a projection onto the sum of the generalized eigenspaces of $K_1(\xi)$ corresponding to eigenvalues of positive real part, while $I - E_0(\xi)$ similarly captures the spectrum of $K_1(\xi)$ with negative real part. If we set

$$A_1(\xi) = \big(2E_0(\xi) - I\big)K_1(\xi), \tag{11.53}$$

then $A_1(\xi)$ is homogeneous of degree 1 in ξ and its eigenvalues all have positive real part, for $\xi \neq 0$. We want to construct a new inner product on $L^2(\mathbb{R}^{n-1})$ with respect to which $-A_1(D_x)$ is "dissipative."

Lemma 11.4. *Let $\mathcal{M}_K^+$ denote the space of complex $K \times K$ matrices with spectrum in $\operatorname{Re} z > 0$, and let $\mathcal{P}_K^+$ be the space of positive-definite, complex $K \times K$ matrices. There is a smooth map*

$$\Phi : \mathcal{M}_K^+ \longrightarrow \mathcal{P}_K^+, \tag{11.54}$$

homogeneous of degree 0, such that

$$A \in \mathcal{M}_K^+,\ P = \Phi(A) \Longrightarrow PA + A^*P \in \mathcal{P}_K^+. \tag{11.55}$$

Proof. First we observe that if $A_0 \in \mathcal{M}_K^+$ is fixed, there exists $P_0 \in \mathcal{P}_K^+$ such that $P_0A_0 + A_0^*P_0 \in \mathcal{P}_K^+$. To see this, use the Jordan normal form to make A_0 similar to B_ε, where B_ε has the eigenvalues of A_0 on its diagonal, εs and 0s right above the diagonal, and 0s elsewhere. Pick ε small compared to the real part of each eigenvalue. Declaring the basis of $\mathbb{C}^K$ with respect to which A_0 takes the form B_ε to be orthonormal specifies a new Hermitian inner product on $\mathbb{C}^K$, of the form $((u, v)) = (P_0u, v)$, where (u, v) is the standard inner product, and this P_0 works.

Thus, given $A \in \mathcal{M}_K^+$, the set $\mathcal{P}(A)$ of $P \in \mathcal{P}_K^+$ such that $PA + A^*P \in \mathcal{P}_K^+$ is nonempty. One readily verifies that $\mathcal{P}(A)$ is an open convex set. Furthermore, given $P \in \mathcal{P}_K^+$, the set of $A \in \mathcal{M}_K^+$ such that $PA + A^*P \in \mathcal{P}_K^+$ is open. The existence of Φ satisfying the conditions of the lemma now follows by a partition of unity argument.

Corollary 11.5. *Given $A_1(\xi)$ constructed by (11.53), there exists $P_0(\xi)$, smooth on $\mathbb{R}^{n-1} \setminus 0$, homogeneous of degree 0, such that both $P_0(\xi)$ and $P_0(\xi)A_1(\xi) + A_1^*(\xi)P_0(\xi)$ are positive-definite, for all $\xi \neq 0$. In fact, for some $a > 0$,*

$$P_0(\xi)A_1(\xi) + A_1^*(\xi)P_0(\xi) \geq a|\xi|I. \tag{11.56}$$

With (u, v) denoting the inner product in $L^2(\mathbb{R}^{n-1})$, we have

$$\begin{aligned} \frac{d}{dy}(P_0\Lambda^{1/2}E_0v, \Lambda^{1/2}E_0v) &= 2\operatorname{Re}\,(P_0\Lambda^{1/2}E_0\frac{\partial v}{\partial y}, \Lambda^{1/2}E_0v) \\ &= 2\operatorname{Re}\,(P_0E_0K\Lambda^{1/2}v, \Lambda^{1/2}E_0v) + 2\operatorname{Re}\,(P_0\Lambda^{1/2}E_0F, \Lambda^{1/2}E_0v), \end{aligned} \tag{11.57}$$

given $Lu = F$. Now $E_0 = (2E_0 - I)E_0$ implies $P_0E_0K_1 = P_0A_1E_0$, so

$$\begin{aligned} 2\operatorname{Re}\,(P_0E_0K\Lambda^{1/2}v, \Lambda^{1/2}E_0v) &= ([P_0A_1 + A_1^*P_0]E_0\Lambda^{1/2}v, E_0\Lambda^{1/2}v) \\ &\quad + 2\operatorname{Re}\,(P_0\Lambda^{1/2}E_0K_0v, \Lambda^{1/2}v). \end{aligned} \tag{11.58}$$

Thus integrating (11.57) over $0 \le y \le 1$ and using (11.56) yield

$$\begin{aligned}(11.59)\qquad &|E_0 v(1)|^2_{1/2} - |E_0 v(0)|^2_{1/2} \\ &\ge C\|E_0 v\|^2_{(0,1)} - C' \int_0^1 |F(y)|_0 \cdot |v(y)|_1 \, dy - C' \|v\|^2_{(0,1/2)},\end{aligned}$$

where, for simplicity of notation, we have set

$$|w|_s^2 = \|w\|^2_{H^s(\mathbb{R}^{n-1})},$$

and we define

$$(11.60)\qquad \|v\|^2_{(0,s)} = \int_0^1 |v(y)|_s^2 \, dy = \int_0^1 \|\Lambda^s v(y)\|^2_{L^2(R^{n-1})} \, dy.$$

More generally, it will be convenient to define the Sobolev-like spaces $H_{(k,s)}(\Omega)$, for $\Omega = [0,1] \times \mathbb{R}^{n-1}$, by

$$(11.61)\qquad \|v\|^2_{(k,s)} = \sum_{j=0}^{k} \int_0^1 \|D_y^j \Lambda^{k-j+s} v(y)\|^2_{L^2(\mathbb{R}^{n-1})} \, dy.$$

Note that $H_{(k,0)}(\Omega) = H^k(\Omega)$. We also note that the standard trace theorem generalizes naturally to

$$(11.62)\qquad \tau : H_{(k,s)}(\Omega) \longrightarrow H^{k+s-1/2}(\mathbb{R}^{n-1}), \quad \tau u(x') = u(0,x').$$

Changing the constants C and C', we can replace $\|v\|^2_{(0,1/2)}$ in (11.59) by $\|v\|^2_{(0,\sigma)}$, for any $\sigma < 1$ (e.g., $\sigma \le 0$). Also, we can write

$$\int_0^1 |F(y)|_0 \cdot |v(y)|_1 \, dy \le \frac{\varepsilon}{2} \|v\|^2_{(0,1)} + \frac{1}{2\varepsilon} \|F\|^2_{(0,0)},$$

and picking $\varepsilon/2 < C/2C'$, obtain from (11.59) the estimate

$$(11.63)\qquad \|v\|^2_{(0,1)} + |E_0 v(0)|^2_{1/2} \le C\|Lv\|^2_{(0,0)} + C|E_0 v(1)|^2_{1/2} + C\|v\|^2_{(0,\sigma)}.$$

Replacing v by $\Lambda^s v$, we have the estimate

$$\begin{aligned}(11.64)\qquad &\|E_0 v\|^2_{(0,s+1)} + |E_0 v(0)|^2_{s+1/2} \\ &\le C\|Lv\|^2_{(0,s)} + C|E_0 v(1)|^2_{s+1/2} + C\|v\|^2_{(0,\sigma)},\end{aligned}$$

for any $\sigma < s+1$. Similarly, with $E_1 = I - E_0$,

$$\begin{aligned}(11.65)\qquad &\|E_1 v\|^2_{(0,s+1)} + |E_1 v(1)|^2_{s+1/2} \\ &\le C\|Lv\|^2_{(0,s)} + C|E_1 v(0)|^2_{s+1/2} + C\|v\|^2_{(0,\sigma)}.\end{aligned}$$

Summing the last two estimates, we have

$$\begin{aligned}(11.66)\qquad &\|v\|^2_{(0,s+1)} + |E_0 v(0)|^2_{s+1/2} + |E_1 v(1)|^2_{s+1/2} \\ &\le C\|Lv\|^2_{(0,s)} + C|E_0 v(1)|^2_{s+1/2} + C|E_1 v(0)|^2_{s+1/2} + C\|v\|^2_{(0,\sigma)}.\end{aligned}$$

Since $\|v\|^2_{(1,s)} = \|D_y v\|^2_{(0,s)} + \|v\|^2_{(0,s+1)}$, and $\partial v/\partial y = Kv + F$, we hence arrive at the estimate

$$\begin{aligned}&\|v\|^2_{(1,s)} + |E_0 v(0)|^2_{s+1/2} + |E_1 v(1)|^2_{s+1/2} \\ &\le C\|Lv\|^2_{(0,s)} + C|E_0 v(1)|^2_{s+1/2} + C|E_1 v(0)|^2_{s+1/2} + C\|v\|^2_{(0,\sigma)}.\end{aligned} \tag{11.67}$$

We can now give a natural condition for the estimate (11.50) to hold. In fact, (11.50) is the $s = 0$ case of the following.

Proposition 11.6. *For any $k \in \mathbb{Z}^+$, $s \in \mathbb{R}$, $\sigma < s$, there is an estimate*

$$\|v\|^2_{(k,s)} \le C\|Lv\|^2_{(k-1,s)} + C|Bv(0)|^2_{k+s-1/2} + C\|v\|^2_{(0,\sigma)}, \tag{11.68}$$

for all $v \in \mathcal{S}(\overline{\mathbb{R}^n_+})$ vanishing for $y \ge 1$, provided that for all s there is an estimate

$$|g|^2_s \le C_1|Bg|^2_s + C_1|E_0 g|^2_s + C_1|g|^2_{s-1}, \tag{11.69}$$

for all $g \in \mathcal{S}(\mathbb{R}^{n-1})$.

Proof. First take $k = 1$. Since (11.67) holds for $v \in H_{(1,s)}(\Omega)$, substitute $E_0 v$ for v in this estimate. Then $E_1 E_0 v(0) = 0$, and $LE_0 v = (\partial_y - K)E_0 v = E_0 Lv - K_0 E_0 v$. Thus we obtain

$$\|E_0 v\|^2_{(1,s)} + |E_0 v(0)|^2_{s+1/2} \le C\|Lv\|^2_{(0,s)} + C|E_0 v(1)|^2_{s+1/2} + C\|E_0 v\|^2_{(0,s)},$$

and hence

$$|E_0 v(0)|^2_{s+1/2} \le C\|Lv\|^2_{(0,s)} + C|E_0 v(1)|^2_{s+1/2} + C\|v\|^2_{(0,\sigma)}. \tag{11.70}$$

Now use (11.69), with $g = v(0)$ and s replaced by $s + 1/2$, to obtain

$$\begin{aligned}|v(0)|^2_{s+1/2} &\le C|Bv(0)|^2_{s+1/2} + C\|Lv\|^2_{(0,s)} \\ &\quad + C|E_0 v(1)|^2_{s+1/2} + C\|v\|^2_{(0,\sigma)}.\end{aligned} \tag{11.71}$$

We can dominate the term $C|E_1 v(0)|^2_{s+1/2}$ on the right side of (11.67) by (11.71), and if $v(1) = 0$, this yields the $k = 1$ case of (11.68).

Then, making use of $\partial v/\partial y = Kv + Lv$, one gets (11.68) by induction for $k \ge 1$. This completes the proof of the proposition.

Now $B(D_x)$ and $E_0(D_x)$ are Fourier multipliers by functions $B(\xi)$ and $E_0(\xi)$. The latter function is homogeneous of degree 0, while the former belongs to $S^0_1(\mathbb{R}^{n-1})$. In fact, we can write $B(\xi) = b_0(\xi) + b_r(\xi)$, where $b_0(\xi)$ is homogeneous of degree 0 and $|b_r(\xi)| \le C\langle\xi\rangle^{-1}$. By the characterization of $H^s(\mathbb{R}^{n-1})$ in terms of behavior of Fourier transforms, we have the following.

Lemma 11.7. *Suppose (1.1) is a $k \times k$ system, so K, B, and E_0 act on functions with values in $\mathbb{C}^\nu$, $\nu = km$. The estimate (11.69) holds if and only if, for each $\xi \ne 0$, there is no nonzero $v \in \mathbb{C}^\nu$ such that*

$$b_0(\xi)v = 0 \text{ and } E_0(\xi)v = 0. \tag{11.72}$$

Note that this is an "ellipticity condition" for some operators that are not differential operators. This is another point to which we will return in Chapter 7.

We want to make the condition for regularity even more explicit by relating it directly to the symbols of P and B_j. We establish the following.

Proposition 11.8. *For given nonzero $\xi \in \mathbb{R}^{n-1}$, the condition that there is no nonzero $v \in \mathbb{C}^\nu$ satisfying (11.72) is equivalent to each of the following two conditions:*

(i) There is no nonzero bounded solution on $[0, \infty)$ of the ODE

$$\frac{d\varphi}{dy} - K_1(\xi)\varphi = 0, \quad b_0(\xi)\varphi(0) = 0. \tag{11.73}$$

(ii) There is no nonzero bounded solution on $[0, \infty)$ of the ODE

$$\frac{d^m}{dy^m}\Phi + \sum_{j=0}^{m-1} \tilde{A}_j(\xi)\frac{d^j}{dy^j}\Phi = 0, \quad \tilde{B}_j\Big(\xi, d/dy\Big)\Phi(0) = 0, \ 1 \le j \le \ell. \tag{11.74}$$

Here $\tilde{A}_j(\xi)$ is the part of $A_j(\xi)$ of (11.40) homogeneous of degree $m - j$, and $\tilde{B}_j\Big(\xi, d/dy\Big)$ comes from taking the part of (11.47) homogeneous of degree m_j, and replacing D_x by ξ.

Proof. The equivalence of the hypothesis of Lemma 11.7 to (i) comes because the solution to (11.73) has the form

$$\varphi(y) = e^{yK_1(\xi)}\varphi(0),$$

and this is bounded for $y \in [0, \infty)$ if and only if $E_0(\xi)\varphi(0) = 0$. The equivalence of (i) and (ii) arises because (11.74) becomes (11.73) when transformed to a first-order system.

It is also useful to note that we can replace (ii) by:

(ii′) There is no nonzero solution to (11.74) that is rapidly decreasing as $y \to +\infty$,

and make a similar replacement of (i).

Since we want to consider boundary problems for which there will be a reasonable existence result as well as a regularity result for solutions, it is natural to consider a further restriction on the boundary condition. Suppose that $\tilde{B}_j(\xi, d/dy)\Phi(0) \in \mathbb{C}^{\lambda_j}$, so

$$b_0(\xi) : \mathbb{C}^\nu \longrightarrow \mathbb{C}^\lambda, \quad \lambda = \lambda_1 + \cdots + \lambda_\ell.$$

Proposition 11.9. *For given nonzero $\xi \in \mathbb{R}^{n-1}$, the following three conditions are equivalent.*

(i) Given $\eta \in \mathbb{C}^\lambda$, there exists a unique bounded solution on $[0, \infty)$ to the ODE

$$\frac{d\varphi}{dy} - K_1(\xi)\varphi = 0, \quad b_0(\xi)\varphi(0) = \eta. \tag{11.75}$$

(ii) Given $\eta_j \in \mathbb{C}^{\lambda_j}$, there exists a unique bounded solution on $[0, \infty)$ to the ODE

$$\frac{d^m}{dy^m}\Phi + \sum \tilde{A}_i(\xi)\frac{d^i}{dy^i}\Phi - 0, \quad \tilde{B}_j\Big(\xi, \frac{d}{dy}\Big)\Phi(0) = \eta_j. \tag{11.76}$$

(iii) With $V(\xi)$ denoting the null space of $E_0(\xi)$ on $\mathbb{C}^\nu$,

$$b_0(\xi) : V(\xi) \longrightarrow \mathbb{C}^\lambda \text{ isomorphically.} \tag{11.77}$$

Proof. The argument here is the same as in the proof of Proposition 11.8. We also note that if these conditions hold, the unique solutions to (11.75) and (11.76) are rapidly decreasing as $y \to +\infty$.

If the boundary problem $\{P(D), B_j(D), 1 \le j \le \ell\}$ satisfies the conditions of Proposition 11.9, it is called a *regular boundary problem*. More generally, if the variable-coefficient boundary problem (11.28) for an elliptic operator $P(x, D)$ produces frozen coefficient problems that satisfy this condition, it is called a regular boundary problem.

As a useful tool for establishing regularity, note that if

$$V(\xi) = \ker E_0(\xi) \text{ has dimension } \lambda \text{ for each nonzero } \xi, \tag{11.78}$$

then (11.77) holds if and only if no nonzero $v \in \mathbb{C}^\nu$ satisfies (11.72). Thus, given (11.78), the conditions of Proposition 11.9 are equivalent to those of Proposition 11.8. Of course, for (11.77) to hold for all $\xi \neq 0$, it is necessary that (11.78) be true.

We now give some examples of regular elliptic boundary problems. Our list will include those studied in §§1, 7, and 9, as well as others, not readily amenable to the methods developed there.

We begin with operators $P(x, D)$, which are *strongly elliptic*, of order $m = 2\mu$. By definition, this means

$$\operatorname{Re} P_m(x, \xi) \ge C|\xi|^m, \tag{11.79}$$

with $C > 0$. If P is a $k \times k$ system, $\operatorname{Re} P_m(x, \xi)$ stands for the matrix-valued function $\big(P_m(x, \xi) + P_m(x, \xi)^*\big)/2$. The *Dirichlet* boundary condition in this case can be written as follows. Let $\partial/\partial\nu$ denote any vector field defined on a neighborhood of ∂M and everywhere transverse to ∂M. Then the boundary condition is

$$u = g_0, \quad \frac{\partial}{\partial \nu}u = g_1, \dots, \Big(\frac{\partial}{\partial \nu}\Big)^{\mu-1} u = g_{\mu-1} \quad \text{on } \partial M. \tag{11.80}$$

If $\mu = 1$, this reduces to $u = g_0$ on ∂M, as in §1.

Proposition 11.10. *If P is a strongly elliptic $k \times k$ system of order 2μ, then the Dirichlet boundary condition is regular.*

Proof. Since (11.78) holds in this case, it suffices to check the uniqueness, namely, that any solution Φ to (11.74) that is rapidly decreasing as $y \to +\infty$ is 0. To see

this, write
(11.81)

$$\left(P_m\left(\xi, i\frac{d}{dy}\right)\Phi, \Phi\right)_{L^2(\mathbb{R}^+)} = \sum \left(L_j\left(\xi, i\frac{d}{dy}\right)\Phi, M_j\left(\xi, i\frac{d}{dy}\right)\Phi\right)_{L^2(\mathbb{R}^+)},$$

where L_j and M_j are differential operators (with coefficients depending on ξ) in y of order $\leq \mu$. Then, by Fourier analyis, if $\Phi(0) = \Phi'(0) = \cdots = \Phi^{(\mu-1)}(0) = 0$, we have (11.81) equal to

$$\int_{-\infty}^{\infty} P_m(\xi, \eta)|\hat{\Phi}(\eta)|^2 \, d\eta. \tag{11.82}$$

Here $\hat{\Phi}(\eta)$ is defined by extending $\Phi(y)$ to be zero for $y \leq 0$. Since

$$\text{Re } P_m(\xi, \eta) \geq C(|\xi|^m + |\eta|^m),$$

if $P_m(\xi, id/dy)\Phi = 0$, this implies $\Phi = 0$, as desired, proving the proposition.

The Dirichlet problem is regular in many additional cases. For example, if $P(x, D)$ is a scalar elliptic differential operator on M, then the Dirichlet problem is always regular, provided dim $M \geq 3$. See the exercises.

The next result contains the fact that the Neumann boundary problem for the Laplace operator is regular. Let $\overline{M}$ have a smooth Riemannian metric.

Proposition 11.11. *If X is a real vector field on ∂M which is everywhere transversal, then the boundary condition $Xu = g_1$ on ∂M is regular for the Laplace operator Δ.*

Proof. To freeze coefficients at a point $p \in M$, pick normal coordinates centered at p, with $\partial/\partial y$ coinciding at p with the unit normal given by the metric tensor. We have (11.78), with $\lambda = 1$. Checking uniqueness of (11.74) comes down to looking at solutions $\Phi(y)$ to

$$\frac{d^2}{dy^2}\Phi - Q(\xi)\Phi = 0, \quad B\Phi'(0) + iA(\xi)\Phi(0) = 0, \tag{11.83}$$

which are bounded for $y \in [0, \infty)$. Here $Q(\xi)$ is a positive definite quadratic form, B is a nonzero constant, and $A(\xi)$ a real linear form in ξ. For $\xi \neq 0$, any bounded solution must be a multiple of $e^{-y\sqrt{Q(\xi)}}$, which has boundary data

$$-B\sqrt{Q(\xi)} + iA(\xi) \neq 0. \tag{11.84}$$

This proves the proposition.

When X is orthogonal to ∂M, this is the Neuman problem; otherwise it is called an oblique derivative problem. Note that if dim $M = 2$, so $\xi \in \mathbb{R}^1$, then one gets a regular elliptic boundary problem for any real vector field that is nowhere vanishing on ∂M, since then (11.84) holds for all $\xi \neq 0$ as long as either $B \neq 0$ or $A \neq 0$. Compare Exercises 4–9 in §4 of Chapter 4. However, when dim $M \geq 3$, so $\xi \in \mathbb{R}^{n-1}$ with $n - 1 \ge 2$, if $B = 0$, then $A(\xi) = 0$ also for ξ in a hyperplane, so regularity fails then.

We start our next line of analysis with an obvious comment. Namely, the direct sum of two regular elliptic boundary problems of (the same) degree m on $\overline{M}$ is also regular. By the same token, if the frozen-coefficient problems all break up into direct sums of regular problems, then they are regular, and hence so is the variable-coefficient problem that gave rise to them (even though it may not break up into such a direct sum). This applies to the Hodge Laplacian Δ on $\Lambda^k(\overline{M})$, with either relative or absolute boundary conditions. In each case, the frozen-coefficient problems can clearly be seen to break up into direct sums of Dirichlet and Neumann problems. This proves:

Proposition 11.12. *If Δ is the Hodge Laplacian on $\Lambda^k(\overline{M})$, then both the relative boundary problem*

$$\Delta u = f \text{ on } M, \quad j^*u = g_0, \; j^*\delta u = g_1 \text{ on } \partial M, \tag{11.85}$$

and the absolute boundary problem

$$\Delta u = f \text{ on } M, \quad u\rfloor\nu = g_0, \; (du)\rfloor\nu = g_1 \text{ on } \partial M, \tag{11.86}$$

are regular elliptic boundary problems.

In (11.85), g_j are forms on ∂M, of degree $k - j$. In (11.86), g_j are sections of the subbundles of $\Lambda^{k-1+j}(\overline{M})\big|_{\partial M}$, defined by $g_j\rfloor\nu = 0$.

Sometimes the fact that the Dirichlet problem is regular can be used as a tool to determine whether another boundary problem is regular. To illustrate this, suppose $P = P(x, D)$ is a strongly elliptic, $k \times k$ system of order 2. Then the ODE in (11.76) takes the form

$$\frac{d^2}{dy^2}\Phi + \sum_{j=0}^{1} \tilde{A}_j(\xi)\frac{d^j}{dy^j}\Phi = 0. \tag{11.87}$$

Let us consider a boundary problem of the form

$$Pu = f, \quad B_0(x)u = g_0, \quad C_0(x)\frac{\partial u}{\partial y} + C_1(x, D_x)u = g_1. \tag{11.88}$$

Here we use coordinates (y, x) on a collar neighborhood of ∂M,

$$B_0(x) : \mathbb{C}^k \longrightarrow \mathbb{C}^{\lambda_1}, \quad C_0(x) : \mathbb{C}^k \longrightarrow \mathbb{C}^{\lambda_2}, \tag{11.89}$$

and $C_1(x, D_x)$ is a first-order differential operator whose coefficients map $\mathbb{C}^k \to \mathbb{C}^{\lambda_2}$. The hypothesis (11.78) is equivalent to $\lambda_1 + \lambda_2 = k$. To complement (11.87) and reproduce (11.76), we have

$$B_0\Phi(0) = \eta_0, \quad C_0\Phi'(0) + C_1(\xi)\Phi'(0) = \eta_1, \tag{11.90}$$

where $B_0 : \mathbb{C}^k \to \mathbb{C}^{\lambda_1}$, $B_1 : \mathbb{C}^k \to \mathbb{C}^{\lambda_2}$, and $C_1(\xi) = \sum A_j\xi_j$, with $A_j : \mathbb{C}^k \to \mathbb{C}^{\lambda_2}$. These arise from freezing the coefficients of $B_0(x)$ and $C_1(x, D_x)$.

Now, for $x \in \partial M, \xi \in \mathbb{R}^{n-1} \setminus 0$, we define a map

$$\mathcal{B}(x,\xi) : \mathbb{C}^k \longrightarrow \mathbb{C}^k \tag{11.91}$$

as follows. Given $\varphi \in \mathbb{C}^k$, let $\Phi_\xi(y)$ be the unique bounded solution to (11.87) such that $\Phi_\xi(0) = \varphi$, and then set

$$\mathcal{N}(x,\xi)\varphi = \Phi'_\xi(0), \tag{11.92}$$

and define

$$\mathcal{B}(x,\xi)\varphi = \big(B_0(x)\varphi, C_0(x)\mathcal{N}(x,\xi)\varphi + C_1(x,\xi)\varphi\big). \tag{11.93}$$

The following is an immediate consequence of Propositions 11.9 and 11.10 and their proofs.

Proposition 11.13. *If P is a second-order, strongly elliptic $k \times k$ system, then the boundary problem (11.88) is regular provided that, for all $x \in \partial M, \xi \in \mathbb{R}^{n-1} \setminus 0$, the map $\mathcal{B}(x,\xi)$ in (11.91) is an isomorphism.*

Note that the proof of Proposition 11.11 can be regarded as a special case of this argument, with $k = 1$. Then $\mathcal{B}(x,\xi)$ (with x supressed) is given by (11.84). It is appropriate to think of $\mathcal{B}(x,\xi)$ and $\mathcal{N}(x,\xi)$ as defined on $T^*(\partial M) \setminus 0$. In Chapter 7 we will see that $\mathcal{N}(x,\xi)$ is the principal symbol of an important pseudodifferential operator.

To close this section, we say a little more about regularity estimates. There are advantages in using spaces like $H_{(k,s)}(\Omega)$ to formulate regularity results of a more general nature than in Proposition 11.2, for regular elliptic boundary problems. Thus, take a collar neighborhood Ω of ∂M, diffeomorphic to $[0,1) \times \partial M$, sitting inside a larger collar neighborhood $\mathcal{C}$, diffeomorphic to $[0,2) \times \partial M$. We use norms $H_{(k,s)}(\Omega)$, given by

$$\|u\|^2_{(0,s)} = \int_0^1 \|u(y,\cdot)\|^2_{H^s(\partial M)}\, dy, \tag{11.94}$$

and more generally

$$\|u\|^2_{(k,s)} = \sum_{j=0}^{k} \int_0^1 \big\| D_y^j u(y,\cdot)\big\|^2_{H^{k-j+s}(\partial M)}\, dy. \tag{11.95}$$

Norms on $H_{(k,s)}(\mathcal{C})$ are analogously defined. These spaces depend on the choice of collaring, but that will not cause a difficulty. Techniques used above are readily extended to prove the following.

Proposition 11.14. *If $P(x,D)$ has order m and $\{P(x,D), B_j(x,D), 1 \le j \le \ell\}$ defines a regular elliptic boundary problem, then, given that*

$$u \in H_{(m,\sigma)}(\mathcal{C}), \tag{11.96}$$

for some $\sigma \in \mathbb{R}$, and given

$$P(x,D)u = f \in H_{(k,s)}(\mathcal{C}), \quad B_j(x,D)u \in H^{m+k-m_j-\frac{1}{2}+s}(\partial M), \tag{11.97}$$

it follows that

$$u \in H_{(m+k,s)}(\Omega), \tag{11.98}$$

with a corresponding estimate.

Part of the usefulness of this extension of Proposition 11.2 arises from the following fact.

Proposition 11.15. *If P is a differential operator of order m, for which ∂M is noncharacteristic, then, for some $\sigma \in \mathbb{R}$,*

$$u \in L^2(M),\ Pu = f \in L^2(M) \Longrightarrow u \in H_{(m,\sigma)}(\Omega). \tag{11.99}$$

Proof. Using an expansion like (11.40) for P, we have

$$\partial_y^m u = f - \sum_{j=0}^{m-1} A_j(y, x, D_x)\, \partial_y^j u. \tag{11.100}$$

If the hypotheses of (11.99) hold, then

$$A_j(y, x, D_x)\, \partial_y^j u \in H^{-j}(I, H^{-m+j}(\partial M)),$$

where $I = [0, 1]$. A solution v_j to $\partial^m v_j = A_j(y, x, D_x)\, \partial_y^j u$ hence belongs to the space $H^{m-j}(I, H^{-m+j}(\partial M))$, so $u \in H_{(1,1-m)}(\Omega)$. Iterating this argument gives (11.99).

Thus Proposition 11.14 is applicable to such $u \in L^2(M)$. Note that the boundary value $B_j(x, D)u|_{\partial M}$ is well defined when u satisfies the conclusion of (11.99).

We stated that part of the point of putting a further restriction on the boundary conditions, as in Proposition 11.9, to define a regular elliptic boundary problem, is to have an existence result as well as a regularity result. In fact, the following is true.

Proposition 11.16. *If $\{P(x, D), B_j(x, D), 1 \le j \le \ell\}$ defines a regular elliptic boundary problem, with*

$$P(x, D) : C^\infty(\overline{M}, E_0) \longrightarrow C^\infty(\overline{M}, E_1) \text{ elliptic of order } m \tag{11.101}$$

and

$$B_j(x, D) : C^\infty(\overline{M}, E_0) \longrightarrow C^\infty(\partial M, G_j) \text{ of order } m_j, \tag{11.102}$$

then, for each $k \ge 0$, the map

$$T : H^{m+k}(M, E_0) \longrightarrow H^k(M, E_1) \oplus \bigoplus_{j=1}^{\ell} H^{m+k-m_j-1/2}(\partial M, G_j) \tag{11.103}$$

defined by

$$Tu = \big(P(x, D)u;\ B_1(x, D)u, \dots, B_\ell(x, D)u\big) \tag{11.104}$$

is Fredholm.

The estimate (11.29) clearly implies that T has finite dimensional kernel. Also, by Proposition 6.7 of Appendix A, the estimate implies that T has closed range. It remains to show that the range of T has finite codimension.

One way to do this is to construct a right Fredholm inverse of T, by the results of §7 in Appendix A. Pseudodifferential operators, introduced in Chapter 7, form a convenient tool to do this. At this stage it is convenient to make a weaker construction, of something that might be called an "approximate Fredholm inverse" of T. The operator we will construct will be called S:

$$S : H^k(M, E_1) \oplus \bigoplus_{j=1}^{\ell} H^{m+k-m_j-1/2}(\partial M, G_j) \longrightarrow H^{m+k}(E_0). \tag{11.105}$$

The function $u = S(f; g_1, \dots, g_\ell)$ is to be an "approximate solution" to

$$P(x, D)u = f, \quad B_j(x, D)u = g_j.$$

To begin, we ignore the boundary condition. Suppose $\overline{M} \subset \Omega$, on which P is elliptic. Let $\tilde{f} \in H^k(U, E_1)$ be an extension of f. Use a partition of unity to write $\tilde{f}$ as a sum of terms $\tilde{f}_\nu$ with support in coordinate charts V_ν on Ω. Then pick a lattice $\Lambda = \Lambda_\varepsilon$, as in (11.6), and set

$$v_\nu = \chi_\nu \sum_{\lambda \in \Lambda} E_\lambda(D)(\chi_\lambda \tilde{f}), \tag{11.106}$$

where the sum is as in (11.18), and $\chi_\nu \in C_0^\infty(\Omega)$ is equal to 1 on V. Now set $v = \sum v_\nu\big|_M$. Note that $v \in H^{m+k}$. The arguments yielding such estimates as (11.20)–(11.23) also give

$$\|Pv - f\|_{H^k} \le \varepsilon \|f\|_{H^k} + C(\varepsilon)\|f\|_{H^{k-1}}. \tag{11.107}$$

Of course, v depends on ε. Let $h_j = B_j(x, D)v\big|_{\partial M}$.

We want $u = v + w$, where w is an approximate solution to

$$P(x, D)w = 0, \quad B_j(x, D)w = g_j - h_j.$$

Cover a collar neighborhood of ∂M in $\overline{M}$ with coordinate charts $\overline{V}_\nu$, straightened out to be regarded as regions in $\overline{\mathbb{R}^n_+}$. Write $e_j = g_j - h_j$ as a sum of terms $e_{j\nu}$ supported in $V_\nu \cap \partial M$, using a partition of unity. Again pick a lattice $\Lambda = \Lambda_\varepsilon$. If $\lambda \in \Lambda \cap \mathbb{R}^{n-1} = \Lambda_0$, we take $w_{\nu\lambda}$ to be the Fourier transform (with respect to ξ) of the solution to (11.76), with $\eta_j(\xi) = \hat{e}_{j\nu\lambda}(\xi)$, where $e_{j\nu\lambda} = \chi_\lambda e_{j\nu}$. Then set

$$w = \sum_\nu \chi_\nu \sum_{\lambda \in \Lambda_0} w_{\nu\lambda}.$$

Parallel to (11.107), or to (11.35), we have, for $\varphi = (f; g_1, \dots, g_\ell)$,

$$\|(I - TS)\varphi\|_{\mathcal{V}_k} \le \varepsilon \|\varphi\|_{\mathcal{V}_k} + C(\varepsilon)\|\varphi\|_{\mathcal{V}_{k-1}}, \tag{11.108}$$

where $\mathcal{V}_k$ is the range space of T in (11.103), and one obtains $\mathcal{V}_{k-1}$ by replacing the index k by $k-1$ at each occurrence. Now this estimate implies that the norm of $[I - TS] \in \mathcal{L}(\mathcal{V}_k)/\mathcal{K}(\mathcal{V}_k)$ is $\leq \varepsilon$. As long as it is < 1, we have $[TS]$ invertible in this quotient algebra, hence TS is a Fredholm operator, with a two-sided Fredholm inverse F. But then SF is a right Fredholm inverse of T, and the proof of Proposition 11.16 is complete.

Recall that in previous sections we have obtained existence results by different means. Some of these methods will be pushed in the next section, leading to an independent proof of the surjectivity, or "almost" surjectivity, of T in many important cases. In §12, the proof above that T in (11.103) has range of finite codimension will not be used.

Exercises

1. If $P(x, D)$ is a strongly elliptic operator of order $2m$, show that
$$\operatorname{Re}\big(P(x, D)u, u\big) \geq C\|u\|^2_{H^m(M)} - C'\|u\|^2_{L^2(M)}$$
for $u \in C_0^\infty(M)$. (*Hint*: Use cut-offs as in the proof of Theorem 11.1 to reduce to an estimate on the quantity $\operatorname{Re}(P_\lambda(D)u, u)$, where $P_\lambda(D)$ is obtained by freezing coefficients. Analyze this inner product via Fourier analysis.)
2. For strongly elliptic $P(x, D)$, show that, for C_1 sufficiently large,
$$P(x, D) + C_1 : H_0^m(M) \longrightarrow H^{-m}(M), \text{ isomorphically.} \tag{11.109}$$
3. Parallel arguments of §1 to show that
$$P(x, D) + C : H_D^{2m+k}(M) \longrightarrow H^k(M), \text{ isomorphically,} \tag{11.110}$$
where $H_D^\ell(M) = H^\ell(M) \cap H_0^m(M)$. Deduce that
$$P(x, D) : H_D^{2m+k}(M) \longrightarrow H^k(M) \text{ is Fredholm,}$$
of index zero.
4. As an alternative, show that (11.109) leads to (11.110) via Propositions 11.14–11.15.

In Exercises 5–7, let $P(x, D)$ be a *scalar* elliptic operator of order m, on $\Omega = [0, 1] \times \partial M$. Let $P_m(x_0, \xi_0, \eta)$ be the principal symbol at $x_0 \in \partial M$, $\xi_0 \in T^*_{x_0}\partial M \setminus 0$, $\eta \in \mathbb{R}$. Then none of the roots $\eta_1, \dots, \eta_m$ of $P_m(x_0, \xi_0, \eta) = 0$ are real. Let
$$M^+(x_0, \xi_0, \eta) = \prod_{k=1}^{\ell}\big(\eta - \eta_k(x_0, \xi_0)\big),$$
the product being over k such that $\eta_k(x_0, \xi_0)$ have positive imaginary part. Let $\tilde{B}_j(x_0, \xi_0, \eta)$ be the principal symbol of $B_j(x.D)$.

5. Show that the conditions for regularity of (P, B_j) in Proposition 11.9 are equivalent to the condition that the set of polynomials in η
$$\{\tilde{B}_j(x_0, \xi_0, \eta) : 1 \leq j \leq \ell\} \tag{11.111}$$
gives a basis of
$$\mathbb{C}[\eta]/\big(M^+(x_0, \xi_0, \eta)\big), \tag{11.112}$$

the quotient of the ring $\mathbb{C}[\eta]$ of polynomials in η, by the ideal generated by $M^+(x_0, \xi_0, \eta)$. (*Hint*: Show that a solution Φ to (11.76), obtained by freezing coefficients at x_0, is bounded if and only if

$$M^+\Big(x_0, \xi_0, \frac{d}{dy}\Big)\Phi = 0.)$$

We say that $P(x, D)$ is *properly elliptic* provided the degree of $M^+(x_0, \xi_0, \eta)$ in η is independent of $(x_0, \xi_0) \in T^*\partial M \setminus 0$. Evidently, if (11.111) is to provide a basis for all $(x_0, \xi_0) \in T^*\partial M \setminus 0$, then $P(x, D)$ must be properly elliptic, since the quotient (11.112) is a vector space whose dimension is the degree of $M^+(x_0, \xi_0, \eta)$.

6. Show that any scalar elliptic $P(x, D)$ is properly elliptic if $\dim M \geq 3$. (*Hint*: $\mathbb{R}^{k-1} \setminus 0$ is connected for $k \geq 3$.)
   Show that $m = 2\ell$.
7. Show that the Dirichlet problem is regular for any properly elliptic scalar operator $P(x, D)$ of order $m = 2\mu$. (*Hint*: Show that $\{1, \eta, \dots, \eta^{\mu-1}\}$ gives a basis of $\mathbb{C}[\eta]/\big(M^+(x_0, \xi_0, \eta)\big)$ under these circumstances.)
8. Consider the following second-order elliptic operator on $\mathbb{R}^2$:

$$L = \Big(\frac{\partial}{\partial \bar{z}}\Big)^2 = \frac{1}{4}\Big(\frac{\partial}{\partial x} + i\frac{\partial}{\partial y}\Big)^2.$$

   Show that L is not properly elliptic. Verify that the Dirichlet problem on the disk for L is not regular by constructing an infinite-dimensional space of solutions to $Lu = 0$, $u\big|_{S^1} = 0$.
9. Let $D = d + \delta$, acting on the space $\oplus \Lambda^j \overline{M}$ of forms on $\overline{M}$. Let $R_0 u = \nu \wedge u$ and $A_0 u = \iota_\nu u$, as in (9.11). Show that the boundary problems $\{D, R_0\}$ and $\{D, A_0\}$ are both regular. Take another look at Exercise 2 in the first set of exercises for §9.

12. Operator properties of regular boundary problems

We want to extend the existence theory, obtained for the Dirichlet and Neumann problems for the Laplace operator in §§1 and 7 and for relative and absolute boundary problems for the Hodge Laplacian in §9, to further classes of elliptic boundary problems. We also study other properties of an elliptic operator $P = P(x, D)$, regarded as an unbounded operator on $L^2(M, E_0)$, with domain

$$\mathcal{D}(P) = \{u \in H^m(M, E_0) : B_j(x, D)u = 0 \text{ on } \partial M, 1 \leq j \leq \ell\}. \tag{12.1}$$

We begin with strongly elliptic, second-order $k \times k$ systems. Note that, in each case studied in §§1, 7, and 9, we had (up to sign) the form

$$P = D^*D + X, \tag{12.2}$$

where

$$D : C^\infty(\overline{M}, E_0) \longrightarrow C^\infty(\overline{M}, E_1) \text{ has injective symbol,} \tag{12.3}$$

that is, $\sigma_D(x, \xi) : E_{0x} \to E_{1x}$ is injective, for each $x \in \overline{M}$, $\xi \neq 0$. If the bundles E_j are endowed with metrics and $\overline{M}$ has a Riemannian metric, then D^* is defined, and D^*D is elliptic. Set $L = D^*D$, so $P = L + X$.

An important tool in the analysis done in §§1, 7, and 9 was Green's formula, which in this generality can be written as

$$(12.4)\qquad (Lu, v) = (Du, Dv) + \frac{1}{i}\int_{\partial M} \langle \sigma_{D^*}(x, \nu)Du, v\rangle \, dS,$$

for sufficiently regular sections u, v of E_0. The boundary integral vanishes for all $v \in C^\infty(\overline{M}, E_0)$ if and only if

$$(12.5)\qquad \sigma_{D^*}(x, \nu)Du = 0 \text{ on } \partial M.$$

The approach to the Neumann boundary problem in §7 started with the fact that $\|du\|_{L^2}^2 + \|u\|_{L^2}^2$ defines the square $H^1(M)$-norm, to establish Proposition 7.1. There exist first-order differential operators D for which the estimate

$$(12.6)\qquad \|Du\|_{L^2}^2 \geq C\|u\|_{H^1}^2 - C'\|u\|_{L^2}^2, \quad u \in H^1(M),$$

is true, but not straightforward, as $|Du(x)|$ does not pointwise dominate a multiple of $|\nabla u(x)|$. There are also first-order elliptic differential operators for which (12.6) is false. We give here a sufficient criterion for the validity of (12.6).

Proposition 12.1. *If (12.5) is a regular elliptic boundary condition for $L = D^*D$, then the estimate (12.6) holds.*

Proof. It is convenient to give this a functional analytic formulation. Let D_1 be the unbounded operator from $L^2(M, E_0)$ to $L^2(M, E_1)$ with domain

$$(12.7)\qquad \mathcal{D}(D_1) = \{u \in L^2(M, E_0) : Du \in L^2(M, E_1)\},$$

and $D_1 u = Du$ for such u; D_1 is the "maximal" extension of D; it is a closed, densely defined operator. Clearly, $H^1(M, E_0) \subset \mathcal{D}(D_1)$. The estimate (12.6) is equivalent to

$$(12.8)\qquad \mathcal{D}(D_1) = H^1(M, E_0).$$

To establish this, we define an unbounded operator $\mathcal{L}$ on $L^2(M, E_0)$ by

$$(12.9)\qquad \begin{aligned} \mathcal{D}(\mathcal{L}) &= \{u \in H^2(M, E_0) : \sigma_{D^*}(x, \nu)Du = 0 \text{ on } \partial M\}, \\ \mathcal{L}u &= Lu = D^*Du, \quad \text{for } u \in \mathcal{D}(\mathcal{L}). \end{aligned}$$

It is clear that

$$(12.10)\qquad \mathcal{D}(\mathcal{L}) \subset \mathcal{D}(D_1^* D_1).$$

In fact, an element $u \in L^2(M, E_0)$ belongs to $\mathcal{D}(D_1^* D_1)$ if and only if

$$(12.11)\qquad Du \in L^2(M, E_1), \quad D^*Du \in L^2(M, E_0), \quad \text{and } \sigma_{D^*}(x, \nu)Du = 0 \text{ on } \partial M.$$

Note that Proposition 11.15 implies that the boundary condition makes sense for $u \in \mathcal{D}(D_1^* D_1)$. It also implies that the regularity result of Proposition 11.14 is applicable, so $\mathcal{D}(D_1^* D_1) \subset H^2(M, E_0)$. Hence

$$(12.12)\qquad D_1^* D_1 = \mathcal{L}.$$

By von Neumann's theorem, $D_1^* D_1$ is automatically self-adjoint; see §8 in Appendix A. Thus $\mathcal{L}$ is self-adjoint. Furthermore,

$$\mathcal{D}(D_1) = \mathcal{D}(\mathcal{L}^{1/2}); \tag{12.13}$$

a proof of this is given in §1 of Chapter 8. By interpolation, we have

$$\mathcal{D}(\mathcal{L}^{1/2}) \subset H^1(M, E_0), \tag{12.14}$$

establishing $\mathcal{D}(D_1) \subset H^1(M, E_0)$ and hence (12.8).

An important example of this phenomenon is the operator that associates to a vector field X its *deformation tensor*, a tensor field of type (0, 2) defined by

$$(\mathrm{Def}\, X)(Y, Z) = \frac{1}{2}\langle \nabla_Y X, Z\rangle + \frac{1}{2}\langle \nabla_Z X, Y\rangle; \tag{12.15}$$

in coordinate notation,

$$(\mathrm{Def}\, X)_{jk} = \frac{1}{2}(X_{j;k} + X_{k;j}). \tag{12.16}$$

This was introduced in (3.35) of Chapter 2. We have

$$\mathrm{Def} : C^\infty(\overline{M}, T) \longrightarrow C^\infty(\overline{M}, S^2T^*). \tag{12.17}$$

If $\tilde{u} \in T^*$ corresponds to $u \in T$ via the metric tensor, then

$$\frac{1}{i}\sigma_{\mathrm{Def}}(x, \xi)u = \frac{1}{2}(\xi \otimes \tilde{u} + \tilde{u} \otimes \xi) = \xi Ⓢ \tilde{u}. \tag{12.18}$$

We also have

$$-\frac{1}{i}\sigma_{\mathrm{Def}^*}(x, \xi)(v Ⓢ w) = \frac{1}{2}\big(\langle v, \xi\rangle w + \langle w, \xi\rangle v\big), \tag{12.19}$$

and hence, for $L = \mathrm{Def}^*\mathrm{Def}$,

$$\sigma_L(x, \xi)u = \frac{1}{2}\big(|\xi|^2 u + \langle \xi, u\rangle \xi\big) = \frac{1}{2}|\xi|^2\big(I + P_\xi\big)u, \tag{12.20}$$

where P_ξ is the orthogonal projection parallel to ξ, if T and T^* are identified via the metric tensor.

Proposition 12.2. *The boundary condition*

$$\sigma_{\mathrm{Def}^*}(x, \nu) Def\, u = g \tag{12.21}$$

is regular for $L = Def^* Def$.

Proof. We will apply Proposition 11.13. For a point $p_0 \in \partial M$, choose local coordinates so that the normal is $\partial/\partial x_n = \partial/\partial y$. Then the symbol of D^*D is (up to a factor of 1/2)

$$(I + P_n)\eta^2 + (I + P_\xi)|\xi|^2. \tag{12.22}$$

Here we are replacing $\xi \in \mathbb{R}^n$ in (12.20) by (ξ, η), $\xi \in \mathbb{R}^{n-1}$. Thus, referring to the notation of (12.20), P_n stands for $P_{(0,1)}$, and P_ξ here stands for $P_{(\xi,0)}$. Consequently, the quantity $\mathcal{N}(x, \xi)$ used in the proof of Proposition 11.13, and defined by (11.92), is seen to be

$$\mathcal{N}(x, \xi) = \big[\alpha P_n + P_n^\perp (I + \beta P_\xi)\big]|\xi|, \quad \alpha = \frac{1}{\sqrt{2}}, \ \beta = \sqrt{2} - 1. \tag{12.23}$$

Here $P_n^\perp = I - P_n$. Note that the range of P_ξ is contained in that of $P_n^\perp$, and so P_n, $P_n^\perp$, and P_ξ in (12.23) all commute.

In the present case, $\mathcal{B}(x, \xi)$ has the form $C_0(x)\mathcal{N}(x, \xi) + C_1(x, \xi)$. In fact, a calculation gives

$$2\mathcal{B}(x, \xi)\varphi = (I + P_n)\mathcal{N}(x, \xi)\varphi + i \sum_{j=1}^{n-1} \varphi_n \xi_j e_j, \tag{12.24}$$

where $\{e_j : 1 \le j \le n - 1\}$ is the standard basis of $\mathbb{R}^{n-1}$. In matrix form,

$$2\mathcal{B}(x, \xi) = \begin{pmatrix} & & & i\xi_1 \\ & P_n^\perp (I + \beta P_\xi)|\xi| & & \vdots \\ & & & i\xi_{n-1} \\ 0 & \cdots\cdots\cdots & 0 & 2\alpha|\xi| \end{pmatrix}. \tag{12.25}$$

It is clear that the determinant of the right side of (12.25) is $2\alpha(1 + \beta)|\xi|^n$, so the asserted regularity follows by Proposition 11.13.

Therefore, Proposition 12.1 yields the following.

Corollary 12.3. *If $\overline{M}$ is a compact Riemannian manifold with boundary, then*

$$\|X\|_{H^1(M)} \le C\|\mathrm{Def}\, X\|_{L^2(M)} + C\|X\|_{L^2(M)}, \tag{12.26}$$

for all smooth vector fields X on $\overline{M}$.

This is called *Korn's inequality* and is useful in elasticity theory.

We have the following Fredholm result.

Proposition 12.4. *If P is given by (12.2) and if (12.5) is a regular boundary condition for D^*D, hence for P, then for $k \ge 0$, the operator*

$$T : H^{k+2}(M, E_0) \longrightarrow H^k(M, E_0) \oplus H^{k+\frac{1}{2}}(\partial M, E_0) \tag{12.27}$$

given by

$$Tu = \big(Pu, \sigma_{D^*}(x, \nu)Du\big|_{\partial M}\big) \tag{12.28}$$

is Fredholm.

Proof. Let $H_B^{k+2} = \{u \in H^{k+2}(M, E_0) : B_1 u = 0\}$, where $B_1 u = B_1(x, D)u = \sigma_{D^*}(x, \nu)Du\big|_{\partial M}$. From the proof of Proposition 12.1, we know that

$$L + I : H_B^2 \longrightarrow L^2(M, E_0) \text{ is bijective,} \tag{12.29}$$

since $H_B^2 = \mathcal{D}(\mathcal{L})$. By elliptic regularity,

$$L + I : H_B^{k+2} \longrightarrow H^k(M, E_0) \text{ is bijective.} \tag{12.30}$$

Now P differs from $L + I$ by a compact operator $K : H_B^{k+2} \to H^k(M, E_0)$, so

$$P : H_B^{k+2} \longrightarrow H^k(M, E_0) \text{ is Fredholm.} \tag{12.31}$$

The Fredholmness of T is an easy consequence.

To return to the study of (12.2)–(12.4), we have the following solvability result.

Proposition 12.5. *If D satisfies (12.3) and B_1 is given by the left side of (12.5), then, with $L = D^*D$,*

$$(L + I) \oplus B_1 : H^{k+2}(M) \longrightarrow H^k(M) \oplus H^{k+1/2}(\partial M), \tag{12.32}$$

isomorphically, and, if $P = L + X$, X of order 1,

$$P \oplus B_1 : H^{k+2}(M) \longrightarrow H^k(M) \oplus H^{k+1/2}(\partial M) \tag{12.33}$$

is Fredholm, of index zero.

We next look at existence results for oblique derivative problems for the Laplace operator, namely,

$$\Delta u = f \text{ on } M, \quad \Big(\frac{\partial}{\partial \nu} + X\Big)u = g \text{ on } \partial M, \tag{12.34}$$

where X is a first-order differential operator of the form $Xu = Yu + \varphi u$, with Y a real vector field tangent to ∂M, $\varphi \in C^\infty(\partial M)$, real. Here we will take the Green identity

$$(\Delta u, v) = (u, \Delta v) + \beta(u, v), \tag{12.35}$$

with

$$\beta(u, v) = \int_{\partial M} \Big(u\frac{\partial \overline{v}}{\partial \nu} - \frac{\partial u}{\partial \nu}\overline{v}\Big)\, dS, \tag{12.36}$$

and rewrite this boundary term as

$$\beta(u, v) = \int_{\partial M} \Big\{u\Big(\frac{\partial \overline{v}}{\partial \nu} + X^t\overline{v}\Big) - \Big(\frac{\partial u}{\partial \nu} + Xu\Big)\overline{v}\Big\}\, dS, \tag{12.37}$$

where X^t is the formal adjoint of X, with respect to the $L^2(\partial M)$-inner product, that is,

$$X^t u = -Yu + (\varphi - \operatorname{div} Y)u,$$

where the divergence is taken with respect to surface measure dS on ∂M.

We define two unbounded operators on $L^2(M)$, denoted $\mathcal{L}_1, \mathcal{L}_2$. These are defined to be $-\Delta$ on their domains, which we specify to be

$$\begin{aligned}\mathcal{D}(\mathcal{L}_1) &= \left\{u \in H^2(M) : \frac{\partial u}{\partial \nu} + Xu = 0 \text{ on } \partial M\right\},\\ \mathcal{D}(\mathcal{L}_2) &= \left\{u \in H^2(M) : \frac{\partial u}{\partial \nu} + X'u = 0 \text{ on } \partial M\right\}.\end{aligned} \tag{12.38}$$

Proposition 12.6. *The operators $\mathcal{L}_j$ have the relation*

$$\mathcal{L}_1^* = \mathcal{L}_2, \tag{12.39}$$

where $\mathcal{L}_1^*$ is the Hilbert space adjoint of $\mathcal{L}_1$. Furthermore, with $Vu = (\partial u/\partial \nu) + Xu\big|_{\partial M}$, we have

$$-\Delta \oplus V : H^{k+2}(M) \longrightarrow H^k(M) \oplus H^{k+1/2}(\partial M) \tag{12.40}$$

is Fredholm, of index zero, and the annihilator of the image, a priori in $H^k(M)^* \oplus H^{-k-1/2}(\partial M)$, is a finite-dimensional subspace of $C^\infty(\overline{M}) \oplus C^\infty(\partial M)$ which is independent of $k \geq 0$.

Proof. To start, suppose $v \in \mathcal{D}(\mathcal{L}_1^*)$, that is, $v \in L^2(M)$, and the map $\mathcal{D}(\mathcal{L}_1) \ni u \mapsto (\mathcal{L}_1 u, v)$ satisfies an estimate

$$|(\Delta u, v)| \leq C(v)\|u\|_{L^2(M)}. \tag{12.41}$$

By (12.35) and (12.36), this can happen if and only if $\Delta v \in L^2(M)$ (hence its boundary data are well defined), and $\beta(u, v) = 0$ for all $u \in \mathcal{D}(\mathcal{L}_1)$, hence

$$\int_{\partial M} u\left(\frac{\partial \overline{v}}{\partial \nu} + X'\overline{v}\right) dS = 0, \quad \text{for all } u \in \mathcal{D}(\mathcal{L}_1). \tag{12.42}$$

Since there exist $u \in \mathcal{D}(\mathcal{L}_1)$ for which $u\big|_{\partial M}$ is an arbitrary element of $C^\infty(\partial M)$, we see that $(\partial v/\partial \nu) + X'v = 0$ on ∂M. Now this is a regular boundary problem for Δ, and Proposition 11.14 applies, to give $v \in \mathcal{D}(\mathcal{L}_2)$. Clearly, $\mathcal{L}_1^* \supset \mathcal{L}_2$, so this proves (12.39). By the same reasoning, $\mathcal{L}_2^* = \mathcal{L}_1$.

Now consider the map

$$\Delta : \mathcal{D}(\mathcal{L}_1) \longrightarrow L^2(M). \tag{12.43}$$

We know it has closed range, $\mathcal{R}(\mathcal{L}_1)$. Let $V \subset L^2(M)$ be its orthogonal complement. Then, by definition, $V \subset \mathcal{D}(\mathcal{L}_1^*)$ and $\mathcal{L}_1^* = 0$ on V. Since we know $\mathcal{L}_1^* = \mathcal{L}_2$, the regularity estimates on $\mathcal{L}_2$ imply that

$$\mathcal{R}(\mathcal{L}_1)^\perp \text{ is a finite-dimensional subspace of } C^\infty(\overline{M}). \tag{12.44}$$

From this we deduce that, for $k = 0$, the range of $-\Delta \oplus V$ in (12.40), which we know to be closed, has orthogonal complement which is a finite dimensional subspace $W \subset C^\infty(\overline{M}) \oplus C^\infty(\partial M)$. Then elliptic regularity implies that, for any $k \in \mathbb{Z}^+$, the annihilator of W in $H^k(M) \oplus H^{k+1/2}(\partial M)$, which we know is in the range of $-\Delta \oplus V$ acting on $H^2(M)$, must be in the range of this operator acting on $H^{k+2}(M)$. Consequently, the annihilator of the range of $-\Delta \oplus V$ in (12.40) is exactly W.

As for the index in (12.40), note that, if $V_s = \partial/\partial\nu + sX$, $s \in [0, 1]$, then $-\Delta \oplus V_s$ is a continuous family of Fredholm operators, on which the index is constant. At $s = 0$ we have the Neumann boundary condition, which has index zero, so Proposition 12.6 is proved.

The method used above for the oblique derivative problem extends to many other situations. For example, suppose $P(x, D)$ is a scalar elliptic operator, of order m, and $B_1(x, D), \dots, B_\ell(x, D)$ scalar operators defining boundary conditions, each of distinct orders $m_j < m$, and each noncharacteristic on ∂M. As indicated in Exercises 5–7 of §11, $P(x, D)$ must be "properly elliptic," of order $m = 2\ell$, and there is an algebraic characterization of regularity. Let $P^t(x, D)$ denote the formal adjoint of $P(x, D)$.

Proposition 12.7. *If $\{P(x, D), B_j(x, D), 1 \le j \le \ell\}$ is a (scalar) regular elliptic boundary problem of the form above, then there are boundary operators $B'_j(x, D)$ such that $\{P^t(x, D), B'_j(x, D), 1 \le j \le \ell\}$ is a regular elliptic boundary problem, and such that, given $v \in L^2(M)$, $P^t(x, D)v \in L^2(M)$,*

$$\big(P(x, D)u, v\big) = \big(u, P^t(x, D)v\big),$$

for all $u \in C^\infty(\overline{M})$ satisfying $B_j(x, D)u = 0$ on ∂M, $1 \le j \le \ell$, if and only if $B'_j(x, D)v = 0$ on ∂M, $1 \le j \le \ell$.

A proof of this can be found in [Sch], pp. 224–237. A related discussion is given in [Ag], pp. 134–151. The reader can try it as an exercise. Once this result is demonstrated, the arguments used above also establish the following.

Proposition 12.8. *For the regular boundary problem $\{P(x, D), B_j(x, D), 1 \le j \le \ell\}$ above, if we define $\mathcal{P}$ and $\mathcal{P}^t$, closed unbounded operators on $L^2(M)$, to be $P(x, D)$ and $P^t(x, D)$, respectively, on domains*

$$\begin{aligned}\mathcal{D}(\mathcal{P}) &= \{u \in H^m(M) : B_j(x, D)u = 0 \text{ on } \partial M\},\\ \mathcal{D}(\mathcal{P}^t) &= \{u \in H^m(M) : B'_j(x, D)u = 0 \text{ on } \partial M\},\end{aligned} \tag{12.45}$$

then

$$\mathcal{P}^* = \mathcal{P}^t, \tag{12.46}$$

where $\mathcal{P}^*$ is the Hilbert space adjoint of $\mathcal{P}$. Furthermore, with

$$Tu = \big(P(x, D)u; B_1(x, D)u, \dots, B_\ell(x, D)u\big),$$

we have

$$T : H^{k+m}(M) \longrightarrow H^k(M) \oplus \bigoplus_{j=1}^{\ell} H^{k+m-m_j-1/2}(\partial M) \text{ Fredholm.} \tag{12.47}$$

We leave it to the reader to consider extensions of these last results to systems, or elliptic operators on sections of vector bundles. As the examples of relative and absolute boundary problems for the Hodge Laplacian illustrate, one natural variant for the noncharacteristic hypothesis on $B_j(x, D)$ made above is that, for $x_0 \in \partial M$,

$$\sigma_{B_j}(x_0, \nu) : E_{0x_0} \longrightarrow G_{jx_0} \text{ is surjective,}$$

where E_0, G_j are the vector bundles used in (11.101) and (11.102).

We postpone until the beginning of Chapter 12 a treatment of natural boundary conditions arising for "elliptic complexes" other than the deRham complex. As we will see, in other cases one need not get regular boundary problems.

Exercises

In Exercises 1–3, we study the oblique derivative problem

$$\Delta u = f \text{ on } M, \quad \frac{\partial u}{\partial \nu} + Xu = g \text{ on } \partial M,$$

where $Xu = Yu + \varphi u$, as in (12.34), and the associated map

$$T = -\Delta \oplus V : H^{k+2}(M) \longrightarrow H^k(M) \oplus H^{k+1/2}(\partial M)$$

of (12.40). Assume M is connected and $\partial M \neq \emptyset$. This problem was treated via Fourier analysis in the case where M is the disk in $\mathbb{R}^2$, in Exercises 4–9 of §4, Chapter 4.

1. If $\varphi = 0$, show that $\ker T$ is the one-dimensional space of constants.
2. If $\varphi \geq 0$ on ∂M, φ not identically zero, show that $\ker T = 0$. Deduce that T is surjective in this case. (*Hint*: Use Zaremba's principle, from §2.)
   Our convention here is that ν is the *outward*-pointing unit normal to ∂M.
3. Give examples where φ changes sign and $\ker T$ has dimension 1. Can you make $\ker T$ have dimension greater than 1?
4. In linear elasticity, one considers the elliptic operator L on vector fields on $\Omega \subset \mathbb{R}^n$, defined by

$$Lu = \mu \Delta u + (\lambda + \mu) \operatorname{grad} \operatorname{div} u, \tag{12.48}$$

   with boundary condition

$$\sum_j \nu_j \sigma_{jk} = 0 \text{ on } \partial\Omega, \quad \sigma_{jk} = \lambda(\operatorname{div} u)\delta_{jk} + \mu\Big(\frac{\partial u_j}{\partial x_k} + \frac{\partial u_k}{\partial x_j}\Big), \tag{12.49}$$

   where ν_j are the components of the normal vector ν. For what values of $\lambda, \mu \in \mathbb{R}$ is this a regular elliptic boundary problem? Show that, for such values, one gets a self-adjoint operator.
5. Let $\overline{M}$ be a compact Riemannian manifold with boundary. Consider the functional

$$K(u) = \int_M f\big(x, \operatorname{Def} u(x)\big)\, dV, \quad f(x, A) = 2\mu \operatorname{Tr} A^2 + \lambda(\operatorname{Tr} A)^2,$$

   arising in linear elasticity. Show that

$$DK(u)w = 4\mu(\operatorname{Def} u, \operatorname{Def} w) + 2\lambda(\operatorname{div} u, \operatorname{div} w) = -(Lu, w) + \int_{\partial M} \langle \beta u, w\rangle\, dS,$$

where (compare formula (4.26) in Chapter 10 and (4.3)–(4.4) in Chapter 17)

$$Lu = \mu \Delta u + (\lambda + \mu) \text{grad div } u + 2\mu \text{ Ric } u,$$
$$\beta u = \lambda (\text{div } u)\nu + 2\mu \sigma_{\text{Def}^*}(x, \nu) \text{Def } u.$$

Show that this leads to the boundary condition (12.49).

6. Let Ω be a smooth, bounded region in $\mathbb{R}^n$, and let $P_1(\xi), \dots, P_k(\xi)$ be (scalar) polynomials, homogeneous of degree m in ξ. Show that there is an estimate

$$\|u\|^2_{H^m(\Omega)} \le C \sum_j \|P_j(D)u\|^2_{L^2(\Omega)} + C\|u\|^2_{L^2(\Omega)}, \tag{12.50}$$

for all $u \in H^m(\Omega)$, if and only if $P_1(\zeta), \dots, P_k(\zeta)$, as polynomials in $\zeta \in \mathbb{C}^n$, have no common zeros, except for $\zeta = 0$.
Remarks: Under the hypothesis of no commmon zeros for $\{P_j(\zeta)\}$ in $\mathbb{C}^n \setminus 0$, there exists M such that, for each α with $|\alpha| = M$,

$$\xi^\alpha = \sum_j A_{j\alpha}(\xi) P_j(\xi),$$

for some polynomials $A_{j\alpha}(\xi)$, homogeneous of degree $M - m$. To prove that such an estimate holds when $\{P_j(\xi) : j \le k\} = \{\xi^\alpha : |\alpha| = m\}$, an inductive approach can be taken. This would yield a variant of (12.50). See Agmon [Ag] for further discussion.

7. As noted in the remarks after Proposition 11.11, for $P = \Delta$ on $\overline{M}$, of dimension 2, the boundary problem $B_1 u = g$ on ∂M, with $B_1 u = Xu$, X any nowhere-vanishing real vector field, possibly tangent to ∂M at points, is regular. Then the noncharacteristic hypothesis of Proposition 12.7 fails. Can you extend Propositions 12.7 and 12.8 to treat this case?

A. Spaces of generalized functions on manifolds with boundary

Let $\overline{M}$ be a compact manifold with smooth boundary. We will define a one-parameter family of spaces of functions and "generalized functions" on M, analogous to the Sobolev spaces defined when $\partial M = \emptyset$. The spaces will be defined in terms of a Laplace operator Δ on M, and a boundary condition for the Laplace operator. We will explicitly discuss only the Dirichlet boundary condition, though the results given work equally well for other coercive boundary conditions yielding self-adjoint operators, such as the Neumann boundary condition.

Fixing on the Dirichlet boundary condition, let us recall from (1.7) the map

$$T : H^{-1}(M) \longrightarrow H^1(M), \tag{A.1}$$

inverting the Laplace operator

$$\Delta : H^1_0(M) \longrightarrow H^{-1}(M). \tag{A.2}$$

The restriction of T to $L^2(M)$ is compact and self-adjoint, and we have an orthonormal basis of $L^2(M)$ consisting of eigenfunctions:

$$u_j \in H^1_0(M) \cap C^\infty(\overline{M}), \quad Tu_j = -\mu_j u_j, \quad \Delta u_j = -\lambda_j u_j, \tag{A.3}$$

where $\mu_j \searrow 0, 0 < \lambda_j \nearrow \infty$.

For a given $v \in L^2(M)$, set

$$v = \sum_j \hat{v}(j)\, u_j, \quad \hat{v}(j) = (v, u_j). \tag{A.4}$$

Now, for $s \geq 0$, we define

$$\begin{aligned} \mathcal{D}_s &= \Big\{ v \in L^2(M) : \sum_{j\geq 0} |\hat{v}(j)|^2 \lambda_j^s < \infty \Big\} \\ &= \Big\{ v \in L^2(M) : \sum_{j\geq 0} \hat{v}(j) \lambda_j^{s/2} u_j \in L^2(M) \Big\}. \end{aligned} \tag{A.5}$$

In view of (A.3), an equivalent characterization is

$$\mathcal{D}_s = (-T)^{s/2} L^2(M). \tag{A.6}$$

Clearly, we have

$$\mathcal{D}_0 = L^2(M). \tag{A.7}$$

Also, $\mathcal{D}_2 = T\, L^2(M)$, and by Theorem 1.3 we have

$$\mathcal{D}_2 = H^2(M) \cap H_0^1(M). \tag{A.8}$$

Generally, $\mathcal{D}_{s+2} = T\, \mathcal{D}_s$, so Theorem 1.3 also gives, inductively,

$$\mathcal{D}_{2k} \subset H^{2k}(M), \ k = 1, 2, 3, \dots . \tag{A.9}$$

A result perhaps slightly less obvious than (A.7)–(A.9) is that

$$\mathcal{D}_1 = H_0^1(M). \tag{A.10}$$

To see this, note that $\mathcal{D}_s$ is the completion of the space $\mathcal{F}$ of finite linear combinations of the eigenfunctions $\{u_j\}$, with respect to the $\mathcal{D}_s$-norm, defined by

$$\|v\|_{\mathcal{D}_s}^2 = \sum_j |\hat{v}(j)|^2 \lambda_j^s. \tag{A.11}$$

Now, if $v \in \mathcal{F}$, then

$$(dv, dv) = (v, -\Delta v) = \sum (v, u_j)(u_j, -\Delta v) = \sum |\hat{v}(j)|^2 \lambda_j, \tag{A.12}$$

so

$$\|v\|_{\mathcal{D}_1}^2 = \|dv\|_{L^2(M)}^2, \tag{A.13}$$

for $v \in \mathcal{F}$. In fact, $\mathcal{D}_s$ is the completion of $\mathcal{D}_\sigma$ in the $\mathcal{D}_s$-norm for any $\sigma > s$. We see that (A.13) holds for all $v \in \mathcal{D}_2$, and, with $\mathcal{D}_2$ characterized by (A.8), it is clear that the completion in the norm (A.13) is described by (A.10).

If the Neumann boundary condition were considered, we would replace λ_j by $\langle \lambda_j \rangle$ to take care of $\lambda_0 = 0$. In such a case, we would have

$$\mathcal{D}_2 = \Big\{ u \in H^2(M) : \frac{\partial u}{\partial \nu} = 0 \text{ on } \partial M \Big\}, \quad \mathcal{D}_1 = H^1(M).$$

Now, for $s < 0$, we define $\mathcal{D}_s$ to be the dual of $\mathcal{D}_{-s}$:

$$\mathcal{D}_s = \mathcal{D}^*_{-s}. \tag{A.14}$$

In particular, for any $v \in \mathcal{D}_s$, and any $s \in \mathbb{R}$, $(v, u_j) = \hat{v}(j)$ is defined, and we see that the characterizations involving the sums in (A.5) continue to hold for all $s \in \mathbb{R}$. Also the norm (A.11) provides a Hilbert space structure on $\mathcal{D}_s$ for all $s \in \mathbb{R}$. By (A.10) we have (for Dirichlet boundary conditions)

$$\mathcal{D}_{-1} = H^{-1}(M). \tag{A.15}$$

Also, we have the interpolation identity

$$[\mathcal{D}_s, \mathcal{D}_\sigma]_\theta = \mathcal{D}_{\theta\sigma+(1-\theta)s}, \tag{A.16}$$

for all $s, \sigma \in \mathbb{R}, \theta \in [0, 1]$, where the interpolation spaces are as defined in Chapter 4.

The isomorphism

$$\Delta : \mathcal{D}_{s+2} \longrightarrow \mathcal{D}_s, \text{ with inverse } T : \mathcal{D}_s \longrightarrow \mathcal{D}_{s+2}, \tag{A.17}$$

obviously valid for $s \geq 0$, extends by duality to an isomorphism $\Delta : \mathcal{D}_{-s} \to \mathcal{D}_{-s-2}$ for $s \geq 0$, so (A.17) also holds for $s \leq -2$. By interpolation, it holds for all real s.

By interpolation, (A.9) implies

$$\mathcal{D}_s \subset H^s(M), \quad \text{for } s \geq 0. \tag{A.18}$$

The natural map $\mathcal{D}_s \hookrightarrow H^s(M)$ is injective, for $s \geq 0$, but it is not generally onto, and the transpose $H^{-s}(M) \to \mathcal{D}_{-s}$ is not generally injective. However, the natural map

$$H^{-s}_{\text{comp}}(M) \longrightarrow \mathcal{D}_{-s} \tag{A.19}$$

is injective, where $H^{-s}_{\text{comp}}(M)$ denotes the space of elements of $H^{-s}(N)$ (N being the double of M) with support in the interior of M. In particular, for any interior point $p \in M$,

$$\delta_p \in \mathcal{D}_{-s}, \quad \text{for } s > \frac{n}{2} \quad (n = \dim M). \tag{A.20}$$

Note that as $p \to \partial M$, $\delta_p \to 0$ in any of these spaces. From the isomorphism in (A.17), we have

$$G_p = \Delta^{-1}\delta_p = T\delta_p \tag{A.21}$$

well defined, and

$$G_p \in \mathcal{D}_{-n/2+2-\varepsilon}, \quad \text{for all } \varepsilon > 0. \tag{A.22}$$

This object is equivalent to the Green function studied in this chapter.

We can write any $v \in \mathcal{D}_s$, even for $s < 0$, as a Fourier series with respect to the eigenfunctions u_j. In fact, defining $\hat{v}(j) = (v, u_j)$, as before, the series

$\sum_j \hat{v}(j)u_j$ is convergent in the space $\mathcal{D}_s$ to v, provided $v \in \mathcal{D}_s$, so we are justified in writing

$$v = \sum_j \hat{v}(j)u_j, \quad v \in \mathcal{D}_s, \quad \text{for any } s \in \mathbb{R}. \tag{A.23}$$

Note that $-\Delta : \mathcal{D}_s \to \mathcal{D}_{s-2}$ is given by

$$-\Delta v = \sum_j \lambda_j \hat{v}(j)u_j, \tag{A.24}$$

for any $s \in \mathbb{R}$. We can define

$$(-\Delta)^{(\sigma+i\tau)} : \mathcal{D}_s \longrightarrow \mathcal{D}_{s-2\sigma}, \tag{A.25}$$

for any $\sigma, \tau \in \mathbb{R}$, by

$$(-\Delta)^{(\sigma+i\tau)} v = \sum \lambda_j^{(\sigma+i\tau)} \hat{v}(j)u_j, \tag{A.26}$$

where $v \in \mathcal{D}_s$ is given by (A.23). The maps (A.25) are all isomorphisms. Note that we can write the $\mathcal{D}_s$-inner product coming from (A.11) as

$$(v, w)_{\mathcal{D}_s} = \big(v, (-\Delta)^s w\big), \tag{A.27}$$

where on the right side the pairing arises from the natural $\mathcal{D}_s : \mathcal{D}_{-s}$ duality.

B. The Mayer-Vietoris sequence in deRham cohomology

Here we establish a useful complement to the long exact sequence (9.67) and illustrate some of its implications. Let X be a smooth manifold, and suppose X is the union of two open sets, M_1 and M_2. Let $U = M_1 \cap M_2$. The Mayer-Vietoris sequence has the form

$$\cdots \to \mathcal{H}^{k-1}(U) \xrightarrow{\delta} \mathcal{H}^k(X) \xrightarrow{\rho} \mathcal{H}^k(M_1) \oplus \mathcal{H}^k(M_2) \xrightarrow{\gamma} \mathcal{H}^k(U) \to \cdots . \tag{B.1}$$

These maps are defined as follows. A closed form $\alpha \in \Lambda^k(X)$ restricts to a pair of closed forms on M_1 and M_2, yielding ρ in a natural fashion. The map γ also comes from restriction; if $\iota_\nu : U \hookrightarrow M_\nu$, a pair of closed forms $\alpha_\nu \in \Lambda^k(M_\nu)$ goes to $\iota_1^*\alpha_1 - \iota_2^*\alpha_2$, defining γ. Clearly, $\iota_1^*(\alpha|_{M_1}) = \iota_2^*(\alpha|_{M_2})$ if $\alpha \in \Lambda^k(X)$, so $\gamma \circ \rho = 0$.

To define the "coboundary map" δ on a class $[\alpha]$, with $\alpha \in \Lambda^k(U)$ closed, pick $\beta_\nu \in \Lambda^k(M_\nu)$ such that $\alpha = \beta_1 - \beta_2$. Thus $d\beta_1 = d\beta_2$ on U. Set

$$\delta[\alpha] = [\sigma] \ \text{ with } \sigma = d\beta_\nu \text{ on } M_\nu. \tag{B.2}$$

To show that (B.2) is well defined, suppose $\beta_\nu \in \Lambda^k(M_\nu)$ and $\beta_1 - \beta_2 = d\omega$ on U. Let $\{\varphi_\nu\}$ be a smooth partition of unity supported on $\{M_\nu\}$, and consider $\psi = \varphi_1\beta_1 + \varphi_2\beta_2$, where $\varphi_\nu\beta_\nu$ is extended by 0 off M_ν. We have $d\psi = \varphi_1 d\beta_1 + \varphi_2 d\beta_2 + d\varphi_1 \wedge (\beta_1 - \beta_2) = \sigma + d\varphi_1 \wedge (\beta_1 - \beta_2)$. Since $d\varphi_1$ is supported on U, we can write

$$\sigma = d\psi - d(d\varphi_1 \wedge \omega),$$

an exact form on X, so (B.2) makes δ well defined. Obviously, the restriction of σ to each M_ν is always exact, so $\rho \circ \delta = 0$. Also, if $\alpha = \iota_1^*\alpha_1 - \iota_2^*\alpha_2$ on U, we can pick $\beta_\nu = \alpha_\nu$ to define $\delta[\alpha]$. Then $d\beta_\nu = d\alpha_\nu = 0$, so $\delta \circ \gamma = 0$.

In fact, the sequence (B.1) is exact, that is,

$$\text{(B.3)} \qquad \operatorname{im} \delta = \ker \rho, \quad \operatorname{im} \rho = \ker \gamma, \quad \operatorname{im} \gamma = \ker \delta.$$

We leave the verification of this as an exercise, which can be done with arguments similar to those sketched in Exercises 11–13 in the exercises on cohomology after §9.

If M_ν are the interiors of compact manifolds with smooth boundary, and $\overline{U} = \overline{M}_1 \cap \overline{M}_2$ has smooth boundary, the argument above extends directly to produce an exact sequence

$$\text{(B.4)} \quad \cdots \to \mathcal{H}^{k-1}(\overline{U}) \xrightarrow{\delta} \mathcal{H}^k(X) \xrightarrow{\rho} \mathcal{H}^k(\overline{M}_1) \oplus \mathcal{H}^k(\overline{M}_2) \xrightarrow{\gamma} \mathcal{H}^k(\overline{U}) \to \cdots .$$

Furthermore, suppose that instead $X = \overline{M}_1 \cup \overline{M}_2$ and $\overline{M}_1 \cap \overline{M}_2 = Y$ is a smooth hypersurface in X. One also has an exact sequence

$$\text{(B.5)} \quad \cdots \to \mathcal{H}^{k-1}(Y) \xrightarrow{\delta} \mathcal{H}^k(X) \xrightarrow{\rho} \mathcal{H}^k(\overline{M}_1) \oplus \mathcal{H}^k(\overline{M}_2) \xrightarrow{\gamma} \mathcal{H}^k(Y) \to \cdots .$$

To relate (B.4) and (B.5), let U be a collar neighborhood of Y, and form (B.4) with $\overline{M}_\nu$ replaced by $\overline{M}_\nu \cup \overline{U}$. There is a map $\pi : \overline{U} \to Y$, collapsing orbits of a vector field transversal to Y, and π^* induces an isomorphism of cohomology groups, $\pi^* : \mathcal{H}^k(\overline{U}) \approx \mathcal{H}^k(Y)$.

To illustrate the use of (B.5), suppose $X = S^n$, $Y = S^{n-1}$ is the equator, and $\overline{M}_\nu$ are the upper and lower hemispheres, each diffeomorphic to the ball $\overline{B^n}$. Then we have an exact sequence

$$\text{(B.6)} \quad \begin{aligned} \cdots \to \mathcal{H}^{k-1}(\overline{B^n}) \oplus \mathcal{H}^{k-1}(\overline{B^n}) \xrightarrow{\gamma} \mathcal{H}^{k-1}(S^{n-1}) \xrightarrow{\delta} \mathcal{H}^k(S^n) \\ \xrightarrow{\rho} \mathcal{H}^k(\overline{B^n}) \oplus \mathcal{H}^k(\overline{B^n}) \to \cdots . \end{aligned}$$

As in (9.70), $\mathcal{H}^k(\overline{B^n}) = 0$ except for $k = 0$, when you get $\mathbb{R}$. Thus

$$\text{(B.7)} \qquad \delta : \mathcal{H}^{k-1}(S^{n-1}) \xrightarrow{\approx} \mathcal{H}^k(S^n), \quad \text{for } k > 1.$$

Granted that the computation $\mathcal{H}^1(S^1) \approx \mathbb{R}$ is elementary, this implies $\mathcal{H}^n(S^n) \approx \mathbb{R}$, for $n \geq 1$. Looking at the segment

$$0 \to \mathcal{H}^0(S^n) \xrightarrow{\rho} \mathcal{H}^0(\overline{B^n}) \oplus \mathcal{H}^0(\overline{B^n}) \xrightarrow{\gamma} \mathcal{H}^0(S^{n-1}) \xrightarrow{\delta} \mathcal{H}^1(S^n) \to 0,$$

we see that if $n \geq 2$, then $\ker \gamma \approx \mathbb{R}$, so γ is surjective, hence $\delta = 0$, so $\mathcal{H}^1(S^n) = 0$, for $n \geq 2$. Also, if $0 < k < n$, we see by iterating (B.7) that $\mathcal{H}^k(S^n) \approx \mathcal{H}^1(S^{n-k+1})$, so $\mathcal{H}^k(S^n) = 0$, for $0 < k < n$. Since obviously $\mathcal{H}^0(S^n) = \mathbb{R}$ for $n \geq 1$, we have a fourth computation of $\mathcal{H}^k(S^n)$, distinct from those sketched in Exercise 10 of §8 and in Exercises 10 and 14 of the set of exercises on cohomology after §9.

We note an application of (B.5) to the computation of Euler characteristics, namely

$$\chi(\overline{M}_1) + \chi(\overline{M}_2) = \chi(X) + \chi(Y). \tag{B.8}$$

Note that this result contains some of the implications of Exercises 17 and 18 in the exercises on cohomology, in §9.

Using this, it is an exercise to show that if one two-dimensional surface X_1 is obtained from another X_0 by adding a handle, then $\chi(X_1) = \chi(X_0) - 2$. In particular, if M^g is obtained from S^2 by adding g handles, then $\chi(M^g) = 2 - 2g$. Thus, if M^g is orientable, since $\mathcal{H}^0(M^g) \approx \mathcal{H}^2(M^g) \approx \mathbb{R}$, we have

$$\mathcal{H}^1(M^g) \approx \mathbb{R}^{2g}. \tag{B.9}$$

It is useful to examine the beginning of the sequence (B.5):

$$0 \to \mathcal{H}^0(X) \xrightarrow{\rho} \mathcal{H}^0(\overline{M}_1) \oplus \mathcal{H}^0(\overline{M}_2) \xrightarrow{\gamma} \mathcal{H}^0(Y) \xrightarrow{\delta} \mathcal{H}^1(X) \to \cdots. \tag{B.10}$$

Suppose C is a smooth, closed curve in S^2. Apply (B.10) with $M_1 = C$, a collar neighborhood of C, and $\overline{M}_2 = \overline{\Omega}$, the complement of C. Since ∂C is diffeomorphic to two copies of C, and since $\mathcal{H}^1(S^2) = 0$, (B.10) becomes

$$0 \to \mathbb{R} \xrightarrow{\rho} \mathbb{R} \oplus \mathcal{H}^0(\overline{\Omega}) \xrightarrow{\gamma} \mathbb{R} \oplus \mathbb{R} \xrightarrow{\delta} 0. \tag{B.11}$$

Thus γ is surjective while $\ker \gamma = \operatorname{im} \rho \approx \mathbb{R}$. This forces

$$\mathcal{H}^0(\overline{\Omega}) \approx \mathbb{R} \oplus \mathbb{R}. \tag{B.12}$$

In other words, $\overline{\Omega}$ has exactly two connected components. This is the smooth case of the Jordan curve theorem. Jordan's theorem holds when C is a homeomorphic image of S^1, but the trick of putting a collar about C does not extend to this case.

More generally, if X is a compact, connected, smooth, oriented manifold such that $\mathcal{H}^1(X) = 0$, and if Y is a smooth, compact, connected, oriented hypersurface, then letting C be a collar neighborhood of Y and $\overline{\Omega} = X \setminus C$, we again obtain the sequence (B.11) and hence the conclusion (B.12). The orientability ensures that ∂C is diffeomorphic to two copies of Y. This produces the following variant of (the smooth case of) the Jordan-Brouwer separation theorem.

Theorem B.1. *If X is a smooth manifold, Y is a smooth submanifold of codimension 1, both are*

compact, connected, and oriented,

and

$$\mathcal{H}^1(X) = 0,$$

then $X \setminus Y$ has precisely two connected components.

If all these conditions hold, *except* that Y is *not* orientable, then we replace $\mathbb{R} \oplus \mathbb{R}$ by $\mathbb{R}$ in (B.11) and conclude that $X \setminus Y$ is connected, in that case. As an example, the real projective space $\mathbb{RP}^2$ sits in $\mathbb{RP}^3$ in such a fashion.

Recall from §19 of Chapter 1 the elementary proof of Theorem B.1 when $X = \mathbb{R}^{n+1}$, in particular the argument using degree theory that if Y is a compact, oriented surface in $\mathbb{R}^{n+1}$ (hence, in S^{n+1}), then its complement has at least two connected components. One can extend the degree-theory argument to the nonorientable case, as follows.

There is a notion of degree mod 2 of a map $F : Y \to S^n$, which is well defined whether or not Y is orientable. For one approach, see [Mil]. This is also invariant under homotopy. Now, if in the proof of Theorem 19.11 of Chapter 1, one drops the hypothesis that the hypersurface Y (denoted X there) is orientable, it still follows that the mod 2 degree of F_p must jump by ± 1 when p crosses Y, so $\mathbb{R}^{n+1} \setminus Y$ still must have at least two connected components. In view of the result noted after Theorem B.1, this situation cannot arise. This establishes the following.

Proposition B.2. *If Y is a compact hypersurface of $\mathbb{R}^{n+1}$ (or S^{n+1}), then Y is orientable.*

References

[Ad] R. Adams, *Sobolev Spaces*, Academic Press, New York, 1975.

[Ag] S. Agmon, *Lectures on Elliptic Boundary Problems*, Van Nostrand, New York, 1964.

[ADN] S. Agmon, A. Douglis, and L. Nirenberg, Estimates near the boundary for solutions of elliptic differential equations satisfying general boundary conditions, *CPAM* 12(1959), 623–727; II, *CPAM* 17(1964), 35–92.

[Ahl] L. Ahlfors, *Complex Analysis*, McGraw-Hill, New York, 1979.

[Au] T. Aubin, *Nonlinear Analysis on Manifolds. Monge-Ampere Equations*, Springer-Verlag, New York, 1982.

[Ba] H. Bateman, *Partial Differential Equations of Mathematical Physics*, Dover, New York, 1944.

[BGM] M. Berger, P. Gauduchon, and E. Mazet, *Le Spectre d'une Variété Riemannienne*, LNM no. 194, Springer-Verlag, New York, 1971.

[BJS] L. Bers, F. John, and M. Schechter, *Partial Differential Equations*, Wiley, New York, 1964.

[BoT] R. Bott and L. Tu, *Differential Forms in Algebraic Topology*, Springer-Verlag, New York, 1982.

[Br] F. Browder, A priori estimates for elliptic and parabolic equations, *AMS Proc. Symp. Pure Math.* IV(1961), 73–81.

[Cor] H. O. Cordes, *Elliptic Pseudodifferential Operators - an Abstract Theory*, LNM no. 756, Springer-Verlag, New York, 1979.

[CH] R. Courant and D. Hilbert, *Methods of Mathematical Physics II*, J. Wiley, New York, 1966.

[DK] B. Dahlberg and C. Kenig, Hardy spaces and the Neumann problem in L^p for Laplace's equation in Lipschitz domains, *Ann. of Math.* 125(1987), 437–465.

[Fol] G. Folland, *Introduction to Partial Differential Equations*, Princeton Univ. Press, Princeton, N. J., 1976.

[FK] G. Folland and J. J. Kohn, *The Neumann Problem for the Cauchy-Riemann Complex*, Princeton Univ. Press, Princeton, N. J., 1972.

[Frd] A. Friedman, *Generalized Functions and Partial Differential Equations*, Prentice-Hall, Englewood Cliffs, N. J., 1963.

[Fri] K. Friedrichs, On the differentiability of the solutions of linear elliptic equations, *CPAM* 6(1953), 299–326.

[Ga] P. Garabedian, *Partial Differential Equations*, Wiley, New York, 1964.

[Gå] L. Gårding, Dirichlet's problem for linear elliptic partial differential equations, *Math. Scand.* 1(1953), 55–72.

[GT] D. Gilbarg and N. Trudinger, *Elliptic Partial Differential Equations of Second Order*, 2nd ed., Springer-Verlag, New York, 1983.

[Gri] P. Grisvard, *Elliptic Problems in Nonsmooth Domains*, Pitman, Boston, 1985.

[Hel] L. Helms, *Introduction to Potential Theory*, Wiley, New York, 1969.

[Hil] E. Hille, *Analytic Function Theory*, Chelsea, New York, 1977.

[Hod] W. Hodge, *The Theory and Applications of Harmonic Integrals*, Cambridge Univ. Press, Cambridge, 1952.

[Ho1] L. Hörmander, *The Analysis of Linear Partial Differential Operators*, Vols. 3 and 4, Springer-Verlag, New York, 1985.

[Ho2] L. Hörmander, *Introduction to Complex Analysis in Several Variables*, Van Nostrand, Princeton, N. J. 1966.

[Jo] F. John, *Partial Differential Equations*, Springer-Verlag, New York, 1975.

[Keg] O. Kellogg, *Foundations of Potential Theory*, Dover, New York, 1954.

[Kic] S. Kichenassamy, Compactness theorems for differential forms, *CPAM* 42(1989), 47–53.

[LU] O. Ladyzhenskaya and N. Ural'tseva, *Linear and Quasilinear Elliptic Equations*, Academic Press, New York, 1968.

[Leb] H. Lebesgue, Sur le problème de Dirichlet, *Rend. Circ. Mat. Palermo* 24(1907), 371–402.

[LM] J. Lions and E. Magenes, *Non-homogeneous Boundary Problems and Applications I, II*, Springer-Verlag, New York, 1972.

[Mil] J. Milnor, *Topology from the Differentiable Viewpoint*, Univ. Press of Virginia, Charlottesville, Va., 1965.

[Mir] C. Miranda, *Partial Differential Equations of Elliptic Type*, Springer-Verlag, New York, 1970.

[Miz] S. Mizohata, *The Theory of Partial Differential Equations*, Cambridge Univ. Press, Cambridge, 1973.

[Mor] C. B. Morrey, *Multiple Integrals in the Calculus of Variations*, Springer-Verlag, New York, 1966.

[Mur] F. Murat, Compacité par compensation, *Ann. Scuola Norm. Sup. Pisa* 5(1978), 485–507.

[Ni] L. Nirenberg, On elliptic partial differential equations, *Ann. Scuola Norm. Sup. Pisa* 13(1959), 116–162.

[Ni2] L. Nirenberg, Estimates and existence of solutions of elliptic equations, *CPAM* 9(1956), 509–530.

[Ni3] L. Nirenberg, *Lectures on Linear Partial Differential Equations*, Reg. Conf. Ser. in Math., no. 17, AMS, 1972.

[PrW] M. Protter and H. Weinberger, *Maximum Principles in Differential Equations*, Springer-Verlag, New York, 1984.

[Ra2] J. Rauch, *Partial Differential Equations*, Springer-Verlag, New York, 1992.

[RT] J. Rauch and M. Taylor, Potential and scattering theory on wildly perturbed domains, *Jour. Funct. Anal.* 18(1975), 27–59.

[RS] M. Reed and B. Simon, *Methods of Mathematical Physics*, Academic Press, New York, Vols. 1,2, 1975; Vols. 3,4, 1978.

[RRT] J. Robbin, R. Rogers, and B. Temple, On weak continuity and the Hodge decomposition, *Trans. AMS* 303(1987), 609–618.

[Ru] W. Rudin, *Real and Complex Analysis*, McGraw-Hill, New York, 1976.

[Sch] M. Schechter, *Modern Methods in Partial Differential Equations, an Introduction*, McGraw-Hill, New York, 1977.

[Se] J. Serrin, The problem of Dirichlet for quasilinear elliptic differential equations with many independent variables, *Phil. Trans. Royal Soc. London Ser. A* 264(1969), 413–496.

[SiT] I. M. Singer and J. Thorpe, *Lecture Notes on Elementary Topology and Geometry*, Springer-Verlag, New York, 1976.

[So2] S. Sobolev, *Partial Differential Equations of Mathematical Physics*, Dover, New York, 1964.

[Stk] I. Stakgold, *Boundary Value Problems of Mathematical Physics*, Macmillan, New York, 1968.

[Sto] J. J. Stoker, *Differential Geometry*, Wiley-Interscience, New York, 1969.

[Tar] L. Tartar, Compensated compactness and applications to partial differential equations, in *Nonlinear Analysis and Mechanics* (R.Knops, ed.), Research Notes in Math., Pitman, Boston, 1979.

[Tre1] F. Treves, *Basic Linear Partial Differential Equations*, Academic Press, New York, 1975.

[Ts] M. Tsuji, *Potential Theory*, Chelsea, New York, 1975.

[Wa] F. Warner, *Foundations of Differentiable Manifolds and Lie Groups*, Scott, Foresman, Glenview, Ill., 1971.

[Wil] C. Wilcox, *Scattering Theory for the d'Alembert Equation in Exterior Domains*, LNM no. 442, Springer-Verlag, New York, 1975.

[Yo] K. Yosida, *Functional Analysis*, Springer-Verlag, New York, 1965.

6

Linear Evolution Equations

Introduction

Here we study linear PDE for which one poses an initial-value problem, also called a "Cauchy problem," say at time $t = t_0$. The emphasis is on the wave and heat equations:

$$\frac{\partial^2 u}{\partial t^2} - \Delta u = 0, \qquad \frac{\partial u}{\partial t} - \Delta u = 0, \tag{0.1}$$

though some other sorts of PDE, such as symmetric hyperbolic systems, are also discussed.

Sections 1 and 2 in particular treat (0.1), for $u = u(t, x)$, where x is in a compact Riemannian manifold, or a noncompact but complete Riemannian manifold (perhaps with boundary), respectively. We make essential use of finite propagation speed for solutions to the wave equation to pass from the compact to the noncompact case. In §3 we treat Maxwell's equations, for the electromagnetic field, by converting this system to the wave equation, where Δ is the Hodge Laplacian, and the boundary conditions are of the form studied in §10 of Chapter 5.

Section 4 establishes the Cauchy-Kowalewsky theorem, for linear PDE with real analytic coefficients and real analytic initial data. We show that the solution $u(t, x)$ is given as a convergent power series $\sum u_j(x)t^j/j!$, whose coefficients $u_j(x)$ belong to certain Banach spaces of holomorphic functions. The argument here differs from the classical method of majorants. While it is straightforward, it does not generalize easily to nonlinear analytic PDE. We will give a treatment of the Cauchy-Kowalewsky theorem in the nonlinear case in Chapter 16.

In §5 we use energy estimates for general second-order, scalar, hyperbolic PDE, derived in Chapter 2, to establish the existence of solutions to the Cauchy problem. We also provide a parallel study of first-order, symmetric, hyperbolic systems. The technique we use involves approximating the coefficients (and initial data) by real analytic functions and using the Cauchy-Kowalewsky theorem. A different technique will be presented in §7 of Chapter 7.

Section 6 discusses geometrical optics, a technique for constructing approximate solutions to certain types of initial-value problems for hyperbolic equations.

We continue this discussion in §7, illustrating the simplest situation where the eikonal equation of geometrical optics breaks down and caustics are formed. We study the geometry behind the formation of the simplest sort of caustics, and we study a class of oscillatory integrals, whose relation to solutions to the wave equation will follow from material developed in the next chapter.

There are two appendices at the end of this chapter. Appendix A establishes estimates for $\partial/\partial x_j$ acting on certain spaces of harmonic functions on the ball, of use in the proof of the linear Cauchy-Kowalewsky theorem in §4. Appendix B establishes the multidimensional case of the stationary phase method, whose one-dimensional case arose in §7. The stationary phase method has other uses; in Chapter 9, we will apply it to some problems in scattering theory.

1. The heat equation and the wave equation on bounded domains

Let $\overline{M}$ be a compact, Riemannian manifold with boundary (which might be empty). On $C^\infty(\overline{M})$ is defined the Laplace operator, as usual. We consider here existence and regularity of solutions to the heat equation, and the wave equation. The heat equation is

$$\frac{\partial u}{\partial t} = \Delta u, \tag{1.1}$$

for $u = u(t, x)$, $t \in \mathbb{R}^+$, $x \in M$. Here, we use $\mathbb{R}^+$ to denote $[0, \infty)$. We set the initial condition

$$u(0, x) = f(x). \tag{1.2}$$

If $\partial M \neq \emptyset$, we also pose a boundary condition. The Dirichlet condition is

$$u(t, x) = 0, \quad x \in \partial M. \tag{1.3}$$

The same methods apply to the Neumann boundary problem, $\partial u/\partial \nu = 0$, for $x \in \partial M$, $t \in \mathbb{R}^+$, and a number of other boundary problems.

Solutions to (1.1)–(1.3) can be constructed with the aid of the eigenfunctions of Δ, which arose in (1.11)–(1.13) of Chapter 5. Recall the orthonormal basis $\{u_j\}$ of $L^2(M)$ satisfying

$$u_j \in H_0^1(M) \cap C^\infty(\overline{M}), \quad \Delta u_j = -\lambda_j u_j, \quad 0 \le \lambda_j \nearrow \infty. \tag{1.4}$$

Given $f \in L^2(M)$, we can write

$$f = \sum_j \hat{f}(j) u_j, \quad \hat{f}(j) = (f, u_j). \tag{1.5}$$

Then set

$$u(t, x) = \sum_j \hat{f}(j) e^{-t\lambda_j} u_j(x), \quad t > 0. \tag{1.6}$$

Recalling the spaces $\mathcal{D}_s$ defined in §A of Chapter 5, we see that

$$f \in \mathcal{D}_s \Longrightarrow u \in C(\mathbb{R}^+, \mathcal{D}_s); \quad \partial_t^j u \in C(\mathbb{R}^+, \mathcal{D}_{s-2j}). \tag{1.7}$$

It is clear that $\partial_t u = \Delta u$, for $t > 0$. If $f \in \mathcal{D}_s$ with $s > n/2$, then $u \in C([0, \infty) \times \overline{M})$, and $u(t, x)$ satisfies (1.2) and (1.3) in the ordinary sense.

The uniqueness of solutions to (1.1)–(1.3) within the class

$$C(\mathbb{R}^+, \mathcal{D}_s) \cap C^1(\mathbb{R}^+, \mathcal{D}_{s-2}) \tag{1.8}$$

is easy to obtain, either by showing that the coefficients in the eigenfunction expansion in terms of the u_j must be given by (1.6), or from the simple energy estimate

$$\frac{d}{dt}\|u(t)\|^2_{\mathcal{D}_{s-2}} = 2 \operatorname{Re}\left(\frac{\partial u}{\partial t}, u(t)\right)_{\mathcal{D}_{s-2}} = -2\|u(t)\|^2_{\mathcal{D}_{s-1}} \le 0, \tag{1.9}$$

for a solution to (1.1) belonging to (1.8). We denote the solution to (1.1)–(1.3) as

$$u(t, x) = e^{t\Delta} f(x). \tag{1.10}$$

Let us note that, by (1.6),

$$u \in C^\infty((0, \infty), \mathcal{D}_\sigma), \quad \text{for all } \sigma \in \mathbb{R}. \tag{1.11}$$

In particular, for the solution u to (1.1)–(1.3), for any $f \in \mathcal{D}_s$,

$$u \in C^\infty((0, \infty) \times \overline{M}). \tag{1.12}$$

There is a maximum principle for solutions to the heat equation (1.1) similar to that for the Laplace equation, discussed in §2 of Chapter 5, namely the following.

Proposition 1.1. *If $u \in C([0, a) \times \overline{M}) \cap C^2((0, a) \times M)$ and u solves (1.1) in $(0, a) \times M$, then*

$$\sup_{[0,a)\times\overline{M}} u(t, x) = \max\left\{\sup_{x\in\overline{M}} u(0, x), \quad \sup_{x\in\partial M, t\in[0,a)} u(t, x)\right\}. \tag{1.13}$$

In particular, if (1.2) and (1.3) hold, then

$$\sup_{[0,a)\times\overline{M}} u(t, x) = \sup_{\overline{M}} f(x). \tag{1.14}$$

Proof. It suffices to show that

$$u > 0 \text{ on } \{0\} \times \overline{M} \cup [0, a) \times \partial M \Longrightarrow u \ge 0 \text{ on } [0, a) \times M.$$

In turn, if we set $u_\varepsilon(t, x) = u(t, x) + \varepsilon t$, it suffices to show that, for any $\varepsilon > 0$, the hypothesis above on u implies $u_\varepsilon > 0$ on $[0, a) \times M$. Indeed, if this implication is false for some u, then, since $\overline{M}$ is compact, there must be a smallest $t_0 \in (0, a)$ such that $u_\varepsilon(t_0, x_0) = 0$, for some $x_0 \in M$. We must have $\partial_t u_\varepsilon(t_0, x_0) \le 0$ and $\Delta u_\varepsilon(t_0, x_0) \ge 0$. However, u_ε satisfies the equation $\partial_t u_\varepsilon = \Delta u_\varepsilon + \varepsilon$, so this yields a contradiction, proving the proposition.

There are sharper versions of the maximum principle, analogous to the Hopf maximum principle for elliptic equations proved in Chapter 5. See [J] and [PW] for more on this.

One corollary of (1.14) is that the map (1.10) extends uniquely from a map of $f \in \mathcal{D}_s \mapsto u \in C(\mathbb{R}^+, \mathcal{D}_s)$ (say for some $s > n/2$) to a mapping

$$f \in C_o(M) \mapsto u \in C([0,\infty) \times \overline{M}), \tag{1.15}$$

where $C_o(M)$ is the space of continuous functions on $\overline{M}$ vanishing on ∂M, that is, the sup norm closure of $C_0^\infty(M)$.

Recall from §A of Chapter 5 that $\delta_p \in \mathcal{D}_{-n/2-\varepsilon}$ for all $\varepsilon > 0$. The "fundamental solution" to the heat equation is

$$H(t,x,p) = e^{t\Delta}\delta_p(x). \tag{1.16}$$

By (1.12), $H(t,x,p)$ is smooth in (t,x), for $t > 0$. Since δ_p is a limit in $\mathcal{D}_{-n/2-\varepsilon}$ of elements of $C_0^\infty(M)$ that are ≥ 0, it follows that

$$H(t,x,p) \geq 0, \quad \text{for } t \in (0,\infty), x \in \overline{M}, p \in M. \tag{1.17}$$

In fact, there is a variant of the strong maximum principle, which strengthens (1.17) to $H(t,x,p) > 0$ for $t > 0$, $x, p \in M$. We refer to [J] and [PW] for details.

Next we look at the wave equation

$$\frac{\partial^2 u}{\partial t^2} - \Delta u = 0, \tag{1.18}$$

for $u = u(t,x)$, $t \in \mathbb{R}$, $x \in M$. The initial conditions are

$$u(0,x) = f(x), \quad u_t(0,x) = g(x), \tag{1.19}$$

and if ∂M is not empty, we impose the Dirichet boundary condition

$$u(t,x) = 0, \quad x \in \partial M. \tag{1.20}$$

If we write $u(t,x)$ as

$$u(t,x) = \sum_j a_j(t) u_j(x), \tag{1.21}$$

with u_j the eigenfunctions (1.4), then the coefficients $a_j(t)$ satisfy

$$a_j''(t) + \lambda_j a_j(t) = 0, \quad a_j(0) = \hat{f}(j), a_j'(0) = \hat{g}(j), \tag{1.22}$$

where $\hat{f}(j) = (f, u_j)$, $\hat{g}(j) = (g, u_j)$, and hence

$$a_j(t) = \hat{f}(j) \cos \lambda_j^{1/2} t + \hat{g}(j) \lambda_j^{-1/2} \sin \lambda_j^{1/2} t. \tag{1.23}$$

If $\partial M = \emptyset$ (and M is connected), then 0 is an eigenvalue of multiplicity one; $\lambda_0 = 0$. In that case, (1.23) is replaced by

$$a_0(t) = \hat{f}(0) + \hat{g}(0)t.$$

For simplicity in writing formulas, we will ignore that case.

Thus, assuming all λ_j are nonzero, a solution to (1.18)–(1.20) is given by

$$u(t,x) = \sum_j \big[\hat{f}(j)\cos\lambda_j^{1/2}t + \hat{g}(j)\lambda_j^{-1/2}\sin\lambda_j^{1/2}t\big]u_j(x). \tag{1.24}$$

This is equivalent to the operator expression

$$u(t,x) = \cos t\sqrt{-\Delta}\, f + \frac{\sin t\sqrt{-\Delta}}{\sqrt{-\Delta}}\, g. \tag{1.25}$$

We see that

$$f \in \mathcal{D}_s,\ g \in \mathcal{D}_{s-1} \Longrightarrow u \in C(\mathbb{R}, \mathcal{D}_s), \quad \partial_t^j u \in C(\mathbb{R}, \mathcal{D}_{s-j}), \tag{1.26}$$

if u is given by (1.24). If $s > n/2$, then $u \in C(\mathbb{R}\times\overline{M})$, and the boundary condition (1.20) is satisfied in the ordinary sense.

If we use the "energy norm," whose square is

$$E_s(t) = \|u(t)\|^2_{\mathcal{D}_s} + \|u_t(t)\|^2_{\mathcal{D}_{s-1}}, \tag{1.27}$$

where $\|v\|_{\mathcal{D}_s} = \|(-\Delta)^{s/2}v\|_{L^2(M)}$, we see that if

$$u \in C^1(\mathbb{R}, \mathcal{D}_s) \cap C^2(\mathbb{R}, \mathcal{D}_{s-1}), \tag{1.28}$$

then

$$\begin{aligned}\frac{dE_s}{dt} &= 2\,\mathrm{Re}\,\big(u_t(t), u(t)\big)_{\mathcal{D}_s} + 2\,\mathrm{Re}\,\big(u_t(t), u_{tt}(t)\big)_{\mathcal{D}_{s-1}} \\ &= 2\,\mathrm{Re}\,\big(u_t(t), (-\Delta)^s u(t)\big) + 2\,\mathrm{Re}\,\big(u_t(t), \Delta(-\Delta)^{s-1}u(t)\big) \\ &= 0,\end{aligned} \tag{1.29}$$

provided u solves the wave equation (1.18). Thus we have the energy identity

$$E_s(t) = E_s(0), \tag{1.30}$$

for all $t \in \mathbb{R}$. In the case $\lambda_0 = 0$, (1.27) annihilates constants, so we don't quite get a norm.

We saw in Chapter 2 that solutions to the wave equation that are sufficiently smooth satisfy the finite propagation speed property. We now show that this holds for general solutions, with initial data f, g as in (1.26). Thus we need to define the support of an element $f \in \mathcal{D}_s$. Consider

$$\mathcal{D}_\infty = \bigcap_j \mathcal{D}_j. \tag{1.31}$$

We know that $\mathcal{D}_\infty \subset C^\infty(\overline{M})$, and we use the usual notion of support of an element of this space. If $K \subset \overline{M}$ is closed, $s \in \mathbb{R}$, we will say $f \in \mathcal{D}_s$ is "$\mathcal{D}$-supported" in K if and only if

$$(v, f) = 0, \quad \text{for all } v \in \mathcal{D}_\infty \text{ such that } \operatorname{supp} v \subset \overline{M} \setminus K. \tag{1.32}$$

Soon we will just say f is supported in K, but a distinct term will be useful until a few points are clarified. We show right away that this notion coincides with the familiar notion of support when $s \geq 0$.

Lemma 1.2. *Let $K \subset \overline{M}$ be closed, $s \in [0, \infty)$, $v \in \mathcal{D}_s \subset L^2(M)$. Then v is $\mathcal{D}$-supported in $K \Longleftrightarrow v$ is supported in K in the usual sense, that is, $v(x) = 0$ for almost all $x \in \overline{M} \setminus K$.*

Proof. Let $w \in \mathcal{D}_\infty$ have support (in the usual sense) in a closed set $L \subset \overline{M} \setminus K$. If $v \in \mathcal{D}_0$ vanishes pointwise a.e. on $\overline{M} \setminus K$, then certainly $(v, w) = \int_M v(x)\overline{w(x)}dV = 0$. This establishes the implication $\Leftarrow$.

Suppose conversely that $(v, w) = 0$ for all $w \in \mathcal{D}_\infty$ that vanish pointwise on a neighborhood of K. In particular, $(v, w) = 0$ for all $w \in C_0^\infty(M \setminus K)$, so v vanishes pointwise a.e. on the open set $U = M \setminus K \subset M$, hence on the closure of U in $\overline{M} \setminus K$. This completes the proof.

It is useful to draw attention to one point related to the proof above, namely

$$\text{for } s \leq 0,\ C_0^\infty(M) \text{ is dense in } \mathcal{D}_s. \tag{1.33}$$

To illustrate the notion of "$\mathcal{D}$-supported" for $s < 0$, we note that, given $p \in \partial M$, there is a nonzero $\nu_p \in \mathcal{D}_s$, for any $s < -n/2-1$, defined by $(u, \nu_p) = \partial u(p)/\partial \nu$, and ν_p is $\mathcal{D}$-supported on $\{p\}$.

We now state the result on finite propagation speed.

Proposition 1.3. *If $K \subset \overline{M}$ is closed, and*

$$K_d = \{x \in \overline{M} :\ dist(x, K) \leq d\}, \tag{1.34}$$

then if $f \in \mathcal{D}_s$, $g \in \mathcal{D}_{s-1}$ are $\mathcal{D}$-supported in K, it follows that

$$\cos t\sqrt{-\Delta}\, f \ \textit{and}\ \frac{\sin t\sqrt{-\Delta}}{\sqrt{-\Delta}}\, g \ \textit{are } \mathcal{D}\textit{-supported in } K_d, \tag{1.35}$$

for $|t| \leq d$.

Proof. Let $v \in \mathcal{D}_\infty$ be supported in $\overline{M} \setminus K_d$. We have

$$(\cos t\sqrt{-\Delta}\, f, v) = (f, \cos t\sqrt{-\Delta}\, v). \tag{1.36}$$

But the results of Chapter 2 apply to $\cos t\sqrt{-\Delta}\, v$, which is smooth, so the right side of (1.36) vanishes for $|t| \leq d$. The same sort of analysis applies to $(-\Delta)^{-1/2} \sin t\sqrt{-\Delta}\, g$, to complete the proof.

The next result should justify one's simply saying that $f \in \mathcal{D}_s$ is supported in a closed set K when it is $\mathcal{D}$-supported in K.

Proposition 1.4. *If $s \in \mathbb{R}$ and $f \in \mathcal{D}_s$ is $\mathcal{D}$-supported in a closed set $K \subset \overline{M}$, then for any neighborhood K_d of K, there exists a sequence $f_j \in \mathcal{D}_\infty$, all supported in K_d, such that $f_j \to f$ in $\mathcal{D}_s$.*

Proof. Pick $\varphi \in C_0^\infty(-d,d)$, $\int \varphi(t)\,dt = 1$, and consider

$$f_j = \int \varphi_j(t) \cos t\sqrt{-\Delta}\, f\, dt, \quad \varphi_j(t) = j\varphi(jt). \tag{1.37}$$

Integration by parts shows that

$$(-\Delta)^k f_j = \int \varphi_j^{(2k)}(t) \cos t\sqrt{-\Delta}\, f\, dt \in \mathcal{D}_s,$$

for all k, so $f_j \in \mathcal{D}_\infty$. That $f_j \to f$ in $\mathcal{D}_s$ is clear. Finally, by Proposition 1.3, each f_j is $\mathcal{D}$-supported in K_d, and so by Lemma 1.2 each f_j is supported in K_d.

Exercises

1. Let $f \in C(\mathbb{R}, \mathcal{D}_s)$. Show that the unique solution $u \in C(\mathbb{R}, \mathcal{D}_{s+1})$ to

$$\frac{\partial^2 u}{\partial t^2} - \Delta u = f, \quad u(0) = u_t(0) = 0$$

is given by

$$u(t) = \int_0^t \frac{\sin(t-\tau)\sqrt{-\Delta}}{\sqrt{-\Delta}} f(\tau)\, d\tau, \tag{1.38}$$

suitably interpreted in case $0 \in \operatorname{Spec}(-\Delta)$. Show that

$$\|u(t)\|_{\mathcal{D}_{s+1}} + \|D_t u(t)\|_{\mathcal{D}_s} \le C \int_0^t \|f(\tau)\|_{\mathcal{D}_s}\, d\tau. \tag{1.39}$$

2. Let $u \in C(\mathbb{R}, L^2(M))$ satisfy

$$\frac{\partial^2 u}{\partial t^2} - \Delta u = f \text{ on } \mathbb{R} \times M, \quad u = 0 \text{ on } \mathbb{R} \times \partial M, \quad u = 0, \text{ for } t < 0.$$

Assume $f \in C^k(\mathbb{R} \times \overline{M})$. Show that $u \in H^{k+1}\big([0,T] \times M\big)$ for any $T < \infty$. (*Hint*: Apply a variant of the $s = 0$ case of Exercise 1 to $D_t^j u$, $0 \le j \le k$. Once you have $\partial_t^2 u = g \in C\big(\mathbb{R}, H_0^1(M)\big) \cap C^1\big(\mathbb{R}, L^2(M)\big)$, apply the PDE to write

$$(\partial_t^2 + \Delta)u = 2g - f, \quad u = 0 \text{ on } \mathbb{R} \times \partial M,$$

and use elliptic regularity. Continue this argument.)

3. Adapt the proof of Hopf's maximum principle, given in §2 of Chapter 5, to the case of the heat equation, proving a stronger version of Proposition 1.1. Establish a version that treats $u(t,x) = e^{t(\Delta - V)} f(x)$, given $V \in C^\infty(\overline{M})$ real-valued and ≥ 0. Using $e^{-\alpha t} u(t,x) = e^{t(\Delta - V - \alpha)} f(x)$, remove the hypothesis that $V \ge 0$.

Exercises 4–10 deal with regularity of solutions to the PDE

$$\frac{\partial u}{\partial t} - \Delta u = f, \quad u\big|_{\mathbb{R}^+ \times \partial M} = 0. \tag{1.40}$$

We assume that $u \in C(\mathbb{R}^+, \mathcal{D}_1)$. Let $I = [0, T]$.

4. Suppose that (1.40) holds and $u(0) = 0$. Show that

$$f \in L^2(I \times M) \Longrightarrow \Delta u \text{ and } \partial_t u \in L^2(I \times M),$$

and $\|\Delta u\|_{L^2(I\times M)} \leq \|f\|_{L^2(I\times M)}$. (*Hint*: If u is sufficiently smooth, compute $(d/dt)\|(-\Delta)^{1/2}u\|^2_{L^2(M)}$ to show that

$$\|(-\Delta)^{1/2}u(T)\|^2_{L^2(M)} + 2\int_0^T \|\Delta u(t)\|^2_{L^2(M)}\,dt = -2\int_0^T \big(f(t), \Delta u(t)\big)_{L^2}\,dt$$
$$\leq \int_0^T \Big[\|f\|^2_{L^2} + \|\Delta u(t)\|^2_{L^2}\Big]\,dt.)$$

5. Omit the hypothesis that $u(0) = 0$. If $I' = [t_0, T]$ for some $t_0 > 0$, show that

$$f \in L^2(I \times M) \Longrightarrow \Delta u \text{ and } \partial_t u \in L^2(I' \times M).$$

(*Hint*: Set $v = \varphi(t)u$, where $\varphi \in C^\infty(\mathbb{R})$, $\varphi(t) = 1$ for $t \geq t_0$, 0 for $t \leq t_0/2$. Then $\partial_t v - \Delta v = \varphi(t)f + \varphi'(t)u$.)

6. Show that

$$\partial_t f \in L^2(I \times M) \Longrightarrow \partial_t^2 u \in L^2(I' \times M), \text{ and } \partial_t u \in C(I', \mathcal{D}_1).$$

(*Hint*: If $\delta_h u(t,x) = h^{-1}\big[u(t+h,x) - u(t,x)\big]$, consider estimates for $\partial_t(\delta_h u)$, using the PDE $\partial_t(\delta_h u) - \Delta(\delta_h u) = \delta_h f$, and let $h \to 0$.)

7. Deduce that if $\partial_t^j f \in L^2(I \times M)$, for $0 \leq j \leq k$, then

$$\partial_t^{j+1} u \in L^2(I' \times M), \text{ and } \partial_t^j u \in C(I', \mathcal{D}_1), \quad 0 \leq j \leq k.$$

8. Assume now that

$$\partial_t f \in L^2(I \times M), \text{ and } f \in L^2\big(I, H^2(M)\big). \tag{1.41}$$

Show that $\Delta u \in L^2\big(I', H^2(M)\big)$, and hence $u \in L^2\big(I, H^4(M)\big)$.
(*Hint*: Note that $\Delta(\Delta u) = \partial_t^2 u - \partial_t f - \Delta f$, while $\Delta u\big|_{I\times\partial M} = -f\big|_{I\times\partial M}$. The term $\partial_t^2 u$ is controlled by Exercise 6. For fixed t, apply elliptic estimates.)

9. Now assume that

$$\partial_t^j f \in L^2\big(I, H^{2k-2j}(M)\big), \quad 0 \leq j \leq k. \tag{1.42}$$

Show that

$$\partial_t^j u \in L^2\big(I', H^{2k+2-2j}(M)\big), \quad 0 \leq j \leq k+1. \tag{1.43}$$

(*Hint*: Reason inductively. Note that $\Delta^j u$ satisfies

$$\Delta(\Delta^j u) = \partial_t^{j+1} u - (\partial_t^j + \partial_t^{j-1}\Delta + \cdots + \partial_t \Delta^{j-1} + \Delta^j)f$$
$$\Delta^j u\big|_{I\times\partial M} = -(\partial_t^{j-1} + \partial_t^{j-2}\Delta + \cdots + \partial_t\Delta^{j-2} + \Delta^{j-1})f\big|_{I\times\partial M}.)$$

10. Deduce in particular that

$$f \in C^\infty\big([0,T] \times \overline{M}\big) \Longrightarrow u \in C^\infty\big((0,T] \times \overline{M}\big). \tag{1.44}$$

11. Parallel the results of Exercisess 4–10 for solutions to $\partial u/\partial t - \Delta u = f$, given
a) Neumann boundary condition, $\quad \partial_\nu u\big|_{\mathbb{R}^+\times\partial M} = 0$,
b) Robin boundary condition, $\quad \partial_\nu u - a(x)u\big|_{\mathbb{R}^+\times\partial M} = 0$.

2. The heat equation and wave equation on unbounded domains

Here we look at the heat and wave equations on $\mathbb{R} \times M$ when M is a noncompact Riemannian manifold.

First we assume that M is complete and without boundary. We construct the solution to the wave equation

$$\frac{\partial^2 u}{\partial t^2} - \Delta u = 0 \text{ on } \mathbb{R} \times M, \quad u(0, x) = f(x), \quad u_t(0, x) = g(x), \tag{2.1}$$

first under the hypothesis that

$$f \in H_0^1(M), \ g \in L^2(M), \quad \text{supp } f, g \subset K, \tag{2.2}$$

where $K \subset M$ is compact. We produce the unique solution

$$u \in C(\mathbb{R}, H^1(M)) \cap C^1(\mathbb{R}, L^2(M)) \tag{2.3}$$

having the property that

$$\text{supp } u(t) \text{ is compact in } M, \quad \forall\, t \in \mathbb{R}. \tag{2.4}$$

To do this, let $\overline{\mathcal{O}}_j \subset M$ be compact subsets with smooth boundary, such that $\mathcal{O}_1 \subset\subset \mathcal{O}_2 \subset\subset \cdots \subset\subset \mathcal{O}_j \subset\subset \nearrow M$. Given supp $f, g \subset K$ and $s > 0$, pick N so large that $K_s \subset \mathcal{O}_N$, where $K_s = \{x \in M : \text{dist}(x, K) \le s\}$. Now let Δ_j be the Laplace operator on $\mathcal{O}_j$, with Dirichlet boundary condition, so that $\cos t\sqrt{-\Delta_j}$ and $(-\Delta_j)^{-1/2} \sin t\sqrt{-\Delta_j}$ are defined on $L^2(\mathcal{O}_j)$, $H_0^1(\mathcal{O}_j)$, and so forth, as in §1. By finite propagation speed, we see that

$$u(t) = \cos t\sqrt{-\Delta_j}\, f + \frac{\sin t\sqrt{-\Delta_j}}{\sqrt{-\Delta_j}}\, g, \quad \text{for } |t| < s, \ j \ge N, \tag{2.5}$$

has support in $\mathcal{O}_N$ and is independent of $j \ge N$. This specifies the solution to (2.1), given (2.2).

We can define

$$U(t)\{f, g\} = \{u(t), \partial_t u(t)\}, \tag{2.6}$$

obtaining a one-parameter family of maps

$$U(t) : C_0^\infty(M) \oplus C_0^\infty(M), \tag{2.7}$$

satisfying the group property

$$U(0) = I, \quad U(t_1 + t_2) = U(t_1)U(t_2). \tag{2.8}$$

Also, if $f, g \in C_0^\infty(M)$, the proof of energy conservation given in Chapter 2 works:

$$\|df\|_{L^2(M)}^2 + \|g\|_{L^2(M)}^2 = \|d_x u(t)\|_{L^2(M)}^2 + \|\partial_t u(t)\|_{L^2(M)}^2, \tag{2.9}$$

for each $t \in \mathbb{R}$. Let us set

$$\mathcal{H} = \text{ completion of } C_0^\infty(M) \text{ in the norm } \|f\|_{\mathcal{H}} = \|df\|_{L^2(M)}. \tag{2.10}$$

We have the following proposition.

Proposition 2.1. *The family of maps $U(t)$ in (2.6) has a unique extension to a unitary group*

$$U(t) : \mathcal{H} \oplus L^2(M) \longrightarrow \mathcal{H} \oplus L^2(M). \tag{2.11}$$

We move on to the heat equation

$$\frac{\partial u}{\partial t} = \Delta u, \quad u(0, x) = f(x), \tag{2.12}$$

first assuming that $f \in L^2(M)$ has support in a compact set K. As with the wave equation, if $K \subset \mathcal{O}_j$, then $e^{t\Delta_j} f$ is defined by §1. Note that, in that case,

$$e^{t\Delta_j} f = \frac{1}{\sqrt{4\pi t}} \int_{-\infty}^{\infty} e^{-s^2/4t} \cos s\sqrt{-\Delta_j}\, f\, ds. \tag{2.13}$$

This suggests considering

$$H(t)f(x) = \frac{1}{\sqrt{4\pi t}} \int_{-\infty}^{\infty} e^{-s^2/4t}\, W(s)f(x)\, ds, \tag{2.14}$$

where $W(s)f(x) = v(t, x)$ solves (2.1), with $g = 0$. Thus, if f is supported on K,

$$W(s)f(x) = \cos s\sqrt{-\Delta_j}\, f(x) \ \text{ if } K_{|s|} \subset \mathcal{O}_j. \tag{2.15}$$

Then

$$H(t)f(x) = e^{t\Delta_j} f(x) + \frac{1}{\sqrt{4\pi t}} \int_{T_j} e^{-s^2/4t} \Big[W(s)f(x) - \cos s\sqrt{-\Delta_j}\, f(x)\Big]\, ds, \tag{2.16}$$

where, assuming $K \subset \mathcal{O}_j$, we set

$$T_j = \big\{s \in \mathbb{R} : \text{dist}(K, \partial\mathcal{O}_j) < |s|\big\}. \tag{2.17}$$

Since $\cos s\sqrt{-\Delta_j}$ and (by (2.15)) $W(s)$ have L^2-operator norms ≤ 1, we have

$$H(t)f = \lim_{j\to\infty} e^{t\Delta_j} f \text{ in } L^2(M), \tag{2.18}$$

given $f \in L^2(M)$ with compact support. Here, $e^{t\Delta_j} f(x)$ is set equal to zero on $M \setminus \mathcal{O}_j$. Thus $H(t)$ extends uniquely to an operator on $L^2(M)$, of norm ≤ 1, and we have

$$H(t)f = \lim_{j\to\infty} e^{t\Delta_j} P_j f \text{ in } L^2(M), \quad \forall f \in L^2(M), \tag{2.19}$$

where $P_j f(x) = \chi_{\mathcal{O}_j}(x) f(x)$. Material in Chapter 8, §2, will show that $H(t)$ is a semigroup, whose infinitesimal generator is the unique self-adjoint extension of Δ from $C_0^\infty(M)$, when M is a complete Riemannian manifold.

We will show that, for $t > 0$, the operator $H(t)$ has a smooth integral kernel:

$$H(t)f(x) = \int_M h(t, x, y)\, f(y)\, dV(y). \tag{2.20}$$

Furthermore, under certain hypotheses on $M, h(t, x, y)$ will be shown to decrease rapidly as $\text{dist}(x, y) \to \infty$, for fixed $t > 0$. Let U_j be open sets in M, containing points x_j, and suppose $\rho = \text{dist}(U_1, U_2) = \inf\{\text{dist}(y_1, y_2) : y_j \in U_j\}$. Assume f is supported in U_1. Then finite propagation speed implies that

$$H(t)f(x) = \frac{1}{\sqrt{4\pi t}} \int_{|s|\ge\rho} e^{-s^2/4t}\, W(s)f(x)\, ds, \quad \text{for } x \in U_2. \tag{2.21}$$

Thus, if $R_j f(x) = \chi_{U_j}(x) f(x)$, we have

$$\|R_2 H(t) R_1\|_{\mathcal{L}(L^2)} \le \frac{1}{\sqrt{4\pi t}} \int_{|s|\ge\rho} e^{-s^2/4t}\, ds \le e^{-\rho^2/4t}, \tag{2.22}$$

since, for $\tau > 0$,

$$\int_\tau^\infty e^{-s^2/4}\, ds = e^{-\tau^2/4} \int_0^\infty e^{-(s^2+2s\tau)/4}\, ds \le \sqrt{\pi}\, e^{-\tau^2/4}. \tag{2.23}$$

To estimate derivatives, we can use the equation $\partial_s^2 W(s) = \Delta W(s)$ and integrate by parts, to write

$$\Delta^k H(t)f(x) = \frac{1}{\sqrt{4\pi t}} \int_{|s|\ge\rho} \left(\partial_s^{2k} e^{-s^2/4t}\right) W(s)f(x)\, ds, \tag{2.24}$$

given $x \in U_2$, $\text{supp}\, f \subset U_1$. Now there are estimates of the form

$$\left|\partial_s^{2k} e^{-s^2/4t}\right| \le C_k t^{-k} \langle (4t)^{-1} s^2 \rangle^k\, e^{-s^2/4t}. \tag{2.25}$$

Hence

$$\begin{aligned} &\|R_2 \Delta^k H(t) R_1\|_{\mathcal{L}(L^2)} \\ &\le C_k t^{-k} \int_{\rho/\sqrt{t}}^\infty (1+s^2)^k\, e^{-s^2/4}\, ds \le C_k t^{-k} \langle t^{-1}\rho^2 \rangle^k\, e^{-\rho^2/4t}, \end{aligned} \tag{2.26}$$

the last inequality following by an appropriate variant of (2.23). Pick $k > n/4$, where $n = \dim M$. There is a Sobolev estimate of the form

$$|f(x_2)| \le C(U_2)\Big[\|\Delta^k f\|_{L^2(U_2)} + \|f\|_{L^2(U_2)}\Big], \tag{2.27}$$

so we have

$$\|h(t, x_2, \cdot)\|_{L^2(U_1)} \le C' C(U_2) \big(1 + t^{-k} \langle t^{-1}\rho^2 \rangle^k\big) e^{-\rho^2/4t}. \tag{2.28}$$

By symmetry and another application of the argument above, we have

$$|h(t, x_2, x_1)| \le C' C(U_1) C(U_2) \big(1 + t^{-k} \langle t^{-1}\rho^2 \rangle^k\big)^2 e^{-\rho^2/4t}. \tag{2.29}$$

Similarly, one can estimate higher derivatives. We have the following.

Proposition 2.2. *If M is a complete Riemannian manifold of dimension n, the operator $H(t)$ given by (2.14) and (2.19) has integral kernel $h(t,x,y)$, smooth on $(0,\infty)\times M\times M$, and satisfying an estimate*

$$0 \le h(t,x,y) \le C\kappa(x,\delta)\kappa(y,\delta)\big(1+t^{-k}\langle t^{-1}\rho^2\rangle^k\big)^2 e^{-\rho^2/4t}, \tag{2.30}$$

where $\operatorname{dist}(x,y)=\rho+2\delta$, $\kappa(x,\delta)=C(U)$, for U the ball of radius δ, centered at x, and $k>n/4$.

The positivity in (2.30) follows from the positivity of the heat kernels $h_j(t,x,y)$ of $e^{t\Delta_j}$ (set equal to zero if x or y is in $M\setminus\mathcal{O}_j$). In fact, using the maximum principle for the heat equation on $\mathbb{R}^+\times\mathcal{O}_j$, we obtain

$$0 \le h_j(t,x,y) \nearrow h(t,x,y) \quad \text{as } j\to\infty. \tag{2.31}$$

In some cases, such as when M is a homogeneous space, perhaps with a compactly supported perturbation in the metric, one has a uniform estimate $\kappa(x,\delta)\le c(\delta)$, independent of $x\in M$. Then (2.30) implies the very rapid decay of $h(t,x,y)$ as $\operatorname{dist}(x,y)\to\infty$. Estimates of the form (2.30) will be of occasional use later, for example, in Chapter 8, §3. Somewhat sharper estimates are proved in [CGT]. Other approaches to heat kernel estimates can be found in [CLY] and [Dav].

The results above can be extended to the case of $\overline{M}$, a complete Riemannian manifold with (smooth) boundary. On ∂M, one could place one of a number of boundary conditions, such as Dirichlet or Neumann. Of course, it is no longer true that the solution operator $U(t)$ as in (2.6) preserves $C_0^\infty(M)\oplus C_0^\infty(M)$, but we do have such results as

$$U(t): H^1_{0,\text{comp}}(M)\oplus L^2_{\text{comp}}(M) \longrightarrow H^1_{0,\text{comp}}(M)\oplus L^2_{\text{comp}}(M), \tag{2.32}$$

in the case of the Dirichlet boundary condition, where $H^1_{0,\text{comp}}(M)$ consists of functions $u\in H^1_0(M)$ such that u is supported on a compact subset of $\overline{M}$. We leave further details on such extension of the results above to the reader.

We now discuss when the heat kernel $h(t,x,y)$ satisfies

$$\int_M h(t,x,y)\,dV(x) = 1, \tag{2.33}$$

for all $t>0$, $y\in M$. This has probabilistic significance. If M is compact (without boundary), (2.33) is clear. If M has boundary, and one uses the Dirichlet boundary condition, then (2.33) fails, but it continues to hold if the Neumann boundary condition is used.

If M is a complete Riemannian manifold (without boundary), then we always have

$$\int_M h(t,x,y)\,dV(x) \le 1, \tag{2.34}$$

as a consequence of (2.31), but (2.33) may fail in some cases; some examples are given in [Az]. In "nice" cases, such as when M has bounded geometry, (2.33) does not fail, as we will now show.

Note that (2.33) holds if and only if

$$\int_M e^{t\Delta} f(x)\, dV(x) = \int_M f(x)\, dV(x), \quad \text{for all } f \in C_0^\infty(M), \tag{2.35}$$

given that M is complete. Our approach to specifying a class of M for which (2.35) holds will use the identity (2.14). Given $f \in L^2(M)$, the integral on the right side of (2.14) is convergent in $L^2(M)$. As long as M is complete, we have

$$\int_M \cos s\sqrt{-\Delta}\, f(x)\, dV(x) = \int_M f(x)\, dV(x), \tag{2.36}$$

for all $s \in \mathbb{R}$, $f \in C_0^\infty(M)$. Consequently, (2.35) holds, for a given $t > 0$, $f \in C_0^\infty(M)$, whenever it can be shown that, for some $\beta < 1/4t$,

$$\left\| \cos s\sqrt{-\Delta}\, f \right\|_{L^1(M)} \le C\, e^{\beta s^2}. \tag{2.37}$$

Three ingredients go into the estimate of this L^1-norm. Two are Cauchy's inequality plus the fact that the L^2-operator norm of $\cos s\sqrt{-\Delta}$ is ≤ 1, yielding

$$\left\| \cos s\sqrt{-\Delta}\, f \right\|_{L^1(B_{s+\sigma}(p))} \le \left\| \cos s\sqrt{-\Delta}\, f \right\|_{L^2(M)} \Big(\text{vol } B_{s+\sigma}(p) \Big)^{1/2}, \tag{2.38}$$

where $B_{s+\sigma}(p) = \{x \in M : \text{dist}(x, p) \le s + \sigma\}$. The third ingredient is finite propagation speed; if f is supported on $B_\sigma(p)$, then $\cos s\sqrt{-\Delta}\, f$ is supported on $B_{s+\sigma}(p)$, so the left side of (2.38) is all of $\| \cos s\sqrt{-\Delta}\, f \|_{L^1(M)}$. Consequently, given $t > 0$, (2.35) holds provided that, for some $\beta < 1/4t$, we have a volume estimate:

$$\text{vol } B_{s+\sigma}(p) \le C(\sigma)\, e^{\beta s^2}, \quad \forall\, s > 0, \sigma > 0. \tag{2.39}$$

In other words, if (2.39) holds, then (2.35) holds for all $t < 1/4\beta$. Then (2.35) extends to all $f \in L^1(M)$, for $t < 1/4\beta$. Consequently, it holds for all $t > 0$, and so does (2.33), as long as (2.39) holds, for some β. Note that (2.39) follows from the estimate $\text{vol } B_s(p) \le C\, e^{\beta s^2/2}$. Relabeling β, we summarize what has been shown.

Proposition 2.3. *If M is a complete Riemannian manifold satisfying, for some $\beta < \infty$, the volume estimate*

$$\text{vol } B_s(p) \le C_p\, e^{\beta s^2}, \tag{2.40}$$

then (2.33) and (2.35) hold.

Exercises

1. Let $V(t)$ denote the solution operator to the following variant of (2.1):

$$\frac{\partial^2 u}{\partial t^2} - (\Delta - 1)u = 0, \quad u(0) = f, \quad \partial_t u(0) = g.$$

Assume M is complete. Show that $V(t)$ preserves $C_0^\infty(M) \oplus C_0^\infty(M)$ and has a unique extension to a unitary group on $H_0^1(M) \oplus L^2(M)$, where $H_0^1(M)$ is, as usual, the completion of $C_0^\infty(M)$ in the norm defined by $\|f\|^2_{H_0^1} = \|df\|^2_{L^2(M)} + \|f\|^2_{L^2(M)}$.

2. Verify the estimates of the s-derivatives of $e^{-s^2/4t}$, given in (2.25).

3. Maxwell's equations

Maxwell's equations for the propagation of electromagnetic waves in a vacuum are written as follows, as seen in §11 of Chapter 2:

$$\frac{\partial E}{\partial t} = \text{curl } B, \quad \frac{\partial B}{\partial t} = -\text{ curl E}, \tag{3.1}$$

and

$$\text{div } E = 0, \quad \text{div } B = 0. \tag{3.2}$$

Here, E is the electric field and B is the magnetic field, both vector fields in a region of $\mathbb{R}^3$, and both varying with time t. Units are chosen so the speed of light c is 1. If the region M in $\mathbb{R}^3$ is bounded by a "perfect conductor," one sets the boundary conditions

$$\nu \times E = 0, \quad \nu \cdot B = 0 \text{ on } \partial M. \tag{3.3}$$

We will investigate the initial-value problem for (3.1)–(3.3), where $E(0, x)$ and $B(0, x)$ are specified, subject to the condition (3.3).

We will transform (3.1)–(3.3) into a system of equations for 1-forms on M rather than vector fields on M, using the metric tensor to identify these, and then make contact with material developed in §10 of Chapter 5. If we let $\tilde{E}$, $\tilde{B}$ be 1-forms on M corresponding to the vector fields E and B, then the equations above become, respectively,

$$\frac{\partial \tilde{E}}{\partial t} = *d\tilde{B}, \quad \frac{\partial \tilde{B}}{\partial t} = - * d\tilde{E}, \tag{3.4}$$

$$\delta \tilde{E} = 0, \quad \delta \tilde{B} = 0, \tag{3.5}$$

and

$$\nu \wedge \tilde{E} = 0, \quad \iota_\nu \tilde{B} = 0 \text{ on } \partial M. \tag{3.6}$$

Here, $\tilde{E} = \tilde{E}(t, x)$, $\tilde{B} = \tilde{B}(t, x)$, and of course d and δ involve only differentiation in the x-variables. The identity div $E = -\delta \tilde{E}$ was demonstrated in Chapter 2; see (10.25) of Chapter 2. Note that we are using forms to describe the electromagnetic field in a completely different way than that used in §11 of Chapter 2. Here we are considering functions of t taking values in spaces of forms on a 3-fold, rather than forms on a 4-fold.

We can define the *energy* of the field $(\tilde{E}, \tilde{B})$ to be

$$\mathcal{E}(t) = \|\tilde{E}(t)\|^2_{L^2(M)} + \|\tilde{B}(t)\|^2_{L^2(M)}. \tag{3.7}$$

The following result expresses conservation of energy and of course also gives a uniqueness result.

Proposition 3.1. *If $\tilde{E}, \tilde{B} \in H^1([T_1, T_2] \times M)$ satisfy (3.4) and the first part of (3.6), then*

$$\frac{d\mathcal{E}}{dt} = 0, \quad \textit{for } t \in (T_1, T_2), \tag{3.8}$$

so

$$\mathcal{E}(t_1) = \mathcal{E}(t_2), \quad \textit{for } t_j \in (T_1, T_2). \tag{3.9}$$

Proof. We have

$$\begin{aligned}\frac{d\mathcal{E}}{dt} &= 2\,\mathrm{Re}\left(\frac{\partial \tilde{E}}{\partial t}, \tilde{E}\right)_{L^2} + 2\,\mathrm{Re}\left(\frac{\partial \tilde{B}}{\partial t}, \tilde{B}\right)_{L^2} \\ &= 2\,\mathrm{Re}\,(*d\tilde{B}, \tilde{E}) - 2\,\mathrm{Re}\,(*d\tilde{E}, \tilde{B}).\end{aligned}$$

Now

$$(*d\tilde{B}, \tilde{E}) = (\delta * \tilde{B}, \tilde{E}) = (*\tilde{B}, d\tilde{E}) = (\tilde{B}, *d\tilde{E}),$$

where the second identity uses the hypothesis that $\nu \wedge \tilde{E} = 0$ on ∂M. Thus (3.8) is proved.

In order to establish the existence of solutions to (3.4)–(3.6), we will produce a second-order wave equation satisfied by $(\tilde{E}, \tilde{B})$, which can be solved by the methods of §1. Note that if we set

$$v = \frac{\partial \tilde{E}}{\partial t} - *d\tilde{B}, \quad w = \frac{\partial \tilde{B}}{\partial t} + *d\tilde{E}, \tag{3.10}$$

then a short computation gives

$$\begin{aligned}\frac{\partial v}{\partial t} + *dw &= \frac{\partial^2 \tilde{E}}{\partial t^2} + \delta d\tilde{E}, \\ \frac{\partial w}{\partial t} - *dv &= \frac{\partial^2 \tilde{B}}{\partial t^2} + \delta d\tilde{B}.\end{aligned} \tag{3.11}$$

If (3.5) holds, then we can replace $\delta d\tilde{E}$ and $\delta d\tilde{B}$ in (3.11) by $-\Delta\tilde{E}$ and $-\Delta\tilde{B}$, respectively. Now, if (3.4) holds, then $v = w = 0$, and hence (3.11) implies

$$\frac{\partial^2 \tilde{E}}{\partial t^2} - \Delta\tilde{E} = 0, \quad \frac{\partial^2 \tilde{B}}{\partial t^2} - \Delta\tilde{B} = 0. \tag{3.12}$$

The appropriate boundary conditions for $\tilde{E}$ and $\tilde{B}$ are

$$\nu \wedge \tilde{E} = 0, \quad \nu \wedge \delta\tilde{E} = 0 \text{ on } \partial M \text{ (relative boundary conditions)}, \tag{3.13}$$

and

$$\iota_\nu \tilde{B} = 0, \quad \iota_\nu d\tilde{B} = 0 \text{ on } \partial M \text{ (absolute boundary conditions)}, \tag{3.14}$$

where the last boundary condition is derived from (3.4) and (3.13), together with the fact that

$$\nu \wedge *d\tilde{B} = 0 \Longleftrightarrow \iota_\nu d\tilde{B} = 0.$$

Now the existence of solutions to the initial value problem

$$\begin{aligned} \tilde{E}(0,x) &= \tilde{E}_0(x), \quad \tilde{E}_t(0,x) = \tilde{E}_1(x), \\ \tilde{B}(0,x) &= \tilde{B}_0(x), \quad \tilde{B}_t(0,x) = \tilde{B}_1(x), \end{aligned} \tag{3.15}$$

follows from the methods of §1, given the material on the Hodge Laplacian with relative or absolute boundary conditions in §9 of Chapter 5.

It remains to show that solving (3.12)–(3.15) produces solutions to the initial-value problem for (3.4)–(3.6). We have the following result.

Proposition 3.2. *Let $(\tilde{E}, \tilde{B})$ solve (3.12)–(3.15), and suppose the initial data in (3.15) satisfy*

$$\delta\tilde{E}_0 = 0, \quad \delta\tilde{B}_0 = 0 \tag{3.16}$$

and

$$\tilde{E}_1 = *d\tilde{B}_0, \quad \tilde{B}_1 = -*d\tilde{E}_0. \tag{3.17}$$

Then $(\tilde{E}, \tilde{B})$ satisfies Maxwell's equations (3.4) and (3.5).

Proof. To see that solving the wave equations, (3.12)–(3.14), preserves the property of being annihilated by δ, note that the eigenfunctions of Δ with either of these boundary conditions can be arranged to belong to one of the terms in the Hodge decomposition; see Exercise 5 of §9, Chapter 5. Thus (3.16) yields (3.5). It remains to prove (3.4). For this, define v, w by (3.10), so (3.17) implies

$$v(0,x) = 0, \quad w(0,x) = 0. \tag{3.18}$$

On the other hand, (3.12) plus $\delta\tilde{E} = \delta\tilde{B} = 0$ implies that (3.11) vanishes, that is,

$$\frac{\partial v}{\partial t} + *dw = 0, \quad \frac{\partial w}{\partial t} - *dv = 0. \tag{3.19}$$

Furthermore, the boundary conditions (3.13) and (3.14) imply

$$\nu \wedge v = 0 \text{ and } \iota_\nu = 0 \text{ on } \partial M. \tag{3.20}$$

Consequently, Proposition 3.1 applies to the pair $(v, -w)$, so v and w are identically zero. This finishes the proof.

Exercises

1. Suppose (E, B) solve Maxwell's equations (3.1)–(3.2) and the boundary condition

$$\nu \times E = 0, \quad \text{for } (t, x) \in \mathbb{R} \times \partial M.$$

Suppose that $\nu \cdot B = 0$ on ∂M at $t = 0$. Show that $\nu \cdot B = 0$ on ∂M for all t.
(*Hint*: Compare (E, B) to the solution discussed in Proposition 3.2.)
What can you say if you drop the hypothesis that $\nu \cdot B = 0$ on ∂M at $t = 0$?

4. The Cauchy-Kowalewsky theorem

The Cauchy-Kowalewsky theorem, in the linear case, asserts the local existence of a real analytic solution to the "Cauchy problem"

$$\frac{\partial^m u}{\partial t^m} + \sum_{j=0}^{m-1} \sum_{|\alpha| \le m-j} A_{j\alpha}(t, x) \frac{\partial^\alpha}{\partial x^\alpha} \frac{\partial^j u}{\partial t^j} = f(t, x), \tag{4.1}$$

$$u(t_0, x) = g_0(x), \ldots, \partial_t^{m-1} u(t_0, x) = g_{m-1}(x),$$

given that $A_{j\alpha}(t, x)$ and $f(t, x)$ are real analytic on a neighborhood of (t_0, x_0) in $\mathbb{R}^{n+1}$ and $g_0, \ldots, g_{m-1}$ are real analytic on a neighborhood of x_0 in $\mathbb{R}^n$. There is no loss of generality in taking $t_0 = 0$, $x_0 = 0$.

Any system of the form (4.1) can be converted to a first-order system

$$\frac{\partial u}{\partial t} = L(t, x)\partial_x u + L_0(t, x)u + f, \quad u(0, x) = g(x), \tag{4.2}$$

where $L(t, x)\partial_x = \sum_{j=1}^n L_j(t, x)\, \partial/\partial x_j$. Here we assume that $L_j(t, x)$ are real analytic, $K \times K$, matrix-valued functions, and f and g are real analytic, with values in $\mathbb{C}^K$. Note that if (4.2) holds, then

$$\partial_t^{j+1} u = \sum_{\ell=0}^{j} \binom{j}{\ell} \Big[(\partial_t^{j-\ell} L)\, \partial_x \partial_t^\ell u + (\partial_t^{j-\ell} L_0)\, \partial_t^\ell u \Big] + \partial_t^j f. \tag{4.3}$$

In particular, we inductively have $\partial_t^{j+1} u(0, x)$ uniquely determined. Thus (4.2) has at most one real analytic, local solution u.

On the other hand, if we can use (4.3) to get sufficiently good estimates on $\partial_t^{j+1} u\big|_{t=0} = u_{j+1}(x)$ that the power series

$$u(t, x) = \sum_{j=0}^{\infty} \frac{1}{j!} u_j(x) t^j \tag{4.4}$$

is shown to converge, for t in some neighborhood of 0, then (4.4) furnishes the solution to (4.2). To be more precise, we set $u_0(x) = g(x)$ and define $u_{j+1}(x)$ inductively by

$$u_{j+1}(x) = \sum_{\ell=0}^{j} \sum_{\nu} \binom{j}{\ell} \partial_t^{j-\ell} L_\nu(0, x) \cdot \partial_\nu u_\ell(x) + \partial_t^j f(0, x). \tag{4.5}$$

We sum over $0 \le \nu \le n$ and make the convention that $\partial_\nu = \partial/\partial x_\nu$ for $\nu \ge 1$, while $\partial_0 u = u$. Our goal will be to get estimates on $u_{j+1}(x)$ guaranteeing the local convergence of (4.4).

As illustrated in results on vector fields with real analytic coefficients (say on an open set $U \subset \mathbb{R}^n$) in Chapter 1, it is often useful to extend the real analytic coefficients and other data to holomorphic functions, defined on a neighborhood of U in $\mathbb{C}^n$. Here we will similarly extend $L(t, x)$, $f(t, x)$, and $g(x)$ as functions holomorphic in x, in a neighborhood of $0 \in \mathbb{C}^n$. We keep t real, for now. Without loss of generality, we can suppose that $L(t, z)$, $f(t, z)$, and $g(z)$ are all holomorphic for z in a neighborhood of the closed unit ball $\overline{B} \subset \mathbb{C}^n$, with real analytic dependence on t, for $|t| \le 1$.

We will use the Banach spaces $\mathfrak{H}_j$ of functions f, holomorphic on B, having the property that

$$N_j(f) = \sup_{z \in B} \delta(z)^j \, |f(z)| \tag{4.6}$$

is finite, where $\delta(z) = 1 - |z|$ is the distance of z from ∂B. We will inductively obtain estimates for $N_j(u_j)$. From (4.5), we have

$$N_{j+1}(u_{j+1}) \le \sum_{\ell=0}^{j} \sum_{\nu} \binom{j}{\ell} \|\partial_t^{j-\ell} L_\nu(0)\|_{L^\infty(B)} N_{j+1}(\partial_\nu u_\ell) + N_{j+1}(\partial_t^j f). \tag{4.7}$$

A key estimate is that, for a certain constant γ, depending only on n, we have

$$N_{j+1}(\partial_{x_\nu} u_\ell) \le \gamma (j+1) N_j(u_\ell). \tag{4.8}$$

In order not to interrupt the flow of the argument, we establish this in Appendix A; see (A.8). Since

$$N_j(v) \le N_\ell(v), \quad \text{for } \ell \le j, \tag{4.9}$$

we have

$$N_{j+1}(u_{j+1}) \le \gamma (j+1) \sum_{\ell=0}^{j} \sum_{\nu} \binom{j}{\ell} \|\partial_t^{j-\ell} L_\nu(0)\|_{L^\infty} N_\ell(u_\ell) + N_{j+1}(\partial_t^j f). \tag{4.10}$$

Given the hypothesis on L, we can assume there are estimates of the form

$$\sum_{\nu} \|\partial_t^m L_\nu(0)\|_{L^\infty(B)} \le C_1 \lambda^m \, m!, \tag{4.11}$$

for certain constants C_1 and λ. Now, our inductive hypothesis on u_ℓ is that there exist constants C_2 and μ such that

$$N_\ell(u_\ell) \le C_2 \, \mu^\ell \, \ell!, \quad 0 \le \ell \le j. \tag{4.12}$$

The $\ell = 0$ case follows from our hypothesis on $g(x)$. We can also assume that, for all j,

$$N_{j+1}(\partial_t^j f) \le C_2 \, \mu^j \, (j+1)!. \tag{4.13}$$

Substitution of these estimates into (4.10) yields

$$N_{j+1}(u_{j+1}) \le \gamma C_1 C_2 (j+1)! \sum_{\ell=0}^{j} \lambda^{j-\ell} \mu^{\ell} + C_2\, \mu^j\, (j+1)!. \tag{4.14}$$

We are permitted to assume that $\mu = 2\lambda$ and $\mu \ge 2\gamma C_1 + 1$. Then $\sum_{\ell=0}^{j} \lambda^{j-\ell}\mu^{\ell} \le 2\mu^j$, so we have

$$N_{j+1}(u_{j+1}) \le C_2(j+1)!\,(2\gamma C_1)\mu^j + C_2\,\mu^j\,(j+1)! \le C_2\,\mu^{j+1}\,(j+1)!. \tag{4.15}$$

This completes the induction; in other words

$$N_j(u_j) \le C_2\,\mu^j\, j!, \quad \text{for all } j. \tag{4.16}$$

We hence have the following proposition.

Proposition 4.1. *Given the real analyticity hypotheses on (4.1), there is a unique real analytic solution $u(t,x)$ on a neighborhood of (t_0, x_0) in $\mathbb{R}^{n+1}$. The size of the region on which $u(t,x)$ is defined and analytic depends on the size of the regions to which the coefficients and data of (4.1) have holomorphic extensions, in a fashion determined by (4.11), (4.12), and (4.16).*

Another approach to the use of estimates of the form (4.8) to prove the linear Cauchy-Kowalewsky theorem can be found in [Ho].

We restate the Cauchy-Kowalewsky theorem in a coordinate-invariant fashion. Let S be a smooth hypersurface in an open set $\mathcal{O} \subset \mathbb{R}^n$. We say that S is noncharacteristic for a differential operator $P = p(x, D)$ of order m if, for each $x \in S$, $\sigma_P(x, \nu) = p_m(x, \nu)$ is invertible, where ν is a nonvanishing normal to S at x. Now assume that $p(x, D)$ has real analytic coefficients and S is a real analytic hypersurface. Let Y be a real analytic vector field transverse to S. We consider the following Cauchy problem:

$$p(x, D)u = f, \quad u\big|_S = g_0, \ \ Yu\big|_S = g_1, \dots, Y^{m-1}u\big|_S = g_{m-1}. \tag{4.17}$$

Then, on a neighborhood of any given $x_0 \in S$, we can make a real analytic change of variable such that, for some real analytic invertible $A(x)$, $Q = A(x)^{-1}p(x, D)$ has the form of the operator in (4.1) and S is given by $t = 0$. We do not claim that $Y = \partial/\partial t$, but clearly $\partial_t^j u\big|_S$ can be determined inductively from $u\big|_S, \dots, Y^j u\big|_S$, and vice versa. Then, with new f and g_j, (4.17) acquires the form (4.1), so we have:

Proposition 4.2. *If $p(x, D)$ is a differential operator of order m with real analytic coefficients on $\mathcal{O}$, S is a real analytic hypersurface in $\mathcal{O}$, Y is a real analytic vector field transverse to S, and f and g_j are real analytic, then there exists a unique real analytic solution to (4.7), on some neighborhood of S.*

Given the linear Cauchy-Kowalewsky theorem, we proceed to a uniqueness result of Holmgren.

Proposition 4.3. *Let $P = p(x, D)$ be a differential operator of order m, with real analytic coefficients on an open set $\mathcal{O} \subset \mathbb{R}^n$, and let $S \subset \mathcal{O}$ be a smooth, noncharacteristic hypersurface. Suppose that $u \in H^m(\mathcal{O})$ solves*

$$(4.18) \qquad p(x, D)u = 0 \text{ on } \mathcal{O}, \quad u\big|_S = 0, \ Yu\big|_S = 0, \dots, Y^{m-1}u\big|_S = 0,$$

where Y is a smooth vector field transverse to S. Then $u = 0$ on a neighborhood of S.

Proof. We can assume that $\mathcal{O} \setminus S$ has two connected components, $\mathcal{O}^+$ and $\mathcal{O}^-$. Alter u to produce $v(x)$, equal to $u(x)$ for $x \in \mathcal{O}^+$ and to 0 for $x \in \mathcal{O}^-$. Then the hypothesis (4.16) implies

$$(4.19) \qquad v \in H^m(\mathcal{O}), \quad p(x, D) = 0 \text{ on } \mathcal{O}.$$

Pick $x_0 \in S$. If S is noncharacteristic at x_0, then there exists a real analytic hypersurface Σ_0, tangent to S at x_0. Cutting down $\mathcal{O}$ if necessary, we can make a real analytic change of variable so that $Q = A(x)^{-1}p(x, D)$ has the form (4.1), for some invertible, real analytic $A(x)$, and Σ_0 is given by $\{t = 0\}$, as illustrated in Fig. 4.1. (Say $t = x_n$.) Picking Σ_0 appropriately, we can arrange that S is given by $t = \varphi(x') \geq |x'|^2$, where $x' = (x_1, \dots, x_{n-1})$. The adjoint operator Q^* also has real analytic coefficients on $\mathcal{O}$. Let $\Sigma_\tau = \mathcal{O} \cap \{t = \tau\}$.

Now, according to the Cauchy-Kowalewsky theorem, together with the estimates on the size of domains of existence discussed above, we have the following. There exists $\delta > 0$ such that, for any $\tau \in (-\delta, \delta)$ and any polynomial $a(x)$ on $\mathbb{R}^n$, the Cauchy problem

$$(4.20) \qquad Q^*w = a, \quad w = \partial_t w = \cdots = \partial_t^{m-1}w = 0 \text{ on } \Sigma_\tau$$

has a solution w, real analytic on $\{x \in \mathcal{O} : |x - x_0| < \delta + \sqrt{\delta}\}$. Thus, if we pick $\tau \in (0, \delta)$ and let $\mathfrak{A}_\tau$ be the set bounded by Σ_τ and S (so $\mathfrak{A}_\tau \subset \mathcal{O}^+$),

$$(4.21) \qquad (u, a)_{L^2(\mathfrak{A}_\tau)} = (v, Q^*w)_{L^2(\mathfrak{A}_\tau)} = (Qv, w)_{L^2(\mathfrak{A}_\tau)} = 0.$$

Since, by the Stone-Weierstrass theorem, the set of polynomials is dense in $C(\mathfrak{A}_\tau)$, this implies $u = 0$ on $\mathfrak{A}_\tau$. Similarly, one establishes that $u = 0$ near x_0 in $\mathcal{O}^-$, and the proposition is proved.

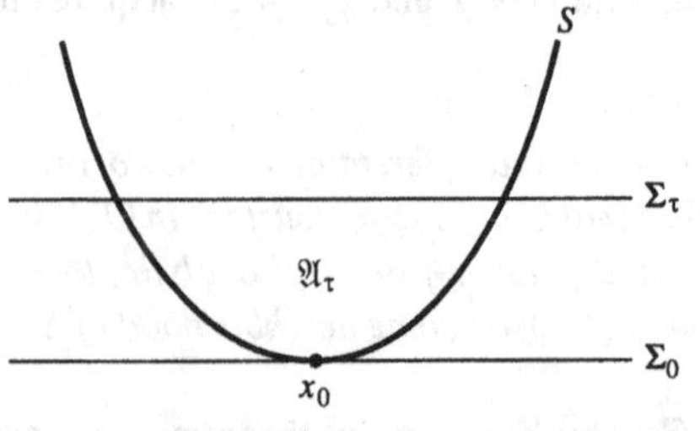

FIGURE 4.1

Exercises

1. Show that the conclusion (4.16), leading to the Cauchy-Kowalewsky theorem, still holds if the hypothesis (4.11) on $\partial_t^m L_\nu(0,x)$ is weakened to

$$\sum_\nu N_m\big(\partial_t^m L_\nu(0)\big) \le C_1\, \lambda^m\, m!. \tag{4.11a}$$

2. In the estimation of (4.14), we took $\mu = 2\lambda$. More generally, work out the analogue of (4.15) when $\lambda = \mu/K$, $K > 1$. What happens if you try to take $\lambda = \mu$? How does this affect your ability to generalize (4.2)–(4.16) to the quasilinear case:

$$\frac{\partial u}{\partial t} = \sum_{j=1}^n L_j(t,x,u)\frac{\partial u}{\partial x_j} + L_0(t,x,u)?$$

   For a proof of the Cauchy-Kowalewsky theorem for nonlinear PDE, see §4 of Chapter 16.
3. If P, $\mathcal{O}$, S, and Y are as in Proposition 4.3, show that whenever $u \in \mathcal{D}'(\mathcal{O})$ satisfies $Pu = 0$ on $\mathcal{O}$, then $Y^j u\big|_S$ is a well-defined element of $\mathcal{D}'(S)$, for all j. Extend Proposition 4.3 to the case of all $u \in \mathcal{D}'(\mathcal{O})$ satisfying (4.18).

5. Hyperbolic systems

We will use energy estimates and Sobolev space theory to establish the existence of solutions to linear hyperbolic equations of a more general form than considered in §1. To begin, let us examine second-order hyperbolic equations, of the form

$$Lu = \Box u + Xu = f, \quad u\big|_{S_0} = g_0, \quad Yu\big|_{S_0} = g_1, \tag{5.1}$$

where $\Box$ is the wave operator on a Lorentz manifold Ω, assumed to be foliated by compact, spacelike hypersurfaces S_τ, the operator X is a first-order differential operator, and Y is a vector field transverse to S_0. The operator $\Box = \partial_t^2 - \Delta$ on $\mathbb{R} \times M$, with $S_\tau = \{(t,x) : t = \tau\}$, dealt with in §1, is a special case, provided $\partial M = \emptyset$.

Energy estimates for (5.1) were established in §8 of Chapter 2. In particular, if $\mathcal{O}$ is the region in Ω swept out by S_τ, $0 \le \tau \le T$, then, by (8.19) of Chapter 2,

$$\|u\|^2_{H^1(\mathcal{O})} \le C\|Lu\|^2_{L^2(\mathcal{O})} + C\|g_0\|^2_{H^1(S_0)} + C\|g_1\|^2_{L^2(S_0)}. \tag{5.2}$$

The argument of Chapter 2 applies as long as $u \in H^2(\mathcal{O})$. If L has formal adjoint $L^* = \Box + X_1$, we similarly have

$$\|v\|_{H^1(\mathcal{O})} \le C\|L^* v\|_{L^2(\mathcal{O})}, \quad \text{for } v \in V_T(\mathcal{O}), \tag{5.3}$$

where

$$V_T(\mathcal{O}) = \{w \in C^\infty(\overline{\mathcal{O}}) : w = dw = 0 \text{ on } S_T\}. \tag{5.4}$$

Now, to solve (4.1), when $g_0 = g_1 = 0$, given $f \in L^2(\mathcal{O})$, it suffices to obtain u such that

$$(u, L^* v) = (f, v), \quad \text{for all } v \in V_T(\mathcal{O}). \tag{5.5}$$

However, by (5.3), given $f \in L^2(\mathcal{O})$, we have

$$|(f, v)| \le C\|f\|_{L^2(\mathcal{O})} \cdot \|L^* v\|_{L^2(\mathcal{O})}, \tag{5.6}$$

so by the Riesz representation theorem, the existence of $u \in L^2(\mathcal{O})$ such that (5.5) holds is guaranteed. In fact, more generally, given $f \in H^1(\mathcal{O})^*$, we have

$$|(f, v)| \le C\|f\|_{H^1(\mathcal{O})^*} \|L^* v\|_{L^2(\mathcal{O})}, \tag{5.7}$$

so we have a solution $u \in L^2(\mathcal{O})$ to (5.5) for all $f \in H^1(\mathcal{O})^*$.

Note that if $u \in L^2(\mathcal{O})$ and $Lu \in L^2(\mathcal{O})$, then $u\big|_{S_0}$ and $Yu\big|_{S_0}$ always exist, in $H^{-2}(S_0)$. If u satisfies (5.7), these Cauchy data vanish. Also, in this case, if we set $f = 0$ and $u = 0$ on $\bigcup_{\tau \le 0} S_\tau = \mathcal{O}^b$, we have $Lu = f$ on $\mathcal{O} \cup \mathcal{O}^b = \mathcal{O}^\#$.

Moving to the nonhomogeneous initial-value problem (5.1), if one has $g_0 \in H^{3/2}(S_0)$ and $g_1 \in H^{1/2}(S_0)$, one can construct $U \in H^2(\mathcal{O})$ with such Cauchy data and subtract this off. Thus the argument above yields a solution $u \in L^2(\mathcal{O})$ to (5.1), given

$$f \in L^2(\mathcal{O}), \quad g_0 \in H^{3/2}(S_0), \quad g_1 \in H^{1/2}(S_0). \tag{5.8}$$

This existence result is not at all satisfactory and will be improved below.

We can extend (5.2) to higher-order a priori estimates for sufficiently smooth solutions to (5.1), as follows. Suppose $u \in H^{k+1}(\mathcal{O})$, which is more than adequate to imply that $f \in H^{k-1}(\mathcal{O})$, $g_0 \in H^k(S_0)$, and $g_1 \in H^{k-1}(S_0)$. For simplicity, take $\Omega = \mathbb{R} \times \mathbb{T}^n$, $S_\tau = \{\tau\} \times \mathbb{T}^n$, so we have natural coordinate systems making D^α meaningful. Then define

$$u_\alpha = D^\alpha u. \tag{5.9}$$

We produce a system of PDE satisfied by $(u_\alpha : |\alpha| \le k-1)$, as follows. There exist first-order differential operators X_β on Ω such that

$$LD^\alpha = D^\alpha L + \sum_{|\beta| \le |\alpha|} X_\beta D^\beta. \tag{5.10}$$

Then

$$Lu_\alpha - \sum_{|\beta| \le |\alpha|} X_\beta u_\beta = D^\beta f. \tag{5.11}$$

We can also determine $u_\alpha\big|_{S_0}$ and $Yu_\alpha\big|_{S_0}$, in terms of derivatives of f, g_0, and g_1, and we have

$$\begin{aligned} u_\alpha\big|_{S_0} &= g_{0\alpha} \in H^{k-|\alpha|}(S_0) \subset H^1(S_0), \\ Yu_\alpha\big|_{S_0} &= g_{1\alpha} \in H^{k-1-|\alpha|}(S_0) \subset L^2(S_0). \end{aligned} \tag{5.12}$$

Now the energy estimate (5.2) applies to the system (5.11)–(5.12), so we have

$$\sum_{|\alpha| \le k-1} \|u_\alpha\|^2_{H^1(\mathcal{O})} \le C \sum_\alpha \Big[\|D^\alpha f\|^2_{L^2(\mathcal{O})} + \|g_{0\alpha}\|^2_{H^1(S_0)} + \|g_{1\alpha}\|^2_{L^2(S_0)} \Big], \tag{5.13}$$

and hence

$$\|u\|^2_{H^k(\mathcal{O})} \le C\|Lu\|^2_{H^{k-1}(\mathcal{O})} + C\|g_0\|^2_{H^k(S_0)} + C\|g_1\|^2_{H^{k-1}(S_0)}. \tag{5.14}$$

We want to show that, given $f \in H^k(\mathcal{O})$, $g_0 \in H^k(S_0)$, $g_1 \in H^{k-1}(S_0)$, with $k \geq 1$, there exists a unique solution u to (5.1) and that $u \in H^k(\mathcal{O})$. We will establish this by obtaining u as a limit of solutions to approximating hyperbolic equations, having analytic coefficients and data, for which a solution is guaranteed by the Cauchy-Kowalewsky theorem. A different sort of existence argument can be found in Chapter 7, §7.

Let us assume S_τ is given by $t = \tau$ in $\Omega = \mathbb{R} \times \mathbb{T}^n$, and $Y = \partial/\partial t$. Now we can approximate the coefficients of L in $C^\infty(\mathbb{R} \times \mathbb{T}^n)$ by functions that are real analytic on $\mathbb{R} \times \mathbb{T}^n$. We can think of these functions as being defined on $\mathbb{R} \times \mathbb{R}^n$, and $\mathbb{Z}^n$-periodic, and can arrange that the coefficients have entire holomorphic extensions to $\mathbb{C} \times \mathbb{C}^n$. Denote the resulting operators by L_ν. Given $k \in \mathbb{Z}^+$, let us assume that

$$f \in H^{k-1}(\Omega), \quad g_0 \in H^k(S_0), \quad g_1 \in H^{k-1}(S_0). \tag{5.15}$$

We approximate these functions, in the appropriate norms, by real analytic functions f_ν, $g_{0\nu}$, $g_{1\nu}$, having entire holomorphic extensions in the sense mentioned above. Consider the initial-value problems

$$L_\nu u_\nu = f_\nu, \quad u_\nu\big|_{S_0} = g_{0\nu}, \quad \partial_t u_\nu\big|_{S_0} = g_{1\nu}. \tag{5.16}$$

The Cauchy-Kowalewsky theorem applies to (5.16), and results of §4 imply that, for each ν, there is a unique solution $u_\nu(t, x)$ that is real analytic on all of $\mathbb{R} \times \mathbb{T}^n$. The energy estimates of the form (5.14) hold uniformly in ν, for any given $\mathcal{O} = [-T, T] \times \mathbb{T}^n$. In other words, given (5.15), $\{u_\nu\}$ is bounded in $H^k(\mathcal{O})$. Thus there is a subsequence $u_{\nu_j} \to u$ weakly in $H^k(\mathcal{O})$. It is clear that such a limit u solves (5.1). We have the following result.

Proposition 5.1. *Given f, g_0, g_1 satisfying (5.15), there is a unique solution $u \in H^k(\Omega)$ to (5.1).*

The final point to discuss in this result is uniqueness. If $k \geq 2$, this is immediate from the energy estimate (5.2). In fact, we can derive a more general uniqueness result by a duality argument. Namely, suppose

$$u \in \mathcal{D}'(\Omega), \quad u = 0 \text{ on } \bigcup_{\tau<0} S_\tau, \quad Lu = 0 \text{ on } \Omega. \tag{5.17}$$

Given $f \in C_0^\infty(\Omega)$, we can apply the existence part of the proposition, with L replaced by L^*, and with time reversed, to produce, for arbitrarily large $k > 0$,

$$v \in H^k(\Omega), \quad v = 0, \quad \text{for } t >> 0, \quad L^*v = f. \tag{5.18}$$

Pick k so large that $u \in H^{2-k}$, on a neighborhood of the support of f. Then

$$(u, f) = (u, L^*v) = (Lu, v) = 0, \tag{5.19}$$

which implies $u = 0$. This finishes the proof of Proposition 5.1 and also establishes the uniqueness of the solution u in (5.8), which can consequently be seen to belong to $H^1(\mathcal{O})$.

We now look at a class of first-order $N \times N$ systems, of the form

$$\frac{\partial u}{\partial t} + \sum_{j=1}^{n} A_j(t,x)\frac{\partial u}{\partial x_j} + B(t,x)u = f(t,x), \quad u(0,x) = g(x). \tag{5.20}$$

Let us suppose the various functions, $f(t,x)$, and so on, are defined on $\Omega = \mathbb{R}\times\mathbb{T}^n$. The system (5.20) is said to be symmetric hyperbolic provided each $N \times N$ matrix A_j satisfies

$$A_j(t,x)^* = A_j(t,x). \tag{5.21}$$

We will derive energy estimates for solutions to (5.21) in a fashion similar to that used in §8 of Chapter 2. Suppose $\mathcal{O} \subset \mathbb{R} \times \mathbb{T}^n$ is bounded by two surfaces, Σ_1 and Σ_2, as illustrated in Fig. 5.1. If we denote the left side of (5.20) by Lu, then, by the Gauss-Green formula, in the form established in (9.17) of Chapter 2,

$$(Lu, u) - (u, L^*u) = \frac{1}{i}\int_{\partial\mathcal{O}} \langle \sigma_L(t,x,\nu)u, u\rangle\, dS, \tag{5.22}$$

where the inner products on the left are inner products in $L^2(\mathcal{O})$, and ν is the inward normal to $\partial\mathcal{O}$, as illustrated in Fig. 5.1. Note that

$$L^*u = -Lu + Cu, \quad Cu = -\sum \frac{\partial A_j}{\partial x_j}u + Bu, \tag{5.23}$$

provided (5.21) holds. Thus we have

$$2\,\mathrm{Re}(Lu, u) - (u, Cu) = \frac{1}{i}\int_{\partial\mathcal{O}} \langle \sigma_L(t,x,\nu)u, u\rangle\, dS. \tag{5.24}$$

Note that if $\nu = (\nu_0, \nu_1, \dots, \nu_n) \in T^*(\mathbb{R} \times \mathbb{T}^n)$, then

$$\frac{1}{i}\sigma_L(t,x,\nu) = \nu_0 I + \sum_{j=1}^{n} A_j(t,x)\nu_j. \tag{5.25}$$

Thus $(1/i)\sigma_L(t,x,\nu)$ is positive-definite on Σ_1 and negative-definite on Σ_2 if these surfaces are close enough to horizontal, that is, if ν is close enough to $(1, 0, \dots, 0)$ on Σ_1 and to $(-1, 0, \dots, 0)$ on Σ_2. If this definiteness condition holds on Σ_j, we say Σ_j is *spacelike*, for the operator L. Compare the notion of a spacelike surface

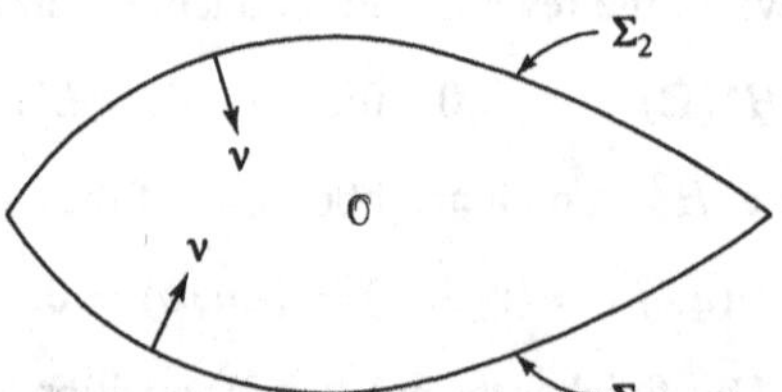

FIGURE 5.1

for $\square$, given in Chapter 2. Also, we say $\zeta \in T^*_z(\Omega) \setminus 0$ is (forward or backward) timelike if $(1/i)\sigma_L(z,\zeta) = \lambda(z,\zeta)$ is (positive- or negative-) definite.

Suppose Σ_1 and Σ_2, bounding $\mathcal{O}$ as above, are both spacelike. Also, suppose that $\mathcal{O}$ is swept out by spacelike surfaces. To be precise, suppose that there is a smooth function φ on a neighborhood of $\overline{\mathcal{O}}$ such that $d\varphi$ is timelike, and set

$$\mathcal{O}(s) = \overline{\mathcal{O}} \cap \{\varphi \le s\}, \quad \Sigma_2(s) = \overline{\mathcal{O}} \cap \{\varphi = s\}. \tag{5.26}$$

We suppose $\mathcal{O}$ is swept out by $\Sigma_2(s)$, $s_0 \le s \le s_1$, as illustrated in Fig. 5.2, with $\Sigma_2 = \Sigma_2(s_1)$. Also set

$$\Sigma_1^b(s) = \Sigma_1 \cap \{\varphi \le s\}. \tag{5.27}$$

As in (5.22), we have (with $\nu_2 = d\varphi/|d\varphi|$):

(5.28)
$$\begin{aligned}\int_{\Sigma_2(s)} \langle \lambda(z,\nu_2)u,u\rangle\, dS &= \int_{\Sigma_1^b(s)} \langle \lambda(z,\nu)u,u\rangle\, dS - 2\,\mathrm{Re}(Lu,u) + (u,Cu)\\ &\le \int_{\Sigma_1^b(s)} \langle \lambda(z,\nu)u,u\rangle\, dS + K\int_{\mathcal{O}(s)} \left[|Lu|^2 + |u|^2\right] dV.\end{aligned}$$

Now, parallel to (8.13) of Chapter 2, we set

$$E(s) = \int_{\mathcal{O}(s)} \langle \lambda(z,\nu_2)u,u\rangle\, dV \tag{5.29}$$

and estimate the rate of change of $E(s)$. Clearly,

$$\frac{dE}{ds} \le C \int_{\Sigma_2(s)} \langle \lambda(z,\nu_2)u,u\rangle\, dS, \tag{5.30}$$

so, by (5.28), we have an estimate of the form

$$\frac{dE}{ds} \le CE(s) + F(s), \tag{5.31}$$

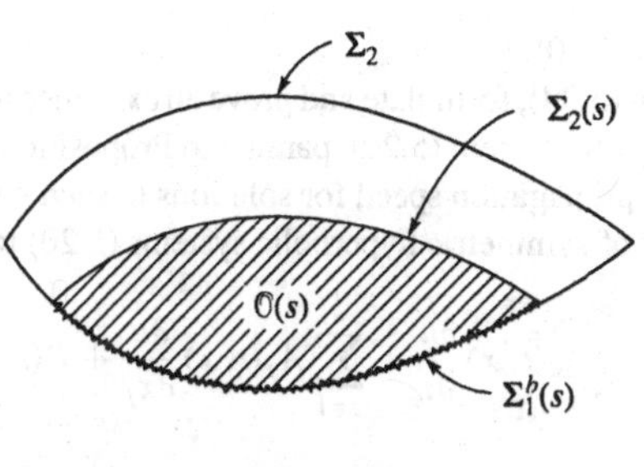

FIGURE 5.2

where

$$F(s) = C \int_{\Sigma_1} |u|^2 \, dS + C \int_{\mathcal{O}(s)} |Lu|^2 \, dV. \tag{5.32}$$

This differential inequality yields

$$E(s) \le \int_{s_0}^{s} e^{C(s-r)} F(r) \, dr. \tag{5.33}$$

Consequently,

$$\int_{\mathcal{O}(s)} |u|^2 \, dV \le C(s - s_0) \int_{\Sigma_1} |u|^2 \, dS + C \int_{\mathcal{O}(s)} |Lu|^2 \, dV. \tag{5.34}$$

From here, the existence and uniqueness of solutions to (5.20), as well as the finite propagation speed, follow by arguments parallel to those used for $L = \Box + X$. We leave the formulation of such results to the reader.

Exercises

1. Supplement (5.2) with

$$\|u\|^2_{H^1(S_T)} + \|Yu\|^2_{L^2(S_T)} \le C\|Lu\|^2_{L^2(\mathcal{O})} + C\|g_0\|^2_{H^1(S_0)} + C\|g_1\|^2_{L^2(S_0)},$$

when L has the form (5.1). More generally, supplement (5.14) with

$$\|u\|^2_{H^k(S_T)} + \|Yu\|^2_{H^{k-1}(S_T)} \le C\|Lu\|^2_{H^{k-1}(\mathcal{O})} + C\|g_0\|^2_{H^k(S_0)} + C\|g_1\|^2_{H^{k-1}(S_0)}.$$

(*Hint*: Look at (8.20), in Chapter 2.)

2. Show that the part of Maxwell's equations given by (3.1) forms a symmetric hyperbolic system.
3. Supplement (5.34) with

$$\|u\|^2_{L^2(\Sigma_2)} \le C\|Lu\|^2_{L^2(\mathcal{O})} + C\|u\|^2_{L^2(\Sigma_1)},$$

when L has the form (5.20).

4. Making use of (5.22)–(5.34), formulate and prove an existence and uniqueness result for the symmetric hyperbolic system (5.20), parallel to Proposition 5.1. Also give a precise formulation of finite propagation speed for solutions to such a system.
5. Generalize the study of symmetric hyperbolic systems (5.20) to include

$$Lu = A_0(t,x)\frac{\partial u}{\partial t} + \sum_{j=1}^{n} A_j(t,x)\frac{\partial u}{\partial x_j} + B(t,x), \tag{5.35}$$

where A_j satisfy (5.21), and in addition $A_0(t,x)$ is positive-definite.

6. Geometrical optics

In this section we look at solutions to the wave equation

$$\frac{\partial^2 u}{\partial t^2} - \Delta u = 0, \tag{6.1}$$

on $\mathbb{R} \times M$, where M is a Riemannian manifold, having either initial data with a simple jump across a smooth surface, of the form

$$u(0, x) = a(x)H\big(\varphi(x)\big), \tag{6.2}$$

or highly oscillatory initial data:

$$u(0, x) = a(x)F\big(\lambda\varphi(x)\big). \tag{6.3}$$

Here, $H(s)$ is the Heaviside function; $H(s) = 1$ for $s > 0$, $H(s) = 0$ for $s < 0$, while $F \in C^\infty(\mathbb{R})$ is bounded, together with all its derivatives, as well as an infinite sequence of antiderivatives. We imagine that λ is large. We assume $a \in C_0^\infty(M)$ and $\nabla\varphi \neq 0$ on a neighborhood U of supp a. For simplicity, we complete the set of initial conditions with

$$u_t(0, x) = 0, \tag{6.4}$$

though the methods developed below extend to more general cases. We will show that, at least for $|t| < T$, with T small enough, $u(t, x)$ has an asymptotic behavior

$$u(t, x) \sim \sum_{j \geq 0} u_j(t, x), \tag{6.5}$$

in a sense that will be made precise below, where, in case (6.2),

$$u_j(t, x) = \sum_{\pm} a_j^\pm(t, x) h_j\big(\varphi^\pm(t, x)\big), \tag{6.6}$$

for certain functions $h_j \in C^\infty(\mathbb{R} \setminus 0)$ whose jth derivative jumps at 0, and, in case (6.3),

$$u_j(t, x) = u_j(t, x, \lambda) = \sum_{\pm} \lambda^{-j} a_j^\pm(t, x) F_j\big(\lambda\varphi^\pm(t, x)\big), \tag{6.7}$$

for certain $F_j \in C^\infty(\mathbb{R})$. In both cases, $a_j^\pm, \varphi^\pm \in C^\infty\big((-T, T) \times M\big)$, with

$$\varphi^\pm(0, x) = \varphi(x), \tag{6.8}$$

and $a_0^+(0, x) + a_0^-(0, x) = a(x)$. The functions $\varphi^\pm$ are called "phase functions," and the functions $a_j^\pm$ are called "amplitudes." We take $h_0 = H$ and $F_0 = F$.

The asymptotic relation (6.5) will imply in particular that $u - \sum_{j \leq N} u_j$ is, for large N, relatively smooth, in case (6.2), and also relatively "small" in case (6.3), as $\lambda \to \infty$. We give the details of the construction in the case (6.3) before sketching a similar treatment of the case (6.2).

In order to compute the action of $\partial_t^2 - \Delta$ on (the sum over $0 \le j \le N$ of) the right side of (6.5), when $u_j(t,x)$ has the form (6.7), we recall that

$$\begin{aligned} \Delta(uv) &= (\Delta u)v + 2\nabla u \cdot \nabla v + u(\Delta v), \\ \Delta\big(F(u)\big) &= F'(u)\Delta u + F''(u)|\nabla u|^2. \end{aligned} \tag{6.9}$$

Here, we use the dot product to denote the inner product with respect to the Riemannian metric; $\nabla u \cdot \nabla v = g^{jk}(\partial_j u)(\partial_k v)$. Thus, if u_j has the form (6.7), we obtain

$$\begin{aligned} (\partial_t^2 - \Delta)u_j = \sum_{\pm} \Big[& \lambda^{2-j} a_j^{\pm} F_j''(\lambda\varphi^{\pm})\Big(|\partial_t\varphi^{\pm}|^2 - |\nabla_x\varphi^{\pm}|^2\Big) \\ & + \lambda^{1-j} F_j'(\lambda\varphi^{\pm})\big(2\varphi_t^{\pm}\partial_t a_j^{\pm} - 2\nabla_x\varphi^{\pm}\cdot\nabla_x a_j^{\pm} + a_j^{\pm}\Box\varphi^{\pm}\big) \\ & - \lambda^{-j} F_j(\lambda\varphi^{\pm})\big(\Box a_j^{\pm}\big)\Big]. \end{aligned} \tag{6.10}$$

In particular, the coefficients of $\lambda^{\mu} = \lambda^{1-j}$ in $(\partial_t^2 - \Delta)\sum_{j\ge 0} u_j(t,x)$ are of the following form:

(6.11)
$$\mu = 2\colon \sum_{\pm} a_0^{\pm} F''(\lambda\varphi^{\pm})\Big(|\partial_t\varphi^{\pm}|^2 - |\nabla_x\varphi^{\pm}|^2\Big),$$

$$\begin{aligned} \mu = 1\colon \sum_{\pm} \Big[& a_1^{\pm} F_1''(\lambda\varphi^{\pm})\Big(|\partial_t\varphi^{\pm}|^2 - |\nabla_x\varphi^{\pm}|^2\Big) \\ & + F'(\lambda\varphi^{\pm})\big(2\varphi_t^{\pm}\partial_t a_0^{\pm} - 2\nabla_x\varphi^{\pm}\cdot\nabla_x a_0^{\pm} + a_0^{\pm}\Box\varphi^{\pm}\big)\Big], \end{aligned} \tag{6.12}$$

$$\begin{aligned} \mu \le 0\colon \sum_{\pm} \Big[& a_{j+1}^{\pm} F_{j+1}''(\lambda\varphi^{\pm})\Big(|\partial_t\varphi^{\pm}|^2 - |\nabla_x\varphi^{\pm}|^2\Big) \\ & + F_j'(\lambda\varphi^{\pm})\big(2\varphi_t^{\pm}\partial_t a_j^{\pm} - 2\nabla_x\varphi^{\pm}\cdot\nabla_x a_j^{\pm} + a_j^{\pm}\Box\varphi^{\pm}\big) \\ & + F_{j-1}(\lambda\varphi^{\pm})\big(\Box a_{j-1}^{\pm}\big)\Big]. \end{aligned} \tag{6.13}$$

We will set these terms successively equal to zero. To begin, the term (6.11) vanishes provided $\varphi^{\pm}$ satisfies the *eikonal equation*:

$$|\partial_t\varphi^{\pm}|^2 - |\nabla_x\varphi^{\pm}|^2 = 0. \tag{6.14}$$

If we use (6.8) to specify $\varphi^{\pm}(0,x)$, then the results on this first-order nonlinear PDE obtained in §15 of Chapter 1 apply. There is a neighborhood U of $K =$ supp a and a $T > 0$ such that this initial-value problem has a unique pair of solutions $\varphi^{\pm} \in C^{\infty}\big((-T,T)\times U\big)$, satisfying

$$\varphi^{\pm}(0,x) = \varphi(x), \quad \partial_t\varphi^{\pm}(0,x) = \pm\big|\nabla_x\varphi(x)\big|. \tag{6.15}$$

Having so specified $\varphi^\pm$, we see that the terms (6.12) and (6.13) simplify. The term (6.12) vanishes provided

$$2\varphi_t^\pm \frac{\partial a_0^\pm}{\partial t} = 2\nabla_x \varphi^\pm \cdot \nabla_x a_0^\pm - a_0^\pm(\Box \varphi^\pm). \tag{6.16}$$

By (6.15) we see that $\varphi^\pm \neq 0$ on U (if $|t|$ is small enough). The linear equations (6.16) for $a_0^\pm$ are called the *first transport equations*. The initial conditions for $a_0^\pm$ are deduced from (6.3) and (6.4). We want

$$a_0^+ + a_0^- = a, \quad \varphi_t^+ a_0^+ + \varphi_t^- a_0^- = 0, \quad \text{at } t = 0; \tag{6.17}$$

hence, in light of (6.15),

$$a_0^+(0,x) = a_0^-(0,x) = \frac{1}{2} a(x). \tag{6.18}$$

We have $a_0^\pm \in C^\infty\big((-T,T) \times U\big)$, compactly supported in U for each $t \in (-T,T)$, if T is small enough.

Next, the term (6.13), for $\mu = 1 - j \le 0$ (i.e., $j \ge 1$), vanishes provided that $F_j'(\lambda\varphi^\pm) = F_{j-1}(\lambda\varphi^\pm)$, that is,

$$F_j(s) = \int F_{j-1}(s)\, ds \tag{6.19}$$

and

$$2\varphi_t^\pm \frac{\partial a_j^\pm}{\partial t} = 2\nabla_x \varphi^\pm \cdot \nabla_x a_j^\pm - (\Box\varphi^\pm) a_j^\pm - \Box a_{j-1}^\pm, \quad j \ge 1, \tag{6.20}$$

which are higher-order transport equations. To obtain the initial conditions, note that if $u(t,x)$ is given by (6.5) and (6.7), then

$$\partial_t u_j \sim \sum_\pm \Big[\lambda^{1-j} a_j^\pm F_j'(\lambda\varphi^\pm)\varphi_t^\pm + \lambda^{-j}(\partial_t a_j^\pm) F_j(\lambda\varphi^\pm)\Big]. \tag{6.21}$$

Thus, using (6.4) and also requiring $u_j(0,x) = 0$ for $j \ge 1$, we require

(6.22)
$$a_j^+ + a_j^- = 0, \quad \sum_\pm \Big[a_j^\pm F_j'(\lambda\varphi^\pm)\varphi_t^\pm + (\partial_t a_{j-1}^\pm) F_{j-1}(\lambda\varphi^\pm)\Big] = 0, \quad \text{at } t = 0,$$

or, using (6.19) and (6.15),

$$a_j^+ + a_j^- = 0, \quad \varphi_t^+(a_j^+ - a_j^-) = -\partial_t\big(a_{j-1}^+ + a_{j-1}^-\big), \quad \text{at } t = 0. \tag{6.23}$$

This specifies $a_j^+(0,x)$ and $a_j^-(0,x)$. Then the transport equations (6.20) have unique solutions $a_j^\pm \in C^\infty\big((-T,T) \times U\big)$, compactly supported in U for each $t \in (-T,T)$.

The construction described above, via the eikonal and transport equations, is the basic case of the method of geometrical optics. We now obtain some estimates on the degree to which such a construction approximates the solution to (6.1), (6.3), and (6.4). If we set

$$v_N = \sum_{j=1}^{N} u_j, \tag{6.24}$$

then v_N satisfies

$$\begin{gathered}\frac{\partial^2 v_N}{\partial t^2} - \Delta v_N = r_N(t,x), \\ v_N(0,x) = a(x)F\big(\lambda\varphi(x)\big), \quad \partial_t v_N(0,x) = \rho_N(x),\end{gathered} \tag{6.25}$$

where

$$\rho_N(x) = \lambda^{-N} \sum_{\pm} \partial_t a_N^{\pm}(0,x) \cdot F_N(\lambda\varphi) \tag{6.26}$$

and

$$r_N(t,x) = \lambda^{-N} \sum_{\pm} (\Box a_N^{\pm}) F_N(\lambda\varphi^{\pm}). \tag{6.27}$$

The following result is elementary.

Proposition 6.1. *If $\varphi^{\pm} \in C^{\infty}\big((-T,T)\times M\big)$ and $b \in C_0^{\infty}(M)$, then*

$$\{\lambda^{-\mu} b(x) F_N(\lambda\varphi^{\pm}) : \lambda > 1\} \tag{6.28}$$

is bounded in $C^j\big((-T,T), H^{\mu-j}(M)\big)$, for each $\mu,\ j \geq 0$, provided $F_N(s)$ and all its derivatives are bounded.

Now, $u - v_N$ satisfies

$$\begin{gathered}(\partial_t^2 - \Delta)(u - v_N) = -r_N, \\ (u - v_N)(0,x) = 0, \quad \partial_t(u - v_N)(0,x) = -\rho_N(x),\end{gathered} \tag{6.29}$$

so we have the following. (Compare with Exercise 1 of §1.)

Proposition 6.2. *The geometrical optics construction of v_N produces an approximation to the solution u to (6.1), (6.3), and (6.4), satisfying*

$$u - v_N \text{ is } O(\lambda^{-\nu}) \text{ in } C^j\big((-T,T), H^{N+1-\nu-j}(M)\big), \tag{6.30}$$

for $0 \leq \nu \leq N$, $j \geq 0$, as long as, for each N, $F_N(s)$ and all its derivatives are bounded.

The most common function to take for $F(s) = F_0(s)$ is $F(s) = e^{is}$, in which case $F_N(s) = i^{-N}e^{is}$. Other equally good functions include $F(s) = \cos s$ and $F(s) = \sin s$.

Let us note that (6.28) is not sharp. We can improve it to

$$\{\lambda^{-\mu} b(x) F_N(\lambda\varphi^{\pm}) : \lambda > 1\} \text{ is bounded in } C^{\mu}\big((-T,T)\times M\big). \tag{6.31}$$

Consequently, we can say that if $N' > N$,

$$\text{(6.32)} \qquad v_{N'} - v_N \text{ is } O(\lambda^{-\nu}) \text{ in } C^{N+1-\nu}\big((-T,T)\times M\big),$$

for $0 \le \nu \le N$. Then if we apply (6.30) to $u - v_{N'}$, with N' very large, we conclude that

$$\text{(6.33)} \qquad u - v_N \text{ is } O(\lambda^{-\nu}) \text{ in } C^{N+1-\nu}\big((-T,T)\times M\big),$$

for $0 \le \nu \le N$.

There is a construction analogous to (6.10)–(6.23) for the initial-value problem (6.1), (6.2), and (6.4), whose initial data have a simple jump discontinuity. As mentioned above, the form (6.5)–(6.6) furnishes an approximate solution. The phase functions $\varphi^\pm$ also satisfy the eikonal equation (6.14), and the amplitudes $a_j^\pm(t,x)$ satisfy transport equations similar to (6.16), and (6.20). Parallel to the relation (6.19) between $F_{j-1}(s)$ and $F_j(s)$, we have $h_j(s) = \int h_{j-1}(s)\,ds$, with $h_0(s) = H(s)$, the Heaviside function. Thus, for $j \ge 1$, we can take

$$\text{(6.34)} \qquad \begin{aligned} h_j(s) = 0, &\qquad \text{for } s < 0, \\ \frac{s^j}{j!}, &\qquad \text{for } s > 0. \end{aligned}$$

Having constructed the terms $u_j(t,x)$ of the form (6.6), we can again use energy estimates for the wave equation to show that $u - \sum_{j\le N} u_j$ has high-order Sobolev regularity if N is large. Comparison with the sum $\sum_{j\le N'} u_j$ for $N' >> N$, parallel to (6.32)–(6.33), then shows that

$$\text{(6.35)} \qquad u - \sum_{j\le N} u_j \in C^{(N,1)}\big((-T,T)\times M\big),$$

i.e., Nth order derivatives are Lipschitz continuous.

Note that the singular support of $\sum_{j\le N} u_j$, hence of u, in $(-T,T)\times M$ is contained in the union of the level sets $\varphi^\pm(t,x) = 0$, each of which is a characteristic surface for $\square$. This phenomenon is a special case of a general result about the "propagation of singularities" of a solution to a PDE, which will be treated in Chapter 7, §9.

Let us mention a geometrical characterization of the level surfaces

$$\text{(6.36)} \qquad S_\beta = \{(t,x) : \varphi(t,x) = \beta\}$$

of a solution φ to the eikonal equation (6.14). Namely, each S_β is swept out by "light rays" $\gamma(t) = (t, x(t))$ passing orthogonally over the level set $\Sigma_\beta = \{x : \varphi(x) = \beta\}$ at $t = 0$, where a light ray is a null geodesic for the Lorentz metric $-dt^2 + \sum g_{jk}\,dx_j\,dx_k$ on $\mathbb{R}\times M$. Equivalently, $x(t)$ is a unit-speed geodesic on M, such that $\dot{x}(0)$ is orthogonal to Σ_β. This follows by arguments used to establish Proposition 15.4 and Corollary 15.5 in Chapter 1.

So far we have looked at approximate solutions to the wave equation whose supports do not intersect a boundary. We now consider the reflection of such waves. Thus, let Ω be an open subset of M, with smooth boundary. Suppose the function $a(x)$ in (6.2)–(6.3) belongs to $C_0^\infty(\Omega)$, and we want to solve (6.1) on $\mathbb{R} \times \Omega$, with the Dirichlet boundary condition,

$$u(t, x) = 0, \quad x \in \partial\Omega, \tag{6.37}$$

plus an initial condition: either (6.2) or (6.3), and (6.4). Suppose that the geometrical optics construction above works, for $t \in (-T, T)$, if we make the construction on $(-T, T) \times M$, and that the associated $u_j(t, x)$, of the form (6.6) or (6.7), have supports intersecting $\partial\Omega$. In that case, we want to construct u, to satisfy (6.36), by subtracting w, the solution to

$$\begin{gathered}\frac{\partial^2 w}{\partial t^2} - \Delta w = 0 \text{ on } \mathbb{R} \times \Omega, \quad w(0, x) = w_t(0, x) = 0, \\ w = v \text{ on } \mathbb{R} \times \partial\Omega,\end{gathered} \tag{6.38}$$

where $v = \sum_{j \le N} u_j$.

Let us restrict attention to $t \in (0, T)$. Suppose our wave has the form (6.5)–(6.6), so it has singularities on the surfaces $\varphi^{\pm}(t, x) = 0$. By the superposition principle, we can consider just one of the terms in the sum over j and $\pm$, so let us drop the $\pm$ superscript and suppose

$$v = a_j(t, x) h_j\big(\varphi(t, x)\big) \tag{6.39}$$

in (6.38). Then we will construct an approximate solution to (6.38), in the form

$$w(t, x) \sim \sum_{\ell \ge j} b_\ell(t, x) h_\ell\big(\psi(t, x)\big) = \sum_{\ell \ge j} w_\ell(t, x), \tag{6.40}$$

granted a geometrical restriction, which we describe below. To do this, we have computations parallel to (6.10)–(6.13). Thus, as in (6.14), we have for $\psi(t, x)$ the eikonal equation:

$$|\partial_t \psi|^2 - |\nabla_x \psi|^2 = 0. \tag{6.41}$$

We want $w_j = v$ on $(0, T) \times \partial\Omega$, so we set

$$\psi(t, x) = \varphi(t, x) \text{ on } (0, T) \times \partial\Omega. \tag{6.42}$$

There are several ways to describe our geometrical hypothesis. One is that the surface $(0, T) \times \partial\Omega$ in $(0, T) \times M$ is noncharacteristic for the eikonal equation (6.41), at the data (6.42) (on the support of $a_j(t, x)$). An equivalent formulation is that if we set

$$C_\beta = S_\beta \cap \{(0, T) \times \partial\Omega\}, \tag{6.43}$$

where S_β is the level set (6.36), then C_β is a spacelike hypersurface of $(0, T) \times \partial\Omega$, with its induced Lorentz metric. Recall that S_β is a union of light rays. Another equivalent hypothesis is that each of these light rays that hits $(0, T) \times \partial\Omega$ does so transversally. Let us assume in addition that each such light ray (inside some S_β, issuing from supp a at $t = 0$) hits $(0, T) \times \partial\Omega$ exactly once.

To continue our construction of the transversally reflected wave, we want to solve (6.41)–(6.42). In fact, under the geometrical hypothesis just stated, this has exactly two solutions. One of them is $\varphi(t, x)$ itself. The level sets $\{\varphi(t, x) = \beta\}$ are swept out by light rays issuing from C_β which point in the negative t-direction as they go into Ω. The solution of current interest to us is the other one; its level sets $\{\psi(t, x) = \beta\}$ are swept out by light rays issuing from C_β which point in the positive t-direction as they go into Ω. See Fig. 6.1.

Having $\psi(t, x)$, we construct the amplitudes $b_\ell(t, x)$ by solving transport equations, parallel to (6.16) and (6.20). We take

$$b_j(t, x) = a_j(t, x), \quad b_\ell(t, x) = 0, \quad x \in \partial\Omega,\ \ell > j. \tag{6.44}$$

In particular, each $b_\ell(t, x)$, hence each $w_\ell(t, x)$, vanishes on $[0, T_1) \times \overline{\Omega}$, for some $T_1 \in (0, T)$. Now if $W_N = \sum_{j \le \ell \le N} w_\ell$, we have $w - W_N$ satisfying:

$$\begin{aligned} &\Big(\frac{\partial^2}{\partial t^2} - \Delta\Big)(w - W_N) = -\widetilde{r}_n, \\ &(w - W_N)(0, x) = 0, \quad \partial_t(w - W_N)(0, x) = 0, \\ &(w - W_N)(t, x) = \widetilde{\rho}_N(t, x), \quad x \in \partial\Omega, \end{aligned} \tag{6.45}$$

similar to (6.29), where $\widetilde{r}_N$ and $\widetilde{\rho}_N$ are fairly smooth, on $(0, T) \times \overline{\Omega}$ and on $(0, T) \times \partial\Omega$, respectively, if N is large, and both vanish for $0 \le t < T_1$, for some $T_1 > 0$. It follows from the results of Exercise 2 in §1 that $w - W_N$ is arbitrarily smooth, for N sufficiently large, so such a construction succeeds in approximating the reflected wave, granted the transversal reflection hypothesis made above.

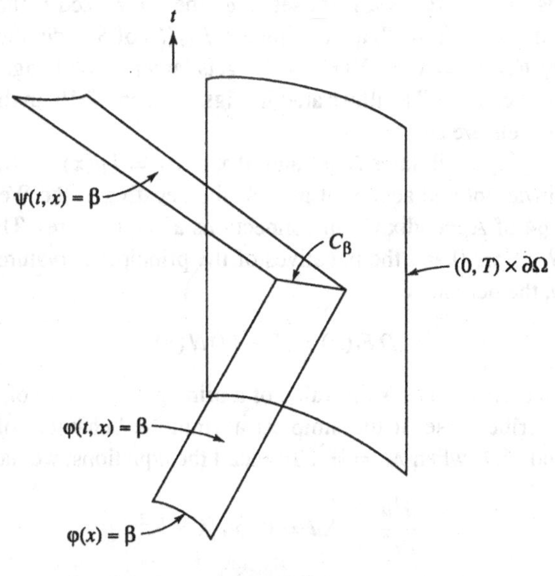

FIGURE 6.1

When the transversality hypothesis made above is violated, the reflected wave can have a much more complicated structure. Some of the basic cases of this phenomenon are dealt with in detail in [Tay], Vol. 3 of [Ho], and [MeT], to which we refer for citations of the original papers.

Exercises

1. Extend the geometrical optics construction of approximate solutions to (6.1) and (6.3), with (6.4) replaced by
$$u_t(0,x) = b(x)\lambda F'\big(\lambda\varphi(x)\big).$$

2. Work out geometrical optics approximations for solutions to hyperbolic systems, of the form (5.20), assuming strict hyperbolicity, that is, for each $\xi \in \mathbb{R}^n \setminus 0$, $\sum A_j(t,x)\xi_j$ has n eigenvalues $\lambda_\nu(x,\xi)$, all real and distinct.

7. The formation of caustics

The geometrical optics construction of §6 breaks down when the eikonal equation (6.14) does not have a global solution, which is a typical state of affairs. We can see this happen in the case where M is $\mathbb{R}^n$, with its flat Euclidean metric. In such a case, for small t, the solution to (6.14) is given implicitly by

$$\varphi^{\pm}(t,y) = \varphi(x), \quad y = x \pm tN(x), \quad N(x) = |\nabla\varphi(x)|^{-1}\nabla\varphi(x). \tag{7.1}$$

In other words, if $S \subset \mathbb{R}^n$ is a level set of φ, then, for fixed t, the level sets of $\varphi^{\pm}(t,\cdot)$ (i.e., the "wavefronts") are the images $F_{\pm t}(S)$ of S under the maps $F_{\pm t}$ on $\mathbb{R}^n$, defined by $F_{\pm t}(x) = x \pm tN(x)$. As $|t|$ gets larger, these images can develop singularities, or "caustics," as illustrated in Figs. 7.1a and 7.1b, in the case $n = 2$, where the level sets are curves.

Note that $DN(x)$ annihilates $N(x)$ and, if $x \in \Sigma_\beta = \{\varphi(x) = \beta\}$, then $DN(x)$ leaves $T_x\Sigma_\beta$ invariant and acts on it as $-A$, the negative of the Weingarten map (discussed in §4 of Appendix C, on connections and curvature). Thus the eigenvalues of $DN(x)$ are 0 and the negatives of the principal curvatures of Σ_β at x. Consequently, the derivative

$$DF_t(x) = I + tDN(x) \tag{7.2}$$

is singular if and only if $1/t$ is the value of a principal curvature of Σ_β at x.

We will describe some of the simplest asymptotic behaviors of solutions to (6.1), (6.3), and (6.4) when $M = \mathbb{R}^2$. To recall the equations, we have

$$\begin{gathered}\frac{\partial^2 u}{\partial t^2} - \Delta u = 0 \ \text{ on } \mathbb{R}\times\mathbb{R}^2,\\ u(0,x) = a(x)F\big(\lambda\varphi(x)\big), \quad u_t(0,x) = 0.\end{gathered} \tag{7.3}$$

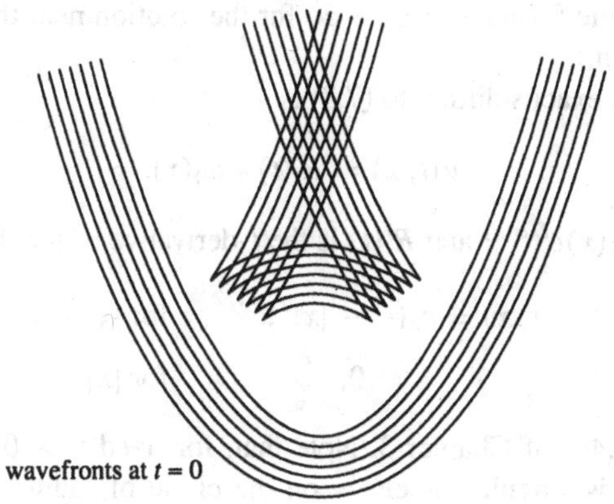

FIGURE 7.1A

We will take

$$F(s) = e^{is}. \tag{7.4}$$

As before, $a \in C_0^\infty(\mathbb{R}^2)$. As shown in §6, there is a short-time approximate solution of the form

$$u(t,x) \sim \sum_{\pm} \sum_{j \ge 0} \lambda^{-j} \, a_j^{\pm}(t,x) \, e^{i\lambda\varphi^{\pm}(t,x)}, \tag{7.5}$$

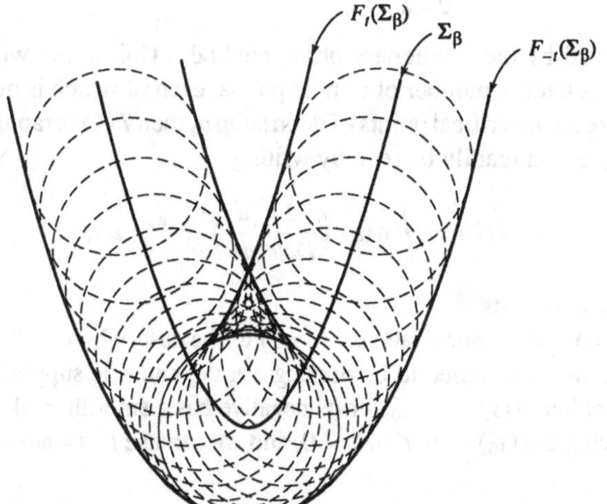

FIGURE 7.1B

where this time we have absorbed the factors i^{-j} into the amplitudes. We now want an asymptotic formula as $\lambda \to \infty$ for the solution near the caustics, where (7.5) breaks down.

Recall that the exact solution to (7.3) is

$$u(t, x) = R'(t) * u_0(x), \tag{7.6}$$

where $u_0(x) = a(x)e^{i\lambda\varphi(x)}$ and $R'(t)$ is the t-derivative of the Riemann function

$$R(t, x) = \begin{cases} c_2\big(t^2 - |x|^2\big)^{-1/2}, & \text{for } |x| < t, \\ 0, & \text{for } |x| > t \end{cases} \tag{7.7}$$

if $t > 0$; see (5.46) of Chapter 3. Note that, for fixed $t > 0$, $R'(t)$ is a radial distribution that is singular precisely on the circle of radius t, centered at the origin. We expect $u(t, x)$ in (7.6) to have qualitative features similar to

$$\begin{aligned} v(t, x) &= \frac{1}{t} \int\limits_{|y-x|=t} u_0(y)\, ds(y) \\ &= \frac{1}{2\pi} \int_{-\pi}^{\pi} a\big(x + t\,\mathrm{cis}(s)\big)\, e^{i\lambda\varphi(x+t\,\mathrm{cis}(s))}\, ds, \end{aligned} \tag{7.8}$$

where $\mathrm{cis}(s) = (\cos s, \sin s)$. The precise relation between $u(t, x)$ and $v(t, x)$ is most easily analyzed using techniques to be developed in the next chapter; see the exercises after §9 of Chapter 7. At this point, we will concentrate on an asymptotic analysis of (7.8).

In the simplest cases, an integral of the form

$$I(\lambda) = \int_{-\infty}^{\infty} a(s)\, e^{i\lambda\psi(s)}\, ds, \quad a \in C_0^\infty(\mathbb{R}), \tag{7.9}$$

can be analyzed by the "stationary phase method." This works when ψ is real-valued and has a finite number of critical points, each of which is nondegenerate. In fact, if there are no critical points of ψ on supp a, then $I(\lambda)$ is rapidly decreasing as $|\lambda| \to \infty$, as can readily be seen by writing

$$I(\lambda) = \int a(s) \left(\frac{1}{i\lambda\psi'} \frac{d}{ds}\right)^k e^{i\lambda\psi(s)}\, ds,$$

and integrating by parts.

Thus we can reduce our analysis of (7.9) to the case where ψ has exactly one critical point, at s_0, assumed to be nondegenerate, and a is supported near s_0. In such a case, either $\psi(s) - \psi(s_0)$ or its negative has a smooth, real-valued square root $t(s)$, such that $t(s_0) = 0$, $t'(s_0) > 0$, and we can use t as a new coordinate, to write

$$I(\lambda) = e^{i\lambda\varphi(s_0)} \int b(t)\, e^{i\alpha\lambda t^2}\, dt, \quad b \in C_0^\infty(\mathbb{R}), \tag{7.10}$$

where $\alpha = \pm 1$. There are several ways to evaluate (7.10) asymptotically; one is to set $x = t^2$, so

$$(7.11)\qquad \begin{aligned} I(\lambda) &= \frac{1}{2} e^{i\lambda\varphi(s_0)} \int_0^\infty \bigl[b(x^{1/2}) + b(-x^{1/2})\bigr] x^{-1/2} e^{i\alpha\lambda x}\, dx \\ &\sim e^{i\lambda\varphi(s_0)}\, \lambda^{-1/2}\bigl[\alpha_0 + \alpha_1\lambda^{-1} + \cdots\bigr], \end{aligned}$$

in view of results on Fourier transforms of singular functions in §8 of Chapter 3. Another method, in the context of the multidimensional stationary phase method, will be given in Appendix B at the end of this chapter.

More generally, if there are a finite number of critical points s_j of $\psi(s)$, all nondegenerate, then

$$(7.12)\qquad I(\lambda) \sim \sum_j A_j(\lambda)\lambda^{-\frac{1}{2}} e^{i\lambda\psi(s_j)}, \quad A_j(\lambda) \sim \alpha_{0j} + \alpha_{1j}\lambda^{-1} + \cdots .$$

If $a(s) = a(y, s)$ and $\psi(s) = \psi(y, s)$ in (7.9) depend smoothly on the parameters y, then we have (7.12) for $I(\lambda) = I(y, \lambda)$, with $\alpha_{kj} = \alpha_{kj}(y)$ and $\psi(s_j) = \psi(y, s_j(y))$ depending smoothly on y, as long as the critical points of $\psi(y, s)$, as a function of s, are all nondegenerate and consequently depend smoothly on y.

Let's return to (7.8). We are assuming that $\nabla\varphi(y) \neq 0$ for $y \in \operatorname{supp} a$. Now, given $x \in \mathbb{R}^2$, $t > 0$, let us denote by $S_t(x)$ the circle of radius t centered at x. The way in which $S_t(x)$ is tangent to various level curves Σ_β of φ determines the nature of the stationary points of the phase in the last integral in (7.8). Clearly, if $1/t$ is bigger than the largest curvature of any Σ_β, then $S_t(x)$ will have only simple tangencies with such level curves, so only nondegenerate stationary points of the phase will appear in (7.8). If $y \in \Sigma_\beta$ (so $\varphi(y) = \beta$) is such a point of intersection, then its contribution to the asymptotic behavior of $v(t, x)$ as $\lambda \to \infty$ is an amplitude times $e^{i\lambda\varphi(y)}$, in agreement with the geometrical optics construction given in §6, since in this case $\varphi(t, x) = \varphi(y)$. This is illustrated in Fig. 7.2.

On the other hand, suppose $y \in \Sigma_\beta$ and $1/t = \kappa(y)$, the curvature of Σ_β at y. Let $x = y + tN(y)$, as illustrated in Fig. 7.3. Then $S_t(x)$ has higher-order tangency with Σ_β at y. Let us assume that y is not a stationary point for κ on Σ_β, that is, if one travels on Σ_β at unit speed, κ is monotonically increasing (or decreasing) at a nonzero rate at y. In such a case, Fig. 7.3 captures the behavior of the image of Σ_γ (for γ close to β) under F_t, by our analysis of (7.2). In this case, the phase function in (7.8) has a simply degenerate critical point at y, so we have an integral of the form (7.9) with $\psi(s_0) = \beta$, $\psi'(s_0) = \psi''(s_0) = 0$, and $\psi'''(s_0) \neq 0$ (say it is > 0). We can treat this in a fashion similar to the nondegenerate case. This time, $\psi(s) - \beta$ has a smooth *cube* root near $s = s_0$, call it $t(s)$, such that $t(s_0) = 0$, $t'(s_0) > 0$, and we can take t as a new coordinate to write

$$(7.13)\qquad I(\lambda) = e^{i\lambda\varphi(s_0)} \int b(t)\, e^{i\lambda t^3}\, dt, \quad b \in C_0^\infty(\mathbb{R}).$$

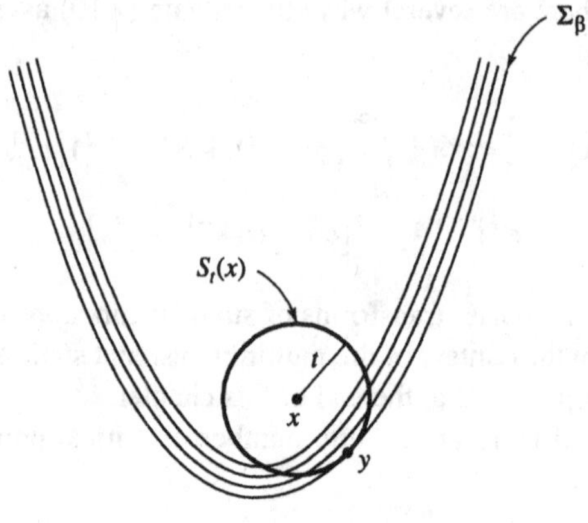

FIGURE 7.2

Parallel to (7.10)–(7.11), we can set $x = t^3$ and write

$$(7.14)\quad \begin{aligned} I(\lambda) &= \frac{1}{3}\, e^{i\lambda\varphi(s_0)} \int b(x^{1/3})x^{-1/3}\, e^{i\lambda x}\, dx \\ &\sim e^{i\lambda\varphi(s_0)}\, \lambda^{-2/3} \left[\alpha_0 + \alpha_1\lambda^{-1} + \cdots\right]. \end{aligned}$$

Note that the exponent in $\lambda^{-2/3}$ here differs from the exponent in $\lambda^{-1/2}$, which appears in (7.11).

Now we want to examine the uniform asymptotic behavior of (7.8), as $\lambda \to \infty$, for x in a neighborhood of a caustic point x_0. We will retain the hypothesis on the curvature made above, namely $x_0 = F_t(y_0)$ with $\kappa(y_0) = 1/t$, $y_0 \in \Sigma_\beta$, and κ not stationary on Σ_β at y_0, so the geometry of $F_t(\Sigma_\gamma)$ for γ near β is as illustrated in

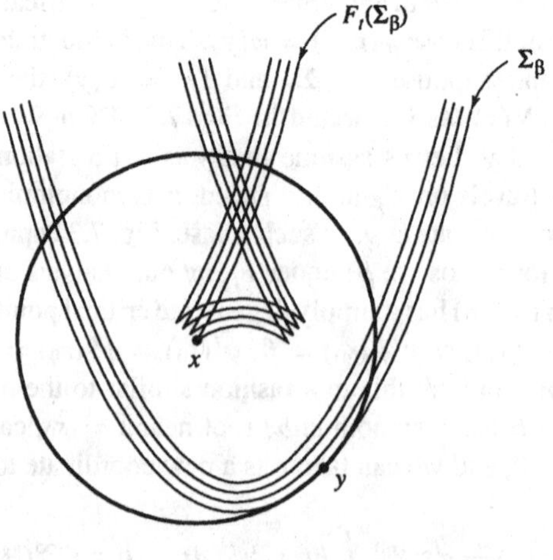

FIGURE 7.3

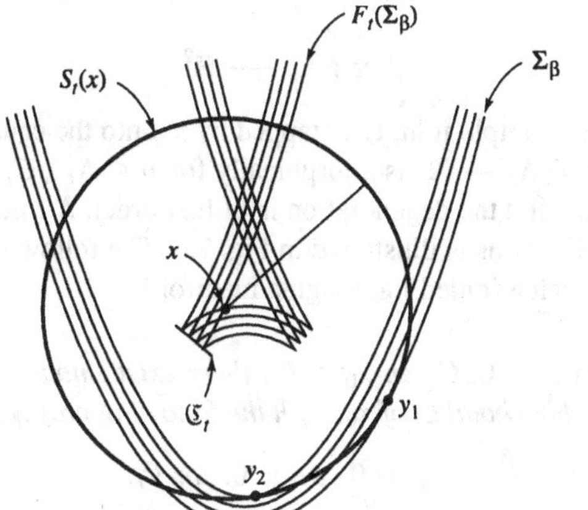

FIGURE 7.4

Fig. 7.3. Thus portions of $F_t(\Sigma_y)$ lie on one side of the caustic set $\mathcal{C}_t$, namely, the image of the critical set of F_t.

Take a point x on this side of $\mathcal{C}_t$, as illustrated in Fig. 7.4. For such x, the circle $S_t(x)$ is simply tangent to two level sets of φ, at points y_1 and y_2, as indicated in Figs. 7.4 and 7.5, and as x approaches $\mathcal{C}_t$, the points y_1 and y_2 coalesce, to a point y such as depicted in Fig. 7.3. Consequently, if $\psi(s) = \psi(t, x, s) = \varphi\big(x + t \operatorname{cis}(s)\big)$, then for x on one side of $\mathcal{C}_t$, $\psi(s)$ has two nondegenerate critical points, s_1 and s_2, which coalesce to a single degenerate critical point s_0 as x approaches $\mathcal{C}_t$.

The side of $\mathcal{C}_t$ on which such x lies is foliated in two ways, by level sets of $\varphi(t, \cdot)$. This arises because the graph of $d\varphi$, a Lagrangian submanifold of $T^*\mathbb{R}^2$, is mapped by the time-t geodesic flow to a Lagrangian manifold $\Lambda_t \subset T^*\mathbb{R}^2$, whose

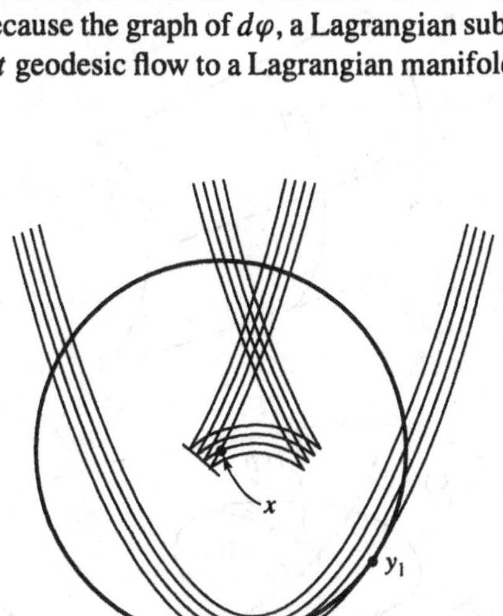

FIGURE 7.5

projection

$$\pi : \Lambda_t \longrightarrow \mathbb{R}^2 \tag{7.15}$$

onto x-space has a simple fold, C_t, mapped by π onto the caustic set $\mathcal{C}_t$. In other words, $D\pi(p) : T_p\Lambda_t \to \mathbb{R}^2$ isomorphically for $p \in \Lambda_t \setminus C_t$, while $D\pi(p)$ has rank 1 for $p \in C_t$, and the degeneration is of first order. A fold map between two two-dimensional regions is illustrated in Fig. 7.6. The following result elucidates the structure of such a folded Lagrangian manifold.

Lemma 7.1. *Fix $t > 0$. Given $x_0 \in \mathcal{C}_t$, there exist smooth functions θ and ρ, defined on a neighborhood U of x_0, with the following properties.*

$$\rho = 0,\ d\rho \neq 0,\ \text{ on } \mathcal{C}_t. \tag{7.16}$$

If $U^{\pm} = \{x \in U : \pm\rho(x) > 0\}$, then Λ_t projects onto U^+ and is the graph of $d\varphi^{\pm}$, where $\varphi^{\pm}$ is the "double-valued" function

$$\varphi^{\pm}(x) = \theta(x) \pm \frac{2}{3}\rho(x)^{3/2}. \tag{7.17}$$

Proof. That (7.15) is a fold implies that, over U^+, Λ_t is the graph of a "double-valued" closed 1-form, i.e., of $d\varphi^{\pm}$. We can put $\varphi^{\pm}$ in the form (7.17) by taking

$$\theta(x) = \frac{1}{2}\big(\varphi^+(x) + \varphi^-(x)\big), \quad \rho(x) = \Big[\frac{3}{4}\big(\varphi^+(x) - \varphi^-(x)\big)\Big]^{2/3}. \tag{7.18}$$

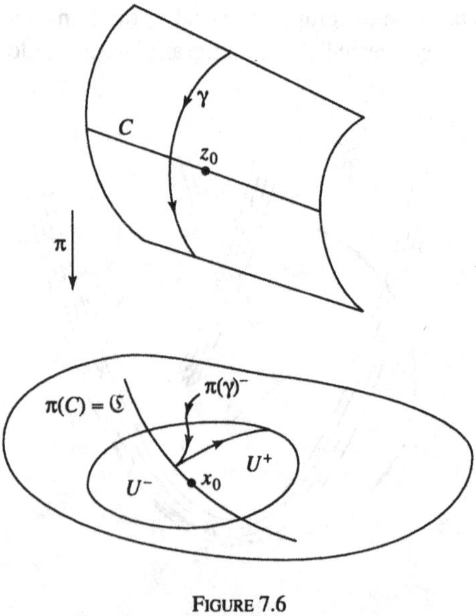

FIGURE 7.6

We need to establish that θ and ρ are smooth on the closure of U^+ in U, in particular at $\mathcal{C}_t$. This is best seen by constructing a function $\Phi \in C^\infty(\Lambda_t)$ such that $\varphi^\pm = \Phi \circ \pi^{-1}$. In fact, if $\kappa = \sum \xi_j \, dx_j$ is the contact form on $T^*\mathbb{R}^2$ and $\iota : \Lambda_t \hookrightarrow T^*\mathbb{R}^2$, then $\iota^*\kappa$ is closed, hence locally exact, and we take Φ such that $d\Phi = \iota^*\kappa$. Compare Exercise 5 in §15 of Chapter 1. There is a smooth involution j on Λ_t, interchanging points with the same image under π, and we can set

$$\Theta = \frac{1}{2}\left(\Phi + \Phi \circ j\right), \quad R = \left[\frac{3}{4}\left(\Phi - \Phi \circ j\right)\right]^{2/3}. \tag{7.19}$$

These formulas define Θ and R as functions on Λ_t that are invariant under j, related to (7.18) by

$$\Theta = \theta \circ \pi, \quad R = \rho \circ \pi. \tag{7.20}$$

It is clear that Θ is smooth, and hence the desired smoothness of θ is established. To examine the smoothness of R and ρ, we reason as follows. Since (7.15) is assumed to be a fold, we know that, for $x \in U^+$ close to x_0,

$$C_1\delta(x)^{1/2} \le |d\varphi^+(x) - d\varphi^-(x)| \le C_2\delta(x)^{1/2},$$

for some $C_j \in (0, \infty)$, where $\delta(x) = \operatorname{dist}(x, \mathcal{C}_t)$. This implies that

$$C_3\delta^{3/2}(x) \le |\varphi^+(x) - \varphi^-(x)| \le C_4\delta(x)^{3/2},$$

and hence, for $z \in \Lambda_t$, close to $z_0 = \pi^{-1}(x_0)$,

$$C_5\delta(z)^3 \le \left|\Phi(z) - \Phi\big(j(z)\big)\right| \le C_6\delta(z)^3,$$

where $\delta(z) = \operatorname{dist}(z, \mathcal{C}_t)$. This implies that R, defined in (7.19), is smooth on Λ_t, which in turn yields the desired smoothness of ρ, and also that $d\rho(x) \neq 0$ on $\mathcal{C}_t$.

We now establish a result that puts the phase function in (7.8) into a normal form, near $\mathcal{C}_t$.

Proposition 7.2. *Fix $t > 0$ and take $x_0 \in \mathcal{C}_t$. For x near x_0, there is a family of diffeomorphisms of $\mathbb{R}$, depending smoothly on the parameter x, transforming $\psi(s) = \psi(t, x, s) = \varphi\big(x + t \operatorname{cis}(s)\big)$, for s near the stationary point s_0 of $\psi_0(s) = \psi(t, x_0, s)$, to*

$$\tilde{\psi}(s) = \frac{1}{3}s^3 - \rho(x)s + \theta(x), \tag{7.21}$$

near $s = 0$.

Proof. We first note that at $x = x_0$ (so $\rho = 0$), $\psi(s)$ can be transformed to $s^3/3 + \theta(x_0)$, as the argument leading to (7.13) shows. We can therefore consider the following situation. Suppose $\psi(\tau, s)$ is smooth,

$$\psi(0, s) = \frac{1}{3}s^3, \quad \frac{\partial}{\partial \tau}\frac{\partial}{\partial s}\psi(0, 0) < 0. \tag{7.22}$$

We want a smooth map, of the form

$$(\tau, s) \mapsto \big(\tau, f(\tau, s)\big), \tag{7.23}$$

transforming ψ to $\tilde{\psi}(\tau, s) = s^3/3 - \rho(\tau)s + \theta(\tau)$, where θ is determined as follows, for $\tau > 0$. By (7.22), for τ small and positive, $\psi(\tau, \cdot)$ has two critical points, close to 0, at $s = s_1(\tau), s_2(\tau)$, and we take $\theta(\tau) = \big(\psi(\tau, s_1)+\psi(\tau, s_2)\big)/2$. The set

$$\Gamma = \big\{(\tau, s) : \partial_s \psi(\tau, s) = 0\big\} \tag{7.24}$$

is a curve tangent to the s-axis at (0, 0), as pictured in Fig. 7.7. There is a smooth involution of Γ, interchanging the points with the same τ-coordinate, and $\theta(\tau)$ is the value of the symmetrization of $\psi\big|_\Gamma$ with respect to this involution, so θ is easily seen to be a smooth function of τ.

We may as well subtract $\theta(\tau)$ and try to achieve the form $\tilde{\psi}(\tau, s) = s^3/3 - \rho(\tau)s$. Note that, in this model case, the analogue of (7.24) is

$$\widetilde{\Gamma} = \{(\tau, s) : s = \pm\sqrt{\rho}\}, \quad \tilde{\psi}(\tau, \pm\sqrt{\rho}) = \mp\frac{2}{3}\rho^{3/2}. \tag{7.25}$$

So $\rho(\tau)$ is uniquely defined for $\tau \geq 0$ by the requirement that $\rho(\tau) > 0$ for $\tau > 0$ and

$$\psi = \mp\frac{2}{3}\rho(\tau)^{3/2}, \quad \text{on } \Gamma. \tag{7.26}$$

To put it simply, $\mp(2/3)\rho(\tau)^{3/2}$ are the critical values of $\psi(\tau, s)$, as a function of s (once $\theta(\tau)$ has been subtracted). Given that now $\psi\big|_\Gamma$ has been arranged to be odd with respect to the involution of Γ described above, it is easy to show that $\rho(\tau)$ is a smooth function of τ, via the sort of argument used in the proof of Lemma 7.1. Also, $d\rho \neq 0$ at $\tau = 0$.

Having specified $\rho(\tau)$, we start to construct the diffeomorphism, of the form (7.23). We want $f(0, s) = s$. For $\tau > 0$, the fact that $\psi(\tau, \cdot)$ and $\tilde{\psi}(\tau, \cdot)$ have identical ranges, for $s \leq -\sqrt{\rho(\tau)}$, for $-\sqrt{\rho(\tau)} \leq s \leq \sqrt{\rho(\tau)}$, and for $s \geq \sqrt{\rho(\tau)}$, implies that there is a ***unique*** homeomorphism $s \mapsto f(\tau, s)$ transforming $\psi(\tau, \cdot)$ to $\tilde{\psi}(\tau, \cdot)$. This homeomorphism is clearly a diffeomorphism (as a function of s), away from $s = \mp\sqrt{\rho(\tau)}$, and, by the sort of argument leading to (7.10), we

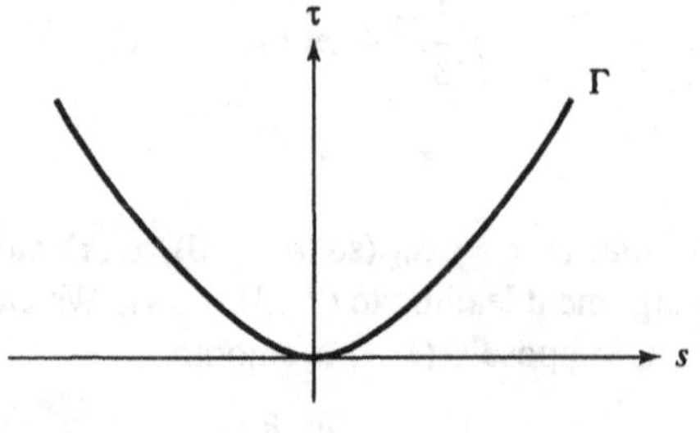

FIGURE 7.7

see that, for each fixed $\tau > 0$, it is a diffeomorphism in a neighborhood of these points too. For $\tau < 0$, both $\psi(\tau, \cdot)$ and $\tilde{\psi}(\tau, \cdot)$ have no critical points (near 0), so the existence of a unique diffeomorphism $s \mapsto f(\tau, s)$ transforming ψ to $\tilde{\psi}$ is easy.

The continuous dependence of $f(\tau, s)$ on τ is easy to establish, but the smooth dependence on τ, at $\tau = 0$, is a bit more subtle, so we finally turn to that point. We will use a device similar to that used in the proof of the Morse lemma (given in Appendix C, §8).

We may as well use ρ instead of τ as a coordinate, so we assume we have a smooth function $\psi(\rho, s)$ satisfying

$$\psi(0, s) = \frac{1}{3}s^3, \quad \partial_\rho \partial_s \psi(0, 0) < 0, \tag{7.27}$$

and, for $\rho > 0$,

$$\partial_s \psi(\rho, \pm\sqrt{\rho}) = 0, \quad \psi(\rho, \pm\sqrt{\rho}) = \mp \rho^{3/2}. \tag{7.28}$$

We want to produce a diffeomorphism of the form $(\rho, s) \mapsto \big(\rho, f(\rho, s)\big)$, such that $f(0, s) = s$, transforming ψ to

$$\tilde{\psi}(\rho, s) = \frac{1}{3}s^3 - \rho s, \tag{7.29}$$

a function that also satisfies (7.28). Now consider the family of functions connecting ψ and $\tilde{\psi}$:

$$\Psi(\sigma, \rho, s) = (1 - \sigma)\psi(\rho, s) + \sigma\Big(\frac{1}{3}s^3 - \rho s\Big). \tag{7.30}$$

Thus $\Psi(0, \rho, s) = \psi(\rho, s)$, $\Psi(1, \rho, s) = s^3/3 - \rho s$, and, for any fixed $\sigma \in \mathbb{R}$, $\Psi(\sigma, \rho, s)$ satisfies (7.27) and (7.28). We will construct a family of diffeomorphisms $s \mapsto F(\sigma, \rho, s) = F_{\sigma,\rho}(s)$, transforming $\psi(\rho, \cdot)$ to $\Psi(\sigma, \rho, \cdot)$, generated by a smooth family of vector fields on a neighborhood of 0 in $\mathbb{R}$:

$$X(\sigma, \rho, s) = \xi(\sigma, \rho, s)\frac{\partial}{\partial s}. \tag{7.31}$$

Given $X(\sigma, \rho, s)$, F is defined by $F(0, \rho, s) = s$, and

$$\frac{\partial}{\partial \sigma} F(\sigma, \rho, s) = \xi(\sigma, \rho, F). \tag{7.32}$$

If $F^*_{\sigma,\rho} g(s) = g\big(F_{\sigma,\rho}(s)\big)$, then

$$\frac{d}{d\sigma} F^*_{\sigma,\rho} g_\sigma(s) = F^*_{\sigma,\rho} \mathcal{L}_{X_{\sigma,\rho}} g_\sigma + F^*_{\sigma,\rho}\Big(\frac{d}{d\sigma} g_\sigma\Big), \tag{7.33}$$

and this quantity vanishes, provided

$$\xi(\sigma, \rho, s)\frac{\partial}{\partial s} g_\sigma = \frac{\partial}{\partial \sigma} g_\sigma. \tag{7.34}$$

Applying this to $g_\sigma(s) = g_{\sigma,\rho}(s) = \Psi(\sigma, \rho, s) = \psi_{\sigma,\rho}(s)$, we have

$$F^*_{\sigma,\rho} \psi_{\sigma,\rho}(s) = \psi_{0,\rho}(s), \quad \forall\, \sigma, \tag{7.35}$$

provided

$$\xi(\sigma, \rho, s) = \frac{s^3/3 - \rho s - \psi(\rho, s)}{(\partial/\partial s)\Psi(\sigma, \rho, s)}. \tag{7.36}$$

Now the denominator of this fraction vanishes on Γ, but by (7.27) the gradient of the denominator does not vanish on Γ. Meanwhile, the numerator vanishes to second order on Γ, so the quotient ξ is C^∞ and vanishes on Γ. Thus (7.31) generates a smooth flow (which leaves Γ invariant, of course), and the proof of Proposition 7.2 is complete.

In order to analyze (7.8), we are now led to discuss the asymptotic evaluation, as $\lambda \to \infty$, of integrals of the form

$$\mathcal{I}(a; \mu, \lambda) = \frac{1}{2\pi} \int_{-\infty}^{\infty} a(s)\, e^{i\lambda(s^3/3 - \mu s)}\, ds, \tag{7.37}$$

given $a \in C_0^\infty(\mathbb{R})$. Such integrals are called *Airy integrals*. We fix $K < \infty$ and assume $|\mu| \le K^2$. The phase function $\varphi_\mu(s) = s^3/3 - \mu s$ has derivative $\varphi_\mu'(s) = s^2 - \mu$, with roots $s = \pm\sqrt{\mu}$, which are the stationary points of the phase when $\mu \ge 0$.

Our first goal is to show simultaneously that the uniform asymptotic behavior of (7.37), as $\lambda \to \infty$, $|\mu| \le K^2$, depends only on $a(s)$ for $-2K \le s \le 2K$, and that (7.37) makes sense for a wider class of amplitudes $a(s)$; namely we allow $a(s) \in S_1^m(\mathbb{R})$, for some $m \in \mathbb{R}$ (i.e., $|D_s^j a(s)| \le C_j \langle s\rangle^{m-j}$). A general $a \in S_1^m(\mathbb{R})$ can be written as a sum of a term in $C_0^\infty(-2K, 2K)$ and a term in $S_1^m(\mathbb{R})$ which vanishes on $[-(3/2)K, (3/2)K]$. If $a_2(s)$ has the latter property, then we can make a change of variable, $y = \varphi_\mu(s)$, and write

$$\mathcal{I}(a_2; \mu, \lambda) = \frac{1}{2\pi} \int b_\mu(y) e^{i\lambda y}\, dy, \quad b_\mu(y) \in S_1^{m/3}(\mathbb{R}), \tag{7.38}$$

where $b_\mu(y)$ depends smoothly on μ. We know that $\hat{b}_\mu(\lambda)$ is an element of $\mathcal{S}'(\mathbb{R})$ that is smooth on $\mathbb{R} \setminus 0$ and rapidly decreasing as $|\lambda| \to \infty$, from material in §8 of Chapter 3. Thus we can take $a \in S_1^m(\mathbb{R})$ in (7.37).

In particular, we can take $a(s) = 1$, obtaining

$$\mathcal{I}(1; \mu, \lambda) = \frac{1}{2\pi} \int_{-\infty}^{\infty} e^{i\lambda(s^3/3 - \mu s)}\, ds = \lambda^{-1/3} Ai\big(-\mu\lambda^{2/3}\big), \quad \lambda > 0, \tag{7.39}$$

where $Ai(x)$ is the Airy function

$$Ai(x) = \frac{1}{2\pi} \int_{-\infty}^{\infty} e^{i(s^3/3 + xs)}\, ds = \mathcal{I}(1; -x, 1), \tag{7.40}$$

for $x \in \mathbb{R}$. If we set $\mu = \pm 1$, $\lambda = x^{3/2}$ in (7.39), we have

$$\begin{aligned} Ai(x) &= x^{1/2}\, \mathcal{I}(1; -1, x^{3/2}), \\ Ai(-x) &= x^{1/2}\, \mathcal{I}(1; 1, x^{3/2}), \end{aligned} \tag{7.41}$$

for $x > 0$. In these cases, μ is a fixed, nonzero quantity, and we can apply the stationary phase method, to get

$$Ai(x) = O(x^{-\infty}), \quad Ai(-x) \sim \frac{1}{\sqrt{\pi}} x^{-1/4} \cos\Big(\frac{2}{3}x^{3/2} - \frac{\pi}{4}\Big), \tag{7.42}$$

as $x \to +\infty$. Let us also note that, since $Ai(x)$ (as an element of $\mathcal{S}'(\mathbb{R})$) is the inverse Fourier transform of $e^{is^3/3}$, which satisfies an obvious first-order, linear ODE, then $Ai(x)$ satisfies the differential equation

$$Ai''(x) - x\, Ai(x) = 0, \tag{7.43}$$

known as *Airy's equation*. It follows that $Ai(x)$ continues to an entire holomorphic function on the complex plane. The graph of $Ai(x)$ is shown in Fig. 7.8. It was constructed by numerically integrating (7.43), using initial data

$$Ai(0) = \frac{3^{-2/3}}{\Gamma(\frac{2}{3})}, \quad Ai'(0) = -\frac{3^{-1/3}}{\Gamma(\frac{1}{3})}. \tag{7.44}$$

Note that $Ai(x)$ is real for $x \in \mathbb{R}$. In fact, (7.40) can be written as

$$Ai(x) = \frac{1}{2\pi} \int_{-\infty}^{\infty} \cos\Big(\frac{1}{3}s^3 + xs\Big)\, ds. \tag{7.45}$$

Taking the Airy function as a basic special function, we see that (7.39) gives the uniform asymptotic behavior of (7.37), for μ in any bounded interval, in the case $a = 1$. We now seek a uniform asymptotic expansion of (7.37), of a similar form, for general $a \in S_1^m(\mathbb{R})$. In fact, the general case will involve both the Airy function and its derivative:

$$Ai'(x) = \frac{i}{2\pi} \int_{-\infty}^{\infty} s\, e^{i(s^3/3 + xs)}\, ds. \tag{7.46}$$

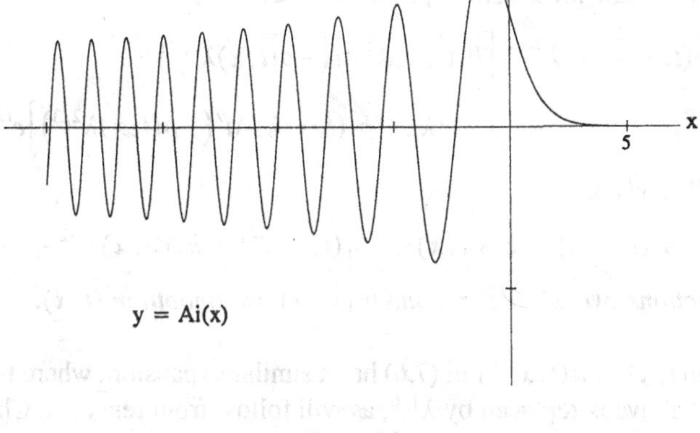

Figure 7.8

To obtain it, write

$$a(s) = a_0 + a_1 s + b_\mu(s)(\mu - s^2), \tag{7.47}$$

where $b_\mu(s) \in S_1^\ell(\mathbb{R})$, $\ell = \max(m/2, 1/2)$, with smooth dependence on μ. Then, for $\lambda > 0$,

(7.48)
$$\mathcal{I}(a; \mu, \lambda) = a_0 \lambda^{-1/3} Ai\big(-\mu\lambda^{2/3}\big) - i a_1 \lambda^{-2/3} Ai'\big(-\mu\lambda^{2/3}\big) - \frac{1}{i\lambda} \mathcal{I}(b'_\mu; \mu, \lambda),$$

where we have used

$$(\mu - s^2) e^{i\lambda(s^3/3 - \mu s)} = \frac{1}{i\lambda} \frac{d}{ds} e^{i\lambda(s^3/3 - \mu s)} \tag{7.49}$$

and integration by parts to evaluate

$$\int b_\mu(s)\, (\mu - s^2)\, e^{i\lambda(s^3/3 - \mu s)}\, ds. \tag{7.50}$$

Now we can apply the same transformation to $\mathcal{I}(b'_\mu; \mu, \lambda)$ and iterate this argument arbitrarily often, to establish the following result:

Proposition 7.3. *Given $a \in S_1^m(\mathbb{R})$, as $\lambda \to +\infty$, we have*

$$\mathcal{I}(a; \mu, \lambda) \sim b_0(\mu, \lambda)\lambda^{-1/3} Ai\big(-\mu\lambda^{2/3}\big) - i b_1(\mu, \lambda)\lambda^{-2/3} Ai'\big(-\mu\lambda^{2/3}\big), \tag{7.51}$$

where

$$b_j(\mu, \lambda) \sim b_{j0}(\mu) + b_{j1}(\mu)\lambda^{-1} + b_{j2}(\mu)\lambda^{-2} + \cdots, \tag{7.52}$$

and where $b_{j\nu}(\mu)$ are smooth in μ. The expansion (7.51) is valid uniformly for $|\mu| \le K^2$, for any fixed $K < \infty$.

When we combine this with an application of Proposition 7.2, we have the following result on the behavior of (7.8) near a caustic set.

Proposition 7.4. *Granted the geometric hypotheses made on the formation of the caustic set C_t before Lemma 7.1, the oscillatory integral (7.8) has the following asymptotic behavior for x near C_t, as $\lambda \to +\infty$:*

$$\begin{aligned} v(t, x, \lambda) = \lambda^{-1/3}\Big[& b_0(t, x, \lambda) Ai\big(-\rho(t, x)\lambda^{2/3}\big) \\ & - i\lambda^{-1/3} b_1(t, x, \lambda) Ai'\big(-\rho(t, x)\lambda^{2/3}\big)\Big] e^{i\lambda\theta(t,x)}, \end{aligned} \tag{7.53}$$

mod $O(\lambda^{-\infty})$, where

$$b_j(t, x, \lambda) \sim b_{j0}(t, x) + b_{j1}(t, x)\lambda^{-1} + b_{j2}(t, x)\lambda^{-2} + \cdots, \tag{7.54}$$

and the functions $\rho(t, x)$, $\theta(t, x)$, and $b_{j\nu}(t, x)$ are smooth in (t, x).

Finally, $u(t, x) = u(t, x, \lambda)$ in (7.6) has a similar expansion, where the leading factor $\lambda^{-1/3}$ above is replaced by $\lambda^{1/6}$, as will follow from results in Chapter 7.

The next order of complexity of a caustic is illustrated in Fig. 7.9. It arises when we alter our hypothesis on the curvature of level curves of φ. In this case, z is a point on a level curve at which κ is stationary, in fact a (nondegenerate) local maximum, such that $\kappa(z) = 1/t$. On nearby curves, this is not a locally maximum value of κ; the set where $\kappa = 1/t$ is denoted $\mathcal{K}_t$ and is mapped by F_t onto the caustic set $\mathcal{C}_t$, which is singular at the "cusp" $v = F_t(z)$. The asymptotic behavior of the functions (7.6) and (7.8) on a neighborhood of v is more complicated than (7.53). A discussion of this (and more complicated caustics) can be found in the last chapter of [GS]. See also [AVG] and [Dui].

In the last chapter of [GS] one can also find an analysis of the wave equation near a caustic of the fold type considered above, making use of a result similar to Lemma 7.1, but replacing the use of Proposition 7.2 by results in "microlocal analysis." The next chapter of this work includes a brief introduction to this area; other applications of microlocal analysis to topics in wave propagation can be found in Vols. 3–4 of [Hol], and in [Tay]. For other approaches to the type of caustic considered here, see [Lud] and references given therein.

Exercises

1. Fix $r > 0$. Let $\gamma_r \in \mathcal{E}'(\mathbb{R}^2)$ denote the unit-mass density on the circle of radius r:

$$\langle u, \gamma_r \rangle = \frac{1}{2\pi} \int_{-\pi}^{\pi} u(r\cos\theta, r\sin\theta)\, d\theta.$$

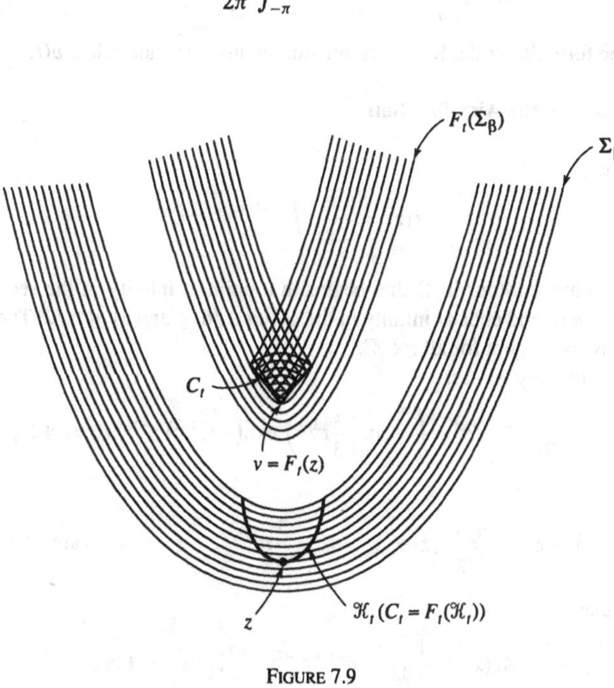

FIGURE 7.9

Show that there exist

$$\alpha_r(\lambda) \sim \lambda^{-1/2}(\alpha_{0r} + \alpha_{1r}\lambda^{-1} + \cdots),$$
$$\beta_r(\lambda) \sim \lambda^{1/2}(\beta_{0r} + \beta_{1r}\lambda^{-1} + \cdots),$$

such that, modulo $O(|\xi|^{-\infty})$,

$$\hat{\gamma}_r(\xi) = \alpha_r(|\xi|)\cos r|\xi| + \beta_r(|\xi|)\frac{\sin r|\xi|}{|\xi|}, \quad |\xi| \to \infty. \tag{7.55}$$

(*Hint*: Use the stationary phase method.)
Compare with formula (6.56) in Chapter 3, in the case $\nu = 0$, in view of the identity

$$\hat{\gamma}_r(\xi) = cJ_0(r|\xi|).$$

2. Give a proof that if $f \in C^\infty(\mathbb{R})$ and $f(x) = f(-x)$, then there exists $g \in C^\infty(\mathbb{R})$ such that $f(x) = g(x^2)$. (*Hint*: For fixed but large k, compare $f(\sqrt{x})$ with $\sum_{j\le k} f^{(2j)}(0)x^j/(2j)!$. Show that if $F \in C^\infty(\mathbb{R})$ vanishes at $x = 0$ to order $2k+1$, then $F(\sqrt{x})$ belongs to $C^k([0,\infty))$.)
3. Extend the result of Exercise 2 to show that if (7.15) is a fold and $f \in C^\infty(\Lambda_t)$ is invariant under the involution j of Λ_t, which interchanges points with the same image under π, then there exists $g \in C^\infty(\mathbb{R}^2)$ such that $f(x) = g(\pi(x))$. This result is used in the proof of Lemma 7.1 and that of Proposition 7.2.
For more material on folds, see [GoG].
4. Suppose $\mathbb{R}^2$ is replaced by $\mathbb{R}^3$ in (7.3). Analyze the following variant of (7.8):

$$v(t,x) = \frac{1}{4\pi t}\int_{|y-x|=t} u_0(y)\,dS(y) = \frac{t}{4\pi}\int_{S^2} a(x+y)e^{i\lambda\varphi(x+y)}\,dS(y).$$

Recall the formula for the Riemann function in this case, and relate $v(t,x)$ to $u(t,x)$.

Exercises on the Airy function

1. Show that

$$Ai(z) = \frac{1}{2\pi i}\int_L e^{v^3/3 - zv}\,dv,$$

where L is any contour in $\mathbb{C}$ that begins at a point at infinity in the sector $-\pi/2 \le \arg(v) \le -\pi/6$ and ends at infinity in the sector $\pi/6 \le \arg(v) \le \pi/2$. The integral on the right is convergent for all $z \in \mathbb{C}$.
2. Show that, for $|\arg z| < \pi$,

$$Ai(z) = \frac{1}{2\pi}e^{-(2/3)z^{3/2}}\int_0^\infty \cos\left(\frac{1}{3}t^{3/2}\right)\exp(-tz^{1/2})\,t^{-1/2}\,dt = \Psi(z)\,e^{-(2/3)z^{3/2}}, \tag{7.56}$$

where

$$\Psi(z) \sim z^{-1/4}\sum_{j=0}^\infty a_j z^{-3j/2}, \quad a_0 = \frac{1}{4}\pi^{-3/2}, \quad z \to \infty, \ |\arg z| \le \pi - \delta. \tag{7.57}$$

In particular,

$$Ai(x) \sim \frac{1}{4}\pi^{-3/2}x^{-1/4}e^{-(2/3)x^{3/2}}, \quad x \to +\infty.$$

3. If we set $A_\pm(z) = Ai\left(e^{\mp 2\pi i/3}z\right)$, show that $A_\pm(z)$ also satisfies the Airy equation. Evaluate $A_\pm(x)$ asymptotically as $x \to +\infty$, showing that $|A_\pm(x)| \to \infty$ as $x \to +\infty$. Show that any two of the functions Ai, A_+, A_- form a basis of solutions to Airy's equation $u''(z) - zu(z) = 0$. Show that

$$A_-(z) = \overline{A_+(\overline{z})}, \quad \text{and } Ai(z) = e^{\pi i/3}A_+(z) + e^{-\pi i/3}A_-(z). \tag{7.58}$$

Using Exercise 2, note that, for $x > 0$,

$$A_+(-x) = \Psi(e^{\pi i/3}x)e^{-(2/3)ix^{3/2}},$$

which, in light of (7.57), implies the second part of (7.42).

4. Show that

$$Ai(z) = \frac{1}{\pi}\left(\frac{z}{3}\right)^{1/2} K_{1/3}\left(\frac{2}{3}z^{3/2}\right), \quad |\arg z| < \frac{2\pi}{3}, \tag{7.59}$$

where $K_\nu(r)$ is the modified Bessel function, defined by (6.50) of Chapter 3 (and satisfying the modified Bessel equation (6.52)). (*Hint*: Denoting the right side of (7.59) by $u(z)$, show that $u(z)$ satisfies Airy's equation and has the same asymptotic behavior as $Ai(z)$, as $z \to +\infty$ in $\mathbb{R}^+$. For the behavior of $K_\nu(r)$ as $r \to +\infty$, use Exercise 2 in §6 of Chapter 3.)

5. Show that

$$Ai(0) = \frac{1}{2\pi}3^{-1/6}\Gamma\left(\frac{1}{3}\right) = \frac{3^{-2/3}}{\Gamma(\frac{2}{3})}, \tag{7.60}$$

as asserted in (7.44). (*Hint*: Show that, given $\nu > 0$,

$$K_\nu(r) \sim \Gamma(\nu)2^{\nu-1}r^{-\nu}, \quad \text{as } r \searrow 0.$$

For the last identity in (7.60), use $\Gamma(z)\Gamma(1-z) = \pi/(\sin \pi z)$.)

6. With $A_\pm(z)$ as in Exercise 3, establish the Wronskian relation

$$A'_+(z)A_-(z) - A_+(z)A'_-(z) = \frac{1}{2\pi i}.$$

(*Hint*: Once you show that the right side is a constant c, use the asymptotic behavior of $A_\pm(x)$ as $x \to +\infty$, obtained via Exercise 2, and the corresponding asymptotic behavior of $A'_\pm(x)$, to evaluate c.)

7. Deduce from Exercises 5 and 6 that

$$Ai'(0) = -\frac{1}{2\pi}3^{1/6}\Gamma\left(\frac{2}{3}\right) = -\frac{3^{-1/3}}{\Gamma(\frac{1}{3})}, \tag{7.61}$$

as asserted in (7.44).

8. Show that

$$\frac{\Gamma(\frac{1}{3})}{\Gamma(\frac{2}{3})} = \frac{2^{-2/3}}{\sqrt{\pi}}\Gamma\left(\frac{1}{6}\right), \quad \frac{\Gamma(\frac{2}{3})}{\Gamma(\frac{1}{3})} = \frac{2^{-1/3}}{\sqrt{\pi}}\Gamma\left(\frac{5}{6}\right).$$

Using $\Gamma(1/3)\Gamma(2/3) = 2\pi/\sqrt{3}$, relate $\Gamma(1/3)^2$ and $\Gamma(2/3)^2$ to $\Gamma(1/6)$. (*Hint*: Use the duplication formula for the gamma function, established in Chapter 3.)

9. Consider the problem of deriving numerical approximations to $\Gamma(1/3)$ and $\Gamma(2/3)$. Try to obtain 10-digit approximations to these quantities. Then write a computer program to produce the graph of $y = Ai(x)$, shown in Fig. 7.8, by solving (7.43) numerically.

A. Some Banach spaces of harmonic functions

If B is the unit ball in $\mathbb{R}^k$, consider the space $\mathfrak{X}_j$ of harmonic functions f on B such that

$$N_j(f) = \sup_{x\in B} \delta(x)^j \, |f(x)| \tag{A.1}$$

is finite, where $\delta(x) = 1-|x|$ is the distance of x from ∂B. In case $k = 2n$ and we identify $\mathbb{R}^{2n}$ with $\mathbb{C}^n$, via $z_\ell = x_\ell + ix_{n+\ell}$, the space $\mathfrak{H}_j$ of holomorphic functions on B such that (A.1) is finite is a closed, linear subspace of $\mathfrak{X}_j$. For results in §4, it is useful both to know that

$$\frac{\partial}{\partial z_\ell} : \mathfrak{H}_j \longrightarrow \mathfrak{H}_{j+1} \tag{A.2}$$

and to estimate its norm. It is just as convenient to estimate the norm of

$$\partial_\ell : \mathfrak{X}_j \longrightarrow \mathfrak{X}_{j+1}, \tag{A.3}$$

where $\partial_\ell = \partial/\partial x_\ell$; then the desired estimate on (A.2) will follow from that on (A.3).

Given $x \in B$, let $B_\rho(x)$ be the ball of radius ρ centered at x; take $\rho \in (0, \delta(x))$. Then, as a consequence of the Poisson integral formula for functions harmonic on a ball (see (3.34) of Chapter 5), we have

$$\partial_\ell u(x) = \frac{k-1}{\rho^2} \operatorname{Avg}_{\partial B_\rho(x)} \big\{(y_\ell - x_\ell)u(y)\big\} \tag{A.4}$$

if u is harmonic on B. Now, for $y \in \partial B_\rho(x)$, $|y_\ell - x_\ell| \le \rho$; furthermore, $\delta(y) \ge \delta(x) - \rho$. If we take $\rho = \beta\delta(x)$, $\beta \in (0,1)$, we obtain

$$\begin{aligned} |\partial_\ell u(x)| &\le \frac{k-1}{\rho^2} \cdot \rho \cdot \big[(1-\beta)\delta(x)\big]^{-j} N_j(u) \\ &= \frac{k-1}{\beta(1-\beta)^j}\, \delta(x)^{-(j+1)} N_j(u), \end{aligned} \tag{A.5}$$

and hence

$$N_{j+1}(\partial_\ell u) \le \frac{k-1}{\beta(1-\beta)^j} N_j(u), \tag{A.6}$$

for $u \in \mathfrak{X}_j$. The factor on the right is minimized at $\beta = 1/(j+1)$. Using the power series expansion of $\log(1-\varepsilon)$, one readily verifies that

$$\Big(1 - \frac{1}{j+1}\Big)^{-j} \le e,$$

so, for all $j \ge 0$, $u \in \mathfrak{X}_j$,

$$N_{j+1}(\partial_\ell u) \le \gamma_k\, (j+1) N_j(u), \quad \gamma_k = (k-1)e. \tag{A.7}$$

Since $\partial/\partial z_\ell = (1/2)(\partial_\ell - i\partial_{n+\ell})$, we also have, for all $j \ge 0$, $u \in \mathfrak{H}_j$,

$$N_{j+1}\Big(\frac{\partial u}{\partial z_\ell}\Big) \le \gamma_{2n}\, (j+1) N_j(u). \tag{A.8}$$

Note that repeated application of (A.7) yields

$$N_m(D^\alpha u) \le \gamma_k^m\ (m!)\ N_0(u), \quad |\alpha| = m, \tag{A.9}$$

for $u \in \mathfrak{X}_0$. This estimate of course implies the well-known real analyticity of harmonic functions. In order for such analyticity to follow from (A.7), it is crucial to have linear dependence in j of the factor on the right side of (A.7). The fact that we can establish (A.7) and (A.8) in this form also makes it an effective tool in the proof of the Cauchy-Kowalewsky theorem, in §4.

B. The stationary phase method

The one-dimensional stationary phase method was derived in §7. Here we discuss the multidimensional case. If M is a Riemannian manifold, $F \in C_0^\infty(M)$, and $\psi \in C^\infty(M)$ is real-valued, with only nondegenerate critical points, there is a formula for the asymptotic behavior of

$$I(\tau) = \int F(x)\, e^{i\tau\psi(x)}\, dV(x) \tag{B.1}$$

as $\tau \to \infty$, given by the stationary phase method, which we now derive. First, using a partition of unity supported on coordinate neighborhoods, we can write (B.1) as a finite sum of integrals of the form

$$J(\tau) = \int f(x)\, e^{i\tau\varphi(x)}\, dx, \tag{B.2}$$

where $f \in C_0^\infty(\mathbb{R}^n)$ and φ has either no critical points on supp f or only one critical point, located at $x = 0$.

Lemma B.1. *If φ has no critical point on supp f, then $J(\tau)$ is rapidly decreasing as $\tau \to \infty$.*

Proof. Cover supp f with open sets on which, by a change of variable, $\varphi(x)$ becomes linear, that is, $\varphi(x) = \xi \cdot x + c$, $\xi \neq 0$. Then $J(\tau)$ is converted to a sum of integrals of the form

$$\int f_j(x) e^{i\tau x\cdot\xi + ic\tau}\, dx = e^{ic\tau} \tilde{f}_j(\tau\xi),$$

with $\tilde{f}_j \in \mathcal{S}(\mathbb{R}^n)$. If $\xi \neq 0$, the rapid decrease as $\tau \to \infty$ is clear.

It remains to consider the case of (B.2) when $\varphi(x)$ has a single critical point, at $x = 0$, which is nondegenerate. In such a case, there exists a coordinate chart near 0 such that

$$\varphi(x) = Ax \cdot x + c,$$

where A is the nonsingular, real, symmetric matrix $(A_{jk}) = (1/2)(\partial_j \partial_k \varphi(0))$, and $c \in \mathbb{R}$. We can assume this holds on supp f. That this can be done is known as the

Morse lemma; a proof is given in §8 of Appendix C. Thus it remains to consider

$$e^{ic\tau}\, K(\tau) = e^{ic\tau} \int g(x)\, e^{i\tau Ax\cdot x}\, dx, \tag{B.3}$$

as $\tau \to +\infty$, where $g \in C_0^\infty(\mathbb{R}^n)$. Using a rotation, we could assume $Ax \cdot x = \sum a_j x_j^2$, where the factors a_j are the eigenvalues of A.

Note that if $P(\xi) = B\xi \cdot \xi$, where B is an invertible, symmetric, real matrix, then

$$e^{itP(D)}\delta(x) = (2\pi)^{-n/2}\, \mathcal{F}\big(e^{itP}\big)(x). \tag{B.4}$$

By diagonalizing B and looking at the one-dimensional cases, $e^{itb\xi^2}$, via techniques used in (6.42) of Chapter 3, we obtain

$$e^{-itP(D)}\delta(x) = \det\big(4\pi i B\big)^{-1/2}\, t^{-n/2}\, e^{iAx\cdot x/t}, \quad A = (4B)^{-1}, \tag{B.5}$$

for $t > 0$, where the determinant is calculated as

$$\lim_{\varepsilon \searrow 0} \det\big(4\pi i (B - i\varepsilon)\big)^{-1/2}, \tag{B.6}$$

using analytic continuation, and the convention that $\det(+4\pi\varepsilon I)^{-1/2} > 0$, for real $\varepsilon > 0$.

Thus, for $K(\tau)$ in (B.3), we have

$$K(t^{-1}) = C(A)\, t^{n/2}\, u(t,0), \tag{B.7}$$

where $C(A) = \det(4\pi i B)^{1/2}$, $4B = A^{-1}$, and $u(t,x)$ solves a generalized Schrödinger equation:

$$u(t,x) = e^{-itP(D)} g(x). \tag{B.8}$$

Given $g \in C_0^\infty(\mathbb{R}^n)$, we know from material of Chapter 3, §5, that

$$u \in C^\infty\big([0,\infty), \mathcal{S}(\mathbb{R}^n)\big) \subset C^\infty\big([0,\infty) \times \mathbb{R}^n\big).$$

Thus we have, for $t \searrow 0$, an expansion

$$u(t,0) \sim \sum_{j\ge 0} a_j t^j, \tag{B.9}$$

with

$$a_j = \frac{1}{j!}\Big(\frac{\partial}{\partial t}\Big)^j u(0,0) = \frac{(-i)^j}{j!} P(D)^j g(0). \tag{B.10}$$

Consequently, for (B.3) we have

$$e^{ic\tau}\, K(\tau) \sim C\tau^{-n/2}\Big(a_0 + a_1\tau^{-1} + a_2\tau^{-2} + \cdots\Big) e^{ic\tau}, \quad \tau \to +\infty, \tag{B.11}$$

where $C = C(A)$ is as in (B.7) and the factors a_j are given by (B.10). We can conclude that $I(\tau)$ in (B.1) is asymptotic to a finite sum of such expansions, under the hypotheses made on $F(x)$ and $\psi(x)$. Let us summarize what has been established.

Proposition B.2. *If $F \in C_0^\infty(M)$ and $\psi \in C^\infty(M)$ is real-valued, with only non-degenerate critical points, at $x_1, \dots, x_k$, then, as $\tau \to +\infty$, the integral (B.1) has the asymptotic behavior*

$$I(\tau) \sim \sum_{j=1}^{k} A_j(\tau)\tau^{-n/2}e^{i\tau\psi(x_j)},$$

(B.12)

$$A_j(\tau) \sim a_{j0} + a_{j1}\tau^{-1} + a_{j2}\tau^{-2} + \cdots .$$

References

[AVG] V. Arnold, A. Varchenko, and S. Gusein-Zade, *Singularities of Differentiable Mappings, I, Classification of Critical Points, Caustics, and Wave Fronts*, Birkhauser, Boston, 1985; II, *Monodromy and Asymptotics of Integrals*, Nauka, Moscow 1984.

[Az] R. Azencott, Behavior of diffusion semi-groups at infinity, *Bull. Soc. Math. France* 102(1974), 193–240.

[BW] M. Born and E. Wolf, *Principles of Optics*, Pergammon Press, New York, 1980.

[CGT] J. Cheeger, M. Gromov, and M. Taylor, Finite propagation speed, kernel estimates for functions of the Laplace operator, and the geometry of complete Riemannian manifolds, *J. Diff. Geom.* 17(1982), 15–83.

[CLY] S. Cheng, P. Li, and S.-T. Yau, On the upper estimate of the heat kernel of a complete Riemannian manifold, *Amer. J. Math.* 103(1981), 1021–1063.

[CFU] C. Chester, B. Friedman, and F. Ursell, An extension of the method of steepest descents, *Proc. Camb. Phil. Soc.* 53(1957), 599–611.

[CLx] R. Courant and P. Lax, The propagation of discontinuities in wave motion, *Proc. NAS, USA* 42(1956), 872–876.

[Dav] E. B. Davies, *Heat Kernels and Spectral Theory*, Cambridge Univ. Press, Cambridge, 1989.

[Dui] J. J. Duistermaat, Oscillatory integrals, Lagrange immersions, and unfolding of singularities, *Comm. Pure Appl. Math.* 27(1974), 207–281.

[DS] N. Dunford and J. Schwartz, *Linear Operators*, Wiley, New York, 1958.

[Fedo] M. Fedoryuk, Asymptotic methods in analysis, *Encycl. of Math. Sciences*, Vol. 13, pp. 83–191. Springer-Verlag, New York, 1989.

[Frl] F. G. Friedlander, *Sound Pulses*, Cambridge Univ. Press, London, 1958.

[Frd] A. Friedman, *Partial Differential Equations*, Holt, Rinehart, and Winston, New York, 1969.

[GoG] M. Golubitsky and V. Guillemin, *Stable Mappings and Their Singularities*, Springer-Verlag, New York, 1973.

[GS] V. Guillemin and S. Sternberg, *Geometric Asymptotics*, AMS, Providence, R. I., 1977.

[Had] J. Hadamard, *Le Problème de Cauchy et les Equations aux Derivées Partielles Linéaires Hyperboliques*, Hermann, Paris, 1932.

[HP] E. Hille and R. Phillips, *Functional Analysis and Semi-groups*, Colloq. Publ. AMS, 1957.

[Ho] L. Hörmander, *The Analysis of Linear Partial Differential Operators*, Vols. 1,2, Springer-Verlag, New York, 1983; Vols. 3,4, Springer-Verlag, New York, 1985.

[J] F. John, *Partial Differential Equations*, Springer-Verlag, New York, 1982.

[K] T. Kato, *Perturbation Theory for Linear Operators*, Springer-Verlag, New York, 1966.

[Lad] O. Ladyzhenskaya, *The Boundary Value Problems of Mathematical Physics*, Springer-Verlag, New York, 1985.

[Lx] P. Lax, Asymptotic solutions of oscillatory initial value problems, *Duke Math. J.* 24(1957), 627–646.

[LP] P. Lax and R. Phillips, *Scattering Theory*, Academic Press, New York, 1967.

[Leb] N. Lebedev, *Special Functions and their Applications*, Dover, New York, 1972.

[Ler] J. Leray, *Hyperbolic Differential Equations*, Princeton Univ. Press, Princeton, N. J., 1952.

[Lud] D. Ludwig, Uniform asymptotic expansions at a caustic, *Comm. Pure Appl. Math.* 19(1966), 215–250.

[MeT] R. Melrose and M. Taylor, *Boundary Problems for Wave Equations with Grazing and Gliding Rays*, in preparation.

[Mil] J. C. P. Miller, *The Airy Integral*, British Assn. for Advancement of Science, Math. Tables, Cambridge Univ. Press, Cambridge, 1946.

[Olv] F. W. J. Olver, *Asymptotics and Special Functions*, Academic Press, New York, 1974.

[Ph] R. Phillips, Dissipative operators and hyperbolic systems of partial differential equations, *Trans. AMS* 90(1959), 193–254.

[PW] M. Protter and H. Weinberger, *Maximum Principles in Differential Equations*, Springer-Verlag, New York, 1984.

[RS] M. Reed and B. Simon, *Methods of Mathematical Physics*, Academic Press, New York, Vols. 1,2, 1975; Vols. 3,4, 1978.

[RN] F. Riesz and B. Sz. Nagy, *Functional Analysis*, Ungar, New York, 1955.

[Stk] I. Stakgold, *Boundary Value Problems of Mathematical Physics*, Macmillan, New York, 1968.

[Str] R. Strichartz, Analysis of the Laplacian on a complete Riemannian manifold, *J. Funct. Anal.* 52(1983), 48–79.

[Tay] M. Taylor, *Pseudodifferential Operators*, Princeton Univ. Press, Princeton, N. J., 1981.

[Tr1] F. Treves, *Linear Partial Differential Equations with Constant Coefficients*, Gordon and Breach, New York, 1967.

[Tr2] F. Treves, *Basic Linear Partial Differential Equations*, Academic Press, New York, 1975.

[Yo] K. Yosida, *Functional Analysis*, Springer-Verlag, New York, 1965.

A

Outline of Functional Analysis

Introduction

Problems in PDE have provided a major impetus for the development of functional analysis. Here, we present some basic results, which are useful for the development of such subjects as distribution theory and Sobolev spaces, discussed in Chapters 3 and 4; the spectral theory of compact and of unbounded operators, applied to elliptic PDE in Chapter 5; the theory of Fredholm operators and their indices, needed for the study of the Atiyah-Singer index theorem in Chapter 10; and the theory of semigroups, of particular value in Chapter 9 on scattering theory, and also germane to studies of evolution equations in Chapters 3 and 6. Indeed, what is thought of as the subject of functional analysis naturally encompasses some of the development of these chapters as well as the material presented in this appendix. One particular case of this is the spectral theory of Chapter 8. In fact, it is there that we present a proof of the spectral theorem for general self-adjoint operators. One reason for choosing to do it this way is that my favorite approach to the spectral theorem uses Fourier analysis, which is not applied in this appendix, though some of the exercises make contact with it. Thus in this appendix the spectral theorem is proved only for compact operators, an extremely simple special case. On the other hand, it is hoped that by the time one gets through the Fourier analysis as developed in Chapter 3, the presentation of the general spectral theorem in Chapter 8 will appear to be very simple too.

1. Banach spaces

A *Banach space* is a complete, normed, linear space. A *norm* on a linear space V is a positive function $\|v\|$ having the properties

$$\begin{aligned} &\|av\| = |a| \cdot \|v\| \text{ for } v \in V,\ a \in \mathbb{C} \text{ (or } \mathbb{R}), \\ &\|v + w\| \le \|v\| + \|w\|, \\ &\|v\| > 0 \text{ unless } v = 0. \end{aligned} \tag{1.1}$$

The second of these conditions is called the *triangle inequality*. Given a norm on V, there is a distance function $d(u, v) = \|u - v\|$, making V a metric space.

A metric space is a set X, with distance function $d : X \times X \to \mathbb{R}^+$, satisfying

$$\begin{aligned} &d(u, v) = d(v, u), \\ (1.2) \qquad &d(u, v) \le d(u, w) + d(w, v), \\ &d(u, v) > 0 \text{ unless } u = v. \end{aligned}$$

A sequence (u_j) is Cauchy provided $d(v_n, v_m) \to 0$ as $m, n \to \infty$; completeness is the property that any Cauchy sequence converges. Further background on metric spaces is given in §1 of Appendix B.

We list some examples of Banach spaces. First, let X be any compact metric space, that is, a metric space with the property that any sequence (x_n) has a convergent subsequence. Then $C(X)$, the space of continuous functions on X, is a Banach space, with norm

$$(1.3) \qquad \|u\|_{\sup} = \sup\{|u(x)| : x \in X\}.$$

Also, for any $\alpha \in [0, 1]$, we set

$$(1.4) \quad \operatorname{Lip}^\alpha(X) = \{u \in C(X) : |u(x) - u(y)| \le C\, d(x, y)^\alpha \text{ for all } x, y \in X\}.$$

This is a Banach space, with norm

$$(1.5) \qquad \|u\|_\alpha = \|u\|_{\sup} + \sup_{x,y\in X} \frac{|u(x) - u(y)|}{d(x, y)^\alpha}.$$

$\operatorname{Lip}^0(X) = C(X)$; the space $\operatorname{Lip}^1(X)$ is typically denoted $\operatorname{Lip}(X)$. For $\alpha \in (0, 1)$, $\operatorname{Lip}^\alpha(X)$ is frequently denoted $C^\alpha(X)$. In all these cases, it is straightforward to verify the conditions (1.1) on the proposed norms and to establish completeness.

Related spaces arise when X is specialized to be a compact Riemannian manifold. We have $C^k(M)$, the space of functions whose derivatives of order $\le k$ are continuous on M. Norms on $C^k(M)$ can be constructed as follows. Pick $Z_1, \dots, Z_N$, smooth vector fields on M that span T_pM at each $p \in M$. Then we can set

$$(1.6) \qquad \|u\|_{C^k} = \sum_{\ell \le k} \|Z_{j_1} \cdots Z_{j_\ell} u\|_{\sup}.$$

If one replaces the sup norm on the right by the C^α-norm (1.5), for some $\alpha \in (0, 1)$, one has a norm for the Banach space $C^{k,\alpha}(M)$.

More subtle examples of Banach spaces are the L^p-spaces, defined as follows. First take $p = 1$. Let (X, μ) be a measure space. We say a measurable function f belongs to $\mathcal{L}^1(X, \mu)$ provided

$$(1.7) \qquad \int_X |f(x)|\, d\mu(x) < \infty.$$

Elements of $L^1(X,\mu)$ consist of equivalence classes of elements of $\mathcal{L}^1(X,\mu)$, where we say

$$f \sim \tilde{f} \Leftrightarrow f(x) = \tilde{f}(x), \quad \text{for } \mu\text{-almost every } x. \tag{1.8}$$

With a slight abuse of notation, we denote by f both a measurable function in $\mathcal{L}^1(X,\mu)$ and its equivalence class in $L^1(X,\mu)$. Also, we say that f, defined only almost everywhere on X, belongs to $L^1(X,\mu)$ if there exists $\tilde{f} \in \mathcal{L}^1(X,\mu)$ such that $\tilde{f} = f$ a.e. The norm $\|f\|_{L^1}$ is given by (1.7); it is easy to see that this norm has the properties (1.1).

The proof of completeness of $L^1(X,\mu)$ makes use of the following key convergence results in measure theory.

Monotone convergence theorem. *If $f_j \in \mathcal{L}^1(X,\mu)$, $0 \le f_1(x) \le f_2(x) \le \cdots$, and $\|f_j\|_{L^1} \le C < \infty$, then $\lim_{j\to\infty} f_j(x) = f(x)$, with $f \in L^1(X,\mu)$ and $\|f_j - f\|_{L^1} \to 0$ as $j \to \infty$.*

Dominated convergence theorem. *If $f_j \in \mathcal{L}^1(X,\mu)$, $\lim f_j(x) = f(x)$, μ-a.e., and there is an $F \in \mathcal{L}^1(X,\mu)$ such that $|f_j(x)| \le F(x)$ μ-a.e., for all j, then $f \in \mathcal{L}^1(X,\mu)$ and $\|f_j - f\|_{L^1} \to 0$.*

To show that $L^1(X,\mu)$ is complete, suppose (f_n) is Cauchy in L^1. Passing to a subsequence, we can assume $\|f_{n+1} - f_n\|_{L^1} \le 2^{-n}$. Consider the infinite series

$$f_1(x) + \sum_{n=1}^{\infty} \big[f_{n+1}(x) - f_n(x)\big]. \tag{1.9}$$

Now the partial sums are dominated by

$$G_m(x) = \sum_{n=1}^{m} |f_{n+1}(x) - f_n(x)|,$$

and since $0 \le G_1 \le G_2 \le \cdots$ and $\|G_m\|_{L^1} \le \sum 2^{-n} \le 1$, we deduce from the monotone convergence theorem that $G_m \nearrow G$ μ-a.e. and in L^1-norm. Hence the infinite series (1.9) is convergent a.e., to a limit $f(x)$, and via the dominated convergence theorem we deduce that $f_n \to f$ in L^1-norm. This proves completeness.

Continuing with a description of L^p-spaces, we define $\mathcal{L}^\infty(X,\mu)$ to consist of bounded, measurable functions, $L^\infty(X,\mu)$ to consist of equivalence classes of such functions, via (1.8), and we define $\|f\|_{L^\infty}$ to be the smallest sup of $\tilde{f} \sim f$. It is easy to show that $L^\infty(X,\mu)$ is a Banach space.

For $p \in (1,\infty)$, we define $\mathcal{L}^p(X,\mu)$ to consist of measurable functions f such that

$$\Big[\int_X |f(x)|^p \, d\mu(x)\Big]^{1/p} \tag{1.10}$$

is finite. $L^p(X, \mu)$ consists of equivalence classes, via (1.8), and the L^p-norm $\|f\|_{L^p}$ is given by (1.10). This time it takes a little work to verify the triangle inequality. That this holds is the content of *Minkowski's inequality*:

$$\|f+g\|_{L^p} \le \|f\|_{L^p} + \|g\|_{L^p}. \tag{1.11}$$

One neat way to establish this is by the following characterization of the L^p-norm. Suppose p and q are related by

$$\frac{1}{p} + \frac{1}{q} = 1. \tag{1.12}$$

We claim that if $f \in L^p(X, \mu)$,

$$\|f\|_{L^p} = \sup\big\{\|fh\|_{L^1} : h \in L^q(X, \mu),\ \|h\|_{L^q} = 1\big\}. \tag{1.13}$$

We can apply (1.13) to $f + g$, which belongs to $L^p(X, \mu)$ if f and g do, since $|f+g|^p \le 2^p(|f|^p + |g|^p)$. Given this, (1.11) follows easily from the inequality $\|(f+g)h\|_{L^1} \le \|fh\|_{L^1} + \|gh\|_{L^1}$.

The identity (1.13) can be regarded as two inequalities. The "$\le$" part can be proven by choosing $h(x)$ to be an appropriate multiple $C|f(x)|^{p-1}$. We leave this as an exercise. The converse inequality, "$\ge$," is a consequence of Hölder's inequality:

$$\int |f(x)g(x)|\, d\mu(x) \le \|f\|_{L^p}\|g\|_{L^q}, \quad \frac{1}{p} + \frac{1}{q} = 1. \tag{1.14}$$

Hölder's inequality can be proved via the following inequality for positive numbers:

$$ab \le \frac{a^p}{p} + \frac{b^q}{q}, \quad a, b > 0, \tag{1.15}$$

assuming that $p \in (1, \infty)$ and (1.12) holds; (1.15) is equivalent to

$$x^{1/p}y^{1/q} \le \frac{x}{p} + \frac{y}{q}, \quad x, y > 0. \tag{1.16}$$

Since both sides of this are homogeneous of degree 1 in (x, y), it suffices to prove it for $y = 1$, that is, to prove that $x^{1/p} \le x/p + 1/q$ for $x \in [0, \infty)$. Now $\varphi(x) = x^{1/p} - x/p$ can be maximized by elementary calculus; one finds a unique maximum at $x = 1$, with $\varphi(1) = 1 - 1/p = 1/q$. This establishes (1.16), hence (1.15). Applying this to the integrand in (1.14) gives

$$\int |f(x)g(x)|\, d\mu(x) \le \frac{1}{p}\|f\|_{L^p}^p + \frac{1}{q}\|g\|_{L^q}^q. \tag{1.17}$$

This looks weaker than (1.14), but now replace f by tf and g by $t^{-1}g$, so that the left side of (1.17) is dominated by

$$\frac{t^p}{p}\|f\|_{L^p}^p + \frac{1}{qt^q}\|g\|_{L^q}^q.$$

Minimizing over $t \in (0, \infty)$ then gives Hölder's inequality. Consequently, (1.10) defines a norm on $L^p(X, \mu)$. Completeness follows as in the $p = 1$ case discussed above.

We next give a discussion of one important method of manufacturing new Banach spaces from old. Namely, suppose V is a Banach space, W a closed linear subspace. Consider the linear space $L = V/W$, with norm

$$\|[v]\| = \inf \{\|v - w\| : w \in W\}, \tag{1.18}$$

where $v \in V$, and $[v]$ denotes its class in V/W. It is easy to see that (1.18) defines a norm on V/W. We record a proof of the following.

Proposition 1.1. *If V is a Banach space and W is a closed linear subspace, then V/W, with norm (1.18), is a Banach space.*

It suffices to prove that V/W is complete. We use the following; compare the use of (1.9) in the proof of completeness of $L^1(X, \mu)$.

Lemma 1.2. *A normed linear space L is complete provided the hypothesis*

$$x_j \in L, \quad \sum_{j=1}^{\infty} \|x_j\| < \infty,$$

implies that $\sum_{j=1}^{\infty} x_j$ converges in L.

Proof. If (y_k) is Cauchy in L, take a subsequence so that $\|y_{k+1} - y_k\| \le 2^{-k}$, and consider $y_1 + \sum_{j=1}^{\infty}(y_{j+1} - y_j)$.

To prove Proposition 1.1 now, say $[v_j] \in V/W$, $\sum \|[v_j]\| < \infty$. Then pick $w_j \in W$ such that $\|v_j - w_j\| \le \|[v_j]\| + 2^{-j}$, to get $\sum_{j=1}^{\infty} \|v_j - w_j\| < \infty$. Hence $\sum (v_j - w_j)$ converges in V, to a limit v, and it follows that $\sum [v_j]$ converges to $[v]$ in V/W.

Note that if W is a proper closed, linear subspace of V, given $v \in V \setminus W$, we can pick $w_n \in W$ such that $\|v - w_n\| \to \operatorname{dist}(v, W)$. Normalizing $v - w_n$ produces $v_n \in V$ such that the following holds.

Lemma 1.3. *If W is a proper closed, linear subspace of a Banach space V, there exist $v_n \in V$ such that*

$$\|v_n\| = 1, \quad \mathit{dist}(v_n, W) \nearrow 1. \tag{1.19}$$

In Proposition 2.1 we will produce an important sharpening of this for Hilbert spaces. For now we remark on the following application.

Proposition 1.4. *If V is an infinite-dimensional Banach space, then the closed unit ball $B_1 \subset V$ is not compact.*

Proof. If V_j is an increasing sequence of spaces, of dimension j, by (1.19) we can obtain $v_j \in V_j$, $\|v_j\| = 1$, each pair a distance $\geq 1/2$; thus (v_j) has no convergent subsequence.

It is frequently useful to show that a certain linear subspace L of a Banach space V is *dense*. We give a few important cases of this here.

Proposition 1.5. *If μ is a Borel measure on a compact metric space X, then $C(X)$ is dense in $L^p(X, \mu)$ for each $p \in [1, \infty)$.*

Proof. First, let K be any compact subset of X. The functions

$$f_{K,n}(x) = \big[1 + n \operatorname{dist}(x, K)\big]^{-1} \in C(X) \tag{1.20}$$

are all ≤ 1 and decrease monotonically to the characteristic function χ_K equal to 1 on K, 0 on $X \setminus K$. The monotone convergence theorem gives $f_{K,n} \to \chi_K$ in $L^p(X, \mu)$ for $1 \leq p < \infty$. Now let $A \subset X$ be any measurable set. Any Borel measure on a compact metric space is regular, that is,

$$\mu(A) = \sup\{\mu(K) : K \subset A,\ K \text{ compact}\}. \tag{1.21}$$

Thus there exists an increasing sequence K_j of compact subsets of A such that $\mu(A \setminus \cup_j K_j) = 0$. Again, the monotone convergence theorem implies $\chi_{K_j} \to \chi_A$ in $L^p(X, \mu)$ for $1 \leq p < \infty$. Thus all *simple* functions on X are in the closure of $C(X)$ in $L^p(X, \mu)$ for $p \in [1, \infty)$. The construction of $L^p(X, \mu)$ directly shows that each $f \in L^p(X, \mu)$ is a norm limit of simple functions, so the result is proved.

This result is easily extended to give the following:

Corollary 1.6. *If X is a metric space that is locally compact and a countable union of compact X_j, and μ is a (locally finite) Borel measure on X, then the space $C_{00}(X)$ of compactly supported, continuous functions on X is dense in $L^p(X, \mu)$ for each $p \in [1, \infty)$.*

Further extensions, involving more general locally compact spaces, can be found in [Lo].

The following is known as the Weierstrass approximation theorem.

Theorem 1.7. *If $I = [a, b]$ is an interval in $\mathbb{R}$, the space $\mathcal{P}$ of polynomials in one variable is dense in $C(I)$.*

There are many proofs of this. One close to Weierstrass's original (and my favorite) goes as follows. Given $f \in C(I)$, extend it to be continuous and compactly supported on $\mathbb{R}$; convolve this with a highly peaked Gaussian; and approximate the result by power series. For a more detailed sketch, in the context of other useful applications of highly peaked Gaussians, see Exercises 14 and 15 in §3 of Chapter 3.

The following generalization is known as the *Stone-Weierstrass theorem.*

Theorem 1.8. *Let X be a compact Hausdorff space and $\mathcal{A}$ a subalgebra of $C_{\mathbb{R}}(X)$, the algebra of real-valued, continuous functions on X. Suppose that $1 \in \mathcal{A}$ and that $\mathcal{A}$ separates points of X, that is, for distinct $p, q \in X$, there exists $h_{pq} \in \mathcal{A}$ with $h_{pq}(p) \neq h_{pq}(q)$. Then the closure $\overline{\mathcal{A}}$ is equal to $C_{\mathbb{R}}(X)$.*

We sketch a proof of Theorem 1.8, making use of Theorem 1.7, which implies that if $f \in \overline{\mathcal{A}}$ and $\varphi : \mathbb{R} \to \mathbb{R}$ is continuous, then $\varphi \circ f \in \overline{\mathcal{A}}$. Consequently, if $f_j \in \overline{\mathcal{A}}$, then $\sup(f_1, f_2) = (1/2)|f_1 - f_2| + (1/2)(f_1 + f_2) \in \overline{\mathcal{A}}$.

The hypothesis of separating points implies that, for distinct $p, q \in X$, there exists $f_{pq} \in \overline{\mathcal{A}}$, equal to 1 at p, 0 at q. Applying appropriate φ, we can arrange also that $0 \leq f_{pq}(x) \leq 1$ on X and that f_{pq} is 1 near p and 0 near q. Taking infima, we can obtain $f_{pU} \in \overline{\mathcal{A}}$, equal to 1 on a neighborhood of p and equal to 0 off a given neighborhood U of p. Applying sups to these, we obtain, for each compact $K \subset X$ and open $U \supset K$, a function $g_{KU} \in \overline{\mathcal{A}}$ such that g_{KU} is 1 on K, 0 off U, and $0 \leq g_{KU}(x) \leq 1$ on X. Once we have gotten this far, it is easy to approximate any continuous $u \geq 0$ on X by a sup of (positive constants times) such g_{KU}, and from there it is easy to prove the theorem.

Theorem 1.8 has a complex analogue. In that case, we add the assumption that $f \in \mathcal{A} \Rightarrow \overline{f} \in \mathcal{A}$ and conclude that $\overline{\mathcal{A}} = C(X)$. This is easily reduced to the real case.

Exercises

1. Let $\mathcal{L}$ be the subspace of $C(S^1)$ consisting of finite linear combinations of the exponentials $e^{in\theta}$, $n \in \mathbb{Z}$. Use the Stone-Weierstrass theorem to show that $\mathcal{L}$ is dense in $C(S^1)$.
2. Show that the space of finite linear combinations of the functions
$$E_\zeta(t) = e^{-\zeta t},$$
as ζ ranges over $(0, \infty)$, is dense in $C_0(\mathbb{R}^+)$, the space of continuous functions on $\mathbb{R}^+ = [0, \infty)$, vanishing at infinity. (*Hint*: Make a slight generalization of the Stone-Weierstrass theorem.)
3. Given $f \in L^1(\mathbb{R}^+)$, the Laplace transform
$$(\mathcal{L}f)(\zeta) = \int_0^\infty e^{-\zeta t} f(t)\, dt$$
is defined and holomorphic for $\operatorname{Re} \zeta > 0$. Suppose $(\mathcal{L}f)(\zeta)$ vanishes for ζ on some open subset of $(0, \infty)$. Show that $f = 0$, using Exercise 2. (*Hint*: First show that $(\mathcal{L}f)(\zeta)$ is identically zero.)
4. Let I be a compact interval, V a Banach space, and $f : I \to V$ a continuous function. Show that the Riemann integral $\int_I f(x)\, dx$ is well-defined. Formulate and establish the fundamental theorem of calculus for V-valued functions. Formulate and verify appropriate basic results on multidimensional integrals of V-valued functions.

5. Let $\Omega \subset \mathbb{C}$ be open, V a (complex) Banach space, and $f : \Omega \to V$. We say f is holomorphic if it is a C^1-map and, for each $z \in \Omega$, $Df(z)$ is $\mathbb{C}$-linear. Establish for such V-valued holomorphic functions the Cauchy integral theorem, the Cauchy integral formula, power-series expansions, and the Liouville theorem.

2. Hilbert spaces

A *Hilbert space* is a complete inner-product space. That is to say, first the space H is a linear space provided with an inner product, denoted (u, v), for u and v in H, satisfying the following defining conditions:

$$(2.1) \qquad \begin{aligned} (au_1 + u_2, v) &= a(u_1, v) + (u_2, v), \\ (u, v) &= \overline{(v, u)}, \\ (u, u) &> 0 \text{ unless } u = 0. \end{aligned}$$

To such an inner product is assigned a norm, by

$$(2.2) \qquad \|u\| = \sqrt{(u, u)}.$$

To establish that the triangle inequality holds for $\|u+v\|$, we can expand $\|u+v\|^2 = (u+v, u+v)$ and deduce that this is $\le \big[\|u\|+\|v\|\big]^2$, as a consequence of Cauchy's inequality:

$$(2.3) \qquad |(u, v)| \le \|u\| \cdot \|v\|,$$

a result that can be proved as follows. The fact that $(u - v, u - v) \ge 0$ implies $2 \operatorname{Re}(u, v) \le \|u\|^2 + \|v\|^2$; replacing u by $e^{i\theta}u$ with $e^{i\theta}$ chosen so that $e^{i\theta}(u, v)$ is real and positive, we get

$$(2.4) \qquad |(u, v)| \le \frac{1}{2}\|u\|^2 + \frac{1}{2}\|v\|^2.$$

Now in (2.4) we can replace u by tu and v by $t^{-1}v$, to get $|(u, v)| \le (t/2)\|u\|^2 + (1/2t)\|v\|^2$; minimizing over t gives (2.3). This establishes Cauchy's inequality, so we can deduce the triangle inequality. Thus (2.2) defines a norm, as in §1, and the notion of completeness is as stated there.

Prime examples of Hilbert spaces are the spaces $L^2(X, \mu)$ for a measure space (X, μ), that is, the case of $L^p(X, \mu)$ discussed in §1 with $p = 2$. In this case, the inner product is

$$(2.5) \qquad (u, v) = \int_X u(x)\overline{v(x)}\, d\mu(x).$$

The nice properties of Hilbert spaces arise from their similarity with familiar Euclidean space, so a great deal of geometrical intuition is available. For example, we say u and v are orthogonal, and write $u \perp v$, provided $(u, v) = 0$. Note that the Pythagorean theorem holds on a general Hilbert space:

$$(2.6) \qquad u \perp v \Longrightarrow \|u + v\|^2 = \|u\|^2 + \|v\|^2.$$

This follows directly from expanding $(u + v, u + v)$.

Another useful identity is the following, called the "parallelogram law," valid for all $u, v \in H$:

$$\|u + v\|^2 + \|u - v\|^2 = 2\|u\|^2 + 2\|v\|^2. \tag{2.7}$$

This also follows directly by expanding $(u+v, u+v)+(u-v, u-v)$, observing some cancellations. One important application of this simple identity is to the following existence result.

Let K be any closed, convex subset of H. Convexity implies that $x, y \in K \Rightarrow (x + y)/2 \in K$. Given $x \in H$, we define the distance from x to K to be

$$d = \inf\{\|x - y\| : y \in K\}. \tag{2.8}$$

Proposition 2.1. *If $K \subset H$ is a closed, convex set, there is a unique $z \in K$ such that $d = \|x - z\|$.*

Proof. We can pick $y_n \in K$ such that $\|x - y_n\| \to d$. It will suffice to show that (y_n) must be a Cauchy sequence. Use (2.7) with $u = y_m - x$, $v = x - y_n$, to get

$$\|y_m - y_n\|^2 = 2\|y_n - x\|^2 + 2\|y_m - x\|^2 - 4\Big\|x - \frac{1}{2}(y_n + y_m)\Big\|^2.$$

Since K is convex, $(1/2)(y_n + y_m) \in K$, so $\|x - (1/2)(y_n + y_m)\| \geq d$. Therefore,

$$\limsup_{m,n\to\infty} \|y_n - y_m\|^2 \leq 2d^2 + 2d^2 - 4d^2 \leq 0,$$

which implies convergence.

In particular, this result applies when K is a closed, linear subspace of H. In this case, for $x \in H$, denote by $P_K x$ the point in K closest to x. We have

$$x = P_K x + (x - P_K x). \tag{2.9}$$

We claim that $x - P_K x$ belongs to the linear space $K^\perp$, called the *orthogonal complement* of K, defined by

$$K^\perp = \{u \in H : (u, v) = 0 \text{ for all } v \in K\}. \tag{2.10}$$

Indeed, take any $v \in K$. Then

$$\begin{aligned}\Delta(t) &= \|x - P_K x + tv\|^2 \\ &= \|x - P_K x\|^2 + 2t \operatorname{Re}(x - P_K x, v) + t^2\|v\|^2\end{aligned}$$

is minimal at $t = 0$, so $\Delta'(0) = 0$ (i.e., $\operatorname{Re}(x - P_K x, v) = 0$), for all $v \in K$. Replacing v by iv shows that $(x - P_K x, v)$ also has vanishing imaginary part for any $v \in K$, so our claim is established. The decomposition (2.9) gives

$$x = x_1 + x_2, \quad x_1 \in K, \ x_2 \in K^\perp, \tag{2.11}$$

with $x_1 = P_K x, x_2 = x - P_K x$. Clearly, such a decomposition is unique. It implies that H is an orthogonal direct sum of K and $K^\perp$; we write

$$H = K \oplus K^\perp. \tag{2.12}$$

From this it is clear that

$$(K^\perp)^\perp = K, \tag{2.13}$$

that

$$x - P_K x = P_{K^\perp} x, \tag{2.14}$$

and that P_K and $P_{K^\perp}$ are *linear* maps on H. We call P_K the *orthogonal projection* of H on K. Note that $P_K x$ is uniquely characterized by the condition

$$P_K x \in K, \ (P_K x, v) = (x, v), \quad \text{for all } v \in K. \tag{2.15}$$

We remark that if K is a linear subspace of H which is not closed, then $K^\perp$ coincides with $\overline{K}^\perp$, and (2.13) becomes $(K^\perp)^\perp = \overline{K}$.

Using the orthogonal projection discussed above, we can establish the following result.

Proposition 2.2. *If $\varphi : H \to \mathbb{C}$ is a continuous, linear map, there exists a unique $f \in H$ such that*

$$\varphi(u) = (u, f), \quad \textit{for all } u \in H. \tag{2.16}$$

Proof. Consider $K = \operatorname{Ker} \varphi = \{u \in H : \varphi(u) = 0\}$, a closed, linear subspace of H. If $K = H$, then $\varphi = 0$ and we can take $f = 0$. Otherwise, $K^\perp \neq 0$; select a nonzero $x_0 \in K^\perp$ such that $\varphi(x_0) = 1$. We claim $K^\perp$ is one-dimensional in this case. Indeed, given any $y \in K^\perp$, $y - \varphi(y)x_0$ is annihilated by φ, so it belongs to K as well as to $K^\perp$, so it is zero. The result is now easily proved by setting $f = ax_0$ with $a \in \mathbb{C}$ chosen so that (2.16) works for $u = x_0$, namely $\overline{a}(x_0, x_0) = 1$.

We note that the correspondence $\varphi \mapsto f$ gives a conjugate linear isomorphism

$$H' \to H, \tag{2.17}$$

where H' denotes the space of all continuous linear maps $\varphi : H \to \mathbb{C}$.

We now discuss the existence of an orthonormal basis of a Hilbert space H. A set $\{e_\alpha : \alpha \in A\}$ is called an *orthonormal set* if each $\|e_\alpha\| = 1$ and $e_\alpha \perp e_\beta$ for $\alpha \neq \beta$. If $B \subset A$ is any finite set, it is easy to see via (2.15) that, for all $x \in H$,

$$P_V x = \sum_{\beta \in B} (x, e_\beta) e_\beta, \quad V = \text{span}\,\{e_\beta : \beta \in B\}, \tag{2.18}$$

where P_V is the orthogonal projection on V discussed above. Note that

$$\sum_{\beta \in B} |(x, e_\beta)|^2 = \|P_V x\|^2 \leq \|x\|^2. \tag{2.19}$$

In particular, we have $(x, e_\alpha) \neq 0$ for at most countably many $\alpha \in A$, for any given x. (Sometimes, A can be an uncountable set.) By (2.19) we also deduce that, with $c_\alpha = (x, e_\alpha)$, $\sum_{\alpha \in A} |c_\alpha|^2 < \infty$, and $\sum_{\alpha \in A} c_\alpha e_\alpha$ is a convergent series in the norm topology of H. We can apply (2.15) again to show that

$$\sum_{\alpha \in A} (x, e_\alpha) e_\alpha = P_L x, \tag{2.20}$$

where P_L is the orthogonal projection on

$$L = \text{closure of the linear span of } \{e_\alpha : \alpha \in A\}. \tag{2.21}$$

We call an orthonormal set $\{e_\alpha : \alpha \in A\}$ maximal if it is not contained in any larger orthonormal set. Such a maximal orthonormal set is a basis of H; the term "basis" is justified by the following result.

Proposition 2.3. *An orthonormal set $\{e_\alpha : \alpha \in A\}$ is maximal if and only if its linear span is dense in H, that is, if and only if L in (2.21) is all of H. In such a case, we have, for all $x \in H$,*

$$x = \sum_{\alpha \in A} c_\alpha e_\alpha, \quad c_\alpha = (x, e_\alpha). \tag{2.22}$$

The proof of the first assertion is obvious; the identity (2.22) then follows from (2.20).

The existence of a maximal orthonormal set in any Hilbert space can be inferred from Zorn's lemma; cf. [DS] and [RS]. This existence can be established on elementary logical principles in case H is separable (i.e., has a countable dense set $\{y_j : j = 1, 2, 3, \dots\}$). In this case, let V_n be the linear span of $\{y_j : j \leq n\}$, throwing out any y_n for which V_n is not strictly larger than V_{n-1}. Then pick unit $e_1 \in V_1$, unit $e_2 \in V_2$, orthogonal to V_1, and so on, via the Gramm-Schmidt process, and consider the orthonormal set $\{e_j : j = 1, 2, 3, \dots\}$. The linear span of $\{e_j\}$ coincides with that of $\{y_j\}$, hence is dense in H.

As an example of an orthonormal basis, we mention

$$e^{in\theta}, \quad n \in \mathbb{Z}, \tag{2.23}$$

a basis of $L^2(S^1)$ with square norm $\|u\|^2 = (1/2\pi) \int_{S^1} |u(\theta)|^2 \, d\theta$. See Chapter 3, §3, or the exercises for this section.

Exercises

1. Let $\mathcal{L}$ be the finite, linear span of the functions $e^{in\theta}$, $n \in \mathbb{Z}$, of (2.23). Use Exercise 1 of §1 to show that $\mathcal{L}$ is dense in $L^2(S^1)$ and hence that these exponentials form an orthonormal basis of $L^2(S^1)$.
2. Deduce that the Fourier coefficients

$$\mathcal{F}f(n) = \hat{f}(n) = \frac{1}{2\pi} \int_{-\pi}^{\pi} f(\theta)\, e^{-in\theta} \, d\theta \tag{2.24}$$

give a norm-preserving isomorphism

$$\mathcal{F} : L^2(S^1) \xrightarrow{\approx} \ell^2(\mathbb{Z}), \tag{2.25}$$

where $\ell^2(\mathbb{Z})$ is the set of sequences (c_n), indexed by $\mathbb{Z}$, such that $\sum |c_n|^2 < \infty$. Compare the approach to Fourier series in Chapter 3, §1.

In the next set of exercises, let μ and ν be two finite, positive measures on a space X, equipped with a σ-algebra $\mathcal{B}$. Let $\alpha = \mu + 2\nu$ and $\omega = 2\mu + \nu$.

3. On the Hilbert space $H = L^2(X, \alpha)$, consider the linear functional $\varphi : H \to \mathbb{C}$ given by $\varphi(f) = \int_X f(x)\, d\omega(x)$. Show that there exists $g \in L^2(X, \alpha)$ such that $1/2 \le g(x) \le 2$ and

$$\int_X f(x)\, d\omega(x) = \int_X f(x) g(x)\, d\alpha(x).$$

4. Suppose ν is absolutely continuous with respect to μ (i.e., $\mu(S) = 0 \Rightarrow \nu(S) = 0$). Show that $\{x \in X : g(x) = \frac{1}{2}\}$ has μ-measure zero, that

$$h(x) = \frac{2 - g(x)}{2g(x) - 1} \in L^1(X, \mu),$$

and that, for positive measurable F,

$$\int_X F(x)\, d\nu(x) = \int_X F(x) h(x)\, d\mu(x).$$

5. The conclusion of Exercise 4 is a special case of the Radon-Nikodym theorem, using an approach due to von Neumann. Deduce the more general case. Allow ν to be a signed measure. (You then need the Hahn decomposition of ν.)

3. Fréchet spaces; locally convex spaces

Fréchet spaces form a class more general than Banach spaces. For this structure, we have a linear space V and a countable family of seminorms $p_j : V \to \mathbb{R}^+$, where a seminorm p_j satisfies part of (1.1), namely

$$p_j(av) = |a| p_j(v), \quad p_j(v + w) \le p_j(v) + p_j(w), \tag{3.1}$$

but not necessarily the last hypothesis of (1.1); that is, one is allowed to have $p_j(v) = 0$ but $v \neq 0$. However, we do assume that

$$v \neq 0 \Longrightarrow p_j(v) \neq 0, \quad \text{for some } p_j. \tag{3.2}$$

Then, if we set

$$d(u, v) = \sum_{j=0}^{\infty} 2^{-j} \frac{p_j(u - v)}{1 + p_j(u - v)}, \tag{3.3}$$

we have a distance function. That $d(u, v)$ satisfies the triangle inequality follows from the next lemma, with $\rho(a) = a/(1 + a)$.

Lemma 3.1. *Let* $\delta : X \times X \to \mathbb{R}^+$ *satisfy*

$$\delta(x,z) \le \delta(x,y) + \delta(y,z), \tag{3.4}$$

for all $x, y, z \in X$. *Let* $\rho : \mathbb{R}^+ \to \mathbb{R}^+$ *satisfy*

$$\rho(0) = 0, \quad \rho' \ge 0, \quad \rho'' \le 0,$$

so that $\rho(a+b) \le \rho(a) + \rho(b)$. *Then* $\delta_\rho(x,y) = \rho\big(\delta(x,y)\big)$ *also satisfies (3.4).*

Proof. We have

$$\rho\big(\delta(x,z)\big) \le \rho\big(\delta(x,y) + \delta(y,z)\big) \le \rho\big(\delta(x,y)\big) + \rho\big(\delta(y,z)\big).$$

Thus V, with seminorms as above, gets the structure of a metric space. If it is complete, we call V a Fréchet space. Note that one has convergence $u_n \to u$ in the metric (3.3) if and only if

$$p_j(u_n - u) \to 0 \text{ as } n \to \infty, \quad \text{for each } p_j. \tag{3.5}$$

A paradigm example of a Fréchet space is $C^\infty(M)$, the space of C^∞-functions on a compact Riemannian manifold M. Then one can take $p_k(u) = \|u\|_{C^k}$, defined by (1.6). These seminorms are actually norms, but one encounters real seminorms in the following situation. Suppose M is a noncompact, smooth manifold, a union of an increasing sequence $\overline{M}_k$ of compact manifolds with boundary. Then $C^\infty(M)$ is a Fréchet space with seminorms $p_k(u) = \|u\|_{C^k(M_k)}$. Also, for such M, and for $1 \le p \le \infty$, $L^p_{\text{loc}}(M)$ is a Fréchet space, with seminorms $p_k(u) = \|u\|_{L^p(\overline{M}_k)}$.

Another important Fréchet space is the Schwartz space of rapidly decreasing functions

$$\mathcal{S}(\mathbb{R}^n) = \{u \in C^\infty(\mathbb{R}^n) : |D^\alpha u(x)| \le C_{N\alpha} \langle x \rangle^{-N} \text{ for all } \alpha, N\}, \tag{3.6}$$

with seminorms

$$p_k(u) = \sup_{x \in \mathbb{R}^n, |\alpha| \le k} \langle x \rangle^k |D^\alpha u(x)|. \tag{3.7}$$

This space is particularly useful for Fourier analysis; see Chapter 3.

A still more general class is the class of locally convex spaces. Such a space is a vector space V, equipped with a family of seminorms, satisfying (3.1)–(3.2). But now we drop the requirement that the family of seminorms be countable, that is, j ranges over some possibly uncountable set $\mathcal{J}$, rather than a countable set like $\mathbb{Z}^+$. Thus the construction (3.3) of a metric is not available. Such a space V has a natural topology, defined as follows. A neighborhood basis of a point $x \in V$ is given by

$$\mathcal{O}(x, \varepsilon, q) = \{y \in V : q(x-y) < \varepsilon\}, \quad \varepsilon > 0, \tag{3.8}$$

where q runs over finite sums of seminorms p_j. Then V is a topological vector space, that is, with respect to this topology, the vector operations are continuous. The term "locally convex" arises because the sets (3.8) are all convex.

Examples of such more general, locally convex structures will arise in the next section.

Exercises

1. Let E be a Fréchet space, with topology determined by seminorms p_j, arranged so that $p_1 \le p_2 \le \cdots$. Let F be a closed linear subspace. Form the quotient E/F. Show that E/F is a Fréchet space, with seminorms
$$q_j(x) = \inf \{p_j(y) : y \in E,\ \pi(y) = x\},$$
where $\pi : E \to E/F$ is the natural quotient map. (*Hint*: Extend the proof of Proposition 1.1. To begin, if $q_j(a) = 0$ for all j, pick $b_j \in E$ such that $\pi(b_j) = a$ and $p_j(b_j) \le 2^{-j}$; hence $p_j(b_k) \le 2^{-k}$, for $k \ge j$. Consider $b_1 + (b_2 - b_1) + (b_3 - b_2) + \cdots = b \in E$. Show that $\pi(b) = a$ and that $p_j(b) = 0$ for all j. Once this is done, proceed to establish completeness.)
2. If V is a Fréchet space, with topology given by seminorms $\{p_j\}$, a set $S \subset V$ is called *bounded* if each p_j is bounded on S. Show that every bounded subset of the Schwartz space $\mathcal{S}(\mathbb{R}^n)$ is relatively compact. Show that no infinite-dimensional Banach space can have this property.
3. Let $T : V \to V$ be a continuous, linear map on a locally convex space. Suppose K is a compact, convex subset of V and $T(K) \subset K$. Show that T has a fixed point in K. (*Hint*: Pick any $v_0 \in K$ and set
$$w_n = \frac{1}{n+1} \sum_{j=0}^{n} T^j v_0 \in K.$$
Show that any limit point of $\{w_n\}$ is a fixed point of T. Note that $Tw_n - w_n = (T^{n+1}v_0 - v_0)/(n+1)$.)

4. Duality

Let V be a linear space such as discussed in §§1–3, for example, a Banach space, or more generally a Fréchet space, or even more generally a Hausdorff topological vector space. The dual of V, denoted V', consists of continuous, linear maps

$$\omega : V \longrightarrow \mathbb{C} \tag{4.1}$$

($\omega : V \to \mathbb{R}$ if V is a real vector space). Elements $\omega \in V'$ are called *linear functionals* on V. Sometimes one finds the following notation for the action of $\omega \in V'$ on $v \in V$:

$$\langle v, \omega \rangle = \omega(v). \tag{4.2}$$

If V is a Banach space, with norm $\| \quad \|$, the condition for the map (4.1) to be continuous is the following: The set of $v \in V$ such that $|\omega(v)| \le 1$ must be a neighborhood of $0 \in V$. Thus this set must contain a ball $B_R = \{v \in V : \|v\| \le R\}$, for some $R > 0$. With $C = 1/R$, it follows that ω must satisfy

$$|\omega(v)| \le C\|v\|, \tag{4.3}$$

for some $C < \infty$. The infimum of the C's for which this holds is defined to be $\|\omega\|$; equivalently,

$$\|\omega\| = \sup\{|\omega(v)| : \|v\| \le 1\}. \tag{4.4}$$

It is easy to verify that V', with this norm, is also a Banach space.

More generally, let ω be a continuous, linear functional on a Fréchet space V, equipped with a family $\{p_j : j \ge 0\}$ of seminorms and (complete) metric given by (3.3). For any $\varepsilon > 0$, there exists $\delta > 0$ such that $d(u, 0) \le \delta$ implies $|\omega(u)| \le \varepsilon$. Take $\varepsilon = 1$ and the associated δ; pick N so large that $\sum_{N+1}^{\infty} 2^{-j} < \delta/2$. It follows that $\sum_1^N p_j(u) \le \delta/2$ implies $|\omega(u)| \le 1$. Consequently, we see that the continuity of $\omega : V \to \mathbb{C}$ is equivalent to the validity of an estimate of the form

$$|\omega(u)| \le C \sum_{j=1}^{N} p_j(u). \tag{4.5}$$

For general Fréchet spaces, there is no simple analogue of (4.4); V' is typically not a Fréchet space. We will give a further discussion of topologies on V' later in this section.

Next we consider identification of the duals of some specific Banach spaces mentioned before. First, if H is a Hilbert space, the inner product produces a conjugate linear isomorphism of H' with H, as noted in (2.17). We next identify the dual of $L^p(X, \mu)$.

Proposition 4.1. *Let (X, μ) be a σ-finite measure space. Let $1 \le p < \infty$. Then the dual space $L^p(X, \mu)'$, with norm given by (4.4), is naturally isomorphic to $L^q(X, \mu)$, with $1/p + 1/q = 1$.*

Note that Hölder's inequality and its refinement (1.13) show that there is a natural inclusion $\iota : L^q(X, \mu) \to L^p(X, \mu)'$, which is an isometry. It remains to show that ι is surjective. We sketch a proof in the case when $\mu(X)$ is finite, from which the general case is easily deduced. If $\omega \in L^p(X, \mu)'$, define a set function ν on measurable sets $E \subset X$ by $\nu(E) = \langle \chi_E, \omega \rangle$, where χ_E is the characteristic function of E; ν is readily verified to be countably additive, as long as $p < \infty$. Furthermore, ν annihilates sets of μ-measure zero, so the Radon-Nikodym theorem implies

$$\int f \, d\nu = \int f w \, d\mu,$$

for some measurable function w. A variant of the proof of (1.13) gives $w \in L^q(X, \mu)$, with $\|w\|_{L^q} = \|\omega\|$.

Note that the countable additivity of ν fails for $p = \infty$; in fact, $L^\infty(X, \mu)'$ can be identified with the space of finitely additive set functions on the σ-algebra of μ-measurable sets that annihilate sets of μ-measure zero.

The following complement to Proposition 4.1 is one of the fundamental results of measure theory. For a proof, we refer to [Ru] and [Yo].

Proposition 4.2. *If X is a compact metric space, $C(X)'$ is isometrically isomorphic to the space $\mathcal{M}(X)$ of (complex) Borel measures on X, with the total variation norm.*

In fact, the generalization of this to the case where X is a compact Hausdorff space, not necessarily metrizable, is of interest. In that case, there is a distinction between the Borel σ-algebra, generated by all compact subsets of X, and the Baire σ-algebra, generated by the compact G_δ subsets of X. For $\mathcal{M}(X)$ here one takes the space of Baire measures to give $C(X)'$. It is then an important fact that each Baire measure has a unique extension to a regular Borel measure. For details, see [Hal].

If M is a smooth, compact manifold, the dual of the Fréchet space $C^\infty(M)$ is denoted $\mathcal{D}'(M)$ and is called the space of *distributions* on M. It is discussed in Chapter 3; also discussed there is the space $\mathcal{S}'(\mathbb{R}^n)$ of tempered distributions on $\mathbb{R}^n$, the dual of $\mathcal{S}(\mathbb{R}^n)$.

For a Banach space, since V' is a Banach space, one can construct its dual, V''. Note that the action (4.2) produces a natural linear map

$$\kappa : V \longrightarrow V'', \tag{4.6}$$

and it is obvious that $\|\kappa(v)\| \le \|v\|$. In fact, $\|\kappa(v)\| = \|v\|$, that is, κ is an isometry. In other words, for any $v \in V$, there exists $\omega \in V'$, $\|\omega\| = 1$, such that $\omega(v) = \|v\|$. This is a special case of the Hahn-Banach theorem, stated below in Proposition 4.3.

Sometimes κ in (4.6) is surjective, so it gives an isometric isomorphism of V with V''. In this case, we say V is *reflexive*. Clearly, any Hilbert space is reflexive. Also, in view of Proposition 4.1, we see that $L^p(X,\mu)$ is reflexive, provided $1 < p < \infty$. On the other hand, $L^1(X,\mu)$ is not reflexive; $L^\infty(X,\mu)'$ is strictly larger than $L^1(X,\mu)$, except for the trivial cases where $L^1(X,\mu)$ is finite-dimensional.

We now state the Hahn-Banach theorem, referred to above. It has a fairly general formulation, useful also for Fréchet spaces and more general, locally convex spaces.

Proposition 4.3. *Let V be a linear space (real or complex), W a linear subspace. Let p be a seminorm on V. Suppose ω is a linear functional on W satisfying $|\omega(v)| \le p(v)$, for $v \in W$. Then there exists an extension of ω to a linear functional Ω on V ($\Omega = \omega$ on W), such that $|\Omega(v)| \le p(v)$ for $v \in V$.*

Note that in case V is a Hilbert space and p the associated norm, this result follows readily from the orthogonal decomposition established in (2.9)–(2.10).

The key to the proof in general is to show that ω can be extended to V when V is spanned by W and one element $z \in V \setminus W$. So one looks for a constant c so that the prescription $\Omega(v + az) = \omega(v) + ac$ works; c is to be picked so that

$$|\omega(v) + ac| \le p(v + az), \quad \text{for } v \in W,\ a \in \mathbb{R} \text{ (or } \mathbb{C}). \tag{4.7}$$

First consider the case of a real vector space. Then (4.7) holds provided $\omega(v)+ac \le p(v+az)$, for all $v \in W, a \in \mathbb{R}$, or equivalently provided

$$\begin{aligned} c &\le a^{-1}\big[p(v+az) - \omega(v)\big], \\ -c &\le a^{-1}\big[p(v-az) - \omega(v)\big], \end{aligned} \tag{4.8}$$

for $v \in W, a > 0$. Such a constant will exist provided

$$\begin{aligned} &\sup_{v_1 \in W, a_1 > 0} a_1^{-1}\big[\omega(v_1) - p(v_1 - a_1 z)\big] \\ &\quad \le \inf_{v_2 \in W, a_2 > 0} a_2^{-1}\big[p(v_2 + a_2 z) - \omega(v_2)\big]. \end{aligned} \tag{4.9}$$

Equivalently, for such v_j and a_j, one must have

$$\omega(a_2 v_1 + a_1 v_2) \le a_1 p(v_2 + a_2 z) + a_2 p(v_1 - a_1 z). \tag{4.10}$$

We know that the left side is dominated by

$$p(a_2 v_1 + a_1 v_2) = p(a_2 v_1 - a_2 a_1 z + a_1 a_2 z + a_1 v_2),$$

which is readily dominated by the right side of (4.10). Hence such a number c exists to make (4.7) work.

A Zorn's lemma argument will then work to show that ω can be extended to all of V in general (i.e., it has a "maximal" extension). In case V is a separable Fréchet space and p a continuous seminorm on V, an elementary inductive argument provides an extension from W to a space dense in V, and hence by continuity to V.

The complex case can be deduced from the real case as follows. Define $\gamma : W \to \mathbb{R}$ as $\gamma(v) = \operatorname{Re} \omega(v)$. Then $\omega(v) = \gamma(v) - i\gamma(iv)$. If $\Gamma : V \to \mathbb{R}$ is a desired real, linear extension of γ to V, then one can set $\Omega(v) = \Gamma(v) - i\Gamma(iv)$.

We now make note of some further topologies on the dual space V'. The first is called the *weak*-topology*. It is the topology of pointwise convergence and is specified by the family of seminorms

$$p_v(\omega) = |\omega(v)|, \tag{4.11}$$

as v varies over V. The following result, called *Alaoglu's theorem*, is useful.

Proposition 4.4. *If V is a Banach space, then the closed unit ball $B \subset V'$ is compact in the weak*-topology.*

This result is readily deduced from the following fundamental result in topology:

Tychonov's Theorem. *If $\{X_\alpha : \alpha \in A\}$ is any family of compact Hausdorff spaces, then the Cartesian product $\prod_\alpha X_\alpha$, with the product topology, is a compact Hausdorff space.*

Indeed, the space $B \subset V'$ above, with the weak*-topology, is homeomorphic to a closed subset of the Cartesian product $\prod\{X_v : v \in B_1\}$, where $B_1 \subset V$ is the unit ball in V, each X_v is a copy of the unit disk in $\mathbb{C}$, and $\kappa : B \to \prod X_v$ is given by $\kappa(\omega) = \{\omega(v) : v \in B_1\}$. For a proof of Tychonov's theorem, see [Dug] and [RS].

We remark that if V is separable, then B is a compact metric space. In fact, if $\{v_j : j \in \mathbb{Z}^+\}$ is a dense subset of $B_1 \subset V$, the weak*-topology on B is given by the metric

$$d(\omega, \sigma) = \sum_{j \geq 0} 2^{-j} |\langle v_j, \omega - \sigma \rangle|. \tag{4.12}$$

Conversely, on V there is the weak topology, the topology of pointwise convergence on V', with seminorms

$$p_\omega(v) = |\omega(v)|, \quad \omega \in V'. \tag{4.13}$$

When V is a reflexive Banach space, $V = V''$, then the weak topology of V coincides with its weak*-topology, as the dual of V'; thus Proposition 4.4 applies to the unit ball in V in this case.

More generally, we say two vector spaces V and W have a dual pairing if there is a bilinear form $\langle v, w \rangle$, defined for $v \in V$, $w \in W$, such that for each $v \neq 0$, $\langle v, w \rangle \neq 0$ for some $w \in W$, and for each $w \neq 0$, this form is nonzero for some $v \in V$. Then the seminorms $p_w(v) = |\langle v, w \rangle|$ on V define a Hausdorff topology called the $\sigma(V, W)$-topology, and symmetrically we have the $\sigma(W, V)$ topology on W. Thus the weak topology on V defined above is the $\sigma(V, V')$-topology, and the weak*-topology on V' is the $\sigma(V', V)$-topology.

We define another topology on the dual space V' of a locally convex space V, called the strong topology. This is the topology of uniform convergence on bounded subsets of V. A set $Y \subset V$ is bounded provided each seminorm p_j defining the topology of V is bounded on Y. The strong topology on V' is defined by the seminorms

$$p_Y(\omega) = \sup\{|\omega(y)| : y \in Y\}, \quad Y \subset V \text{ bounded}. \tag{4.14}$$

In case V is a Banach space, $Y \subset V$ is bounded if and only if it is contained in some ball of finite radius, and then each seminorm (4.14) is dominated by some multiple of the norm on V', given by (4.3). Thus in this case the strong topology and the norm topology on V' coincide. For more general Fréchet spaces, such as $V = C^\infty(M)$, the strong topology on V' does not make V' a normed space, or even a Fréchet space.

There are many interesting results in the subject of duality, concerning the topologies discussed above and other topologies, such as the Mackey topology, which we will not describe here. For further material, see [S].

Exercises

1. Suppose $\{u_j : j \in \mathbb{Z}^+\}$ is an orthonormal set in a Hilbert space H. Show that $u_j \to 0$ in the weak* topology as $j \to \infty$.

2. In the setting of Exercise 1, suppose $H = L^2(X, \mu)$, and the u_j also satisfy uniform bounds: $|u_j(x)| \le M$. Show that $u_j \to 0$ in the weak* topology of $L^\infty(X, \mu)$, as the dual to $L^1(X, \mu)$.
3. Deduce that if $f \in L^1(S^1)$, with Fourier coefficients $\hat{f}(n)$ given by (2.24), then $\hat{f}(n) \to 0$ as $n \to \infty$.
4. Prove the assertion made in the text that, when V is a separable Banach space, then the unit ball B in V', with the weak* topology, is metrizable. (*Hint*: To show that (4.12) defines a topology coinciding with the weak* topoogy, use the fact that if $\varphi : X \to Y$ is continuous and bijective, with X compact and Y Hausdorff, then φ is a homeomorphism.)
5. On a Hilbert space H, suppose $f_j \to f$ weakly. Show that if

$$\|f\| \ge \limsup_{j\to\infty} \|f_j\|,$$

then $f_j \to f$ in *norm*. (*Hint*: Expand $(f - f_j, f - f_j)$.)
6. Suppose X is a closed, linear subspace of a reflexive Banach space V. Show that X is reflexive. (*Hint*: Use the Hahn-Banach theorem. First show that $X' \approx V'/X^\perp$, where $X^\perp = \{\omega \in V' : \omega(v) = 0, \forall\, v \in X\}$. Thus, a bounded linear functional on X' gives rise to a bounded linear functional on V', annihilating $X^\perp$.)

5. Linear operators

If V and W are two Banach spaces, or more generally two locally convex spaces, we denote by $\mathcal{L}(V, W)$ the space of continuous, linear transformations from V to W. As in the derivation of (4.4), it is easy to see that, when V and W are Banach spaces, a linear map $T : V \to W$ is continuous if and only if there exists a constant $C < \infty$ such that

$$\|Tv\| \le C\|v\| \tag{5.1}$$

for all $v \in V$. Thus we call T a *bounded* operator. The infimum of all the C's for which this holds is defined to be $\|T\|$; equivalently,

$$\|T\| = \sup\{\|Tv\| : \|v\| \le 1\}. \tag{5.2}$$

It is clear that $\mathcal{L}(V, W)$ is a linear space. If V and W are Banach spaces and $T_j \in \mathcal{L}(V, W)$, then $\|T_1 + T_2\| \le \|T_1\| + \|T_2\|$; completeness is also easy to establish in this case, so $\mathcal{L}(V, W)$ is also a Banach space. If X is a third Banach space and $S \in \mathcal{L}(W, X)$, it is clear that $ST \in \mathcal{L}(V, X)$, and

$$\|ST\| \le \|S\| \cdot \|T\|. \tag{5.3}$$

The space $\mathcal{L}(V) = \mathcal{L}(V, V)$, with norm (5.2), is a Banach algebra for any Banach space V. Generally, a Banach algebra is defined to be a Banach space B with the structure of an algebra, so that, for any $S, T \in B$, the inequality (5.3) holds. Another example of a Banach algebra is the space $C(X)$, for compact X, with norm (1.3), the product being given by the pointwise product of functions.

If V and W are Banach spaces and $T \in \mathcal{L}(V, W)$, then the adjoint $T' \in \mathcal{L}(W', V')$ is uniquely defined to satisfy

$$\langle Tv, w\rangle = \langle v, T'w\rangle, \quad v \in V, \ w \in W'. \tag{5.4}$$

Using the Hahn-Banach theorem, it is easy to see that

$$\|T\| = \|T'\|, \tag{5.5}$$

both norms being the sup of the absolute value of (5.4) over $\|v\| = 1$, $\|w\| = 1$. When V and W are reflexive, it is clear that $T'' = T$. We remark that (5.4) also defines T' for general locally convex V and W.

In case V and W are Hilbert spaces and $T \in \mathcal{L}(V, W)$, then we also have an adjoint $T^* \in \mathcal{L}(W, V)$, given by

$$(Tv, w) = (v, T^*w), \quad v \in V, \ w \in W, \tag{5.6}$$

using the inner products on W and V, respectively. As in (5.5) we have $\|T\| = \|T^*\|$. Also it is clear that $T^{**} = T$.

When H is a Hilbert space, the Banach algebra $\mathcal{L}(H)$ is a C*-algebra. Generally, a C*-algebra B is a Banach algebra, equipped with a conjugate linear involution $T \mapsto T^*$, satisfying $\|T^*\| = \|T\|$ and

$$\|T^*T\| = \|T\|^2. \tag{5.7}$$

To see that (5.7) holds for $T \in \mathcal{L}(H)$, note that both sides are equal to the sup of the absolute value, over $\|v_1\| \le 1$, $\|v_2\| \le 1$, of

$$(T^*Tv_1, v_2) = (Tv_1, Tv_2), \tag{5.8}$$

such a supremum necessarily being obtained over the set of pairs satisfying $v_1 = v_2$. Note that $C(X)$, considered above, is also a C*-algebra. However, for a general Banach space V, $\mathcal{L}(V)$ will not have the structure of a C*-algebra.

We consider some simple examples of bounded linear operators. If (X, μ) is a measure space, $f \in L^\infty(X, \mu)$, then the multiplication operator M_f, defined by $M_f u = fu$, is bounded on $L^p(X, \mu)$ for each $p \in [1, \infty]$, with $\|M_f\| = \|f\|_{L^\infty}$. If X is a compact Hausdorff space and $f \in C(X)$, then $M_f \in \mathcal{L}(C(X))$, with $\|M_f\| = \|f\|_{\sup}$. In case X is a compact Riemannian manifold and P is a differential operator of order k on X, with smooth coefficients, then P does not give a bounded operator on $C(X)$, but one has $P \in \mathcal{L}(C^k(X), C(X))$, and more generally $P \in \mathcal{L}(C^{k+m}(X), C^m(X))$, for $m \ge 0$. For related results on Sobolev spaces, see Chapter 4.

Another class of examples, a little more elaborate than those just mentioned, is given by integral operators, of the form

$$Ku(x) = \int_X k(x, y)\, u(y)\, d\mu(y), \tag{5.9}$$

where (X, μ) is a measure space. We have the following result:

Proposition 5.1. *Suppose k is measurable on $X \times X$ and*

$$\int_X |k(x,y)|\,d\mu(x) \le C_1, \quad \int_X |k(x,y)|\,d\mu(y) \le C_2, \tag{5.10}$$

for all y and for all x, respectively. Then (5.9) defines K as a bounded operator on $L^p(X,\mu)$, for each $p \in [1,\infty]$, with

$$\|K\| \le C_1^{1/p}\, C_2^{1/q}, \quad \frac{1}{p} + \frac{1}{q} = 1. \tag{5.11}$$

Proof. For $p \in (1,\infty)$, we estimate

$$\Bigl|\int_X \int_X k(x,y) f(y) g(x)\,d\mu(x)\,d\mu(y)\Bigr| \tag{5.12}$$

via the estimate $ab \le a^p/p + b^q/q$ of (1.15), used to prove Hölder's inequality. Apply this to $|f(y)g(x)|$. Then (5.12) is dominated by

$$\frac{C_1}{p}\|f\|_{L^p}^p + \frac{C_2}{q}\|g\|_{L^q}^q \tag{5.13}$$

provided (5.10) holds. Replacing f, g by $tf, t^{-1}g$, we see that (5.12) is dominated by $(C_1 t^p/p)\|f\|_{L^p}^p + (C_2/qt^q)\|g\|_{L^q}^q$; minimizing over $t \in (0,\infty)$, via elementary calculus, we see that (5.12) is dominated by

$$C_1^{1/p} C_2^{1/q} \|f\|_{L^p}\|g\|_{L^q}, \tag{5.14}$$

proving the result. The exceptional cases $p = 1$ and $p = \infty$ are easily handled.

We call $k(x,y)$ the *integral kernel* of K. Note that K' is an integral operator, with kernel $k'(x,y) = k(y,x)$. In the case of the Hilbert space $L^2(X,\mu)$, K^* is an integral operator, with kernel $k^*(x,y) = \overline{k(y,x)}$.

Chapter 7 includes a study of a much more subtle class of operators called *singular integral operators*, or *pseudodifferential operators of order zero*; L^p-estimates for this class are made in Chapter 13.

We next consider some results about linear transformations on Banach spaces which use the following general result about complete metric spaces, known as the *Baire category theorem*.

Proposition 5.2. *Let X be a complete metric space, and X_j, $j \in \mathbb{Z}^+$, nowhere-dense subsets; that is, the closure $\overline{X}_j$ contains no nonempty open set. Then $\bigcup_j X_j \ne X$.*

Proof. The hypothesis on X_1 implies there is a closed ball $B_{r_1}(p_1) \subset X \setminus X_1$, for some $p_1 \in X$, $r_1 > 0$. Then the hypothesis on X_2 gives a ball $B_{r_2}(p_2) \subset B_{r_1}(p_1) \setminus X_2$, $0 < r_2 \le r_1/2$. Continue, getting balls

$$B_{r_j}(p_j) \subset B_{r_{j-1}}(p_{j-1}) \setminus X_j, \quad 0 < r_j \le 2^{-j+1} r_1. \tag{5.15}$$

Then (p_j) is Cauchy; it must converge to a point $p \notin \cup_j X_j$, as p belongs to each $B_{r_j}(p_j)$.

Our first application is to a result called the *uniform boundedness principle.*

Proposition 5.3. *Let V, W be Banach spaces, $T_j \in \mathcal{L}(V, W)$, $j \in \mathbb{Z}^+$. Assume that for each $v \in V$, $\{T_j v\}$ is bounded in W. Then $\{\|T_j\|\}$ is bounded.*

Proof. Let $X = V$. Let $X_n = \{v \in X : \|T_j v\| \le n \text{ for all } j\}$. The hypothesis implies $\cup_n X_n = X$. Clearly, each X_n is closed. The Baire category theorem implies that some X_N has nonempty interior, so there exists $v_0, r > 0$ such that $\|v\| \le r \Rightarrow \|T_j(v_0 + v)\| \le N$, for all j. Hence

$$\|v\| \le r \Rightarrow \|T_j v\| \le N + \|T_j v_0\| \le R \quad \forall\, j, \tag{5.16}$$

using the boundedness of $\{T_j v_0\}$. This implies $\|T_j\| \le R/r$, completing the proof.

The next result is known as the *open mapping theorem.*

Proposition 5.4. *If V and W are Banach spaces and $T \in \mathcal{L}(V, W)$ is onto, then any neighborhood of 0 in V is mapped onto a neighborhood of 0 in V.*

Proof. Let B_1 denote the unit ball in V, $X_n = T(nB_1) = nT(B_1)$. The hypothesis implies $\bigcup_{n\ge1} X_n = W$. The Baire category theorem implies that some $\overline{X}_N$ has nonempty interior, hence contains a ball $B_r(w_0)$; symmetry under sign change implies $\overline{X}_N$ also contains $B_r(-w_0)$. Hence $\overline{X}_{2N} = 2\overline{X}_N$ contains $B_{2r}(0)$. By scaling, $\overline{X}_1$ contains a ball $B_\varepsilon(0)$. Our goal now is to show that X_1 itself contains a ball. This will follow if we can show that $\overline{X}_1 \subset X_2$.

So let $y \in \overline{X}_1 = \overline{T(B_1)}$. Thus there is an $x_1 \in B_1$ such that $y - Tx_1 \in B_{\varepsilon/2}(0) \subset \overline{X}_{1/2}$. For the same reason, there is an $x_2 \in B_{\frac{1}{2}}$ such that $(y - Tx_1) - Tx_2 \in B_{\varepsilon/4}(0) \subset \overline{X}_{1/4}$. Continue, getting $x_n \in B_{2^{1-n}}$ such that

$$y - \sum_{j=1}^{n} Tx_j \in B_{\varepsilon/2^n}(0).$$

Then $x = \sum_{j=1}^{\infty} x_j$ is in B_2 and $Tx = y$. This completes the proof.

Corollary 5.5. *If V and W are Banach spaces and $T : V \to W$ is continuous and bijective, then $T^{-1} : W \to V$ is continuous.*

In such a situation, we say that T is a topological isomorphism.

The third basic application of the Baire category theorem is called the *closed-graph theorem.* For a given linear map $T : V \to W$, its graph is defined to be

$$G_T = \{(v, Tv) \in V \oplus W : v \in V\}. \tag{5.17}$$

It is easy to see that, whenever V and W are topological vector spaces, then if T is continuous, G_T is closed. The following is a converse.

Proposition 5.6. *Let V and W be Banach spaces, $T : V \to W$ a linear map. If G_T is closed in $V \oplus W$, then T is continuous.*

Proof. The hypothesis implies that G_T is a Banach space, with norm $\|(v, Tv)\| = \|v\| + \|Tv\|$. Now the maps $J : G_T \to V, K : G_T \to W$, given by $J(v, Tv) = v$, $K(v, Tv) = Tv$, are clearly continuous, and J is bijective. Hence J^{-1} is continuous, and so $T = KJ^{-1}$ is also continuous.

Propositions 5.3–5.6 have extensions to Fréchet spaces, since they are also complete metric spaces. For example, let V be a Fréchet space in Proposition 5.3 (keep W a Banach space). In this case, the hypothesis that $\{T_j v\}$ is bounded in W for each $v \in V$ implies that there exists a neighborhood $\mathcal{O}$ of the origin in V, of the form (3.8), such that $v \in \mathcal{O} \Rightarrow \|T_j v\| \le 1$ for all j, that is, for some finite sum q of seminorms defining the Fréchet space structure of V,

$$\|T_j v\| \le K\, q(v), \quad \text{for all } j, \tag{5.18}$$

with K independent of j.

Propositions 5.4–5.6 extend directly to the case where V and W are Fréchet spaces, with only slight extra complications in the proofs.

We now give an important application of the open mapping theorem, to a result known as the *closed-range theorem*. If W is a Banach space and $L \subset W$ is a linear subspace, we denote by $L^\perp$ the subspace of W' consisting of linear functionals on W that annihilate L.

Proposition 5.7. *If V and W are Banach spaces and $T \in \mathcal{L}(V, W)$, then*

$$\text{Ker}\, T' = T(V)^\perp. \tag{5.19}$$

If, in addition, $T(V)$ is closed in W, then $T'(W')$ is closed in V' and

$$T'(W') = (\text{Ker}\, T)^\perp. \tag{5.20}$$

Proof. For the first identity, by $\langle Tv, w\rangle = \langle v, T'w\rangle$, it is obvious that $T(V)^\perp = \text{Ker}\, T'$. If $T(V)$ is closed, it follows from Corollary 5.5 that $\widetilde{T} : V/\text{Ker}\, T \to T(V)$ is a topological isomorphism. Thus we have a topological isomorphism

$$\widetilde{T}' : T(V)' \overset{\approx}{\to} (V/\text{Ker}\, T)'. \tag{5.21}$$

Meanwhile, there is a natural isomorphism of Banach spaces

$$(V/\text{Ker}\, T)' \approx (\text{Ker}\, T)^\perp, \tag{5.22}$$

and, by the Hahn-Banach theorem, there is a natural surjection $W' \to T(V)'$. (See Exercise 4 below.) Composing these operators yields T'. Thus we have (5.20).

In the Hilbert space case, we have the same result for T^*.

Since one frequently looks at equations $Tu = v$, it is important to consider the notion of invertibility. An operator $T \in \mathcal{L}(V, W)$ is invertible if there is an $S \in \mathcal{L}(W, V)$ such that ST and TS are identity operators. One useful fact is that all operators close to the identity in $\mathcal{L}(V)$ are invertible.

Proposition 5.8. *Let V be a Banach space, $T \in \mathcal{L}(V)$, with $\|T\| < 1$. Then $I - T$ is invertible.*

Proof. The power series $\sum_{n=0}^{\infty} T^n$ converges to $(I - T)^{-1}$.

When V is a Banach space, we say $\zeta \in \mathbb{C}$ belongs to the resolvent set of an operator $T \in \mathcal{L}(V)$ (denoted $\rho(T)$) provided $\zeta I - T$ is invertible; then the resolvent of T is

$$R_\zeta = (\zeta I - T)^{-1}. \tag{5.23}$$

It easily follows from the method of proof of Proposition 5.8 that the resolvent set of any $T \in \mathcal{L}(V)$ is open in $\mathbb{C}$. Furthermore, R_ζ is a holomorphic function of $\zeta \in \rho(T)$. In fact, if $\zeta_0 \in \rho(T)$, then we can write $\zeta - T = (\zeta_0 - T)\big(I - (\zeta_0 - \zeta)R_{\zeta_0}\big)$, and hence, for ζ close to ζ_0,

$$R_\zeta = R_{\zeta_0} \sum_{n=0}^{\infty} R_{\zeta_0}^n (\zeta_0 - \zeta)^n.$$

It is also clear that ζ belongs to the resolvent set whenever $|\zeta| > \|T\|$, since

$$(\zeta - T)^{-1} = \zeta^{-1}(I - \zeta^{-1}T)^{-1}. \tag{5.24}$$

The complement of the resolvent set is called the *spectrum* of T. Thus, for any $T \in \mathcal{L}(V)$, the spectrum of T (denoted $\sigma(T)$) is a compact set in $\mathbb{C}$. By (5.24), $\|R_\zeta\| \to 0$ as $|\zeta| \to \infty$. Since R_ζ is holomorphic on $\rho(T)$, it follows by Liouville's theorem that, for any $T \in \mathcal{L}(V)$, $\rho(T)$ cannot be all of $\mathbb{C}$, so $\sigma(T)$ is nonempty.

Using the resolvent as a tool, we now discuss a holomorphic functional calculus for an operator $T \in \mathcal{L}(V)$, and applications to spectral theory. Let Ω be a bounded region in $\mathbb{C}$, with smooth boundary, containing the spectrum $\sigma(T)$ in its interior. If f is holomorphic on a neighborhood of Ω, we set

$$f(T) = \frac{1}{2\pi i} \int_\gamma f(\zeta)\,(\zeta - T)^{-1}\,d\zeta, \tag{5.25}$$

where $\gamma = \partial\Omega$. Note that if T were a complex number in Ω, this would be Cauchy's formula. Here are a couple of very basic facts.

Lemma 5.9. *If $f(z) = 1$, then $f(T) = I$, and if $f(z) = z$, then $f(T) = T$.*

Proof. Deform γ to be a large circle and use (5.24), plus

$$(I - \zeta^{-1}T)^{-1} = I + \sum_{n=1}^{\infty} (\zeta^{-1}T)^n. \tag{5.26}$$

We next derive a multiplicative property of this functional calculus, making use of the following result, known as the *resolvent identity*.

Lemma 5.10. *If $z, \zeta \in \rho(T)$, then*

$$R_z - R_\zeta = (\zeta - z) R_z R_\zeta. \tag{5.27}$$

Proof. For any $\zeta \in \rho(T)$, R_ζ commutes with $\zeta - T$, hence with T, hence with any $z - T$. If, in addition, $z \in \rho(T)$, we have both $R_\zeta R_z (z - T) = R_\zeta$ and $R_z R_\zeta (z - T) = R_z (z - T) R_\zeta = R_\zeta$, hence

$$R_z R_\zeta = R_\zeta R_z. \tag{5.28}$$

Thus

$$\begin{aligned} R_z - R_\zeta &= (\zeta - T) R_\zeta R_z - (z - T) R_z R_\zeta \\ &= (\zeta - z) R_\zeta R_z, \end{aligned}$$

proving (5.27).

Now for our multiplicative property:

Proposition 5.11. *If f and g are holomorphic on a neighborhood of Ω, then*

$$f(T) g(T) = (fg)(T). \tag{5.29}$$

Proof. Let $\gamma = \partial\Omega$, as above, and let γ_1 be the boundary of a slightly larger region, on which f and g are holomorphic. Write

$$g(T) = \frac{1}{2\pi i} \int_{\gamma_1} g(z)(z - T)^{-1} \, dz,$$

and hence, using (5.25), write $f(T)g(T)$ as a double integral. The product $R_\zeta R_z$ of resolvents of T appears in the new integrand. Using the resolvent identity (5.27), we obtain

$$f(T) g(T) = -\frac{1}{4\pi^2} \int_{\gamma_1} \int_{\gamma} (\zeta - z)^{-1} f(\zeta) g(z) (R_z - R_\zeta) \, d\zeta \, dz. \tag{5.30}$$

The term involving R_z as a factor has $d\zeta$-integral equal to zero, by Cauchy's theorem. Doing the dz-integral for the other term, using Cauchy's identity

$$g(\zeta) = \frac{1}{2\pi i} \int_{\gamma_1} (z - \zeta)^{-1} g(z) \, dz,$$

we obtain from (5.30)

$$f(T)g(T) = \frac{1}{2\pi i} \int_\gamma f(\zeta)g(\zeta)R_\zeta \, d\zeta, \tag{5.31}$$

which gives (5.29).

One interesting situation that frequently arises is the following. Ω can have several connected components, $\Omega = \Omega_1 \cup \cdots \cup \Omega_M$, each Ω_j containing different pieces of $\sigma(T)$. Taking a function equal to 1 on Ω_j and 0 on the other components produces operators

$$P_j = \frac{1}{2\pi i} \int_{\gamma_j} (\zeta - T)^{-1} \, d\zeta, \quad \gamma_j = \partial\Omega_j. \tag{5.32}$$

By (5.29) we see that

$$P_j^2 = P_j, \quad P_j P_k = 0, \quad \text{for } j \neq k, \tag{5.33}$$

so $P_1, \dots, P_M$ are mutually disjoint projections. By Lemma 5.9, $P_1 + \cdots + P_M = I$. It follows easily that if T_j denotes the restriction of T to the range of P_j, then

$$\sigma(T_j) = \sigma(T) \cap \Omega_j. \tag{5.34}$$

Exercises

1. Extend the $p = 2$ case of Proposition 5.1 to the following result of Schur. Let (X, μ) and (Y, ν) be measure spaces, and let $k(x, y)$ be measurable on $(X \times Y, \mu \times \nu)$. Assume that there are measurable functions $p(x)$, $q(y)$, positive a.e. on X and Y, respectively, such that

$$\int_X |k(x, y)| p(x) \, d\mu(x) \leq C_1 q(y), \quad \int_Y |k(x, y)| q(y) \, d\nu(y) \leq C_2 p(x). \tag{5.35}$$

   Show that $Ku(x) = \int_Y k(x, y)u(y) \, d\nu(y)$ defines a bounded operator

$$K : L^2(Y, \nu) \longrightarrow L^2(X, \mu), \quad \|K\|^2 \leq C_1 C_2.$$

   Give an appropriate modification of the hypothesis (5.35) in order to obtain an operator bound $K : L^p(Y, \nu) \to L^p(X, \mu)$.

2. Show that $k(x, y)$ is the integral kernel of a bounded map $K : L^2(\mathbb{R}^n) \to L^2(\mathbb{R}^n_+)$ provided it satisfies the estimate

$$|k(x, y)| \leq C\Big(|x' - y'|^2 + x_1^2 + y_1^2\Big)^{-n/2}, \quad x = (x_1, x'), \ y = (y_1, y'). \tag{5.36}$$

   (*Hint*: Construct $p(x)$ and $q(y)$ so that (5.35) holds. Here, $\mathbb{R}^n_+ = \{x \in \mathbb{R}^n : x_1 \geq 0\}$.)

3. Show that $k(x, y)$ is the integral kernel of a bounded map $K : L^p(\mathbb{R}^n_+) \to L^p(\mathbb{R}^n_+)$, for $1 \leq p \leq \infty$, provided it satisfies the estimates

$$|k(x, y)| \leq C x_1 \Big(|x_1 + y_1| + |x' - y'|\Big)^{-(n+1)}$$

and

$$|k(x,y)| \leq C y_1 \Big(|x_1 + y_1| + |x' - y'| \Big)^{-(n+1)}$$

4. Let K be a closed, linear subspace of a Banach space V; consider the natural maps $j : K \hookrightarrow V$ and $\pi : V \to V/K$. Show that $j' : V' \to K'$ is surjective and that $\pi' : (V/K)' \to V'$ has range $K^{\perp}$.
5. Show that the set of invertible, bounded, linear maps on a Banach space V is open in $\mathcal{L}(V)$. (*Hint*: If T^{-1} exists, write $T + R = T(I + T^{-1}R)$.)
6. Let X be a compact metric space and $F : X \to X$ a continuous map. Define $T : C(X) \to C(X)$ by $Tu(x) = u(F(x))$. Show that $T' : \mathcal{M}(X) \to \mathcal{M}(X)$ is given by $(T'\mu)(E) = \mu(F^{-1}(E))$, for any Borel set $E \subset X$. Using Exercise 3 of §3, show that there is a *probability measure* μ on X such that $T'\mu = \mu$.

6. Compact operators

Throughout this section we will restrict attention to operators on Banach spaces. An operator $T \in \mathcal{L}(V, W)$ is said to be compact provided T takes any bounded subset of V to a relatively compact subset of W, that is, a set with compact closure. It suffices to assume that $T(B_1)$ is relatively compact in W, where B_1 is the closed unit ball in V. We denote the space of compact operators by $\mathcal{K}(V, W)$. The following proposition summarizes some elementary facts about $\mathcal{K}(V, W)$.

Proposition 6.1. *$\mathcal{K}(V, W)$ is a closed, linear subspace of $\mathcal{L}(V, W)$. Any T in $\mathcal{L}(V, W)$ with finite-dimensional range is compact. Furthermore, if $T \in \mathcal{K}(V, W)$, $S_1 \in \mathcal{L}(V_1, V)$, and $S_2 \in \mathcal{L}(W, W_2)$, then $S_2 T S_1 \in \mathcal{K}(V_1, W_2)$.*

Most of these assertions are obvious. We show that if $T_j \in \mathcal{K}(V, W)$ is norm convergent to T, then T is compact. Given any sequence (x_n) in B_1, one can pick successive subsequences on which $T_1 x_n$ converges, then $T_2 x_n$ converges, and so on, and by a diagonal argument produce a single subsequence (which we'll still denote (x_n)) such that for each j, $T_j x_n$ converges as $n \to \infty$. It is then easy to show that $T x_n$ converges, giving compactness of T.

A particular case of Proposition 6.1 is that $\mathcal{K}(V) = \mathcal{K}(V, V)$ is a closed, two-sided ideal of $\mathcal{L}(V)$.

The following gives a useful class of compact operators.

Proposition 6.2. *If X is a compact metric space, then the natural inclusion*

$$\iota : Lip(X) \longrightarrow C(X) \tag{6.1}$$

is compact.

Proof. It is easy to show that any compact metric space has a countable, dense subset; let $\{x_j : j = 1, 2, 3, \dots\}$ be dense in X. Say (f_n) is a bounded sequence in $\text{Lip}(X)$. We want to prove that a subsequence converges in $C(X)$. Since bounded subsets of $\mathbb{C}$ are relatively compact, we can pick a subsequence of (f_n) converging at x_1; then we can pick a further subsequence of this subsequence, converging at x_2, and so forth. The standard diagonal argument then produces a subsequence (which we continue to denote (f_n)) converging at each x_j. We claim that (f_n) converges uniformly on X, as a consequence of the uniform estimate

$$|f_n(x) - f_n(y)| \le K\, d(x, y), \tag{6.2}$$

with K independent of n. Indeed, pick $\varepsilon > 0$. Then pick $\delta > 0$ such that $K\delta < \varepsilon/3$. Since X is compact, we can select from $\{x_j\}$ finitely many points, say $\{x_1, \dots, x_N\}$, such that any $x \in X$ is of distance $\le \delta$ from one of these. Then pick M so large that $f_n(x_j)$ is within $\varepsilon/3$ of its limit for $1 \le j \le N$, for all $n \ge M$. Now, for any $x \in X$, picking $\ell \in \{1, \dots, N\}$ such that $d(x, x_\ell) \le \delta$, we have, for $k \ge 0$, $n \ge M$,

$$\begin{aligned} |f_{n+k}(x) - f_n(x)| &\le |f_{n+k}(x) - f_{n+k}(x_\ell)| \\ &\quad + |f_{n+k}(x_\ell) - f_n(x_\ell)| + |f_n(x_\ell) - f_n(x)| \\ &\le K\delta + \frac{\varepsilon}{3} + K\delta < \varepsilon, \end{aligned} \tag{6.3}$$

proving the proposition.

The argument given above is easily modified to show that $\iota : \mathrm{Lip}^\alpha(X) \to C(X)$ is compact, for any $\alpha > 0$. Indeed, there is the following more general result. Let $\omega : X \times X \to [0, \infty)$ be any continuous function, vanishing on the diagonal $\Delta = \{(x, x) : x \in X\}$. Fix $K \in \mathbb{R}^+$. Let $\mathcal{F}$ be any subset of $C(X)$ satisfying

$$|u(x)| \le K, \quad |u(x) - u(y)| \le K\, \omega(x, y). \tag{6.4}$$

The latter condition is called *equicontinuity*. Ascoli's theorem states that such a set $\mathcal{F}$ is relatively compact in $C(X)$ whenever X is a compact Hausdorff space. The proof is a further extension of the argument given above.

We note another refinement of Proposition 6.2, namely that the inclusion ι : $\mathrm{Lip}^\alpha(X) \to \mathrm{Lip}^\beta(X)$ is compact whenever $0 \le \beta < \alpha \le 1$, X a compact metric space. Compare results on inclusions of Sobolev spaces given in Chapter 4.

We next look at persistence of compactness upon taking adjoints.

Proposition 6.3. *If $T \in \mathcal{K}(V, W)$, then T' is also compact.*

Proof. Let (ω_n) be sequence in B_1', the closed unit ball in W'. Consider (ω_n) as a sequence of continuous functions on the compact space $X = \overline{T(B_1)}$, B_1 being the unit ball in V. Ascoli's theorem, indeed its special case, Proposition 6.2, applies; there exists a subsequence (ω_{n_k}) converging uniformly on X. Thus $(T'\omega_{n_k})$ is a sequence in V' converging uniformly on B_1, hence in the V'-norm. This completes the proof.

The following provides a useful improvement over the a priori statement that, for $T \in \mathcal{K}(V, W)$, the image $T(B_1)$ of the closed unit ball $B_1 \subset V$ is relatively compact in W.

Proposition 6.4. *Assume V is separable and reflexive. If $T : V \to W$ is compact, then the image of the closed unit ball $B_1 \subset V$ under T is compact.*

Proof. From Proposition 4.4 and the remark following its proof, B_1, with the weak*-topology (the $\sigma(V, V')$-topology, since $V = V''$), is a compact metric space, granted that V' is also separable, which we now demonstrate. Indeed, for any Banach space Y, it is a simple consequence of the Hahn-Banach theorem that Y is separable provided Y' is separable; if Y is reflexive, this implication can be reversed.

Consequently, given a sequence $v_n \in B_1$, possessing a subsequence $v_n^{(1)}$ such that $Tv_n^{(1)}$ converges in W, say to w, you can pass to a further subsequence $v_n^{(2)}$, which is weak*-convergent in V, with limit $v \in B_1$. It follows that $Tv_n^{(2)}$ is weakly convergent to Tv; for any $\omega \in W'$, $Tv_n^{(2)}(\omega) = v_n^{(2)}(T'\omega) \to v(T'\omega) = (Tv)(\omega)$. Hence $Tv = w$. This shows that $T(B_1)$ is closed in W, and hence completes the proof.

Remark: It is possible to drop the assumption that V is separable, via an argument replacing sequences by nets in order to construct the weak* limit point v.

We next derive some results on the spectral theory of a compact operator A on a Hilbert space H that is self-adjoint, so $A = A^*$. For simplicity, we will assume that H is separable, though that hypothesis can easily be dropped.

Proposition 6.5. *If $A \in \mathcal{L}(H)$ is compact and self-adjoint, then either $\|A\|$ or $-\|A\|$ is an eigenvalue of A, that is, there exists $u \neq 0$ in H such that*

$$Au = \lambda u, \tag{6.5}$$

with $\lambda = \pm\|A\|$.

Proof. By Proposition 6.4, we know that the image under A of the closed unit ball in H is compact, so the norm assumes a maximum on this image. Thus there exists $u \in H$ such that

$$\|u\| = 1, \quad \|Au\| = \|A\|. \tag{6.6}$$

Pick any unit $w \perp u$. Self-adjointness implies $\|Ax\|^2 = (A^2x, x)$, so we have, for all real s,

$$\big(A^2(u + sw), u + sw\big) \le \|A\|^2(1 + s^2), \tag{6.7}$$

equality holding at $s = 0$. Since the left side is equal to

$$\|A\|^2 + 2s \operatorname{Re}(A^2u, w) + s^2\|Aw\|^2,$$

this inequality for $s \to 0$ implies $\text{Re}(A^2u, w) = 0$; replacing w by iw gives $(A^2u, w) = 0$ whenever $w \perp u$. Thus A^2u is parallel to u, that is, $A^2u = cu$ for some scalar c; (6.6) implies $c = \|A\|^2$. Now, assuming $A \neq 0$, set $v = \|A\|u + Au$. If $v = 0$, then u satisfies (6.5) with $\lambda = -\|A\|$. If $v \neq 0$, then v is an eigenvector of A with eigenvalue $\lambda = \|A\|$.

The space of $u \in H$ satisfying (6.5) is called the λ-*eigenspace* of A. Clearly, if A is compact and $\lambda \neq 0$, such a λ-eigenspace must be finite-dimensional. If $Au_j = \lambda_j u_j$, $A = A^*$, then

$$\lambda_1(u_1, u_2) = (Au_1, u_2) = (u_1, Au_2) = \overline{\lambda}_2(u_1, u_2). \tag{6.8}$$

With $\lambda_1 = \lambda_2$ and $u_1 = u_2$, this implies that each eigenvalue of $A = A^*$ is *real*. With $\lambda_1 \neq \lambda_2$, it then yields $(u_1, u_2) = 0$, so any distinct eigenspaces of $A = A^*$ are orthogonal. We also note that if $Au_1 = \lambda_1 u_1$ and $v \perp u_1$, then $(u_1, Av) = (Au_1, v) = \lambda_1(u_1, v) = 0$, so $A = A^*$ leaves invariant the orthogonal complement of any of its eigenspaces.

Now if A is compact and self-adjoint on H, we can apply Proposition 6.5, restrict A to the orthogonal complement of its $\pm\|A\|$-eigenspaces (where its norm must be strictly smaller, as a consequence of Proposition 6.5), apply the proposition again, to this restriction, and continue. In this fashion we arrive at the following result, known as the *spectral theorem* for compact, self-adjoint operators.

Proposition 6.6. *If $A \in \mathcal{L}(H)$ is a compact, self-adjoint operator on a Hilbert space H, then H has an orthonormal basis u_j of eigenvectors of A. With $Au_j = \lambda_j u_j$, (λ_j) is a sequence of real numbers with only 0 as an accumulation point.*

The spectral theorem has a more elaborate formulation for general self-adjoint operators. It is proved in Chapter 8.

We next give a result that will be useful in the study of spectral theory of compact operators that are not self-adjoint. It will also be useful in §7. Let V, W and Y be Banach spaces.

Proposition 6.7. *Let $T \in \mathcal{L}(V, W)$. Suppose $K \in \mathcal{K}(V, Y)$ and*

$$\|u\|_V \leq C\|Tu\|_W + C\|Ku\|_Y, \tag{6.9}$$

for all $u \in V$. Then T has closed range.

Proof. Let $Tu_n \to f$ in W. We need $v \in V$ with $Tv = f$. Let $L = \text{Ker}\, T$. We divide the argument into two cases.

If $\text{dist}(u_n, L) \leq a < \infty$, take $v_n = u_n \bmod L$, $\|v_n\| \leq 2a$; then $Tv_n = Tu_n \to f$. Passing to a subsequence, we have $Kv_n \to g$ in Y. Then (6.9), applied to $u = v_n - v_m$, implies that (v_n) is Cauchy, so $v_n \to v$ and $Tv = f$.

If $\text{dist}(u_n, L) \to \infty$, we can assume that $\text{dist}(u_n, L) \geq 2$ for all n. Pick $v_n = u_n \bmod L$ such that $\text{dist}(u_n, L) \leq \|v_n\| \leq \text{dist}(u_n, L) + 1$, and set $w_n = v_n/\|v_n\|$. Note that $\text{dist}(w_n, L) \geq 1/2$. Since $\|w_n\| = 1$, we can take a subsequence and assume $Kw_n \to g$ in Y. Since $Tw_n \to 0$, (6.9) applied to $w_n - w_m$ implies (w_n) is Cauchy. Thus $w_n \to w$ in V, and we see that simultaneously $\text{dist}(w, L) \geq 1/2$ and $Tw = 0$, a contradiction. Hence this latter case is impossible, and the proposition is proved.

Note that Proposition 6.7 applies to the case $V = W = Y$ and $T = \zeta I - K$, for $K \in \mathcal{K}(V)$ and ζ a nonzero scalar. Such an operator therefore has closed range. The next result is called the *Fredholm alternative*.

Proposition 6.8. *For $\zeta \neq 0$, $K \in \mathcal{K}(V)$, the operator $T = \zeta I - K$ is surjective if and only if it is injective.*

Proof. Assume T is injective. Then $T : V \to \mathcal{R}(T)$ is bijective. By Proposition 6.7, $\mathcal{R}(T)$ is a Banach space, so the open mapping theorem implies that $T : V \to \mathcal{R}(T)$ is a topological isomorphism. If $\mathcal{R}(T) = V_1$ is not all of V, then $V_2 = T(V_1)$, $V_3 = T(V_2)$, and so on, form a strictly decreasing family of closed subspaces. By Lemma 1.3, we can pick $v_\nu \in V_n$ with $\|v_n\| = 1$, $\text{dist}(v_n, V_{n+1}) \geq 1/2$. Thus, for $n > m$,

$$Kv_m - Kv_n = \zeta v_m + [-\zeta v_n - (Tv_m - Tv_n)] \\ = \zeta v_m + w_{mn}, \tag{6.10}$$

with $w_{mn} \in V_{m+1}$. Hence $\|Kv_n - Kv_m\| \geq |\zeta|/2$, contradicting compactness of K. Consequently, T is surjective if it is injective.

For the converse, we use Proposition 5.7. If T is surjective, (5.19) implies $T' = \zeta I - K'$ is injective on V'. Since K' is compact, the argument above implies T' is surjective, and hence, by (5.20), T is injective.

A substantial generalization of this last result will be contained in Proposition 7.4 and Corollary 7.5.

It follows that every $\zeta \neq 0$ in the spectrum of a compact K is an eigenvalue of K. We hence derive the following result on $\sigma(K)$.

Proposition 6.9. *If $K \in \mathcal{K}(V)$, the spectrum $\sigma(K)$ has only 0 as an accumulation point.*

Proof. Suppose we have linearly independent $v_n \in V$, $\|v_n\| = 1$, with $Kv_n = \lambda_n v_n$, $\lambda_n \to \lambda \neq 0$. Let V_n be the linear span of $\{v_1, \dots, v_n\}$. By Lemma 1.3, there exist $y_n \in V_n$, $\|y_n\| = 1$, such that $\text{dist}(y_n, V_{n-1}) \geq 1/2$. With $T_\lambda = \lambda I - K$, we have, for $n > m$,

$$\lambda_n^{-1}Ky_n - \lambda_m^{-1}Ky_m = y_n + \left[-y_m + \lambda_n^{-1}T_{\lambda_n}y_n + \lambda_m^{-1}T_{\lambda_m}y_m\right] \\ = y_n + z_{nm}, \tag{6.11}$$

where $z_{nm} \in V_{n-1}$ since $T_{\lambda_n}y_n \in V_{n-1}$. Hence $\|\lambda_n^{-1}Ky_n - \lambda_m^{-1}Ky_m\| \geq 1/2$, which contradicts compactness of K.

Note that if $\lambda_j \neq 0$ is such an isolated point in the spectrum $\sigma(K)$ of a compact operator K, and we take γ_j to be a small circle enclosing λ_j but no other points of $\sigma(K)$, then, as in (5.32), the operator

$$P_j = \frac{1}{2\pi i} \int_{\gamma_j} (\zeta - K)^{-1} \, d\zeta$$

is a projection onto a closed subspace V_j of V with the property that the restriction of K to V_j (equal to $P_j K P_j$) has spectrum consisting of the one point $\{\lambda_j\}$. Thus V_j must be finite-dimensional. $K|_{V_j}$ may perhaps not be scalar; it might have a Jordan normal form with λ_j down the diagonal and some ones directly above the diagonal.

Having established a number of general facts about compact operators, we take a look at an important class of compact operators on Hilbert spaces: the Hilbert-Schmidt operators, defined as follows. Let H be a separable Hilbert space and $A \in \mathcal{L}(H)$. Let $\{u_j\}$ be an orthonormal basis of H. We say A is a *Hilbert-Schmidt operator*, or an HS operator for short, provided

$$\sum_j \|Au_j\|^2 < \infty, \tag{6.12}$$

or equivalently, if

$$\sum_{j,k} |a_{jk}|^2 < \infty, \quad a_{jk} = (Au_k, u_j). \tag{6.13}$$

The class of HS operators on H will be denoted $\text{HS}(H)$. The first characterization makes it clear that if A is HS and B is bounded, then BA is HS. The second makes it clear that A^* is HS if A is; hence $AB = (B^*A^*)^*$ is HS if A is HS and B is bounded. Thus (6.12) is independent of the choice of orthonormal basis $\{u_j\}$. We also define the Hilbert-Schmidt norm of an HS operator:

$$\|A\|_{\text{HS}}^2 = \sum_j \|Au_j\|^2 = \sum_{j,k} |a_{jk}|^2. \tag{6.14}$$

The first identity makes it clear that $\|BA\|_{\text{HS}} \le \|B\| \cdot \|A\|_{\text{HS}}$ if A is HS and B is bounded, and in particular

$$\|UA\|_{\text{HS}} = \|A\|_{\text{HS}}$$

when U is unitary. The second identity in (6.14) shows that

$$\|A^*\|_{\text{HS}} = \|A\|_{\text{HS}}.$$

Using $AU = (U^*A^*)^*$, we deduce that

$$\|AU\|_{\text{HS}} = \|A\|_{\text{HS}}$$

when U is unitary. Thus, for U unitary, $\|UAU^{-1}\|_{\text{HS}} = \|A\|_{\text{HS}}$, so the HS-norm in (6.14) is independent of the choice of orthonormal basis for H.

From (6.12) it follows that an HS operator A is a norm limit of finite-rank operators, hence compact. If $A = A^*$, and we choose an orthonormal basis of eigenvectors of A, with eigenvalues μ_j, then

(6.15) $$\sum_j |\mu_j|^2 = \|A\|_{\mathrm{HS}}^2.$$

A compact, self-adjoint operator A is HS if and only if the left side of (6.15) is finite.

If $A : H_1 \to H_2$ is a bounded operator, we can say it is HS provided AV is HS for some unitary map $V : H_2 \to H_1$, with obvious adjustments when either H_1 or H_2 is finite-dimensional.

The following classical result might be called the Hilbert-Schmidt kernel theorem. In Chapter 4 it is used as an ingredient in the proof of the celebrated Schwartz kernel theorem.

Proposition 6.10. *If $T : L^2(X_1, \mu_1) \to L^2(X_2, \mu_2)$ is HS, then there exists a function $K \in L^2(X_1 \times X_2, \mu_1 \times \mu_2)$ such that*

(6.16) $$(Tu, v)_{L^2} = \iint K(x_1, x_2) u(x_1) \overline{v(x_2)}\, d\mu_1(x_1)\, d\mu_2(x_2).$$

Proof. Pick orthonormal bases $\{f_j\}$ for $L^2(X_1)$ and $\{g_k\}$ for $L^2(X_2)$, and set

$$K(x_1, x_2) = \sum_{j,k} a_{jk} \overline{f_j(x_1)} g_k(x_2),$$

where $a_{jk} = (Tf_j, g_k)$. The hypothesis that T is HS is precisely what is necessary to guarantee that $K \in L^2(X_1 \times X_2)$, and then (6.16) is obvious. It is also clear that

(6.17) $$\|T\|_{\mathrm{HS}}^2 = \|K\|_{L^2}^2.$$

Also of interest is the converse, proved simply by reversing the argument:

Proposition 6.11. *If $K \in L^2(X_1 \times X_2, \mu_1 \times \mu_2)$, then (6.16) defines an HS operator T, satisfying (6.17).*

We note that the HS-square norm polarizes to a Hilbert space inner product on $\mathrm{HS}(H)$:

(6.18) $$(A, B)_{\mathrm{HS}} = \sum_{j,k} a_{jk} \overline{b}_{jk}$$

if, parallel to (6.13), $b_{jk} = (Bu_k, u_j)$, given an orthonormal basis $\{u_j\}$. Since the norm uniquely determines the inner product, we have without further calculation the independence of $(A, B)_{\mathrm{HS}}$ under change of orthonormal basis; more generally, $(A, B)_{\mathrm{HS}} = (UAV, UBV)_{\mathrm{HS}}$ for unitary U and V on H.

Note that $\sum_k a_{jk} \overline{b}_{\ell k} = c_{j\ell}$ form the matrix coefficients of $C = AB^*$, and (6.18) is the sum of the diagonal elements of C; we write

(6.19) $$(A, B)_{\mathrm{HS}} = \mathrm{Tr}\, AB^*.$$

Generally, we say an operator $C \in \mathcal{L}(H)$ is *trace class* if it can be written as a product of two HS operators; call them A and B^*, and then $\operatorname{Tr} C$ is defined to be given by (6.19). It is not clear at first glance that TR, the set of trace class operators, is a linear space, but this can be seen as follows. If $C_j = A_j B_j^*$, then

$$C_1 + C_2 = (A_1 \quad A_2)\begin{pmatrix} B_1^* \\ B_2^* \end{pmatrix}. \tag{6.20}$$

Note that a given $C \in \mathrm{TR}$ may be written as a product of two HS operators in many different ways, but the computation of $\operatorname{Tr} C$ is unaffected, since as we have already seen, the definition (6.19) leads to the computation

$$\operatorname{Tr} C = \sum_j c_{jj}, \quad c_{jj} = (Cu_j, u_j). \tag{6.21}$$

This formula shows that $\operatorname{Tr}:\mathrm{TR} \to \mathbb{C}$ is a linear map. Furthermore, by our previous remarks on $(\ ,\ )_{\mathrm{HS}}$, the trace formula (6.21) is independent of the choice of orthonormal basis of H.

There is an intrinsic characterization of trace class operators:

Proposition 6.12. *An operator $C \in \mathcal{L}(H)$ is trace class if and only if C is compact and the operator $(C^*C)^{1/2}$ has the property that its set of eigenvalues $\{\lambda_j\}$ is summable; $\sum \lambda_j < \infty$.*

Proof. Given C compact, let $\{u_j\}$ be an orthonormal basis of H consisting of eigenvectors of C^*C, which is compact and self-adjoint. Say $C^*Cu_j = \lambda_j^2 u_j$, $\lambda_j \geq 0$. Then the identity $(C^*C)^{1/2}u_j = \lambda_j u_j$ defines $(C^*C)^{1/2}$.

Note that, for all $v \in H$,

$$\|(C^*C)^{1/2}v\|^2 = (C^*Cv, v) = \|Cv\|^2. \tag{6.22}$$

Thus $Cv \mapsto (C^*C)^{1/2}v$ extends to an isometric isomorphism between the ranges of C and of $(C^*C)^{1/2}$, yielding in turn operators V and W of norm 1 such that

$$C = V(C^*C)^{1/2}, \quad (C^*C)^{1/2} = WC. \tag{6.23}$$

Now, if $\sum \lambda_j < \infty$, define $A \in \mathcal{L}(H)$ by $Au_j = \lambda_j^{1/2}u_j$. Hence A is Hilbert-Schmidt, and $C = VA \cdot A$, so C is trace class. Conversely, if $C = AB^*$ with $A, B \in \mathrm{HS}$, then $(C^*C)^{1/2} = WA \cdot B^*$ is a product of HS operators, hence of trace class. The computation (6.21), using the basis of eigenvectors of C^*C, then yields $\sum \lambda_j = \operatorname{Tr}(C^*C)^{1/2} < \infty$, and the proof is complete.

It is desirable to establish some results about TR as a linear space. Given $C \in \mathrm{TR}$, we define

$$\|C\|_{\mathrm{TR}} = \inf\{\|A\|_{\mathrm{HS}}\|B\|_{\mathrm{HS}} : C = AB^*\}. \tag{6.24}$$

This is a norm; in particular,

$$\|C_1 + C_2\|_{\mathrm{TR}} \leq \|C_1\|_{\mathrm{TR}} + \|C_2\|_{\mathrm{TR}}. \tag{6.25}$$

This can be seen by using (6.20), with A_2 replaced by tA_2 and B_2^* by $t^{-1}B_2^*$, and minimizing over $t \in (0,\infty)$ the quantity

$$\begin{aligned}\|(A_1, tA_2)\|_{HS}^2 \cdot \|(B_1, t^{-1}B_2)^t\|_{HS}^2 \\ = \big(\|A_1\|_{HS}^2 + t^2\|A_2\|_{HS}^2\big)\cdot\big(\|B_1\|_{HS}^2 + t^{-2}\|B_2\|_{HS}^2\big).\end{aligned}$$

Next, we note that (6.24) easily yields

$$\|C^*\|_{TR} = \|C\|_{TR} \tag{6.26}$$

and, for bounded S_j,

$$\|S_1 C S_2\|_{TR} \le \|S_1\| \cdot \|C\|_{TR} \cdot \|S_2\|, \tag{6.27}$$

with equality if S_1 and S_2 are unitary. Also, using (6.23), we have

$$\|C\|_{TR} = \|(C^*C)^{1/2}\|_{TR}. \tag{6.28}$$

Using (6.24) with C replaced by $D = (C^*C)^{1/2}$, the choice $A = B = D^{1/2}$ yields

$$\|(C^*C)^{1/2}\|_{TR} \le \|(C^*C)^{1/4}\|_{HS}^2 = \operatorname{Tr}(C^*C)^{1/2}. \tag{6.29}$$

On the other hand, we have, by (6.19) and Cauchy's inequality,

$$|\operatorname{Tr}(AB^*)| \le \|A\|_{HS}\|B\|_{HS}, \tag{6.30}$$

and hence, for $C \in$ TR,

$$|\operatorname{Tr} C| \le \|C\|_{TR}. \tag{6.31}$$

If we apply this, with C replaced by $(C^*C)^{1/2}$, and compare with (6.28)–(6.29), we have

$$\|C\|_{TR} = \operatorname{Tr}(C^*C)^{1/2}. \tag{6.32}$$

Either directly or as a simple consequence of this, we have

$$\|C\|_{TR} \ge \|C\|_{HS} \ge \|C\|. \tag{6.33}$$

We can now establish:

Proposition 6.13. *Given a Hilbert space H, the space TR of trace class operators on H is a Banach space, with norm (6.24).*

Proof. It suffices to prove completeness. Thus let (C_j) be Cauchy in TR. Passing to a subsequence, we can assume $\|C_{j+1} - C_j\|_{TR} \le 8^{-j}$. Then write $C = \sum \widetilde{C}_j$, where $\widetilde{C}_1 = C_1$ and, for $j \ge 2$, $\widetilde{C}_j = C_j - C_{j-1}$. By (6.33), C is a bounded operator on H. Write

$$\widetilde{C}_j = \widetilde{A}_j \widetilde{B}_j^*, \quad \|\widetilde{A}\|_{HS}, \|\widetilde{B}\|_{HS} \le 2^{-j}.$$

Then we can form

$$A = \widetilde{A}_1 \oplus \widetilde{A}_2 \oplus \cdots, \quad B = \widetilde{B}_1 \oplus \widetilde{B}_2 \oplus \cdots \in \mathcal{L}(\mathcal{H}, H),$$

where $\mathcal{H} = H \oplus H \oplus \cdots$, check that A and B are Hilbert-Schmidt, and note that $C = AB^*$. Hence $C \in \mathrm{TR}$ and $C_j \to C$ in TR-norm.

The classes HS and TR are the most important cases of a continuum of ideals $\mathcal{I}_p \subset \mathcal{L}(H)$, $1 \le p < \infty$. One says $C \in \mathcal{K}(H)$ belongs to $\mathcal{I}_p$ if and only if $(C^*C)^{p/2}$ is trace class. Then $\mathrm{TR} = \mathcal{I}_1$ and $\mathrm{HS} = \mathcal{I}_2$. For more on this topic, see [Si].

We next discuss the trace of an integral operator. Let A and B be two HS operators on $L^2(X,\mu)$, with integral kernels $K_A, K_B \in L^2(X \times X, \mu \times \mu)$. Then $C = AB$ is given by

$$Cu(x) = \iint K_A(x,z) K_B(z,y) u(y)\, d\mu(y)\, d\mu(z), \tag{6.34}$$

and we have, by (6.17) and (6.19),

$$\operatorname{Tr} C = \iint K_A(x,z) K_B(z,x)\, d\mu(z)\, d\mu(x). \tag{6.35}$$

Now C has an integral kernel $K_C \in L^2(X \times X, \mu \times \mu)$:

$$K_C(x,y) = \int K_A(x,z) K_B(z,y)\, d\mu(z), \tag{6.36}$$

which strongly suggests the trace formula

$$\operatorname{Tr} C = \int K_C(x,x)\, d\mu(x). \tag{6.37}$$

The only sticky point is that the diagonal $\{(x,x) : x \in X\}$ may have measure 0 in $X \times X$, so one needs to define $K_C(x,y)$ carefully. The formula (6.35) implies, via Fubini's theorem, that

$$K_C(x,x) = \int K_A(x,z) K_B(z,x)\, d\mu(z)$$

exists for μ-almost every $x \in X$, and for this function, the identity (6.37) holds. In many cases of interest, X is a locally compact space and $K_C(x,y)$ is continuous, and then passing from (6.35) to (6.37) is straightforward.

We next give a treatment of the determinant of $I + A$, for trace class A. This is particularly useful for results on trace formulas and the scattering phase, in Chapter 9. Our treatment largely follows [Si]; another approach can be found in Chapter 11 of [DS].

With $\Lambda^j C$ the operator induced by C on $\Lambda^j H$, we define

$$\det(I + C) = 1 + \sum_{j \ge 1} \operatorname{Tr} \Lambda^j C. \tag{6.38}$$

It is not hard to show that if $C_j = \Lambda^j C$ and $D_j = (C_j^* C_j)^{1/2}$, then $D_j = \Lambda^j (C^*C)^{1/2}$, so

$$\|C_j\|_{\mathrm{TR}} = \operatorname{Tr} D_j = \sum_{i_1 < \cdots < i_j} \mu_{i_1} \cdots \mu_{i_j}, \tag{6.39}$$

where μ_i, $i \geq 1$, are the positive eigenvalues of the compact, positive operator $(C^*C)^{1/2}$, counted with multiplicity. In particular,

$$\|C_j\|_{\mathrm{TR}} \leq \frac{1}{j!}\|C\|_{\mathrm{TR}}^j, \tag{6.40}$$

so (6.38) is absolutely convergent for any $C \in \mathrm{TR}$. Note that in the finite-dimensional case, (6.38) is simply the well-known expansion of the characteristic polynomial. Replacing C by zC, $z \in \mathbb{C}$, we obtain an entire holomorphic function of z:

$$\det(I + zC) = 1 + \sum_{j\geq 1} z^j \operatorname{Tr} \Lambda^j C. \tag{6.41}$$

This replacement causes D_j to be replaced by $|z|^j D_j$, and (6.39) implies

$$|\det(I + zC)| \leq \det(I + |z|D) = \prod_{i\geq 1}(1 + \mu_i|z|), \tag{6.42}$$

the latter identity following by diagonalization of the compact, self-adjoint operator D. Note that since $1 + r \leq e^r$, for $r \geq 0$,

$$\prod_{i\geq \ell}(1 + \mu_i|z|) \leq e^{\kappa_\ell |z|}, \quad \kappa_\ell = \sum_{i\geq \ell} \mu_i. \tag{6.43}$$

Taking $\ell = 1$, we have

$$|\det(I + zC)| \leq e^{|z|\|C\|_{\mathrm{TR}}}. \tag{6.44}$$

Also,

$$|\det(I + zC)| \leq \Big\{\prod_{i=1}^{\ell-1}(1 + \mu_i|z|)\Big\} e^{\kappa_\ell |z|}, \quad \forall\, \ell. \tag{6.45}$$

Hence, for any $C \in \mathrm{TR}$,

$$|\det(I + zC)| \leq C_\varepsilon e^{\varepsilon|z|}, \quad \forall\, \varepsilon > 0. \tag{6.46}$$

We next establish the continuous dependence of the determinant.

Proposition 6.14. *We have a continuous map $F : TR \to \mathbb{C}$, given by*

$$F(A) = \det(I + A).$$

Proof. For fixed $C, D \in TR$, $g(z) = F(C + zD)$ is holomorphic, as one sees from (6.40) and (6.41). Now consider

$$h(z) = F\Big(\frac{1}{2}(A + B) + z(A - B)\Big). \tag{6.47}$$

Then

$$\begin{aligned} |F(A) - F(B)| = \Big|h\Big(\frac{1}{2}\Big) - h\Big(-\frac{1}{2}\Big)\Big| &\leq \sup\Big\{|h'(t)| : -\frac{1}{2} \leq t \leq \frac{1}{2}\Big\} \\ &\leq R^{-1} \sup_{|z|\leq R+1/2} |h(z)|. \end{aligned} \tag{6.48}$$

In turn, we can estimate $|h(z)|$ using (6.45). If we take $R = \|A - B\|_{\mathrm{TR}}^{-1}$, we get

$$|F(A) - F(B)| \le \|A - B\|_{\mathrm{TR}} \exp\{\|A\|_{\mathrm{TR}} + \|B\|_{\mathrm{TR}} + 1\}, \tag{6.49}$$

which proves the proposition.

One use of Proposition 6.14 is as a tool to prove the following.

Proposition 6.15. *For any $A, B \in TR$,*

$$\det\big((I + A)(I + B)\big) = \det(I + A) \cdot \det(I + B). \tag{6.50}$$

Proof. By Proposition 6.14, it suffices to prove (6.50) when A and B are finite rank operators, in which case it is elementary.

The following is an important consequence of (6.50).

Proposition 6.16. *Given $A \in TR$, we have*

$$I + A \text{ invertible} \Longleftrightarrow \det(I + A) \neq 0. \tag{6.51}$$

Proof. If $I + A$ is invertible, the inverse has the form

$$(I + A)^{-1} = I + B, \quad B = -A(I + A)^{-1} \in \mathrm{TR}. \tag{6.52}$$

Hence (6.50) implies $\det(I + A)\det(I + B) = 1$, so $\det(I + A) \neq 0$.

For the converse, assume $I + A$ is not invertible, so $-1 \in \operatorname{Spec}(A)$. Since A is compact, we can consider the associated spectral projection P of H onto the generalized (-1)-eigenspace of A. Since $(PA)(I - P)A = 0$, we have

$$\det(I + A) = \det(I + AP) \cdot \det(I + A(I - P)). \tag{6.53}$$

It is elementary that $\det(I + AP) = 0$, so the proposition is proved.

As another application of (6.50), we can use the identity

$$I + A + sB = (I + A)\big(I + s(I + A)^{-1}B\big) \tag{6.54}$$

to show that

$$\frac{d}{ds}\det\big(I + A(s)\big) = \det\big(I + A(s)\big) \cdot \operatorname{Tr}\big((I + A(s))^{-1}A'(s)\big), \tag{6.55}$$

when $A(s)$ is a differentiable function of s with values in TR.

Exercises

1. If A is a Hilbert-Schmidt operator, show that

$$\|A\| \le \|A\|_{\mathrm{HS}},$$

where the left side denotes the operator norm. (*Hint*: Pick unit u_1 such that $\|Au_1\| \ge \|A\| - \varepsilon$, and make that part of an orthonormal basis.)

2. Suppose $K \in L^2(X \times X, \mu \times \mu)$ satisfies $K(x, y) = \overline{K(y, x)}$. Show that
$$K(x, y) = \sum c_j \, u_j(x) \, \overline{u_j(y)}$$
with $\{u_j\}$ an orthonormal set in $L^2(X, \mu)$, $c_j \in \mathbb{R}$, and $\sum c_j^2 < \infty$.
(*Hint*: Apply the spectral theorem for compact, self-adjoint operators.)
3. Define $T : L^2(I) \to L^2(I)$, $I = [0, 1]$, by
$$Tf(x) = \int_0^x f(y) \, dy.$$
Show that T has range $\mathcal{R}(T) \subset \{u \in C(I) : u(0) = 0\}$. Show that T is compact, that T has no eigenvectors, and that $\sigma(T) = \{0\}$. Also, show that T is HS, but not trace class.
4. Let K be a closed bounded subset of a Banach space B. Suppose T_j are compact operators on B and $T_j x \to x$ for each $x \in B$. Show that K is compact if and only if $T_j \to I$ uniformly on K.
5. Prove the following result, also known as part of Ascoli's theorem. If X is a compact metric space, B_j are Banach spaces, and $K : B_1 \to B_2$ is a compact operator, then $\kappa f(x) = K(f(x))$ defines a compact map $\kappa : C^\alpha(X, B_1) \to C(X, B_2)$, for any $\alpha > 0$.
6. Let B be a bounded operator on a Hilbert space H, and let A be trace class. Show that
$$\operatorname{Tr}(AB) = \operatorname{Tr}(BA).$$
(*Hint*: Write $A = A_1 A_2$ with $A_j \in \text{HS}$.)
7. Given a Hilbert space H, define $\Lambda^j H$ as a Hilbert space and justify (6.39). Also, check the finite rank case of (6.50).
8. Assume $\{u_j : j \ge 1\}$ is an orthonormal basis of the Hilbert space H, and let P_n denote the orthogonal projection of H onto the span of $\{u_1, \dots, u_n\}$. Show that if $A \in \text{TR}$, then $P_n A P_n \to A$ in TR-norm. (This is used implicitly in the proof of Proposition 6.15.)

7. Fredholm operators

Again in this section we restrict attention to operators on Banach spaces. An operator $T \in \mathcal{L}(V, W)$ is said to be Fredholm provided

$$\text{Ker}\, T \text{ is finite-dimensional} \tag{7.1}$$

and

$$T(V) \text{ is closed in } W, \text{ of finite codimension,} \tag{7.2}$$

that is, $W/T(V)$ is finite-dimensional. We say T belongs to $\text{Fred}(V, W)$. We define the *index* of T to be

$$\text{Ind}\, T = \dim \text{Ker}\, T - \dim W/T(V), \tag{7.3}$$

the last term also denoted $\text{Codim}\, T(V)$. Note the isomorphism $\big(W/T(V)\big)' \approx T(V)^\perp$. By (5.19), $T(V)^\perp = \text{Ker}\, T'$. Consequently,

$$\text{Ind}\, T = \dim \text{Ker}\, T - \dim \text{Ker}\, T'. \tag{7.4}$$

Furthermore, by Proposition 5.7, we deduce that if T is Fredholm, $T' \in \mathcal{L}(W', V')$ is also Fredholm, and

$$\operatorname{Ind} T' = -\operatorname{Ind} T. \tag{7.5}$$

The following is a useful characterization of Fredholm operators.

Proposition 7.1. *Let $T \in \mathcal{L}(V, W)$. Then T is Fredholm if and only if there exist $S_j \in \mathcal{L}(W, V)$ such that*

$$S_1 T = I + K_1 \tag{7.6}$$

and

$$T S_2 = I + K_2, \tag{7.7}$$

with K_1 and K_2 compact.

Proof. The identity (7.6) implies $\operatorname{Ker} T \subset \operatorname{Ker}(I + K_1)$, which is finite-dimensional. Also, by Proposition 6.7, (7.6) implies T has closed range. On the other hand, (7.7) implies $T(V)$ contains the range of $I + K_2$, which has finite codimension in light of the spectral theory of K_2 derived in the last section. The converse result, that $T \in \operatorname{Fred}(V, W)$ has such "Fredholm inverses" S_j, is easy.

Note that, by virtue of the identity

$$S_1(I + K_2) = S_1 T S_2 = (I + K_1) S_2, \tag{7.8}$$

we see that whenever (7.6) and (7.7) hold, S_1 and S_2 must differ by a compact operator. Thus we could take $S_1 = S_2$.

The following result is an immediate consequence of the characterization of the space $\operatorname{Fred}(V, W)$ by (7.6)–(7.7).

Corollary 7.2. *If $T \in \operatorname{Fred}(V, W)$ and $K : V \to W$ is compact, then $T + K \in \operatorname{Fred}(V, W)$. If also $T_2 \in \operatorname{Fred}(W, X)$, then $T_2 T \in \operatorname{Fred}(V, X)$.*

Proposition 7.1 also makes it natural to consider the quotient space $\mathcal{Q}(V) = \mathcal{L}(V)/\mathcal{K}(V)$. Recall that $\mathcal{K}(V)$ is a closed, two-sided ideal of $\mathcal{L}(V)$. Thus the quotient is a Banach space, and in fact a Banach algebra. It is called the *Calkin algebra*. One has the natural algebra homomorphism $\pi : \mathcal{L}(V) \to \mathcal{Q}(V)$, and a consequence of Proposition 7.1 is that $T \in \mathcal{L}(V)$ is Fredholm if and only if $\pi(T)$ is invertible in $\mathcal{Q}(V)$. For general $T \in \operatorname{Fred}(V, W)$, the operators $S_1 T$ and $T S_2$ in (7.6) and (7.7) project to the identity in $\mathcal{Q}(V)$ and $\mathcal{Q}(W)$, respectively. Now the argument made in §5 that the set of invertible elements of $\mathcal{L}(V)$ is open, via Proposition 5.8, applies equally well when $\mathcal{L}(V)$ is replaced by any Banach algebra with unit. Applying it to the Calkin algebra, we have the following:

Proposition 7.3. *$\operatorname{Fred}(V, W)$ is open in $\mathcal{L}(V, W)$.*

We now establish a fundamental result about the index of Fredholm operators.

Proposition 7.4. *The index map*

$$\text{(7.9)} \qquad Ind : Fred(V, W) \longrightarrow \mathbb{Z}$$

defined by (7.3) is constant on each connected component of Fred(V, W).

Proof. Let $T \in \text{Fred}(V, W)$. It suffices to show that if $S \in \mathcal{L}(V, W)$ and if $\|T - S\|$ is small enough, then $\text{Ind}\, S = \text{Ind}\, T$. We can pick a closed subspace $V_1 \subset V$, complementary to $\text{Ker}\, T$ and a (finite-dimensional) $W_0 \subset W$, complementary to $T(V)$, so that

$$\text{(7.10)} \qquad V = V_1 \oplus \text{Ker}\, T, \quad W = T(V) \oplus W_0.$$

Given $S \in \mathcal{L}(V, W)$, define

$$\text{(7.11)} \qquad \tau_S : V_1 \oplus W_0 \to W, \quad \tau_S(v, w) = Sv + w.$$

The map τ_T is an isomorphism of Banach spaces. Thus $\|T - S\|$ small implies τ_S is an isomorphism of $V_1 \oplus W_0$ onto W. We restrict attention to such S, lying in the same component of $\text{Fred}(V, W)$ as T.

Note that $\tau_S(V_1)$ is closed in W, of codimension equal to $\dim W_0$; now $\tau_S(V_1) = S(V_1)$, so we have the semicontinuity property

$$\text{(7.12)} \qquad \text{Codim}\, S(V) \le \text{Codim}\, T(V).$$

We also see that $\text{Ker}\, S \cap V_1 = 0$. Thus we can write

$$V = \text{Ker}\, S \oplus Z \oplus V_1,$$

for a finite-dimensional $Z \subset V$. S is injective on $Z \oplus V_1$, taking it to $S(V) = S(Z) \oplus S(V_1)$, closed in W, of finite codimension. It follows that

$$\text{(7.13)} \qquad \text{Codim}\, S(V) = \text{Codim}\, T(V) - \dim S(Z),$$

while

$$\text{(7.14)} \qquad \dim \text{Ker}\, S + \dim Z = \dim \text{Ker}\, T.$$

Since $S(Z)$ and Z have the same dimension, this gives the desired identity, namely $\text{Ind}\, S = \text{Ind}\, T$.

Corollary 7.5. *If $T \in Fred(V, W)$ and $K \in \mathcal{K}(V, W)$, then $T + K$ and T have the same index.*

Proof. For $s \in [0, 1]$, $T + sK \in \text{Fred}(V, W)$.

The next result rounds out a useful collection of tools in the study of index theory.

Proposition 7.6. *If $T \in Fred(V, W)$ and $S \in Fred(W, X)$, then*

$$\text{Ind } ST = \text{Ind } S + \text{Ind } T. \tag{7.15}$$

Proof. Consider the following family of operators in $\mathcal{L}(V \oplus W, W \oplus X)$:

$$\begin{pmatrix} I & 0 \\ 0 & S \end{pmatrix} \begin{pmatrix} \cos t & \sin t \\ -\sin t & \cos t \end{pmatrix} \begin{pmatrix} T & 0 \\ 0 & I \end{pmatrix}, \tag{7.16}$$

the middle factor belonging to $\mathcal{L}(W \oplus W)$. For each $t \in \mathbb{R}$, this is Fredholm. For $t = 0$, it is

$$\begin{pmatrix} T & 0 \\ 0 & S \end{pmatrix},$$

of index Ind $T+$ Ind S, while for $t = -\pi/2$, it is

$$\begin{pmatrix} 0 & -I \\ ST & 0 \end{pmatrix},$$

of index Ind ST. The identity of these two quantities now follows from Proposition 7.4.

Exercises

Exercises 1–4 may be compared to Exercises 3–7 in Chapter 4, §3. Let H denote the subspace of $L^2(S^1)$ that is the range of the projection P:

$$Pf(\theta) = \sum_{n=0}^{\infty} \hat{f}(n) e^{in\theta}.$$

Given $\varphi \in C(S^1)$, define the "Toeplitz operator" $T_\varphi : H \to H$ by $T_\varphi u = P(\varphi u)$. Clearly, $\|T_\varphi\| \le \|\varphi\|_{\text{sup}}$.

1. By explicit calculation, for $\varphi(\theta) = E_k(\theta) = e^{ik\theta}$, show that
$$T_{E_k} T_{E_\ell} - T_{E_k E_\ell} \text{ is compact on } H.$$
2. Show that, for any $\varphi, \psi \in C(S^1)$, $T_\varphi T_\psi - T_{\varphi\psi}$ is compact on H. (*Hint*: Approximate φ and ψ by linear combinations of exponentials.)
3. Show that if $\varphi \in C(S^1)$ is nowhere vanishing, then $T_\varphi : H \to H$ is Fredholm. (*Hint*: Show that a Fredholm inverse is given by T_ψ, $\psi(\theta) = \varphi(\theta)^{-1}$.)
4. A nowhere-vanishing $\varphi \in C(S^1)$ is said to have degree $k \in \mathbb{Z}$ if φ is homotopic to $E_k(\theta) = e^{ik\theta}$, through continuous maps of S^1 to $\mathbb{C} \setminus 0$. Show that this implies
$$\text{Index } T_\varphi = \text{Index } T_{E_k}.$$
Compute this index by explicitly describing Ker T_{E_k} and Ker $T_{E_k}^*$. Show that the calculation can be reduced to the case $k = 1$.

8. Unbounded operators

Here we consider unbounded linear operators on Banach spaces. Such an operator T between Banach spaces V and W will not be defined on all of V, though for simplicity we write $T : V \to W$. The domain of T, denoted $\mathcal{D}(T)$, will be some linear subspace of T. Generalizing (5.17), we consider the graph of T:

$$G_T = \{(v, Tv) \in V \oplus W : v \in \mathcal{D}(T)\}. \tag{8.1}$$

Then G_T is a linear subspace of $V \oplus W$; if G_T is closed in $V \oplus W$, we say T is a closed operator. By the closed-graph theorem, if T is closed and $\mathcal{D}(T) = V$, then T is bounded. If T is a linear operator, the closure of its graph $\overline{G}_T$ may or may not be the graph of an operator. If it is, we write $\overline{G}_T = G_{\overline{T}}$ and call $\overline{T}$ the closure of T.

For a linear operator $T : V \to W$ with dense domain $\mathcal{D}(T)$, we define the adjoint $T' : W' \to V'$ as follows. There is the identity

$$\langle Tv, w' \rangle = \langle v, T'w' \rangle, \tag{8.2}$$

for $v \in \mathcal{D}(T)$, $w' \in \mathcal{D}(T') \subset W'$. We define $\mathcal{D}(T')$ to be the set of $w' \in W'$ such that the map $v \mapsto \langle Tv, w' \rangle$ extends from $\mathcal{D}(T) \to \mathbb{C}$ to a continuous, linear functional $V \to \mathbb{C}$. For such w', the identity (8.2) uniquely determines $T'w' \in V'$.

It is useful to note the following relation between the graphs of T and T'. The graph G_T has annihilator $G_T^{\perp} \subset V' \oplus W'$ given by

$$G_T^{\perp} = \{(v', w') \in V' \oplus W' : \langle Tv, w' \rangle = -\langle v, v' \rangle \text{ for all } v \in \mathcal{D}(T)\}. \tag{8.3}$$

Comparing the definition of T', we see that, with

$$\mathcal{J} : V' \oplus W' \to W' \oplus V', \quad \mathcal{J}(v', w') = (w', -v'),$$

we have

$$G_{T'} = \mathcal{J}\, G_T^{\perp}. \tag{8.4}$$

We remark that $\mathcal{D}(T)$ is dense if and only if the right side of (8.5) is the graph of a (single-valued) transformation. Using $X^{\perp\perp} = \overline{X}$ for a linear subspace of a reflexive Banach space, we have the following.

Proposition 8.1. *A densely defined linear operator $T : V \to W$ between reflexive Banach spaces has a closure $\overline{T}$ if and only if T' is densely defined. T' is always closed, and $T'' = \overline{T}$.*

If H_0 and H_1 are Hilbert spaces and $T : H_0 \to H_1$, with dense domain $\mathcal{D}(T)$, we define the adjoint $T^* : H_1 \to H_0$ by replacing the dual pairings in (8.2) by the Hilbert space inner products. Parallel to (8.4), we have

$$G_{T^*} = \mathcal{J}\, G_T^{\perp}, \tag{8.5}$$

where $\mathcal{J} : H_0 \oplus H_1 \to H_1 \oplus H_0$, $\mathcal{J}(v, w) = (w, -v)$, and one takes Hilbert space orthogonal complements. Again, T has a closure if and only if T^* is densely defined, T^* is always closed, and $T^{**} = \overline{T}$. Note that, generally, the range $\mathcal{R}(T)$ of T satisfies

$$\mathcal{R}(T)^\perp = \text{Ker } T^*. \tag{8.6}$$

A densely defined operator $T : H \to H$ on a Hilbert space is said to be symmetric provided T^* is an extension of T (i.e., $\mathcal{D}(T^*) \supset \mathcal{D}(T)$ and $T = T^*$ on $\mathcal{D}(T)$). An equivalent condition is that $\mathcal{D}(T)$ is dense and

$$(Tu, v) = (u, Tv), \quad \text{for } u, v \in \mathcal{D}(T). \tag{8.7}$$

If $T^* = T$ (so $\mathcal{D}(T^*) = \mathcal{D}(T)$), we say T is *self-adjoint*. In light of (8.5), T is self-adjoint if and only if $\mathcal{D}(T)$ is dense and

$$G_T^\perp = \mathcal{J}\, G_T. \tag{8.8}$$

Note that if T is symmetric and $\mathcal{D}(T) = H$, then T^* cannot be a proper extension of T, so we must have $T^* = T$; hence T is closed. By the closed graph theorem, T must be bounded in this case; this result is called the *Hellinger-Toeplitz theorem.*

For a bounded operator defined on all of H, being symmetric is equivalent to being self-adjoint; in the case of unbounded operators, self-adjointness is a stronger and much more useful property. We discuss some results on self-adjointness. In preparation for this, it will be useful to note that if $T : H_0 \to H_1$ has range $\mathcal{R}(T)$, and if T is injective on $\mathcal{D}(T)$, then $T^{-1} : H_1 \to H_0$ is defined, with domain $\mathcal{D}(T^{-1}) = \mathcal{R}(T)$, and we have

$$G_{T^{-1}} = \mathcal{J}\, G_{-T}. \tag{8.9}$$

Since generally $\mathcal{R}(T)^\perp = \text{Ker } T^*$, the following is an immediate consequence.

Proposition 8.2. *If T is self-adjoint on H and injective, then T^{-1}, with dense domain $\mathcal{R}(T)$, is self-adjoint.*

From this easy result we obtain the following more substantial conclusion.

Proposition 8.3. *If $T : H \to H$ is symmetric and $\mathcal{R}(T) = H$, then T is self-adjoint.*

Proof. The identity (8.6) implies Ker $T = 0$ if $\mathcal{R}(T) = H$, so T^{-1} is defined. Writing $f, g \in H$ as $f = Tu$, $g = Tv$, and using

$$(T^{-1}f, g) = (T^{-1}Tu, Tv) = (u, Tv) = (Tu, v) = (f, T^{-1}g),$$

we see that T^{-1} is symmetric. Since $\mathcal{D}(T^{-1}) = H$, the Hellinger-Toeplitz theorem implies that T^{-1} is bounded and self-adjoint, so Proposition 8.2 applies to T^{-1}.

Whenever $T : H_0 \to H_1$ is a closed, densely defined operator between Hilbert spaces, the spaces G_T and $\mathcal{J}G_{T^*}$ provide an orthogonal decomposition of $H_0 \oplus H_1$; that is,

$$H_0 \oplus H_1 = \{(v, Tv) + (-T^*u, u) : v \in \mathcal{D}(T), u \in \mathcal{D}(T^*)\}, \tag{8.10}$$

where the terms in the sum are mutually orthogonal. Using this observation, we will be able to prove the following important result, due to J. von Neumann.

Proposition 8.4. *If $T: H_0 \to H_1$ is closed and densely defined, then T^*T is self-adjoint, and $I + T^*T$ has a bounded inverse.*

Proof. Pick $f \in H_0$. Applying the decomposition (8.10) to $(f, 0) \in H_0 \oplus H_1$, we obtain unique $v \in \mathcal{D}(T)$, $u \in \mathcal{D}(T^*)$, such that

$$f = v - T^*u, \quad u = -Tv. \tag{8.11}$$

Hence

$$v \in \mathcal{D}(T^*T) \text{ and } (I + T^*T)v = f. \tag{8.12}$$

Consequently, $I + T^*T : \mathcal{D}(T^*T) \to H_0$ is bijective, with inverse $(I + T^*T)^{-1} : H_0 \to H_0$ having range $\mathcal{D}(T^*T)$. Now, with $u = (I + T^*T)^{-1}f$ and $v = (I + T^*T)^{-1}g$, we easily compute

$$\begin{aligned} \big(f, (I + T^*T)^{-1}g\big) &= \big((I + T^*T)u, v\big) \\ &= (u, v) + (Tu, Tv) = \big((I + T^*T)^{-1}f, g\big), \end{aligned} \tag{8.13}$$

so $(I + T^*T)^{-1}$ is a symmetric operator on H. Since its domain is H, we have $(I+T^*T)^{-1}$ bounded and self-adjoint, and thus Proposition 8.2 finishes the proof.

If T is symmetric, note that

$$\|(T \pm i)u\|^2 = \|Tu\|^2 + \|u\|^2, \quad \text{for } u \in \mathcal{D}(T). \tag{8.14}$$

If T is closed, it follows that the ranges $\mathcal{R}(T \pm i)$ are closed. The following result provides an important criterion for self-adjointness.

Proposition 8.5. *Let $T : H \to H$ be symmetric. The following three conditions are equivalent:*

$$T \text{ is self-adjoint}, \tag{8.15}$$

$$T \text{ is closed and Ker } (T^* \pm i) = 0, \tag{8.16}$$

$$\mathcal{R}(T \pm i) = H. \tag{8.17}$$

Proof. Assume (8.17) holds, that is, both ranges are all of H. Let $u \in \mathcal{D}(T^*)$; we want to show that $u \in \mathcal{D}(T)$. $\mathcal{R}(T - i) = H$ implies there exists $v \in \mathcal{D}(T)$ such that $(T - i)v = (T^* - i)u$. Since $\mathcal{D}(T) \subset \mathcal{D}(T^*)$, this implies $u - v \in \mathcal{D}(T^*)$ and $(T^* - i)(u - v) = 0$. Now the implication (8.17) $\Rightarrow$ $\text{Ker}(T^* \mp i) = 0$ is clear from (8.6), so we have $u = v$; hence $u \in \mathcal{D}(T)$, as desired. The other implications of the proposition are straightforward.

In particular, if T is self-adjoint on H, $T \pm i : \mathcal{D}(T) \to H$ bijectively. Hence

$$U = (T - i)(T + i)^{-1} : H \longrightarrow H, \tag{8.18}$$

bijectively. By (8.14) this map preserves norms; we say U is unitary. The association of such a unitary operator (necessarily bounded) with any self-adjoint operator (perhaps unbounded) is J. von Neumann's unitary trick. Note that $I - U = 2i(T+i)^{-1}$, with range equal to $\mathcal{D}(T)$. We can hence recover T from U as

$$T = i(I+U)(I-U)^{-1}, \tag{8.19}$$

both sides having domain $\mathcal{D}(T)$.

We next give a construction of a self-adjoint operator due to K. O. Friedrichs, which is particularly useful in PDE. One begins with the following set-up. There are two Hilbert spaces H_0 and H_1, with inner products $(\ ,\ )_0$ and $(\ ,\ )_1$, respectively, and a continuous injection

$$J : H_1 \longrightarrow H_0, \tag{8.20}$$

with dense range. We think of J as identifying H_1 with a dense linear subspace of H_0; given $v \in H_1$, we will often write v for $Jv \in H_0$. A linear operator $A : H_0 \to H_0$ is defined by the identity

$$(Au, v)_0 = (u, v)_1, \tag{8.21}$$

for all $v \in H_1$, with domain

$$\begin{aligned} \mathcal{D}(A) = \{u \in H_1 \subset H_0 : v \mapsto (u, v)_1 \text{ extends from } H_1 \to \mathbb{C} \text{ to a} \\ \text{continuous, conjugate-linear functional } H_0 \to \mathbb{C}\}. \end{aligned} \tag{8.22}$$

Thus the graph of A is described as

$$\begin{aligned} G_A = \{(u, w) \in H_0 \oplus H_0 : u \in H_1 \text{ and} \\ (u, v)_1 = (w, v)_0 \text{ for all } v \in H_1\}. \end{aligned} \tag{8.23}$$

We claim that G_A is closed in $H_0 \oplus H_0$; this comes down to establishing the following.

Lemma 8.6. *If $(u_n, w_n) \in G_A$, $u_n \to u$, $w_n \to w$ in H_0, then $u \in H_1$ and $u_n \to u$ in H_1.*

Proof. Let $u_{mn} = u_m - u_n$, $w_{mn} = w_m - w_n$. We know that $(u_{mn}, v)_1 = (w_{mn}, v)_0$, for each $v \in H_1$. Taking $v = u_{mn}$ gives $\|u_{mn}\|_1^2 = (w_{mn}, u_{mn})_0 \to 0$ as $m, n \to \infty$. This implies that (u_n) is Cauchy in H_1, and the rest follows.

Actually, we could have avoided writing down this last short proof, as it will not be needed to establish our main result:

Proposition 8.7. *The operator A defined above is a self-adjoint operator on H_0.*

Proof. Consider the adjoint of J, $J^* : H_0 \to H_1$. This is also injective with dense range, and the operator JJ^* is a bounded, self-adjoint operator on H_0, that is injective with dense range. To restate (8.22), $\mathcal{D}(A)$ consists of elements $u = J\tilde{u}$ such that $v \mapsto (\tilde{u}, v)_1$ is continuous in Jv, in the H_0-norm, that is, there exists $w \in H_0$ such that $(\tilde{u}, v)_1 = (w, Jv)_0$, hence $\tilde{u} = J^* w$. We conclude that

$$\mathcal{D}(A) = \mathcal{R}(JJ^*) \tag{8.24}$$

and, for $u \in H_0$, $v \in H_1$,

$$(AJJ^*u, Jv)_0 = (J^*u, v)_1 = (u, Jv)_0. \tag{8.25}$$

It follows that

$$A = (JJ^*)^{-1}, \tag{8.26}$$

and Proposition 8.2 finishes the proof.

We remark that, given a closed, densely defined operator T on H_0, one can make $\mathcal{D}(T) = H_1$ a Hilbert space with inner product $(u, v)_1 = (Tu, Tv)_0 + (u, v)_0$. Thus Friedrichs' result, Proposition 8.7, contains von Neumann's result, Proposition 8.4. This construction of Friedrichs is used to good effect in Chapter 5.

We next discuss the resolvent and spectrum of a general closed, densely defined operator $T : V \to V$. By definition, $\zeta \in \mathbb{C}$ belongs to the resolvent set $\rho(T)$ if and only if $\zeta - T : \mathcal{D}(T) \to V$, bijectively. Then the inverse

$$R_\zeta = (\zeta - T)^{-1} : V \longrightarrow \mathcal{D}(T) \subset V \tag{8.27}$$

is called the resolvent of T; clearly, $R_\zeta \in \mathcal{L}(V)$. As in §5, the complement of $\rho(T)$ is called the spectrum of T and denoted $\sigma(T)$.

Such an operator may have an empty resolvent set. For example, the unbounded operator on $L^2(\mathbb{R}^2)$ defined by multiplication by $x_1 + ix_2$, with domain consisting of all $u \in L^2(\mathbb{R}^2)$ such that $(x_1 + ix_2)u \in L^2(\mathbb{R}^2)$, has this property. There are also examples of unbounded operators with empty spectrum. Note that Proposition 8.5 implies that $\pm i \in \rho(T)$ whenever T is self-adjoint. The same argument shows that any $\zeta \in \mathbb{C} \setminus \mathbb{R}$ belongs to $\rho(T)$, hence $\sigma(T)$ is contained in $\mathbb{R}$, when T is self-adjoint.

We note some relations between $\sigma(T)$ and $\sigma(R_\zeta)$, given that $\zeta \in \rho(T)$. Clearly, 0 belongs to $\rho(R_\zeta)$ if and only if $\mathcal{D}(T) = V$. Since R_ζ is bounded, we know that its spectrum is a nonempty, compact subset of $\mathbb{C}$. If $\lambda \in \rho(R_\zeta)$, write $S_\lambda = (\lambda - R_\zeta)^{-1}$. It follows easily that S_λ and R_ζ commute, and both preserve $\mathcal{D}(T)$. A computation gives

$$\begin{aligned} I = (\lambda - R_\zeta)S_\lambda &= \lambda(\zeta - T)S_\lambda(\zeta - T)^{-1} - S_\lambda(\zeta - T)^{-1} \\ &= \lambda(\zeta - \lambda^{-1} - T)S_\lambda(\zeta - T)^{-1} \text{ on } V, \end{aligned} \tag{8.28}$$

and similarly,

$$\begin{aligned} I &= \lambda(\zeta - T)^{-1}S_\lambda(\zeta - T) - (\zeta - T)^{-1}S_\lambda \\ &= \lambda S_\lambda(\zeta - T)^{-1}(\zeta - \lambda^{-1} - T) \text{ on } \mathcal{D}(T). \end{aligned} \tag{8.29}$$

This establishes the following:

Proposition 8.8. *Given* $\zeta \in \rho(T)$, *if* $\lambda \in \rho(R_\zeta)$ *and* $\lambda \neq 0$, *then* $\zeta - \lambda^{-1} \in \rho(T)$. *Hence* $\rho(T)$ *is open in* $\mathbb{C}$. *We have, for such* λ,

$$(\zeta - \lambda^{-1} - T)^{-1} = \lambda(\lambda - R_\zeta)^{-1}(\zeta - T)^{-1}. \tag{8.30}$$

The second assertion follows from the fact that $\lambda \in \rho(R_\zeta)$ provided $|\lambda| > \|R_\zeta\|$.

If there exists $\zeta \in \rho(T)$ such that R_ζ is compact, we say T has compact resolvent. By Proposition 8.8 it follows that when T has compact resolvent, then $\sigma(T)$ is a discrete subset of $\mathbb{C}$. Every resolvent in (8.30) is compact in this case. If T is self-adjoint on H with compact resolvent, there exists $z \in \rho(T) \cap \mathbb{R}$, and $(z - T)^{-1}$ is a compact, self-adjoint operator, to which Proposition 6.6 applies. Thus H has an orthonormal basis of eigenvectors of T:

$$v_j \in \mathcal{D}(T), \quad Tv_j = \lambda_j v_j, \tag{8.31}$$

where $\{\lambda_j\}$ is a sequence of real numbers with no finite accumulation point. Important examples of unbounded operators with compact resolvent arise amongst differential operators; cf. Chapter 5.

Exercises

1. Consider the following operator, which is densely defined on $L^2(\mathbb{R})$:

$$Tf(x) = f(0)e^{-x^2}, \quad \mathcal{D} = C_0^\infty(\mathbb{R}).$$

Show that T is unbounded and also that T has no closure.

9. Semigroups

If V is a Banach space, a one-parameter semigroup of operators on V is a set of bounded operators

$$P(t) : V \longrightarrow V, \quad t \in [0, \infty), \tag{9.1}$$

satisfying

$$P(s + t) = P(s)P(t), \tag{9.2}$$

for all $s, t \in \mathbb{R}^+$, and

$$P(0) = I. \tag{9.3}$$

We also require strong continuity, that is,

$$t_j \to t \Longrightarrow P(t_j)v \to P(t)v, \tag{9.4}$$

for each $v \in V$, the convergence being in the V-norm. A semigroup of operators will by definition satisfy (9.1)–(9.4). If $P(t)$ is defined for all $t \in \mathbb{R}$ and satisfies these conditions, we say it is a one-parameter group of operators.

A simple example is the translation group

$$T_p(t) : L^p(\mathbb{R}) \longrightarrow L^p(\mathbb{R}), \quad 1 \le p < \infty, \tag{9.5}$$

defined by

$$T_p(t)f(x) = f(x-t). \tag{9.6}$$

The properties (9.1)–(9.3) are clear in this case. Note that $\|T_p(t)\| = 1$ for each t. Also, $\|T_p(t) - T_p(t')\| = 2$ if $t \neq t'$; to see this, apply the difference to a function f with support in an interval of length $|t-t'|/2$. To verify the strong continuity (9.4), we make the following observation. As noted in §1, the space $C_{00}(\mathbb{R})$ of compactly supported, continuous functions on $\mathbb{R}$ is dense in $L^p(\mathbb{R})$ for $p \in [1,\infty)$. Now, if $f \in C_{00}(\mathbb{R})$, $t_j \to t$, then $T_p(t_j)f(x) = f(x-t_j)$ have support in a fixed compact set and converge uniformly to $f(x-t)$, so clearly we have convergence in (9.4) in L^p-norm for each $f \in C_{00}(\mathbb{R})$. The folowing simple but useful lemma completes the proof of (9.4) for T_p.

Lemma 9.1. *Let $T_j \in \mathcal{L}(V,W)$ be uniformly bounded. Let L be a dense, linear subspace of V, and suppose*

$$T_j v \to T_0 v, \quad \text{as } j \to \infty, \tag{9.7}$$

in the W-norm, for each $v \in L$. Then (9.7) holds for all $v \in V$.

Proof. Given $v \in V$ and $\varepsilon > 0$, pick $w \in L$ such that $\|v - w\| < \varepsilon$. Suppose $\|T_j\| \le M$ for all j. Then

$$\begin{aligned} \|T_j v - T_0 v\| &\le \|T_j v - T_j w\| + \|T_j w - T_0 w\| + \|T_0 w - T_0 v\| \\ &\le \|T_j w - T_0 w\| + 2M\|v - w\|. \end{aligned}$$

Thus

$$\limsup_{j\to\infty} \|T_j v - T_0 v\| \le 2M\varepsilon,$$

which proves the lemma.

Many examples of semigroups appear in the main text, particularly in Chapters 3, 6, and 9, so we will not present further examples here.

We note that a uniform bound on the norm

$$\|P(t)\| \le M, \quad \text{for } |t| \le 1 \tag{9.8}$$

for some $M \in [1,\infty)$, holds for any strongly continuous semigroup, as a consequence of the uniform boundedness principle. From (9.8) we deduce that, for all $t \in \mathbb{R}^+$,

$$\|P(t)\| \le M\, e^{Kt}, \tag{9.9}$$

for some K; for a group, one would use $M\,e^{K|t|}$, $t \in \mathbb{R}$.

Of particular interest are unitary groups—strongly continuous groups of operators $U(t)$ on a Hilbert space H such that

$$U(t)^* = U(t)^{-1} = U(-t). \tag{9.10}$$

Clearly, in this case $\|U(t)\| = 1$. The translation group T_2 on $L^2(\mathbb{R})$ is a simple example of a unitary group.

A one-parameter semigroup $P(t)$ of operators on V has an *infinitesimal generator* A, which is an operator on V, often unbounded, defined by

$$Av = \lim_{h\to 0} h^{-1}\big(P(h)v - v\big), \tag{9.11}$$

on the domain

$$\mathcal{D}(A) = \{v \in V : \lim_{h\to 0} h^{-1}\big(P(h)v - v\big) \text{ exists in } V\}. \tag{9.12}$$

The following provides some basic information on the generator.

Proposition 9.2. *The infinitesimal generator A of $P(t)$ is a closed, densely defined operator. We have*

$$P(t)\mathcal{D}(A) \subset \mathcal{D}(A), \tag{9.13}$$

for all $t \in \mathbb{R}^+$, and

$$AP(t)v = P(t)Av = \frac{d}{dt}P(t)v, \quad \textit{for } v \in \mathcal{D}(A). \tag{9.14}$$

If (9.9) holds and $\operatorname{Re}\zeta > K$, then ζ belongs to the resolvent set of A, and

$$(\zeta - A)^{-1}v = \int_0^\infty e^{-\zeta t}\, P(t)v\, dt, \quad v \in V. \tag{9.15}$$

Proof. First, if $v \in \mathcal{D}(A)$, then for $t \in \mathbb{R}^+$,

$$h^{-1}\big(P(h)P(t)v - P(t)v\big) = P(t)\,h^{-1}\big(P(h)v - v\big), \tag{9.16}$$

which gives (9.13), and also (9.14), if we replace $P(h)P(t)$ by $P(t+h)$ in (9.16). To show that $\mathcal{D}(A)$ is dense in V, let $v \in V$, and consider

$$v_\varepsilon = \varepsilon^{-1}\int_0^\varepsilon P(t)v\, dt.$$

Then

$$\begin{aligned} h^{-1}\big(P(h)v_\varepsilon - v_\varepsilon\big) &= \varepsilon^{-1}\Big[h^{-1}\int_\varepsilon^{\varepsilon+h} P(t)v\,dt - h^{-1}\int_0^h P(t)v\,dt\Big] \\ &\to \varepsilon^{-1}\big(P(\varepsilon)v - v\big), \text{ as } h \to 0, \end{aligned}$$

so $v_\varepsilon \in \mathcal{D}(A)$ for each $\varepsilon > 0$. But $v_\varepsilon \to v$ in V as $\varepsilon \to 0$, by (9.4), so $\mathcal{D}(A)$ is dense in V.

Next we prove (9.15). Denote the right side of (9.15) by R_ζ, clearly a bounded operator on V. First we show that

$$R_\zeta(\zeta - A)v = v, \quad \text{for } v \in \mathcal{D}(A). \tag{9.17}$$

In fact, by (9.14) we have

$$\begin{aligned} R_\zeta(\zeta - A)v &= \int_0^\infty e^{-\zeta t} P(t)(\zeta v - Av)\, dt \\ &= \int_0^\infty \zeta e^{-\zeta t} P(t)v\, dt - \int_0^\infty e^{-\zeta t} \frac{d}{dt} P(t)v\, dt, \end{aligned}$$

and integrating the last term by parts gives (9.17). The same sort of argument shows that $R_\zeta : V \to \mathcal{D}(A)$, that $(\zeta - A)R_\zeta$ is bounded on V, and that

$$(\zeta - A)R_\zeta v = v, \tag{9.18}$$

for $v \in \mathcal{D}(A)$. Since $(\zeta - A)R_\zeta$ is bounded on V and $\mathcal{D}(A)$ is dense in V, (9.18) holds for all $v \in V$. This proves (9.15). Finally, since the resolvent set of A is nonempty, and $(\zeta - A)^{-1}$, being continuous and everywhere defined, is closed, so is A. The proof of the proposition is complete.

We write, symbolically,

$$P(t) = e^{tA}. \tag{9.19}$$

In view of the following proposition, the infinitesimal generator determines the one-parameter semigroup with which it is associated uniquely. Hence we are justified in saying "A generates $P(t)$."

Proposition 9.3. *If $P(t)$ and $Q(t)$ are one-parameter semigroups with the same infinitesimal generator, then $P(t) = Q(t)$ for all $t \in \mathbb{R}^+$.*

Proof. Let $v \in V$ and $w \in V'$. Then, for Re ζ large enough,

$$\begin{aligned} \int_0^\infty e^{-\zeta t} \langle P(t)v, w\rangle\, dt &= \langle (\zeta - A)^{-1} v, w\rangle \\ &= \int_0^\infty e^{-\zeta t} \langle Q(t)v, w\rangle\, dt. \end{aligned} \tag{9.20}$$

Uniqueness for the Laplace transform of a scalar function implies $\langle P(t)v, w\rangle = \langle Q(t)v, w\rangle$ for all $t \in \mathbb{R}^+$ and for any $v \in V$ and $w \in V'$. Then the Hahn-Banach theorem implies $P(t)v = Q(t)v$, as desired.

We note that if $P(t)$ is a semigroup satisfying (9.9) and if we have a function $\varphi \in L^1(\mathbb{R}^+, e^{Kt} dt)$, we can define $P(\varphi) \in \mathcal{L}(V)$ by

$$P(\varphi)v = \int_0^\infty \varphi(t)\, P(t)v\, dt. \tag{9.21}$$

In particular, this works if $\varphi \in C_0^\infty(0,\infty)$. In such a case, it is easy to verify that, for all $v \in V$, $P(\varphi)v$ belongs to the domain of all powers of A and

$$A^k P(\varphi)v = (-1)^k \int_0^\infty \varphi^{(k)}(t)P(t)v\,dt. \tag{9.22}$$

This shows that all the domains $\mathcal{D}(A^k)$ are dense in V, refining the proof of denseness of $\mathcal{D}(A)$ in V given in Proposition 9.2.

A general characterization of generators of semigroups, due to Hille and Yosida, is briefly discussed in the exercises. Here we mention two important special cases, which follow from the spectral theorem, established in Chapter 8.

Proposition 9.4. *If A is self-adjoint and positive (i.e., $(Au,u) \geq 0$ for $u \in \mathcal{D}(A)$), then $-A$ generates a semigroup $P(t) = e^{-tA}$ consisting of positive, self-adjoint operators of norm ≤ 1.*

Proposition 9.5. *If A is self-adjoint, then iA generates a unitary group, $U(t) = e^{itA}$.*

In both cases it is easy to show that the generator of such (semi)groups must be of the form hypothesized. For example, if $U(t)$ is a unitary group and we denote by iA the generator, the identity

$$h^{-1}\big([U(h)-I]u, v\big) = h^{-1}\big(u, [U(-h)-I]v\big) \tag{9.23}$$

shows that A must be symmetric. By Proposition 9.2, all $\zeta \in \mathbb{C} \setminus \mathbb{R}$ belong to the resolvent set of A, so by Proposition 8.5, A is self-adjoint. If A is self-adjoint, iA is said to be skew-adjoint.

We now give a criterion for a symmetric operator to be essentially self-adjoint, that is, to have self-adjoint closure. This is quite useful in PDE; see Chapter 8 for some applications.

Proposition 9.6. *Let A_0 be a linear operator on a Hilbert space H, with domain $\mathcal{D}$, assumed dense in H. Let $U(t)$ be a unitary group, with infinitesimal generator iA, so A is self-adjoint, $U(t) = e^{itA}$. Suppose $\mathcal{D} \subset \mathcal{D}(A)$ and $A_0 u = Au$ for $u \in \mathcal{D}$, or equivalently*

$$\lim_{h\to 0} h^{-1}\big(U(h)u - u\big) = A_0 u, \quad \textit{for all } u \in \mathcal{D}. \tag{9.24}$$

Also suppose $\mathcal{D}$ is invariant under $U(t)$:

$$U(t)\mathcal{D} \subset \mathcal{D}. \tag{9.25}$$

Then A_0 is essentially self-adjoint, with closure A. Suppose, furthermore, that

$$A_0 : \mathcal{D} \longrightarrow \mathcal{D}. \tag{9.26}$$

Then A_0^k, with domain $\mathcal{D}$, is essentially self-adjoint for each positive integer k.

Proof. It follows from Proposition 8.5 that A_0 is essentially self-adjoint if and only if the range of $i + A_0$ and the range of $i - A_0$ are dense in H. So suppose $v \in H$ and (for one choice of sign)

$$((i \pm A_0)u, v) = 0, \quad \text{for all } u \in \mathcal{D}. \tag{9.27}$$

Using (9.25) together with the fact that $A_0 = A$ on $\mathcal{D}$, we have

$$((i \pm A_0)u, U(t)v) = 0, \quad \text{for all } t \in \mathbb{R},\ u \in \mathcal{D}. \tag{9.28}$$

Consequently, $\int \rho(t)U(t)v\,dt$ is orthogonal to the range of $i \pm A_0$, for any $\rho \in L^1(\mathbb{R}^+)$. Choosing $\rho \in C_0^\infty(0, \infty)$ an approximate identity, we can approximate v by elements of $\mathcal{D}(A)$, indeed of $\mathcal{D}(A^k)$ for all k. Thus we can suppose in (9.27) that $v \in \mathcal{D}(A)$. Hence, taking adjoints, we have

$$(u, (-i \pm A)v) = 0, \quad \text{for all } u \in \mathcal{D}. \tag{9.29}$$

Since $\mathcal{D}$ is dense in H and $\text{Ker}(-i \pm A) = 0$, this implies $v = 0$. This yields the first part of the proposition. Granted (9.26), the same proof works with A_0 replaced by A_0^k (but $U(t)$ unaltered), so the proposition is proved.

This result has an extension to general semigroups which is of interest.

Proposition 9.7. *Let $P(t)$ be a semigroup of operators on a Banach space B, with generator A. Let $\mathcal{L} \subset \mathcal{D}(A)$ be a dense, linear subspace of B, and suppose $P(t)\mathcal{L} \subset \mathcal{L}$ for all $t \geq 0$. Then A is the closure of its restriction to $\mathcal{L}$.*

Proof. By Proposition 9.2, it suffices to show that $(\lambda - A)(\mathcal{L})$ is dense in B provided $\text{Re}\,\lambda$ is sufficiently large, namely, $\text{Re}\,\lambda > K$ with $\|P(t)\| \leq Me^{Kt}$. If $w \in B'$ annihilates this range and $w \neq 0$, pick $u \in \mathcal{L}$ such that $\langle u, w\rangle \neq 0$. Now

$$\frac{d}{dt}\langle P(t)u, w\rangle = \langle AP(t)u, w\rangle = \langle \lambda P(t)u, w\rangle$$

since $P(t)u \in \mathcal{L}$. Thus $\langle P(t)u, w\rangle = e^{\lambda t}\langle u, w\rangle$. But if $\text{Re}\,\lambda > K$ as above, this is impossible unless $\langle u, w\rangle = 0$. This completes the proof.

We illustrate some of the preceding results by looking at the infinitesimal generator A_p of the group T_p given by (9.5)–(9.6). By definition, $f \in L^p(\mathbb{R})$ belongs to $\mathcal{D}(A_p)$ if and only if

$$h^{-1}\big(f(x-h) - f(x)\big) \tag{9.30}$$

converges in L^p-norm as $h \to 0$, to some limit. Now the limit of (9.30) always exists in the space of distributions $\mathcal{D}'(\mathbb{R})$ and is equal to $-(d/dx)u$, where d/dx is applied in the sense of distributions. In fact, we have the following.

Proposition 9.8. *For $p \in [1, \infty)$, the group T_p given by (9.5)–(9.8) has infinitesimal generator A_p given by*

$$A_p f = -\frac{df}{dx}, \tag{9.31}$$

for $f \in \mathcal{D}(A_p)$, with

$$\mathcal{D}(A_p) = \{f \in L^p(\mathbb{R}) : f' \in L^p(\mathbb{R})\}, \tag{9.32}$$

where $f' = df/dx$ is considered a priori as a distribution.

Proof. The argument above shows that (9.31) holds, with $\mathcal{D}(A_p)$ contained in the right side of (9.32). The reverse containment can be derived as a consequence of the following simple result, taking $\mathcal{L} = C_0^\infty(\mathbb{R})$.

Lemma 9.9. *Let $P(t)$ be a one-parameter semigroup on B, with infinitesimal generator A. Let $\mathcal{L}$ be a weak*-dense, linear subspace of B'. Suppose that $u, v \in B$ and that*

$$\lim_{h\to 0} h^{-1}\langle P(h)u - u, w\rangle = \langle v, w\rangle, \quad \forall\, w \in \mathcal{L}. \tag{9.33}$$

Then $u \in \mathcal{D}(A)$ and $Au = v$.

Proof. The hypothesis (9.33) implies that $\langle P(t)u, w\rangle$ is differentiable and that

$$\frac{d}{dt}\langle P(t)u, w\rangle = \langle P(t)v, w\rangle, \quad \forall\, w \in \mathcal{L}.$$

Hence $\langle P(t)u - u, w\rangle = \int_0^t \langle P(s)v, w\rangle\, ds$, for all $w \in \mathcal{L}$. The weak* denseness of $\mathcal{L}$ implies $P(t)u - u = \int_0^t P(s)v\, ds$, and the convergence in the B-norm of $h^{-1}\big(P(h)u - u\big) = h^{-1}\int_0^h P(s)v\, ds$ to v as $h \to 0$ follows.

The space (9.32) is the Sobolev space $H^{1,p}(\mathbb{R})$ studied in Chapter 13; in case $p = 2$, it is the Sobolev space $H^1(\mathbb{R})$ introduced in Chapter 4.

Note that if we define

$$A_0 : C_0^\infty(\mathbb{R}) \longrightarrow C_0^\infty(\mathbb{R}), \quad A_0 f = -\frac{df}{dx}, \tag{9.34}$$

then Proposition 9.7 applies to T_p, $p \in [1, \infty)$, with $B = L^p(\mathbb{R})$, $\mathcal{L} = C_0^\infty(\mathbb{R})$, to show that, as a closed operator on $L^p(\mathbb{R})$,

$$A_p \text{ is the closure of } A_0, \quad \text{for } p \in [1, \infty). \tag{9.35}$$

This amounts to saying that $C_0^\infty(\mathbb{R})$ is dense in $H^{1,p}(\mathbb{R})$ for $p \in [1, \infty)$, which can easily be verified directly.

The fact that a semigroup $P(t)$ satisfies the operator differential equation (9.14) is central. We now establish the following converse.

Proposition 9.10. *Let A be the infinitesimal generator of a semigroup. If a function $u \in C\big([0, T), \mathcal{D}(A)\big) \cap C^1\big([0, T), V\big)$ satisfies*

$$\frac{du}{dt} = Au, \quad u(0) = f, \tag{9.36}$$

then $u(t) = e^{tA} f$, for $t \in [0, T)$.

Proof. Set $v(s,t) = e^{sA}u(t) \in C^1(Q, V)$, $Q = [0,\infty) \times [0,T)$. Then (9.36) implies that $(\partial_s - \partial_t)v = e^{sA}Au(t) - e^{sA}Au(t) = 0$, so $u(t) = v(0,t) = v(t,0) = e^{tA}f$.

We can thus deduce that, given $g \in C\big([0,T), \mathcal{D}(A)\big)$, $f \in \mathcal{D}(A)$, the solution $u(t)$ to

$$\frac{\partial u}{\partial t} = Au + g(t), \quad u(0) = f, \tag{9.37}$$

is unique and is given by

$$u(t) = e^{tA}f + \int_0^t e^{(t-s)A}g(s)\,ds. \tag{9.38}$$

This is a variant of Duhamel's principle.

We can also define a notion of a "weak solution" of (9.37) as follows. If A generates a semigroup, then $\mathcal{D}(A')$ is a dense, linear subspace of V'. Suppose that, for every $\psi \in \mathcal{D}(A')$, $\langle u(t), \psi\rangle \in C^1([0,T))$; if $f \in V$, $g \in C([0,T), V)$, and

$$\frac{d}{dt}\langle u(t), \psi\rangle = \langle u(t), A'\psi\rangle + \langle g(t), \psi\rangle, \quad u(0) = f, \tag{9.39}$$

we say $u(t)$ is a weak solution to (9.37).

Proposition 9.11. *Given $f \in V$ and $g \in C\big([0,T), V\big)$, (9.37) has a unique weak solution, given by (9.38).*

Proof. First, consider (9.38), with $f \in V$, $g \in C(J, V)$, and $J = [0,T)$. Let $f_j \to f$ in V and $g_j \to g$ in $C(J,V)$, where $f_j \in \mathcal{D}(A)$ and $g_j \in C^1(J,V) \cap C(J, \mathcal{D}(A))$. Then, by Proposition 9.10,

$$u_j(t) = e^{tA}f_j + \int_0^t e^{(t-s)A}g_j(s)\,ds \tag{9.40}$$

is the unique solution in $C^1(J,V) \cap C(J, \mathcal{D}(A))$ to

$$\frac{\partial u_j}{\partial t} = Au_j + g_j, \quad u_j(0) = f_j.$$

Thus, for any $\psi \in \mathcal{D}(A')$, u_j solves (9.39), with g and f replaced by g_j and f_j, respectively, and hence

$$\langle u_j(t), \psi\rangle = \langle f_j, \psi\rangle + \int_0^t \langle u_j(s), A'\psi\rangle\,ds + \int_0^t \langle g_j(s), \psi\rangle\,ds. \tag{9.41}$$

Passing to the limit, we have

$$\langle u(t), \psi\rangle = \langle f, \psi\rangle + \int_0^t \langle u(s), A'\psi\rangle\,ds + \int_0^t \langle g(s), \psi\rangle\,ds, \tag{9.42}$$

which implies (9.39).

For the converse, suppose that $u \in C(J, V)$ is a weak solution, satisfying (9.39), or equivalently, that (9.42) holds. Set $\varphi(t) = j$ for $0 \le t \le 1/j$, 0 elsewhere, and consider $P(\varphi_j)$, defined by (9.21). We see that $\langle Av, P(\varphi_j)'\psi\rangle = \langle AP(\varphi_j)v, \psi\rangle$. Hence $P(\varphi_j)' : V' \to \mathcal{D}(A')$, and also $\langle v, A'P(\varphi_j)'\psi\rangle = \langle AP(\varphi_j)v, \psi\rangle$ for $v \in \mathcal{D}(A), \psi \in V'$. If you replace ψ by $P(\varphi_j)'\psi$ in (9.41), then $u_j(t) = P(\varphi_j)u(t)$ satisfies (9.41), with $f_j = P(\varphi_j)f$, $g_j(t) = P(\varphi_j)g(t)$; hence $u_j \in C^1(J, V) \cap C(J, \mathcal{D}(A))$ is given by (9.40), and passing to the limit gives (9.38) for u.

We close this section with a brief discussion of when we can deduce that, given a generator A of a semigroup and another operator B, then $A + B$ also generates a semigroup. There are a number of results on this, to the effect that $A+B$ works if B is "small" in some sense, compared to A. These results are part of the "perturbation theory" of semigroups. The following simple case is useful.

Proposition 9.12. *If A generates a semigroup e^{tA} on V and B is bounded on V, then $A + B$ also generates a semigroup.*

Proof. The idea is to solve the equation

$$\frac{\partial u}{\partial t} = Au + Bu, \quad u(0) = f, \tag{9.43}$$

by solving the integral equation

$$u(t) = e^{tA} f + \int_0^t e^{(t-s)A} Bu(s)\, ds. \tag{9.44}$$

In other words, we want to solve

$$(I - \mathcal{N})u(t) = e^{tA} f \in C\big([0, \infty), V\big), \tag{9.45}$$

where

$$\mathcal{N}u(t) = \int_0^t e^{(t-s)A} Bu(s)\, ds, \quad \mathcal{N} : C(\mathbb{R}^+, V) \to C(\mathbb{R}^+, V). \tag{9.46}$$

Note that

$$\begin{aligned}\mathcal{N}^k u(t) = \int_0^t \int_0^{t_{k-1}} \cdots \int_0^{t_1} & e^{(t-t_{k-1})A} Be^{(t_{k-1}-t_{k-2})A} \cdots \\ & \cdots Be^{(t_1-t_0)A} Bu(t_0)\, dt_0 \cdots dt_{k-1}.\end{aligned} \tag{9.47}$$

Hence, if e^{tA} satisfies the estimate (9.9),

$$\sup_{0 \le t \le T} \|\mathcal{N}^k u(t)\| \le \big(M\|B\|\big)^k e^{tK} \cdot \big(\text{vol } S_k^T\big) \cdot \sup_{0 \le t \le T} \|u(t)\|, \tag{9.48}$$

where vol S_k^T is the volume of the k-simplex

$$S_k^T = \{(t_0, \ldots, t_{k-1}) : 0 \le t_0 \le \cdots \le t_{k-1} \le T\}.$$

Looking at the case $A = 0$, $B = b$ (scalar) of (9.43), with solution $u(t) = e^{tb} f$, we see that

$$\text{vol } S_k^T = \frac{T^k}{k!}. \tag{9.49}$$

It follows that

$$\mathcal{S}g(t) = g(t) + \sum_{k=1}^{\infty} \mathcal{N}^k g(t) \tag{9.50}$$

is convergent in $C(\mathbb{R}^+, V)$, given $g(t) \in C(\mathbb{R}^+, V)$. Now consider

$$Q(t)f = e^{tA} f + \sum_{k=1}^{\infty} \mathcal{N}^k (e^{tA} f). \tag{9.51}$$

It is straightforward to verify that $Q(t)$ is a strongly continuous semigroup on V, with generator $A + B$.

An extension of Proposition 9.12—part of the perturbation theory of R. Phillips—is given in the exercises. We mention another perturbation result, due to T. Kato. A semigroup $P(t)$ is called a *contraction semigroup* on V if $\|P(t)\| \le 1$ for all $t \ge 0$.

Proposition 9.13. *If A generates a contraction semigroup on V, then $A + B$ generates a contraction semigroup, provided $\mathcal{D}(B) \supset \mathcal{D}(A)$, B is "dissipative," and*

$$\|Bf\| \le \vartheta \|Af\| + C_1 \|f\|, \tag{9.52}$$

for some $C_1 < \infty$ and $\vartheta < 1/2$. If V is a Hilbert space, we can allow any $\vartheta < 1$.

To say that B is dissipative means that if $u \in \mathcal{D}(B) \subset V$ and $u^\# \in V'$ satisfies $\langle u, u^\# \rangle = \|u\|^2$, then

$$\text{Re}\, \langle Bu, u^\# \rangle \le 0. \tag{9.53}$$

If V is a Hilbert space with inner product (,), this is equivalent to

$$\text{Re}\, (Bu, u) \le 0, \quad \text{for } u \in \mathcal{D}(B). \tag{9.54}$$

Proofs of Proposition 9.13 typically use the Hille-Yosida characterization of which A generate a contraction semigroup. See the exercises for further discussion.

Exercises

In Exercises 1–3, define, for $I = (0, 1)$,

$$A_0 : C_0^\infty(I) \longrightarrow C_0^\infty(I), \quad A_0 f = -\frac{df}{dx}. \tag{9.55}$$

1. Given $f \in L^2(I)$, define Ef on $\mathbb{R}$ to be equal to f on I and to be periodic of period 1, and define $U(t) : L^2(I) \to L^2(I)$ by

$$U(t)f(x) = (Ef)(x-t)|_I. \tag{9.56}$$

Show that $U(t)$ is a unitary group whose generator D is a skew-adjoint extension of A_0. Describe the domain of D.

2. More generally, for $e^{i\theta} \in S^1$, define Ef on $\mathbb{R}$ to equal f on I and to satisfy

$$(Ef)(x+1) = e^{i\theta} f(x).$$

Then define $U_\theta(t) : L^2(I) \to L^2(I)$ by (9.56), with this E. Show that $U_\theta(t)$ is a unitary group whose generator D_θ is a skew-adjoint extension of A_0. Describe the domain of D_θ.

3. This time, define Ef on $\mathbb{R}$ to equal f on I and zero elsewhere. For $t \geq 0$, define $P(t) : L^2(I) \to L^2(I)$ by (9.56) with this E. Show that $P(t)$ is a strongly continuous semigroup. Show that $P(t) = 0$ for $t \geq 1$. Show that the infinitesimal generator B of $P(t)$ is a closed extension of A_0 which has empty spectrum. Describe the domain of B.

4. Let P^t be a strongly continuous semigroup on the Banach space X, with infinitesimal generator A. Suppose A has compact resolvent. If K is a closed bounded subset of X, show that K is compact if and only if $P^t \to I$ uniformly on K. (*Hint*: Let $T_j = h^{-1} \int_0^h P^t \, dt$, $h = 1/j$, and use Exercise 4 of §6.)

Exercises 5–8 deal with the case where $P(t)$ satisfies (9.1)–(9.3) but the strong continuity of $P(t)$ is replaced by weak continuity, that is, convergence in (9.4) holds in the $\sigma(V, V')$-topology on V. We restrict attention to the case where V is reflexive.

5. If $\varphi \in C_0^\infty(\mathbb{R}^+)$, show that $P(\varphi)v$ is well defined in V, satisfying

$$\langle P(\varphi)v, \omega\rangle = \int_0^\infty \varphi(t) \langle P(\varphi)v, \omega\rangle \, dt, \quad v \in V, \ \omega \in V'.$$

6. Show that $V_0 = \operatorname{span}\{P(\varphi)v : v \in V, \varphi \in C_0^\infty(\mathbb{R}^+)\}$ is *dense* in V. (*Hint*: Suppose $\omega \in V'$ annihilates V_0.)

7. Show that $P(t_j)P(\varphi)v = P(\varphi_j)v$, where $\varphi_j(\tau) = \varphi(\tau - t_j)$ for $\tau \geq t_j$, 0 for $\tau < t_j$. Deduce that as $t_j \to t$,

$$P(t_j)P(\varphi)v \to P(t)P(\varphi)v, \quad \text{in } V\text{-norm},$$

for $v \in V$, $\varphi \in C_0^\infty(\mathbb{R}^+)$. (*Hint*: Estimate $\|P(\varphi_j - \varphi_0)v\|$, with $\varphi_0(\tau) = \varphi(\tau - t)$. To do this, show that (9.9) continues to hold.)

8. Deduce that the hypotheses on $P(t)$ in Exercises 5–7 imply the strong continuity (9.4). (*Hint*: Use Lemma 9.1.)

9. If $P(t)$ is a strongly continuous semigroup on V, then $Q(t) = P(t)'$, acting on V', satisfies (9.1)–(9.3), with weak* continuity in place of (9.4). Deduce that if V is reflexive, $Q(t)$ is a strongly continuous semigroup on V'. Give an example of $P(t)$ on a (nonreflexive) Banach space V for which $P(t)'$ is *not* strongly continuous in $t \in [0, \infty)$.

10. Extend Proposition 9.12 to show that if A generates a semigroup e^{tA} on V and if $\mathcal{D}(B) \supset \mathcal{D}(A)$ is such that Be^{tA} is bounded for $t > 0$, satisfying

$$\|Be^{tA}\|_{\mathcal{L}(V)} \leq C_0 t^{-\alpha}, \quad t \in (0, 1],$$

for some $\alpha < 1$, then $A + B$ also generates a semigroup.
(*Hint*: Show that (9.51) still works. Note that the integrand in the formula (9.57) for $\mathcal{N}^k(e^{tA} f)$ is of the form $\cdots Be^{(t_1-t_0)A}Be^{t_0 A} f$.)

11. Recall that $P(t)$ is a contraction semigroup if it satisfies (9.1)–(9.4) and $\|P(t)\| \le 1$ for all $t \ge 0$. Show that the infinitesimal generator A of a contraction semigroup has the following property:

$$\lambda > 0 \Longrightarrow \lambda \in \rho(A), \text{ and } \|(\lambda - A)^{-1}\| \le \frac{1}{\lambda}. \tag{9.57}$$

12. The Hille-Yosida theorem states that whenever $\mathcal{D}(A)$ is dense in V and there exist $\lambda_j > 0$ such that

$$\lambda_j \nearrow +\infty, \quad \lambda_j \in \rho(A), \quad \|(\lambda_j - A)^{-1}\| \le \frac{1}{\lambda_j}, \tag{9.58}$$

then A generates a contraction semigroup. Try to prove this. (*Hint*: With $\lambda = \lambda_j$, set $A_\lambda = \lambda A(\lambda - A)^{-1}$, which is in $\mathcal{L}(V)$. Define $P_\lambda(t) = e^{tA_\lambda}$ by the power-series expansion. Show that

$$\|P_\lambda(t)\| \le 1, \quad \big\|\big(P_\lambda(t) - P_\mu(t)\big)f\big\| \le t\big\|(A_\lambda - A_\mu)f\big\|, \tag{9.59}$$

and construct $P(t)$ as the limit of $P_{\lambda_j}(t)$.)

13. If $P(t)$ satisfies (9.9), set $Q(t) = e^{-Kt}P(t)$, so $\|Q(t)\| \le M$ for $t \ge 0$. Show that $|||f||| = \sup_{t\ge 0} \|Q(t)f\|$ defines an equivalent norm on V, for which $Q(t)$ is a contraction semigroup. Then, using Exercisess 11 and 12, produce a characterization of generators of semigroups.
14. Show that if $P(t)$ is a contraction semigroup, its generator A is dissipative, in the sense of (9.53).
15. Show that if $\mathcal{D}(A)$ is dense, if $\lambda_0 \in \rho(A)$ for some λ_0 such that Re $\lambda_0 > 0$, and if A is dissipative, then A generates a contraction semigroup. (*Hint*: First show that the hypotheses imply $\lambda \in \rho(A)$ whenever Re $\lambda > 0$. Then apply the Hille-Yosida theorem.)
Deduce Propositions 9.4 and 9.5 from this result.
16. Prove Proposition 9.13. (*Hint*: Show that $\lambda \in \rho(A + B)$ for some $\lambda > 0$, and apply Exercise 15. To get this, show that when A is dissipative and $\lambda > 0$, $\lambda \in \rho(A)$, then

$$\|A(\lambda - A)^{-1}\| \le \kappa,$$

where $\kappa = 2$ for a general Banach space V, while $\kappa = 1$ if V is a Hilbert space.)

References

[Ad] R. Adams, *Sobolev Spaces*, Academic Press, New York, 1975.
[Do] W. Donoghue, *Distributions and Fourier Transforms*, Academic Press, New York, 1969.
[Dug] J. Dugundji, *Topology*, Allyn and Bacon, New York, 1966.
[DS] N. Dunford and J. Schwartz, *Linear Operators*, Wiley, New York, 1958.
[Hal] P. Halmos, *Measure Theory*, van Nostrand, New York, 1950.
[HP] E. Hille and R. Phillips, *Functional Analysis and Semi-groups*, Colloq. Publ. AMS, 1957.

[K] T. Kato, *Perturbation theory for Linear Operators*, Springer-Verlag, New York, 1966.

[LP] P. Lax and R. Phillips, *Scattering Theory*, Academic Press, New York, 1967.

[Lo] L. Loomis, *Abstract Harmonic Analysis*, van Nostrand, New York, 1953.

[Nel] E. Nelson, *Operator Differential Equations*, Graduate Lecture Notes, Princeton Univ., Princeton, N. J., 1965.

[Pal] R. Palais, ed., *Seminar on the Atiyah-Singer Index Theorem*, Princeton Univ. Press, Princeton, N. J., 1963.

[RS] M. Reed and B. Simon, *Methods of Mathematical Physics*, Academic Press, New York, Vols. 1, 2, 1975; Vols. 3, 4, 1978.

[RN] F. Riesz and B. Sz. Nagy, *Functional Analysis*, Ungar, New York, 1955.

[Ru] W. Rudin, *Real and Complex Analysis*, McGraw-Hill, New York, 1976.

[S] H. Schaefer, *Topological Vector Spaces*, MacMillan, New York, 1966.

[Sch] L. Schwartz, *Théorie des Distributions*, Hermann, Paris, 1950.

[Si] B. Simon, *Trace Ideals and Their Applications*, London Math. Soc. Lecture Notes, no. 35, Cambridge Univ. Press, Cambridge, 1979.

[Tri] H. Triebel, *Theory of Function Spaces*, Birkhauser, Boston, 1983.

[Yo] K. Yosida, *Functional Analysis*, Springer-Verlag, New York, 1965.

B

Manifolds, Vector Bundles, and Lie Groups

Introduction

This appendix provides background material on manifolds, vector bundles, and Lie groups, which are used throughout the book. We begin with a section on metric spaces and topological spaces, defining some terms that are necessary for the concept of a manifold, defined in §2, and for that of a vector bundle, defined in §3. These sections contain mostly definitions; however, a few results about compactness are proved.

In §4 we establish the easy case of a theorem of Sard, a useful result in manifold theory. It is used only once in the text, in the development of degree theory in Chapter 1, §19.

In §5 we introduce the concept of a Lie group G and its Lie algebra $\mathfrak{g}$ and establish the correspondence between Lie subgroups of G and Lie subalgebras of $\mathfrak{g}$. We also define a Haar measure on a Lie group. In §6 we establish an important relation between Lie groups and Lie algebras, known as the Campbell-Hausdorff formula.

In §7 we discuss representations of a Lie group and associated representations of its Lie algebra. Some basic results on representations of compact Lie groups are given in §8, and in §9 we specialize to the groups SU(2) and SO(3) and to some related groups, such as SO(4). Material in §9 is useful in Chapter 8, Spectral Theory, particularly in its study of the simplest quantum mechanical model of the hydrogen atom.

1. Metric spaces and topological spaces

A metric space is a set X together with a distance function $d : X \times X \to [0, \infty)$, having the properties that

$$
\begin{aligned}
& d(x, y) = 0 \Longleftrightarrow x = y, \\
(1.1) \qquad & d(x, y) = d(y, x), \\
& d(x, y) \le d(x, z) + d(y, z).
\end{aligned}
$$

The third of these properties is called the *triangle inequality*. An example of a metric space is the set of rational numbers $\mathbb{Q}$, with $d(x, y) = |x - y|$. Another example is $X = \mathbb{R}^n$, with $d(x, y) = \sqrt{(x_1 - y_1)^2 + \cdots + (x_n - y_n)^2}$.

If (x_ν) is a sequence in X, indexed by $\nu = 1, 2, 3, \ldots$ (i.e., by $\nu \in \mathbb{Z}^+$), one says $x_\nu \to y$ if $d(x_\nu, y) \to 0$, as $\nu \to \infty$. One says (x_ν) is a Cauchy sequence if $d(x_\nu, x_\mu) \to 0$ as $\mu, \nu \to \infty$. One says X is a complete metric space if every Cauchy sequence converges to a limit in X. Some metric spaces are not complete; for example, $\mathbb{Q}$ is not complete. One can take a sequence (x_ν) of rational numbers such that $x_\nu \to \sqrt{2}$, which is not rational. Then (x_ν) is Cauchy in $\mathbb{Q}$, but it has no limit in $\mathbb{Q}$.

If a metric space X is not complete, one can construct its completion $\widehat{X}$ as follows. Let an element ξ of $\widehat{X}$ consist of an equivalence class of Cauchy sequences in X, where we say $(x_\nu) \sim (y_\nu)$, provided $d(x_\nu, y_\nu) \to 0$. We write the equivalence class containing (x_ν) as $[x_\nu]$. If $\xi = [x_\nu]$ and $\eta = [y_\nu]$, we can set $d(\xi, \eta) = \lim_{\nu \to \infty} d(x_\nu, y_\nu)$ and verify that this is well defined and makes $\widehat{X}$ a complete metric space.

If the completion of $\mathbb{Q}$ is constructed by this process, you get $\mathbb{R}$, the set of real numbers.

A metric space X is said to be compact provided any sequence (x_ν) in X has a convergent subsequence. Clearly, every compact metric space is complete. There are two useful conditions, each equivalent to the characterization of compactness just stated, on a metric space. The reader can establish the equivalence, as an exercise.

(i) If $S \subset X$ is a set with infinitely many elements, then there is an *accumulation point*, that is, a point $p \in X$ such that every neighborhood U of p contains infinitely many points in S.

Here, a neighborhood of $p \in X$ is a set containing the ball

$$
B_\varepsilon(p) = \{x \in X : d(x, p) < \varepsilon\}, \tag{1.2}
$$

for some $\varepsilon > 0$.

(ii) Every open cover $\{U_\alpha\}$ of X has a finite subcover.

Here, a set $U \subset X$ is called open if it contains a neighborhood of each of its points. The complement of an open set is said to be closed. Equivalently, $K \subset X$

is closed provided that

$$(1.3) \qquad x_\nu \in K,\ x_\nu \to p \in X \Longrightarrow p \in K.$$

It is clear that any closed subset of a compact metric space is also compact.

If X_j, $1 \le j \le m$, is a finite collection of metric spaces, with metrics d_j, we can define a product metric space

$$(1.4) \qquad X = \prod_{j=1}^{m} X_j, \quad d(x,y) = d_1(x_1,y_1) + \cdots + d_m(x_m,y_m).$$

Another choice of metric is $\delta(x,y) = \sqrt{d_1(x_1,y_1)^2 + \cdots + d_m(x_m,y_m)^2}$. The metrics d and δ are equivalent; that is, there exist constants $C_0, C_1 \in (0,\infty)$ such that

$$(1.5) \qquad C_0\delta(x,y) \le d(x,y) \le C_1\delta(x,y), \quad \forall\, x,y \in X.$$

We describe some useful classes of compact spaces.

Proposition 1.1. *If X_j are compact metric spaces, $1 \le j \le m$, so is the product space $X = \prod_{j=1}^{m} X_j$.*

Proof. Suppose (x_ν) is an infinite sequence of points in X; let us write $x_\nu = (x_{1\nu}, \dots, x_{m\nu})$. Pick a convergent subsequence $(x_{1\nu})$ in X_1, and consider the corresponding subsequence of (x_ν), which we relabel (x_ν). Using this, pick a convergent subsequence $(x_{2\nu})$ in X_2. Continue. Having a subsequence such that $x_{j\nu} \to y_j$ in X_j for each $j = 1, \dots, m$, we then have a convergent subsequence in X.

The following result is called the *Heine-Borel theorem*:

Proposition 1.2. *If K is a closed bounded subset of $\mathbb{R}^n$, then K is compact.*

Proof. The discussion above reduces the problem to showing that any closed interval $I = [a,b]$ in $\mathbb{R}$ is compact. Suppose S is a subset of I with infinitely many elements. Divide I into two equal subintervals, $I_1 = [a, b_1]$, $I_2 = [b_1, b]$, $b_1 = (a+b)/2$. Then either I_1 or I_2 must contain infinitely many elements of S. Say I_j does. Let x_1 be any element of S lying in I_j. Now divide I_j in two equal pieces, $I_j = I_{j1} \cup I_{j2}$. One of these intervals (say I_{jk}) contains infinitely many points of S. Pick $x_2 \in I_{jk}$ to be one such point (different from x_1). Then subdivide I_{jk} into two equal subintervals, and continue. We get an infinite sequence of distinct points $x_\nu \in S$, and $|x_\nu - x_{\nu+k}| \le 2^{-\nu}(b-a)$, for $k \ge 1$. Since $\mathbb{R}$ is complete, (x_ν) converges, say to $y \in I$. Any neighborhood of y contains infinitely many points in S, so we are done.

If X and Y are metric spaces, a function $f : X \to Y$ is said to be continuous provided $x_\nu \to x$ in X implies $f(x_\nu) \to f(x)$ in Y.

Proposition 1.3. *If X and Y are metric spaces, $f : X \to Y$ continuous, and $K \subset X$ compact, then $f(K)$ is a compact subset of Y.*

Proof. If (y_ν) is an infinite sequence of points in $f(K)$, pick $x_\nu \in K$ such that $f(x_\nu) = y_\nu$. If K is compact, we have a subsequence $x_{\nu_j} \to p$ in X, and then $y_{\nu_j} \to f(p)$ in Y.

If $F : X \to \mathbb{R}$ is continuous, we say $f \in C(X)$. A corollary of Proposition 1.3 is the following:

Proposition 1.4. *If X is a compact metric space and $f \in C(X)$, then f assumes a maximum and a minimum value on X.*

A function $f \in C(X)$ is said to be uniformly continuous provided that, for any $\varepsilon > 0$, there exists $\delta > 0$ such that

$$x, y \in X,\ d(x, y) \le \delta \Longrightarrow |f(x) - f(y)| \le \varepsilon. \tag{1.6}$$

An equivalent condition is that f have a modulus of continuity, in other words, a monotonic function $\omega : [0, 1) \to [0, \infty)$ such that $\delta \searrow 0 \Rightarrow \omega(\delta) \searrow 0$ and such that

$$x, y \in X,\ d(x, y) \le \delta \le 1 \Longrightarrow |f(x) - f(y)| \le \omega(\delta). \tag{1.7}$$

Not all continuous functions are uniformly continuous. For example, if $X = (0, 1) \subset \mathbb{R}$, then $f(x) = \sin(1/x)$ is continuous, but not uniformly continuous, on X. There *is* a case where continuity implies uniform continuity:

Proposition 1.5. *If X is a compact metric space and $f \in C(X)$, then f is uniformly continuous.*

Proof. If not, there exist $x_\nu, y_\nu \in X$ and $\varepsilon > 0$ such that $d(x_\nu, y_\nu) \le 2^{-\nu}$ but

$$|f(x_\nu) - f(y_\nu)| \ge \varepsilon. \tag{1.8}$$

Taking a convergent subsequence $x_{\nu_j} \to p$, we also have $y_{\nu_j} \to p$. Now continuity of f at p implies $f(x_{\nu_j}) \to f(p)$ and $f(y_{\nu_j}) \to f(p)$, contradicting (1.8).

If X and Y are metric spaces, the space $C(X, Y)$ of continuous maps $f : X \to Y$ has a natural metric structure, under some additional hypotheses. We use

$$D(f, g) = \sup_{x \in X} d\big(f(x), g(x)\big). \tag{1.9}$$

This sup exists provided $f(X)$ and $g(X)$ are bounded subsets of Y, where to say $B \subset Y$ is bounded is to say $d : B \times B \to [0, \infty)$ has bounded image. In particular, this supremum exists if X is compact. The following result is useful in the proof of the fundamental local existence theorem for ODE, in Chapter 1.

Proposition 1.6. *If X is a compact metric space and Y is a complete metric space, then C(X, Y), with the metric (1.9), is complete.*

We leave the proof as an exercise.

The following extension of Proposition 1.1 is a special case of Tychonov's theorem.

Proposition 1.7. *If $\{X_j : j \in \mathbb{Z}^+\}$ are compact metric spaces, so is the product $X = \prod_{j=1}^{\infty} X_j$.*

Here, we can make X a metric space by setting

$$d(x, y) = \sum_{j=1}^{\infty} 2^{-j} \frac{d_j(p_j(x), p_j(y))}{1 + d_j(p_j(x), p_j(y))}, \tag{1.10}$$

where $p_j : X \to X_j$ is the projection onto the jth factor. It is easy to verify that if $x_\nu \in X$, then $x_\nu \to y$ in X, as $\nu \to \infty$, if and only if, for each j, $p_j(x_\nu) \to p_j(y)$ in X_j.

Proof. Following the argument in Proposition 1.1, if (x_ν) is an infinite sequence of points in X, we obtain a nested family of subsequences

$$(x_\nu) \supset (x^1{}_\nu) \supset (x^2{}_\nu) \supset \cdots \supset (x^j{}_\nu) \supset \cdots \tag{1.11}$$

such that $p_\ell(x^j{}_\nu)$ converges in X_ℓ, for $1 \le \ell \le j$. The next step is a "diagonal construction." We set

$$\xi_\nu = x^\nu{}_\nu \in X. \tag{1.12}$$

Then, for each j, after throwing away a finite number $N(j)$ of elements, one obtains from (ξ_ν) a subsequence of the sequence $(x^j{}_\nu)$ in (1.11), so $p_\ell(\xi_\nu)$ converges in X_ℓ for all ℓ. Hence (ξ_ν) is a convergent subsequence of (x_ν).

We turn now to the notion of a topological space. This is a set X, together with a family $\mathcal{O}$ of subsets, called "open," satisfying the following conditions:

$$\begin{gathered} X, \emptyset \text{ open}, \\ U_j \text{ open}, 1 \le j \le N \Rightarrow \bigcap_{j=1}^{N} U_j \text{ open}, \\ U_\alpha \text{ open}, \alpha \in A \Rightarrow \bigcup_{\alpha \in A} U_\alpha \text{ open}, \end{gathered} \tag{1.13}$$

where A is *any* index set. It is obvious that the collection of open subsets of a metric space, defined above, satisfies these conditions. As before, a set $S \subset X$ is closed provided $X \setminus S$ is open. Also, we say a subset $N \subset X$ containing p is a neighborhood of p provided N contains an open set U that in turn contains p.

If X is a topological space and S is a subset, S gets a topology as follows. For each U open in X, $U \cap S$ is declared to be open in S. This is called the *induced topology*.

A topological space X is said to be Hausdorff provided that any distinct $p, q \in X$ have disjoint neighborhoods. Clearly, any metric space is Hausdorff. Most important topological spaces are Hausdorff.

A Hausdorff topological space is said to be compact provided the following condition holds. If $\{U_\alpha : \alpha \in A\}$ is any family of open subsets of X, covering X (i.e., $X = \bigcup_{\alpha \in A} U_\alpha$), then there is a finite subcover, that is, a finite subset $\{U_{\alpha_1}, \dots, U_{\alpha_N} : \alpha_j \in A\}$ such that $X = U_{\alpha_1} \cup \cdots \cup U_{\alpha_N}$. An equivalent formulation is the following, known as the *finite intersection property*. Let $\{S_\alpha : \alpha \in A\}$ be any collection of closed subsets of X. If each finite collection of these closed sets has nonempty intersection, then the complete intersection $\bigcap_{\alpha \in A} S_\alpha$ is nonempty. It is not hard to show that any compact metric space satisfies this condition.

Any closed subset of a compact space is compact. Furthermore, any compact subset of a Hausdorff space is necessarily closed.

Most of the propositions stated above for compact metric spaces have extensions to compact Hausdorff spaces. We mention one nontrivial result, which is the general form of Tychonov's theorem; for a proof, see [Dug].

Theorem 1.8. *If S is any nonempty set (possibly uncountable) and if, for any $\alpha \in S$, X_α is a compact Hausdorff space, then so is $X = \prod_{\alpha \in S} X_\alpha$.*

A Hausdorff space X is said to be locally compact provided every $p \in X$ has a neighborhood N that is compact (with the induced topology).

A Hausdorff space is said to be paracompact provided every open cover $\{U_\alpha : \alpha \in A\}$ has a locally finite refinement, that is, an open cover $\{V_\beta : \beta \in B\}$ such that each V_β is contained in some U_α and each $p \in X$ has a neighborhood N_p such that $N_p \cap V_\beta$ is nonempty for only finitely many $\beta \in B$. A typical example of a paracompact space is a locally compact Hausdorff space X that is also σ-compact (i.e., $X = \bigcup_{n=1}^\infty X_n$ with X_n compact). Paracompactness is a natural condition under which to construct partitions of unity, as will be illustrated in the next two sections.

A map $F : X \to Y$ between two topological spaces is said to be continuous provided $F^{-1}(U)$ is open in X whenever U is open in Y. If $F : X \to Y$ is one-to-one and onto, and both F and F^{-1} are continuous, F is said to be a homeomorphism. For a bijective map $F : X \to Y$, the continuity of F^{-1} is equivalent to the statement that $F(V)$ is open in Y whenever V is open in X; another equivalent statement is that $F(S)$ is closed in Y whenever S is closed in X.

If X and Y are Hausdorff, and $F : X \to Y$ is continuous, then $F(K)$ is compact in Y whenever K is compact in X. In view of the discussion above, there arises the following useful sufficient condition for a continuous map $F : X \to Y$ to be a homeomorphism. Namely, if X is compact, Y is Hausdorff, and F is one-to-one and onto, then F is a homeomorphism.

2. Manifolds

A *manifold* is a Hausdorff topological space with an "atlas," that is, a covering by open sets U_j together with homeomorphisms $\varphi_j : U_j \to V_j$, V_j open in $\mathbb{R}^n$. The number n is called the dimension of M. We say that M is a smooth manifold provided the atlas has the following property. If $U_{jk} = U_j \cap U_k \neq \emptyset$, then the map

$$\psi_{jk} : \varphi_j(U_{jk}) \to \varphi_k(U_{jk}), \tag{2.1}$$

given by $\varphi_k \circ \varphi_j^{-1}$, is a smooth diffeomorphism from the open set $\varphi_j(U_{jk})$ to the open set $\varphi_k(U_{jk})$ in $\mathbb{R}^n$. By this, we mean that ψ_{jk} is C^∞, with a C^∞-inverse. If the ψ_{jk} are all C^ℓ-smooth, M is said to be C^ℓ-smooth. The pairs (U_j, φ_j) are called local coordinate charts.

A continuous map from M to another smooth manifold N is said to be smooth if it is smooth in local coordinates. Two different atlases on M, giving a priori two structures of M as a smooth manifold, are said to be equivalent if the identity map on M is smooth from each one of these two manifolds to the other. Actually, a smooth manifold is considered to be defined by equivalence classes of such atlases, under this equivalence relation.

One way manifolds arise is the following. Let $f_1, \dots, f_k$ be smooth functions on an open set $U \subset \mathbb{R}^n$. Let $M = \{x \in U : f_j(x) = c_j\}$, for a given $(c_1, \dots, c_k) \in \mathbb{R}^k$. Suppose that $M \neq \emptyset$ and, for each $x \in M$, the gradients ∇f_j are linearly independent at x. It follows easily from the implicit function theorem that M has a natural structure of a smooth manifold of dimension $n - k$. We say M is a submanifold of U. More generally, let $F : X \to Y$ be a smooth map between smooth manifolds, $c \in Y$, $M = F^{-1}(c)$, and assume that $M \neq \emptyset$ and that, at each point $x \in M$, there is a coordinate neighborhood U of x and V of c such that the derivative DF at x has rank k. More pedantically, (U, φ) and (V, ψ) are the coordinate charts, and we assume the derivative of $\psi \circ F \circ \varphi^{-1}$ has rank k at $\varphi(x)$; there is a natural notion of $DF(x) : T_xX \to T_cY$, which will be defined in the next section. In such a case, again the implicit function theorem gives M the structure of a smooth manifold.

We mention a couple of other methods for producing manifolds. For one, given any connected smooth manifold M, its universal covering space $\tilde{M}$ has the natural structure of a smooth manifold. $\tilde{M}$ can be described as follows. Pick a base point $p \in M$. For $x \in M$, consider smooth paths from p to x, $\gamma : [0, 1] \to M$. We say two such paths γ_0 and γ_1 are equivalent if they are homotopic, that is, if there is a smooth map $\sigma : I \times I \to M (I = [0, 1])$ such that $\sigma(0, t) = \gamma_0(t)$, $\sigma(1, t) = \gamma_1(t)$, $\sigma(s, 0) = p$, and $\sigma(s, 1) = x$. Points in $\tilde{M}$ lying over any given $x \in M$ consist of such equivalence classes.

Another construction produces quotient manifolds. In this situation, we have a smooth manifold M and a discrete group Γ of diffeomorphisms on M. The quotient space $\Gamma \setminus M$ consists of equivalence classes of points of M, where we set $x \sim \gamma(x)$ for each $x \in M$, $\gamma \in \Gamma$. If we assume that each $x \in M$ has a neighborhood U containing no $\gamma(x)$, for $\gamma \neq e$, the identity element of Γ, then $\Gamma \setminus M$ has a natural smooth manifold structure.

We next discuss partitions of unity. Supose M is paracompact. In this case, using a locally finite covering of M by coordinate neighborhoods, we can construct $\psi_j \in C_0^\infty(M)$ such that, for any compact $K \subset M$, only finitely many ψ_j are nonzero on K (we say the sequence ψ_j is locally finite) and such that, for any $p \in M$, some $\psi_j(p) \neq 0$. Then

$$\varphi_j(x) = \Big(\sum_k \psi_k(x)^2\Big)^{-1} \psi_j(x)^2 \tag{2.2}$$

is a locally finite sequence of functions in $C_0^\infty(M)$, satisfying $\sum_j \varphi_j(x) = 1$. Such a sequence is called a *partition of unity*. It has many uses.

Using local coordinates plus such cut-offs as appear in (2.2), one can easily prove that any smooth, compact manifold M can be smoothly imbedded in some Euclidean space $\mathbb{R}^N$, though one does not obtain so easily Whitney's optimal value of N ($N = 2 \dim M$, valid for paracompact M, not just compact M), proved in [Wh].

A more general notion than manifold is that of a smooth *manifold with boundary*. In this case, $\overline{M}$ is again a Hausdorff topological space, and there are two types of coordinate charts (U_j, φ_j). Either φ_j takes U_j to an open subset V_j of $\mathbb{R}^n$ as before, or φ_j maps U_j homeomorphically onto an open subset of $\mathbb{R}^n_+ = \{(x_1, \dots, x_n) \in \mathbb{R}^n : x_n \geq 0\}$. Again appropriate transition maps are required to be smooth. In case $\overline{M}$ is paracompact, there is again the construction of partitions of unity. For one simple but effective application of this construction, see the proof of the Stokes formula in §13 of Chapter I.

3. Vector bundles

We begin with an intrinsic definition of a tangent vector to a smooth manifold M, at a point $p \in M$. It is an equivalence class of smooth curves through p, that is, of smooth maps $\gamma : I \to M$, I an interval containing 0, such that $\gamma(0) = p$. The equivalence relation is $\gamma \sim \gamma_1$ provided that, for some coordinate chart (U, φ) about p, $\varphi : U \to V \subset \mathbb{R}^n$, we have

$$\frac{d}{dt}(\varphi \circ \gamma)(0) = \frac{d}{dt}(\varphi \circ \gamma_1)(0). \tag{3.1}$$

This equivalence is independent of the choice of coordinate chart about p.

If $V \subset \mathbb{R}^n$ is open, we have a natural identification of the set of tangent vectors to V at $p \in V$ with $\mathbb{R}^n$. In general, the set of tangent vectors to M at p is denoted T_pM. A coordinate cover of M induces a coordinate cover of TM, the disjoint union of T_pM as p runs over M, making TM a smooth manifold. TM is called the *tangent bundle* of M. Note that each T_pM has the natural structure of a vector space of dimension n, if n is the dimension of M. If $F : X \to M$ is a smooth map between manifolds, $x \in X$, there is a natural linear map $DF(x) : T_xX \to T_pM$, $p = F(x)$, which agrees with the derivative as defined in §1 of Chapter 1, in local coordinates. $DF(x)$ takes the equivalence class of a smooth curve γ through x to that of the curve $F \circ \gamma$ through p.

The tangent bundle TM of a smooth manifold M is a special case of a vector bundle. Generally, a smooth vector bundle $E \to M$ is a smooth manifold E, together with a smooth map $\pi : E \to M$ with the following properties. For each $p \in M$, the "fiber" $E_p = \pi^{-1}(p)$ has the structure of a vector space, of dimension k, independent of p. Furthermore, there exists a cover of M by open sets U_j, and diffeomorphisms $\Phi_j : \pi^{-1}(U_j) \to U_j \times \mathbb{R}^k$ with the property that, for each $p \in U_j$, $\Phi_j : E_p \to \{p\} \times \mathbb{R}^k \to \mathbb{R}^k$ is a linear isomorphism, and if $U_{j\ell} = U_j \cap U_\ell \neq \emptyset$, we have smooth "transition functions"

$$\Phi_\ell \circ \Phi_j^{-1} = \Psi_{j\ell} : U_{j\ell} \times \mathbb{R}^k \to U_{j\ell} \times \mathbb{R}^k, \tag{3.2}$$

which are the identity on the first factor and such that for each $p \in U_{j\ell}$, $\Psi_{j\ell}(p)$ is a linear isomorphism on $\mathbb{R}^k$. In the case of complex vector bundles, we systematically replace $\mathbb{R}^k$ by $\mathbb{C}^k$ in the discussion above.

The structure above arises for the tangent bundle as follows. Let (U_j, φ_j) be a coordinate cover of M, $\varphi_j : U_j \to V_j \subset \mathbb{R}^n$. Then $\Phi_j : TU_j \to U_j \times \mathbb{R}^n$ takes the equivalence class of smooth curves through $p \in U_j$ containing an element γ to the pair $\big(p, (\varphi_j \circ \gamma)'(0)\big) \in U_j \times \mathbb{R}^n$.

A *section* of a vector bundle $E \to M$ is a smooth map $\beta : M \to E$ such that $\pi(\beta(p)) = p$ for all $p \in M$. For example, a section of the tangent bundle $TM \to M$ is a vector field on M. If X is a vector field on M, generating a flow $\mathcal{F}^t$, then $X(p) \in T_pM$ coincides with the equivalence class of $\gamma(t) = \mathcal{F}^t p$.

Any smooth vector bundle $E \to M$ has associated a vector bundle $E^* \to M$, the "dual bundle" with the property that there is a natural duality of E_p and E_p^* for each $p \in M$. In case E is the tangent bundle TM, this dual bundle is called the *cotangent bundle* and is denoted T^*M.

More generally, given a vector bundle $E \to M$, other natural constructions involving vector spaces yield other vector bundles over M, such as tensor bundles $\otimes^j E \to M$ with fiber $\otimes^j E_p$, mixed tensor bundles with fiber $(\otimes^j E_p) \otimes (\otimes^k E_p^*)$, exterior algebra bundles with fiber ΛE_p, and so forth. Note that a k-form, as defined in Chapter 1, is a section of $\Lambda^k T^*M$. A section of $(\otimes^j T) \otimes (\otimes^k T^*)M$ is called a tensor field of type (j, k).

A Riemannian metric tensor on a smooth manifold M is a smooth, symmetric section g of $\otimes^2 T^*M$ that is positive-definite at each point $p \in M$; that is, $g_p(X, X) > 0$ for each nonzero $X \in T_pM$. For any fixed $p \in M$, using a local coordinate patch (U, φ) containing p, one can construct a positive, symmetric section of $\otimes^2 T^*U$. Using a partition of unity, we can hence construct a Riemannian metric tensor on any smooth, paracompact manifold M. If we define the length of a path $\gamma : [0, 1] \to M$ to be

$$L(\gamma) = \int_0^1 g\big(\gamma'(t), \gamma'(t)\big)^{1/2} dt,$$

then

$$d(p,q) = \inf\{L(\gamma) : \gamma(0) = p, \gamma(1) = q\} \tag{3.3}$$

is a distance function making M a metric space, provided M is connected.

The notion of vector bundle often aids in making intrinsic definitions of important mathematical concepts. As an illustration, we note the following intrinsic characterization of the contact form κ on T^*M, which was specified in local coordinates in (14.17) of Chapter 1. Let $z \in T^*M$; if $\pi : T^*M \to M$ is the natural projection, let $p = \pi(z)$, so $z \in T_p^*M$. To define κ at z, as $\kappa(z) \in T_z^*(T^*M)$, we specify how it acts on a tangent vector $v \in T_z(T^*M)$. The specification is

$$\langle v, \kappa(z)\rangle = \langle (D\pi)v, z\rangle, \tag{3.4}$$

where $D\pi : T_z(T^*M) \to T_pM$ is the derivative of π, and the right side of (3.4) is defined by the usual dual pairing of T_pM and T_p^*M. It is routine to check that this agrees with (14.17) of Chapter 1 in any coordinate system on M. This establishes again the result of §14 of Chapter 1, that the symplectic form $\sigma = d\kappa$ is well defined on a cotangent bundle T^*M.

4. Sard's theorem

Let $F : \Omega \to \mathbb{R}^n$ be a C^1-map, with Ω open in $\mathbb{R}^n$. If $p \in \Omega$ and $DF(p) : \mathbb{R}^n \to \mathbb{R}^n$ is not surjective, then p is said to be a critical point and $F(p)$ a critical value. The set C of critical points can be a large subset of Ω, even all of it, but the set of critical values $F(C)$ must be small in $\mathbb{R}^n$. This is part of Sard's theorem.

Theorem 4.1. *If $F : \Omega \to \mathbb{R}^n$ is a C^1-map, then the set of critical values of F has measure 0 in $\mathbb{R}^n$.*

Proof. If $K \subset \Omega$ is compact, cover $K \cap C$ with m-dimensional cubes Q_j, with disjoint interiors, of side δ_j. Pick $p_j \in C \cap Q_j$, so $L_j = DF(p_j)$ has rank $\le n-1$. Then, for $x \in Q_j$,

$$F(p_j + x) = F(p_j) + L_j x + R_j(x), \quad \|R_j(x)\| \le \rho_j = \eta_j \delta_j,$$

where $\eta_j \to 0$ as $\delta_j \to 0$. Now $L_j(Q_j)$ is certainly contained in an $(n-1)$-dimensional cube of side $C_0\delta_j$, where C_0 is an upper bound for $\sqrt{m}\|DF\|$ on K. Since all points of $F(Q_j)$ are a distance $\le \rho_j$ from (a translate of) $L_j(Q_j)$, this implies

$$\text{meas } F(Q_j) \le 2\rho_j (C_0\delta_j + 2\rho_j)^{n-1} \le C_1 \eta_j \delta_j^n,$$

provided δ_j is sufficiently small that $\rho_j \le \delta_j$. Now $\sum_j \delta_j^n$ is the volume of the cover of $K \cap C$. For fixed K, this can be assumed to be bounded. Hence

$$\text{meas } F(C \cap K) \le C_K\, \eta,$$

where $\eta = \max \{\eta_j\}$. Picking a cover by small cubes, we make η arbitrarily small, so meas $F(C \cap K) = 0$. Letting $K_j \nearrow \Omega$, we complete the proof.

Sard's theorem also treats the more difficult case when Ω is open in $\mathbb{R}^m$, $m > n$. Then a more elaborate argument is needed, and one requires more differentiability, namely that F is class C^k, with $k = m - n + 1$. A proof can be found in [Stb]. The theorem also clearly extends to smooth mappings between separable manifolds.

Theorem 4.1 is applied in Chapter 1, in the study of degree theory. We give another application of Theorem 4.1, to the existence of lots of Morse functions. This application gives the typical flavor of how one uses Sard's theorem, and it is used in a Morse theory argument in Appendix C. The proof here is adapted from one in [GP]. We begin with a special case:

Proposition 4.2. *Let $\Omega \subset \mathbb{R}^n$ be open, $f \in C^\infty(\Omega)$. For $a \in \mathbb{R}^n$, set $f_a(x) = f(x) - a \cdot x$. Then, for almost every $a \in \mathbb{R}^n$, f_a is a Morse function, that is, it has only nondegenerate critical points.*

Proof. Consider $F(x) = \nabla f(x)$; $F : \Omega \to \mathbb{R}^n$. A point $x \in \Omega$ is a critical point of f_a if and only if $F(x) = a$, and this critical point is degenerate only if, in addition, a is a critical value of F. Hence the desired conclusion holds for all $a \in \mathbb{R}^n$ that are not critical values of F.

Now for the result on manifolds:

Proposition 4.3. *Let M be an n-dimensional manifold, imbedded in $\mathbb{R}^K$. Let $f \in C^\infty(M)$, and, for $a \in \mathbb{R}^K$, let $f_a(x) = f(x) - a \cdot x$, for $x \in M \subset \mathbb{R}^K$. Then, for almost all $a \in \mathbb{R}^K$, f_a is a Morse function.*

Proof. Each $p \in M$ has a neighborhood Ω_p such that some n of the coordinates x_ν on $\mathbb{R}^K$ produce coordinates on Ω_p. Let's say $x_1, \dots, x_n$ do it. Let $(a_{n+1}, \dots, a_K)$ be fixed, but arbitrary. Then, by Proposition 4.2, for almost every $(a_1, \dots, a_n) \in \mathbb{R}^n$, f_a has only nondegenerate critical points on Ω_p. By Fubini's theorem, we deduce that, for almost every $a \in \mathbb{R}^K$, f_a has only nondegenerate critical points on Ω_p. (The set of bad $a \in \mathbb{R}^K$ is readily seen to be a countable union of closed sets, hence measurable.) Covering M by a countable family of such sets Ω_p, we finish the proof.

5. Lie groups

A *Lie group* G is a group that is also a smooth manifold, such that the group operations $G \times G \to G$ and $G \to G$ given by $(g, h) \mapsto gh$ and $g \mapsto g^{-1}$ are smooth maps. Let e denote the identity element of G. For each $g \in G$, we have left and right translations, L_g and R_g, diffeomorphisms on G, defined by

$$L_g(h) = gh, \quad R_g(h) = hg. \tag{5.1}$$

The set of left-invariant vector fields X on G, that is, vector fields satisfying

$$(DL_g)X(h) = X(gh), \tag{5.2}$$

is called the *Lie algebra* of G, and is denoted $\mathfrak{g}$. If $X, Y \in \mathfrak{g}$, then the Lie bracket $[X, Y]$ belongs to $\mathfrak{g}$. Evaluation of $X \in \mathfrak{g}$ at e provides a linear isomorphism of $\mathfrak{g}$ with T_eG.

A vector field X on G belongs to $\mathfrak{g}$ if and only if the flow $\mathcal{F}_X^t$ it generates commutes with L_g for all $g \in G$, that is, $g(\mathcal{F}_X^t h) = \mathcal{F}_X^t(gh)$ for all $g, h \in G$. If we set

$$\gamma_X(t) = \mathcal{F}_X^t e, \tag{5.3}$$

we obtain $\gamma_X(t+s) = \mathcal{F}_X^s(\mathcal{F}_X^t e) \cdot e = (\mathcal{F}_X^t e)(\mathcal{F}_X^s e)$, and hence

$$\gamma_X(s+t) = \gamma_X(s)\gamma_X(t), \tag{5.4}$$

for $s, t \in \mathbb{R}$; we say γ_X is a smooth, one-parameter subgroup of G. Clearly,

$$\gamma_X'(0) = X(e). \tag{5.5}$$

Conversely, if γ is any smooth, one-parameter group satisfying $\gamma'(0) = X(e)$, then $\mathcal{F}^t g = g \cdot \gamma(t)$ defines a flow generated by the vector field $X \in \mathfrak{g}$ coinciding with $X(e)$ at e.

The exponential map

$$\operatorname{Exp} : \mathfrak{g} \longrightarrow G \tag{5.6}$$

is defined by

$$\operatorname{Exp}(X) = \gamma_X(1). \tag{5.7}$$

Note that $\gamma_{sX}(t) = \gamma_X(st)$, so $\operatorname{Exp}(tX) = \gamma_X(t)$. In particular, under the identification $\mathfrak{g} \to T_eG$,

$$D\operatorname{Exp}(0) : T_eG \longrightarrow T_eG \text{ is the identity map.} \tag{5.8}$$

The fact that each element $X \in \mathfrak{g}$ generates a one-parameter group has the following generalization, to a fundamental result of S. Lie. Let $\mathfrak{h} \subset \mathfrak{g}$ be a Lie subalgebra, that is, $\mathfrak{h}$ is a linear subspace and $X_j \in \mathfrak{h} \Rightarrow [X_1, X_2] \in \mathfrak{h}$. By Frobenius's theorem (established in §9 of Chapter 1), through each point p of G there is a smooth manifold M_p of dimension $k = \dim \mathfrak{h}$, which is an integral manifold for $\mathfrak{h}$ (i.e., $\mathfrak{h}$ spans the tangent space of M_p at each $q \in M_p$). We can take M_p to be the maximal such (connected) manifold, and then it is unique. Let H be the maximal integral manifold of $\mathfrak{h}$ containing the identity element e.

Proposition 5.1. *H is a subgroup of G.*

Proof. Take $h_0 \in H$ and consider $H_0 = h_0^{-1}H$; clearly, $e \in H_0$. By left invariance, H_0 is also an integral manifold of $\mathfrak{h}$, so $H_0 = H$. This shows that $h_0, h_1 \in H \Rightarrow h_0^{-1}h_1 \in H$, so H is a group.

In addition to left-invariant vector fields on G, one can consider all left-invariant differential operators on G. This is an algebra, isomorphic to the "universal enveloping algebra" $\mathfrak{U}(\mathfrak{g})$, which can be defined as

$$\mathfrak{U}(\mathfrak{g}) = \bigotimes \mathfrak{g}_{\mathbb{C}}/J, \tag{5.9}$$

where $\mathfrak{g}_{\mathbb{C}}$ is the complexification of $\mathfrak{g}$ and J is the two-sided ideal in the tensor algebra $\bigotimes \mathfrak{g}_{\mathbb{C}}$ generated by $\{XY - YX - [X, Y] : X, Y \in \mathfrak{g}\}$.

There are other classes of objects whose left-invariant elements are of particular interest, such as tensor fields (particularly metric tensors) and differential forms.

Given any $\alpha_0 \in \Lambda^k T_e^* G$, there is a unique k-form α on G, invariant under L_g, that is, satisfying $L_g^* \alpha = \alpha$ for all $g \in G$, equal to α_0 at e. In case $k = n = \dim G$, if ω_0 is a nonzero element of $\Lambda^n T_e^* G$, the corresponding left-invariant n-form ω on G defines also an orientation on G, and hence a left-invariant volume form on G, called a (left) Haar measure. It is uniquely defined up to a constant multiple. Similarly one has a right Haar measure. It is very important to be able to integrate over a Lie group using Haar measure.

In many but not all cases left Haar measure is also right Haar measure; then G is said to be *unimodular*. Note that if $\omega \in \Lambda^n(G)$ gives a left Haar measure, then, for each $g \in G$, $R_g^* \omega$ is also a left Haar measure, so we must have

$$R_g^* \omega = \mu(g)\omega, \quad \mu : G \to (0, \infty). \tag{5.10}$$

Furthermore, $\mu(gg') = \mu(g)\mu(g')$. If G is compact, this implies $\mu(g) = 1$ for all g, so all compact Lie groups are unimodular.

There are some particular Lie groups that we want to mention. Let $n \in \mathbb{Z}^+$ and $F = \mathbb{R}$ or $\mathbb{C}$. Then $\mathrm{Gl}(n, F)$ is the group of all invertible $n \times n$ matrices with entries in F. We set

$$\mathrm{Sl}(n, F) = \{A \in \mathrm{Gl}(n, F) : \det A = 1\}. \tag{5.11}$$

We also set

$$\begin{aligned} \mathrm{O}(n) &= \{A \in \mathrm{Gl}(n, \mathbb{R}) : A^t = A^{-1}\}, \\ \mathrm{SO}(n) &= \{A \in \mathrm{O}(n) : \det A = 1\}, \end{aligned} \tag{5.12}$$

and

$$\begin{aligned} \mathrm{U}(n) &= \{A \in \mathrm{Gl}(n, \mathbb{C}) : A^* = A^{-1}\}, \\ \mathrm{SU}(n) &= \{A \in \mathrm{U}(n) : \det A = 1\}. \end{aligned} \tag{5.13}$$

The Lie algebras of the groups listed above also have special names. We have $\mathrm{gl}(n, F) = M(n, F)$, the set of $n \times n$ matrices with entries in F. Also,

$$\begin{aligned} \mathrm{sl}(n, F) &= \{A \in M(n, F) : \mathrm{Tr}\, A = 0\}, \\ \mathrm{o}(n) = \mathrm{so}(n) &= \{A \in M(n, \mathbb{R}) : A^t = -A\}, \\ \mathrm{u}(n) &= \{A \in M(n, \mathbb{C}) : A^* = -A\}, \\ \mathrm{su}(n) &= \{A \in \mathrm{u}(n) : \mathrm{Tr}\, A = 0\}. \end{aligned} \tag{5.14}$$

There are many other important matrix Lie groups and Lie algebras with special names, but we will not list any more here. See [Helg], [T], or [Var1] for such lists.

6. The Campbell-Hausdorff formula

The Campbell-Hausdorff formula has the form

$$\text{Exp}(X)\,\text{Exp}(Y) = \text{Exp}\big(C(X,Y)\big), \tag{6.1}$$

where G is any Lie group, with Lie algebra $\mathfrak{g}$, and Exp: $\mathfrak{g} \to G$ is the exponential map defined by (5.7); X and Y are elements of $\mathfrak{g}$ in a sufficiently small neighborhood U of zero. The map $C : U \times U \to \mathfrak{g}$ has a universal form, independent of $\mathfrak{g}$. We give a demonstration similar to one in [HS], which was also independently discovered by [Str].

We begin with the case $G = \text{Gl}(n, \mathbb{C})$ and produce an explicit formula for the matrix-valued analytic function $X(s)$ of s in the identity

$$e^{X(s)} = e^X e^{sY}, \tag{6.2}$$

near $s = 0$. Note that this function satisfies the ODE

$$\frac{d}{ds} e^{X(s)} = e^{X(s)} Y. \tag{6.3}$$

We can produce an ODE for $X(s)$ by using the following formula, derived in Exercises 7–10 of §4, Chapter 1:

$$\frac{d}{ds} e^{X(s)} = e^{X(s)} \int_0^1 e^{-\tau X(s)} X'(s) e^{\tau X(s)}\, d\tau. \tag{6.4}$$

As shown there, we can rewrite this as

$$\frac{d}{ds} e^{X(s)} = e^{X(s)}\, \Xi\big(\text{ad}\, X(s)\big) X'(s). \tag{6.5}$$

Here, ad is defined as a linear operator on the space of $n \times n$ matrices by

$$\text{ad}\, X(Y) = XY - YX; \tag{6.6}$$

the function Ξ is

$$\Xi(z) = \int_0^1 e^{-\tau z}\, d\tau = \frac{1 - e^{-z}}{z}, \tag{6.7}$$

an entire holomorphic function of z; and a holomorphic function of an operator is defined either as in Exercise 10 of that set, or as in §5 of Appendix A. Comparing (6.3) and (6.5), we obtain

$$\Xi\big(\text{ad}\, X(s)\big) X'(s) = Y, \quad X(0) = X. \tag{6.8}$$

We can obtain a more convenient ODE for $X(s)$ as follows. Note that

$$e^{\text{ad}\, X(s)} = \text{Ad}\, e^{X(s)} = \text{Ad}\, e^X \cdot \text{Ad}\, e^{sY} = e^{\text{ad}\, X}\, e^{s\, \text{ad}\, Y}. \tag{6.9}$$

Now let $\Psi(\zeta)$ be holomorphic near $\zeta = 1$ and satisfy

$$\Psi(e^a) = \frac{1}{\Xi(a)} = \frac{a}{1-e^{-a}}, \tag{6.10}$$

explicitly,

$$\Psi(\zeta) = \frac{\zeta \log \zeta}{\zeta - 1}. \tag{6.11}$$

It follows that

$$\Psi\big(e^{\mathrm{ad}\, X} e^{s\, \mathrm{ad}\, Y}\big)\Xi\big(\mathrm{ad}\, X(s)\big) = I, \tag{6.12}$$

so we can transform (6.8) to

$$X'(s) = \Psi\big(e^{\mathrm{ad}\, X} e^{s\, \mathrm{ad}\, Y}\big)Y, \quad X(0) = X. \tag{6.13}$$

Integrating gives the Campbell-Hausdorff formula for $X(s)$ in (6.2):

$$X(s) = X + \int_0^s \Psi\big(e^{\mathrm{ad}\, X} e^{t\, \mathrm{ad}\, Y}\big)Y\, dt. \tag{6.14}$$

This is valid for $\|sY\|$ small enough, if also X is close enough to 0.

Taking the $s = 1$ case, we can rewrite this formula as

$$e^X e^Y = e^{C(X,Y)}, \quad C(X,Y) = X + \int_0^1 \Psi\big(e^{\mathrm{ad}\, X} e^{t\, \mathrm{ad}\, Y}\big)Y\, dt. \tag{6.15}$$

The formula (6.15) gives a power series in ad X and ad Y which is norm-summable provided

$$\|\mathrm{ad}\, X\| \le x, \quad \|\mathrm{ad}\, Y\| \le y, \tag{6.16}$$

with $e^{x+y} - 1 < 1$, that is,

$$x + y < \log 2. \tag{6.17}$$

We can extend the analysis above to the case where X and Y are vector fields on a manifold M, asking for a vector field $X(s)$ such that

$$\mathcal{F}^1_{X(s)} = \mathcal{F}^1_X \mathcal{F}^s_Y, \tag{6.18}$$

where $\mathcal{F}^t_X$ is the flow generated by X, evaluated at time t. If there is such a family $X(s)$, depending smoothly on s, material in §6 of Chapter 1, in place of material in §4 cited above, leads to a formula parallel to (6.4), and hence to (6.8), in this context. However, we cannot always solve (6.8), because ad $X(s)$ tends not to act as a bounded operator on a Banach space of vector fields, and in fact one cannot always solve (6.18) for $X(s)$ is this case. However, if there is a *finite-dimensional* Lie algebra $\mathfrak{g}$ of vector fields containing X and Y, then the analysis (6.9)–(6.17) extends. We have

$$\mathcal{F}^t_X \mathcal{F}^t_Y = \mathcal{F}^t_{C(t,X,Y)}, \tag{6.19}$$

with

$$C(t, X, Y) = X + \int_0^1 \Psi\big(e^{\mathrm{ad}\, tX} e^{\mathrm{ad}\, stY}\big) Y\, ds, \tag{6.20}$$

provided $\|\mathrm{ad}\, tX\| + \|\mathrm{ad}\, tY\| < \log 2$, the operator norm $\|\mathrm{ad}\, X\|$ being computed using any convenient norm on $\mathfrak{g}$. In particular, if $M = G$ is a Lie group with Lie algebra $\mathfrak{g}$, and $X, Y \in \mathfrak{g}$, this analysis applies to yield the Campbell-Hausdorff formula for general Lie groups.

7. Representations of Lie groups and Lie algebras

We define a representation of a Lie group G on a finite-dimensional vector space V to be a smooth map

$$\pi : G \longrightarrow \mathrm{End}(V) \tag{7.1}$$

such that

$$\pi(e) = I, \quad \pi(gg') = \pi(g)\pi(g'), \quad g, g' \in G. \tag{7.2}$$

If $F \in C_0(G)$, that is, if F is continuous with compact support, we can define $\pi(F) \in \mathrm{End}(V)$ by

$$\pi(F)v = \int_G F(g)\pi(g)v\, dg. \tag{7.3}$$

We get different results depending on whether left or right Haar measure is used. Right now, let us use *right* Haar measure. Then, for $g \in G$, we have

$$\pi(F)\pi(g)v = \int_G F(x)\pi(xg)v\, dx = \int_G F(xg^{-1})\pi(x)v\, dx. \tag{7.4}$$

We also define the derived representation

$$d\pi : \mathfrak{g} \longrightarrow \mathrm{End}(V) \tag{7.5}$$

by

$$d\pi = D\pi(e) : T_eG \longrightarrow \mathrm{End}(V), \tag{7.6}$$

using the identification $\mathfrak{g} \approx T_eG$. Thus, for $X \in \mathfrak{g}$,

$$d\pi(X)v = \lim_{t\to 0} \frac{1}{t}\big[\pi(\mathrm{Exp}\, tX)v - v\big]. \tag{7.7}$$

The following result states that $d\pi$ is a Lie algebra homomorphism.

Proposition 7.1. *For $X, Y \in \mathfrak{g}$, we have*

$$\big[d\pi(X), d\pi(Y)\big] = d\pi\big([X, Y]\big). \tag{7.8}$$

Proof. We will first produce a formula for $\pi(F)d\pi(X)$, given $F \in C_0^\infty(G)$. In fact, making use of (7.4), we have

$$\begin{aligned}
\pi(F)d\pi(X)v &= \lim_{t\to 0} \frac{1}{t} \int_G \big[F(g)\pi(g)\pi(\operatorname{Exp} tX) - F(g)\pi(g)\big]v\, dg \\
&= \lim_{t\to 0} \frac{1}{t} \int_G \big[F\big(g\cdot \operatorname{Exp}(-tX)\big) - F(g)\big]\pi(g)v\, dg \\
&= -\pi(XF)v,
\end{aligned} \tag{7.9}$$

where XF denotes the left-invariant vector field X applied to F. It follows that

$$\begin{aligned}
&\pi(F)\big[d\pi(X)d\pi(Y) - d\pi(Y)d\pi(X)\big]v \\
&\qquad = \pi(YXF - XYF)v = -\pi\big([X,Y]F\big)v,
\end{aligned} \tag{7.10}$$

which by (7.9) is equal to $\pi(F)d\pi\big([X,Y]\big)v$. Now, if F is supported near $e \in G$ and integrates to 1, is is easily seen that $\pi(F)$ is close to the identity I, so this implies (7.8).

There is a representation of G on $\mathfrak{g}$, called the *adjoint representation*, defined as follows. Consider

$$K_g : G \longrightarrow G, \quad K_g(h) = ghg^{-1}. \tag{7.11}$$

Then $K_g(e) = e$, and we set

$$\operatorname{Ad}(g) = DK_g(e) : T_eG \longrightarrow T_eG, \tag{7.12}$$

identifying $T_eG \approx \mathfrak{g}$. Note that $K_g \circ K_{g'} = K_{gg'}$, so the chain rule implies $\operatorname{Ad}(g)\operatorname{Ad}(g') = \operatorname{Ad}(gg')$.

Note that $\gamma(t) = g\operatorname{Exp}(tX)g^{-1}$ is a one-parameter subgroup of G satisfying $\gamma'(0) = \operatorname{Ad}(g)X$. Hence

$$\operatorname{Exp}(t\operatorname{Ad}(g)X) = g\operatorname{Exp}(tX)\,g^{-1}. \tag{7.13}$$

In particular,

$$\operatorname{Exp}\big((\operatorname{Ad}\operatorname{Exp} sY)tX\big) = \operatorname{Exp}(sY)\operatorname{Exp}(tX)\operatorname{Exp}(-sY). \tag{7.14}$$

Now, the right side of (7.15) is equal to $\mathcal{F}_Y^{-s} \circ \mathcal{F}_X^t \circ \mathcal{F}_Y^s(e)$, so by results on the Lie derivative of a vector field given in (8.1)–(8.3) of Chapter 1, we have

$$\operatorname{Ad}(\operatorname{Exp} sY)X = \mathcal{F}^s_{Y\#}X. \tag{7.15}$$

If we take the s-derivative at $s = 0$, we get a formula for the derived representation of Ad, which is denoted ad, rather than d Ad. Using (8.3)–(8.5) of Chapter 1, we have

$$\operatorname{ad}(Y)X = [Y, X]. \tag{7.16}$$

In other words, the adjoint representation of $\mathfrak{g}$ on $\mathfrak{g}$ is given by the Lie bracket. We mention that Jacobi's identity for Lie algebras is equivalent to the statement that

$$\mathrm{ad}([X, Y]) = [\mathrm{ad}(X), \mathrm{ad}(Y)], \quad \forall\, X, Y \in \mathfrak{g}. \tag{7.17}$$

If V has a positive-definite inner product, we say that the representation (7.1) is unitary provided $\pi(g)$ is a unitary operator on V, for each $g \in G$ (i.e., $\pi(g)^{-1} = \pi(g)^*$).

We say the representation (7.1) is irreducible if V has no proper linear subspace invariant under $\pi(g)$ for all $g \in G$. Irreducible unitary representations are particularly important. The following version of Schur's lemma is useful.

Proposition 7.2. *A unitary representation π of G on V is irreducible if and only if, for any $A \in End(V)$,*

$$\pi(g)A = A\pi(g), \quad \forall\, g \in G \Longrightarrow A = \lambda I. \tag{7.18}$$

Proof. First, suppose π is irreducible and A commutes with $\pi(g)$ for all g. Then so does A^*, hence $A + A^*$ and $(1/i)(A - A^*)$, so we may as well suppose $A = A^*$. Now, any polynomial $p(A)$ commutes with $\pi(g)$ for all g, so it follows that each projection P_λ onto an eigenspace of A commutes with all $\pi(g)$. Hence the range of P_λ is invariant under π, so if $P_\lambda \neq 0$, it must be I, and $A = \lambda I$.

Conversely, suppose the implication (7.18) holds. Then if $W \subset V$ is invariant under π, the orthogonal projection P of V onto W must commute with all $\pi(g)$, so P is a scalar multiple of I, hence either 0 or I. This completes the proof.

Corollary 7.3. *Assume G is connected. Then a unitary representation of G on V is irreducible if and only if, for any $A \in End(V)$,*

$$d\pi(X)A = A\, d\pi(X), \quad \forall\, X \in \mathfrak{g} \Longrightarrow A = \lambda I. \tag{7.19}$$

Proof. We mention that

$$\pi(\mathrm{Exp}\, tX) = e^{t\, d\pi(X)} \tag{7.20}$$

and leave the details to the reader.

Given a representation π of G on V, there is also a representation of the universal enveloping algebra $\mathfrak{U}(\mathfrak{g})$, defined as follows. If

$$P = \sum_{\mu \le m} c_{i_1 \cdots i_\mu} X_{i_1} \cdots X_{i_\mu}, \quad X_j \in \mathfrak{g}, \tag{7.21}$$

with $c_{i_1 \cdots i_\mu} \in \mathbb{C}$, we have

$$d\pi(P) = \sum_{\mu \le m} c_{i_1 \cdots i_\mu} d\pi(X_{i_1}) \cdots d\pi(X_{i_\mu}). \tag{7.22}$$

Proposition 7.4. *Suppose G is connected. Let $P \in \mathfrak{U}(\mathfrak{g})$, and assume*

$$PX = XP, \quad \forall\, X \in \mathfrak{g}. \tag{7.23}$$

If π is an irreducible unitary representation of G on V, then $d\pi(P)$ is a scalar multiple of the identity, that is,

$$d\pi(P) = \lambda I.$$

Proof. Immediate from Corollary 7.3.

So far in this section we have concentrated on finite-dimensional representations. It is also of interest to consider infinite-dimensional representations. One example is the right-regular representation of G on $L^2(G)$:

$$R(g)f(x) = f(xg). \tag{7.24}$$

If G has right-invariant Haar measure, then $R(g)$ is a unitary operator on $L^2(G)$ for each $g \in G$, and one readily verifies that $R(g)R(g') = R(gg')$. However, the smoothness hypothesis made on π in (7.1) does not hold here. When working with an infinite-dimensional representation π of G on a Banach space V, one makes instead the hypothesis of strong continuity: For each $v \in V$, the map $g \mapsto \pi(g)v$ is continuous from G to V, with its norm topology. If the map is C^∞, one says v is a smooth vector for the representation v. For example, each $f \in C_0^\infty(G)$ is a smooth vector for the representation (7.24). Of course, $C_0^\infty(G)$ is dense in $L^2(G)$. More generally, the set of smooth vectors for any strongly continuous representation π of G on a Banach space V is dense in V. In fact, for $F \in C_0^\infty(G)$, $\pi(F)$ is still well defined by (7.3), and the space

$$\mathcal{G}_\pi = \{\pi(F)v : F \in C_0^\infty(G), v \in V\} \tag{7.25}$$

is readily verified to be a dense subspace of V consisting of smooth vectors. If V is finite dimensional, this implies that $\mathcal{G}_\pi = V$, so any strongly continuous, finite-dimensional representation of a Lie group automatically possesses the smoothness property used above.

The occasional use made of Lie group representations in this book will not require much development of the theory of infinite-dimensional representations, so we will not go further into it here. One can find treatments in many places, including [HT], [Kn], [T], [Var2], and [Wal1].

8. Representations of compact Lie groups

Throughout this section, G will be a compact Lie group. If π is a representation of G on a finite-dimensional complex vector space V, we can always put an inner product on V so that π is unitary. Indeed, let $((u, v))$ be any Hermitian inner product on V, and set

$$(u, v) = \int_G ((\pi(g)u, \pi(g)v))\, dg. \tag{8.1}$$

Note that if V_1 is a subspace of V invariant under $\pi(g)$ for all $g \in G$, and if π is unitary, then the orthogonal complement of V_1 is also invariant. Thus, if π is not irreducible on V, we can decompose it, and we can obviously continue this process only a finite number of times if $\dim V$ is finite. Thus π breaks up into a direct sum of irreducible unitary representations of G.

Let π and λ be two representations of G, on V and W, respectively. We say they are equivalent if there is $A \in \mathcal{L}(V, W)$, invertible, such that

$$\pi(g) = A^{-1}\lambda(g)A, \quad \forall\, g \in G. \tag{8.2}$$

If these representations are unitary, we say they are unitarily equivalent if A can be taken to be unitary.

Suppose that π and λ are irreducible and unitary, and (8.2) holds. Then $\pi(g)^* = A^*\lambda(g)^*(A^{-1})^*$, for all $g \in G$, so $\pi(g) = (A^*A)\pi(g)(A^*A)^{-1}$. By Schur's lemma, A^*A must be a (positive) scalar, say b^2. Replacing A by $b^{-1}A$ makes it unitary. Breaking up a general π into irreducible representations, we deduce that whenever π and λ are finite-dimensional, unitary representations, if they are equivalent, then they are unitarily equivalent.

We now derive some results known as *Weyl orthogonality relations*, which play an important role in the study of representations of compact Lie groups. To begin, let π and λ be two irreducible representations of a compact group G, on finite-dimensional spaces V and W, respectively. Consider the representation $\nu = \pi \otimes \overline{\lambda}$ on $V \otimes W' \approx \mathcal{L}(W, V)$, defined by

$$\nu(g)(A) = \pi(g)A\lambda(g)^{-1}, \quad g \in G,\ A \in \mathcal{L}(W, V). \tag{8.3}$$

Let Z be the linear subspace of $\mathcal{L}(V, W)$ on which ν acts trivially. We want to specify Z. Note that $A_0 \in Z$ if and only if

$$\pi(g)A_0 = A_0\pi(g), \quad \forall\, g \in G. \tag{8.4}$$

Since this implies that the range of A_0 is invariant under π and $\operatorname{Ker} A_0$ is invariant under λ, we see that either $A_0 = 0$ or A_0 is an isomorphism from W to V. In the latter case, we have $\pi(g) = A_0\lambda(g)A_0^{-1}$, so the representations π and λ would have to be equivalent. In this case, for arbitrary $A \in Z$, we would have

$$\pi(g)A = A\lambda(g) = AA_0^{-1}\pi(g)A_0,$$

or $\pi(g)AA_0^{-1} = AA_0^{-1}\pi(g)$, so Schur's lemma implies that AA_0^{-1} is a scalar. We have proved the following result:

Proposition 8.1. *If π and λ are finite-dimensional, irreducible representations of G and if $\nu = \pi \otimes \overline{\lambda}$, then the trivial representation occurs not at all in ν if π and λ are not equivalent, and it occurs acting on a one-dimensional subspace of $V \otimes W'$ if π and λ are equivalent.*

The next ingredient for the orthogonality relation is the study of the operator

$$P = \int_G \pi(g)\,dg. \tag{8.5}$$

Here π is a finite-dimensional representation of the compact group G, not necessarily irreducible, and dg denotes Haar measure, with total mass 1. Note that

$$\pi(y)P = \int_G \pi(yg)\,dg = P = P\pi(y), \tag{8.6}$$

for all $y \in G$. Hence

$$P^2 = P \int_G \pi(g)\,dg = \int_G P\pi(g)\,dg = P, \tag{8.7}$$

so P is a projection. Also, if π is unitary, we see that $P = P^*$.

Now, if π is unitary, it gives a representation both on the range $\mathcal{R}(P)$ and on the kernel Ker P. It is clear from (8.5) that, given $v \in V$, $\|Pv\| < \|v\|$ unless $\pi(g)v = v$, for all $g \in G$. Consequently, π operates like the identity on $\mathcal{R}(P)$, but we do not have $\pi(g)v = v$ for all $g \in G$, for any nonzero $v \in$ Ker P. We have proved:

Proposition 8.2. *If π is a unitary representation of G on V, then P, given by (8.5), is the orthogonal projection onto the subspace of V on which π acts trivially.*

The following is a special case:

Corollary 8.3. *If π is a nontrivial, irreducible, unitary representation, and P is given by (8.5), then $P = 0$.*

We apply Proposition 8.2 to

$$Q = \int_G \pi(g) \otimes \overline{\lambda}(g)\,dg, \tag{8.8}$$

with π and λ irreducible. By Proposition 8.1, we see that

$$Q = 0 \ \text{ if } \pi \text{ and } \lambda \text{ are not equivalent.} \tag{8.9}$$

On the other hand, if $\lambda = \pi$, then Q has as its range the set of scalar multiples of the identity operator on V (if π acts on V). Note that $\pi \otimes \overline{\pi}$ leaves invariant the space of elements $A \in \mathcal{L}(V, V)$ of trace zero, which is the orthogonal complement (with respect to the Hilbert-Schmidt inner product) of the space of scalars, so Q must annihilate this space. Thus Q is given by

$$Q(A) = (d^{-1}\,\mathrm{Tr}\,A)I, \quad \pi = \lambda,\ d = \dim V. \tag{8.10}$$

The identities (8.9) and (8.10) are equivalent to the Weyl orthogonality relations. If we express π and λ as matrices, with respect to some orthonormal bases, we get the following theorem:

Theorem 8.4. *Let π and λ be inequivalent irreducible, unitary representations of G, on V and W, with matrix entries π_{ij} and $\lambda_{k\ell}$, respectively. Then*

$$\int_G \pi_{ij}(g)\lambda_{k\ell}(g)\,dg = 0. \tag{8.11}$$

Also,

$$\int_G \pi_{ij}(g)\overline{\pi_{k\ell}(g)}\,dg = 0, \quad \text{unless } i = k \text{ and } j = \ell. \tag{8.12}$$

Furthermore,

$$\int_G |\pi_{ij}(g)|^2\,dg = d^{-1}, \tag{8.13}$$

where $d = \dim V = \operatorname{Tr} \pi(e)$.

Hence, if $\{\pi^k\}$ is a complete set of inequivalent, irreducible, unitary representations of G on spaces V_k, of dimension d_k, then

$$d_k^{1/2}\,\pi_{ij}^k(g) \tag{8.14}$$

forms an orthonormal set in $L^2(G)$. The following is the Peter-Weyl theorem:

Theorem 8.5. *The orthonormal set (8.14) is complete.*

In other words, the linear span of (8.14) is dense. If G is given as a group of unitary $N \times N$ matrices, this result is elementary. In fact, the linear span of (8.14) is an algebra (take tensor products of π^k and π^ℓ and decompose into irreducibles), and is closed under complex conjugates (pass from π to $\overline{\pi}$), so if we know it separates points (which is clear if $G \subset U(N)$), the Stone-Weierstrass theorem applies.

If we do not know a priori that $G \subset U(N)$, we can prove the theorem by considering the right-regular representation of G on $L^2(G)$:

$$R(g)f(x) = f(xg). \tag{8.15}$$

If we endow G with a bi-invariant Riemannian metric and consider the associated Laplace operator Δ, which is then a bi-invariant differential operator, we see that the representation R leaves invariant each eigenspace E_ℓ of Δ. Now, E_ℓ is finite-dimensional, and the restriction R_ℓ of R to E_ℓ splits into irreducibles:

$$E_\ell = E_{\ell 1} \oplus \cdots \oplus E_{\ell N}, \quad N = N(\ell), \tag{8.16}$$

say $R_\ell\big|_{E_{\ell m}} = R_{\ell m}$. Thus there is a unitary map $A : E_{\ell m} \to V_k$, for some $k = k(\ell, m)$, such that $R_{\ell m} = A\pi^k A^{-1}$. If $\{e_i\}$ is an orthonormal basis of V_k with respect to which the matrix of $\pi^k(g)$ is $\big(\pi^k_{ij}(g)\big)$, then $u_i = A^{-1}e_i$ gives an orthonormal basis of $E_{\ell m}$, and we have

$$u_i(xg) = \sum_j \pi^k_{ij}(g) u_j(x). \tag{8.17}$$

In particular, taking $x = e$,

$$u_i(g) = \sum_j c_j \pi^k_{ij}(g), \quad c_j = u_j(e). \tag{8.18}$$

This shows that each space $E_{\ell m}$ consists of finite linear combinations of the functions in (8.14). Since

$$L^2(G) = \bigoplus_\ell E_\ell = \bigoplus_\ell \bigoplus_m E_{\ell m},$$

this proves Theorem 8.5.

The following corollary will be useful in the next section.

Corollary 8.6. *If G_1 and G_2 are two compact Lie groups, then the irreducible, unitary representations of $G = G_1 \times G_2$ are, up to unitary equivalence, precisely those of the form*

$$\pi(g) = \pi_1(g_1) \otimes \pi_2(g_2), \tag{8.19}$$

where $g = (g_1, g_2) \in G$, and $\pi_j \in \widehat{G}_j$ is a general, irreducible, unitary representation of G_j.

Proof. Given irreducible, unitary representations π_j of G_j, the irreducibility and unitarity of (8.19) are clear. It remains to prove the completeness of the set of such representations. For this, it suffices to show that the matrix entries of such representations have dense linear span in $L^2(G_1 \times G_2)$. This follows from the general elementary fact that tensor products of orthonormal bases of $L^2(G_1)$ and $L^2(G_2)$ form an orthonormal basis of $L^2(G_1 \times G_2)$.

9. Representations of SU(2) and related groups

The group SU(2) is the group of 2×2, complex, unitary matrices of determinant 1, that is,

$$\mathrm{SU}(2) = \left\{ \begin{pmatrix} z_1 & z_2 \\ -\overline{z}_2 & \overline{z}_1 \end{pmatrix} : |z_1|^2 + |z_2|^2 = 1,\ z_j \in \mathbb{C} \right\}. \tag{9.1}$$

As a set, SU(2) is naturally identified with the unit sphere S^3 in $\mathbb{C}^2$. Its Lie algebra su(2) consists of 2×2, complex, skew-adjoint matrices of trace zero. A basis of su(2) is formed by

$$X_1 = \frac{1}{2}\begin{pmatrix} i & 0 \\ 0 & -i \end{pmatrix}, \quad X_2 = \frac{1}{2}\begin{pmatrix} 0 & 1 \\ -1 & 0 \end{pmatrix}, \quad X_3 = \frac{1}{2}\begin{pmatrix} 0 & i \\ i & 0 \end{pmatrix}. \tag{9.2}$$

Note the commutation relations

$$[X_1, X_2] = X_3, \quad [X_2, X_3] = X_1, \quad [X_3, X_1] = X_2. \tag{9.3}$$

The group SO(3) is the group of linear isometries of $\mathbb{R}^3$ with determinant 1. Its Lie algebra so(3) is spanned by elements J_ℓ, $\ell = 1, 2, 3$, which generate rotations about the x_ℓ-axis. One readily verifies that these satisfy the same commutation relations as in (9.3). Thus SU(2) and SO(3) have isomorphic Lie algebras. There is an explicit homomorphism

$$p : \mathrm{SU}(2) \longrightarrow \mathrm{SO}(3), \tag{9.4}$$

which exhibits SU(2) as a double cover of SO(3). One way to construct p is the following. The linear span $\mathfrak{g}$ of (9.2) over $\mathbb{R}$ is a three-dimensional, real vector space, with an inner product given by $(X, Y) = -\operatorname{Tr} XY$. It is clear that the representation p of SU(2) by a group of linear transformations on $\mathfrak{g}$ given by $p(g) = gXg^{-1}$ preserves this inner product and gives (9.4). Note that Ker $p = \{I, -I\}$.

If we regard X_j as left-invariant vector fields on SU(2), set

$$\Delta = X_1^2 + X_2^2 + X_3^2, \tag{9.5}$$

a second-order, left-invariant differential operator. It follows easily from (9.3) that X_j and Δ commute:

$$\Delta X_j = X_j \Delta, \quad 1 \le j \le 3. \tag{9.6}$$

Suppose π is an irreducible unitary representation of SU(2) on V. Then π induces a skew-adjoint representation $d\pi$ of the Lie algebra su(2) and an algebraic representation of the universal enveloping algebra. By (9.6), $d\pi(\Delta)$ commutes with $d\pi(X_j)$, $j = 1, \dots, 3$. Thus, if π is irreducible, Proposition 7.4 implies

$$d\pi(\Delta) = -\lambda^2 I, \tag{9.7}$$

for some $\lambda \in \mathbb{R}$. (Since $d\pi(\Delta)$ is a sum of squares of skew-adjoint operators, it must be negative.) Let

$$L_j = d\pi(X_j). \tag{9.8}$$

Now we will diagonalize L_1 on V. Set

$$V_\mu = \{v \in V : L_1 v = i\mu v\}, \quad V = \bigoplus_{i\mu \in \operatorname{spec} L_1} V_\mu. \tag{9.9}$$

The structure of π is defined by how L_2 and L_3 behave on V_μ. It is convenient to set

$$L_\pm = L_2 \mp i L_3. \tag{9.10}$$

We have the following key identity, as a direct consequence of (9.3):

$$[L_1, L_\pm] = \pm i L_\pm. \tag{9.11}$$

Using this, we can establish the following:

Lemma 9.1. *We have*

$$L_\pm : V_\mu \longrightarrow V_{\mu\pm 1}. \tag{9.12}$$

In particular, if $i\mu \in$ spec L_1, then either $L_+ = 0$ on V_μ or $\mu + 1 \in$ spec L_1, and also either $L_- = 0$ on V_μ or $\mu - 1 \in$ spec L_1.

Proof. Let $v \in V_\mu$. By (9.11) we have

$$L_1 L_\pm v = L_\pm L_1 v \pm i L_\pm v = i(\mu \pm 1) L_\pm v,$$

which establishes the lemma. The operators $L_\pm$ are called *ladder operators*.

To continue, if π is irreducible on V, we claim that spec L_1 must consist of a sequence

$$\text{spec } L_1 = \{\mu_0, \mu_0 + 1, \dots, \mu_0 + k = \mu_1\}, \tag{9.13}$$

with

$$L_+ : V_{\mu_0+j} \to V_{\mu_0+j+1} \quad \text{isomorphism, for } 0 \le j \le k-1, \tag{9.14}$$

and

$$L_- : V_{\mu_1-j} \to V_{\mu_1-j-1} \quad \text{isomorphism, for } 0 \le j \le k-1. \tag{9.15}$$

In fact, we can compute

$$L_- L_+ = L_2^2 + L_3^2 + i[L_3, L_2] = -\lambda^2 - L_1^2 - iL_1 \tag{9.16}$$

on V, and

$$L_+ L_- = -\lambda^2 - L_1^2 + iL_1 \tag{9.17}$$

on V, so

$$\begin{aligned} L_- L_+ &= \mu(\mu+1) - \lambda^2 \text{ on } V_\mu, \\ L_+ L_- &= \mu(\mu-1) - \lambda^2 \text{ on } V_\mu. \end{aligned} \tag{9.18}$$

Note that since L_1 and L_2 are skew-adjoint, $L_+ = -L_-^*$, so

$$L_+ L_- = -L_-^* L_-, \quad L_- L_+ = -L_+^* L_+.$$

Thus

$$\text{Ker } L_+ = \text{ Ker } L_- L_+, \quad \text{Ker } L_- = \text{ Ker } L_+ L_-.$$

These observations establish (9.13)–(9.15).

Considering that $d\pi$ acts on the linear span of $\{v, L_+v, \dots L_+^{\mu_1-\mu_0}v\}$ for any nonzero $v \in V_{\mu_0}$, and that irreducibility implies this must be all of V, we have

$$\dim V_\mu = 1, \quad \mu_0 \le \mu \le \mu_1. \tag{9.19}$$

From (9.18) we see that $\mu_1(\mu_1+1) = \lambda^2 = \mu_0(\mu_0 - 1)$. Hence,

$$\mu_1 - \mu_0 = k \Longrightarrow \mu_0 = -\frac{k}{2}, \ \mu_1 = \frac{k}{2}, \tag{9.20}$$

and we have

$$\dim V = k+1, \quad \lambda^2 = \frac{1}{4}k(k+2) = \frac{1}{4}(\dim V^2 - 1). \tag{9.21}$$

A nonzero element $v \in V$ such that $L_+v = 0$ is called a "highest-weight vector" for the representation π of SU(2) on V. It follows from the analysis above that all highest-weight vectors for an irreducible representation on V belong to the one-dimensional space V_{μ_1}.

The calculations above establish that an irreducible, unitary representation π of SU(2) on V is determined uniquely up to equivalence by $\dim V$. We are ready to prove the following:

Proposition 9.2. *There is precisely one equivalence class of irreducible, unitary representations of SU(2) on $\mathbb{C}^{k+1}$, for each $k = 0, 1, 2, \dots$.*

We will realize each such representation, which is denoted $D_{k/2}$, on the space

$$\mathcal{P}_k = \{p(z) : p \text{ homogeneous polynomial of degree } k \text{ on } \mathbb{C}^2\}, \tag{9.22}$$

with SU(2) acting on $\mathcal{P}_k$ by

$$D_{k/2}(g)f(z) = f(g^{-1}z), \quad g \in SU(2),\ z \in \mathbb{C}^2. \tag{9.23}$$

Note that, for $X \in \mathrm{su}(2)$,

$$dD_{k/2}(X)f(z) = \frac{d}{dt}f(e^{-tX}z)\Big|_{t=0} = -(\partial_1 f, \partial_2 f)\cdot X\begin{pmatrix} z_1 \\ z_2 \end{pmatrix}, \tag{9.24}$$

where $\partial_j f = \partial f/\partial z_j$. A calculation gives

$$\begin{aligned} L_1 f(z) &= -\frac{i}{2}(z_1\partial_1 f - z_2\partial_2 f), \\ L_2 f(z) &= -\frac{1}{2}(z_2\partial_1 f - z_1\partial_2 f), \\ L_3 f(z) &= -\frac{i}{2}(z_2\partial_1 f + z_1\partial_2 f). \end{aligned} \tag{9.25}$$

In particular, for

$$\varphi_{kj}(z) = z_1^{k-j} z_2^j \in \mathcal{P}_k, \quad 0 \le j \le k, \tag{9.26}$$

we have

$$L_1\varphi_{kj} = i\Big(-\frac{k}{2} + j\Big)\varphi_{kj}, \tag{9.27}$$

so

$$V = \mathcal{P}_k \Longrightarrow \text{span } \varphi_{kj} = V_{-k/2+j}, \quad 0 \le j \le k. \tag{9.28}$$

Note that

$$L_+ f(z) = -z_2 \partial_1 f(z), \quad L_- f(z) = z_1 \partial_2 f(z), \tag{9.29}$$

so

$$L_+ \varphi_{kj} = -(k-j)\varphi_{k,j+1}, \quad L_- \varphi_{kj} = j\varphi_{k,j-1}. \tag{9.30}$$

We see that the structure of the representation $D_{k/2}$ of SU(2) on $\mathcal{P}_k$ is as described in (9.12)–(9.21). The last detail is to show that $D_{k/2}$ is irreducible. If not, then $\mathcal{P}_k$ splits into a direct sum of several irreducible subspaces, each of which has a one-dimensional space of highest-weight vectors, annihilated by L_+. But as seen above, within $\mathcal{P}_k$, only multiples of z_2^k are annihilated by L_+, so the representaiton $D_{k/2}$ of SU(2) on $\mathcal{P}_k$ is irreducible.

We can deduce the classification of irreducible, unitary representations of SO(3) from the result above as follows. We have the covering homomorphism (9.4), and Ker $p = \{\pm I\}$. Now each irreducible representation d_j of SO(3) defines an irreducible representation $d_j \circ p$ of SU(2), which must be equivalent to one of the representations $D_{k/2}$ described above. On the other hand, $D_{k/2}$ factors through to yield a representation of SO(3) if and only if $D_{k/2}$ is the identity on Ker p, that is, if and only if $D_{k/2}(-I) = I$. Clearly, this holds if and only if k is even. Thus all the irreducible, unitary representations of SO(3) are given by representations $\widetilde{D}_j$ on $\mathcal{P}_{2j}$, uniquely defined by

$$\widetilde{D}_j\big(p(g)\big) = D_j(g), \quad g \in \text{SU}(2). \tag{9.31}$$

It is conventional to use D_j instead of $\widetilde{D}_j$ to denote such a representation of SO(3). Note that D_j represents SO(3) on a space of dimension $2j+1$, and

$$dD_j(\Delta) = -j(j+1). \tag{9.32}$$

Also, we can classify the irreducible representations of U(2), using the results on SU(2). To do this, use the exact sequence

$$1 \to K \to S^1 \times \text{SU}(2) \to \text{U}(2) \to 1, \tag{9.33}$$

where "1" denotes the trivial multiplicative group, and

$$K = \{(\omega, g) \in S^1 \times \text{SU}(2) : g = \omega^{-1} I, \omega^2 = 1\}. \tag{9.34}$$

The irreducible representations of $S^1 \times$ SU(2) are given by

$$\pi_{mk}(\omega, g) = \omega^m D_{k/2}(g) \ \text{ on } \mathcal{P}_k, \tag{9.35}$$

with $m, k \in \mathbb{Z}, k \ge 0$. Those giving a complete set of irreducible representations of U(2) are those for which $\pi_{mk}(K) = I$, that is, those for which $(-1)^m D_{k/2}(-I) = I$. Since $D_{k/2}(-I) = (-1)^k I$, we see the condition is that $m+k$ be an even integer.

We now consider the representations of SO(4). First note that SO(4) is covered by SU(2)×SU(2). To see this, equate the unit sphere $S^3 \subset \mathbb{R}^4$, with its standard metric, to SU(2), with a bi-invariant metric. Then SO(4) is the connected component of the identity in the isometry group of S^3. Meanwhile, SU(2)×SU(2) acts as a group of isometries, by

$$(g_1, g_2) \cdot x = g_1 x g_2^{-1}, \quad g_j \in \mathrm{SU}(2). \tag{9.36}$$

Thus we have a map

$$\tau : \mathrm{SU}(2) \times \mathrm{SU}(2) \longrightarrow \mathrm{SO}(4). \tag{9.37}$$

This is a group homomorphism. Note that $(g_1, g_2) \in \operatorname{Ker} \tau$ implies $g_1 = g_2 = \pm I$. Furthermore, a dimension count shows τ must be surjective, so

$$\mathrm{SO}(4) \approx \mathrm{SU}(2) \times \mathrm{SU}(2)/\{\pm(I, I)\}. \tag{9.38}$$

As shown in §8, if G_1 and G_2 are compact Lie groups, and $G = G_1 \times G_2$, then the set of all irreducible, unitary representations of G, up to unitary equivalence, is given by

$$\{\pi(g) = \pi_1(g_1) \otimes \pi_2(g_2) : \pi_j \in \widehat{G}_j\}, \tag{9.39}$$

where $g = (g_1, g_2) \in G$ and $\widehat{G}_j$ parameterizes the irreducible, unitary representations of G_j. In particular, the irreducible unitary representations of SU(2)×SU(2), up to equivalence, are precisely the representations of the form

$$\gamma_{k\ell}(g) = D_{k/2}(g_1) \otimes D_{\ell/2}(g_2), \quad k, \ell \in \{0, 1, 2, \dots\}, \tag{9.40}$$

acting on $\mathcal{P}_k \otimes \mathcal{P}_\ell \approx \mathbb{C}^{k+1} \otimes \mathbb{C}^{\ell+1}$. By (9.38), the irreducible, unitary representations of SO(4) are given by all $\gamma_{k\ell}$ such that $k + \ell$ is even, since, for $p_0 = (-I, -I) \in \mathrm{SU}(2)\times\mathrm{SU}(2)$, $\gamma_{k\ell}(p_0) = (-1)^{k+\ell} I$.

We next consider the problem of decomposing the tensor-product representations $D_{k/2} \otimes D_{\ell/2}$ of SU(2) (i.e., the composition of (9.40) with the diagonal map SU(2)↪SU(2)×SU(2)) into irreducible representations. We may as well assume that $\ell \le k$. Note that $\pi_{k\ell} = D_{k/2} \otimes D_{\ell/2}$ acts on

$$\begin{aligned} \mathcal{P}_{k\ell} = \{&f(z, w) : \text{polynomial on } \mathbb{C}^2 \times \mathbb{C}^2, \\ &\text{homogeneous of degree } k \text{ in } z,\ \ell \text{ in } w\}, \end{aligned} \tag{9.41}$$

as

$$\pi_{k\ell}(g) f(z, w) = f(g^{-1} z, g^{-1} w). \tag{9.42}$$

Parallel to (9.25) and (9.29), we have, on $\mathcal{P}_{k\ell}$,

$$\begin{aligned} L_1 f &= -\frac{i}{2}(z_1 \partial_{z_1} f - z_2 \partial_{z_2} f + w_1 \partial_{w_1} f - w_2 \partial_{w_2} f), \\ L_+ f &= -z_2 \partial_{z_1} f - w_2 \partial_{w_1} f, \quad L_- f = z_1 \partial_{z_2} f + w_1 \partial_{w_2} f. \end{aligned} \tag{9.43}$$

To decompose $\mathcal{P}_{k\ell}$ into irreducible subspaces, we specify Ker L_+. In fact, a holomorphic function $f(z, w)$ annihilated by L_+ is of the form

$$f(z, w) = g(z_2, w_2, w_2 z_1 - z_2 w_1), \tag{9.44}$$

and the kernel of L_+ in $\mathcal{P}_{k\ell}$ is the linear span of

$$\psi_{k\ell\mu}(z, w) = z_2^{k-\mu} w_2^{\ell-\mu} (w_2 z_1 - z_2 w_1)^{\mu}, \quad 0 \leq \mu \leq \ell. \tag{9.45}$$

A calculation gives

$$L_1 \psi_{k\ell\mu} = \frac{i}{2}(k + \ell - 2\mu)\psi_{k\ell\mu}. \tag{9.46}$$

It follows that, for fixed $k, \ell, 0 \leq \ell \leq k$, and for each $\mu = 0, \ldots, \ell$, $\psi_{k\ell\mu}$ is the highest-weight vector of a representation equivalent to $D_{(k+\ell-2\mu)/2}$, so we have

$$D_{k/2} \otimes D_{\ell/2} \approx \bigoplus_{\mu=0}^{\ell} D_{(k+\ell-2\mu)/2} = D_{(k-\ell)/2} \oplus D_{(k-\ell)/2+1} \oplus \cdots \oplus D_{(k+\ell)/2}. \tag{9.47}$$

This is called the *Clebsch-Gordon series.*

Extensions of the results presented here to more general compact Lie groups, due mainly to E. Cartan and H. Weyl, can be found in a number of places, including [T], [Var1], and [Wal1].

References

[Dug] J. Dugundji, *Topology*, Allyn and Bacon, New York, 1966.

[GP] V. Guillemin and A. Pollack, *Differential Topology*, Prentice-Hall, Englewood-Cliffs, New Jersey, 1974.

[HS] M. Hausner and J. Schwartz, *Lie Groups; Lie Algebras*, Gordon and Breach, London, 1968.

[Helg] S. Helgason, *Differential Geometry, Lie Groups, and Symmetric Spaces*, Academic Press, New York, 1978.

[HT] R. Howe and E. Tan, *Non-Abelian Harmonic Analysis*, Springer-Verlag, New York, 1992.

[Hus] D. Husemuller, *Fibre Bundles*, McGraw-Hill, New York, 1966.

[Kn] A. Knapp, *Representation Theory of Semisimple Groups*, Princeton Univ. Press, Princeton, N. J., 1986.

[Stb] S. Sternberg, *Lectures on Differential Geometry*, Prentice Hall, Englewood Cliffs, N. J., 1964.

[Str] R. Strichartz, The Campbell-Hausdorff formula and DuHamel's principle, Preprint.

[T] M. Taylor, *Noncommutative Harmonic Analysis*, AMS, Providence, R. I., 1986.

[Var1] V. S. Varadarajan, *Lie Groups, Lie Algebras, and Their Representations*, Springer-Verlag, New York, 1984.

[Var2] V. S. Varadarajan, *An Introduction to Harmonic Analysis on Semisimple Lie Groups*, Cambridge Univ. Press, Cambridge, 1986.

[Wal1] N. Wallach, *Harmonic Analysis on Homogeneous Spaces*, Marcel Dekker, New York, 1973.

[Wal2] N. Wallach, *Real Reductive Groups, I*, Academic Press, New York, 1988.

[Wh] H. Whitney, Sphere spaces, *Proc. NAS, USA* 21(1939), 462–468.

Index

Texts in Applied Mathematics

(continued from page ii)

31. *Brémaud:* Markov Chains: Gibbs Fields, Monte Carlo Simulation, and Queues.
32. *Durran:* Numerical Methods for Wave Equations in Geophysical Fluid Dynamics.
33. *Thomas:* Numerical Partial Differential Equations: Conservation Laws and Elliptic Equations.
34. *Chicone:* Ordinary Differential Equations with Applications.

9 780387 946542